Hand Book of Processed Functional Meat Products

Sajad A. Rather • F. A. Masoodi
Editors

Hand Book of Processed Functional Meat Products

Springer

Editors
Sajad A. Rather
Food Science & Technology
University of Kashmir
Srinagar, Jammu and Kashmir, India

F. A. Masoodi
Food Science & Technology
University of Kashmir
Srinagar, Jammu and Kashmir, India

ISBN 978-3-031-69870-5　　ISBN 978-3-031-69868-2　(eBook)
https://doi.org/10.1007/978-3-031-69868-2

This Springer imprint is published by the registered company Springer Nature Switzerland AG
The registered company address is: Gewerbestrasse 11, 6330 Cham, Switzerland

Preface

Functional food technology is an impending applied science that claims to boost health or well-being by providing health benefits beyond that of fundamental nutrition. Meat and meat products have multifarious disease-preventing and health-promoting benefits. However, the meat industry has met new challenges since the World Health Organization has classified processed meat in carcinogenic group. In addition consumers often associate meat with a contravening health image. The deplorable image of meat is mainly due to fat content – saturated fatty acids, and cholesterol and process-induced toxicants like N-nitroso compounds and polycyclic aromatic hydrocarbons (PAHs) and the alliance of these with chronic diseases. In this context, the functional food concept in the meat processing industry has gained importance, especially by reduction/replacement of fat, sodium, synthetic preservatives such as nitrites and antioxidants, reduction of process-induced toxicants, or addition of beneficial components such as probiotics, bioactive compounds, dietary fibre, etc.

This book has 24 chapters and is a professional reference book. It covers the following major areas: meat processing industry; post-mortem muscle chemistry; functional meat products design and development; bioactive compounds in functional meat products; herbs, spices and condiments for processed meat products; probiotic meat products; cultured meat; reduction of carcinogenic compounds in processed meat products; application of enzymes in processed meat products; nanotechnology in meat processing industry; innovation in the sensory assessment of processed meat products; authentication, market, regulatory and consumer challenges of functional meat products; 3D printing in the development of meat-based products; valorization of meat waste and by-products; and environmental issues of meat. This book differs from others in the market in several aspects. It offers a comprehensive coverage and discussion of the most recent science, technology and applications of functional meat products. Many meat science and technology books available in the market describe meat chemistry, properties and basic science with only rudimentary understanding of meat processing, functional meat products development and applications. Thus, we hope this reference book will be read by and helpful to researchers, students, teachers, nutritionists, dieticians, industry and government policy makers. This book *Handbook of Processed Functional Meat Products* is the first of its kind on the compilation of processed functional meat products and their quality evaluation.

The editors wish to thank all the contributing authors and co-authors for their hard work and excellent results on the chapters of this book. We were able to bring all of them on the same platform even though they have been working on various aspects of processed functional meat products. We are also grateful to Springer Nature for publishing this comprehensive book on processed functional meat products.

Srinagar, Jammu and Kashmir, India

Sajad A. Rather
F. A. Masoodi

Contents

Contributors

Martin Aduah Africa Centre of Excellence for Mycotoxin and Food Safety, Federal University of Technology, Minna, Nigeria

Frederick Adzitey Department of Animal Science, University for Development Studies, Tamale, Ghana

Sheikh Rafeh Ahmad Division of Livestock Products Technology, FVSc & AH, SKUAST-K, Srinagar, Jammu and Kashmir, India

Tanbir Ahmad Division of Livestock Products Technology, ICAR-Indian Veterinary Research Institute, Izatnagar, Bareilly, Uttar Pradesh, India

Najmeenah Akhter Department of Food Technology, Islamic University of Science and Technology, Pulwama, India

Shazia Akhter Division of Food Science and Technology, SKUAST-Kashmir, Srinagar, Jammu and Kashmir, India

Tawheed Amin Division of Food Science and Technology, SKUAST, Srinagar, Jammu and Kashmir, India

Nadira Anjum Division of Food Science and Technology, Sher-e-Kashmir University of Agricultural Sciences and Technology of Kashmir, Srinagar, India

Tajalli Asad Division of Food Science and Technology, Sher-e-Kashmir University of Agricultural Sciences and Technology of Kashmir, Srinagar, India

Barjees Ashaq Division of Food Science and Technology, Sher-e-Kashmir University of Agricultural Sciences and Technology of Kashmir, Srinagar, India

Kitipong Assatarakul Department of Food Technology, Faculty of Science, Chulalongkorn University, Bangkok, Thailand

Qudsiya Ayaz Division of Food Science and Technology, Sher-e-Kashmir University of Agricultural Sciences and Technology of Kashmir, Srinagar, India

Iqra Bashir Division of Food Science and Technology, Sher-e-Kashmir University of Agricultural Sciences and Technology of Kashmir, Srinagar, India

B. N. Dar Department of Food Technology, Islamic University of Science and Technology, Pulwama, India

I. Prince Devadason Division of Livestock Products Technology, ICAR-Indian Veterinary Research Institute, Izatnagar, Bareilly, Uttar Pradesh, India

Mahdi Ebrahimi Department of Veterinary Preclinical Sciences, Faculty of Veterinary Medicine, University Putra Malaysia, Serdang, Malaysia

Humaira Fayaz Division of Livestock Products Technology, FVSc & AH, SKUAST-K, Srinagar, Jammu and Kashmir, India

Tariq Ahmad Ganie Department of Food Technology, Islamic University of Science and Technology, Pulwama, Jammu and Kashmir, India

Saqib Gulzar Department of Food Technology, Engineering and Science, Agrotecnio – CERCA Center, University of Lleida, Lleida, Spain

Nurul Huda Department of Livestock Production, Faculty of Sustainable Agriculture, Universiti Malaysia Sabah, Sandakan, Malaysia

Syed Arshad Hussain Division of Livestock Products Technology, FVSc & AH, SKUAST-K, Srinagar, Jammu and Kashmir, India

Ishamri Ismail School of Food Industry, Faculty of Bioresources and Food Industry, Universiti Sultan Zainal Abidin, Kampus Besut, Terengganu, Malaysia

Saeid Jafari Department of Food Technology, Faculty of Science, Chulalongkorn University, Bangkok, Thailand

Samar Jan Division of Food Science and Technology, Sher-e-Kashmir University of Agricultural Sciences and Technology of Kashmir, Srinagar, India

Javed-ur-Rehman Central Sericulture Research and Training Institute, Berhampore, West Bengal, India

Adamu Yusuf Kabiru Department of Biochemistry, Federal University of Technology, Minna, Nigeria

Devendra Kumar Division of Livestock Products Technology, ICAR-Indian Veterinary Research Institute, Izatnagar, Bareilly, Uttar Pradesh, India

Pavan Kumar Department of Livestock Products Technology, College of Veterinary Science, Guru Angad Dev Veterinary and Animal Sciences University, Ludhiana, Punjab, India

Mansha Manzoor Division of Food Science and Technology, Sher-e-Kashmir University of Agricultural Sciences and Technology of Kashmir, Srinagar, India

Neelofar Manzoor Department of Food Technology, Islamic University of Science and Technology, Pulwama, India

F. A. Masoodi Department of Food Science and Technology, University of Kashmir, Srinagar, India

Sajad Ahmad Mir Department of Life Science (Microbiology and Food Science and Technology), GITAM (Deemed to be University), Visakhapatnam, Andhra Pradesh, India

Tanveer Ahmad Mir Department of Food Technology, Islamic University of Science and Technology, Pulwama, Jammu and Kashmir, India

A. R. Muhammad Sayyaf Division of Livestock Products Technology, ICAR-Indian Veterinary Research Institute, Izatnagar, Bareilly, Uttar Pradesh, India

Haseeba Muzaffar Department of Food Science and Technology, University of Kashmir, Srinagar, India

Syed Muzaffar Department of Chemistry AAAM Degree College Bemina, Srinagar, Jammu and Kashmir, India

Shahida Naseem Department of Environmental Science, University of Kashmir, Srinagar, India

Zahida Naseem Division of Food Science and Technology, Sher-e-Kashmir University of Agricultural Sciences and Technology of Kashmir, Srinagar, India

Tahir Nazir Division of Livestock Products Technology, FVSc & AH, SKUAST-K, Srinagar, Jammu and Kashmir, India

Naseh Nisar Division of Food Science and Technology, Sher-e-Kashmir University of Agricultural Sciences and Technology of Kashmir, Srinagar, India

Tina Nurkhoeriyati Food Technology Department, Faculty of Engineering, Bina Nusantara University, Jakarta, Indonesia

Hilal Ahmad Punoo Department of Food Science and Technology, University of Kashmir, Srinagar, India

Asma Irshad Qureshi Division of Livestock Products Technology, FVSc & AH, SKUAST-K, Srinagar, Jammu and Kashmir, India

Kurnia Ramadhan Department of Food Science and Technology, Universitas Bakrie, Jakarta, Indonesia

Shakir Rashid Division of Livestock Products Technology, FVSc & AH, SKUAST-K, Srinagar, India

Jahangir Ahmad Rather Department of Food Science and Technology, Islamic University of Science and Technology, Awantipora, Jammu and Kashmir, India

Department of Food Science and Technology, Islamic University of Science and Technology, Awantipora, Jammu and Kashmir, India

Sajad A. Rather Department of Food Science and Technology, University of Kashmir, Srinagar, India

Sadiya Sajad Division of Livestock Products Technology, FVSc & AH, SKUAST-Kashmir, Srinagar, India

Samira Division of Food Science and Technology, Sher-e-Kashmir University of Agricultural Sciences and Technology of Kashmir, Srinagar, India

Awis Qurni Sazili Department of Animal Science, Faculty of Agriculture, Universiti Putra Malaysia, Serdang, Selangor, Malaysia

Halal Products Research Institute, Universiti Putra Malaysia, Serdang, Selangor, Malaysia

A. R. Sen Division of Livestock Products Technology, ICAR-Indian Veterinary Research Institute, Izatnagar, Bareilly, Uttar Pradesh, India

Khursheed Ahmad Shiekh Department of Food Technology, Faculty of Science, Chulalongkorn University, Bangkok, Thailand

Department of Food Science, College of Agriculture and Life Sciences, Cornell University, Ithaca, NY, USA

Haamiyah Sidiq Division of Food Science and Technology, Sher-e-Kashmir University of Agricultural Sciences and Technology of Kashmir, Srinagar, India

Aaruba Maqbool Sofi Division of Food Science and Technology, SKUAST, Srinagar, Jammu and Kashmir, India

Asif H. Sofi Division of Livestock Products Technology, FVSc & AH, SKUAST-K, Srinagar, Jammu and Kashmir, India

Aisha Sultana Division of Food Science and Technology, Sher-e-Kashmir University of Agricultural Sciences and Technology of Kashmir, Srinagar, India

Mohamed Tagrida Department of Food Technology, Engineering and Science, University of Lleida—Agrotecnio CERCA Center, Lleida, Spain

Suman Talukder Division of Livestock Products Technology, ICAR-Indian Veterinary Research Institute, Izatnagar, Bareilly, Uttar Pradesh, India

Ruqaya Tariq Division of Food Science and Technology, SKUAST, Srinagar, Jammu and Kashmir, India

Sajad Mohd Wani Division of Food Science and Technology, Sher-e-Kashmir University of Agricultural Sciences and Technology of Kashmir, Srinagar, India

About the Editors

Sajad A. Rather obtained his Master's Degree and Doctorate in Food Technology in 2011 and 2017, respectively, with specialization in meat processing from the University of Kashmir, India. He qualified ICAR NET Exam in Food Technology through ICAR-ASRB in 2011. Sajad A. Rather has worked as Junior Research Fellow (JRF) for a period of 3 years and Senior Research Fellow (SRF) for a period of 1 year. He has published 70 research papers and review articles in reputed peer-reviewed journals. He has more than 15 book chapters to his credit. Sajad A. Rather is on the editorial board of some reputed journals besides being reviewer of many others. He has a strong linkage with the meat industries of the Union territory of Jammu and Kashmir, India. Sajad A. Rather is a life member of the American Society for Microbiology (ASM) and the Association of Food Scientists and Technologists (India) [AFST (I)]. He is currently the Food Safety Officer at Food and Drugs Administration, Jammu and Kashmir, India.

F. A. Masoodi currently serves as Professor of Food Technology at the University of Kashmir, India. He has a degree in Agricultural Sciences and obtained his Master's Degree from Punjab Agriculture University (PAU), Ludhiana, India, and his Doctorate in Food Technology from G. B. Pant University of Agriculture and Technology, Uttarakhand, India. Before joining the University of Kashmir, Prof. Masoodi worked in some reputed universities of India like CCS HAU Hisar. He has guided about 30 Ph.D. scholars and published about 360 research papers and review articles in peer-reviewed journals. He has also authored a few books and more than 50 book chapters and has been on the editorial board of some journals. Prof. Masoodi has been on the expert panel of some prestigious research institutions in India, which include the National Dairy Research Institute (NDRI), Karnal; the Institute of Chemical Technology (ICT) Mumbai; and the National Agri-Food Biotechnology Institute (NABI) Mohali, Punjab. He has worked on various research projects as Principal Investigator and generated significant extramural grants from

FCP	Free Choice Profiling
FDA	Food and Drug Administration
FP	Flash Profiling
GHP	Good Hygiene Practices
GMP	Good Manufacturing Practices
GRAS	Generally Recognized as Safe
HAAs	Heterocyclic Aromatic Amines
HACCP	Hazard Analysis Critical Control Point
HDL	High-Density Lipoprotein
HPP	High Pressure Processing
IARC	International Agency for Research on Cancer
ICMR	Indian Council of Medical Research
IMP	Inosine 5'-Monophosphate
KCl	Potassium Chloride
KDa	Kilodalton
LAB	Lactic Acid Bacteria
LBG	Locust Bean Gum
LCA	Life Cycle Evaluation
LDL	Low-Density Lipoproteins
MAP	Modified Atmosphere Packaging
Mb	Myoglobin
MCT	Medium Chain Triglycerides
MDA	Malondialdehyde
MetMb	Methemyoglobin
MPa	Megapascal
MS	Mass Spectrometry
MSG	Monosodium Glutamate
MWP	Microparticulated Whey Protein
NHLBI	National Heart, Lung and Blood Institute
NO	Nitric Oxide
NOCs	N-Nitroso Compounds
NPIP	N-Nitrosopiperidine
OUOD	'One Up One Down'
OWP	Overwrapped Packaging
OxyMb	Oxymyoglobin
PAHs	Polycylic Aromatic Hydrocarbons
PCR	Polymerase Chain Reaction
PCR-RFLP	Polymerase Chain Reaction-Restriction Fragment Length Polymorphism
PEF	Pulsed Electric Field
PER	Protein Efficiency Ratio
PG	Propyl-Gallate
PNCs	Polymer Nanocomposites
PSE	Pale, Soft and Exudative

PUFAs	Polyunsaturated Fatty Acids
QR	Quick Response
QSRs	Quick Service Restaurants
RATA	Rate-All-That-Apply
RFID	Radio Frequency Identification
RFLPs	Restriction Fragment Length Polymorphisms
ROOH	Hydroperoxides
ROS	Reactive Oxygen Species
RTE	Ready-to-Eat
RT-PCR	Reverse Transcription Polymerase Chain Reaction
SARS-CoV-2	Severe Acute Respiratory Syndrome Coronavirus 2
SDS-PAGE	Sodium Dodecyl Sulphate-Polyacrylamide Gel Electrophoresis
SFA	Saturated Fatty Acid
SNPs	Single Nucleotide Polymorphisms
TGase	Transglutaminase
VSP	Vacuum Skin Packaging
WHC	Water-Holding Capacity
WHO	World Health Organization

Chapter 1
Introduction to Meat Processing Industry

Asma Irshad Qureshi, Syed Arshad Hussain, Sheikh Rafeh Ahmad, and Humaira Fayaz

1.1 Introduction

Processed meats are the result of the need to preserve meat in ancient times. The knowledge to preserve meat by making fermented sausages was already known in ancient times (Pederson 1979). Such type of products was found in old Greek, Roman and even Babylonic scripts. In Northern and Central Europe "meat animals" were slaughtered before winter (November). It is said that Roman army success in conquering nearly all the territories of the "Old World" was partly due to the knowledge of preserving meat (dry cured ham and fermented sausage), which made the long distance shipment and supply of the troops feasible (Vandendriessche 2008).

Preservation of meat by means of a_w reduction combined with increase in acidity, can be considered as the oldest technology. a_w reduction was achieved by salting and/or drying. In any case, meat processing had its origin before the dawn of civilization. The ancient Egyptians recorded the preservation of meat products by salting and sun-drying (Vandendriessche 2008). If fermentation was involved in preservation of meat, this was mainly due to the 'meat borne' microorganisms. The early Romans are credited with being the first to use ice and snow as a means of preservation. Thus, meat processing and preservation was considered being an art, a craft.

Modern food processing and preservation traces its origin to the invention of canning for which Nicholas Appert, a chef, received an official commendation from the French government in 1809 (Christensen 2023). Since that time, advances in technology have continued to change processing methods. The newer methods of meat preservation include (1) use of chemical additives, (2) development of

A. I. Qureshi (✉) · S. A. Hussain (✉) · S. R. Ahmad (✉) · H. Fayaz
Division of Livestock Products Technology, FVSc & AH, SKUAST-K, Srinagar, India
e-mail: sh.rafeh@skuastkashmir.ac.in

S. A. Rather, F. A. Masoodi (eds.), *Hand Book of Processed Functional Meat Products*, https://doi.org/10.1007/978-3-031-69868-2_1

intermediate-moisture foods, (3) freeze dehydration, (4) enzyme treatments, (5) use of biopreservatives and (6) irradiation and other novel treatments. Increased scientific information and technological advances have greatly changed many processing methods such as developments that have led to commercial freezing and new techniques of curing and smoking. Currently, meat processing is categorized into two groups: (1) those contributing to the preservation and production of meat products and (2) those involved in modifying the form of fresh meat in preparation of healthier meat products for consumption. Today meat processing is a period of emphasis on quality, safety and nutrition/health aspects.

1.2 History of Meat Processing

The history of meat processing is nothing short of a narrative that unfolded over millennia, reflecting the ingenuity of human societies in handling and preserving meat for sustenance marking the evolution of human societies through different eras through a continuous journey of innovation and adaptation to meet the changing needs of societies throughout different historical periods. Following is a brief overview of the historical progression (Chiles and Fitzgerald 2017; Thompson 2009; MacLachlan 2006):

1.2.1 Primitive Techniques (Prehistoric Era)

In ancient times, humans were primarily hunters and gatherers. Basic meat processing involved techniques such as skinning, butchering and roasting over open fire. The use of bones as tools for cutting and extracting marrow is evident in archaeological findings.

1.2.2 Domestication of Animals (Neolithic Era)

The shift from a nomadic lifestyle to settled agriculture marked the Neolithic Era. Humans began domesticating animals, leading to the development of herding practices. This necessitated more sophisticated meat processing techniques, including the use of bones for tools and the utilization of various cooking methods.

1.2.3 Development of Preservation Techniques (Ancient Civilizations)

In ancient civilizations like Mesopotamia, Egypt and China, people developed methods to preserve meat, such as salting, smoking and drying. These techniques helped extend the shelf life of meat and enabled trade over long distances.

1.2.4 Roman Culinary Innovations (Roman Era)

The Romans are credited with introducing sausages and improving meat preservation techniques. They mixed ground meat with spices and stuffed it into casings made from intestines, pioneering an early form of processed meat. Additionally, Romans built elaborate structures for smoking and curing meats.

1.2.5 Medieval and Renaissance Periods

During the middle ages, meat processing became more organized, with the establishment of butcher guilds which regulated the trade, ensuring quality and safety standards. Smoking and salting continued to be important methods for preserving meat. The Renaissance period saw advancements in kitchen technology, such as the invention of the sausage grinder that facilitated the mass production of sausages and other processed meat products.

1.2.6 Industrial Revolution (Eighteenth to Nineteenth Centuries)

The industrial revolution brought significant changes to meat processing. With the invention of machinery, mass production became possible. Meat packing plants emerged and innovations like refrigeration and canning further revolutionized the industry. In fact some 50 years before Louis Pasteur demonstrated the role of microbes in food spoilage and developed pasteurization, Nicholas Appert, a French confectioner, developed a process of preserving food sealed and cooked inside a jar in 1809 for which he received an official commendation from the French government. The modern food processing traces its origin to the development of canning.

water (Pal and Devrani 2018; Lawrie and Ledward 2006; Teodorovic et al. 2015). A new approach, which is a variation of super-fast chilling, is stepwise chilling. This process includes three steps: fast chilling, tempering and fast chilling again. This approach leads to 5% energy cost savings compared to super-fast chilling, but it also results in meat tenderization because of meat enzyme activation during the tempering phase (Kristensen et al. 2014). A future perspective could be so-called pad chilling, based on the direct contact of the animal carcass on a cold surface, whereby the temperature is lowered through the conduction process and not evaporation like in conventional chilling. This process is 30% faster, chill loss is very low (0.1–0.2%) and the energy cost saving is about 50% compared to conventional chilling methods (Kristensen et al. 2014). As a newer method that is between chilling and freezing, super chilling involves initial freezing of the surface, then meat is exposed to equalization of ice crystals at only 1–2 °C below the freezing temperature. The shelf life of such meat could be up to four times longer than that of chilled meat. Furthermore, because there is no need for any thawing process, this method reduces costs in labour, energy and weight loss, as well as resulting in meat with better sensory characteristics than frozen meat has after thawing. However, this method requires more precise temperature control and monitoring systems because of the narrow range between super chilling and chilling temperatures (Zhou et al. 2010). Freezing of meat expanded at the end of nineteenth century, as a method that provided long shelf life of meat and long distance transport from one part of the world to another. Although meat storage below −18 °C stops all microbial growth and cellular metabolism, lipid oxidation remains the main problem during storage. Adequate packaging films could help in lowering the oxidation processes and the incidence of freezer burns on meat surfaces (Pal and Devrani 2018; Lawrie and Ledward 2006; Teodorovic et al. 2015). A relatively newer method, cryogenic freezing is faster than air freezing and requires only tanks or sprays for cryogen application. However, it is suitable only for smaller and packed meat pieces and the cost of the cryogenic liquid makes this method relatively limited in commercial use (Zhou et al. 2010).

1.7.1.2 Salting and Curing

Salting was recognized very early in human civilization as a method of prolonging meat's shelf life without chilling (Lawrie and Ledward 2006). Curing also arose long ago in history, when ancient civilizations incorporated nitrates in meat via contaminated salt, but it took a long time until the actual chemical processes in cured meat were completely understood (Teodorovic et al. 2015; Sindelar and Milkowski 2011). At present, nitrite plays a very important role in the meat industry, in the first place because of its strong antimicrobial activity, especially against *Clostridium botulinum*, as well as it being a colouring agent. Recent studies, however, proved the carcinogenic potential of N-nitrosamines that can be formed in cured meat products, which provoked a series of investigation in order to find nitrite substitutes. Promising results were obtained with essential oils, polyphenols, lactic acid bacteria and acid whey, etc. (Vasilev et al. 2017b), but these single ingredient solutions could provide only one of the main functions of nitrite in meat products. As it is also known that endogenous production of nitric compounds in the human body occurs physiologically from other sources of nitrate and nitrosamines not from meat products, the exclusion of the use of nitrites is not yet in sight (Sindelar and Milkowski 2011). Additionally, although some traditional meat products are produced without the use of curing salts, many of these products contain nitrate and nitrite residues which are indirectly incorporated in the products via spices (Vuković et al. 2011, 2014). Such products are often labelled as "uncured" or "without preservatives" which could be misleading for consumers in terms of nitrite presence in the product (Sindelar and Milkowski 2011) and in terms of product labelling (Henderikx 2017). As for the use of NaCl, some meat products are a significant source of high sodium intake in the human diet, which is recognized as cardiovascular disease promoting factor. Partial replacement of NaCl with K, Ca and Mg salts (chloride, lactate etc.) could reduce the sodium content in meat products without adverse effects on products' properties or safety (Vasilev et al. 2016, 2017b).

1.7.1.3 Smoking

Smoking of foods, including meat, originated thousands of years ago and is a preservation method based on chemical compounds from smoke that influence the sensory properties of meat products and have antimicrobial and antioxidative roles on the meat surface (Teodorovic et al. 2015; McDonald 2015). Recently, emphasis has been given to the investigation of harmful smoke compounds such as polycyclic aromatic hydrocarbons (PAHs), because of their carcinogenic potential. There are several approaches in order to produce smoked meat products with less or with no PAHs, which include measures to obtain smoke with less PAHs by the controlled process of wood pyrolysis, smoke purification, or even to reduce the content of smoke in meat products by means of starter cultures or some spices (Vasilev et al. 2017). The other approach to obtain smoked meat products free of PAHs is to use smoke flavourings. As the toxic compounds are not soluble in the water phase of the liquid smoke, the compounds are easily removed, producing smoke flavourings with no detectable amounts of PAHs (McDonald 2015).

1.7.1.4 Drying

Along with salting and smoking, drying is one of the oldest meat processing methods. Until the industrial revolution, meat was dried under the influence of natural climate conditions, but during the last century, the use of climate chambers became irreplaceable in the meat industry. However, as climate chambers are significant energy consumers, this fact provoked investigations on shortening the drying process or introducing new air drying systems (Kvalsvik 2015). Some approaches to shorten the drying process are microwave vacuum drying, which is a process combining microwave heating and vacuum drying (Pawlak et al. 2018) and ultrasonic vacuum drying (Başlar et al. 2014). Drying in a vacuum chamber is more than three times faster than drying under usual conditions and the meat's sensory properties are not affected, while microbiological stability is achieved more rapidly (Vukovic and Klettner 2004). As for air drying systems, a heat pump dryer is more efficient than hot air drying systems. Among heat pump drying systems, the best appear to be CO_2-heat pumps, followed by glycol heat pumps, while ammonia heat pumps were less effective (Kvalsvik 2015).

1.7.1.5 Heat Treatment

Although early humans used fire to prepare meat for consumption, the commercial thermal processing of meat started at the beginning of nineteenth century with Nicolas Appert and design of the first canned products (Lawrie and Ledward 2006). Heat treatment provides safety, shelf life, sensory properties and better digestibility of meat and depending on the chosen temperature intensity, pasteurization, cooking or sterilization can result (Raseta et al. 2018). Hermetically sealed cans are a good environment for growth of anaerobic bacteria, with special attention paid to toxigenic *Clostridium botulinum*. Therefore, process control of canning gained its importance, which resulted in development of Fo-values. These precisely measure whether the canning process ensures meat safety with respect to *C. botulinum* (Fo $\geq$ 3) (Vasilev and Vukovic 2008; Lawrie and Ledward 2006; Teodorovic et al. 2015). Sterilized cans are usually overheated in practice (Fo $>$ 8) in order to ensure product safety, but such processing significantly affects the nutritive value of the canned products. Such unnecessarily strong heat treatment could be optimized by Fo-value and Co-value (cooking value) determinations, in order to provide safety and preserve the nutritive value as much as possible (Raseta et al. 2018). Furthermore, it is possible to produce shelf stable meat products by mild heat treatment and which can be stored at room temperature, by means of the hurdle technology concept based on the simultaneous action of several antimicrobial parameters (pH, water activity, redox potential, Fo-value, temperature). This approach led to the development of the so-called refrigerated processed foods of extended durability, where the accent is on mild heat treatment and prolonged shelf life (Vasilev and Vukovic 2008).

1.7.1.6 Packaging

The main purpose of packaging was firstly just to mechanically protect products from contamination, over-drying, oxidation, weight and nutrient losses, adsorbing strange odours from the environment, etc., simply by overwrapping with packaging material. With the development of polymer materials with low gas permeability and the technology to remove air from the packaging, vacuum packaging emerged, which additionally provides reduced oxidative changes and limited growth of aerobic bacteria. This led further to the development of modified atmosphere packaging, which provides antimicrobial effects due to the high CO_2 concentration in the packs. Another novel method is active packaging, where active components are purposefully incorporated into the packaging material. There are several variations of active packaging: antimicrobial packaging which releases antimicrobial substances during storage (silver ions, bacteriocins, essential oils, nisin, etc.), antioxidative packaging containing oxygen scavengers and finally, intelligent packaging which contains sensors and indicators for detection of gas, temperature abuse, package integrity, etc., and which utilise colour changes to inform the consumer about the state of the products (Zhou et al. 2010; Dimitrijevic et al. 2015). As active packaging is mostly based on nanotechnology, there are still issues concerning the lack of knowledge about the impact of nanomaterials on human health and their occurrence and destiny in the environment, which should be further, studied (Dimitrijevic et al. 2015).

1.7.2 Novel Meat Processing Methods

1.7.2.1 Irradiation

Food irradiation was developed during the second half of the twentieth century (Yousefi and Razdari 2015) and involves exposure of food to ionizing irradiation (Gamma rays, Electron beams) in order to promote safety and prolong the food's shelf life, but also refers to product quality examination using computer tomography (CT) based on X rays. Gamma rays are obtained from radioactive isotopes Co_{60} or Cs_{137} with relatively short half-lives (5.27 and 30.19 years respectively), with electron beams produced in an electron accelerator and X-rays produced by slamming electrons into metal (tantalum or platinum). Irradiation dose refers to the amount of absorbed energy in Greys (Gy) (Yousefi and Razdari 2015; Ibrahiim 2013). Irradiation is a highly effective, cold, penetrative and relatively easy to control process, which does not affect the sensory or nutritive properties of food if properly used. One problem can be lipid oxidation in high fat foods, but simultaneous use of antioxidants can prevent this (Yousefi and Razdari 2015). Although food irradiation has been approved since 1989 by the USDA and FDA, nowadays, about 30% of countries worldwide have adopted it as a commercial method of food preservation, led by the USA, China, The Netherlands, Belgium, Brazil, Thailand and Australia. The types of foods irradiated are mostly spices, crops, vegetables, fruits, etc. and to a smaller extent, meat, mostly ground beef, poultry, fish and seafood. The main obstacle for wider use is acceptance by consumers but this could be overcome by their education and proper labelling of irradiated food (Maherani et al. 2016).

1.7.2.2 High Pressure Processing

The idea of high pressure implemented as a food preservation method started at the end of the nineteenth century with the first studies, but it gained importance only during recent decades because of the rising demand for food with preserved nutritive properties (Campus 2010). High pressure processing (HPP) includes the use of a pressure chamber at 100–600 MPa, transmitted via water as the pressure transfer medium and conducted at room temperature. The preservative effect is based on the destruction of microorganisms as well enzyme inactivation, while nutritive value is not affected. In fresh meat, pressure above 150 MPa starts to change the colour by affecting myoglobin, so HPP is more appropriate for cured meat products such as cooked and dry cured ham. Because of the differences in bacterial resistances to high pressure, hurdle technology using a combination of other antimicrobial parameters could provide a stronger preservation effect at low pressure (Hugas et al. 2002). HPP could have special importance as a post packaging processing step in the control of *Listeria monocytogenes* in ready-to-eat products.

Furthermore, the fact that this technology is environmental friendly and waste free could contribute to its greater importance in the future (Campus 2010). However, it is considered that the possibilities of HPP are still underestimated because of the lack of activity from large companies and limited commercial research as well as consumer awareness of its benefits (Abera 2019).

1.7.2.3 Pulsed Electric Field

Pulsed electric field (PEF) is based on the discharge of high voltage short electric pulses that leads to permeabilization of cell membranes of microorganisms without negative impact on the nutritive value of food. It is mainly used for liquid foods and has no wide use in the meat industry (Stoica et al. 2013), but some studies indicate some interesting approaches concerning the use of PEF (1–3 kV/cm in 100 pulses) as a support for drying, marinating and salting of meat. Specifically, PEF could increase both the speed of salt diffusion and drying during production of dry cured ham, provide faster fermentation in fermented sausages and improve brine distribution in cooked ham production (Toepfl et al. 2006).

1.7.2.4 Pulsed Light

Pulsed light (PL) is based on short time light pulses which are capable of inactivating microorganisms on food surfaces. The antimicrobial effect is provided through the UV spectra of the light, which damages the microbial DNA. PL treatment includes 1–20 light flashes per second, applying energy from 0.01 to 50 J/cm^2 (Mahendran et al. 2019). There are several possible uses in the meat industry, such as decontamination of carcasses (skin and meat) (Koch et al. 2019), improved safety of fresh food products such as beef and tuna carpaccio (Hierro et al. 2012), decontamination of equipment (knives) after being in contact with meat and meat products (Rajkovic et al. 2010) and sliced fermented sausages (Rajkovic et al. 2017). As too intensively pulsed light could lead to changes in sensory properties of meat, especially colour and aroma, there are some limitations concerning the intensity of the treatment which should be optimised according to the number of flashes, voltage, spectral range and the distance between the product and the light source. Another concern is the economic aspect because of the need for specific equipment (Mahendran et al. 2019).

1.7.2.5 Cold Plasma

Cold plasma (CP) consists of an ionized gas that contains ions (+ and −), electrons, free radicals and photons. Such accumulated charged particles act destructively on microorganisms, providing the antimicrobial effect. CP could serve for decontamination of surfaces of equipment and meat and meat products (Stoica et al. 2013), but as some adverse effects on fresh meat colour were observed and as lipid oxidation and a certain degree of off-flavour developed, a balance must be found between the antimicrobial effect and sensory qualities of the product (Misra and Jo 2017). An interesting approach could be application of plasma-treated water (PTW) as a curing agent in emulsion type sausages, because plasma treatment of water generates NO^{2-} ions which could give colour to the sausages as if they were treated with nitrite (Jung et al. 2015).

1.8 Functional Meat Products

The concept of functional food has gained interest with consumers' awareness of the influence of nutrition on their health and includes the approach of food modification to reduce the amount of potential harmful components and enrich food with health promoting components. As meat products are often described as potentially harmful because of their saturated fat and salt contents, as well as the moieties formed through smoking and curing, most research in this area deals with these issues (Vasilev et al. 2017). Saturated fat reduction is based on the replacement of animal fat with prebiotics, gums, starches, plant proteins, etc. (Vasilev et al. 2017, 2009) and with emulsion systems containing oils rich in polyunsaturated fatty acids (PUFAs) (Vasilev et al. 2011; Glisic et al. 2019). As for enrichment of meat products with functional ingredients, probiotics could be successfully implemented in fermented sausages, prebiotics both in fermented and heat treated products simultaneously serving as animal fat replacers, minerals such as salts of Ca, Mg or K could be used as partial NaCl replacers and at the same time, would serve as functional ingredients (Vasilev et al. 2016; Vasilev et al. 2009). The main problems concerning functional meat products enriched with PUFAs are their proneness to lipid oxidation

Mahendran R, Ramanan KR, Barba FJ, Lorenzo JM, López-Fernández O, Munekata PES, Roohinejad S, Sant'Ana AS, Tiwari BK (2019) Recent advances in the application of pulsed light processing for improving food safety and increasing shelf life. Trends Food Sci Technol 88:67–79
Maherani B, Hossain F, Criado P, Ben-Fadhel Y, Salmieri S, Lacroix M (2016) World market development and consumer acceptance of irradiation technology. Foods 5:79
McDonald ST (2015) Comparison of health risks of smoked foods as compared to smoke flavorings: are smoke flavors "healthier"? Adv Food Technol Nutr Sci 1:130–134
Meixner HW, Karnitzschky I (2001) Warmfleisch richtig verpackt. Fleischwirtsch 81:30–31
Misra NN, Jo C (2017) Applications of cold plasma technology for microbiological safety in meat industry. Trends Food Sci Technol 64:74–86
Mottram DS (1998) Flavour formation in meat and meat products: a review. Food Chem 62:415–424
OECD/FAO (2023) OECD-FAO agricultural outlook 2023–2032 185.
Pal M, Devrani M (2018) Application of various techniques for meat preservation. J Exp Food Chem 4:134
Pawlak T, Gawałek J, Ryniecki A, Stangierski J, Siatkowski I, Peplińska B, Pospiech E (2018) Microwave vacuum drying and puffing of the meat tissue – process analysis. Dry Technol. https://doi.org/10.1080/07373937.2018.1444635
Pederson CS (1979) Microbiology of food fermentations. The Avi Publishing Company, Inc, Westport/ Connecticut (USA)
Pisula A, Tyburcy A (1996) Hot processing of meat. Meat Sci 43:125–134
Rajkovic A, Tomasevic I, Smigic N, Uyttendaele M, Radovanovic R, Devileghere F (2010) Pulsed UV light as an intervention strategy against Listeria monocytogenes and Escherichia coli O157:H7 on the surface a meat slicing knife. J Food Eng 100:446–451
Rajkovic A, Tomasevic I, De Meulenaer B, Devlieghere F (2017) The effect of pulsed UV light on Escherichia coli O157:H7, Listeria monocytogenes, Salmonella Typhimurium, Staphylococcus aureus and staphylococcal enterotoxin A on sliced fermented salami and its chemical quality. Food Control 73:829–837
Raseta M, Mrdovic B, Djordjevic V, Polacek V, Becskei Z, Brankovic-Lazic I, Vasilev D (2018) Determination of co-value as an indicator of nutritive value of pâté sterilised by regular and optimized regime. Vet Glas 72:101–111
Savell JW, Mueller SL, Baird BE (2005) The chilling of carcasses. Meat Sci 70:449–459
Shahbandeh M (2023) Production of meat worldwide from 2016 to 2023 (in million metric tons). Statista Agriculture/Farming-2023
Simmons NJ, Daly CC, Cummings TL, Morgan SK, Johnson NV, Lombard A (2008) Reassessing the principles of electrical stimulation. Meat Sci 80:110–122
Sindelar JJ, Milkowski AL (2011) Sodium nitrite in processed meat and poultry meats: a review of curing and examining the risk/benefit of its use AMSA. White Paper Series 3:1–14
Sørheim O, Hildrum KI (2002) Muscle stretching techniques for improving meat tenderness. Trends Food Sci Technol 13:127–135
Sorinmade SO, Cross HR, Ono K, Wergin (1982) Mechanisms of ultrastructural changes in electrically stimulated beef longissimus. Meat Sci 6:71
Stoica M, Mihalcea L, Borda D, Alexe P (2013) Non-thermal novel food processing technologies: an overview. J Agroaliment Processes Technol 19:212–217
Teodorovic V, Karabasil N, Dimitrijevic M, Vasilev D (2015) Higijena i Tehnologija Mesa (Meat hygiene and technology). Naucna KMD, Belgrade
Thompson J (2009) Origins of the meat industry. In: Meat technology, pp 1–10
Toepfl S, Heinz V and Knorr D (2006) Pulsed Electric Fields (PEF) Processing of Meat (IUFoST) 20,060,591.
Troy DJ (1995) Modern methods to improve and control meat quality International Developments in Process Efficiency and Quality in the Meat Industry Dublin Castle, Ireland (Proceedings) 7–72.

Troy DJ (2006) Hot-boning of meat: a new perspective. In: Leo Nollet LM, Toldrá F (eds) Advanced technologies for meat processing. Publishers CRC Press, Baton Rouge, pp 73–86
Troy D, Kerry J (2010) Consumer perception and the role of science in the meat industry. Meat Sci 86(1):214–226
USDA (United States Department of Agriculture). (2005) www.nal.usda.gov/fnic/foodcomp
Vandendriessche F (2008) Meat products in the past, today and in the future. Meat Sci 78:104–113
Vasilev D, Vukovic I (2008) Hazard analysis and possibilities for preventing botulism originating from meat products. Vet Glas 62:317–328
Vasilev D, Vuković I, Tomović V, Jokanović M, Vasiljević N, Milanović-Stevanović M, Tubić M (2009) Some important physical, physico-chemical and sensory properties of functional fermented sausages. Meat Technol 50:342–350
Vasilev D, Vuković I, Saičić S (2011) Some quality parameters of functional fermented, cooked and liver sausages. Meat Technol 52:141–153
Vasilev D, Jovetić M, Vranić D, Tomović V, Jokanović M, Dimitrijevic M, Karabasil N, Vasiljević N (2016) Qualität und Mikroflora von funktionellen Rohwürsten. Fleiscwirtschaft 96:96–102
Vasilev D, Djordjevic V, Karabasil N, Dimitrijevic M, Petrovic Z, Velebit B, Teodorovic V (2017a) Inulin as a prebiotic and fat replacer in meat products. Theory Pract Meat Process 2:4–13
Vasilev D, Glisic M, Jankovic V, Dimitrijevic M, Karabasil N, Suvajdzic B, Teodorovic V (2017b) Perspectives in production of functional meat products. IOP Conf Ser: Earth Environ Sci 85: 012033
Vukovic I and Klettner PG (2004) Trocknung von Rohschinken in der Klimakammer und im Vakuum mit anschließender Reifung Mitteilungsblatt BAFF 42: 145–150.
Vuković I, Petrović LJ, Vasilev D, Saičić S (2011) Untersuchung von Rohdauerwürsten aus Serbien. Fleischwirtchaft 91:118–122
Vuković I, Vasilev D, Saičić S (2014) Mikroflora und Qualitä Traditioneller Rohwurst Lemeški Kulen. Fleischwirtchaft 94:114–118
West RL (1983) Functional characteristics of hot-boned meat. Food Technol 37:57–66
Yousefi MR, Razdari AM (2015) Irradiation and its potential to food preservation. Int J Adv Biol Biomed Res 3:51–54
Zhang W, Xiao S, Samaraweera H, Lee EJ, Ahn DU (2010) Improving functional value of meat products. Meat Sci 86:15–31
Zhou GH, Xu XL, Liu Y (2010) Preservation technologies for fresh meat – a review. Meat Sci 86: 119–128

the sarcomere and measuring over one μm in length, the titin molecule is anchored with its N-terminus in the Z-disk and its C-terminus in the M-line. Titin features an elastic component in the I-band region, establishing a connection between the end of the myosin filament and the Z-line. The myosin filament portion of titin remains fixed, while the size and structure in the I-band region vary due to different isoforms of the molecule. These isoforms exhibit a range of molecular weights, reported to be approximately 2.4 MDa and up to around 4 MDa in vertebrate skeletal muscle (Lindstedt and Nishikawa 2017). The sarcomeric structural component nebulin may play a regulatory role in muscle contraction. Obscurin, identified as the most recently discovered giant protein in the sarcomere, poses challenges in characterization due to its substantial size and insolubility (Young et al. 2001). With a molecular weight ranging from 720 to 900 kDa, obscurin is not localized within the sarcomere but rather surrounds it at the Z-disk and M-line levels. Obscurin exhibits interactions with myosin, titin, and proteins present in the sarcoplasmic reticulum membranes.

2.3 Composition of Post-mortem Muscle

2.3.1 Water Content

Water is the most essential constituent of meat comprising around 75% of weight and has an important influence on colour, texture and surface appearance (Toldra 2006). According to Juárez et al. (2012), the water content changes inversely with the fat content, while the protein content is constantly maintained inside the muscle. Water is a critical component in meat influencing its quality characteristics. Post-mortem changes in water content play a crucial role in determining meat quality.

1. Post-mortem changes in water content:
   - (a) Rigor mortis and water loss: Rigor mortis is the stiffening of muscles after death. It contributes to initial water loss. The impact of rigor mortis on water content varies with species and muscle types.
   - (b) Protein denaturation and water holding capacity: Changes in protein conformation affect water holding capacity (WHC). Denaturation processes post-mortem influence the ability of meat to retain water.
   - (c) Evaporation and drip loss: Loss of water through evaporation and drip loss occurs during post-mortem storage. Factors such as temperature and humidity influence the extent of water loss.
2. Impact on meat quality
   - (a) Texture and tenderness: Changes in water content affect the texture and tenderness of meat. Proper water retention is crucial for achieving desired sensory attributes.

(b) Colour and appearance: Water content influences the colour stability of meat. Post-mortem changes in water content contribute to variations in meat appearance. Water content is linked to the nutritional composition of meat. Understanding water dynamics aids in assessing the nutritional value of meat.

2.3.2 Protein Composition

Protein is the most precious ingredient in meat from the nutritional and processing meat point of view. It serves as the structural component of the muscular tissue. Amino acids are the basic chemical molecules that make up the complex molecule known as protein. There are 20 amino acids that are essential for human development and metabolic processes. Eight of these amino acids cannot be synthesized in our body and are deemed as essential, indicating that it is necessary to be present in our diets (Geletu et al. 2021). The remaining amino acids are inessential ones. Leucine, isoleucine, lysine, methionine, cystine, phenylalanine, threonine, tryptophan, valine, arginine, and histidine are among the necessary amino acids found in beef proteins; the latter two are thought to be particularly important for newborns (Olaoye 2011). The two major categories of meat proteins: myofibrillar proteins and sarcoplasmic proteins.

(a) Myofibrillar Proteins: Myofibrillar proteins constitute a significant portion of the muscle tissue, comprising approximately 50–55% of the total protein content. These proteins are primarily involved in muscle contraction and contribute to the structural integrity of the meat. Actin and myosin are the major myofibrillar Proteins (Huff-Lonergan and Lonergan 2005).

    Myosin is a motor protein that forms thick filaments and plays a crucial role in muscle contraction. Its interaction with actin generates the force required for muscle movement. Actin, a globular protein, forms thin filaments and is vital for the regulation of muscle contraction. It interacts with myosin to produce the sliding motion observed during contraction. Tropomyosin and Troponin are the regulatory proteins that work in concert to control the interaction between myosin and actin, modulating the contraction-relaxation cycle. Myofibrillar proteins greatly influence the texture, tenderness, and overall quality of meat. Changes in myofibrillar protein composition post-mortem impact the meat's structural integrity, influencing its sensory attributes.

(b) Sarcoplasmic Proteins: Sarcoplasmic proteins represent approximately 30–35% of the total protein content in meat. These proteins are located in the sarcoplasm, the cytoplasm of muscle cells, and contribute to various cellular functions. Sarcoplasmic proteins play a key role in maintaining cellular function and influencing meat colour and flavour. Enzymatic activities post-mortem, influenced by sarcoplasmic proteins, contribute to the maturation process and subsequent changes in meat quality. Myoglobin is major sarcoplasmic heme-containing protein responsible for meat colour. Its colour changes during the emerges as a rich source of polyunsaturated fatty acids, while the brain uniquely contains high concentrations of C-22 polyunsaturated fatty acids. Notably, offal tissues, especially the brain, exhibit higher cholesterol concentrations than muscle tissues (Soladoye et al. 2015).

2.4.1.3 Carbohydrates

The primary reserve of carbohydrates within an animal's body resides in the liver, containing roughly half of the total carbohydrate content. These carbohydrates are primarily stored as "glycogen" in the liver and muscles, albeit in smaller quantities within glands and other organs. Substantial amounts also exist in the bloodstream in the form of glucose. Glycogen significantly influences meat characteristics such as colour, texture, tenderness, and water retention. The intricate process of converting stored glycogen into glucose, and subsequently into lactic acid, is regulated by hormones and enzymes (Kamruzzaman et al. 2016). As muscles age initially, the lactic acid levels increase, leading to a decrease in pH.The pH level strongly impacts muscle texture, tenderness, colour, and water retention. Normally, muscle pH is around 5.6. However, if an animal undergoes severe stress or strenuous exercise prior to slaughter without sufficient time to restore its glycogen levels, only minimal glycogen remains to convert into lactic acid. This situation results in an elevated pH (e.g., 6.5), causing the meat to become dark, firm, and dry (DFD). This condition arises from glycogen depletion due to exhaustion before slaughter (Jensen et al. 2011; Adzitey 2011).

2.4.2 Micronutrients

2.4.2.1 Minerals

Minerals are essential nutrients found in food that lack carbon and play a vital role in the proper development, growth, and maintenance of the human body. They are categorized into two groups: macro- and micro-minerals, based on the body's varying needs. Macro-minerals, such as sodium, calcium, phosphorus, magnesium, chloride, potassium, and sulphur, are required in larger quantities. Conversely, micro-minerals, including iron, zinc, iodine, copper, cobalt, manganese, selenium, and fluoride, are needed in smaller amounts (Yu et al. 2017). Potassium stands out as the most abundant mineral, followed by phosphorus, sodium, and magnesium. Iron, zinc, and selenium are notably abundant in meat. These minerals serve diverse functions in supporting human growth, development, and maintenance, as detailed below (Wood et al. 2008).

2.4.2.2 Vitamins

Vitamins encompass a group of organic compounds that play multifaceted roles within the human body. Despite their requirement in small quantities, they are pivotal for proper growth, development, and maintenance. Particularly crucial during early childhood, these elements actively participate in a multitude of metabolic processes, driving various chemical and biochemical reactions. A defining characteristic is their inability to be synthesized by mammalian cells, necessitating their intake through dietary sources (Grunert et al. 2004). Vitamins are commonly categorized into water-soluble (such as B-complex vitamins like thiamin, riboflavin, nicotinic acid, pyridoxine, choline, biotin, folic acid, cyanocobalamin, inositol, vitamin-B6, vitamin-B12, and vitamin C) and fat-soluble types (like vitamin A, vitamin D, and vitamin K). In the context of meat, it serves as a significant source for five B-complex vitamins—thiamin, riboflavin, nicotinic acid, vitamin B6, and vitamin B12. Additionally, it contains pantothenic acid and biotin, albeit with a limited folacin content (Woods and Fearon 2009).

2.4.3 *Energy Content of Post-mortem Muscle*

2.4.3.1 Caloric Value

Lean meat primarily comprises three key components: water (constituting 75%), protein (19%), and fat (2.5%). The residual 3.5% consists of minor components like carbohydrates (approximately 1%), soluble nitrogenous substances, minerals, and vitamins (Lawrie 1998). These composition values are subject to variation due to multiple factors, both internal (such as muscle type, breed, sex, and species) and external (like feeding patterns). Particularly in processed meat, the main components can exhibit even greater variability. Carbohydrates are generally present in minimal quantities, usually less than 1%, across most meat products, contributing insignificantly to their overall calorific value.

In fresh meat, proteins, offering 4 kcal/g, contribute roughly 80 kcal/100 g, yet in many processed meat products, the protein content diminishes to 15% or lower. Fat content in fresh meat typically ranges between 1% and 2%, while in certain meat products, it can surge to 25% or exceed 30% (Keeton 1994) Consequently, the fat's calorific contribution to a 100 g serving of meat, estimated at 8 kcal/g of adipose tissue (considering it's approximately 90% fat and 10% water), may fluctuate from around 10–15% up to over 80%. This implies total calories ranging from approximately 100 to over 300 kcal.

Assessing the calorific value of meats necessitates considering meat consumption patterns within affluent societies' average diets. Per capita meat consumption is often calculated from meat production data that encompass bones, fatty tissues, organs, trimmings, and waste of diverse kinds. If inedible portions are excluded, roughly 60% of meat production is actually used for human consumption (Honikel 1992;

Klurfeld 1994). For example, in a country like Germany, real meat consumption averages around 178 g/day, inclusive of meat products. The average composition would consist of approximately 17% protein and 20% fat, with fresh meat accounting for 8.6% of the fat and meat products for 24.3%. Thus, meats contribute around 240 kcal/100 g or slightly over 400 kcal/day (Rourke et al. 1992). Similar statistics have been reported for the United States. (Klurfeld 1994).

While total meat consumption varies among countries, the average composition of meat remains relatively consistent. Consequently, meat's contribution to caloric intake in most developed countries is estimated to range from 10% to 20% of total calories. Other researchers have also reported comparable findings regarding meat consumption and its caloric contribution (Klurfeld 1994).

2.4.3.2 Impact on Dietary Consideration

The alterations observed in meat after death, particularly in muscle tissue, play a crucial role in its dietary implications. These modifications encompass changes in nutritional content, texture, flavour, and overall quality.

- Protein Denaturation and Structural Modifications: After death, muscle proteins undergo denaturation and structural alterations, influencing meat texture (Offer and Trinick 1983).
- Metmyoglobin Formation: The transformation of myoglobin into metmyoglobin post-mortem results in the discoloration (browning) of meat (Faustman and Cassens 1990).
- Lipid Oxidation: Following death, meat lipids are prone to oxidation, causing shifts in flavour, aroma, and nutritional value (Decker and Park 2010).
- Alterations in Nutrient Composition: Post-mortem changes can impact meat's nutritional profile, encompassing variations in moisture, protein, fat, and other essential nutrients (Savell et al. 1986).
- Tenderness Enhancement: Post-mortem aging, or conditioning, improves meat tenderness through natural enzymatic processes (Shackelford et al. 2001).
- Enzymatic Influence on Texture and Flavour: Enzymatic activities after death significantly influence the development of meat's texture and flavour.

2.5 Biochemistry of Post-mortem Muscle

2.5.1 Rigor Mortis

The quality of meat heavily depends on muscle-to-meat conversion, impacted by antemortem and postmortem changes affecting tenderness, juiciness, and colour. Optimal conditions during this conversion enhance the likelihood of delivering high-quality products. Harvesting triggers cellular-level changes, and minimizing the time between restraint, immobilization, and exsanguination aids the transition from ante mortem to post-mortem metabolism.

During exsanguination, about 50% of blood is lost, affecting oxygen supply and leading to a shift from aerobic respiration to anaerobic glycolysis, causing a gradual decrease in muscle pH. Lactic acid accumulation leads to ATP depletion, resulting in rigor mortis. Factors like increased lactic acid levels and ante mortem stress influence rigor onset, affecting tissue beyond.

The living muscle's functional state provides insights into transformations during harvesting. Normal muscle contraction relies on a balance of chemical matrices regulated by the endocrine and nervous systems. ATP, generated through metabolic pathways, sustains muscle contraction. Post-mortem, changes in metabolic pathways impact meat quality (Leyton et al. 1975).

Exsanguination removes 50% of blood, depleting muscle oxygen. Lack of blood flow hinders heat dissipation, potentially accelerating rigor. Muscle fiber type influences metabolism, with faster twitch fibers entering rigor faster. Elimination of the circulatory system impacts temperature control and affects postmortem muscle changes. Muscle flexibility's return, post-rigor, is not fully understood but is presumed to involve damage to muscle proteins. The process of rigor mortis can be explained in several steps:

1. Depletion of ATP (Adenosine Triphosphate): ATP is the energy currency of cells, and it is required for muscle contraction. After death, the production of ATP ceases, and existing ATP is rapidly used.
2. No ATP for cross-bridge detachment: Muscle contraction involves the interaction between actin and myosin filaments, forming cross-bridges. ATP is necessary for the detachment of myosin heads from actin filaments during muscle relaxation. In the absence of ATP, these cross-bridges cannot detach, resulting in the stiffening of muscles.
3. Increased calcium ion concentration: In living muscle cells, calcium ions are stored in the sarcoplasmic reticulum and are released during muscle contraction. After death, the cell membrane becomes permeable, allowing calcium ions to flood the cytoplasm. Elevated calcium levels activate enzymes that break down ATP, further depleting the available ATP.
4. Formation of cross-bridges: In the absence of ATP, cross-bridges between myosin and actin filaments form and cannot be broken. This cross-bridge formation leads to the stiffening of the muscle tissue. The process of rigor mortis begins within a few hours after death, peaks at around 12 h, and gradually diminishes over the next 48–72 h as autolysis and decay set in (Visschers et al. 1975).

2.5.2 *Proteolysis*

Proteolysis constitutes an important group of reactions during the processing of meat and meat products. In fact, proteolysis has a high impact on texture, and thus meat tenderness, because it contributes to the breakdown of the myofibrillar proteins responsible for muscle network, but proteolysis also generates peptides and free amino acids that have a direct influence on taste and also act as substrates for further reactions contributing to aroma (Toldrá 2002). In general, proteolysis has several consecutive stages as follows:

(1) action of calpains and cathepsins on major myofibrillar proteins, generating protein fragments and intermediate size polypeptides;
(2) these generated fragments and polypeptides are further hydrolzyed to small peptides by DPPs and TPPs; and
(3) dipeptidases, aminopeptidases, and carboxypeptidases are the last proteolytic enzymes that act on previous polypeptides and peptides to generate free amino acids. The progress of proteolysis varies depending on the processing conditions, the type of muscle, and the amount of endogenous proteolytic enzymes.

2.5.2.1 Proteolysis in Aged Meat and Cooked Meat Products

During meat aging, there is proteolysis of important myofibrillar proteins such as troponin T by calpains, with the associated release of a characteristic 30 kDa fragment, which is associated with meat tenderness, nebulin, desmin, titin, troponin I, myosin heavy chain, and proteins at the Z-line level (Toldrá 2012). A 95 kDa fragment is also characteristically generated (Koohmaraie 1994). The fastest aging rate is observed in chicken, followed by pork, lamb, and beef. For instance, it has been reported that chicken myofibrils are easily damaged by cathepsin L, whereas beef myofibrils are much more resistant. The reasons for this difference are the differences between species in enzymatic activity, inhibitor content, and susceptibility to proteolysis of the myofibrillar structure. These factors are also strongly linked to the type of muscle metabolism, which has a strong influence on the aging rate (Toldrá 2006a). In fact, proteolysis is faster in fast-twitch white fibers (which contract rapidly) than in slow-twitch red fibers (which contract slowly), that is, aging rate increases with increasing speed of contraction but decreases with the level of heme iron.

Even though the effect of muscle type is estimated to be tenfold lower than the effect of temperature, it is threefold higher than animal effects. Of these conditions, the most significant are pH, which decreases once the animal is slaughtered, and osmolality, which increases from 300 to around 550 mOsm within 2 days, due to the release of ions to the cytosol. The osmolality has been observed to be higher in fast-twitch muscle, which also experiences a faster aging rate (Valin and Ouali 1992). Age of the animal also decreases the aging rate, because collagen content as well as the crosslinks that make the muscle more heat stable and mechanically resistant, increase with age.

2.5.3 *Lipolysis*

Lipolysis makes an important contribution to the quality of meat products by the generation of free fatty acids, some of which have a direct influence on flavour, and others that, with poly-unsaturations, may be oxidized to volatile aromatic compounds, acting as flavour precursors. In addition, the breakdown of triacylglycerols affects the texture of the adipose tissue, and an excess of lipolysis/oxidation may contribute to the development of rancid aromas or yellowish colors in fat (Toldrá and Flores 1998).

2.5.3.1 Role of Lipid Oxidation in Flavour or Off-Flavour Development

Lipids are susceptible to oxidation, which is believed to play the main non-microbial role in the quality degradation of food and food products including meat, fish, and oil-based products. Usually, off-flavours are linked very closely to lipids compared to proteins and carbohydrates. Lipid oxidation not only lessens the nutritional benefits of products due to the changes in essential fatty acids and vitamins but also negatively affects the sensory qualities such as flavour, texture, and colour, which impacts on the overall consumer acceptance (Shahidi and Abad 2019; Hossain et al. 2020). However, in some cases, lipid oxidation promotes the development of pleasant aromas, mainly during the ripening or dry-cured stages of meat products, which is one of the quality parameters of these products (Domínguez et al. 2019). Lipid oxidation is a very complex phenomenon that undergoes a variety of reactions and produces a wide range of compounds. In a nutshell, fatty acids, mainly PUFAs, react with molecular oxygen and produce primary oxidation products (hydroperoxides) via free radical mechanisms. These compounds are highly unstable due to a weak oxygen–oxygen bond and are believed to make no contribution to odour and aroma development, but they break down rapidly and produce a wide variety of secondary constituents, which are responsible for the formation of off-flavours (Shahidi 2000). However, not all of these constituents similarly contribute to the flavour profiles, since the overall aroma perception is mainly dependent on the olfactory threshold, concentration, and type of products. Among these compounds, aldehydes seem to be the major compounds contributing to flavour development due to their low odour threshold and higher concentration. The most common aldehydes generated from lipid oxidation are *n*-alkanals, 4-hydroxy-2-alkenals, 2-alkenals, and malondialdehyde (MDA) (Domínguez et al. 2019). Hence, lipid oxidation can occur in various pathways including autoxidation, photooxidation, thermal oxidation, and enzyme-catalyzed oxidation. Autoxidation is a very common lipid oxidation pathway, occurring through a continuous free-radical chain reaction. However, hydroperoxides are formed during the initiation stage of autoxidation, which is the main difference between the mechanisms of photo and enzymatic oxidation (Shahidi and Zhong 2015).

2.5.4 *Factors Influencing Post-mortem Muscle Characteristics*

2.5.4.1 Pre slaughter Factors

Stress and Its Effects

Stress occurs when an animal is exposed to intrinsic or extrinsic stressors, which may disturb the normal physiological equilibrium, thereby eliciting a threatened homeostatic state (Wiepkema and Koolhaas 1993). The transportation of animals to or from a farm, along with the mixing of groups and confinement before slaughter, unavoidably induces stress. This stress response, in turn, affects the physiological and metabolic functions of the animals, influencing postmortem biochemical changes (Cassens et al. 1975). Therefore, the type, duration, and intensity of the stressors and the susceptibility of the animal to them are of great importance for animal welfare and meat quality development.

Intense pre-slaughter stress has been associated with significant meat quality defects, such as bruises and hemorrhages, as well as an elevated risk of pathogen spread (Barbut et al. 2008; Bianchi et al. 2006). The stress response is recognized as complex and multivariate, which might be affected by the interactions among stressors, which leads to unpredictable responses in animals (Bianchi et al. 2006; Dokmanovic et al. 2015). Animals may undergo various challenges, such as fear, dehydration, hunger, heightened fatigue, and physical injuries, depending on the source of stress. These factors can lead to potential alterations in their energy balance, intracellular ion levels, protease system, and skeletal muscle proteins (Ouali et al. 2006; Scheffler and Gerrard 2007). These factors are all significant, affecting the conversion of muscle into meat.

Stress Factors That Affect Meat Quality

From the farm to abattoir, multiple stressors, including environmental, nutritional, and pre-slaughter handling stress, impact the quality of meat. Several research studies have shown that individual or combined stressors can induce alterations in both the behavioural and physiological condition of animals before slaughter. These changes, in turn, influence muscle metabolites, subsequently affecting meat characteristics and the overall welfare of the animals (Xing et al. 2019).

Heat Stress

For all species, responses to environmental changes linked to temperature, particularly heat stress, play a vital role. Instances of mortality related to heat stress have been documented in various livestock, such as deaths in deer during pre-slaughter transportation, sheep exposed to sunlight, and elevated death rates upon arrival for cattle and poultry transported under high ambient temperatures. Furthermore, heat stress has implications for both the quality and safety of meat. Elevated temperatures, particularly during warmer seasons, heighten the risk of heat stress, leading to the development of pale, soft, and exudative (PSE) pork (dalla Costa et al. 2007).

Pre-slaughter Handling Stress

As large-scale production and intensified animal husbandry continue to evolve, the livestock and poultry industries are increasingly focused on understanding the effects of pre-slaughter conditions on both meat quality and animal welfare. The pre-slaughter handling process involves a sequence of events that can impose significant physical and psychological stress on animals, resulting in heightened respiratory rates, elevated heart rates, anxiety, and irritability. Additionally, there is an excessive increase in nutrient and water consumption, along with the release of hormones, leading to shifts in plasma biochemical indicators and disruptions in energy metabolism. These changes have negative implications for animal welfare, production efficiency, and meat quality (Adzitey 2011; Ferguson and Warner 2008). Recent research has explored the impact of pre-slaughter handling practices, encompassing aspects such as food/water deprivation, transportation, lairage, stunning, crating, shackling, and other handling procedures, on the stress levels and meat quality of animals (Xing et al. 2019).

Oxidative Stress

The various pre-slaughter stress factors contribute to oxidative stress, resulting in the degradation of meat quality. For instance, transport-related stress elevates serum MDA concentrations, which are associated with instances of bovine respiratory disease and mortality in calves (Chirase et al. 2004). The transportation of broilers in high ambient temperatures results in the overproduction of reactive oxygen species (ROS) and subsequent excessive oxidation, contributing to the development of meat with characteristics similar to pale, soft, and exudative (PSE) meat (Xing et al. 2017).

Animal Age and Type

As used in meat grading standards, maturity is defined as the physiological age of the animals or birds from which carcasses are derived. The maturity or age at slaughter is very closely related to meat tenderness. In general, tenderness decreases as age increases. Consequently, meat from a young animal is generally more tender than that from a mature animal of the same species and breed. This phenomenon is primarily attributed to alterations in inter-molecular cross-linkages within collagen fibers (Cianciullo 2013). In young animals, there are fewer cross-linkages, and they are more susceptible to breaking. As the animal matures, the number of cross-linkages rises, transforming them into more stable forms. As a result, collagen in meat from young animals is more soluble during cooking, while it becomes less soluble as the animal ages. The age of an animal significantly influences the colour and flavour of its meat (Mashele 2017). As the animal ages, the muscle colour tends to darken, primarily because of the rising concentration of myoglobin (Mb). When utilized in processed meats, muscles with higher myoglobin concentration, such as those from leg muscles, play a positive role in achieving desirable product colour. Leg muscles exhibit a deeper red hue compared to loin, and myoglobin concentration tends to increase with age. Consequently, veal and lamb meats appear paler in comparison to beef. Mb concentration is low in rapidly respiring animal like rabbit (0.02%). Myoglobin (Mb) constitutes 0.7% or more of the muscle weight in horses and reaches levels of 5–8% in seals and whales. As animals age, there is an increase in meat flavour intensity. This change in flavour is primarily attributed to the elevated concentration of nucleotides in the muscle, which undergo degradation to form inosinic acid and hypoxanthine post mortem (Abhijith et al. 2021).

Breed

Meat quality varies among different breeds within the same species. Traditionally, superior quality beef is frequently derived from specific beef breeds such as Hereford and Aberdeen Angus. These breeds, known for early maturation, tend to accumulate subcutaneous and intra-muscular fats, including marbling fat, in carcasses at an earlier age compared to dairy breeds such as Friesian, Holstein, and Jersey. Beef breeds are known for yielding a higher meat-to-bone ratio, resulting in meat that is more tender, juicy, and flavorful, primarily due to a greater degree of marbling. Inherent amounts of intra-muscular fat vary among different breeds (Martín et al. 2022). Notable exotic beef breeds, in addition to Hereford and Aberdeen Angus, include Belgian Blue, Limousin, Chamlais, and others. Meat from humped indigenous cattle (*Bos indicus*) is generally tougher compared to that from European breeds (*Bos taurus*) such as Hereford and Aberdeen Angus. In the context of Indian goat breeds, they are categorized into milch type, meat type, and fibre type based on their production potential. Notable meat type goat breeds include Bengal, Barbari, Gohilwadi, Kutchi, Malabari (Tellicherry), Marwari, Mehasana, Osmanabadi, Sangamneri, and Zalawadi. Likewise, significant sheep breeds for mutton production include Bikaneri, Magra, Malpura, Mandya, Muzaffarnagari, Madras red, Nellore, and others (Rodrigues et al. 2017).

2.5.4.2 Post-slaughter Factors

Temperature and Storage Conditions

The rate of post-mortem chemical reactions in muscles is significantly influenced by temperature. Consequently, the temperatures at which freshly dressed carcasses/meats are stored have a considerable impact on the functional and eating qualities of meat. Swiftly reducing muscle temperature after slaughter is always preferred to minimize protein denaturation and inhibit microbial growth in meat. However, two unfavorable conditions, referred to as 'cold shortening' and 'thaw rigor,' can occur in muscles due to low temperatures before the onset of rigor mortis (Maribo et al. 1998). In addition to the mentioned alterations, storing meat at low temperatures leads to enhancements in tenderness, juiciness, and flavour, a phenomenon known as 'ageing' or 'conditioning.' This is a crucial process that significantly influences the palatability of meat, especially in the case of beef. The primary factor behind meat 'ageing' is the degradation or proteolysis of muscle proteins. This process in post-mortem muscles is mainly facilitated by calcium-activated sarcoplasmic factors (CASF), specifically calpains, and the level of their inhibitor, calpastatin, in the muscles (Álvarez et al. 2021).

Hot Processing/Accelerated Processing

The traditional method of beef processing involves chilling carcasses to temperatures of 7 °C or below before proceeding with the division into smaller cuts and subsequent processing. In this approach, not only muscles but also bones and fats are cooled, resulting in inefficiencies since a significant portion of bones and fats is typically removed during the preparation of carcasses for retail sale. This result in a significant waste of energy required for carcass cooling. Additionally, due to the large size and irregular shape of the whole carcass, uneven cooling occurs. To address this issue, a processing method known as hot boning, hot deboning, or hot cutting has been developed. In this method, deboning (removal of muscle from the bones) is done when the carcass is still hot (before any cooling). Hot processing has both positive and negative effects on the eating quality of meat. On the positive side, hot boning enhances the functional properties of meat, especially water-holding capacity (WHC). However, it also promotes rapid microbial growth unless rapid chilling is implemented. Meat processed through hot boning yields better results, with approximately 2% more meat, improved water-holding capacity (WHC), leading to reduced drip during storage, and a more uniform colour. Additionally, this method reduces the requirements for refrigeration space, energy, labour costs, and shortens the time needed to produce marketable meat. There are some disadvantages, which include the inability to maintain the characteristic shapes of joints, and the resulting meat may be slightly tougher. The toughness in hot-processed meats is attributed to both 'heat shortening' from prolonged exposure to high temperatures and 'cold shortening' due to rapid cooling compared to whole carcasses. It is essential to exercise caution in maintaining hygiene during the process of hot deboning, as the conditions are highly favourable for microbial contamination and growth. This is attributed to the extended handling of carcasses at relatively high temperatures (Neto et al. 2013).

2.5.5 *Methods of Studying Post-mortem Muscle*

Microscopic and molecular techniques are employed to study post-mortem muscle.

2.5.5.1 Microscopic Techniques

(a) Histology is the microscopic study of animal and plant cell and tissues through staining and sectioning and examining them under a microscope (electron or light microscope) (Black and Black 2018). It plays a fundamental role in studying post-mortem muscle tissue, offering insights into the structural integrity and abnormalities within muscle fibers. Histological techniques are used in studying post-mortem Muscle fiber morphology, Staining techniques, Nuclei distribution, pathological changes and Quantitative analysis (Demirci et al. 2022).

In muscle fibre morphology histological examination is used for visualization of muscle fiber type's viz. skeletal, smooth and cardiac their arrangement, and any structural abnormalities. This includes assessing the size, shape, and distribution of muscle fibers, which can indicate conditions like atrophy, hypertrophy, or degeneration (Talbot and Maves 2016).

In staining techniques various staining methods are employed to highlight specific components of post-mortem muscle tissue. Hematoxylin and eosin (H&E) staining is commonly used to visualize cellular structures, whereas specialized stains like Gomori trichrome or Masson's trichrome stain can reveal collagen content and differentiate muscle fibers (Alturkistani et al. 2016). In Nuclei distribution histology helps in assessing the distribution and appearance of nuclei within muscle fibers. Changes in the nuclei distribution can indicate muscle regeneration, repair, or pathological conditions (Ripolone et al. 2022).

In pathological changes histological examination is used in identifying pathological changes in post-mortem muscle tissue. This includes detecting signs of inflammation, necrosis, fibrosis, or other structural alterations that could indicate underlying myopathies or injuries (Swash and Schwartz 2013).

In Quantitative Analysis advanced histological techniques allow for quantitative analysis of various parameters such as fiber size, fiber type distribution, or the presence of infiltrating cells. Image analysis software assists in quantifying these parameters, providing quantitative data for research purposes (Moscalu et al. 2023).

(b) Electron microscopy (EM) is a powerful microscopic technique that allows for the visualization of post-mortem muscle tissue at an ultrastructural level. EM is used in Ultrastructural analysis, Sarcomere organization, Myofibrillar abnormalities, Connective tissue and Extracellular matrix (McDonald and Auer 2006).

Ultrastructural Analysis provides high-resolution images of post-mortem muscle tissue, allowing researchers to examine cellular structures in detail. This includes visualizing sarcomeres, mitochondria, sarcoplasmic reticulum, and other organelles at a level not achievable with light microscopy (Vincent et al. 2019).

Sarcomere Organization reveals the precise organization of sarcomeres within muscle fibers. This detailed view aids in understanding how muscle fibers contract and the structural changes that occur post-mortem (Lonergan et al. 2010).

Myofibrillar Abnormalities can detect myofibrillar disarray, Z-line streaming, or other structural abnormalities within muscle fibers. These abnormalities might indicate various myopathies or pathological conditions affecting the muscle tissue (Selcen and Engel 2011).

Connective tissue and extracellular matrix allows the visualization of collagen fibers, proteoglycans, and other components of the extracellular matrix within post-mortem muscle. Changes in the matrix might suggest tissue damage or remodeling (Roy and Bruce 2023).

2.5.5.2 Molecular Techniques

Proteomic and genomic studies on post-mortem muscle tissue offer a comprehensive understanding of molecular changes, protein profiles, genetic variations, and their implications in muscle physiology, diseases, and forensic investigations.

(a) Proteomic analysis using mass spectrometry (MS) allows researchers to identify and quantify proteins within post-mortem muscle tissue. Various approaches like shotgun proteomics or targeted proteomics reveal protein profiles, modifications, and interactions. Mass spectrometry-based proteomics identifies proteins present in the tissue, offering insights into their abundance, post-translational modifications (PTMs), and structural variations (Zhai et al. 2020).

    This aids in understanding muscle physiology and pathology. Proteomic analysis helps uncover post-mortem changes in protein phosphorylation, acetylation, or methylation, shedding light on altered signaling pathways and their role in muscle function or disease progression. Comparing protein profiles between healthy and diseased post-mortem muscle tissue enables the identification of disease-specific markers, potential therapeutic targets, or biomarkers for muscle disorders. Proteomic analysis assists in forensic investigations by identifying protein signatures related to traumatic injuries, time of death estimation, or differentiating between post-mortem changes and pathological conditions (Zhong et al. 2023).

(b) Genomic studies involve sequencing the entire genome of post-mortem muscle tissue. Targeted genomic studies focus on specific regions of interest within the genome, investigating candidate genes related to muscle function, development, or diseases in post-mortem samples. Genomic studies aid in identifying mutations or genetic variants contributing to inherited muscle disorders, enabling better understanding, diagnosis, and potential therapeutic interventions. Analysing genomic data from post-mortem muscle tissues helps in population studies, identifying genetic variations prevalent in certain demographics or regions that might impact muscle health or diseases. Genomic studies on post-mortem tissues raise ethical considerations regarding consent, privacy, and proper utilization of genetic data for research purpose (Martínez-Barrios et al. 2023).

2.6 Applications and Future Perspectives

Post-mortem meat, also known as post-slaughter meat, refers to the meat obtained from an animal after it has been slaughtered. The study and utilization of post-mortem meat have various applications across the meat industry, research advancements, and future directions in understanding post-mortem muscle. Here are some aspects of its applications:

2.6.1 Impact on Meat Industry

(a) **Meat quality improvement:** Understanding the post-mortem changes in meat helps in developing strategies to improve meat quality. Factors such as muscle pH, temperature, and aging processes affect the tenderness, flavour, and colour of meat.
(b) **Processing Techniques:** Knowledge of post-mortem changes aids in developing efficient processing techniques. For example, understanding rigor mortis and its resolution can impact the processing time and methods, influencing the texture of the final product.
(c) **Quality Control:** Post-mortem studies contribute to the establishment of quality control standards in the meat industry, ensuring consistency and safety in meat products.

2.6.2 Research Advancements

(a) **Molecular understanding:** Post-mortem meat research involves molecular studies to understand the biochemical and genetic changes occurring in muscle tissues after slaughter. This includes exploring gene expression, proteomic changes, and metabolic pathways.
(b) **Enzyme activity:** Investigating post-mortem enzyme activity provides insights into the processes affecting meat quality. For example, studying the activity of proteolytic enzymes helps in understanding the mechanisms behind meat tenderization.
(c) **Nutritional value**: Research on post-mortem meat contributes to understanding the nutritional composition and changes in the meat during storage and processing.

2.6.3 Future Directions in Understanding Post-mortem Muscle

(a) **Genomic Approaches:** Advancements in genomics can provide a deeper understanding of the genetic factors influencing post-mortem muscle characteristics, leading to selective breeding for desired meat quality traits.
(b) **Integrative Systems Biology:** Combining various omics approaches (genomics, transcriptomics, proteomics, metabolomics) can offer a holistic view of post-mortem muscle changes, aiding in comprehensive understanding and targeted interventions.
(c) **Sustainable Practices:** Future research may focus on sustainable post-mortem practices, considering factors such as environmental impact, waste reduction, and ethical considerations in meat production.

2.6.4 Future Perspectives of Post-mortem Meat

Future perspectives may involve a stronger focus on sustainable and ethical meat production. This could include innovations in alternative protein sources, reducing waste in the meat industry, and implementing eco-friendly practices. The meat industry might see increased adoption of advanced processing technologies, such as cultured or lab-grown meat production, which could have implications for post-mortem meat quality and processing methods. Moreover, the rise of cellular agriculture and alternative protein sources may reshape the meat industry. Cultivated meat, produced from animal cells without traditional slaughter, could impact post-mortem meat production methods and lead to more sustainable and ethical practices.

2.7 Conclusion

The examination of post-mortem muscle provides valuable insights into the structure, nutritive value, and biochemistry of this tissue after death. An essential aspect in the development of meat quality is the phenomenon of protein oxidation. When myosin heavy chain undergoes oxidation, it can lead to a reduction in meat tenderness. Additionally, the cross-linking of myosin heavy chain with titin may decrease protein solubility, potentially affecting meat tenderness. The oxidation of calpain is recognized for inhibiting calpain-catalysed protein degradation, thereby raising concerns about pro-oxidants negatively impacting meat tenderness by promoting protein aggregation and inhibiting protein degradation. Understanding the structural alterations in post-mortem muscle, including rigor mortis and proteolysis is crucial for comprehending meat quality and its potential applications in the food industry. The nutritive value of post-mortem muscle is influenced by factors such as protein content, amino acid composition, and lipid profile, crucial considerations for nutritionists and consumers alike. Biochemically, post-mortem muscle undergoes significant transformations, encompassing enzymatic and non-enzymatic reactions that impact texture, flavour, and overall palatability. Understanding these processes is vital for the meat industry to develop strategies enhancing meat quality and extending shelf life.

In conclusion, exploring post-mortem muscle involves a multidisciplinary approach, integrating structural, nutritive, and biochemical aspects. This knowledge contributes not only to advancements in food science and technology but also aids in developing sustainable and nutritious meat products for global consumers.

References

Abhijith A, Warner RD, Ha M, Dunshea FR, Leury BJ, Zhang M, Joy A, Osei-Amponsah R, Chauhan SS (2021) Effect of slaughter age and post-mortem days on meat quality of longissimus and semimembranosus muscles of Boer goats. Meat Sci 175:108466

Adzitey F (2011) Effect of pre-slaughter animal handling on carcass and meat quality. Int Food Res J 18(2):1–9

Alturkistani HA, Tashkandi FM, Mohammedsaleh ZM (2016) Histological stains: a literature review and case study. Global J Health Sci 8(3):72

Álvarez S, Mullen AM, Hamill R, O'Neill E, Álvarez C (2021) Dry-aging of beef as a tool to improve meat quality. Impact of processing conditions on the technical and organoleptic meat properties. In: Advances in food and nutrition research, vol 95. Academic, pp 97–130

Bai X, Yin F, Ru A, Tian W, Li J, Zhang G, Chen Q, Chai R, Xiao K, Zhu C, Zhao G (2022) Muscle fiber composition affects the postmortem redox characteristics of yak beef. Food Chem 397: 133797

Bailey AJ, Light ND (1989) Connective tissue in meat and meat products. Elsevier Applied Science, London, pp 195–224

Bakhsh A, Hwang YH, Joo ST (2019) Effect of slaughter age on muscle fiber composition, intramuscular connective tissue, and tenderness of goat meat during post-mortem time. Food Secur 8(11):571

Barbut S, Sosnicki AA, Lonergan SM, Knapp T, Ciobanu DC, Gatcliffe LJ, Huff-Lonergan E, Wilson EW (2008) Progress in reducing the pale, soft and exudative (PSE) problem in pork and poultry meat. Meat Sci 79(1):46–63

Barrón-Hoyos JM, Archuleta AR, del Refugio F-VM, Canett-Romero R, Cinco-Moroyoqui FJ, Romero-Barancini AL, Rueda-Puente EO (2013) Protein quality evaluation of animal food proteins by in-vitro methodologies. Food Nutr Sci 04(04): 376–384

Bianchi M, Petracci M, Cavani C (2006) The influence of genotype, market live weight, transportation, and holding conditions prior to slaughter on broiler breast meat colour. Poult Sci 85(1): 123–128

Black JG, Black LJ (2018) Microbiology principles and explorations. John wiley and sons Hoboken, New Jersey

Bottinelli RY, Reggiani C (2000) Human skeletal muscle fibres: molecular and functional diversity. Prog Biophys Mol Biol 73(2–4):195–262

Cassens RG, Marple DN, Eikelenboom (1975) Animal physiology and meat quality. Adv Food Res 21:71–155

Chauhan SS, England EM (2018) Postmortem glycolysis and glycogenolysis: insights from species comparisons. Meat Sci 144:118–126

Chirase NK, Greene LW, Purdy CW, Loan RW, Auvermann BW, Parker DB, Walborg EF, Stevenson DE, Xu Y, Klaunig JE (2004) Effect of transport stress on respiratory disease, serum antioxidant status, and serum concentrations of lipid peroxidation biomarkers in beef cattle. Am J Vet Res 65(6):860–864

Choe JH, Choi YM, Lee SH, Shin HG, Ryu YC, Hong KC, Kim BC (2008) The relation between glycogen, lactate content and muscle fiber type composition, and their influence on postmortem glycolytic rate and pork quality. Meat Sci 80(2):355–362

Choi YM, Oh HK (2016) Carcass performance, muscle fiber, meat quality, and sensory quality characteristics of crossbred pigs with different live weights. Korean J Food Sci Anim Resour 36: 389–396

Christensen M, Henckel P, Purslow PP (2004) Effect of muscle type on the rate of post-mortem proteolysis in pigs. Meat Sci 66(3):595–601

Cianciullo D (2013) Effect of prebiotic and synbiotic injected in ovo on performance, meat quality and hystopathological changes in muscle of broiler chickens. Doctoral thesis submitted by University of Molise from the Department of Agricultural, Environmental and Food Sciences

Culioli J, Berri C, Mourot J (2003) Muscle foods: consumption, composition and quality. Sci Aliment 23(1):13–34
dalla Costa OA, Faucitano L, Coldebella A, Ludke JV, Peloso JV, dalla Roza D, da Costa MP (2007) Effects of the season of the year, truck type and location on truck on skin bruises and meat quality in pigs. Livest Sci 107(1):29–36
Decker EA, Park Y (2010) Healthier meat products as functional foods. Meat Sci 86(1):49–55
Demirci T, Sener E, Kok AN, Sener MT (2022) Postmortem histochemical and electron microscopic changes in skeletal muscle fibers of fatal hypothermia: an experimental study. Aust J Forensic Sci 54(4):464–479
Dokmanovic M, Baltic MZ, Duric J, Ivanovic J, Popovic L, Todorovic M, Markovic R, Pantic S (2015) Correlations among stress parameters, meat and carcass quality parameters in pigs. Asian Australas J Anim Sci 28(3):435–441
Domínguez R, Pateiro M, Gagaoua M, Barba FJ, Zhang W, Lorenzo JM (2019) A comprehensive review on lipid oxidation in meat and meat products. Antioxidants 8(10):429
Ertbjerg P, Puolanne E (2017) Muscle structure, sarcomere length and influences on meat quality: a review. Meat Sci 132:139–152
Faustman C, Cassens RG (1990) The biochemical basis for discoloration in fresh meat: a review. J Muscle Foods 1(3):217–243
Ferguson DM, Warner RD (2008) Have we underestimated the impact of pre-slaughter stress on meat quality in ruminants? Meat Sci 80(1):12–19
Fraterman S, Zeiger U, Khurana TS, Wilm M, Rubinstein NA (2007) Quantitative proteomics profiling of sarcomere associated proteins in limb and extraocular muscle allotypes. Mol Cell Proteomics 6:728–737
Geletu US, Usmael MA, Mummed YY, Ibrahim AM (2021) Quality of cattle meat and its compositional constituents. Vet Med Int 2021:1–9
Grunert KG, Bredahl L, Brunsø K (2004) Consumer perception of meat quality and implications for product development in the meat sector – a review. Meat Sci 66(2):259–272
Honikel KO (1992) Fleisch-und Fleischfettverzehr. Fleischwirtschaft (Frankfurt) 72(8):1145–1148
Hossain A, Dave D, Shahidi F (2020) Northern Sea cucumber (*Cucumaria frondosa*): a potential candidate for functional food, nutraceutical, and pharmaceutical sector. Mar Drugs 18(5):274
Huff-Lonergan E, Lonergan SM (2005) Mechanisms of water-holding capacity of meat: the role of postmortem biochemical and structural changes. Meat Sci 71(1):194–204
Hwang IH, Thompson JM (2001) The interaction between pH and temperature decline early postmortem on the calpain system and objective tenderness in electrically stimulated beef longissimus dorsi muscle. Meat Sci 58(2):167–174
Jensen J, Rustad PI, Kolnes AJ, Lai YC (2011) The role of skeletal muscle glycogen breakdown for regulation of insulin sensitivity by exercise. Front Physiol 2:112
Joo ST, Kim GD, Hwang YH, Ryu YC (2013) Control of fresh meat quality through manipulation of muscle fiber characteristics. Meat Sci 95:828–836
Juárez M, Aldai N, López-Campos Ó, Dugan ME, Uttaro B, Aalhus JL (2012) Beef texture and juiciness. In: Hui YH (ed) Handbook of meat and meat processing, 2nd edn. CRC Press, Boca Raton
Kamruzzaman M, Makino Y, Oshita S (2016) Parsimonious model development for real-time monitoring of moisture in red meat using hyperspectral imaging. Food Chem 196:1084–1091
Karunanayaka DS, Jayasena DD, Jo C (2016) Prevalence of pale, soft, and exudative (PSE) condition in chicken meat used for commercial meat processing and its effect on roasted chicken breast. J Anim Sci Technol 58:1–8
Keeton JT (1994) Low-fat meat products technological problems with processing. Meat Sci 36(1–2):261–276
Klurfeld DM (1994) Human nutrition and health implications of meat with more muscle and less fat. In: Low-fat meats. Academic, pp 35–51, Cambridge, UK
Kontrogianni-Konstantopoulos A, Ackermann MA, Bowman AL, Yap SV, Bloch RJ (2009) Muscle giants: molecular scaffolds in sarcomerogenisis. Physiol Rev 89:1217–1267

Koohmaraie M (1994) Muscle proteinases and meat aging. Meat Sci 36(1–2):93–104
Koohmaraie M, Geesink GH (2006) Contribution of postmortem muscle biochemistry to the delivery of consistent meat quality with particular focus on the calpain system. Meat Sci 74(1):34–43
Lawrie RA (1998) Meat science, 6th edn. Woodhead Publishing, Cambridge
Lawrie RA, Ledward D (2006) Meat and human nutrition. R. AB T.-L. MS. In: Lawrie SE (ed) Lawrie's meat science. Woodhead Publishing, Cambridge, UK, pp 342–357
Leyton A, Ostor JD, Overton RDP (1975) ATP levels in post-mortem muscle. J Sci Food Agric 26(3):395–399
Lindstedt S, Nishikawa K (2017) Huxleys missing filament: form and function of titin in vertebrate striated muscle. Annu Rev Physiol 79:145–466
Lonergan EH, Zhang W, Lonergan SM (2010) Biochemistry of postmortem muscle – lessons on mechanisms of meat tenderization. Meat Sci 86(1):184–195
Mahan DC, Shields RG Jr (1998) Essential and nonessential amino acid composition of pigs from birth to 145 kilograms of body weight, and comparison to other studies. J Anim Sci 76(2): 513–521
Marangoni F, Corsello G, Cricelli C, Ferrara N, Ghiselli A, Lucchin L, Poli (2015) Role of poultry meat in a balanced diet aimed at maintaining health and wellbeing: an Italian consensus document. Food Nutr Res 59(1):27606
Maribo H, Olsen EV, Barton-Gade P, Møller AJ, Karlsson A (1998) Effect of early post-mortem cooling on temperature, pH fall and meat quality in pigs. Meat Sci 50(1):115–129
Martín NP, Schreurs NM, Morris ST, López-Villalobos N, McDade J, Hickson RE (2022) Meat quality of beef-cross-dairy cattle from Angus or Hereford sires: A case study in a pasture-based system in New Zealand. Meat Sci 190:108840
Martínez-Barrios E, Grassi S, Brión M, Toro R, Cesar S, Cruzalegui J, Coll M, Alcalde M, Brugada R, Greco A, Ortega-Sánchez ML (2023) Molecular autopsy: twenty years of post-mortem diagnosis in sudden cardiac death. Front Med 10:1118585
Mashele GA (2017) Effect of slaughter age between 5 and 14 months on the quality of sheep meat with specific focus on collagen concentration and solubility: a thesis presented in partial fulfilment of the requirements for the degree of masters of science in animal science at Massey University, Manawatu, New Zealand. Doctoral dissertation, Massey University
McDonald KL, Auer M (2006) High-pressure freezing, cellular tomography, and structural cell biology. BioTechniques 41(2):137–143
Meyer LC, Wright NT (2013) Structure of giant muscle proteins. Front Physiol 4:368–380
Miroshnichenko NS, Balanuk IV, Nozdrenko DN (2000) Packing of myosin molecules in muscle thick filaments. Cell Biol Int 24:327–333
Moscalu M, Moscalu R, Dascălu CG, Ţarcă V, Cojocaru E, Costin IM, Ţarcă E, Şerban IL (2023) Histopathological images analysis and predictive modeling implemented in digital pathology – current affairs and perspectives. Diagnostics 13(14):2379
Mourot J (2009) Optimising the nutritional and sensorial profile of pork. In: Kerry JP, Ledward D (eds) Improving the sensory and nutritional quality of fresh meat, Woodhead Publishing in food science, technology and nutrition. Woodhead Publishing, Saint Gilles, pp 342–355
Nathanailides C, Panopoulos S, Kakali F, Karipoglou C, Lenas D (2011) Antemortem and postmortem biochemistry, drip loss and lipid oxidation of European sea bass muscle tissue. Procedia Food Sci 1:1099–1104
Neto MP, Beraquet NJ, Cardoso S (2013) Effects of chilling methods and hot-boning on quality parameters of *M. longissimus* lumborum from Bos indicus Nelore steer. Meat Sci 93(2): 201–206
Nishimura T (2010) The role of intramuscular connective tissue in meat texture. Anim Sci J 81:21–27
Offer G, Trinick J (1983) On the mechanism of water holding in meat: the swelling and shrinking of myofibrils. Meat Sci 8(4):245–281
Olaoye OA (2011) Meat: an overview of its composition, biochemical changes and associated microbial agents. Int Food Res J 18(3):877–885

Ouali A, Herrera-Mendez CH, Coulis G, Becila S, Boudjellal A, Aubry L, Sentandreu MA (2006) Revisiting the conversion of muscle into meat and the underlying mechanisms. Meat Sci 74(1): 44–58

Pette D, Staron RS (2000) Myosin isoforms, muscle fiber types, and transitions. Microsc Res Tech 50(6):500–509

Purslow PP (2005) Intramuscular connective tissue and its role in meat quality. Meat Sci 70:435–447

Ripolone M, Velardo D, Mondello S, Zanotti S, Magri F, Minuti E, Cazzaniga S, Fortunato F, Ciscato P, Tiberio F, Sciacco M (2022) Muscle histological changes in a large cohort of patients affected with Becker muscular dystrophy. Acta Neuropathol Commun 10(1):48

Rodrigues RT, Chizzotti ML, Vital CE, Baracat-Pereira MC, Barros E, Busato KC, Gomes RA, Ladeira MM, Martins TD (2017) Differences in beef quality between Angus (Bos taurus taurus) and Nellore (Bos taurus indicus) cattle through a proteomic and phosphoproteomic approach. PLoS One 12(1):e0170294

Rourke TJ, Ramsey CB, Davis GW, Clark JR, Wheeler TL (1992) Consumption and plate waste of regular and total trim roasted pork loins. J Muscle Foods 3(4):341–354

Roy BC, Bruce HL (2023) Contribution of intramuscular connective tissue and its structural components on meat tenderness-revisited: a review. Crit Rev Food Sci Nutr 63:1–31

Sakomura NK, Ekmay RD, Mei SJ, Coon CN (2015) Lysine, methionine, phenylalanine, arginine, valine, isoleucine, leucine, and threonine maintenance requirements of broiler breeders. Poult Sci 94(11):2715–2721

Savell JW, Cross HR, Smith GC (1986) Percentage ether extractable fat and moisture content of beef longissimus muscle as related to USDA marbling score. J Food Sci 51(3):838–839

Scheffler TL, Gerrard DE (2007) Mechanisms controlling pork quality development: the biochemistry controlling postmortem energy metabolism. Meat Sci 77(1):7–16

Schweigert BS, Payne BJ (1956) A summary of the nutrient content of meat. American Meat Institute Foundation, Chicago

Selcen D, Engel AG (2011) Myofibrillar myopathies. Handb Clin Neurol 101:143–154

Shackelford SD, Wheeler TL, Meade MK, Reagan JO, Byrnes BL, Koohmaraie M (2001) Consumer impressions of tender select beef. J Anim Sci 79(10):2605–2614

Shahidi F (2000) Lipids in flavor formation. In: Risch SJ, Ho C-T (eds) Flavor chemistry: industrial and academic research. American Chemical Society, Oxford, pp 24–43

Shahidi F, Abad A (2019) Lipid-derived flavours and off-flavours in food. Encycl Food Chem 2: 182–192

Shahidi F, Zhong Y (2015) Measurement of antioxidant activity. J Funct Foods 18:757–781

Soladoye OP, Juárez ML, Aalhus JL, Shand P, Estévez M (2015) Protein oxidation in processed meat: mechanisms and potential implications on human health. Compr Rev Food Sci Food Saf 14(2):106–122

Squire J (2019) The actin-myosin interaction in muscle: background and overview. Int J Mol Sci 20(22):5715

Swash M, Schwartz MS (2013) Biopsy pathology of muscle. Springer: Cambridge, UK

Talbot J, Maves L (2016) Skeletal muscle fiber type: using insights from muscle developmental biology to dissect targets for susceptibility and resistance to muscle disease. Wiley Interdiscip Rev Dev Biol 5(4):518–534

Toldrá F (2002) The role of muscle enzymes in dry-cured meat products with different drying conditions. Trends Food Sci Technol 17(4):164–168

Toldrá F (2006) Meat: chemistry and biochemistry. In: Hui Y (ed) Handbook of food science, technology and engineering. CRC Press, Taylor & Francis Group, Boca Raton

Toldrá F (2012) Biochemistry of processing meat and poultry. In: Food biochemistry and food processing. Wiley-Blackwell, Ames, pp 303–316

Toldrá F, Flores M (1998) The role of muscle proteases and lipases in flavor development during the processing of dry-cured ham. Crit Rev Food Sci 38(4):331–352

Tskhovrebova L, Trinick J (2010) Roles of Titin in the structure and elasticity of the sarcomere. J Biomed Biotechnol 2010:1–7. https://doi.org/10.1155/2010/612482
Valin C, Ouali A (1992) Proteolytic muscle enzymes and post mortem meat tenderisation. In: New technologies for meat and meat products. ECCEAMST, Audet Tijdschriften, Nijmegen, pp 163–179
Vincent AE, White K, Davey T, Philips J, Ogden RT, Lawless C, Warren C, Hall MG, Ng YS, Falkous G, Holden T (2019) Quantitative 3D mapping of the human skeletal muscle mitochondrial network. Cell Rep 26(4):996–1009
Visschers PM, Slaughter PE, Rust RS (1975) The mechanism of rigor mortis in the broiler chicken. Poult Sci 54(1):13–20
Weaver AD, Bowker BC, Gerrard DE (2009) Sarcomere length influences mu-calpain-mediated proteolysis of bovine myofibrils. J Anim Sci 87(6):2096–2103
Wiepkema PR, Koolhaas JM (1993) Stress and animal welfare. Anim Welf 2(3):195–218
Wong GC, Lin A, Tang JX, Li Y, Janmey PA, Safinya CR (2003) Lamellar phase of stacked two-dimensional rafts of actin filaments. Phys Rev Lett 91(1):018103
Wood JD, Enser M, Fisher AV, Nute GR, Sheard PR, Richardson RI, Hughes SI, Whittington FM (2008) Fat deposition, fatty acid composition and meat quality: A review. Meat Sci 78(4): 343–358
Woods VB, Fearon AM (2009) Dietary sources of unsaturated fatty acids for animals and their transfer into meat, milk and eggs: A review. Livest Sci 126(1–3):1–20
Xing T, Wang C, Zhao X, Dai C, Zhou G, Xu X (2017) Proteome analysis using isobaric tags for relative and absolute analysis quantitation (iTRAQ) reveals alterations in stress-induced dysfunctional chicken muscle. J Agric Food Chem 65(13):2913–2922
Xing T, Gao F, Tume RK, Zhou G, Xu X (2019) Stress effects on meat quality: A mechanistic perspective. Compr Rev Food Sci Food Saf 18(2):380–401
Young P, Ehler E, Gautel M (2001) Obscurin, a giant sarcomeric rho guanine nucleotide exchange factor protein involved in sarcomere assembly. J Cell Biol 154:123–136
Yu TY, Morton JD, Clerens S, Dyer JM (2017) Cooking-induced protein modifications in meat. Compr Rev Food Sci Food Saf 16(1):141–159
Zhai C, Djimsa BA, Prenni JE, Woerner DR, Belk KE, Nair MN (2020) Tandem mass tag labeling to characterize muscle-specific proteome changes in beef during early postmortem period. J Proteome 222:103794
Zhong Q, Xiao X, Qiu Y, Xu Z, Chen C, Chong B, Zhao X, Hai S, Li S, An Z, Dai L (2023) Protein posttranslational modifications in health and diseases: functions, regulatory mechanisms, and therapeutic implications. Med Comment 4(3):e261

Chapter 3
Quality Requirements of Meat for Processing

Syed Arshad Hussain, Sheikh Rafeh Ahmad, Shakir Rashid, Humaira Fayaz, and Asma Irshad Qureshi

3.1 Introduction

The quality of meat and meat products plays a role in the food industry influencing aspects like taste, texture, safety and nutrition. Meat quality refers to the features that make meat appealing for consumption including factors such as color, tenderness, juiciness, flavour and safety standards. It encompasses both intrinsic qualities of the meat itself-like animal genetics, age, diet and muscle composition as well as external factors such as environmental conditions, handling practices and processing methods in the supply chain. The Food and Agriculture Organization (FAO) defines meat quality as the combination of attributes that determine its acceptability to consumers and other users. This definition highlights the importance of meeting consumer expectations by ensuring that meat is not only safe but meets desired criteria in terms of taste, appearance and overall sensory experience. Stringent quality standards are imposed on meat intended for processing purposes based on its intended use in the product. For example, lean cuts of meat with minimal connective tissue may be preferred for processed products like sausages or deli meats as they contribute to a more desirable texture and mouthfeel. Furthermore, the absence of pathogens and foreign substances is imperative to ensure the safety and integrity of processed meat products. Stringent hygiene practices and adherence to food safety regulations are essential at every stage of processing from slaughter and fabrication to packaging and distribution. In short, the quality of meat and meat products is a multifaceted concept that encompasses various attributes and considerations. By prioritizing important factors like animal welfare, handling practices and safety standards, producers can ensure that consumers receive meat products that meet their expectations for taste, texture as well as overall quality.

S. A. Hussain (✉) · S. R. Ahmad (✉) · S. Rashid · H. Fayaz · A. I. Qureshi
Division of Livestock Products Technology, FVSc & AH, SKUAST-K, Srinagar, India
e-mail: sh.rafeh@skuastkashmir.ac.in

S. A. Rather, F. A. Masoodi (eds.), *Hand Book of Processed Functional Meat Products*, https://doi.org/10.1007/978-3-031-69868-2_3

3.2 Animal Breed and Meat Quality

The quality of meat can vary depending on the breed of livestock used. This can be due to natural adaptations of the breed to their environment or deliberate crossbreeding efforts aimed at enhancing desirable traits. However, these modifications often result in improvements in some characteristics while compromising others. Brahman cattle normally thrive in the tougher climates but often yield tougher meat due to high levels of calpastatin in their muscles (Ibrahim et al. 2008). Conversely, Wagyu cattle produce tender, marbled meat. By crossbreeding Brahman with Wagyu, resulting in Waguli cattle, a blend of high marbling and low calpastatin activity breed, yielding tender meat post-slaughter. While Brahman meat eventually tenderizes after about 14 days of aging, initial tenderness varies between beef from Holstein and Simmental cattle, with differing aging responses as reported by Schone et al. (2006). According to Thomas et al. (2008), beef from medium-framed, early maturing animals exhibited the highest marbling scores and contained the highest concentration of total n-3 fatty acids, with the lowest n-6/n-3 ratio. In contrast, Lynch et al. (2002) found that meat from Hereford cattle contained higher levels of C14:0, C16:1 and C18:0 in the phospholipid fraction compared to Friesian and Charolais cattle. Similarly, the influence of breed on beef flavor is also significant, with variations observed in nitrogen and sulfur compounds, free amino acids, alcohols, aldehydes and ketones in flavor volatiles across different cattle breeds (Sato et al. 1995; Insausti et al. 2005). Beef from Friesian cattle exhibits a distinct fatty flavor and aftertaste along with a different volatile profile compared to Pirenaica cattle (Gorraiz et al. 2002). Enzymes like μ- and m-calpain, primarily associated with textural changes, can also impact flavor by generating peptides that contribute significantly to the taste profile. Interestingly, meat from Bos taurus and Bos indicus cattle inheriting specific genotypes at the calpastatin and μ-calpain genes tends to produce steaks with more intense flavor (Casas et al. 2006). These genes are linked to increased rancid, sour and salty flavors as well as decreased umami flavor (Toldrá and Flores 2000). Moreover, variations in volatile compounds such as hexane and 2,2,4,6,6-pentamethylheptane are observed between Friesian and Pirenaica cattle (Gorraiz et al. 2002). Additionally, breed influences beef color, with Holstein cattle showing higher a* values (redness) compared to Simmental and Simmental x Limousin cattle (Frickh and Solkner 1997). Similarly, swine genetics also play a significant role in shaping the quality of pork. Pigs homozygous for the Halothane gene (nn) tend to have leaner carcasses and higher weight gain: feed ratios compared to Halothane negative (NN) and heterozygous (Nn) pigs (Leach et al. 1996). However, pigs carrying one or two copies of the Halothane gene may produce pork with higher lean content but are prone to developing pale, soft and exudative (PSE) meat due to rapid pH decline post-slaughter (Sather et al. 1990). Fernandez et al. (2004) found that NN and Nn pigs displayed similar postmortem changes, with differences observed in visual color intensity, meat tenderness and fat content. Duroc pigs are known for producing highly marbled meat with good eating quality (Ellis et al. 1996), while pork from Danish Landrace Duroc pigs is noted for its tenderness compared to Landrace, Duroc and various Yorkshire crosses (Brewer et al. 2004). Additionally, breed influences the fatty acid composition of intramuscular neutral lipid, with Berkshire and Tamworth pigs yielding fattier carcasses and higher saturated fatty acid content, while Duroc and Large White pigs produce leaner meat with more polyunsaturated fatty acids, including 20:5n-3 and 22:6n-3 (Wood et al. 2004). Numerous studies have delved into how breed impacts carcass quality and characteristics, aiming to breed animals tailored to specific market demands while enhancing the profitability of the production system (Crouse et al. 1981). Santos-Silva et al. (2002) compared 'Merino Branco' lambs with those resulting from crossbreeding 'San Merino' and 'Île de France', noting that the crossbred lambs exhibited better carcass yields and muscle-to-bone ratios. However, they concluded that these differences held limited significance as they didn't greatly affect the carcass commercial value. In contrast, Rodrigues et al. (2006) observed disparities in the carcasses of 'Churra Galega Bragancana' and 'Suffolk Down' lambs, with the latter yielding a higher proportion of relevant cuts, such as the leg, shoulder and breast. Barone et al. (2007) conducted a comparative analysis of 'Gentile di Puglia', 'Île de France' sheep breeds and their crossbred lambs, uncovering notable differences in carcass characteristics between the pure breeds, with more pronounced distinctions seen in comparison to the crossbred lambs. Santos-Silva and Vaz (2001) also identified significant differences between the carcasses of 'Serra da Estrela' and 'Merino Branco' breeds across various variables, except carcass yield.

3.3 Animal Diet and Meat Quality

The composition of diet of an animal can impact meat quality in two distinct ways: directly, through the direct deposition of compounds from the feed source into the meat or indirectly by increasing fat content (Brewer 2010), thus affecting both the carcass composition (Lewis et al. 2002) as well as the degree of fattening. Grazing animals typically have lower fat levels, making them more appealing to consumers (Sañudo et al. 2000). Additionally, the fat of grass-fed animals tends to exhibit a yellow hue (Priolo et al. 2002). Grass-fed lambs yield lighter carcasses compared to lambs fed on concentrates of the same age (Priolo et al. 2002). Consequently, animals primarily fed concentrates require less time to reach slaughter readiness (Mustafa et al. 2008). Feeding practices significantly influence meat quality. Diets containing fish byproducts, raw soybeans, canola oil and meal may impart undesirable flavors to meat (Melton 1990). Pigs are particularly susceptible to changes in dietary fat source as they lack the capacity to biohydrogenate unsaturated fats effectively leading to softer, oilier carcass fat when fed high levels of polyunsaturated fatty acids (PUFA) (Whitney et al. 2006). The timing of dietary changes also plays a role with alterations closer to slaughter having a more pronounced effect on pork fat (Lampe et al. 2006). Finishing diets can impact pork quality differently; while low-protein diets may improve tenderness and juiciness, they can decrease flavor quality (Wood et al. 2004). Similarly, diets low in digestible carbohydrates can affect muscle glycogen stores and proteolysis, influencing meat tenderness (Rosenvold et al. 2001). Supplements such as plant-based diets containing conjugated linoleic acid, selenium and vitamin E may slightly affect meat quality (Janz et al. 2008). In beef animals, grain feeding tends to increase carcass weight, intramuscular fat content and flavor intensity compared to forage diets (Melton 1990). Beta-agonists like ractopamine hydrochloride and zilpaterol can improve feedlot performance and carcass characteristics but may decrease meat tenderness (Quinn et al. 2008; Avendaño-Reyes et al. 2006). Implants with anabolic steroids can increase carcass weight and muscle area but have minimal effects on other characteristics (Smith et al. 2007). Supplementation with vitamin D_3 may enhance tenderness by affecting muscle calcium concentrations and proteolysis (Montgomery et al. 2004). Additionally, vitamin E supplementation can stabilize meat colour and decrease fluid loss in swine (Asghar and Pearson 1980). Ramírez-Retamal et al. (2013) conducted a comparative analysis of Chilote lamb carcasses, examining those raised on naturalized pasture versus natural rangeland. They found no discernible differences based on the types of grassland utilized. Overall, dietary interventions can significantly impact meat quality but their effects vary depending on various factors such as timing, composition and species. While both breed and feed type playing significant roles in determining carcass quality individually, there is a possibility that the interaction between multiple factors also influences the quality and characteristics of lamb carcasses as suggested by Díaz et al. (2002) and Mustafa et al. (2008).

3.4 Effect of Meat Composition on the Quality of Processed Meat Products

Meat industry strives to deliver meat products that offer high eating satisfaction while remaining affordable to consumers. Consumer satisfaction and enjoyment while eating meat are influenced by specific factors known as eating quality parameters. Pearson (1968) proposed that quality encompasses attributes like flavor, juiciness, texture, tenderness and appearance, all of which contribute to the overall enjoyment and desirability of the product. Processed meat products also offer convenience, flavor and versatility. The quality of these products is influenced by various factors, with meat composition playing a pivotal role. The composition of meat, including its protein, fat, moisture, collagen and other components, directly impacts the sensory attributes, texture, safety and shelf life of processed meats.

Protein Content and Composition The protein level in meat significantly correlates with its ability to retain water, its capacity to emulsify fat and its overall nutritional value. The protein content and composition, particularly myofibrillar proteins such as myosin and actin, influence the texture and binding properties of processed meats. Typically, consumers consider tenderness to be the primary factor influencing the overall eating experience of meat (Brewer and Novakofski 2008). Tenderness encompasses various tactile sensations perceived before, during and after chewing, such as particle size, oiliness, juiciness and mouth coating (Bourne 1992). It comprises mechanical, particulate and chemical elements including hardness, cohesiveness, elasticity, grittiness, fibrousness, juiciness and oiliness (Bourne 1992). Factors affecting meat tenderness include myofibrillar structure, connective tissue and composition, with influences ranging from cold shortening and proteolytic degradation to animal age, activity level, mechanical tenderization and composition (Pearson and Young 1989). Lawrie (1966) suggests that the tenderness of meat can be attributed to three protein categories found within the muscle tissue which include proteins within the myofibril (such as actin, myosin and tropomyosin), proteins within the connective tissue (like collagen, elastin, reticulin and the mucopolysaccharides of the matrix) and proteins within the sarcoplasm (such as sarcoplasmic proteins and sarcoplasmic reticulum). These proteins contribute to the formation of a stable protein matrix during processing affecting the firmness, cohesiveness and juiciness of the final products. Muscle-based foods inherently possess specific textural qualities determined by the nature of the muscle tissue. Typically, meat from less utilized muscles tends to be more tender as compared to the more utilized ones, which tends to be tougher. Nevertheless, both genetic factors and age play significant roles in determining tenderness. Meat from 2-year-old Angus/Wagyu heifers exhibited similar tenderness and juiciness to that of younger animals. However, meat from 2-year-old pure Angus breeds was less tender and juicy compared to meat from yearlings or crossbred Angus/Wagyu animals (Rentfrow et al. 2004). While connective tissue plays a role in the toughness of meat, the overall concentrations of connective tissue components like collagen and elastin did not strongly correlate with sensory panel ratings for tenderness. However, the percentage of soluble collagen, indicating susceptibility of collagen to heat, showed a significant relationship with the contribution of connective tissue to toughness, whereas elastin concentration did not exhibit the same correlation (Cross et al. 1973). Proteins also play a crucial role in emulsification and gelation processes in processed meats. They interact with fats and water to form stable emulsions and gels enhancing the moisture retention, succulence and mouthfeel of processed meat products. The salt soluble myofibrillar proteins play a pivotal in meat processing, driving structural development and texture formation in meat products (Culioli et al. 1993; Xiong 1997; Zayas 1997). Constituting over half of muscle proteins (Cheftel et al. 1985), they exhibit crucial functionalities by forming viscoelastic gels, binding water and creating cohesive interfaces in emulsions. Actin, myosin and actomyosin predominantly govern functional attributes of meat, contributing to about 95% of its water-holding capacity and 75–90% of its emulsifying capacity (Li Chan et al. 1985). Myosin, particularly, plays a vital role in fat emulsification and water retention (Galluzzo and Regenstein 1978; Xiong et al. 2000), owing to its distinctive composition featuring hydrophobic residues in the head region and hydrophilic groups in the tail region, fostering protein-protein interaction and molecular flexibility at interfaces (Xiong 1997). Another protein, myoglobin is the primary pigment responsible for meat color, although hemoglobin, cytochromes and ribonucleases also contribute in smaller amounts. Myoglobin comprises of a peptide portion and an oxygen-binding prosthetic group called heme, containing an iron atom. There are three main forms of myoglobin in muscles: deoxymyoglobin, oxymyoglobin, and metmyoglobin. Any chemical process that denatures or destabilizes proteins, including myoglobins, can influence the formation of pigment forms, altering meat color. The ultimate pH and its rate of decrease, along with factors like temperature, affect chemical reactions and muscle structure, influencing meat color. Generally, lower pH levels result in a paler appearance, while higher pH levels lead to a darker color. Maintaining a red or pink color during cooking, known as pinkness or pinking, depends on preserving myoglobin in its reduced state, which is heat-stable. Meats with higher pigment content may retain a pink-red color for longer periods.

Fat Content and Composition The fat content and composition significantly contribute to the flavor, aroma and juiciness of processed meats. Higher fat content can impart richness and succulence to the products, enhancing their palatability. Of particular importance is the intramuscular fat termed as marbling which include fat cells situated amidst muscle fiber bundles, and intracellular fat, primarily composed of lipids within cell membranes and droplets in vesicles (Cassens and Cooper 1971; Gandemer 1999). Concerning meat texture, an increase in fat content may dilute the protein content, leading to reduced bulk density, while fat within connective tissue can decrease the force needed to slice the meat. The composition of fat is crucial too, as increased unsaturation tends to soften fat, consequently lowering its melting point influencing the mouthfeel and perceived greasiness of processed meats. Dikeman (1987) conducted a comprehensive review of literature examining the correlation between marbling and tenderness, revealing that marbling contributes to only 5–10% of the variability in beef palatability. Despite this, a minimum muscle lipid content of 3% has been suggested for achieving acceptable eating quality (Savell et al. 1986). While studies such as those by Smith et al. (1985) and Aalhus et al. (2004) have yielded inconsistent findings regarding the relationship between fat content and texture, many countries, including Canada and the United States, incentivize producers with premiums for higher levels of marbling in beef due to the perceived associations with flavor, tenderness and juiciness. Furthermore, fat acts as a lubricant between muscle fibers, potentially enhancing the perception of tenderness and juiciness during chewing (Calkins and Sullivan 2007). This lubricating effect, particularly from marbling, stimulates salivation and contributes to the perceived juiciness of meat. Earlier studies by Cross et al. (1980), Penfield et al. (1989) and Savell et al. (1989) highlighted the significant role of varying fat levels in enhancing juiciness and acceptability. While intramuscular fat, predominantly in the form of marbling, has been linked to meat juiciness (Gaddis et al. 1950; Thompson 2004), the precise mechanism underlying this phenomenon remains largely unknown. The complexity of studying the impact of intramuscular fat on texture and juiciness is compounded by interactive effects with aging, endpoint temperature and psychological factors influencing the eating experience.

Moisture Content The moisture content of meat directly impacts its juiciness, tenderness and succulence in processed products. Optimal moisture levels are necessary to maintain the desired sensory attributes and prevent textural defects such as dryness or sponginess. It also affects the cooking loss and yield of processed meats, with excessive moisture loss leading to shrinkage and reduced product yield. Proper moisture management during processing is crucial to ensure consistent product quality and minimize economic losses. The moisture content within meat plays a pivotal role in its binding capacity with additives like binders and fillers during the production of processed products. This moisture content is closely associated with the fat content of the meat and influences the overall shelf stability of processed products by affecting the water activity, thereby impacting microbial growth. Lawrie (1966) observed that variations in pH, water retention, fat content and texture directly influenced the juiciness of cooked meats. The water-holding capacity of meat is particularly crucial as it affects the appearance of meat, its behavior during cooking and juiciness upon chewing (Lawrie 1966). Reduction in water retention is evident through fluid exudation, known as weep/purge in uncooked unfrozen meat, drip in thawed uncooked meat, and shrink/cooking loss in cooked meat, from both aqueous and fatty components (Lawrie 1966). Weir (1960) further noted that juiciness results from both initial fluid release and sustained juiciness influenced by the stimulating effect of fat on saliva production. One of the most important reasons which significantly impacts the juiciness of meat, especially regarding the initial release of fluid is the type of cooking method employed. Cooking techniques that preserve more meat fluid, whether water or lipid, resulting in lower cooking losses, tend to produce a juicier final product (Smith 1972). Generally, meat cooked "rare" retains more juices compared to "well-done" meat, with beef being juicier than pork, lamb and veal when cooked to the same degree (Weir 1960). High-quality meats typically experience lower cooking losses compared to poor-quality meats, despite losing more fat due to their higher fat content. This could be attributed to structural changes, such as marbling, enhancing their water-holding capacity (Saffle and Bratzler 1957). Shrinkage during cooking is influenced by various factors, including the loss of fluid fat due to high temperatures and protein denaturation, which reduces water-holding capacity, leading to the loss of moisture (Lawrie 1966). Increased shrinkage during cooking is directly linked to a decrease in juiciness when consumed.

In conclusion, the composition of meat significantly influences the quality of processed meat products encompassing sensory attributes, texture, safety and shelf life. Understanding the intricate interplay between meat components, processing techniques and additives is essential for producing high-quality processed meats that meet consumer expectations for taste, texture and overall satisfaction.

3.5 Influence of Postmortem Changes in Muscle on Quality of Processed Meat Products

Upon death, muscles undergo a series of physical and chemical transformations over several hours before fully transitioning into meat. These changes involve the initiation of rigor mortis and proteolytic processes after death. Following the cessation of blood circulation and cardiac activity post-exsanguination, there is a metabolic shift from aerobic to anaerobic pathway leading to the production of lactic acid causing a decrease in pH from approximately 7.0 to 5.3, ultimately stabilizing at 5.9. This transition is accompanied by the depletion of glycogen and creatine phosphate stores, as well as a drop in body temperature from 39 to 36 °C. The specific levels of these changes are influenced by the animal's metabolic rate, pre-slaughter handling, slaughter techniques and post-slaughter conditioning (Cruz-Monterrosa and Guerrero-Legarreta 2012). Water, which accounts for approximately 75% of muscle tissue, is predominantly found surrounding polar molecules and between layers of cellular structures. The intermolecular spaces between salt-soluble proteins like actin and myosin harbour most of the water content, which can fluctuate due to various internal and external factors. The movement of water within these spaces is regulated primarily by the arrangement of myofilaments. Factors such as changes in pH, interaction with ions like magnesium and calcium, denaturation caused by rapid pH drop at high carcass temperatures and the presence of substances like ATP and enzymes influence the spatial organization of myofilaments and consequently water distribution within muscle tissue (Brewer 2010). Before rigor mortis sets in, magnesium adenosine triphosphate (Mg-ATP) acts as a barrier against the formation of connections between the contractile proteins actin and myosin. This function of Mg-ATP complex helps to uphold the interfilamental space, allowing water to permeate through the muscle tissue (Siegel and Schmidt 1979). In the immediate post mortem period, glycogen in tissues is metabolized through anaerobic pathways, leading to a decrease in pH. Once ATP is exhausted to ADP and IMP, there is insufficient energy for rephosphorylation, thus the plasticizing effect of ATP diminishes, preventing muscle relaxation. Consequently, actin and myosin bind together, causing the sarcomere to contract, resulting in muscle fibers contracting and rigor mortis setting in. At this stage, meat exhibits minimum tenderness, resulting in almost nonexistent physicochemical properties such as water-holding capacity, emulsifying capacity and cohesiveness (Cruz-Monterrosa and Guerrero-Legarreta 2012). Sensory evaluation reveals meat that is utterly unacceptable due to its toughness, bland flavor and dark coloration (Young et al. 2001; Pérez-Alvarez 2006). This process of contraction at rigor involves both longitudinal and lateral contraction of muscle cells, typically occurring within a 24-h period. This leads to a reduction in water-holding capacity (WHC) along with other meat quality attributes. The lowest WHC of meat is observed at pH 5.0, aligning with the isoelectric point of actomyosin (Brewer 2010). Moreover, toughness shows an inverse relationship with the initial pH level and the rate at which pH declines. Research by Zamora et al. (1996) suggests that two-thirds of WHC reductions during rigor mortis are attributable to ATP depletion, while the remaining one-third is linked to pH decrease. Various factors impacting the rapidity of pH decline and residual glycogen levels in tissues, influence tenderness, WHC and color. Additionally, factors influencing the ultimate pH, such as the ryanodine gene status and stress-induced alterations in muscle glycogen content also impact these characteristics (Brewer 2010). In addition to changes in pH levels, variations in carcass temperature can also exert notable impacts on meat quality, particularly in terms of tenderness and water-holding capacity (WHC). Rapid postmortem glycolysis is linked with elevated drip loss, diminished WHC and the paler hue characteristic of pale soft exudative (PSE) pork, as observed in the study by Bendall and Wismer-Pedersen (1962). Additionally, when the ultimate pH level of muscle remains elevated ($\geq$6.0) due to stress-induced glycogen depletion before slaughter, it leads to the formation of dark firm dry (DFD) muscle. While DFD meat exhibits increased water retention, its elevated pH level shortens its microbial shelf life. However, due to its exceptional water-binding capabilities, it is well-suited for processed meat products. Although it's notably dark color might discourage consumers from purchasing it at retail meat counters.

3.6 Aging and Quality of Processed Meat Products

Following slaughter, meat undergoes a post-rigor phase, commonly referred to as the aging/conditioning process. During postmortem conditioning, rigor resolution gradually takes place, leading to partial restoration of tenderness and elasticity in muscles. This process is facilitated by enzymatic activity that targets various myofibrillar proteins, including desmin, troponin T, viculin, titin, nebulin, actin and myosin, resulting in tension reduction and alterations in myofibril structure. Desmin fragmentation weakens the Z-line, while phospholipid liberation induced by Ca^{2+} release from the sarcoplasmic reticulum degrades Z-disks. Endogenous proteases, including cathepsins and calpains, act on myofibrillar proteins. Cathepsins, such as A, B, C, D, H, L and J, primarily target myosin and actin. Calpains, represented by m-calpains and μ-calpains (calpains I and II), require different calcium concentrations for activation and exhibit peak activity between pH 6.6 and 6.8. Another calpain, P94-calpain, has also been identified in this process (Cruz-Monterrosa and Guerrero-Legarreta 2012). Throughout this period, the tenderness of the meat typically improves. Additionally, the taste and flavor of the meat are enhanced. Variation in meat tenderness is predominantly influenced by factors such as sarcomere length, muscle composition, connective tissue proteins and proteolytic degradation (Koohmaraie et al. 2002). During growth, the accelerated turnover of proteins results in increased proteolytic activity, contributing to the aging process (Zgur et al. 2003). This heightened proteolytic activity aids aging by breaking down certain structural proteins, allowing sarcomeres to relax (Kristensen and Purslow 2001). Consequently, this relaxation facilitates the influx of water previously expelled during rigor, a process potentially influenced by differences in protein concentration between intra and extracellular compartments of muscle cells. There is substantial diversity in the enhancement of tenderness during aging, observed both among animals and different muscles within an animal, which may be associated with the initial tenderness of the meat (Novakofski and Brewer 2006; Stolowski et al. 2006). Wicklund et al. (2005) found that it took 14 days of aging for strip steaks to exhibit noticeable changes in tenderness. Novakofski and Brewer (2006) observed that the extent of shear improvement during the first week of aging varied depending on the initial shear value, but no further differences were noted between 7 and 14 days. Rentfrow et al. (2004) noted that Warner Bratzler shear values decreased and tenderness increased in beef from 1 and 2-year-old heifers with aging, with maximum improvement seen after just 7 days. Bruce et al. (2005) suggested that aging for up to 14 days resulted in increased tenderness. The process of aging is usually accomplished through two popular methods: wet aging and dry aging. While wet aging involves storing fresh beef in vacuum-sealed packages at refrigerated temperatures, typically below 5 °C, often resulting in a tangier and more pronounced bloody or serum-like flavor compared to dry-aged beef (Laster et al. 2008). During wet aging, nucleotides degrade into inosine 5′-monophosphate (IMP), leading to the accumulation of hypoxanthine, inosine and ribose, which contribute to both the umami and bitter tastes of the meat (Aaslyng and Meinert 2017). The umami taste is attributed to elevated levels of glutamate. On the other hand, dry aging entails exposing beef to controlled airflow, temperature and humidity without packaging, resulting in a more robust, beefy and brown-roasted flavor profile compared to wet-aged beef (Laster et al. 2008). Previous studies have reported that dry aging, particularly in low-marbled beef, improves tenderness and sensory properties (Lee et al. 2017; Lepper-Blilie et al. 2016). Proteolysis of myofibrillar proteins and degradation of structural proteins during both wet and dry aging contribute to improved tenderness, while changes in peptide and free amino acid concentrations enhance flavor (Aaslyng and Meinert 2017; Laville et al. 2009). Both wet and dry aging methods have been demonstrated to enhance the overall eating experience of beef through distinct processes (Kim et al. 2017). However, consumer feedback has favored sensory aspects of dry-aged meat, noting superior tenderness and juiciness compared to vacuum-aged alternatives. Dry aging significantly enhances the texture and succulence of both beef and pork.

3.7 Factors Affecting Texture of Processed Meat Products

The texture/tenderness of meat is probably the most important attribute in determining its overall eating quality. Consumers consistently prioritize meat tenderness, often willing to pay more for cuts that promise a softer texture. Extensive research has delved into understanding, managing, and anticipating this attribute. Meat texture encompasses various elements, from initial tenderness upon biting to overall chewability, along with nuanced sensory experiences like fiber cohesiveness, adhesion and chew count. These factors, combined with the intricate interplay between meat properties, oral sensations and brain interpretation during chewing, form the foundation of psychorheology. Moreover, textural properties in the mouth are intertwined with other meat characteristics like water-holding capacity and fat content, which aid in necessary lubrication during chewing (Juárez et al. 2012). The texture of meat is an intricate attribute shaped by factors like muscle structure, cell integrity, enzyme activity and extracellular matrix composition (McCormick 2009). For example, the arrangement of muscle fibers within fascicles confers specific mechanical properties, influencing the tissue's resistance to shear. Intramuscular fat levels (Nishimura et al. 1999) and postmortem energy metabolism (Thompson et al. 2006) also play roles in determining meat tenderness. The interaction of these factors complicates the process, making it challenging to isolate individual effects. In a broader sense, various pre and postmortem factors combine together and collectively play a crucial role to define the texture of meat which can be categorized into four key factors; the background toughness (established before slaughter), the toughening phase (during rigor mortis), the tenderization phase (during postmortem aging) and the denaturation/solubilization of proteins (during cooking). Antemortem factors affecting meat moisture or fat content, as well as the effects of toughening, tenderization, and cooking phases on proteins, also impact meat juiciness (Winger and Hagyard 1999).

3.7.1 Ante-mortem Factors

Ante-mortem factors which determine the texture of the meat include type of animal species/domestication status; type of breed; feeding status, growth rate and age of animal at slaughter; condition of animal at slaughter.

3.7.1.1 Type of Animal Species/Domestication Status

The species effect can be gauged by the fact that as the proportion of *Bos indicus* heritage increases, tenderness tends to decline (Barkhouse et al. 1996). This decrease in tenderness is attributed to the reduced proteolytic capacity in *Bos indicus*, primarily due to higher levels of calpastatin activity as evidenced by various studies (Shackelford et al. 1991). Additionally, differences in overall carcass leanness between *Bos taurus* and *Bos indicus* cattle contribute to a minor extent to the overall variation in tenderness (Wheeler et al. 1994).

As far as the effect of domestication is concerned, there has been a general trend towards increased muscle mass, primarily achieved by enlarging fiber size rather than increasing fiber number (Rehfeldt et al. 2008). This shift has also involved a transition from oxidative to glycolytic fibers, resulting in an increase in cross-sectional area (Essén-Gustavsson and Lindhölm 1984; Weiler et al. 1995). Out of the different fiber types, the cross-sectional area of muscle fibers, particularly higher in white fibers, has been noted to have a negative correlation with beef tenderness

(Calkins et al. 1981; Seideman and Crouse 1986). Selection for lean growth affects muscles differently, with propulsive muscles experiencing more pronounced changes compared to postural muscles due to variations in fiber composition (Ashmore et al. 1972). It has been suggested that these alterations may influence meat quality, with shifts towards larger white fibers potentially leading to coarser fiber grain perception and reduced susceptibility to cold shortening (Goldstein et al. 1985; Xu and Becker 1998), attributed to their extensive sarcoplasmic reticulum and lower mitochondrial density (Essén-Gustavsson et al. 1992). Recent research has also identified differences in the content of total and insoluble connective tissue between highly selected and less selected beef breeds (Christensen et al. 2011).

3.7.1.2 Type of Breed

A wide range of cattle breeds contribute to beef production, each with its unique genetic makeup, leading to diverse meat qualities (Albertí et al. 2008). Studies have highlighted variations in meat texture among different beef breeds (Mandell et al. 1997; Dikeman et al. 2005; Christensen et al. 2011). Certain breeds are believed to exhibit finer muscle grain, which is often preferred by consumers, although Albrecht et al. (2006) found differences in fiber bundle size and connective tissue among breeds without a confirmed link to tenderness. Recent advancements in genomic analysis have identified gene associations with tenderness. Single-nucleotide polymorphisms in genes such as μ-calpain (CAPN1), Calpain 3 (CAPN3), and calpastatin (CAST) have been linked to beef tenderness (Vieira et al. 2007; Warner et al. 2010). Downregulation of the DNAJA1 gene, crucial in apoptosis, has also been correlated with tenderness variability in Charolais bulls (Bernard et al. 2007). Moreover, the double-muscled syndrome, stemming from mutations in the myostatin gene (MSTN or GDF8), results in more tender meat due to lower collagen content and increased cathepsin activity (Allais et al. 2010; Caballero et al. 2007; Sierra et al. 2010). Genotype further influences tenderness by affecting glycolytic rate modulated by variations in muscle fiber type across different genotypes (Thompson et al. 2006).

3.7.1.3 Feeding Status, Growth Rate and Age of Animal at Slaughter

Factors like breed, diet and exercise can influence growth rates, leading to varying maturity stages among animals of similar chronological age. Each muscle possesses a distinct connective tissue framework, with variations based on its role in movement. Muscles involved in locomotion typically contain higher collagen levels compared to those involved in posture, and the function of each muscle influences how collagen is distributed within it. Connective tissue forms the primary structure of the epimysium (encasing the entire muscle), perimysium (encasing muscle bundles or fascicles), and endomysium (encasing individual muscle fibers), extending from the tendons (Harper 1999). Collagens, found primarily in endomysial and perimysial connective tissues, play a vital role in meat tenderness. They contribute to tenderness through both their overall content and the formation of heat-stable cross-links. The arrangement of perimysium, which defines muscle fascicle size or "grain" size also influences tenderness. As animals mature, the tenderness of their meat often declines, primarily due to changes in collagen composition (Reagan et al. 1976). While collagen cross-linking in older animals is considered a background factor affecting meat texture, establishing direct links between collagen cross-link content in aged animals and cooked meat toughness remains challenging (Purslow 2005). Interestingly, some studies have found no significant tenderness differences among animals of different ages, highlighting the complex interplay of factors such as breed, diet, gender and production practices (Riley et al. 1986; Shackelford et al. 1995). While postmortem changes may alter the spatial distribution of connective tissue, their impact on tenderness is not well understood. Although enzymes capable of breaking down connective tissue have been identified, structural weakening of connective tissue is typically observed only during extended aging. Therefore, the proteolytic breakdown of connective tissue during aging may not significantly affect meat tenderness (Purslow 2005; Lepetit et al. 2000; Nishimura et al. 1998). Variations in fat content can influence the texture of meat. Higher fat levels may dilute the protein content, reducing bulk density, while fat within the connective tissue can ease the cutting process. Additionally, the development of intra and intermuscular fat occurs at a slower pace compared to bone or muscle growth, contributing to tenderness improvements observed between 9 and 18 months establishing a link between marbling and texture (Lopez 1977; Calkins and Sullivan 2007). Moreover, the increase in fatness as animals age plays a role in carcass chilling, providing additional insulation against rapid temperature decline postmortem (Lawrie 1998), thereby impacting the development of rigor mortis and the activity of proteolytic enzymes and thus texture. The degree of fat unsaturation affects its softness and melting point. According to Dikeman (1987), marbling accounts for only 5–10% of the variability in beef palatability. Savell et al. (1986) reported a minimum muscle lipid content of 3% for acceptable eating quality or palatability. In this context, while studies by Smith et al. (1985) and Aalhus et al. (2004) have shown inconsistent results linking fat content and texture, premiums are often paid for beef with greater levels of marbling, particularly in countries like Canada and the United States, as it is associated with flavor, tenderness and juiciness. Furthermore, as already discussed, fat acts as a lubricant between muscle fibers, potentially enhancing tenderness and juiciness perception. The role of feeding status/diet of an animal on various quality parameters particularly the amount of fat content of carcass has been elaborated in the previous sections. Moreover, when comparing high-energy concentrate diets to forage-based diets in cattle, there is a known increase in marbling levels. However, the correlation between marbling and juiciness has not consistently surfaced in research (French et al. 2000). Therefore, to discern a noticeable difference, substantial variations in feed quality and quantity are typically required, impacting factors such as growth rates, carcass fatness, and marbling levels (Basarab et al. 2007). Additionally, the aging process in cattle also contributes to increased fatness, particularly in older animals. This heightened fat content plays a crucial role during carcass chilling, acting as insulation and safeguarding against rapid temperature declines during postmortem stages (Lawrie 1998). The rates of protein synthesis and degradation while the animal is alive can impact the tenderization process after death. The amount of stored glycogen can affect the speed and extent of postmortem glycolysis. Compensatory growth strategies have shown promise in enhancing beef tenderness without altering shear force significantly (Therkildsen et al. 2008). Yet another human intervention to affect the growth of meat animals has been the use of hormonal growth promotants (HGPs) which are widely employed in various countries to enhance the growth rate and feed conversion efficiency of beef cattle while reducing carcass fat content. However, the impact of HGPs on beef tenderness remains a topic of debate. Some studies suggest that HGPs tend to increase beef toughness (Thompson et al. 2008). Koohmaraie et al. (2002) proposed a mechanism linking reduced tenderization with HGPs, suggesting that these substances may inhibit protein degradation, potentially due to heightened calpastatin activity in live animals. Elevated calpastatin levels could slow down the aging process post-slaughter, resulting in less tender meat, especially in muscles with rapid aging rates or high calpain-to-calpastatin ratios (Thompson et al. 2008). Furthermore, implanted animals tend to display advanced physiological maturity at slaughter, indicated by increased ossification (Morgan 1997), which may lead to higher levels of mature collagen cross-links (Bailey and Shimokomaki 1971). However, the specific effects of different types of HGP treatments, such as single or combined regimens, or variations in hormone types (estrogenic and androgenic) on beef texture require further investigation, thus underscoring the need for continued research to elucidate the complex relationship between HGPs and beef quality attributes.

3.7.1.4 Condition of Animal at Slaughter

The condition of the animal at the time of slaughter particularly the level of stress and glycogen depletion, directly impacts the postmortem biochemical changes in the muscle tissue and consequently the meat texture and tenderness. This is true particularly in the cases of DFD and PSE meat. These changes affect the protein structure and water-binding capacity of the meat, ultimately influencing its texture and tenderness. DFD meat typically occurs in animals that have experienced prolonged stress or agitation before slaughter, leading to high levels of glycogen depletion in the muscles and higher postmortem pH. This condition results in meat that is dark in color, firm to the touch with dry appearance and improved tenderness and juiciness owing to higher water-holding capacity. On the other hand, PSE meat is characterized by its pale color, soft texture and low water-holding capacity. It is typically found in animals that undergo extreme stress or excitement before slaughter, leading to a rapid decline in muscle pH. The low pH level in PSE meat can cause protein denaturation, resulting in a soft, mushy texture and reduced water-binding capacity. PSE meat also tends to lose more moisture during cooking, leading to dry and unpalatable meat products. Therefore, minimizing stress and ensuring proper handling practices before slaughter are essential to prevent the occurrence of DFD and PSE meat and maintain desirable meat quality. Proper pre-slaughter management can help preserve muscle glycogen levels, pH balance and protein integrity, resulting in meat with optimal texture, tenderness, and juiciness.

3.7.2 Post-mortem Factors

In addition to the ante-mortem factors, which exert a significant impact in deciding the overall texture of meat, there are several post-mortem factors having a definite bearing on the texture and tenderness of the meat. These include carcass aging, carcass hanging, electrical stimulation, carcass chilling and freezing, artificial tenderization and method of cooking as discussed below.

3.7.2.1 Carcass Aging

The attainment of optimal meat tenderness heavily relies on the extent of modification and weakening of myofibrillar structures, elucidating the tenderness enhancement observed in aged meat compared to its unaged counterpart. Disruptions occurring at the junction of the I-band and Z-disk, alongside the degradation of myofibrillar and cytoskeletal proteins like titin, desmin, troponin-T or nebulin, consistently correlate with meat tenderization. Traditionally, three endogenous proteolytic enzyme systems, the lysosomal cathepsins, the calpain system and the multicatalytic proteinase complex, have been implicated in facilitating such myofibrillar degradation (Koohmaraie and Geesink 2006). However, it has been found that enzyme systems like calpain and caspase system are majorly accountable for the breakdown of crucial myofibrillar proteins postmortem (Koohmaraie and Geesink 2006). The calpain system encompasses a minimum of three calcium-activated proteases: μ-calpain (calpain-1), m-calpain (calpain-2) and P94-calpain (calpain-3). All these calpains along with an inhibitor of μ and m-calpain known as calpastatin. Calpastatin serves as a specific endogenous inhibitor of μ and m-calpain, requiring calcium to bind and effectively inhibit calpains at concentrations close to or below the levels needed to activate calpain. Variations in calpastatin levels and its inhibition of calpain during the initial postmortem hours have been frequently linked to meat tenderness, with higher calpastatin levels often associated with increased toughness in meat. The interaction between μ-calpain and calpastatin within 3 days post-slaughter appears pivotal in postmortem proteolysis and meat tenderization (Juárez et al. 2012). Although the calpain system is traditionally deemed crucial for ultimate meat tenderness, emerging research suggests that additional processes and systems such as apoptosis mediated by the caspase protease family could also contribute (Ouali et al. 2006). For a given duration of aging, the variations in texture between muscles can be attributed to several factors including differences in fat and moisture content, shape, sarcomere length, fiber type and connective tissue composition. However, the changes observed in texture parameters over time during aging are primarily influenced by the rate of postmortem rigor shortening, proteolysis of myofibrillar proteins, moisture loss, and breakdown of collagen, along with the complex interactions among these factors. (Aalhus et al. 2004; Moreno et al. 2007; Janz et al. 2006; Willems and Purslow 1996; Lefaucheur 2010; Light et al. 1985; Olsson et al. 1994; Koohmaraie et al. 1996; Purslow 2005).

3.7.2.2 Carcass Hanging

Carcass hanging methods are used to prevent excessive rigor shortening during chilling. While various techniques of carcass hanging have been explored like horizontal hanging, neck-tied, hip-tied and hip-free methods using the obturator foramen in the aitch bone, studies like (López-Campos et al. 2010) suggest that the latter approach, known as Aitch bone suspension or Tenderstretch, leads to elongated sarcomeres and enhanced tenderness in muscles like the longissimus and posterior muscles of the hind limb (semimembranosus, semitendinosus). Another method, the Tendercut, involves cutting bones and connective tissues in the mid-loin and the sirloin junction of beef carcass sides. According to Claus et al. (1997), this technique results in increased sarcomere lengths and tenderness across various muscles in the loin and round.

3.7.2.3 Electrical Stimulation

Electrical stimulation has emerged as a commercially viable method for enhancing beef tenderness, although originally it was developed to prevent cold shortening. It has been reported to reduce shear force values by 15–30% showing greater efficacy in lower-grade carcasses (Smith 1985). Sensory evaluations conducted by taste panels indicate that electrically stimulated meat is perceived as 10–50% more tender compared to non-stimulated meat (Chrystall and Devine 1985). Researchs comparing the effectiveness of high and low-voltage stimulation has found that the greatest improvements in tenderness are linked to high-voltage stimulation (Bouton et al. 1980; Aalhus et al. 1994). Several mechanisms have been proposed to explain the effects of electrical stimulation including physical damage caused by contractions, prevention of cold-induced shortening and acceleration of proteolysis, either through non-enzymatic disruption of myofibrillar structure or mediated by calpains or lysosomal enzymes (Aalhus (1995). Recent literature on apoptosis suggests that caspase enzymes may also be involved (Kemp et al. 2010).

3.7.2.4 Carcass Chilling and Freezing

Although lower temperatures ensure prevention of microbial spoilage in meat, however, excessive cooling may lead to certain negative effects like cold shortening. As such, several strategies have been employed to prevent scarcomere shortening during rigor deveoplment and thus enhance the texture of meat. Delayed chilling has been found to have a softening effect on beef, as demonstrated by Fields et al. (1976) and Smith et al. (1979). When carcasses are chilled at temperatures exceeding 15 °C for 15–20 h, there is a reduction in sarcomere shortening and an increase in proteolysis (Juárez et al. 2012). However, due to the heightened risk of microbial growth, this method is not suitable for commercial applications. To prevent cold shortening, it is essential to ensure that the muscle temperature does not drop below 11 °C before the pH reaches 6.1–6.3 (Juárez et al. 2012). Consequently, slower chilling protocols have been developed, extending the chilling process to 36–48 h instead of the traditional 24-h cycle. The commercial chilling procedures have been adopted to prevent carcass meat temperatures from falling below 11 °C within the initial 10 h post-mortem. This adaptation has been shown to improve tenderness acceptability by 10% to 40% across different muscle types (Aalhus 2002). Conversely, there are indications from various studies that extremely rapid chilling, achieving temperatures of −1 °C within approximately 5 h postmortem, may enhance tenderness in pork and lamb but not beef (Davey and Garnett 1980; Jones et al. 1993; Joseph 1996). The swift reduction in temperature during rapid chilling prompts a substantial release of calcium from the sarcoplasmic reticulum into the myofibrils (Jaime et al. 1992). This early influx of free calcium coupled with a high muscle pH could lead to an accelerated and heightened activation of calpains. Also, it has been reported that thawed beef which is aged before freezing has lower shear force values compared to chilled meat aged the same time (Shanks et al. 2002; Enfält et al. 2004).

3.7.2.5 Artificial Tenderization

Given the significance of texture in determining meat quality, various methods like carcass infusion post-slaughter, blade and needle tenderization, marination, application of proteolytic enzymes and high-pressure processing have been explored to enhance the tenderness of beef. Injection of calcium chloride into hip muscles before rigor mortis accelerates the tenderization process, reducing the required aging time to just one day postmortem (Wheeler et al. 1991), which can be attributed to the early activation of the calpain system, leading to increased proteolysis rates and consequently, a more tender meat texture. However, its commercial application is limited owing to development of off flavors.

Another method is the use of blade/needle tenderization to mechanically disrupt the structural integrity of myofibrils, muscle fibers and collagen, reticulin, and elastin fibrils in beef cuts (Smith et al. 2008). Several studies have observed a reduction in shear force following blade tenderization across various muscles (Seideman et al. 1977; Pietrasik and Shand 2004).

Marination is another popular method of augmenting meat tenderness. A variety of substances have been utilized to enhance beef tenderness including acids like vinegar and lemon juice which disrupt hydrogen bonds within collagen fibrils, causing connective tissue to expand (Lawrie 1974). Additionally, using salt (sodium chloride) as a pre-cooking seasoning at a concentration of 0.3% w/w improved tenderness, juiciness and browning (Pérez et al. 1998; Aalhus et al. 2009). Phosphate salts have also been shown to significantly enhance beef tenderness and juiciness (Baublits et al. 2005) which is partly attributed to increased water-holding capacity. Throughout history, marination with proteolytic enzymes has been a traditional method used by various cultures to enhance beef tenderness. These enzyme formulations typically break down meat proteins resulting in tenderization and an increase in peptide and free amino acid concentrations (Weir et al. 1958). Fungal or bacterial enzymes (such as fungal amylase and protease) target muscle fiber proteins, while tropical plant enzymes (like bromelain, papain and ficin) act on collagen and elastin. However, careful attention to time and temperature is necessary to avoid excessive tenderization and a mushy texture. The impact of these tenderizers, along with other marination ingredients, is primarily observed on the surface of the meat unless the cuts are thin or somehow penetrate the interior of meat (Rao et al. 1989).

High pressure processing (HPP) is a popular alternative to conventional heat-based methods for preserving food by applying intense isostatic pressures ranging from 100 to 600 MPa at near-room temperatures. Its primary goal is to extend shelf life and ensure safety by deactivating microbes while preserving the fresh attributes of food products. Unlike thermal processing, HPP primarily affects non-covalent bonds (Bak et al. 2009). This unique mechanism can yield favorable structural changes in meat products leading to reduced cook loss and improved texture (Sikes et al. 2009). Additionally, hydrodynamic pressure (HDP) technology, employing high-pressure shock waves in water for brief intervals, has shown promise in enhancing tenderness and minimizing variations in beef texture (Spanier et al. 2000; Claus et al. 2001).

3.7.2.6 Method of Cooking

Cooking marks the final stage in processing after slaughter, just before consumption, where heat transforms muscle tissue, giving meat its distinct flavor, texture and aroma. This heat-induced process alters myofibrillar proteins and connective tissue, influencing water-holding capacity (WHC) and ultimately affecting the texture and juiciness of the cooked product. Initially, as heat is applied, meat toughens significantly, increasing three to four times between 40 °C and 50 °C. However, between 50 °C and 60 °C, toughness decreases, followed by another doubling beyond 60 °C (Davey & Gilbert 1974). Christensen et al. (2000) observed that the initial toughening phase, occurring below 60 °C, is attributed to strengthened perimysium, likely due to collagen fiber concentration changes induced by denaturation (Lewis and Purslow 1989). Conversely, the subsequent toughening phase above 60 °C correlates with increased individual muscle fiber strength coinciding with actin and sarcoplasmic protein denaturation. This period also marks the onset of collagen solubilization possibly contributing to tenderization (Lewis and Purslow 1989). Okitani et al. (2009) suggested irreversible actin dissociation from myosin within this temperature range as another tenderization factor. Between 40 °C and approximately 55 °C, myosin denatures and coagulates, leading to myofilament shrinkage, microstructure tightening and sarcomere shortening (Cheng and Parrish 1976; Hostetler and Landmann 1968). Heating beyond 60 °C induces collagen shrinkage and sheath contracture (Davey & Gilbert 1974). Actin denaturation occurs between 70 °C and 80 °C, while collagen gelatinization begins above 75 °C (Cheng and Parrish 1976 Bejerholm & Aaslyng 2004). Gelatinizing collagen forms channels facilitating intramuscular fat movement, reducing moisture loss during cooking which is particularly beneficial for highly marbled meat (Cross 1987). Meats with significant connective tissue become tender when cooked with moist heat for prolonged periods, while those with little connective tissue toughen due to myofibrillar protein denaturation (Cross 1987). Overall, these changes lead to decreased immobilized water and increased evaporative and drip losses during cooking, with cooking losses starting around 30 °C, peaking between 50 °C and 70 °C and slowing thereafter. Extended cooking resulting in reduced juiciness is also associated with decreased tenderness (Cross 1987). Thus, the final tenderness and juiciness of cooked meat reflect a blend of intrinsic (collagen content, myofibrillar protein denaturation, and water loss) and extrinsic (temperature and cooking duration) factors. These multifaceted influences underscore the complexity of factors affecting meat texture and highlight the need for comprehensive approaches in understanding and optimizing meat quality.

3.8 Conclusion

Muscle, as a living tissue, undergoes continuous changes influenced by its immediate environment before, during and after the animal slaughter in its journey of conversion to meat. These responses wield significant influences, potentially shaping the quality and worth of the ensuing meat product, either positively or adversely. The quality of processed meat products is thus a result of various pre and post-slaughter factors which ultimately determine its acceptance or rejection. Ensuring optimal animal management practices throughout the process of animal rearing as well as during slaughter and post-slaughter periods, including controlled carcass chilling and aging and other strategic processing techniques collectively contribute to the production of high-quality meat products. By understanding and optimizing these factors meat producers can consistently deliver products with superior sensory attributes, nutritional value and safety, meeting the diverse preferences and expectations of consumers worldwide.

References

Aalhus JL (1995) Postmortem effects on the appearance and eating quality of meat. In: Morgan-Jones SD (ed) Quality and grading of meat animals. CRC Press, Inc., Boca Raton, pp 97–114

Aalhus JL (2002) Carcass cooling, suspension and ageing. In: Beef palatability enhancement workshop, Calgary, pp 1–6
Aalhus JL, Jones SDM, Lutz S, Best DR, Robertson WM (1994) The efficacy of high and low voltage electrical stimulation under different chilling regimes. Can J Anim Sci 74:433–442
Aalhus JL, Jeremiah LE, Dugan MER, Larsen IL, Gibson LL (2004) Establishment of consumer thresholds for beef quality attributes. Can J Anim Sci 84(4):631–638
Aalhus JL, Juárez M, Aldai N, Uttaro B, Dugan MER (2009) Meat preparation and eating quality. In: 55th international congress of meat science and technology, Copenhagen, pp 1058–1063
Aaslyng MD, Meinert L (2017) Meat flavour in pork and beef-from animal to meal. Meat Sci 132: 112–117
Albertí P, Panea B, Sañudo C, Olleta JL, Ripoll G, Ertbjerg P, Christensen M et al (2008) Live weight, body size and carcass characteristics of young bulls of fifteen European breeds. Livest Sci 114(1):19–30
Albrecht E, Teuscher F, Ender K, Wegner J (2006) Growth- and breed-related changes of muscle bundle structure in cattle. J Anim Sci 84(11):2959–2964. https://doi.org/10.2527/jas.2006-345
Allais S, Levéziel H, Payet-Duprat N, Hocquette JF, Lepetit J, Rousset S, Denoyelle C et al (2010) The two mutations, Q204X and nt821, of the myostatin gene affect carcass and meat quality in young heterozygous bulls of French beef breeds. J Anim Sci 88(2):446–454
Asghar A, Pearson A (1980) Influence of ante- and post-mortem treatments on muscle composition and meat quality. Adv Food Res 26:53–61
Ashmore CR, Tompkins G, Doerr L (1972) Postnatal development of muscle fiber types in domestic animals. J Anim Sci 34(1):37–41
Avendaño-Reyes L, Torres-Rodríguez V, Meraz-Murillo FJ, Pérez-Linares C, Figueroa-Saavedra F, Robinson PH (2006) Effects of two beta – adrenergic agonists on finishing performance, carcass characteristics, and meat quality of feedlot steers. J Anim Sci 84(12):3259–3265
Bailey AJ, Shimokomaki MS (1971) Age related changes in the reducible cross-links of collagen. FEBS Lett 16(2):86–88
Bak KH, Orlien V, Karlsson AH, Lindahl G (2009) Effect of high pressure processing, temperature, and storage on the color of pork longissimus dorsi. In: 55th international conference of meat science and technology, Denmark, pp 581–584
Barkhouse KL, Van Vleck LD, Cundiff LV, Koohmaraie M, Lunstra DD, Crouse JD (1996) Prediction of breeding values for tenderness of market animals from measurements on bulls. J Anim Sci 74(11):2612–2621
Barone C, Colatruglio P, Girolami A, Matassino D, Zullo A (2007) Genetic type, sex, age at slaughter and feeding system effects on carcass and cut composition in lambs. Livest Sci 112: 133–142
Basarab JA, Aalhus JL, Shah MA, Mir PS, Baron VS, Dugan M, Okine EK, Robertson WM (2007) Effect of feeding sunflower seeds on the performance, carcass characteristics, meat quality, retail stability and sensory characteristics of pasture-fed and feedlot finished beef. Can J Anim Sci 87(1):15–27
Baublits RT, Pohlman FW, Brown JAH, Johnson ZB (2005) Effects of sodium chloride, phosphate type and concentration, and pump rate on beef biceps femoris quality and sensory characteristics. Meat Sci 70(2):205–214
Bejerholm C, Aaslyng MD (2004) Cooking of meat. In: Werner Klinth J (ed) Encyclopedia of meat sciences. Elsevier, Oxford, pp 343–349
Bendall JR, Wismer-Pedersen J (1962) Some properties of the fibrillar proteins of normal and watery pork muscle. J Food Sci 27:144–159
Bernard C, Cassar-Malek I, Le Cunff M, Dubroeucq H, Renand G, Hocquette J-F (2007) New indicators of beef sensory quality revealed by expression of specific genes. J Agric Food Chem 55(13):5229–5237
Bourne M (1992) Calibration of rheological techniques used for foods. J Food Eng 16:151–163

Bouton PE, Ford AL, Harris PV, Shaw FD (1980) Electrical stimulation of beef sides. Meat Sci 4: 145–155

Brewer S (2010) Technological quality of meat for processing. In: Toldrá F (ed) Handbook of meat processing. Wiley-Blackwell Publishing, Iowa, p 27

Brewer S, Novakofski J (2008) Consumer quality evaluation of aging of beef. J Food Sci 73(1): S78–S82

Brewer SK, Ryan J, Jensen C, Prestat L, Zhu A, Sosnicki B, Field R, Hankes MKF (2004) Enhancement effects on quality characteristics of pork derived from pigs of various commercial genetic backgrounds. J Food Sci 69(1):5–10

Bruce HL, Beilken SL, Leppard P (2005) Textural descriptions of cooked steaks from bovine M. Longissimus thoracis et lumborum from different production and aging regimes. J Food Sci 70:S309–S316

Caballero B, Sierra V, Oliván M, Vega-Naredo I, Tomás-Zapico C, Alvarez-García Ó, Tolivia D, Hardeland R, Rodríguez-Colunga MJ, Coto-Montes A (2007) Activity of cathepsins during beef aging related to mutations in the myostatin gene. J Sci Food Agric 87(2):192–199

Calkins CR, Sullivan G (2007) Ranking of beef muscles for tenderness. Beef facts: Product enhancement. National Cattlemen's Beef Association. http://www.beefresearch.org

Calkins CR, Dutson TR, Smith GC, Carpenter ZL, Davis GW (1981) Relationship of fiber type composition to marbling and tenderness of bovine muscle. J Food Sci 46(3):708–710

Casas E, White SN, Wheeler T, Shackelford SD, Koohmaraie M, Riley DG, Chase CC, Johnson D, Smith TPL (2006) Effects of calpastatin and o-calpain markers in beef cattle on tenderness traits. J Anim Sci 84(3):520–525

Cassens R, Cooper C (1971) Red and white muscle. Adv Food Res 19:1–74

Cheftel JC, Cuq JL, Lorient D (1985) Le système protéique musculaire. In: Protéines alimentaires. Ed. Lavoisier Tec et Doc, Paris, pp 123–146

Cheng CS, Parrish FC (1976) Scanning electron microscopy of bovine muscle: effect of heating on ultrastructure. J Food Sci 41(6):1449–1454

Christensen M, Purslow PP, Larsen LM (2000) The effect of cooking temperature on mechanical properties of whole meat, single muscle fibres and perimysial connective tissue. Meat Sci 55(3): 301–307

Christensen M, Ertbjerg P, Failla S, Sañudo C, Richardson RI, Nute GR, Olleta JL et al (2011) Relationship between collagen characteristics, lipid content and raw and cooked texture of meat from young bulls of fifteen European breeds. Meat Sci 87(1):61–65

Chrystall BB, Devine CE (1985) Electrical stimulation: its early development in New Zealand. In: Pearson AM, Dutson TR (eds) Advances in meat research: electrical stimulation. Westport, AVI, pp 73–90

Claus JR, Wang H, Marriott NG (1997) Prerigor carcass muscle stretching effects on tenderness of grain-fed beef under commercial conditions. J Food Sci 62(6):1231–1234

Claus JR, Schilling JK, Marriott NG, Duncan SE, Solomon MB, Wang H (2001) Hydrodynamic shockwave tenderization effects using a cylinder processor on early deboned broiler breasts. Meat Sci 58(3):287–292

Cross HR (1987) Sensory characteristics of meat. Part 1. Sensory factors and evaluation. In: Price JF, Schweigert BS (eds) The science of meat and meat products, 3rd edn. Food & Nutrition Press, Inc., Westport, pp 307–330

Cross HR, Carpenter ZL, Smith GC (1973) Effects of intramuscular collagen and elastin on bovine muscle tenderness. J Food Sci 38:998–1003

Cross H, Berry B, Wells L (1980) Effects of fat level and source on the chemical, sensory and cooking properties of ground beef patties. J Food Sci 45(4):791–794

Crouse J, Busboom J, Field R, Ferrell C (1981) The effects of breed, diet, sex, location and slaughter weight on lamb growth, carcass composition and meat flavor. J Anim Sci 53:376–386

Culioli J, Boyer C, Vignon X, Ouali A (1993) Heat-induced gelation properties of myosin – Influence of purification and muscle type. Sci Aliment 13:249–260

Davey CL, Garnett KJ (1980) Rapid freezing, frozen storage and the tenderness of lamb. Meat Sci 4(4):319–322
Davey CL, Gilbert KV (1974) Temperature-dependent cooking toughness in beef. J Sci Food Agric 25(8):931–938
Díaz M, Velasco S, Cañeque V, Lauzurica S, de Huidobro FR, Pérez C et al (2002) Use of concentrate or pasture for fattening lambs and its effect on carcass and meat quality. Small Rumin Res 43:257–268
Dikeman M (1987) Fat reduction in animals and the effects on palatability and consumer acceptance of meat products. Proc Reciprocal Meat Conf 40:93–103
Dikeman ME, Pollak EJ, Zhang Z, Moser DW, Gill CA, Dressler EA (2005) Phenotypic ranges and relationships among carcass and meat palatability traits for fourteen cattle breeds, and heritabilities and expected progeny differences for Warner–Bratzlershear force in three beef cattle breeds. J Anim Sci 83(10):2461–2467
Ellis M, Webb A, Avery P, Brown I (1996) The influence of terminal sire genotype, sex, slaughter weight, feeding regime and slaughter – house on growth performance and carcass and meat quality in pigs and on the organoleptic properties of fresh pork. J Anim Sci 62(3):521–530
Enfält AC, Lundesjö Ahnström M, Svensson K, Hansson I, Lundström K (2004) Tenderness in M. Longissimus dorsi from cows – effect of pelvic suspension and ageing time. In: 50th international congress of meat science and technology, Helsinki, p 82
Essén-Gustavsson B, Lindhölm A (1984) Fiber types and metabolic characteristics in muscles of wild boars, normal and halothane sensitive Swedish landrace pigs. Comp Biochem Physiol A Physiol 78(1):67–71
Essén-Gustavsson B, Karlström K, Lundström K (1992) Muscle fibre characteristics and metabolic response at slaughter in pigs of different halothane genotypes and their relation to meat quality. Meat Sci 31(1):1–11
Fernandez X, Neyraud E, Astruc T, Sante V (2004) Effects of halothane genotype and pre-slaughter treatment on pig meat quality. 1. Postmortem metabolism, meat quality indicators and sensory traits of Longissimus lumborum. Meat Sci 62(4):429–437
Fields PA, Carpenter ZL, Smith GC (1976) Effects of elevated temperature conditioning on youthful and mature beef carcasses. J Anim Sci 42(1):72–83
French P, O'Riordan EG, Monahan FJ, Caffrey PJ, Vidal M, Mooney MT, Troy DJ, Moloney AP (2000) Meat quality of steers finished on autumn grass, grass silage or concentrate-based diets. Meat Sci 56(2):173–180
Frickh J, Solkner J (1997) Colour as a beef quality trait – results of a breed comparison. Zuchtungskunde 69(3):163–180
Gaddis A, Hankins O, Hiner R (1950) Relationships between the amount and composition of press fluid, palatability and other factors of meat. Food Technol 4:498–503
Galluzzo SJ, Regenstein JM (1978) Emulsion capacity and timed emulsification of chicken breast muscle myosin. J Food Sci 43:1757–1760
Gandemer G (1999) Lipids and meat quality: lipolysis, oxidation, Maillard reaction and flavour. Sci Aliment 19:439–458
Goldstein MA, Murphy DL, van Winkle WB, Entman ML (1985) Cytochemical studies of a glycogen-sarcoplasmic reticulum complex. J Muscle Res Cell Motil 6(2):177–187
Gorraiz C, Beriain MJ, Chasco J, Insausti K (2002) Effect of aging time on volatile compounds, odor, and flavor of cooked beef from Pirenaica and Friesian bulls and heifers. J Food Sci 67: 916–922
Harper GS (1999) Trends in skeletal muscle biology and the understanding of toughness in beef. Aust J Agric Res 50(7):1105–1129
Hostetler RL, Landmann WA (1968) Photomicrographic studies of dynamic changes in muscle fiber fragments. 1. Effect of various heat treatments on length, width and birefringence. J Food Sci 33(5):468–470

Ibrahim RM, Goll DE, Marchello JA, Duff GC, Thompson VF, Mares SW, Ahmad HA (2008) Effect of two dietary concentrate levels on tenderness, calpain and calpastatin activities, and carcass merit in Waguli and Brahman steers. J Anim Sci 86(6):1426–1433

Insausti K, Goni V, Petri E, Gorraiz C, Beriain MJ (2005) Effect of weight at slaughter on the volatile compounds of cooked beef from Spanish cattle breeds. Meat Sci 70:83–90

Jaime I, Beltrán JA, Ceña P, López-Lorenzo P, Roncalés P (1992) Tenderisation of lamb meat: effect of rapid postmortem temperature drop on muscle conditioning and aging. Meat Sci 32(4): 357–366

Janz JAM, Aalhus JL, Price MA (2006) The effect of epimysial connective tissue on factors related to tenderness of beef semitendinosus. J Muscle Foods 17(1):43–55

Janz J, Morel P, Purchas R, Corrigan V, Cumarasamy S, Wilkinson B, Hendriks W (2008) The influence of diets supplemented with conjugated linoleic acid, selenium, and vitamin E, with or without animal protein, on the quality of pork from female pigs. J Anim Sci 86(6):1402–1409

Jones SDM, Jeremiah LE, Robertson WM (1993) The effects of spray and blast-chilling on carcass shrinkage and pork muscle quality. Meat Sci 34(3):351–362

Joseph RL (1996) Very fast chilling of beef and tenderness – a report from an EU concerted action. Meat Sci 43(Suppl. 1):217–227

Juárez et al (2012) Beef texture and juiciness. In: Hui YH (ed) Handbook of meat and meat processing, 2nd edn. CRC Press, Taylor & Francis Group, Florida, pp 177–206

Kemp CM, Sensky PL, Bardsley RG, Buttery PJ, Parr T (2010) Tenderness – an enzymatic view. Meat Sci 84(2):248–256

Kim JH, Kim DH, Ji DS, Lee HJ, Yoon DK, Lee CH (2017) Effect of aging process and time on physicochemical and sensory evaluation of raw beef top round and shank muscles using an electronic tongue. Korean J Food Sci Anim Resour 37:823–832

Koohmaraie M, Geesink GH (2006) Contribution of postmortem muscle biochemistry to the delivery of consistent meat quality with particular focus on the calpain system. Meat Sci 74(1):34–43

Koohmaraie M, Doumit ME, Wheeler TL (1996) Meat toughening does not occur when rigor shortening is prevented. J Anim Sci 74(12):2935–2942

Koohmaraie M, Kent MP, Shackelford SD, Veiseth E, Wheeler T (2002) Meat tenderness and muscle growth: is there any relationship? Meat Sci 62:345–352

Kristensen L, Purslow PP (2001) The effect of aging on the water-holding capacity of pork: role of cytoskeletal proteins. Meat Sci 58:17–23

Lampe JF, Baas TJ, Mabry JW (2006) Comparison of grain sources for swine diets and their effect on meat and fat quality traits. J Anim Sci 84(4):1022–1029

Laster MA, Smith RD, Nicholson KL, Nicholson JDW, Miller RK, Griffin DB, Harris KB, Savell JW (2008) Dry versus wet aging of beef: retail cutting yields and consumer sensory attribute evaluations of steaks from ribeyes, strip loins, and top sirloins from two quality grade groups. Meat Sci 80:795–804

Laville E, Sayd T, Morzel M, Blinet S, Chambon C, Lepetit J, Renand G, Hocquette JF (2009) Proteome changes during meat aging in tough and tender beef suggest the importance of apoptosis and protein solubility for beef aging and tenderization. J Agric Food Chem 57: 10755–10764

Lawrie RA (1966) The eating quality of meat. In: Meat science. Pergamon Press, London

Lawrie RA (1974) Meat science, 2nd edn. Pergamon Press, Oxford

Lawrie RA (1998) The eating quality of meat. In: Lawrie RA (ed) Meat science. Woodhead Publishing Ltd, Cambridge, pp 212–228

Leach L, Ellis M, Sutton D, McKeith F, Wilson E (1996) Influence of slaughter weight on growth and carcass characteristics, commercial cutting and curing yields, and meat quality of barrows and gilts from two genotypes. J Anim Sci 75:934–941

Lee HJ, Choe J, Kim KT, Oh J, Lee DG, Kwon KM, Choi YI, Jo C (2017) Analysis of low-marbled Hanwoo cow meat aged with different dry-aging methods. Asian Australas J Anim Sci 30: 1733–1738. https://doi.org/10.5713/ajas.17.0318

Lefaucheur L (2010) A second look into fibre typing – relation to meat quality. Meat Sci 84(2): 257–270
Lepetit J, Grajales A, Favier R (2000) Modelling the effect of sarcomere length on collagen thermal shortening in cooked meat: consequence on meat toughness. Meat Sci 54(3):239–250
Lepper-Blilie AN, Berg EP, Buchanan DS, Berg PT (2016) Effects of post-mortem aging time and type of aging on palatability of low marbled beef loins. Meat Sci 112:63–68. https://doi.org/10.1016/j.meatsci.2015.10.017
Lewis GJ, Purslow PP (1989) The strength and stiffness of perimysial connective tissue isolated from cooked beef muscle. Meat Sci 26(4):255–269
Lewis R, Emmans G, Simm G (2002) Effects of index selection on the carcass composition of sheep given either ad libitum or controlled amounts of food. Anim Sci 75:185–195
Li Chan E, Nakai S, Wood DF (1985) Relationship between functional (fat binding, emulsifying) and physicochemical properties of muscle proteins. Effects of heating, freezing, pH and species. J Food Sci 50(4):1034–1040
Light N, Champion AE, Voyle C, Bailey AJ (1985) The role of epimysial, perimysial and endomysial collagen in determining texture in six bovine muscles. Meat Sci 13(3):137–149
Lopez AL (1977) Effects of age, breed and method of carcass suspension on bullock beef palatability, carcass traits, histological measurements and certain age and yield predictions. Diss Abstr Int Sec B 37:5891
López-Campos Ó, Aalhus JL, Larsen IL, Basarab JA, Robertson WM (2010) Impacts of altered carcass suspension on beef tenderness and grade attributes. Canadian Meat Science Association/ Canadian Meat Research Institute/Canadian Meat Council Associate Members Technical Symposium, Victoria, p 4
Lynch A, Buckley DJ, Galvin K, Mullen AM, Troy DJ, Kerry JP (2002) Evaluation of rib steak colour from Friesian, Hereford and Charolais heifers pastured or overwintered prior to slaughter. Meat Sci 61(3):227–232
Mandell IB, Gullett EA, Wilton JW, Kemp RA, Allen OB (1997) Effects of gender and breed on carcass traits, chemical composition, and palatability attributes in Hereford and Simmental bulls and steers. Livest Prod Sci 49(3):235–248
McCormick RJ (2009) Collagen. In: Du M, McCormick RJ (eds) Applied muscle biology and meat science. CRC Press, Boca Raton, pp 129–148
Melton SL (1990) Effects of feeds on flavor of red meat: a review. J Anim Sci 68:4421–4435
Montgomery JL, King MB, Gentry JG, Barham AR, Barham BL, Hilton GG, Blanton JR Jr, Horst RL, Galyean ML, Morrow J Jr, Wester DB, Miller MF (2004) Supplemental vitamin D3 concentration and biological type of steers. II. Tenderness, quality, and residues of beef. J Anim Sci 82(7):2092–2104
Moreno T, Pérez N, Oliete B, Carballo JA, Franco D, Monserrat L (2007) Effects on quality attributes of commercial veal pieces under different ageing treatments. Int J Food Sci Technol 42(3):373–379
Morgan JB (1997) B-maturity: factors affecting physiological maturity. Range Beef Cow Symposium XV, Rapid City
Mustafa M, Chadwick J, Akhtar P, Ali S, Lateef M, Sultan J (2008) The effect of concentrate- and silage-based finishing diets on the growth performance and carcass characteristics of Suffolk Cross and Scottish blackface lambs. Turk J Vet Anim Sci 32:191–197
Nishimura T, Liu A, Hattori A, Takahashi K (1998) Changes in mechanical strength of intramuscular connective tissue during postmortem aging of beef. J Anim Sci 76:528–532
Nishimura T, Hattori A, Takahashi K (1999) Structural changes in intramuscular connective tissue during the fattening of Japanese black cattle: effect of marbling on beef tenderization. J Anim Sci 77:93–104
Novakofski J, Brewer S (2006) The paradox of toughening during the aging of tender steaks. J Food Sci 71:S473–S479
Okitani A, Ichinose N, Itoh J, Tsuji Y, Oneda Y, Hatae K, Migita K, Matsuishi M (2009) Liberation of actin from actomyosin in meats heated to 65°C. Meat Sci 81(3):446–450

Olsson U, Hertzman C, Tornberg E (1994) The influence of low temperature, type of muscle and electrical stimulation on the course of rigor mortis, ageing and tenderness of beef muscles. Meat Sci 37(1):115–131. https://doi.org/10.1016/0309-1740(94)90149-X

Ouali A, Herrera-Mendez CH, Coulis G, Becila S, Boudjellal A, Aubry L, Sentandreu MA (2006) Revisiting the conversion of muscle into meat and the underlying mechanisms. Meat Sci 74(1): 44–58

Pearson AM (1968) Estimating meat yield and quality in live animals. Proceedings of The Second World Conference on Animal Production, p 139

Pearson A, Young R (1989) Muscle and meat biochemistry. Academic, San Diego

Penfield M, Costello C, McNeil M, Riemann M (1989) Effects of fat level and cooking methods on physical and sensory characteristics of restructured beef steaks. J Food Qual 11(5):349–356

Pérez ML, Escalona H, Guerrero I (1998) Effect of calcium chloride marination on calpain and quality characteristics of meat from chicken, horse, cattle and rabbit. Meat Sci 48(1–2):125–134

Pérez-Alvarez JA (2006) Color. In: Hui YH, Guerrero-Legarreta I, Rosmini MR (eds) Ciencia y Tecnología de Carnes. Noriega Editores, Mexico City, pp 161–197

Pietrasik Z, Shand PJ (2004) Effect of blade tenderization and tumbling time on the processing characteristics and tenderness of injected cooked roast beef. Meat Sci 66(4):871–879

Priolo A, Micol D, Agabriel J, Prache S, Dransfield E (2002) Effect of grass or concentrate feeding systems on lamb carcass and meat quality. Meat Sci 62:179–185

Purslow PP (2005) Intramuscular connective tissue and its role in meat quality. Meat Sci 70(3): 435–447

Quinn MJ, Reinhardt CD, Loe ER, Depenbusch BE, Corrigan ME, May ML, Drouillard JS (2008) The effects of ractopamine-hydrogen chloride (Optaflexx) on performance, carcass characteristics, and meat quality of finishing feedlot heifers. J Anim Sci 86(4):902–908

Ramírez-Retamal J, Morales R, Martínez ME, de la Barra R (2013) Effect of breed and feeding on the carcass characteristics of the Chilote breed lamb. Chil J Agric Res 73:48–54

Rao MV, Gault NFS, Kennedy S (1989) Variations in water-holding capacity due to changes in the fibre diameter, sarcomere length and connective tissue morphology of some beef muscles under acidic conditions below the ultimate pH. Meat Sci 26(1):19–37

Reagan JO, Carpenter ZL, Smith GC (1976) Age-related traits affecting the tenderness of the bovine longissimus muscle. J Anim Sci 43(6):1198–1205

Rehfeldt C, Henning M, Fiedler I (2008) Consequences of pig domestication for skeletal muscle growth and cellularity. Livest Sci 116(1–3):30–41

Rentfrow G, Brewer MS, Carr TR, Berger LA, McKeith FK (2004) The effects of feeding elevated levels of vitamins D3 and E on beef quality. J Muscle Foods 15(3):205–223

Riley RR, Smith GC, Cross HR, Savell JW, Long CR, Cartwright TC (1986) Chronological age and breed-type effects on carcass characteristics and palatability of bull beef. Meat Sci 17(3): 187–198

Rodrigues S, Cadavez V, Teixeira A (2006) Breed and maturity effects on Churra Galega Bragancana and Suffolk lamb carcass characteristics: killing-out proportion and composition. Meat Sci 72:288–293

Rosenvold K, Petersen JS, Laerke HN, Jensen SK, Therkildsen M, Karlsson AH, Moller HS, Andersen HJ (2001) Muscle glycogen stores and meat quality as affected by strategic finishing feeding of slaughter pigs. J Anim Sci 79(2):382–391

Saffle RL, Bratzler LJ (1957) The effect of different degrees of pork carcass fatness on some processing and palatability characteristics. J Anita Sci 16:1071

Santos-Silva J, Vaz A (2001) The effect of weight on carcass and meat quality of Serra da Estrela and Merino Branco lambs fattened with dehydrated lucerne. Anim Res 50:289–298

Santos-Silva J, Mendes I, Bessa R (2002) The effect of genotype, feeding system and slaughter weight on the quality of light lambs. 1. Growth, carcass composition and meat quality. Livest Prod Sci 76:17–25

Sañudo C, Enser M, Campo MM, Nute G, María G, Sierra I et al (2000) Fatty acid composition and sensory characteristics of lamb carcasses from Britain and Spain. Meat Sci 54:339–346

Sather AP, Schaefer AL, Tong AKW (1990) Muscle and rectal temperature response curves to a short-term halothane challenge in eight-week-old piglets with known genotype at the halothane locus. Can J Anim Sci 70(1):9–14
Sato M, Nakamura T, Numata M, Kuwahara K, Homma S, Sato A, Fujimaki M (1995) Study on factors related to beef quality on the flavor and umami taste of Japanese black cattle branded beef. Anim Sci Technol 66:274–282
Savell JW, Cross HR, Smith GC (1986) Percentage ether extractable fat and moisture content of beef longissimus muscle as related to USDA marbling score. J Food Sci 51:838–839
Savell J, Cross H, Francis J, Wise J, Hale D, Wilkes D, Smith G (1989) National consumer retail beef study: interaction of trim level, price and grade on consumer acceptance of beef steaks and roasts. J Food Qual 12(4):251–274
Schone F, Kirchheim U, Kinast C (2006) Quality of the meat of bullocks: 1. Physically-chemical characteristics depending on breed, meat cut and storage. Fleischwirtschaft 86(11):101–107
Seideman SC, Crouse JD (1986) The effect of sex condition, genotype and diet on bovine muscle fiber characteristics. Meat Sci 17(1):55–72
Seideman SC, Smith GC, Carpenter ZL, Marshall WH (1977) Blade tenderization of beef psoas major and semitendinosus muscles. J Food Sci 42(6):1510–1512
Shackelford SD, Koohmaraie M, Miller MF, Crouse JD, Reagan JO (1991) An evaluation of tenderness of the longissimus muscle of Angus by Hereford versus Brahman crossbred heifers. J Anim Sci 69(1):171–177
Shackelford SD, Koohmaraie M, Wheeler TL (1995) Effects of slaughter age on meat tenderness and USDA carcass maturity scores of beef females. J Anim Sci 73(11):3304–3309
Shanks BC, Wulf DM, Maddock RJ (2002) Technical note: the effect of freezing on Warner–Bratzler shear force values of beef longissimus steaks across several postmortem aging periods. J Anim Sci 80(8):2122–2125
Siegel DG, Schmidt G (1979) Ionic, pH, and temperature effects on the binding ability of myosin. J Food Sci 44:1686–1691
Sierra V, Guerrero L, Fernández-Suárez V, Martínez A, Castro P, Osoro K, Rodríguez-Colunga MJ, Coto-Montes A, Oliván M (2010) Eating quality of beef from biotypes included in the PGI “Ternera Asturiana” showing distinct physicochemical characteristics and tenderization pattern. Meat Sci 86(2):343–351
Sikes AL, Tobin AB, Tume RK (2009) Use of high pressure to reduce cook loss and improve texture of low-salt beef sausage batters. Innov Food Sci Emerg Technol 10(4):405–412
Smith GC (1972) Relationship of the cow-calf producer to the consumer. In: Commercial beef cattle production. Lea and Febiger, Philadelphia
Smith GC (1985) Effects of electrical stimulation on meat quality, color, grade, heat ring, and palatability. In: Pearson AM, Dutson TR (eds) Advances in meat research: electrical stimulation. AVI Publishing Company, Inc., Westport, pp 121–158
Smith GC, Carpenter ZL, Dutson TR, Hostetler RL (1979) Singular and combined methods for increasing tenderness of beef. Proc Eur Meet Meat Res 25:5
Smith GC, Carpenter ZL, Cross HR, Murphey CE, Abraham HC, Savell JW, Davis GW, Berry BW, Parrish FC Jr (1985) Relationship of USDA marbling groups to palatability of cooked beef. J Food Qual 7(4):289–308
Smith KR, Duckett SK, Azain MJ, Sonon RN Jr, Pringle TD (2007) The effect of anabolic implants on intramuscular lipid deposition in finished beef cattle. J Anim Sci 85(2):430–440
Smith GC, Tatum JD, Belk KE, Scanga JA (2008) Post-harvest practices for enhancing beef tenderness. In: Research & knowledge management. National Cattlemen’s Beef Association, Centennial
Spanier AM, Berry BW, Solomon MB (2000) Variation in the tenderness of beef strip loins and improvement in tenderness by use of hydrodynamic pressure processing (HDP). J Muscle Foods 11(3):183–196

Stolowski GD, Baird BE, Miller RK, Savell J, Sams A, Taylor J, Sanders JO, Smith SB (2006) Factors influencing the variation in tenderness of seven major beef muscles from three Angus and Brahman breed crosses. Meat Sci 73:475–483

Therkildsen M, Houbak MB, Byrne DV (2008) Feeding strategy for improving tenderness has opposite effects in two different muscles. Meat Sci 80(4):1037–1045

Thomas ML, Brown AH, Kellogg DW (2008) Fatty acids and meat characteristics of different biological types of beef cattle developed under a management-intensive grazing system. J Food Qual 31(2):189–204

Thompson J (2004) The effects of marbling on flavour and juiciness scores of cooked beef, after adjusting to a constant tenderness. Aust J Exp Agric 44(7):645–652

Thompson JM, Perry D, Daly B, Gardner GE, Johnston DJ, Pethick DW (2006) Genetic and environmental effects on the muscle structure response post-mortem. Meat Sci 74(1):59–65

Thompson JM, McIntyre BM, Tudor GD, Pethick DW, Polkinghorne R, Watson R (2008) Effects of hormonal growth promotants (HGP) on growth, carcass characteristics, the palatability of different muscles in the beef carcass and their interaction with aging. Aust J Exp Agric 48(11): 1405–1414

Toldrá F, Flores M (2000) The use of muscle enzymes as predictors of pork meat quality. Food Chem 69:387–395

Vieira C, Cerdeño A, Serrano E, Lavín P, Mantecón AR (2007) Breed and ageing extent on carcass and meat quality of beef from adult steers (oxen). Livest Sci 107(1):62–69

Warner RD, Greenwood PL, Pethick DW, Ferguson DM (2010) Genetic and environmental effects on meat quality. Meat Sci 86(1):171–183

Weiler U, Appell HJ, Kremser M, Hofacker S, Claus R (1995) Consequences of selection on muscle composition. A comparative study on gracilis muscle in wild and domestic pigs. Anat Histol Embryol 24(2):77–80

Weir CE (1960) Palatability characteristics of meat. In: The science of meat and meat products. W. H. Freeman and Company, San Francisco

Weir CE, Wang H, Birkner ML, Parsons J, Ginger B (1958) Studies of enzymatic tenderization of meat. II. Panel and histological analysis of meat treated with liquid tenderizer containing papain. Food Res 23:411–422

Wheeler TL, Koohmaraie M, Crouse JD (1991) Effects of calcium chloride injection and hot boning on the tenderness of round muscles. J Anim Sci 69(12):4871–4875

Wheeler TL, Cundiff LV, Koch RM (1994) Effect of marbling degree on beef palatability in Bos taurus and Bos indicus cattle. J Anim Sci 72:3145–3151

Whitney MH, Shurson GC, Johnston LJ (2006) Growth performance and carcass characteristics of grower-finisher pigs fed high-quality corn distillers dried grain with solubles originating from a modern Midwestern ethanol plant. J Anim Sci 84(12):3356–3363

Wicklund S, Homco-Ryan C, Ryan K, McKeith F, McFarlane B, Brewer S (2005) Aging and enhancement effects on quality characteristics of beef strip steaks. J Food Sci 70:242–248

Willems MET, Purslow PP (1996) Effect of postrigor sarcomere length on mechanical and structural characteristics of raw and heat-denatured single porcine muscle fibres. J Texture Stud 27(2):217–233

Winger R, Hagyard C (1999) Juiciness – its importance and contributing factors. In: Pearson AM, Dutson TR (eds) Quality attributes and their measurement in meat, poultry and fish products. Aspen Publishing, Gaithersburg, pp 94–116

Wood JD, Nute GR, Richardson RI (2004) Effects of breed, diet and muscle on fat deposition and eating quality in pigs. Meat Sci 67(4):651–667

Xiong YL (1997) Structure – function relationships of muscle proteins. In: Damodaran S, Paraf A (eds) Food proteins and their applications. Marcel Dekker, New York

Xiong YL, Lou X, Harmon RJ, Wang C, Moody WG (2000) Salt and pyrophosphate-induced structural changes in myofibrils from chicken red and white muscles. J Sci Food Agric 80:1176–1182

Xu KY, Becker LC (1998) Ultrastructural localization of glycolytic enzymes on sarcoplasmic reticulum vesicles. J Histochem Cytochem 46(4):419–427
Young OA, Lovatt SJ, Simmons NJ, Devine CE (2001) Carcass processing: quality controls. In: Hui YH, Shorthose R, Young O, Koohmaraie M, Rogers P (eds) Meat science and applications. Marcel Dekker, New York, pp 295–318
Zamora F, Debiton E, Lepetit J, Lebert A, Dransfield E, Ouali A (1996) Predicting variability of aging and toughness in beef longissimus lumborum et thoracis. Meat Sci 43(3/4):321–333
Zayas JF (1997) Emulsifying properties of proteins. In: Zayas JF (ed) Functionality of proteins in food. Springer, Berlin
Zgur S, Cepon M, Cepin S (2003) Influence of growth rate in two growth periods on intramuscular connective tissue and palatability traits of beef. Czeh J Anim Sci 48:113–119

Chapter 4
Processed Meat Products

Jahangir Ahmad Rather, Najmeenah Akhter, Shazia Akhter, Hilal Ahmad Punoo, Neelofar Manzoor, Haseeba Muzaffar, and B. N. Dar

4.1 Introduction

In recent years, there is a notable rise in both the consumption and production of meat. The demand for high-quality meat and its products is consistently rising, driven by a growing awareness of the nutritional and palatable qualities of meat as well as the sense of satisfaction derived from its consumption. Factors such as increased purchasing power, evolving socio-economic status, and changing lifestyles have also played a role in the heightened consumption of processed and convenient meat products (Font-i-Furnols 2023).

Meat processing encompasses various treatments that bring about significant chemical and physical alterations to the natural state of meat. This includes processes such as grinding, curing, smoking, cooking, canning, freezing, fermentation, dehydration, and the use of additives like seasonings, chemicals, and enzymes, among others. While processing modifies the properties of fresh meat, it is crucial to ensure that the fundamental quality of "being meat" remains intact even after processing (Shirsath and Henchion 2021). Meat technology involves the application of any or all applied sciences that have practical value or industrial use in the production of meat. It is the utilization of scientific and modern knowledge to streamline various aspects, including slaughter, processing, transportation, storage, and marketing, aiming for more efficient methods that offer practical advantages over traditional

J. A. Rather (✉) · H. A. Punoo · H. Muzaffar
Department of Food Science and Technology, University of Kashmir, Srinagar, India
e-mail: Jahangir.rather@islamicuniversity.edu.in

N. Akhter · N. Manzoor · B. N. Dar (✉)
Department of Food Technology, Islamic University of Science and Technology, Pulwama, India

S. Akhter
Division of Food Science and Technology, SKUAST, Srinagar, Jammu and Kashmir, India

S. A. Rather, F. A. Masoodi (eds.), *Hand Book of Processed Functional Meat Products*, https://doi.org/10.1007/978-3-031-69868-2_4

production approaches. Hence, processing is an integral component of meat technology.

The processing of meat and meat products varies within the countries and regions of the world. In the northern regions Harrisa, Goshtaba, Rista, Tabakhmaz, Natyakhni and Methimaz are most famous processed meat products. The processed meat products are gaining importance and are attracting the tourists of the world owing to the rich flavor of these meat products (Rather et al. 2024). In the Eastern regions the most popular processed meat products are Meat Curry, Mutton Dopiyaza, Mutton Korma and Rapka. In the Southern, Eastern and Northwestern regions, prevalent culinary choices include meat rolls made from shoulder or pork ham, vindaloo crafted from pickled meat chunks deep-fried in fat, and shakudi, a dish prepared by cooking meat in coconut milk or grated coconut along with a blend of spices. Canned meat products are also gaining importance nowadays due to the safety and shelf stability of the canning meat products. The meat, fish and poultry products are being canned and sterilized at 121 °C after canning and transferred to various parts of world (Rather et al. 2021).

Consumers are increasingly interested in fermented meat products due to their health-promoting properties. Traditional fermented and preserved meat options from different countries include European salami, Portuguese alheira, Spanish androlla, Thai nham, Indian and Nepalese kargyong, satchu, and suka ko masu, Indian arjia, jamma and chartayshya, as well as Vietnamese nem chua. Additionally, cured meat and meat products enjoy global popularity, featuring variations in smoking, fermentation, and curing methods, whether dry or wet. Illustrating this trend are prominent instances of dry-cured hams like Parma, Country ham, Prosciutto, Smithfield, and Westphalia and the renowned Jinhua ham from China represents another distinctive style of dry-cured ham (Anagnostopoulos and Tsaltas 2019). Sausages are also important processed meat products of various countries. These are minced meat products may be smoked, un-smoked, fermented or unfermented, cured or un-cured and are being placed in the casings (Carballo 2021).

4.2 Significance of Processed Meat Products

The processing of meat contributes to the creation of value-added, diverse, and convenient meat products that align with lifestyle preferences. It enables the efficient utilization of various components such as cuts, carcasses, and edible byproducts (Brar et al. 2021). The processing of meat enables the integration of non-meat ingredients, leading to enhanced quality and cost-effectiveness. It plays a crucial role in facilitating distribution, transportation, preservation, and marketing to broader populations. Additionally, it supports export efforts, fosters entrepreneurial ventures, competes with imports, and also fosters employment. Value-added products, characterized by increased convenience through reduced preparation time and steps, allow the use of specific parts, the generation of diverse flavors, and extended product shelf life.

These products can be generally classified according to their processing methods, function and variety and convenience (Mora et al. 2019).

4.3 Processed Meat and Meat Products Demand

Muscle foods play a crucial role in the human diet, supplying essential macro and micronutrients necessary for growth and the maintenance of health. On a per capita basis the increase in consumption of animal products is more pronounced in developing countries compared to developed ones (Savarino et al. 2021). In India, per capita animal protein consumption stands at approximately 10.4 g per day, contrasting with the world average of 25 g/day. Considering the recommended minimum daily intake of 20 g per capita for animal protein, comprising 10 g from milk, 4 g from meat, and 2 g from eggs, the anticipated current populations demand is estimated to be 104 million metric tons (mt) for milk, 7.7 mt for meat, and 4.6 mt for eggs (based on a population of 104 billion). This contrasts with the current production levels of 84 mt for milk, 6.04 mt for meat, 5.6 mt for fish, and 30 billion for eggs (wakhungu 2019).

4.4 Processed Meat Products

Striving for technological advancements and efficient processes is crucial to attain reduced production expenses, product diversification, higher yields, byproduct utilization, improved shelf life, and robust quality control and management systems. Product diversification becomes essential to minimize imports and support domestic industries. While there is currently a modest production of processed pork products relative to the abundant and affordable pork available, emphasis should be placed on global competitiveness, socio-ecological friendliness, and energy conservation in processing technology (Arora 2023).

Organized processing, under professional supervision, ensures the delivery of safe and affordable products to consumers. The muscle-based fast-food industry holds significant potential in the country, with multinational food companies already establishing their presence. Globalization has expanded product choices and value for consumers, with an increasing demand for convenient muscle-based fast food driven by industrialization, urbanization, higher standards of living, an uptick in working women, rising literacy, and growing health awareness. The changing trends in food consumption, transitioning from cereals to dairy and meat products, are notably conspicuous among the growing middle class that possesses increased purchasing power.

4.4.1 Processed Meat Products in the Northern Region of India

In India, the state of Jammu and Kashmir (J&K) stands out not only for its picturesque natural landscapes but also for its indigenous meat products, turning it into a culinary heaven. The region boasts a variety of traditional meat products collectively known as 'Wazwan,' once exclusive to royal feasts. The preparation of most Wazwan dishes is a specialized culinary practice primarily undertaken by a specific group of individuals known as "Waza" (chefs who prepare Wazwan for special occasions, such as feasts during marriages). This culinary practice is often a family profession and comes with a high price tag. The Wazwan encompasses flavorful, ready-to-eat meat products that are freshly prepared and served hot as part of elaborate meals. These meat products have gained immense popularity, establishing a niche for themselves and becoming a predominant component of the local market (Rather et al. 2016b). In addition to meeting local demand, these products also cater to the fast-food needs of foreign and domestic tourists, delighting all those who visit the Union territory of J & K. Notable meat delicacies include Harrisa, Rista, Goshtaba, Tabakhmaz, Nate Yakhni, Kabab, Aab-Gosh, Rogan-Josh, Dopiyaza, and more, as shown in Fig. 4.1. These delicacies are typically prepared using lamb or mutton as the primary raw material. While some have Persian origins, others have been developed in the Kashmir Valley over the past few centuries. Two essential Wazwan products, Goshtaba and Rista are emulsion-type meat products created from communated meat and are integral components of the world-famous traditional Kashmiri cuisine (Rather et al., 2024). Both Rista and Goshtaba are vital due to their highly appealing flavor, texture, and other palatability characteristics. In the Northern plains of the country, indigenous products include butter chicken, curry, meat and chicken pickles, tandoori chicken, korma, and a variety of kababs and biryanis. Some of these products are described below:

4.4.1.1 Harrissa

Harrissa, highly enjoyed in J & K and other states of India, especially during winter mornings, is a spice-rich meat product. Typically made from mutton, chevon, or beef. Harrissa has a pasty consistency with a distinct spice-rich flavor. The cooking process involves adding meat, salt, and spice mix to water, cooking it in large vessels overnight for 5–8 h until a clear paste is obtained. Rice paste and condiments are then added, and the mixture is simmered for 15 mins. Preheated refined oil is finally added, and the mixture is simmered until a thick consistency is achieved. Harrisa is commonly served for breakfast in the Kashmir Valley with local bread, topped with hot oil and garnished with methi-maz, saffron, and Kabab pieces.

Fig. 4.1 Ethnic meat products of Northern region meat products, (**a**) Harrisa, (**b**) Tabakhmaz, (**c**) Goshtaba, (**d**) Rista, (**e**) Nate-yakhni, (**f**) Seekh kabab and (**g**) Methi maz

4.4.1.2 Tabaq-Maz

Tabaq-Maz, a main course delicacy in Wazwan, is a ready-to-consume meat item crafted from the rib section of chevon, beef or lamb. The product is braised in water or milk containing spices, condiments, and salt until the meat becomes tender and absorbs the rich spice flavor. The ribs are subsequently fragmented into rectangular pieces and gently fried in ghee over a low flame, using partially charred wood pieces, until the product achieves a semi-dry and crispy texture. Tabaq-Maz is known for its unique crunch and flavor, often served as a main dish in Jammu and Kashmir restaurants.

4.4.1.3 Goshtaba

Goshtaba is a traditional meat product prepared by pounding meat and fat together on a stone slab using a wooden mallet. The resulting smooth emulsion is shaped into large round balls and cooked in a white-colored gravy made of water, dahi, salt, condiments and spices. Goshtaba has a slightly sour flavor and is typically served with rice(Rather et al. 2024).

4.4.1.4 Rista

Similar to Goshtaba in its preparation, Rista is crafted by pounding meat and fat on a stone slab with a wooden mallet. The resultant meat emulsion is then shaped into small balls and cooked in a red gravy infused with condiments, spices, and water (Rather et al. 2016a). The unique red hue of Wazwan gravies is attained through the inclusion of saffron or Celosia flowers.

4.4.1.5 Nate Yakhni

Nate Yakhni, a popular meat product in Jammu and Kashmir, is prepared by Kashmiri people during occasions and festivals, and it is a significant dish in Wazwan. The dish is made using chevon or mutton in white-colored gravy of curd. The meat pieces are first partially cooked in water and then further cooked in a gravy made by stirring curd containing condiments, spices, animal fat, salt, and meat broth (Rather et al., 2016a). Nate Yakhni is frequently garnished with powdered dried mint and typically enjoyed alongside rice.

4.4.1.6 Seekh Kababs

Seekh Kababs, renowned as convenient meat offerings from Wazwan, are commonly served in global restaurants and food establishments either as appetizers or as part of the main course. Typically, they are grilled or broiled over charcoal. Variations can be created using the meat of other animals such as fish and chicken. The meat is ground coarsely, mixed with condiments, spices, and a binder, shaped into a cigar on a skewer, and cooked on charcoal. The product is turned several times until it is cooked and semi-dry, often brushed with fat during cooking (Hakeem et al. 2014).

4.4.1.7 Methi-Maaz

Methi-Maaz, made from lamb offals takes the lead as the initial dish served in Wazwan. The offals are first thoroughly cooked in boiling water with spices and condiments. The chopped offals are then cooked in gravy with condiments, spices, fat, and fenugreek leaves powder. This meat product has a thick, gravy-like consistency, is served with rice, and is known for its rich taste and flavor.

4.4.2 Processed Meat Products in the Eastern Region of India

Several traditional meat delicacies are popularly prepared and enjoyed in the Eastern parts of the country. Important products include:

4.4.2.1 Meat Curry

Crafted from bone-in chevon or mutton chunks, this curry is infused with a combination of salt, spices, and condiments. The chunks can be either marinated in a mixture of salt, spices, condiments, and sour curd or cooked without marination. After frying on a low flame, the meat chunks are cooked until tender.

4.4.2.2 Mutton Dopiyaza

This meat dish consists of mutton, onions, sour curd, condiments, spices and lemon juice, and is typically savored with rice or chapati.

4.4.2.3 Mutton Korma and Rapka

Mutton korma, a popular dish in northern India, involves frying marinated medium-size meat pieces in ghee with spices and onions. The meat is gently cooked in a covered pan, and curd is added, continuing until the meat reaches a tender consistency. While korma can be prepared from chevon, lamb, or chicken, rapka is made from yak and Mithun meat, gaining popularity in Arunachal Pradesh. Prepared within bamboo containers and exposed to a traditional fire source for several weeks, this meat product possesses a semi-dry texture and a smoky flavor.

4.4.3 Meat Products in the Western, Southern and North-Eastern Regions of India

While the western part of the country boasts various traditional meat products, many are undocumented and require scientific attention. Some popular items include meat rolls made from pork ham or shoulder, shakudi, in which meat is cooked with coconut milk or grated coconut along with spices and vindaloo prepared from pickled meat chunks deep-fried in fat. Southern meat products tend to be spicier and hotter than those in other parts of India, featuring fried products, curries and biryanis prepared from mutton, chicken, and chevon. In states like Andhra Pradesh and Tamil Nadu, Biltong, a dried and salted meat product, is particularly popular. The northeastern part of the country sees a diverse range of traditional meat products prepared and preserved using local methods. Products made from chevon, mutton, beef, or chicken can be preserved for months or years. One popular product is Kargyong, a hard-to-soft brown sausage-like product indigenous to Darjeeling, Sikkim, and Ladakh, prepared from pork, beef, or yak.

Cheu, a native pork-based delicacy, is grilled over an open flame using bamboo sticks. Doh kpu, a minced pork product originating from the Meghalaya region, entails creating an emulsion shaped into balls and then deep-frying them in oil. Egadin banum is a pork-based dish made by cooking marinated pork pieces on bamboo sticks over an open flame. In the northeastern region, it's not uncommon to find meat products based on unconventional animals such as frogs, snakes, and lizards.

4.4.4 Other Processed Meat Products of Different Countries

Numerous other meat products are being processed and consumed all around the world. These meat products usually differ on the traditional recipe and processing method applied. The various other processed meat products are as under;

4.4.4.1 Canned Meat Products

The meat industry is directing its attention towards exports, seeking distinctive meat products with extended shelf life suitable for both international and national markets. However, the considerable perishability of meat and its products poses a significant hurdle to their commercialization. Consequently, heat treatment is employed to create meat products with an extended shelf life and ensured microbiological safety (Madian et al., 2023). Canned meat products have been a staple in the food industry for decades, providing convenient and long-lasting options for consumers worldwide. The process of canning meat involves preserving it in a sealed container, typically tin-coated steel or aluminum, through heat treatment to destroy microorganisms that could cause spoilage. Traditional meat products, processed industrially or canned, are provided to consumers with a labelled shelf life exceeding 12 months (Rather et al. 2021). The industrial processing (canning) of these meat products takes place at a temperature of 121 °C. This increases the shelf life of meat and meat products by destroying the spoilage-causing microorganisms and making these products safe for consumers. The canning process begins with the preparation of the meat. It involves cleaning, trimming, and sometimes cooking the meat before placing it in cans. Seasonings and sauces may be added for flavor enhancement. Once prepared, the meat is carefully placed into cans, and any additional liquids, such as broth or sauce, are added. Proper filling is crucial to ensure even heat distribution during the subsequent heat treatment. The filled cans are sealed to create an airtight environment. This step is essential for preventing the entry of microorganisms that could lead to spoilage. The type of seal used depends on the canning method and the specific requirements of the product. The sealed cans undergo heat treatment to destroy bacteria, molds, and enzymes that could cause food spoilage. This process, known as sterilization, typically involves exposing the cans to high temperatures (121 °C) for a specified period. After heat treatment, the cans are cooled to room temperature. This step is crucial for maintaining the quality and safety of canned meat products. Rapid cooling helps prevent overcooking and maintains the texture of the meat. Reports indicate that subjecting meat products to thermal processing at elevated temperatures leads to lipid oxidation, alterations in fatty acid compositions, and the generation of cholesterol oxidation products (COPs) as well as Maillard's reaction (Liu et al., 2022). The generation of COPs (cholesterol oxidation products) is increased during the storage of meat and its products. Generally, factors such as high temperatures, light, pH, oxygen, water activity, and the presence of unsaturated fatty acids play crucial roles in the formation of cholesterol oxidation products during processing or storage. Overconsumption of cholesterol oxidation products has been shown to lead to carcinogenic, cytotoxic, atherogenic and mutagenic effects.

4.4.4.2 Sausages

Recognizing research conducted in other nations, Germany is credited with pioneering scientific knowledge in sausage manufacturing during the mid-decades of the last century. Manuals documenting a substantial body of scientific insights derived from the analysis of various local German sausage varieties are now considered timeless classics and indisputable references in the scientific literature of this field (Carballo, 2021). These manuals also provide guidance and recommendations for producing sausages with consistently high quality, while helping to prevent common defects and manufacturing mishaps. Among various types of sausages, the production of dry-fermented sausages presents particular complexity. This is due to the fact that the sensory characteristics of these products arise from a sequence of changes in the raw materials and ingredients, facilitated by enzymes present in meat tissue and interacting microorganisms. These activities are further influenced by ingredients such as salt and spices, as well as the environmental conditions during the ripening process (Rodríguez-González et al. 2020). As they undergo no heat treatment, dry-fermented sausages pose the highest microbiological risks among all sausage types. The predominant research focus in this field has been on the exploration of these sausages. Commencing in the 1980s, a series of pioneering studies, comprehensively reviewed by Luke (1985) paved the way for subsequent investigations into the biological, chemical and microbial characterization of various local types of air dried fermented sausages. These studies, primarily conducted on French, Italian, Greek, Spanish, and Portuguese sausages utilizing spontaneous fermentation, consistently highlight the presence, albeit with varying intensity of proteolytic, glycolytic, lipolytic, and oxidation processes during maturation. They also underscore the significant roles played by muscle and fat enzymes, as well as microorganisms, in these processes. Concerning microbiological characteristics, LAB and microorganisms from the Micrococci and Staphylococci families emerged as the predominant microbial flora in these local sausage varieties, with some contribution from surface molds and yeasts in specific cases. Lactic acid bacteria exhibit rapid development during the fermentation of raw-cured sausages, reaching counts of 108–109 c.f.u./g, which remain stable until the end of the drying/ripening stage. This microbial group plays a crucial role in ensuring the hygienic/sanitary quality of the sausage by producing organic acids (lactic and acetic), reducing pH values, inhibiting the growth of spoilage bacteria, hastening product dehydration, affecting the development and endurance of color, as well as contributing to the aroma and taste, especially in sausages with short maturation periods. Additionally, lactic acid bacteria can produce bacteriocins, protein compounds that enhance antimicrobial activity, facilitating their establishment and inhibiting the growth of undesirable microorganisms. While their lipolytic activity is considered weak, their proteolytic activity, especially in various Lactobacillus species, has been extensively studied and demonstrated (Carballo 2021).

4.4.5 *Fermented Meat Products*

Fermentation, a technique with a long history, can be seen as a biological method for preserving food, imparting desirable properties such as excellent sensory qualities to the end product (Smid and Hugenholtz 2010). Its popularity has grown in many cultures where fermented foods are integral to daily consumption, owing to their nutritional value and diverse sensory attributes. Overall, fermentation is an environmentally friendly and economically efficient process that is essential for ensuring food safety and creating products with desirable sensory characteristics (Liu et al. 2011). The benefits of fermentation, such as the creation of novel appealing tastes and textures in the final product, are entirely different from those found in the original materials. For instance, table olives only become enjoyable after the microbial (or chemical) induced elimination of bitterness caused by phenolic compounds. Yeast (*Saccharomyces cerevisiae*), either on its own or in conjunction with lactic acid bacteria (LAB), plays a role in leavening dough in bread production or transforming grape must into wine. Despite minor technological advancements, many fermentation processes, including these examples, have largely remained unchanged since ancient times. Meanwhile, sciences such as microbiology, chemistry, and food physics have analytically studied them, and new manufacturing technologies have emerged (Plessas et al. 2011).

Historically conserved and fermented meat items from diverse nations (Table 4.1) include European salami, Portugal's alheira, Spain's androlla, Thailand's nham, India and Nepal's kargyong, satchu, and suka ko masu, as well as India's arjia, chartayshya, and jamma, and Vietnam's nem chua. The key microbes involved in fermentation of meat include LAB species and coagulase-negative cocci, with the presence of yeasts and enterococci in some meat products. The Denaturing Gradient Gel Electrophoresis (DGGE) method, employed for culture-independent

Table 4.1 Fermented meat and fish products with diverse probiotic strains

Product	Microorganisms	References
Balao-balao	*Lactobacillus spp., enterococcus spp., Leuconostoc spp., Pediococcus spp.*	Alexandraki et al. (2013)
Burong	*Leuconostoc spp., lactobacillus spp.*	Dalmacio et al. (2011)
Gulbi	*Aspergillus sp., staphylococcus sp., Candida sp*	Anagnostopoulos and Tsaltas (2019)
Ika-shiokara	*Staphylococcus sp., micrococcus sp., Debaryomyces sp.*	Alexandraki et al. (2013)
Narezushi	*Leuconostoc spp., lactobacillus spp.*	Alexandraki et al. (2013)
Sukuti	*Lactobacillus spp., Lactococcus spp., Candida spp., enterococcus spp., Saccharomycopsis sp.*	Anagnostopoulos and Tsaltas (2019)
Lanhouin	*Bacillus spp.*	Anihouvi et al. (2012)
Momone	*Lactobacillus spp., bacillus spp., Lactococcus spp., Debaryomyces spp.,*	Anagnostopoulos and Tsaltas (2019)
Plaa-som	*Lactococcus spp., lactobacillus spp., streptococcus spp., Pediococcus*	Kopermsub and Yunchalard (2010)
Sausages	*Leuconostoc spp., lactobacillus spp., Weisella spp.*	Rantsiou et al. (2005)
Salame	*Lactobacillus spp., staphylococcus spp., macrococcus spp.,*	Anagnostopoulos and Tsaltas (2019)

identification, has unveiled *L. curvatus* and *L. sakei* as the primary lactic acid bacteria (LAB) species participating in the transformation process. This is complemented by coagulase-negative cocci *Staphylococcus xylosus* during the fermentation and ripening of meat (Anagnostopoulos and Tsaltas 2019). Predominant lactic acid bacteria (LAB) found in fermented meats encompass *L. curvatus, L. brevis, Lb. plantarum, L. casei, L. sakei, Lactobacillus carnis, L. curvatus, Lactobacillus divergens, Lactobacillus sanfranciscensis, Leuconostoc carnosum, Leuconostoc gelidium, Leuconostoc pseudomesenteroides, Leuconostoc citreum, L. mesenteroides, Pediococcus acidilactici, P. pentosaceus, W. cibaria, Weissella viridescens, Bacillus lentus, B. licheniformis, Bacillus mycoides, B. subtilis, B. thuringiensis, Enterococcus cecorum, E. durans, E. faecalis, E. faecium* and *Enterococcus hirae* (Nguyen et al. 2010), and coagulase-negative staphylococci, micrococci, and Enterobacteriaceae (Marty et al. 2012). Fermented sausages of Spanish origin commonly incorporate yeast species, including *Cattleya curvata, Cattleya zeylanoides, C. parapsilosis, Trichosporon ovoides, Citeromyces matritensis*, and *Y. lipolytica* (Anagnostopoulos and Tsaltas 2019).

Preserving fish by fermentation, smoking, sun-drying, and salting is a longstanding tradition in coastal regions or areas near lakes and rivers. The resulting preserved or fermented fish products find use as seasonings, condiments, curries, and side dishes (Salampessy et al. 2010). Microorganisms commonly isolated from fish include Bacillus, Lactobacillus, Debaryomyces, Pediococcus, Pseudomonas, and Hansenula. Examples of fermented fish products from various cultures include hentak and ngari in India, bagoong in the Philippines, terasi in Indonesia, belacan in Malaysia, and ngapi in Myanmar. Halophilic bacteria such as *Lentibacillus jeotgali, Gracilibacillus thailandensis, Paenibacillus tyraminigenes, Piscibacillus salipiscarius* and others participate in the fermentation of fish paste. These ethnic fermented fish products are found worldwide (Table 4.1) include hentak, ngari, and tungtap in India, bordia, karati, and lashim in India, jeotgal or jeot or saeujeot in Korea, plaa-som in Thailand, shiokara in Japan, patis in the Philippines, surströmming in Sweden, and sun-dried or smoked fish products such as gnuchi, sidra, and sukuti in India, Nepal, and Bhutan. Bacteria such as *E. faecalis, Lb. plantarum, L. reuteri, Streptococcus salivarius*, species of Bacillus, Micrococcus, Pediococcus, and yeasts including species of Candida and Saccharomyces are reported in thailand fermented and sun-dried fish products (Anagnostopoulos and Tsaltas 2019). Moreover, *Haloanaerobium praevalens* has been detected in surströmming, the fermented herrings of Sweden, whereas *Haloanaerobium fermentans, Tetragenococcus muriaticus*, and *Tetragenococcus halophilus* have been observed in Japanese puffer fish (Devi et al. 2015). Other microbial species, including B. *subtilis, E. faecalis, E. faecium, Lactococcus plantarum, Bacillus pumilus, Lactobacillus fructosus, Lc. lactis subsp. cremoris, Lc. lactis subsp. lactis, Lactobacillus amylophilus, Lactobacillus confusus, Lactobacillus corynifomis subsp. torquens, Lb. plantarum, L. mesenteroides, Saccharomycopsis spp.*, Micrococcus, as well as yeasts like *Candida bombicola, Candida chiropterorum* and *P. pentosaceus,* were isolated from fermented and sun-dried fish products in India (Anagnostopoulos and Tsaltas 2019).

4.4.5.1 Cured Meat

The various cured meat and meat products are shown in the Fig. 4.2. Curing, or the salting of meat, holds a rich historical legacy as one of the oldest preservation methods, dating back to 3000 BC. Preservation was achieved by applying concentrated salt to the lean surface of the meat, leading to dehydration to a level where bacterial growth and spoilage were inhibited, resulting in a safe and enjoyable product. In modern times, the primary purpose of curing is to impart an appealing flavor and color to cured meat. Two primary methods for applying curing ingredients are pickle or brine cures and dry cures. Dry cures, employed to produce country-style or dry-cured hams and bacon for niche markets, involve hand-rubbing a mixture of salt and nitrate or nitrite onto the meat's surface. This process requires several applications, and the cured products are ready for merchandising after several months of curing under environmental or controlled temperature conditions. Examples of dry-cured hams include Country ham, Westphalia, Smithfield, Parma, and Prosciutto. The renowned Jinhua ham in China represents another style of dry ham. On the other hand, the most commonly used modern method of curing is the pickle or brine cure. In pickle curing, salt is the primary and essential ingredient for achieving the desired flavor in cured meats. Different concentrations of pickles are formed by dissolving varying salt concentrations in water, measured by a salimeter and expressed as degrees of pickle. Pickle cures typically range from 50° to 85°, with

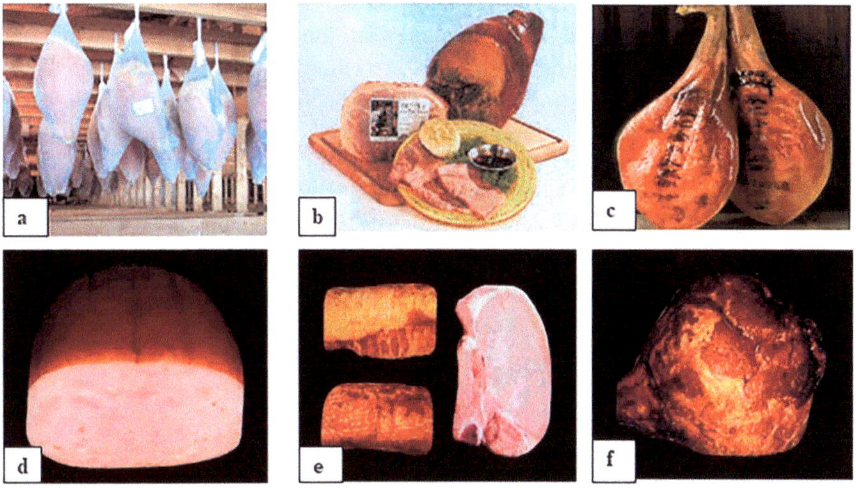

Fig. 4.2 Cured meat products, (**a**) ageing of ham, (**b**) dry cured ham, (**c**) Chinese jinhua ham, (**d**) turkey ham, (**e**) cured pork ham and (**f**) cured and smoked meat with bone

65° pickles being the most common. These pickles result in an acceptable salt level desired by consumers, representing about 2–3% in hams and 2% in bacon. Although these salt levels are lower than those in dry-cured products, they still help inhibit some microbial growth. Due to the lower salt concentrations, mechanical refrigeration is necessary to preserve and extend the shelf life of the products.

Incorporating a pickle or brine into meat is commonly achieved through a low-pressure pump injection system, which speeds up curing, ensures rapid and uniform distribution of curing ingredients, and facilitates the production of marinated or enhanced products. Artery pumping, stitch pumping, and multi needle injection are methods employed for the uniform distribution of the cure throughout the interior of cuts. Artery pumping involves uniformly injecting the cure via the vascular system, while stitch pumping uses a needle with multiple holes inserted into the muscle in various locations. Multineedle injection, a variation of stitch pumping, is a widely used commercial injection system in curing hams and bellies. The optimal internal temperature for meat cuts during the curing process is 34–40 °F, and the pickle or brine temperature should not exceed 40 °F for the best cure retention. Hams are typically pumped to 110% of their green weight, while bacons can vary from 105% to 110% pump. However, bacon must be shrunk back to the original green weight after processing. A typical automated ham processing line can inject 150,000 pounds a day. Massaging or tumbling of boneless hams ensures the uniform distribution of cure, extraction of proteins for binding meat pieces, and the attainment of the desired texture. To reduce curing times, large commercial processors often move products directly from the curing process to thermal processing units for cooking and smoking. After heating and smoking, the cured products are refrigerated and then packaged for retail sale as whole cuts, parts, or specialties such as sliced products. Hams can be bone-in, partially boneless (whole or part), or boneless. Bellies, hams, and picnic shoulders are the primary pork cuts cured and smoked. Some heavy boneless pork loins are cured and smoked and sliced into Canadian bacon, a popular pizza topping. Cured beef is often found in the form of corned beef, with spices, garlic, and whole peppers added during production. Corned beef can be slow-cooked as a roast or sliced thin for sandwiches like the Reuben. Additionally, some beef round muscles are cured, smoked, and dried, sold as dried beef. The boneless beef plate can be cured, smoked, and marketed as beef bacon. The serratus ventralis muscle in the short ribs may be cured, smoked, and spiced for pastrami. Turkey thigh meat is another option for curing, forming, smoking, and cooking to create turkey ham. Various factors influence the quality of cured meats, with high levels of sanitation and low refrigeration temperatures being crucial for producing high-quality products. The curing room should be maintained at refrigerated temperatures, and proper sanitation and temperature control are essential during packaging, transportation, storage, and retail sales of the processed products.

4.5 Conclusions

Various traditional meat products are crafted in diverse regions of India. Some of these items have achieved nationwide recognition and can be found in restaurants across the country and beyond. These indigenous meat products play a dual role in enhancing both nutrition and local community revenue. Urgent scientific attention is warranted to systematically document, standardize and modernize the processing techniques for these traditional meat products. Additionally, there is a need to delve into the packaging requirements and devise suitable packaging systems tailored for these indigenous meat products. Furthermore, comprehensive research efforts are essential to enhance the shelf life of select indigenous meat products, exploring natural and environmentally friendly methods such as the utilization of plant extracts and the development of biodegradable or cellulose-based packaging materials.

References

Alexandraki V, Tsakalidou E, Papadimitriou K, Holzapfel WH (2013) Commission on genetic resources for food and agriculture. In: Status trends conservation of sustainable use and conservation of microorganism. Food Process. FAO Background. Study Paper

Anagnostopoulos DA, Tsaltas D (2019) Fermented foods and beverages. In: Innovations in traditional foods. Woodhead Publishing, pp 257–291

Anihouvi VB, Kindossi JM, Hounhouigan JD (2012) Processing and quality characteristics of some major fermented fish products from Africa: a critical review. Int Res J Biol Sci 1:72–84

Arora K (2023) The status and potential of unorganized sector of food processing in India and its contribution in rural economy. In: Transforming organic agri-produce into processed food products. Apple Academic Press, pp 301–325

Brar PS, Mehta N, Singh A, Sivakumar S, Phand S (2021) Value addition of milk and meat: a push to entrepreneurship [E-book]. Hyderabad Guru Angad Dev Veterinary and Animal Sciences University. Ludhiana National Institute of Agricultural Extension Management, Hyderabad, India

Carballo J (2021) Sausages: nutrition, safety, processing and quality improvement. Food Secur 10: 890

Dalmacio LM, Angeles AK, Larcia LL, Balolong M, Estacio R (2011) Assessment of bacterial diversity in selected Philippine fermented food products through PCR-DGGE. Benefic Microbes 2:273–281

Devi KR, Deka M, Jeyaram K (2015) Bacterial dynamics during yearlong spontaneous fermentation for production of ngari, a dry fermented fish product of Northeast India. Int J Food Microbiol 199:62–71

Font-i-Furnols M (2023) Meat consumption, sustainability and alternatives: an overview of motives and barriers. Food Secur 12:2144

Hakeem HR, Kumar A, Shafat S, Gupta S (2014) Standardization and optimization of the basic formulation and processing conditions for the preparation of chicken seekh kabab from meat of spent hens. Vet Pract 15:154–156

Kopermsub P, Yunchalard S (2010) Identification of lactic acid bacteria associated with the production of plaa-som, a traditional fermented fish product of Thailand. Int J Food Microbiol 138:200–204

Liu S, Han Y, Zhou Z (2011) Lactic acid bacteria in traditional fermented Chinese foods. Food Res Int 44:643–651

Liu J, Han L, Han W, Gui L, Yuan Z, Hou S, Wang Z, Yang B, Raza SHA, Alowais AFS (2022) Effect of different heat treatments on the quality and flavor compounds of black Tibetan sheep meat by HS-GC-IMS coupled with multivariate analysis. Molecules 28:165

Luke FK (1985) Fermented sausages. In: Wood BJB (ed) Microbiology of fermented foods, vol 1. Elsevier Applied Science Publisher, London/New York, pp 41–83

Madian MA, Mousa MM, Ayoub KH (2023) Chemical and microbiological spoilage of canned meat products. Alexandria J Vet Sci 79:73–80

Marty E, Buchs J, Eugster-Meier E, Lacroix C, Meile L (2012) Identification of staphylococci and dominant lactic acid bacteria in spontaneously fermented Swiss meat products using PCR–RFLP. Food Microbiol 29:157–166

Mora L, Toldrá-Reig F, Reig M, Toldrá F (2019) Possible uses of processed slaughter byproducts. In: Sustainable meat production and processing. Elsevier, pp 145–160

Nguyen H, Elegado F, Librojo-Basilio N, Mabesa R, Dizon E (2010) Isolation and characterisation of selected lactic acid bacteria for improved processing of Nem chua, a traditional fermented meat from Vietnam. Benefic Microbes 1:67–74

Plessas S, Alexopoulos A, Mantzourani I, Koutinas A, Voidarou C, Stavropoulou E, Bezirtzoglou E (2011) Application of novel starter cultures for sourdough bread production. Anaerobe 17:486–489

Rantsiou K, Urso R, Iacumin L, Cantoni C, Cattaneo P, Comi G, Cocolin L (2005) Culture-dependent and-independent methods to investigate the microbial ecology of Italian fermented sausages. Appl Environ Microbiol 71:1977–1986

Rather SA, Masoodi FA, Akhter R (2016a) Ethnic meat products of Kashmiri wazwan: a review. J Ethn Foods 3:246–250

Rather SA, Masoodi FA, Akhter R, Gani A, Wani SM, Malik AH (2016b) Effects of guar gum as fat replacer on some quality parameters of mutton goshtaba, a traditional Indian meat product. Small Rumin Res 137:169–176

Rather SA, Masoodi FA, Rather JA, Gani A, Wani SM, Ganaie TA, Akhter R (2021) Impact of thermal processing and storage on fatty acid composition and cholesterol oxidation of canned traditional low-fat meat product of India. LWT 139:110503

Rather JA, Akhter N, Rather SA, Masoodi FA, Dar BN (2024) Effect of xanthan gum treatment on the shelf-life enhancement of retorted meatballs (Goshtaba): a traditional meat product of India. Meas Food 13:100127
Rodríguez-González M, Fonseca S, Centeno JA, Carballo J (2020) Biochemical changes during the manufacture of Galician chorizo sausage as affected by the addition of autochthonous starter cultures. Food Secur 9:1813
Salampessy J, Kailasapathy K, Thapa N (2010) Fermented fish products. In: Fermented foods and beverages of the world, pp 289–307
Savarino G, Corsello A, Corsello G (2021) Macronutrient balance and micronutrient amounts through growth and development. Ital J Pediatr 47:1–14
Shirsath AP, Henchion MM (2021) Bovine and ovine meat co-products valorisation opportunities: a systematic literature review. Trends Food Sci Technol 118:57–70
Smid EJ, Hugenholtz J (2010) Functional genomics for food fermentation processes. Annu Rev Food Sci Technol 1:497–519
Wakhungu HK (2019) Dietary habits, Nutritional status and prevalence of dental caries among 5-year-old school children in Uasin-Gishu County, Kenya

Chapter 5
Herbs, Spices and Condiments for Processed Meat Products

Tanbir Ahmad, A. R. Muhammad Sayyaf, Suman Talukder, Devendra Kumar, A. R. Sen, and I. Prince Devadason

5.1 Introduction

In addition to cultivating plants and animals, the skill of cooking is essential for human survival. The development of taste, along with cooking, suggests that the incorporation of spices into meals may have originated out of necessity. The immediate consumption of a large slaughtered animal was crucial to prevent spoilage and spices might have been introduced to either slow down this process or mask unpleasant flavours, making the food more enjoyable. However, the belief that spices emerged solely out of necessity is questionable, considering that spices were historically seen as a luxury. Those who could afford spices likely had access to fresh meat and food. Many of the spices commonly used today also possess therapeutic qualities, suggesting that their integration into regular cooking might have started as medicinal practices, gradually transitioning into culinary use due to their unique ability to enhance the quality of food. Spices, derived from various plant parts, such as cloves, nutmeg, pepper, mace, cinnamon, and ginger, are known for their aromatic properties. These ingredients, when used in moderation, not only add flavour and colour to dishes but also offer nutritional, medicinal, antibacterial, and antioxidant benefits. The indirect impacts of using spices include the potential to reduce salt and sugar in recipes and influence the cooking or processing methods, showcasing the versatile role of spices in culinary practices.

T. Ahmad (✉) · A. R. Muhammad Sayyaf · S. Talukder · D. Kumar · A. R. Sen · I. P. Devadason
Division of Livestock Products Technology, ICAR-Indian Veterinary Research Institute, Izatnagar, Bareilly, Uttar Pradesh, India

S. A. Rather, F. A. Masoodi (eds.), *Hand Book of Processed Functional Meat Products*, https://doi.org/10.1007/978-3-031-69868-2_5

5.2 What Are Spices?

Spices, as per Farrell's definition in 1990, are characterized as dried, fragrant, aromatic, or pungent vegetable or plant substances, whether in whole, broken, or ground form. Their primary function is to enhance flavour, adding a touch of piquancy or relish to foods and beverages. Unlike nutrient sources, spices primarily serve as seasonings.

5.2.1 *Natural and Soluble Spices*

Natural spices

(a) Spices in their natural form are known as natural spices.
(b) Mostly used in dry or semidry sausages.
(c) High in anthocyanins and flavones than soluble spices.

Soluble spices

Oils and oleoresins are extracted using solvents and transferred to a salt or sugar base. This mixture is called soluble spice.

(a) Mostly used in canned meats.
(b) Low in anthocyanins and flavones than natural spices.
(c) Canned meats made with natural spices may darken when heated due to presence of anthocyanins and flavones.
(d) Soluble spices are low in these compounds and so, give a brighter appearance to canned meat products.

5.2.2 *How Spices Are Distinct from Condiments and Herbs*

Herbs and condiments are often used interchangeably with spices, but there is a notable distinction. While condiments are typically added post-cooking based on personal preferences, spices serve as ingredients during food preparation. Spices, such as coriander and pepper, can also function as condiments when incorporated into dishes during or after preparation. The terms "herbs"and "spices"are commonly used synonymously, mainly referring to fresh or dried leaves of plants for flavouring food and beverages. However, the term "spice" encompasses any plant part used to enhance food quality, including stems, barks, tubers, seeds, and leaves. Many herbal spices are valued for their therapeutic qualities and find use in various recipes, with examples like cinnamon, fennel, and fenugreek, which are employed for both their leaves and other components. Unlike herbs, spices impart a robust, zesty, spicy, hot, and fragrant flavour, contributing to the distinctive and interesting taste of meals (Sastry 2013).

5.3 Roles of Spices in Meat Preparations

Spices play a crucial role in diverse segments of the food industry, including meat, fish, vegetable products, bakery items, and snacks, with the meat industry being a prominent consumer. While households typically rely on a limited selection of spices such as pepper, nutmeg, cinnamon, paprika, and vanilla, the food industry shows a growing demand for lesser-known spices like turmeric, coriander, mace, ginger, chillies, and cardamom. Throughout history, spices have been essential for global food flavouring, with varying spice usage across cultures. An evolving trend is observed in the meat industry, with an increasing diversity of spices used in meat preparations. Beyond flavour enhancement, spices contribute micronutrients to food, augmenting its palatability, and are acknowledged for their potential to combat pathogens, thereby contributing to the overall well-being of those who appreciate the rich flavours spices bring. This enduring tradition emphasizes the significant role of spices in culinary practices and their potential health-related advantages.

5.3.1 Basic Uses of Spices

The fundamental roles of spices encompass a range of direct and indirect effects on food. Directly, spices contribute to food by providing colour, often in shades of red, yellow, and green, imparting various flavours that may include pungency, bitterness, or sweetness. Additionally, they offer valuable properties such as antifungal, antibacterial, or antioxidant characteristics when incorporated into food. Beyond their direct effects, spices exhibit multifaceted impacts, playing a role in masking undesirable food odors, particularly those associated with meats, enhancing food texture, and contributing to food preservation. In culinary applications and confections, spices serve essential functions, including flavouring, colouring, adding pungency, and deodorizing or masking, as outlined in Table 5.1.

5.3.1.1 Flavour and Aroma

Spices derive their range of subtle to intense flavours from a variety of chemical compounds, particularly volatile essential oils. The composition of essential oils in spices varies depending on the spice type (Table 5.2) and factors such as cultivation location, harvest time, and storage conditions contribute to the distinct flavours of spices. The storage duration and temperature are directly correlated with the loss of flavour components from spices.

In the context of meat preparation, marinades incorporate spices and flavourings to enhance the taste and appearance of the final product. Commonly used spices in marinades include basil, cinnamon, cloves, fennel seed, garlic, onion, paprika, rosemary, thyme, and turmeric. As consumer preferences shift towards more

Table 5.4 Pungent compounds in spices (From Hirasa and Takemasa 1998)

Spice	Pungent compound	Basic structure	Sensation
Red pepper capsaicin	Hihydrocapsaicin	Acid amide group	↓
Ginger	Zingerrol, Shogaol	Carbonyl group	
Onion	Diallyl sulfide	Thioether group	
Garlic	Diallyl sulfide		
Mustard	Allyl isothiocyanate, Phydroxybenzyl isothiocyanate	Isothiocyanate group	
Horse radish	Allyl-isothiocyanate	Isothiocyanate group	Sharp

oils. Some spice components exhibit strong aromatic qualities even in small quantities. As identified by Hirasa and Takemasa (1998), the volatile components can be classified into carbonyl, thioether, acid-amide, or isothiocyanate compounds, with Table 5.4 listing the noxious chemicals present in various spices. The mucous membrane of the oral cavity and nose is stimulated by a sharp feeling, while a hot sensation spreads throughout the mouth. Volatile substances primarily evoke a sharp feeling, whereas non-volatile compounds are associated with a hot sensation.

5.3.1.4 Masking Undesirable Flavours and Balancing Flavours

Condiments and spices play a crucial role in harmonizing the overall taste profile of a dish. They serve to balance excessive sweetness, acidity, or blandness, creating a well-rounded and pleasing flavour. Additionally, spices and condiments are employed to mask undesirable tastes that may arise from certain ingredients or processing methods. Particularly robust-flavoured spices can effectively cover and balance less appetizing tastes.

Traditionally, the use of onions and garlic in meat preparation serves a dual purpose of covering or neutralizing unpleasant meat odors. Generally, spices exhibit three distinct deodorizing effects. Chemical deodorization involves molecules with odors undergoing chemical reactions with non-volatile or odorless substances, rendering them scentless. A masking effect occurs when a strong flavour conceals an undesirable odor, or when two different aromas combine to create an odorless blend. Physical deodorization involves odor-absorbing materials absorbing scents, though this is relatively uncommon in the realm of spices. The efficacy of deodorization depends on the timing and method of spice incorporation into the dish. For instance, garlic is less effective when added at the end of meal preparation but produces desired results when introduced early in the cooking process.

Spices, renowned for their preservative properties and the ability to extend the shelf life of certain food items, can obscure changes in the sensory attributes of sous vide meat products. A study comparing two sous vide dishes—one rich in herbs and spices (bolognaise meat sauce) and another with fewer spices (chicken tikka masala)—revealed that the former remained acceptable for a significantly longer duration (Armstrong and McIlveen 2000). This suggests that a higher quantity of herbs and spices contributes to enhanced preservation and prolonged acceptability of sous vide meat products.

5.4 Preservative Properties of Spices in Meat Products

In various sectors of the food industry, particularly the meat industry, microorganisms play a crucial role, and their significance in food processing cannot be overstated. As mentioned earlier, many spices inherently possess antimicrobial and antifungal properties. The awareness of these antimicrobial properties in spices has a rich history, dating back centuries. For example, in ancient Egypt, cinnamon, thyme, and cumin were utilized in the mummification process, showcasing their antimicrobial attributes. Similarly, in ancient Rome and Greece, coriander was known for extending the preservation period of meat, and mint was employed to prevent the spoilage of milk.

Vegetables can serve as valuable alternatives to traditional preservatives like nitrites in food, given that certain spices exhibit antioxidant, bactericidal, and fungicidal effects (Gassara et al. 2016). Numerous studies have delved into the preservative properties of spices and their essential oils, with compounds such as eugenol, thymol, carvacrol, vanillin, allicin, cinnamic aldehyde, and allyl isothiocyanate demonstrating antimicrobial properties.

Many spices and condiments possess natural antimicrobial properties that help inhibit bacteria growth, thereby contributing to the preservation of processed meats. This not only enhances the shelf life of meat products but also eliminates the need for excessive artificial preservatives. While spices are primarily recognized for enhancing the flavour of meat products, they play a crucial role in preserving meat through various mechanisms.

5.4.1 Antimicrobial Properties

For centuries, spices have not only enhanced culinary experiences by providing heightened flavours but have also demonstrated antimicrobial properties, particularly in the realm of meat processing. In meat products, spices play a crucial role in ensuring safety and longevity by inhibiting the proliferation of microorganisms, including bacteria and fungi.

Numerous spices contain inherent compounds known for their antimicrobial efficacy. Notably, allicin in garlic and onion exhibits antibacterial and antifungal effects. Cinnamaldehyde, a component in cinnamon, contributes significantly to meat preservation through its antimicrobial attributes. Unique spices like coriander and rosemary exhibit anti-spoilage effects by restraining bacterial and fungal growth, thus extending the shelf life of meat products. Many spices harbor essential oils with intrinsic antimicrobial capabilities, such as thyme and oregano, which contain thymol and carvacrol known for their antimicrobial and antioxidant attributes.

The combination of organic acids and spices has proven effective in controlling the growth of clostridium. Treatments involving sodium citrate or various spice powders with ingredients like chili, garlic, oregano, curry, and clove have been suggested to significantly reduce the population of *C. perfringens* in cooked ground beef. However, the antimicrobial activity of spices can vary due to factors like form (powders or essential oils), storage conditions affecting oxidation, heat-sensitive compounds impacted by cooking, and the degradation of phenolic compounds. The extended shelf life of sous vide-cooked chicken tikka masala, attributed to the presence of spices and herbs, demonstrates the synergistic effects of diverse spice blends. This synergy enhances the overall antimicrobial efficacy, surpassing the potency of individual components (Gómez et al. 2020).

It's important to note that spices and herbs themselves may contain spores contributing to spore counts in raw formulated cooked or heat-treated meat and poultry products. For example, a survey found *C. perfringens* spores in 80% of 54 different spices and herbs.

5.4.2 Natural Antioxidants in Spice Preservation

Spices often possess antioxidant properties that play a crucial role in preventing the oxidation and spoilage of meat. By mitigating oxidative reactions, spices contribute to preserving the quality and safety of meat products. Contemporary knowledge indicates that certain spices, such as rosemary and sage, function as antioxidants, not only preserving the quality and colour of meats but also offering health benefits to consumers. The antioxidant properties are attributed, in part, to the reduction of fat oxidation (Alirezalu et al. 2020).

The gradual deterioration of stored foods is influenced by various factors, with oxidation of oils or fats being a significant contributor. Spices, acting as antioxidants, have the potential to extend the shelf life of various foods, opening avenues for the development of new food products. The oxidation of fats in meat leads to the formation of peroxides, subsequent oxidation, and degradation into low-molecular-weight alcohol and aldehyde compounds, resulting in rancidity. These "free radicals" are believed to contribute to damage to human DNA, leading to conditions such as cancer and aging.

To combat oxidation in packaged foods, methods like replacing oxygen with an inert gas have been utilized, but the prevalent approach involves the use of antioxidants. Both synthetic and natural antioxidants, including spices, can be employed. While synthetic antioxidants are effective, concerns about their decomposition at high temperatures and potential toxicity have been raised. Tocopherol, a natural antioxidant, is widely used in the food industry, although its antioxidant effects are considered inferior to synthetic counterparts.

Numerous studies on spice antioxidants highlight sage and rosemary as particularly effective in retarding fat or oil oxidation. Additionally, other leafy spices such as oregano and thyme exhibit stronger antioxidant activities compared to many others spices. Cinnamon, containing cinnamic aldehyde, and cloves, containing eugenol, are among the most effective germicidal spices. Containing inherent polyphenol antioxidants, these spices effectively inhibit lipid oxidation in both fresh and cooked meat. The introduction of dried spices resulted in a reduction in malondialdehyde content in meat samples, with ethanol extracts proving more potent than their dried counterparts. Incorporating an ethanolic extract of spices into sodium salt-treated meat patties significantly decreased lipid oxidation, as evident from the lowered malondialdehyde and peroxide value levels.

Utilizing antioxidants sourced from natural origins, such as culinary herbs or other plant materials, has the potential to extend the preservation of meat products for up to 6 months. Polyphenols, recognized as substantial antioxidants following vitamins C, A, and E, represent naturally occurring and non-toxic plant-derived compounds. Meat products treated with these natural antioxidants, when stored properly, have the potential to maintain favorable sensory attributes and elevated nutritive value. Opting for natural antioxidants is more appealing to consumers concerned about food items supplemented with synthetic counterparts. Further research is needed to identify the most suitable method for substituting synthetic antioxidants, aiming to create optimal food choices conducive to optimal health. Table 5.5 provides a summary of the impact of incorporating natural antioxidant ingredients into the formulation of meat and meat products, focusing on their sensory characteristics.

Table 5.5 Effect of the addition of natural antioxidants in the formulation of meat and meat products on their sensory characteristics

Spices and condiments	Meat product	Effects
Rosemary extract	Turkey meat, pork burgers	No effects on sensory characteristics
Oregano extract	Sheep burgers	No effects on sensory characteristics
Garlic	Low-salt sausages	No effects on aroma, flavour and overall appearance
Additional fruit extracts, including blueberries, raspberry pomace, pomegranate peel, and pomegranate juice, are also considered	Pork burgers and cooked pork ham Beef burgers Chicken patties	Sensory quality was not negatively altered
Citrus fibre	Boulogne sausage	The addition of citrus fibre alone did not have any noticeable impact on colour or texture properties. However, when combined with rosemary essential oil, there was an enhancement in sensory parameters
	Rabbit burgers Indian sheep meat product	Turmeric powder contributes to increased yellow colouration, while cumin and cardamom result in elevated overall acceptability scores.

5.4.3 pH Regulation and Its Influence

In meat processing, pH regulation is a crucial aspect that influences the quality, safety, and shelf life of the final product. pH refers to the acidity or alkalinity of a substance and is measured on a scale from 0 to 14, with 7 being neutral. Optimal microbial growth occurs at pH values between 6.5 and 7 although most microorganisms continue to grow within the pH range of 4 and 9.5, for fresh meat, pH varies around 5.0 and 6.5, hence microorganisms can easily grow into the meat and maintaining the appropriate pH is essential for various reasons, including:

(a) Microbial Safety: Controlling pH helps inhibit the growth of pathogenic bacteria, such as Salmonella and *Escherichia coli* (*E. coli*). A lower pH creates an environment that is less favourable for the survival and reproduction of these harmful microorganisms.
(b) Texture and Tenderness: pH affects the texture and tenderness of meat. Meat with an optimal pH range is often more tender and has a desirable texture. This is particularly important in processed meat products where texture plays a significant role in consumer acceptance.
(c) Colour Stability: pH influences the colour of meat. Meat with a lower pH tends to have a brighter red colour, which is generally more appealing to consumers. Colour stability is crucial for the visual appeal of meat products.
(d) Protein Functionality: pH affects the functionality of proteins in meat. Proteins play a key role in the texture and structure of meat products. Proper pH regulation helps maintain the functionality of proteins during processing.

Some spices can influence the pH of meat products. The pH level is critical for microbial growth. Spices like black pepper and paprika, for example, can help control the pH, creating an environment less favourable for the growth of spoilage and pathogenic microorganisms.

5.4.4 Reducing Water Activity (a_w)

Spices can indirectly contribute to antimicrobial properties by reducing the water activity in meat products. Lower water activity creates an environment less conducive to microbial growth and spoilage.

5.5 Spices as Functional Ingredient

Spices, which have long been the basis of traditional medicine in many countries, have also been the subject of study, particularly by the chemical, pharmaceutical, and food industries, because of their potential use for improving health. Elements

Table 5.6 Examples of herbs and spices as functional foods (Kumar et al. 2012)

Herb/ spice	Key compounds	Activities/effects
Rosemary	Phenolic compounds	Potent antioxidant activity—regeneration of endogenous tocopherol—comparable efficacy to BHA/BHT in preserving sausages—enhances colour and freshness—prevents lipid oxidation in cooked products
Green tea	Catechins (epicatechin, epigallocatechin gallate, etc.)	Antioxidant properties—antibacterial, anticarcinogenic, and antiviral benefits—outperforms α-tocopherol and BHA in reducing peroxide formation—inhibits mutagen formation during meat cooking
Clove	Eugenol	Antimicrobial activity—inhibits L. monocytogenes growth in mutton and chicken frankfurters—effective against various foodborne pathogens—potent antioxidant, prevents discolouration and retards lipid oxidation
Garlic	Allicin	Antimicrobial activity against gram-positive and gram-negative bacteria Aqueous garlic extract inhibits microbial contaminants on poultry carcasses
Sage	Carnosol, carnosic acid, rosmadial, rosmanol, epirosmanol, methyl carnosate	Rich in antioxidant compounds—commonly used in sausages for preservation
Oregano essential oil	–	Reduces oxidation levels in pork and beef during refrigerated storage—extends shelf life of fresh chicken breast meat—concentration-dependent impact on sensory quality—delays microbial growth under modified atmosphere

derived from herbs and spices encompass numerous phytochemicals, offering significant reservoirs of natural antioxidants like phenolic diterpenes, flavonoids, tannins, and phenolic acids. These constituents exhibit antioxidant, anti-inflammatory, and anticancer properties (Viuda-Martos et al. 2010). Within food applications, they contribute to enhanced flavour, hinder lipid oxidation-induced deterioration, impede microbial growth, and contribute to reducing the likelihood of certain diseases. Table 5.6 provides some examples.

5.5.1 *Antioxidant Properties*

Herbs and spices have emerged as potent sources of natural antioxidants, capable of retarding lipid oxidation—a principal cause of food spoilage. This oxidative process, occurring during storage, processing, or final product storage, not only leads to a loss of nutritional value but also impacts sensory qualities, shortening shelf life. Traditionally, synthetic antioxidants like BHT, BHA, and PG have been employed, but concerns about their safety have fuelled interest in plant-derived alternatives. Spices, essential oils, and their constituents exhibit antioxidant activities through mechanisms such as scavenging free radicals, hydrogen donation, and chelating metallic ions (Manessis et al. 2020).

5.5.2 *Antibacterial Properties*

As guardians against microbial threats, herbs and spices, particularly in the form of essential oils, play a pivotal role. These natural antimicrobials can extend the shelf life of foods by inhibiting the growth or viability of microorganisms. Studies have showcased their effectiveness against various bacteria and molds. The composition and structure of essential oils are key determinants of their antimicrobial capacity, making them valuable in the quest for natural preservatives (Flores and Toldrá 2021).

5.5.3 *Anti-Inflammatory Activity*

Certain spices have been recognized for their anti-inflammatory potential, offering relief from various discomforts. Flavonoids present in spices, such as galangin, demonstrate the ability to inhibit enzymes associated with inflammation. The mechanisms behind this anti-inflammatory action involve the modulation of prostaglandins and the inhibition of inflammatory enzymes.

5.5.4 *Inhibition of Platelet Aggregation*

Spices and their extracts, rich in natural antioxidants, exhibit antiplatelet activity, potentially contributing to cardiovascular health. Compounds like carnosol and luteolin present in herbs and spices have been linked to the inhibition of platelet aggregation, suggesting a role in thrombosis prevention.

5.5.5 *Antiviral Activity*

Several spices and herbs showcase antiviral properties, with compounds like curcumin and eugenol exhibiting inhibitory effects against viruses such as HIV and herpes simplex. Essential oils derived from various spices also demonstrate strong antiviral activity against specific strains.

5.5.6 *Anti-Carcinogenic Properties*

Certain components of herbs and spices have demonstrated potential as chemoprotective agents against cancer. Compounds like curcumin in turmeric and carnosol in rosemary exhibit anti-carcinogenic properties through various mechanisms, including the induction of apoptosis and the inhibition of tumor angiogenesis.

5.5.7 *Other Health Benefits*

Herbs and spices contribute to digestive health by stimulating saliva and gastric juices, aiding in the digestion process. Additionally, they offer protection against ulcers, with flavonoids playing a crucial role in increasing prostaglandins and preventing oxidative damage. Moreover, some spices exhibit anti-diabetic properties, influencing insulin sensitivity and glucose metabolism.

5.6 Relevant Considerations While Using Spices in Processed Meat

When incorporating herbs and spices into food, it is crucial to acknowledge potential microbiological concerns associated with their natural origin. Since these ingredients grow in natural environments, their cultivation, harvest, and distribution methods can introduce a natural microflora, including microorganisms that may be human pathogens. Herbs and spices intended for use in ready-to-eat foods must themselves be free of pathogenic microorganisms. For those used in cooked foods, the presence of spore-forming bacteria, especially in items with extended shelf life, needs consideration. Spores from bacteria like Clostridium or Bacillus, if not properly addressed during cooking and storage, can survive and multiply, posing a risk of food poisoning. Adequate time-temperature conditions during food preparation and storage are essential to mitigate this risk. Conservation methods such as gamma irradiation (up to 10 kGy), heat, and steam treatments offer viable options to ensure herbs and spices are disinfested and decontaminated (Martínez-Graciá et al. 2015).

The broader utilization of plant-derived when incorporated as antioxidants in meat systems faces significant challenges, including (i) elevated processing and storage expenses associated with plants/fruits and their derived antioxidants, (ii) the consistently squeezed profit margins in the meat trading industry, and (iii) the necessity for increased quantities of these antioxidants to demonstrate effective antioxidant action compared to synthetic antioxidants. This heightened requirement may lead to modifications in the organoleptic characteristics of meat, such as taste, colour, and odors, potentially resulting in reduced market appeal.

Although numerous studies have investigated the efficacy of plant derivatives, the availability of commercially viable natural antimicrobials and antioxidant agents remains limited. This scarcity can be attributed to the variability in phenolic compounds found in plant extracts and essential oils (EOs), along with variations in extraction methods used during production and storage. Addressing this challenge may involve modeling and optimizing extraction processes to ensure the comprehensive extraction of bioactive compounds from plant sources while preserving volatile components during preparation and storage.

In recent decades, advancements in packaging technology have contributed to the development of innovative techniques aimed at enhancing quality and extending the shelf life of various meat products. The adoption of active packaging has emerged as a potential alternative to conventional meat packaging. The incorporation of antimicrobial and antioxidant agents into active packaging has demonstrated the potential to enhance the stability and prolong the shelf life of minimally processed meat products. Consequently, the application of natural antimicrobials, antioxidants, and active packaging not only meets consumer preferences for reducing reliance on chemical preservatives but also offers promising protective effects against microbial growth and lipid oxidation in fresh meat and meat products.

Before integrating plant-based antimicrobials and antioxidants into meat products, additional research is essential to identify the optimal concentration. This process involves assessing the toxicity of these plants to human cells through in vivo studies and clinical trials to gain a comprehensive understanding of their effects on consumers. These investigations should also encompass studies evaluating the bio accessibility and bioavailability of compounds from these plants within the human body.

5.7 Conclusions

In conclusion, the incorporation of spices in meat preparations is a complex and multifaceted process that goes beyond mere flavour enhancement. This book chapter has highlighted the diverse roles of spices, ranging from their historical significance in preserving meats to their contemporary applications as functional ingredients with antimicrobial and antioxidant properties. Despite the wealth of studies demonstrating the effectiveness of plant-derived antioxidants, challenges such as legislative fragmentation, complex matrices in applications, and the need for extensive studies on concentration, toxicity, bioaccessibility, and bioavailability remain.

The utilization of spices in the meat industry offers not only flavour diversity but also potential health benefits. The preservation properties of spices, attributed to natural antimicrobial compounds and antioxidants, contribute to extending the shelf life of meat products. Additionally, spices play a vital role in pH regulation and water activity reduction, influencing the overall quality and safety of processed meats. As the demand for natural alternatives to synthetic additives grows, the chapter suggests that further research is needed to optimize extraction processes, evaluate concentration levels, and address challenges related to organoleptic traits. The potential of active packaging as an alternative preservation method is also highlighted, offering a promising avenue for meeting consumer demands for natural and healthier meat products. The chapter underscores the intricate interplay between culinary traditions, scientific research, and market dynamics in shaping the acceptance and utilization of natural, plant-derived spices in the meat industry. Despite existing challenges, the versatile and valuable contributions of spices to meat preparations position them as indispensable elements in the evolving landscape of food science and technology.

References

Alirezalu K, Pateiro M, Yaghoubi M, Alirezalu A, Peighambardoust SH, Lorenzo JM (2020) Phytochemical constituents, advanced extraction technologies and techno-functional properties of selected Mediterranean plants for use in meat products. A comprehensive review. Trends Food Sci Technol 100:292–306

Armstrong GA, McIlveen H (2000) Effects of prolonged storage on the sensory quality and consumer acceptance of sous vide meat-based recipe dishes. Food Qual Prefer 11(5):377–385

Flores M, Toldrá F (2021) Chemistry, safety, and regulatory considerations in the use of nitrite and nitrate from natural origin in meat products-invited review. Meat Sci 171:108272

Gassara F, Kouassi AP, Brar SK, Belkacemi K (2016) Green alternatives to nitrates and nitrites in meat-based products–a review. Crit Rev Food Sci Nutr 56(13):2133–2148

Gómez I, Janardhanan R, Ibañez FC, Beriain MJ (2020) The effects of processing and preservation technologies on meat quality: sensory and nutritional aspects. Food Secur 9(10):1416

Hirasa K, Takemasa M (1998) Spice Science & Technology. Marcel Dekker, Inc., New York

Kumar A, Kumar S, Sharma BD, Mendiratta SK, Verma OP, Patel AK (2012) Functional meat and meat products: an overview. J Adv Vet Res 2(4):314–324

Manessis G, Kalogianni AI, Lazou T, Moschovas M, Bossis I, Gelasakis AI (2020) Plant-derived natural antioxidants in meat and meat products. Antioxidants 9(12):1215

Martínez-Graciá C, González-Bermúdez CA, Cabellero-Valcárcel AM, Santaella-Pascual M, Frontela-Saseta C (2015) Use of herbs and spices for food preservation: advantages and limitations. Curr Opin Food Sci 6:38–43

Peter KV (2004) Handbook of Herbs and Spices, Vol. 2. CRC Press, New York/Woodhead Publishing Limited, Cambridge, England, pp 77–81

Sastry ED (2013) Understanding spices for processing. Ethiop J Appl Sci Technol 1:81–89

Viuda-Martos M, Ruiz-Navajas Y, Fernández-López J, Pérez-Álvarez JA (2010) Spices as functional foods. Crit Rev Food Sci Nutr 51(1):13–28

Yusop SM, O'Sullivan MG, Kerry JP (2011) Marinating and enhancement of the nutritional content of processed meat products. In: Processed meats. Elsevier, Amsterdam, pp 421–449

Chapter 6
Introduction to Functional Meat Products

Zahida Naseem, Sajad Ahmad Mir, Shahida Naseem, and Shazia Akhter

6.1 Introduction

Food has a variety of purposes for people, including preventing diseases linked to malnutrition, enhancing physical and mental health and satisfying hunger and providing essential nutrients. These days, consumers are more worried about health-related issues and demand food products that are higher in nutrients and have more essential components that promote health. Research on the health effects of food components has been sparked by growing understanding of the effects of diet and particular food elements. Foods that boost health and prevent diseases are therefore needed. Such foods have a positive impact on the body systems by offering more nutrients than is necessary. Accordingly, foods are classified as functional based on how they affect health; hence, foods that have been left unaltered and altered (via technical or biotechnological methods) fall under this category (Nisa et al. 2017).

Functional food is defined as a food that may offer health assistances beyond basic nutrition and resembles conventional food that is meant to be consumed as part of a regular diet but has been altered to fulfill physiological roles other than meeting basic nutrient requirements (Bhat and Hina 2011). Foods that are promoted as functional generally have additional, scientifically advanced ingredients that offer a particular health advantage.

Z. Naseem · S. Akhter
Division of Food Science and Technology, SKUAST-Kashmir, Srinagar, India

S. A. Mir (✉)
Department of Life Science (Microbiology and Food Science and Technology), GITAM (Deemed to be University), Visakhapatnam, Andhra Pradesh, India

S. Naseem
Department of Environmental Science, University of Kashmir, Srinagar, India

S. A. Rather, F. A. Masoodi (eds.), *Hand Book of Processed Functional Meat Products*, https://doi.org/10.1007/978-3-031-69868-2_6

Meat is an essential constituent of any stable diet. Meat contains higher concentration of protein, comprising all important amino acids, fatty acids, and vitamins, primarily B complex vitamins and especially B12. Additionally, meat has a higher degree of bioavailability as compared to other dietary sources (Haque et al. 2023). Meat also contains minor quantity of omega-3 fatty acids, cobalamin, zinc and manganese. Therefore, meat is known as nutritional powerhouse. For the best possible muscular growth and development, meat is very beneficial. Consumption of processed meat products, like bologna, burgers, meat batters, meat patties, kebabs, frankfurters, meatballs, sausages, fermented sausage, and others is increasing and becoming more and more popular. Due to growing consumer demand for their accessibility and superior high quality in terms of texture, taste, flavor and nutritional profile, processed meat products are becoming more and more common in the developed world.

However, the perception that meat, especially red meat, should be avoided has been fueled by the impression that animal fat is high in saturated fatty acids (SFA). Furthermore, the negative consequences connected to meat and processed meat are caused by the deficiency of dietary fiber (DF) in these diets. Regular eating of meat has been associated with numerous health problems, such as cardiovascular diseases, hypertension obesity, and coronary heart disease (Micha et al. 2012). Scientific studies have confirmed that there is an association between specific components in meat and neurological problems also.

The demand for functional meat products has grown in recent years due to consumer trends toward better eating. As a result, several beneficial bioactive ingredients are currently being explored and included in meat-based products (Rashwan et al. 2020). A food product is considered functional if it contains components from other natural sources that have been demonstrated to improve physiological processes of the body, like improved immunity and anti-aging (Elkhalifa et al. 2021). The basic tenet of functional food is to lower the incidence of chronic illnesses by reducing the intake of foods that are regularly consumed. Large food corporations, especially the meat industry, are becoming increasingly interested in formulating foods based on the potential health benefits of its non-nutritional elements. In terms of meat, the changes that can be made to it in order to provide it functional qualities depend on whether the animal's diet is altered or if the carcass is altered after postmortem. The first method allows for the modification of the fatty acid, lipid and vitamin E content, while as the second involves the mechanical removal of fat. When it comes to meat products, the main focus is on reformulating them by changing the amount of fat in them, as well as by adding a variety of useful ingredients like fiber, vitamins, plant proteins, mono- or polyunsaturated fatty acids, phytochemicals, calcium and so on.

6.2 Development of Functional Meat Products

Functional foods are new foods that have been developed to include ingredients or living microorganisms that may improve health or prevent disease, provided that the concentration is safe and high enough to produce the desired effect. Meat and meat products are given high important in the diet of technologically advanced countries. Meat and meat-related items can be reformed by adding components thought to be healthful or by removing or minimizing those deemed detrimental. Creating meat products that are both useful and healthful can be accomplished through a variety of approaches. From animal production to product processing, every step needs to be taken into account. As mentioned earlier, it is possible to improve the composition of animal products (such as conjugated linoleic acid) by altering the feed that animals receive. Furthermore, the functional properties of meat products can be altered by adding or removing food items deemed harmful or advantageous to health. A wide range of functional meat products can be developed by the help of functional ingredients through reformulation method. Techniques for enhancing and changing meat products are summed up as follows.

6.3 Alteration of the Fatty Acid and Cholesterol Levels in Meat

Consumer acceptability of meat products is determined by various factors. These include the quality of end product, the cost and quantity of protein and its nutritional worth. These days consumers who are concerned about their health are also worried about the fat level of meat and its products. Lean meat has a very low lipid content, however processed meat products have a substantially larger percentage of fat. Meat fat mostly consists of saturated and monounsaturated fatty acids. Consuming saturated fat is one of the best indicators of having high blood cholesterol and a major contributing factor to heart disease, cancer and obesity.

One of the main sources of fat in the diet is meat, particularly saturated fatty acids (SFA), which have been linked to a number of ailments in modern society, particularly in industrialized nations. In general, it is thought that the presence of diet-related disorders such cardiac illnesses is connected to the high percentage of animal fat in human diets (Mozaffarian et al. 2010). Since meat has a high fat content and about 40% of it is saturated. It is advisable to alter the amount and quality of the fat to produce novel meat products with useful qualities.

Another risk factor for cancer and coronary heart disease is the ratio of ω-6:ω-3 polyunsaturated fatty acids (PUFA), particularly for the development of blood clots that can result in a heart attack. The primary methods for achieving a decrease in carcass fat content are genetic line selection within breeds, modifications to animal feeding procedures, such as the addition of feed additives like probiotics and antibiotics and manipulation of animal metabolism with tools like growth hormone, anabolic implants and agonists. The firmness or softness of meat fat is significantly influenced by changes in the content of fatty acids, particularly in the subcutaneous and intermuscular.

Utilizing non-fat, low-energy alternatives is the first step in fat reduction in the meat products. The second is using marine or plant oils such as coconut, peanut, sunflower, palm, tea seed, olive oil, hydrogenated coconut, hydrogenated palm or hydrogenated corn, soybean oil that are high in polyunsaturated fatty acids (PUFAs) to enhance the fatty acid profile of the products. The prebiotic inulin is the most suitable non-fat alternative discovered so far because of its white color, neutral scent and ability to dissolve in water to produce a stable gel that resembles fat. According to Šojić et al. (2011), inulin can be added to cooked sausages in powder form up to 5% or as a gel (inulin-water ratio 1:3) up to 8%. Inulin in gel form is more conveniently used in stuffing of sausages than powder because the gel contains more readily disseminated and dissolved inulin. In addition to inulin, carrageenan and gums act as potential fat replacers because of their excellent gelling qualities (Grasso et al. 2014). The most popular application to enhance the fatty acid profile is the use of plant oils. Meat is a rather high source of cholesterol; depending on the animal species, it can contain anywhere from 75 to 100 mg per 100 g. The cholesterol can also be reduced by substituting plant-based ingredients for meat and fat or by utilizing *Eubacterium coprostanoligenes* (cholesterol-reducing bacteria) in fermented meat products (Pogorzelska-Nowicka et al. 2018).

6.4 Addition of Plant Proteins

Meat is a chief source valuable proteins of high biological value and vital amino acids. The nine amino acids like tryptophan, threonine, leucine, isoleucine, lysine, methionine, phenylalanine, valine and histidine are essential for human nutrition. In nutritional assessment of food proteins, tyrosine and cystine are supposed to play important roles in human nutrition. Tyrosine, cystine, and other important amino acids are present in high concentrations in the majority of meat and meat products. It is found that consuming a protein derived from red meat twice a day or more increases the risk of colon cancer. The use of plant proteins in meat products is becoming more and more popular. Meat products contain a variety of plant proteins including lentil, chickpea, sunflower, black eye bean and soy protein. In meat products, plant proteins are employed for nutritional purposes, economic savings and more recently because of their potential to improve health. Plant proteins improve the nutritional value of animal products by lowering cholesterol and energy levels, increasing the content of proteins that promote health and so on. Plant-based proteins, such as those from soy and sunflower can help prevent and treat osteoporosis, cancer, cardiovascular disease and menopausal symptoms.

Certain plant-based proteins, such as sunflower, walnut and other sources, have been included into meat systems to maintain healthy, beneficial lysine/arginine ratios (Naveen et al. 2015). In order to regulate blood cholesterol levels, soy protein has been utilized as a functional ingredient in a variety of meat products, including low-fat sausage. Functional vegetable proteins can enhance the structure of the products, making them easier to slice using contemporary high-speed slicing equipment, in addition to their capacity to bind water into the cooked items, so boosting the final yield and guaranteeing succulence.

Nuts have a high protein content. Despite their high fat content, nuts have a high percentage of unsaturated fats such as monounsaturated fats, which when substituted for dietary fatty acids and/or carbohydrates can help decrease cholesterol. Nuts also include dietary fiber and a variety of bioactive substances, including plant sterols, which have the ability to reduce cholesterol.

6.5 Addition of Dietary Fibre

The edible portion of plants or analogous carbohydrates that are resilient to absorption and digestion in the small intestine and that undergo either full or incomplete fermentation in the large intestine of humans is known as dietary fiber (Pame et al. 2022). Fiber consumption has been shown to lower the risk of many illnesses thus, consuming foods high in this component can increase the amount of fiber in the diet. A sufficient intake of dietary fiber lower blood pressure and LDL and plasma cholesterol, lessen insulin and glycemic response, lowers the risk of obesity, certain types of cancer, type 2 diabetes, cardiovascular disease and gut-related issues. Also, it can cut down calories. Elevated dietary fiber consumption is also linked to decreased blood pressure and serum cholesterol levels.

Furthermore, an appropriate intake of fiber appears to enhance gut health and the interactions between the gut microbiota and fiber, perhaps improving immune function and contributing to weight loss or prevention through the regulation of satiety or fullness. It has been reported that children's who consume more fiber have a decreased chance of being overweight or obese (Choumenkovitch et al. 2013). In addition to their health benefits, fibers are supplemented to meat products for their functional abilities. These help in reducing cooking loss because of their high-water binding capacity, enhance texture and add flavor and mouth feel because of their fat binding ability. For the production of functional meat products, a variety of dietary fiber sources, including sorghum, wheat fiber, carrot, beet extract, orange, apple, peach, olive and oat bran can be utilized (Table 6.1). Functional chicken nuggets can be developed by combining several dietary fiber sources, such as pea hull flour, gram hull flour, apple pulp, and bottle gourd, at a 10% level. Choosing the right sources of dietary fiber and amounts of supplements can make meat products healthier.

6.11 Reduction in Salt Content

Salt contains high amounts of sodium which plays a necessary role in the transmission of nerve impulses, balancing acid-base equilibrium and maintaining plasma volume in human metabolism. Meat has a comparatively low salt content. It only contains 50–90 mg of sodium per 100 g. However, meat products add a substantial quantity of salt to the diet. It is estimated (based on eating habits) that 20–30% coming from meat and meat derivatives. Meat products are known to contain a lot of salt. The primary health concern with sodium chloride is its sodium level, as excessive consumption may lead to harmful effects on human health like hypertension which is associated to cardiovascular disease and stroke.

Regular use of NaCl is a part of meat processing techniques, it helps to provide the necessary technological and sensory properties (flavor, texture, protein solubility, water holding capacity and stuffing stability). It also reduces water activity, which increases product safety (Tunieva and Gorbunova 2017). But NaCl content in meat products can be limited only up to 10%, more decrease would result in undesirable product qualities. A part of table salt can be replaced by other salts which do not contain sodium. For instance, potassium lactate and potassium chloride can replace up to 30% of NaCl and glycine up to 40%. These levels have no effect on the sensory qualities of the products.

Moreover, salt mixes comprising calcium, potassium, magnesium chloride, or lactate could substitute 30–53% of the NaCl. Regarding the antimicrobial effect of salt substitutes, lactates, potassium and calcium chloride all had effects comparable to those of NaCl. This indicates that these substances are able to offer microbial safety comparable to that of conventional meat products in addition to similar sensory qualities.

6.12 Conclusion

Meat-based functional foods are designed with two main goals: boosting the health status of customers by minimizing the risk of degenerative disease development and/or improving the composition and shelf-life of meat products. Modified meat products guarantee stability, health advantages and acceptable quality characteristics in terms of physicochemical and sensory properties. The two most common methods of obtaining functional meat with antioxidant properties are animal feeding and food composition. The nutritional content and adequacy of diets can be enhanced by incorporating meat and meat products as ideal food matrices that deliver bioactive substances (such fiber, ω-3 PUFAs, probiotics, and bioactive peptides) without requiring changes to dietary habits.

References

Ahmad SR, Sharma BD, Irshad A, Kumar RR, Malav OP, Talukder S (2021) Effect of aerobic storage conditions on the quality of functional restructured buffalo meat fillets enriched with natural sources of dietary fibers and antioxidant components. J Food Process Preserv 45(1): e15072

Banerjee R, Verma AK, Das AK, Rajkumar V, Shewalkar AA, Narkhede HP (2012) Antioxidant effects of broccoli powder extract in goat meat nuggets. Meat Sci 91(2):179–184

Bhat ZF, Hina B (2011) Fibre-based functional meat products. Asian J Food Agro-Indus 4(4): 261–273

Choumenkovitch SF, McKeown NM, Tovar A, Hyatt RR, Kraak VI, Hastings AV, Herzog JB, Economos CD (2013) Whole grain consumption is inversely associated with BMI Z-score in rural school-aged children. Public Health Nutr 16(2):212–218

Cruzen SM, Bradkim YH, Lonergan SM, Grubbs JK, Fritchen AN, Huff-Lonergan E (2015) Effect of early postmortem enhancement of calcium lactate/phosphate on quality attributes of beef round muscles under different packaging systems. Meat Sci 101:63–72

Devatkal SK, Narsaiah K, Borah A (2010) Anti-oxidant effect of extracts of kinnow rind, pomegranate rind and seed powders in cooked goat meat patties. Meat Sci 85(1):155–159

Devatkal SK, Kadam DM, Naik PK, Sahoo J (2011) Quality characteristics of gluten-free chicken nuggets extended with sorghum flour. J Food Qual 34(2):88–92

Devatkal SK, Thorat PR, Manjunatha M, Anurag RK (2012) Comparative antioxidant effect of aqueous extracts of curry leaves, fenugreek leaves and butylated hydroxytoluene in raw chicken patties. J Food Sci Technol 49(6):781–785

Eim VS, Simal S, Rossello C, Femenia A, Bon J (2013) Optimisation of the addition of carrot dietary fiber to a dry fermented sausage (Sobrassada) using arficial neutral networks. Meat Sci 94:341–348

Elkhalifa AEO, Al-Shammari E, Adnan M, Alcantara JC, Mehmood K, Eltoum NE, Awadelkareem AM, Khan MA, Ashraf SA (2021) Development and characterization of novel biopolymer derived from *Abelmoschus esculentus* L. Extract and its antidiabetic potential. Molecules 26(12):3609

El-Nashi HB, Abdel Fattah AFAK, Abdel Rahman NR, Abd El-Razik MM (2015) Quality characteristics of beef sausage containing pomegranate peels during refrigerated storage. Ann Agric Sci 60:403–412

Fernandez-Gimes JM, Fernandez-Lopez J, Sayas-Barbera ME, Sendra E, Perez Alvarez JA (2004) Lemon albedo as a new source of dietary fiber application to Bologna sausage. Meat Sci 67:7–13

Food and Agriculture Organization of the United Nations (FAO); World Health Organization (WHO) (2001) Report of a joint FAO/WHO expert consultation on evaluation of health and nutritional properties of probiotics in food including powder milk with live lactic acid bacteria. Food and Agriculture Organization of the United Nations (FAO), Córdoba

Grasso S, Brunton NP, Lyng JG, Lalor F, Monahan FJ (2014) Healthy processed meat product-regulatory, reformulation and consumer challenges. Trends Food Science Tech 39:4–17

Grigelmo-Miguel N, Abadias-Seros MT, Martin-Belloso O (1999) Characterisation of low-fat high density fiber frankfurters. Meat Sci 52(3):247–256

Haque A, Ahmad S, Azad ZRAA, Adnan M, Ashraf SA (2023) Incorporating dietary fiber from fruit and vegetable waste in meat products: a systematic approach for sustainable meat processing and improving the functional, nutritional and health attributes. Peer J 11:e14977

Hegazy AI (2011) Influence of using fenugreek seed flour as antioxidant and antimicrobial agent in the manufacturing of beef burger with emphasis on frozen storage stability. World J Agric Sci 7(4):391–399

Javidipour I, Vural H, Ozboy-Ozbas O, Tekin A (2005) Effects of interesterified vegetable oils and sugar beet fiber on the quality of Turkish-type salami. Int J Food Sci Technol 40:177–185

Jiang J, Xiong YLL (2016) Natural antioxidants as food and feed additives to promote health benefits and quality of meat products: a review. Meat Sci 120:107–117
Jimenez-Colmenero F, Francisco J, Sanchez-Muniz J, Olmedilla Alonso B (2010) Design and development of meat-based functional foods with walnut:Technological, nutritional and health impact. Food Chem 123:959–967
Kausar T, Hanan E, Ayob O, Praween B, Azad ZRAA (2019) A review on functional ingredients in red meat products. Bioinformation 15(5):358–363
Kim YJ, Park WY, Choi IH (2010) Effects of dietary α-tocopherol, selenium, and their different combinations on growth performance and meat quality of broiler chickens. Poult Sci 89:603–608
Kumar D, Tanwar VK (2011) Utilization of clove powder as phytopreservative for chicken nuggets preparation. J Stored Prod Postharvest Res 2(1):11–14
Landete JM (2013) Dietary intake of natural antioxidants: vitamins and polyphenols. Crit Rev Food Sci Nutr 53(7):706–721
Macho-González A, Bastida S, Garcimartín A, López-Oliva ME, González P, Benedí J, Sánchez-Muniz FJ (2021) Functional meat products as oxidative stress modulators: a review. Adv Nutr 12(4):1514–1539
Madane P, Das AK, Pateiro M, Nanda PK, Bandyopadhyay S, Jagtap P, Barba FJ, Shewalkar A, Maity B, Lorenzo JM (2019) Drumstick (*Moringa oleifera*) flower as an antioxidant dietary fibre in chicken meat nuggets. Food Secur 1, 8(8):307
Marco ML, Sanders ME, Gänzle M, Arrieta MC, Cotter PD, De Vuyst L, Hutkins R (2021) The International Scientific Association for Probiotics and Prebiotics (ISAPP) consensus statement on fermented foods. Nat Rev Gastroenterol Hepatol 18(3):196–208
Micha R, Michas G, Mozaffarian D (2012) Unprocessed red and processed meats and risk of coronary artery disease and type 2 diabetes – an updated review of the evidence. Curr Atheroscler Rep 14(6):515–524
Mir SA, Naseem Z, Rizwan D, Masoodi FA (2022) Bioactive peptides from meat. In: Nollet LM, Ötleş S (eds) Bioactive peptides from food: sources, analysis, and functions. CRC Press
Mozaffarian D, Micha R, Wallace S (2010) Effects on coronary heart disease of increasing polyunsaturated fat in place of saturated fat: a systematic review and meta-analysis of randomized controlled trials. PLoS Med 7:e1000252
Naseem Z, Mir SA, Wani SM, Rouf MA, Bashir I, Zehra A (2023) Probiotic-fortified fruit juices: Health benefits, challenges, and future perspective. Nutrition 115:112154
Naveen KTJ, Nayar R, Rajagopal K (2015) Technological and health impacts of meat based functional foods – a review. Int J Sci Res 4(10):475–481
Naveena BM, Sen AR, Kingsly RP, Singh DB, Kondaiah N (2008) Antioxidant activity of pomegranate rind powder extract in cooked chicken patties. Int J Food Sci Technol 43(10): 1807–1812
Nisa AU, Zahra N, Hina S, Masood S, Javed A, ManzarInam S (2017) Development of meat-based functional foods: a review. J Appl Biol Biotechnol 5(03):086–092
Olmedilla-Alonso B, Jiménez-Colmenero F, Sánchez-Muniz FJ (2013) Development and assessment of healthy properties of meat and meat products designed as functional foods. Meat Sci 95(4):919–930
Pame K, Daimary B, Borah S (2022) Utilization of dietary fibre in meat products as functional foods. Int J Ag Env Biotech 15(01):147–151
Pogorzelska-Nowicka E, Atanasov IAG, Horbanczuk J, Wierzbicka A (2018) Bioactive compounds in functional meat products. Molecules 23:307
Rashwan AK, Karim N, Shishir MRI, Bao T, Lu Y, Chen W (2020) Jujube fruit: a potential nutritious fruit for the development of functional food products. J Funct Foods 75(9):104205
Salejda AM, Olender K, Zielinska-Dawidziak M, Mazur M, Szperlik J, Miedzianka J, Zawiślak I, Kolniak-Ostek J, Szmaja A (2022) Frankfurter-type sausage enriched with buckwheat by-product as a source of bioactive compounds. Food Secur 11(5):674

Saricoban C, Yılmaz MT, Karakaya M (2009) Responce surface methodology study on the optimization of effects of fat, wheat bran and salt on chemical, textural and sensory properties of patties. Meat Sci 83:610–619

Sebranek JG, Sewalt VJH, Robbins K, Houser TA (2005) Comparison of a natural rosemary extract and BHA/BHT for relative antioxidant effectiveness in pork sausage. Meat Sci 69(2):289–296

Serdaroğlu M, Kavuşan HS, İpek G, Öztürk B (2018) Evaluation of the quality of beef patties formulated with dried pumpkin pulp and seed. Korean J Food Sci Anim Resour 38:1–13

Šojić BV, Petrović LS, Pešović BM, Tomović VM, Jokanović MR, Džinić NR, Salitrežić PP (2011) The influence of inulin addition on the physico-chemical and sensory characteristics of reduced-fat cooked sausages. Acta Periodica Technologica 42:157–164

Soto AM, Morales P, Haza AI, García ML, Selgas MD (2014) Bioavailability of calcium from enriched meat products using Caco-2 cells. Food Res Int 55:263–270

Talukdar S, Sharma DP (2010) Development of dietary fiber rich chicken meat patties using wheat bran and oat bran. J Food Sci Technol 47(2):224–229

Tunieva EK, Gorbunova NA (2017) Alternative methods of technological processing to reduce salt in meat products. Theory Pract Meat Process 2(1):47–56

Ursachi CŞ, Perţa-Crişan S, Munteanu FD (2020) Strategies to improve meat products' quality. Food Secur 9(12):1883

Varvara M, Bozzo G, Celano G, Disanto C, Pagliarone CN, Celano GV (2016) The use of ascorbic acid as a food additive: technical-legal issues. Ital J Food Saf 5(1):4313

Verma A, Rajkumar V (2021) Antioxidant effect of Amla (Emblica officinalis) fruit and curry (Murraya koenijii) leaf extracts on quality of goat meat nuggets. Indian J Small Rumin 27:105–112

Verma AK, Rajkumar V, Banerjee R, Biswas S, Das AK (2013) Guava (*Psidium guajava* L.) powder as an antioxidant dietary fibre in sheep meat nuggets. Asian Australas J Anim Sci 26(6): 886–895

Verma AK, Rajkumar V, Kumar MS, Jayant SK (2020) Antioxidative effect of drumstick (*Moringa oleifera* L.) flower on the quality and stability of goat meat nuggets. Nutr Food Sci 50(1):84–95

Wenjiao F, Yunchuan C, Junxiu S, Yongkui Z (2014) Effects of tea polyphenol on quality and shelf life of pork sausages. J Food Sci Technol 51:191–195

Yılmaz I (2005) Physicochemical and sensory characteristics of low-fat meatballs with added wheat bran. J Food Eng 69:369–373

Chapter 7
Design and Development of Meat-Based Functional Foods

Sheikh Rafeh Ahmad, Asma Irshad Qureshi, Syed Arshad Hussain, Humaira Fayaz, and Asif H. Sofi

7.1 Introduction

The growing awareness of how diet impacts health has sparked interest in studying food components' effects on our well-being. Consumers are increasingly concerned about health and seek more nutritious food choices that offer added health benefits. Functional foods, known for positively influencing body systems and promoting health while preventing diseases, encompass both unaltered and modified foods, achieved through technological or biotechnological means. Recent research has delved into tertiary functions of foods, focusing on how specific food components regulate physiological systems to reduce disease risks. Functions like anticarcinogenecity, antimutagenecity, antioxidative activity and antiaging effects are examples of these tertiary functions. This has led to an upsurge in the production of such functional foods on a larger scale due to a growing awareness of the link between diet and a healthy lifestyle. For instance, there is considerable attention on dairy products fortified with increased probiotics. Meat, a significant source of food proteins along with valuable nutrients like minerals and vitamins such as iron, vitamin B12, and folic acid, faces health concerns due to certain constituents associated with chronic diseases. This has put pressure on food manufacturers to develop healthier meat products. Functional foods offer a significant opportunity for the meat industry to create products with enhanced health-promoting factors while minimizing components linked to health issues. Various techniques in animal processing and product manufacturing are being considered to design healthier meat products.

Red meat refers to unprocessed muscle meat as beef, veal, pork, lamb, mutton, horse and goat (Bouvard et al. 2015), while processed meat refers to meat that has

S. R. Ahmad (✉) · A. I. Qureshi · S. A. Hussain · H. Fayaz · A. H. Sofi
Division of Livestock Products Technology, FVSc & AH, SKUAST-K, Srinagar, India
e-mail: sh.rafeh@skuastkashmir.ac.in

S. A. Rather, F. A. Masoodi (eds.), *Hand Book of Processed Functional Meat Products*, https://doi.org/10.1007/978-3-031-69868-2_7

been transformed through salting, curing, fermentation, smoking or other processes to increase shelf life and enhance flavour (Bouvard et al. 2015; Jiménez-Colmenero et al. 2017; Salim et al. 2018). Although fresh meat and processed meat products are consumed regularly by the majority of the population, there is an increasing concern based on many studies that report the association of high intake of red and/or processed meat with chronic diseases such as obesity, type 2 diabetes, cardiovascular disease and a variety of cancers (Boada et al. 2016). Besides, there is a growing understanding of the relationship between diet, specific food ingredients and health, leading consumers to look for foods with additional dietary benefits (Khalaf et al. 2021). Some consumer groups believe that meat and meat product consumption is unhealthy, because of their high animal fat, sodium, nitrites, cholesterol and synthetic antioxidants (Grasso et al. 2014; Hygreeva et al. 2014). However, these potentially negative components in meats can be minimised by the selection of lean meat cuts, removal of adipose fat and dietary manipulation. In addition to genetic and nutritional strategies, the reduction of compounds with negative health implications can be also minimised by reformulation and technological strategies (Grasso et al. 2014; Monteiro et al. 2015; Gomes et al. 2016). Therefore, the development of healthier meat and meat products may represent an opportunity for the meat industry to alleviate the growing negative perception on the meat consumption (Canto et al. 2014; Bessa et al. 2016; Guedes-Oliveira et al. 2016; Alvares et al. 2018; Cunha et al. 2018). This chapter provides an overview of the potential strategies used to improve the health implications of meat products.

7.2 Strategies for the Development of Meat Based Functional Foods

Numerous strategies for the development of meat based functional foods encompass aspects ranging from animal farming practices to alterations in meat formulations. Furthermore, factors such as meat storage conditions and consumption methods are also taken into account, as they significantly influence meat composition.

As far as meat is concerned, strategies for functional meat and products development are basically related to increasing the presence of beneficial compounds and limiting those with negative health implications in meat and meat products. Regarding meat products, efforts are mainly directed toward their reformulations which include reduction of fat, modification of fatty acid profile, control and reduction of sodium content, cholesterol, allergens (vegetable and egg proteins), biogenic amines, addition of vegetable oils, fish oils, conjugated linoleic acid, plant-based proteins (soy protein), natural extracts, vitamins C and E, minerals (Ca, Se, Fe, Mg, Mn), plant sterols, phytate, probiotics, dietary fiber, oligosaccharides and L-carnitine (Fernández-Ginés et al. 2004; Jiménez-Colmenero 2007) (Fig. 7.1). For successful functional food development, both consumer needs and opportunities and legislative aspects need to be taken into consideration (Ares and Gambaro 2007).

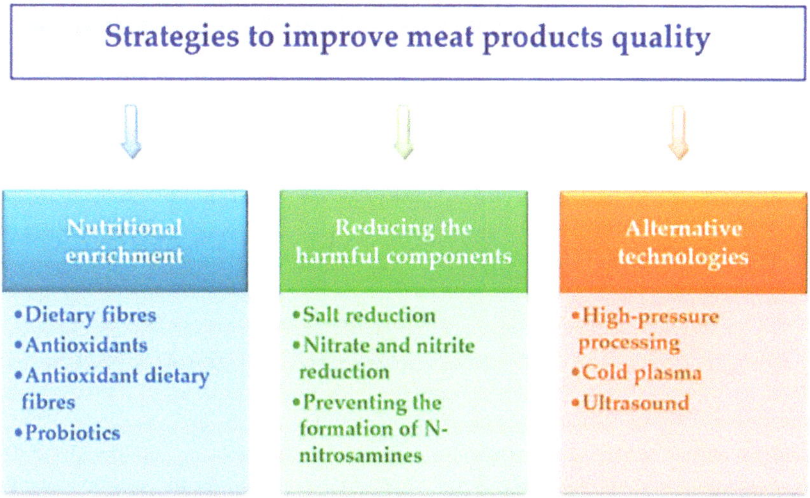

Fig. 7.1 Strategies to improve meat products quality

Genetic selection, nutrition and feeding management, growth-promoting and nutrient partitioning agents, immunisation of animals against target circulation hormones or releasing factors and gene manipulation techniques are available strategies to induce changes in protein, lipid content, vitamin E level and fatty acid composition of meat (Jiménez-Colmenero et al. 2001; Burnett et al. 2020). For the processed meat industry, the first strategy involves reducing ingredients that could be considered unhealthy, for example synthetic antioxidants, nitrites and sodium. The second is to incorporate ingredients that are potentially health-enhancing and may have technological functions, such as fibre, fat substitutes, vegetable proteins, natural antioxidants, mineral and vitamins (Jiménez-Colmenero et al. 2001; Olmedilla-Alonso et al. 2013; Grasso et al. 2014; Guedes-Oliveira et al. 2016). Food scientists have been working on developing various strategies to produce healthier meat and meat products (Fig. 7.2).

7.2.1 Animal Production Practices

The availability of bioactive compounds in meat can be modified through animal production practices. The composition of animal tissue and that of carcasses vary according to type of animals and the other characteristics. In-vivo modification approaches are present to modify fatty acid, minerals and vitamin profile and composition. These compounds have a demonstrated efficacy as bioactive compounds and are easy to modify in the tissues (Jiménez-Colmenero 2007). Animal production practices represent the first opportunity to modify the presence of

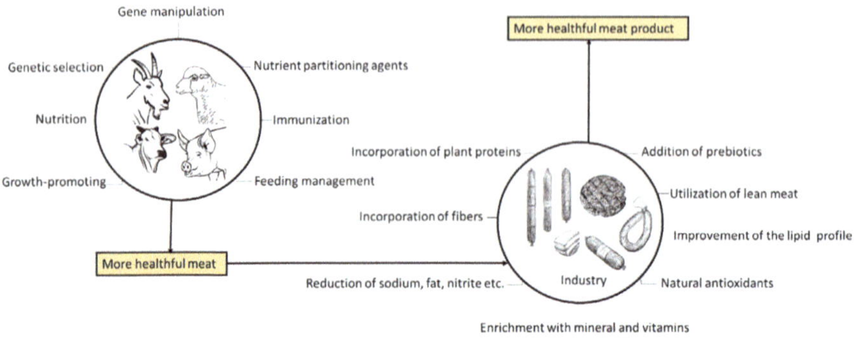

Fig. 7.2 Strategies to achieve potentially more healthful fresh meat and processed meat products

bioactive components. The composition of the animal tissues and hence, that of carcasses and commercial cuts (and meat raw materials), varies not only according to species, but also according to breed, age, sex, feeding type, etc. Several strategies are available for inducing in vivo changes, although these compounds should meet two essential requirements: on the one hand, have a demonstrated efficacy as bioactive compounds and, on the other, be modifiable in the tissues of the animals. These compounds include lipids (fat content, fatty acid composition or cholesterol level), fat soluble vitamins and minerals (Jiménez-Colmenero 2007). Some of the more relevant procedures are described here:

7.2.1.1 Nutrition and Management

Lipids are the bioactive compounds that have gained much importance in designing meat based functional foods due to their effect on health. In animal tissues, lipids can be deposited either by endogenous process i.e. denovo system or by exogenous methods (supplied by feed). Several feeding trials have been done on animals to increase the content of mono (MUFA) and poly-unsaturated (PUFA) fatty acids in animal tissues. By increasing MUFA's in animal feed, the content of MUFA in the meat can be upgraded. Some studies showed that PUFA's in animal meat can be enhanced by adding different PUFA rich ingredients of plant and marine origin (Raes et al. 2004; Scollan et al. 2006; Wood et al. 2008). The level of CLA in chicken, beef and lamb has been augmented by dietary supplementation (Raes et al. 2004; Lynch and Kerry 2000; Schmid et al. 2006). Studies have shown that in poultry, the content of cholesterol can be reduced selectively in muscle tissue of living animal by raising the copper level in feed (Clarke 1997). Also production practices for reducing fat content in farm animals have also been successfully implemented (Bass et al. 1990; Hays and Preston 1994; Moloney 2006). Peroxidation risk in muscle food may increase due to unsaturated fatty acids and cause oxidative damage. This can be prevented by enhancing the muscle antioxidants through animal diet. Recent studies have indicated that when vitamin E was added as a supplement in animal diet, its concentration in animal tissue increased. Mineral contents like Fe, Mg and Se of meat can also be increased by dietary supplementation with these healthy elements.

7.2.1.2 Genetic Strategies

Genetic improvement to produce changes that are favourable from an economic and productive perspective is a practice that offers interesting prospects in healthy food production. Although the objectives of selection have varied over time, they have basically been aimed at reducing fat content and increasing fat unsaturation.

Genetics play a crucial role in the development of farm animals and thus on the quality and nutritional profile of meat obtained from them. There are many breeding strategies that have been proposed in recent years to reduce the fat content of the carcass by reducing the back fat thickness and intramuscular fat with an increment in lean percentage of the carcass. Some of the research studies have been directed towards the production of more tender meat (Johnson et al. 1990; Wheeler et al. 1990). Several strategies for increasing the ratio of lean to fat in farm animals have been proposed which may be based on the repartitioning effects of somatotropin, specific β-adrenergic agonists, some growth promoting agents and gene manipulation techniques (Solomon et al. 2002). Other strategies may be based on different breeding schemes, but most entail a backcross or a three-way or four-way cross. It has been noted that leaner carcasses are generally associated with a more glycolytic muscle fibre type and hence more rapid post-mortem metabolism, pale meat colour and lower water holding capacity (De Semet et al. 2002). Some changes in lipids have been further reported to be dependent on sire genetic type (Armero et al. 2002).

7.2.1.3 Genetic Information

Genetic advances have been opening up new possibilities for genetic improvement selection programs to manipulate carcass composition. Genetic markers have become strategic tools for identification of loci underlying the expression of quantitative traits (quantitative trait loci [QTL]) that have been found to affect carcass characteristics related to fat content and fatty acid profile. Selective breeding and management practices have been used in recent decades to reduce carcass fat content (Goutefongea and Dumont 1990; Higgs 2000).

7.2.1.4 Genetic Modification

Biotechnology in animals (cloning, transgenesis or transgenesis followed by cloning) is a more extreme scientific method. It offers the potential to improve the quality, yield and safety of food products by direct genetic manipulation (Solomon et al. 2008). These possibilities affect, for example, the improvement of fatty acid profiles. Livestock cannot convert n-6 fatty acids into n-3 fatty acids because they lack an n-3 fatty acid desaturase gene. Transgenic pigs have been produced that express a Δ12 fatty acid desaturase gene from spinach (Saeki et al. 2004) or *Caenorhabditis elegans* encoding one n-3 fatty acid desaturase gene (Lai et al. 2006) for synthesis of LNA, or produce high levels of n-3 PUFAs and reduce the n-6/n-3 PUFA ratio.

7.2.2 *In-Vitro Meat Production Systems*

Conventional meat production is considered an inefficient process of converting vegetable proteins into animal proteins by rearing and slaughtering meat animals. With the ever increasing human population crossing the 8 billion mark recently, it become quite imperative to ensure a proper meat supply in the future in a sustainable manner with the already dwindling natural resources (Kumar et al. 2017, 2019, 2021a, b, 2022a). To fulfil the future demand for meat for the burgeoning population, the meat production system needs a major transformation and technological breakthrough as incremental gains by improving the available technologies or scaling up the production of conventional meat production would not be sufficient (Chen et al. 2022). Thus, the present focus of the food industry is to ensure a sustainable supply of meat by harnessing the available resources and improving the technology for the production of meat alternatives such as plant-based meat analogs, cultured meat, insect meat and hybrid meat products (Kumar et al. 2022b, c; Sharma et al. 2022). Among these alternatives, cultured meat/in-vitro meat has attracted significant attention from researchers and policymakers worldwide. As per Reiss et al. (2021) estimations, in-vitro meat production requires 89% less water, 99% less land and 96% lower GHGs (greenhouse gases) as compared to intensive livestock farming.

In-vitro meat as the term refers to producing edible biomass from stem cells by applying cell culturing technology (Stephens et al. 2018). This technology has the potential to supply high-quality animal proteins at low environmental impact with minimal use of animals, later restricted to the harvesting of stem cells through muscle biopsy (Post et al. 2020). The in-vitro meat production process is completed in the 4 basic steps viz., (1) harvesting stem cells from embryos or biopsy, (2) proliferation of these stem cells in the presence of growth factors, and nutrients followed by cell growth on scaffolds/micro-carriers, (3) induced maturation of cells into myofibers, adipocytes, and other mature cells and (4) processing this edible tissue into meat products (Guan et al. 2021). To get desired texture and sensory attributes, co-culturing of multiple cells or 3D/4D bioprinting are applied (Zhang and Wang 2019). There are three approaches for in-vitro myogenesis viz., muscle stem cells, differentiated from pluripotent stem cells and direct reprogramming applying transgenesis (Choi et al. 2019, 2020; Ding et al. 2018; Genovese et al. 2017). During the harvesting of efficient stem cells, animal species, age (age-induced loss of proliferation could be controlled by IGF 1 factor) and sex (effect of sex hormones, males have more satellite cells than females) affect the overall collection of the

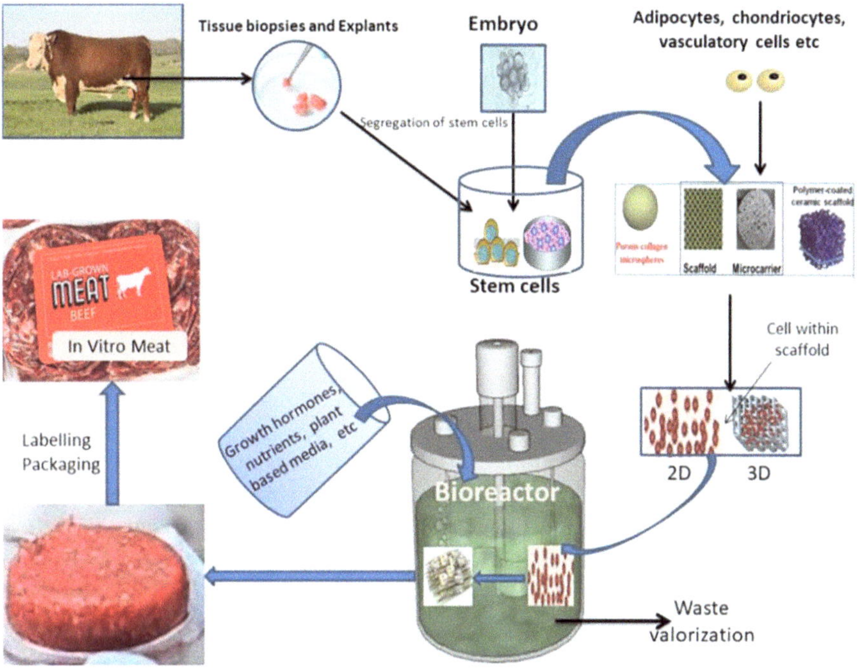

Fig. 7.3 In-vitro meat production process

satellite cells (Choi et al. 2020). For harvesting stem cells from donor animals, tissue biopsy (by using a needle, quick process to collect up to 0.5 g sample, little stress to animals) and a small incision (more invasive, up to 15 g sample) could be used with a frequency of 4 samples collection in every session with one session in every 3 months to facilitate proper recovery to animals (Melzener et al. 2021).

Meat produced in vitro has been proposed as a humane, safe and environmentally beneficial alternative to slaughtered animal flesh as a source of nutritional muscle tissue (Datar and Betti 2010). By co-culture, medium formulation or genetic engineering, it is theoretically possible to create products with different tastes, textures and nutrient profiles, including altering the presence of bioactive compounds such as fatty acids (Datar and Betti 2010) (Fig. 7.3).

Due to its minimal involvement of animals and potential sustainability, in-vitro meat is also known as "cell-based, "cultivated", "clean", "slaughter-free", "lab-grown", and "nano-pastured" meat (Pakseresht et al. 2022). The production techniques involve the collection and purification of cells and storage, transportation, standardization, quality control, and food processing technology are employed on it for the creation of meat products based on in-vitro meat (Post et al. 2020). The in-vitro meat is created under sterile circumstances, thus associated with the additional merits of being a safe, long-lasting meat, with lower food loss. Furuhashi et al. (2021) provided the whole production process follows high-quality GMP (good manufacturing practices), hygiene and sanitation as per the globally recognized standards such as European Union and FDA standards.

The fundamental advantage of "in-vitro meat" is that it potentially allows for mass production from a small number of animal cells, derived either from farm animals or from cell lines that multiply in bioreactors. As a result, this method utilizes fewer/minimum animals to produce a large quantity of meat, hence termed "victimless meat", "slaughter-free meat" or "cruelty-free meat" (Chriki and Hocquette 2020). However, while producing in-vitro meat, suitable technological innovations such as co-culturing should be applied for matching the nutritive value of in-vitro meat to conventional meat such as fatty acid profile, vitamin B12 and trace minerals. It is also desired to produce in-vitro meat owing to organoleptic attributes such as color, appearance, flavor and texture similar to conventional meat. Cultured meat production is thought to be safer for customers, healthier and disease-free as compared to traditional meat production (Arshad et al. 2017). The other associated merits acclaimed to in-vitro meat are a production method complying with animal welfare due to slaughter-free production, better compositional control and nutritive quality of the developed product and better food safety and lower public health risk (zoonoses) (Post et al. 2020). Based on the life cycle assessment (LCA), in-vitro meat had a 7–45% lower energy requirement, 78–96% lower greenhouse gas emission, and 82–96% lower land requirement as compared to conventional meat (Tuomisto and de Mattos 2011). However, the gains in energy and greenhouse gas (GHG) emissions depend upon the specific production process and utilization of clean energy in the production of in-vitro meat (Lynch and Pierrehumber 2019). The increased CO_2 would likely affect agricultural production and food quality with comparatively less effect on photosynthesis in C_4 plants. In general, increased CO_2 results in increased growth, higher photosynthesis, lower stomatal conductance, decreased water consumption and lowering tissue protein contents in plants (Taub 2010; Gamage et al. 2018). In the context of cell culturing, CO_2 via bicarbonate buffer plays a crucial role in pH maintenance and cellular integrity; where CO_2 inhibition causes acidification of medium (Dubey et al. 2021). Further, the CO_2 concentration in the bioreactor also affects the utilization of glucose and production of lactate as up to 40–50% lower glucose utilization at 140 mmHg of partial CO_2 (DeZengotita et al. 1998). In addition to avian and mammalian cell culture for in-vitro meat production, fish cells are regarded as potentially better suitable for tissue culture due to their physiological properties such as tolerance to hypoxia, salt tolerance, buffering capacity, ability to grow at low temperatures (Rubio et al. 2019).

The production process of in-vitro meat has three basic requirements: cell, media, and scaffold (Allan et al. 2019). To realize the acclaimed merits of in-vitro meat, it should be popularized and produced at a mass scale. For example, for the production of 1 ton of in-vitro meat, there are 1014 cells need to be produced, which would need a bioreactor of 10,000 L with 107 cells/mL density in the bioreactor (Guan et al. 2021). Similarly, for producing 1 kg protein from muscle cells, approx. 2.9 × 1011 to 8 × 1012 cells are required (Rosser and Thomas-Vazquez 2018). Thus, for a perceivable contribution to conventional meat production, huge investment and technological breakthrough are needed. As per a report by CE Delft, the availability and cost of cell culture media are the main drivers affecting the mass-scale production of in-vitro meat (CE Delft 2021), with growth factors and recombinant proteins accounts up to 99% of the cost of cell culture media.

Based on information gathered from industry leaders and published literature, Garrison et al. (2022) analyzed the cost of production of in-vitro meat production on a large scale with an annual production capacity of 5,40,000 kg in a scenario assuming several technological advancements/ innovations leading to marked cost reduction in the growth hormones and culture media. The authors reported the cost of 1 kg of in-vitro meat to $63 in the proposed model with cell-culture medium, bioreactors, and labor charges accounting for $55/ kg (more than 80%) of the overall cost of production. Thus, there is an urgent need to develop low-cost culture media, large-scale bioreactors, and automation to make this novel product affordable to the common consumer. Thus, to make in-vitro meat affordable and competitive with conventional meat, readily and cheap availability of these ingredients is crucial. According to one estimate, the reduction of the cost of cell culture media has the potential to decrease the cost of in-vitro meat to $15/kg (Chen et al. 2022). Other important factors are a trained and skilled workforce and the utilization of efficient and clean energy. This warrants the upgradation and scaling up of the supply chain and ensures a proper supply of inputs such as media, stem cells, etc.

7.2.3 *Meat Product Reformulation Techniques*

Development of meat based functional foods by modification in transformation systems of meat is an effective scientific strategy. Content of different bioactive compounds can be altered using several techniques. Meat formulation process allows the use of traditional ingredients and other ingredients to make meat products healthier. Technological strategies used to design and develop functional foods based on changes in meat transformation systems are especially promising. A number of the approaches can be used to remove, reduce, increase, add and/or replace different bioactive components. The modification of the meat formulation process also makes it possible to use traditional and other ingredients specifically designed with certain attributes that confer health-promoting properties.

7.2.3.1 Modifying Fat Content

Utilizing strategies for meat reformulation, fat content can be improved by modifying fatty acid profile, reducing fat and cholesterol contents. Water and other fat replacers (gums, protein based or carbohydrate based) can be added to reduce the fat density. To design meat products with reduced fat, leaner meat is used as raw material. Manufacturing and preparation procedures are very important in meat reformulation strategies (Jiménez-Colmenero 1996; Keeton 1994; Kerry and Kerry

2006). Fatty acid composition of meat is very important in making healthier products because plasma lipids are differently affected by each fatty acid. Strategies are present to alter the fatty acid profile of meat by replacing unhealthy fats with healthy ones. Healthy fats have smaller proportions of SFA's and larger proportions of unsaturated fatty acids and also low cholesterol values. Lipids either from plant origin or from marine sources can be incorporated directly as liquid oils into meat products or by incorporation in encapsulated and emulsified forms (simple and multiple emulsions). Plant sources for lipids include maize, soy, bean, olive, peanut and cottonseed, etc. and marine sources are mainly algae and fish. Successful experiments to inject commercially produced CLA isomers in meat have been conducted (Baublits et al. 2007). It has also been incorporated into meat products such as pate (Martín et al. 2008) or sausage (Juarez et al. 2009) to make these products healthier.

Approaches for cholesterol removal from meat raw materials by physicochemical procedures have been reported (Clarke 1997), through the use of cholesterol-reducing bacteria (*Eubacterium coprostanoligenes*) in fermented meat products (Madden et al. 1999) or by diluting fat and lean meat raw materials (since cholesterol is strictly linked to animal cells) with vegetable oils and plant based proteins (Clarke 1997).

7.2.3.2 Adding Plant Proteins

For past 3–4 years, there is an increased trend in using plant proteins in meat products because of their health enhancing properties and low cost. Plant proteins cut down cholesterol and energy levels and increase content of health promoting proteins and thus upgrade nutritional value of meat products. Studies depict that increasing plant proteins from sources like sunflower, walnut, etc. in meat can balance lysine/arginine ratio and thus can make meat more beneficial. Several meat products have been designed to maintain blood cholesterol level by using soy protein as functional ingredient (Jiménez Colmenero et al. 2012).

Plant proteins have been used in meat products for technological purposes, to reduce costs, for nutritional reasons and recently for their health-promoting properties (Table 7.1). They provide nutritional benefits by lowering energy and cholesterol contents (when used as fat replacers), raising protein levels or balancing amino acid profiles and incorporating certain health-enhancing components (Jiménez Colmenero et al. 2012, Martín et al. 2008). Some proteins from plant sources (sunflower, walnut, etc.) have been used in the meat system to balance (reduce) lysine/arginine ratios and have healthy beneficial effects (Jiménez-Colmenero et al. 2010). Soy protein has been used as functional ingredient in various FOSHU meat products (such as low-fat sausage), with claims informing that an acceptable blood cholesterol level can be maintained by consuming these products (Arihara 2006).

Table 7.1 Examples of healthy and unhealthy components considered in the development of meat-based products

Target condition/ function (markers)	Compounds associated with beneficial effect	Compounds associated with an unhealthy effect
Cardiovascular Lipids Blood pressure Obesity others Cancer	MUFA, n-3 PUFA, CLA, dietary fiber, phytosterols, vitamins C and E, bioactive peptides, histidyl dipeptides, L-carnitine, lycopene. Bioactive peptides, vegetable/plant proteins CLA, dietary fiber, creatine Folic acid, vitamins B6 and B12, lycopene, lutein, selenium, taurine Coenzyme Q10, extracts of fruits, herbs and spices rich in flavonoids and phenolic compounds Dietary fiber, folic acid, vitamin E, selenium, CLA, probiotics, histidyl dipeptides, lycopene, extracts of fruits, herbs and spices rich in flavonoids and phenolic compounds	Fat, SFA, trans-fatty acids, cholesterol Sodium Fat, SFA, trans-fatty acids Nitrite (nitrosamines), lipid oxidation compounds, inorganic phosphate, polycyclic aromatic hydrocarbons, heterocyclic amines, iron
Bone diseases	Calcium, magnesium, L-carnitine	
Immunological status	Probiotics, selenium, iron	
Anemia (iron-deficiency)	Iron	
Migraine and allergy		Biogenic amines Allergens (gluten, lactose)
Growth and development	Iodine	

7.2.4 *Probiotic Meat Products*

Because of their health benefits (Table 7.1), probiotics could be a promising market for the meat industry. Probiotics can be efficiently ingested through fermented meat products which are processed without heating. There are many published studies which show the incorporation of probiotic bacteria (*Lactobacillus acidophilus*, *Lactobacillus casei*, *Lactobacillus rhamnosus*, *Lactobacillus paracasei*, *Lactobacillus plantarum*, etc.) to fermented sausages (Jiménez Colmenero et al. 2012). Functional fermented meat products containing lactic acid bacteria of human gut are now being marketed (Arihara 2006). Probiotic strains are not damaged by various technical treatments during manufacturing process and also not in the gastrointestinal tract. Resistance to technological processing (presence of nitrite and sodium chloride, pH, water activity, mechanical and thermal treatments, etc.) and to the environment in the gastrointestinal tract are among the main characteristics of the microorganisms suitable for use as probiotic strains in this fermented matrix.

Commercial probiotic functional fermented meat products (salami in Germany and meat spread in Japan) have been reported to contain human intestinal lactic acid bacteria such as *L. acidophilus, L. casei, Bifidobacterium spp. and L. rhamnosus* (Arihara 2006; Jiménez-Colmenero 1996).

7.2.5 Addition of Fiber and Prebiotics Contents

Owing to physiological and medicinal properties, prebiotics and dietary fibers are used as important functional ingredients in meat products. Development of meat products using varying quantities of fibers and prebiotics is increasing. Many fruits, vegetables, legumes, cereals and sea weeds are good sources of these compounds (Jiménez-Colmenero and Delgado-Pando 2013; Verma and Banerjee 2010). Different prebiotics and dietary fibers have been used in the manufacture of meat products for their technological qualities (as non-caloric bulking agents, to improve water and oil retention or emulsion stability, to modify texture, to help overcome the effects produced by composition changes, etc.) and physiological properties (reducing diabetes risk, favoring blood lipid regulation, preventing CVD, colon cancer and constipation, regulating intestinal transit). In recent years, there has been an extraordinary increase in the number of studies on the addition of different concentrations (1.5–3%) of prebiotics (inulin), dietary fiber and fiber-rich ingredients from highly diverse sources—cereals (oats, rice, wheat, etc.), fruits (apple, lemon, orange, etc.), legumes (soy, peas, etc.), roots (carrot, sugarbeet, konjac, etc.), tubers (potato) and seaweeds (red and brown algae)—in the formulation of fresh (ground and sausage), cooked and fermented meat products. The use of dietary fiber in the development of fiber-enriched meat products has recently been reviewed (Jiménez-Colmenero and Delgado-Pando 2013; Verma and Banerjee 2010).

7.2.6 Mineral Enrichment

Many beneficial minerals (Fe, Se, Ca, Zn etc) are present in meats. Different techniques have been devised to modify the concentration of these minerals. Meat products enriched with selenium (Garcia-Iniguez et al. 2010), calcium (Caceres et al. 2006) and iron (Navas-Carretero et al. 2011) have been produced. Studies show that non meat ingredients (walnut, sea weed etc.) can also raise the mineral content of processed meat (Jiménez Colmenero et al. 2010; Cofrades et al. 2011a). Dry fermented sausage was enriched with iodine in a study (Garcia-Iniguez et al. 2010). Although meats are sources of many minerals (Fe, Se, Ca, Zn, etc.) with effects on human health (Table 7.2), technological strategies have been employed to modulate their concentrations in these foods. Thus, different meat products have been enriched with selenium (Garcia-Iniguez et al. 2010), calcium (Caceres et al. 2006) or iron (Navas-Carretero et al. 2009). In this regard, a Spanish producer

Table 7.2 Intervention studies with meat based functional products in humans

Food Functional ingredient	Intake	Objective	Design	Subjects	Markers	Results	References
Meat and meat products obtained from nutrition & feeding management							
n-3 PUFA enriched pork products (steak, stir-fry, diced, mince, and sausage) from animals feeding with diet fortified with tuna fishmeal products	1000 g of five fresh cuts/week 1.3 g LC n-3 PUFA/week, from pork meat products	Cardiovascular health benefits to consumers	Parallel, double-blind, randomized, placebo Controlled 12 weeks	Healthy $N = 33$ (17 men, 16 women) 18–65 years	Serum lipids Erythrocyte fatty acid composition TG Thromboxane B2 Body weight	Erythrocyte DHA increased 15% No changes in EPA TG decreased 0.3% Serum thromboxane increased less than in control group	Coates et al. (2009)
Selenium-enriched chicken meat from animal feeding with diet supplemented with sources of selenium	Chicken (800 g/week) selenium (22 µg/day)	Body composition biochemical and antioxidant status	Parallel double-blind 10 weeks	Normo and overweight $N = 24$ 20–45 years	Antioxidant status (selenium, glutathione peroxidase activity) Inflammatory status (homocysteine, uric acid, C-reactive protein) Lipids, glucose and insuline	Weight reduction. Improve of body composition No effect on antioxidant and inflammatory biomarkers	Navas-Carretero et al. (2011)
n-3 PUFA enriched chicken meat from animals feeding with linseed and rapeseed oils	160 g chicken meat/day n-3 PUFA enriched chicken meat (160 g/day)	Blood lipid concentrations.	Double-blind, randomized, Placebocontrolled 4 weeks	Healthy $N = 46$ (11 men, 35 women) 20–29 years	Cholesterol (total, LDL, HDL), TG Phospholipid fatty acid Blood pressure C-reactive protein Body weight	EPA and AA/EPA increased in serum phospholipids DHA increased No difference in other markers, body weight or blood pressure	Haug et al. (2012)

(continued)

Table 7.2 (continued)

Food Functional ingredient	Intake	Objective	Design	Subjects	Markers	Results	References
Beef high in MUFA and low in SFA ground beef: (a) low-MUFA (b) high-MUFA	114 g ground beef patties, five per week	Effect on HDL-C concentration and LDL particle diameter.	Crossover, randomized 5 weeks/ diet	Normocholesterolemic (physical and blood examination) $N = 27$ (men) 23–60 years	Lipoproteins (plasma), VLDL, LDL and HDL-cholesterol HDL2, HDL3 and LDL lipoprotein diameters Insulin and glucose	Both interventions decreased plasma insulin and HDL2 and HDL3 particle diameters and increased plasma 18:0 and 20:4 (n-6) relative to baseline High-MUFA increased the HDL-C concentration from baseline	Gilmore et al. (2011)
Reformulated meat products							
n-3 PUFA enriched meat products (luncheon meat and sausages) fortified with fish oil	Wide range of foodstuffs naturally containing n-3 PUFA and items fortified (among others is luncheon meat and sausages)	Effect on intake and blood and tissue proportions of n-3 PUFA	Dietary intervention (2 points time) 4 weeks	Healthy $N = 16$ (male) 39 ± 4.5 years	Dietary intake—Plasma, platelet and mononuclear cell phospholipid fatty acids	Dietary intake of ALA, EPA and DHA increased, and LA decreased. Proportions of EPA and DHA increased in phospholipids pools Increment in phospholipid pools	Metcalf et al. (2003)
Frankfurters and pâtés: (a) reduced-fat products (RF), (b) n-3-enriched reduced-fat products	200 g frankfurters & 250 g pâtés/ week. RF provided 2 g n-3 fatty acids/ day, 1.5 g of which	Lipids, lipoproteins, atherogenic ratios, oxidized LDL, blood pressure	Crossover non-randomized, placebo controlled 4 weeks	Overweight/obese mild-hypercholesterolemic $N = 22$ 44 ± 12 years	Cholesterol (total, HDL-, LDL-) Oxidized-LDL Blood pressure	n-3 RF vs NF products: LDL-cholesterol, oxid-LDL and LDL-chol/HDL-	Delgado-Pando et al. (2013)

(n-3 RF) fortified with a combination of linseed, olive and fish oils, c) Normal-fat products (NF	were ALA and approx. 0.4 g EPA + DHA					chol., decreased 12%, 17% and 11% respectively RF vs NF products: Oxid-LDL decreased 15%	
Probiotic fermented (raw) sausage (Lactobacillius paracasei LTH 2579)	Sausage (50 g/day)	Blood lipids, Immunological paramaters	Placebo-controlled 5 weeks	Healthy (control) $N = 20$ (10 men, 10 women) 22–55 years	Cholesterol (total, LDL, HDL)-TG Antibodies against oxidized-LDL Immunological parameters	No effect on cholesterol fractions and TG Improved host immunity (higher titer of antibodies against ox.LDL)	Jahreis et al. (2002)
Meat products (pátês) iron-enriched with ferrous sulfate, and ferric pyrophosphate encapsulated in liposomes (with or without added pigment (AproRED))	19 mg iron/day	Bioavailability study	Crossover, double-blind, randomized, postprandial 6 h	Low iron stores $N = 17$ (women) 21–25 years	Serum iron	Serum iron concentration increased	Navas-Carretero et al. (2009)
Low-fat meat products (sausages, meat balls, cold cuts) enriched with plant sterols	Sterols (1.25 or 2.5 or 5 g/day) provided by different foods meat products provide 1/3 de daily plant sterol dose	Reduction of cholesterol and LDL-chol.	Double-blind, randomized, placebo-controlled. 5 weeks	Mild to moderate hypercholesterolemic. $N = 78$ 25–75 years	Total cholesterol, LDL-col, HDL/LDL-col.	Cholesterol 8% and LDL-cholesterol 13% decreased Lowest dose with good cholesterol lowering effect: 0.9 g/d	Tikkanen et al. (2001)

(continued)

Table 7.2 (continued)

Food Functional ingredient	Intake	Objective	Design	Subjects	Markers	Results	References
Soybean phytosterols added to ground beef	Ground beef (hamburger, sloppy joe, meat sauce, tacos and cheeseburger) Soybean phytosterols (2.7 g/portion of ground beef)	Plasma lipid and lipoprotein	Triple-blind randomized, placebo-controlled 4 weeks	Mildly to hypercholesterolemic $N = 34$ (men)		Total cholesterol decreased and LDL-cholesterol 14.6% LDL particle size did not change	Matvienko et al. (2002)
Non-esterified plant sterol enriched, and mineral enriched low-fat and low-salted meat products	Frankfurters and cold cuts enriched with plant sterols (1.2 g/day or 2.1 g/day)	Serum lipids. Blood pressure	Single-blind randomized, placebo-controlled 3 weeks	Midly to hypercholesterolemic $N = 21$ 30–65 years	Cholesterol TG Blood pressure	Reduction in serum total cholesterol (when intake of sterols is 2.1 g/day) No effect on blood pressure	Tapola et al. (2004); Olmedilla-Alonso et al. (2008)
Walnut-enriched (~20%) meat product: Beef steak restructured and frankfurter Midly to hypercholesterolemic and overweight $N = 25$ 45–70 years	136 g walnuts/week provide by 600 g/week of beef steak restructured and by 80 g/week of frankfurters	Reduction of CHD biomarkers	(a) Crossover, single-dose bioavailability study (g-tocopherol) 1 day (b) Crossover, randomized placebo-controlled 5 weeks	Midly to hypercholesterolemic and overweight $N = 25$ 45–70 years	Serum cholesterol (total, HDLand LDL-) TG, homocysteine, vitamins B6 and B12, folic acid, HDL/LDL and platelet function test. Platelet aggregation, plasma thromboxane A2, prostacyclin I2, thrombogenic ratio (TXB2/6-keto-PGF1α)	Serum g-tocopherol increased (exposure marker), 4.5% decrease in total cholesterol from baseline and 3% respect to meat products without walnuts Improvement in thrombogenic status.	Jiménez et al. (2010); Canales et al. (2009, 2010, 2011); Nus et al. (2007); Sánchez-Muniz et al. (2012)

					Quimiotactic and adhesion molecules for monocytes (VCAM, ICAM, LTB4) Antioxidant enzymes (SOD, CAT, GR, GPx, AE, PON-1), GSH and GSH/(GSH + GSSG)	Dietary response dependent on polymorphisms of PON-1 and ApoA4. Decrease in quimiotactic and adhesion molecules Improvement in antioxidant status	
Processing, storage and consumption							
Chargrilled meat (diet enriched with)		Induction of P-glycoprotein	Mechanistic study 7 days	Healthy $N = 10$	CYP1A, CYP13A and P-glycoprotein messenger RNA (duodenal biopsy)	Induction of CYP1A enzymes but not CYP13A or P-glycoproteins	Fontana et al. (1999)
High meat vs a vegetarian (VEG) period (1) fresh haem in red meat (RM) vs vegetarian diet. (2) nitrite-preserved red meat (PM) vs vegetarian (VEG) diet	420 g/day (men) 366 g/d (women)	Colorectal cancer risk (by stimulating nitrosation and DNA damage)	Randomized, crossover 14 days/period	Healthy (physical and blood examination) $N = 12$ (period 1) $N = 16$ (period 2) 20–85 years		Meat cured with nitrite has the same effect as fresh RM on endogenous nitrosation but show increased fecal water-induced oxidative DNA damage	Joosen et al. (2009)

TG triacylglycerol, *AA* arachidonic acid, *SOD* superoxide dismutase, *CAT* catalase, *GR* Glutathione reductase, *GPx* Glutathione peroxidase, *AE* arylesterase, *PON-1* Paraoxonase, *GSH* Reduced glutathione, and *GSSG* Oxidized reductase. Markers analyzed in serum when no other sample matrix is indicated marketed a cooked ham enriched with calcium and magnesium. On the other hand, some non-meat ingredients (walnut, seaweed, etc.) added to processed meat increase the levels of a number of minerals (Cu, Mg, Mn, K, etc.) (Cofrades et al. 2011a; Jiménez-Colmenero et al. 2010). Iodine has been incorporated into dry fermented sausage (Garcia-Iniguez et al. 2010).

7.2.7 Incorporation of Vitamins and Other Antioxidants

Although meats are sources of other vitamins, strategies have been employed to incorporate vitamins in meat to produce healthier products (Jiménez-Colmenero et al. 2012). These compounds have been used in product formulation either in isolated form or may be added from some other sources like honey, walnut, wheat grain, etc. Different functional meat products like vitamin E enriched Frankfurters (Jiménez-Colmenero et al. 2010), vitamin C enriched beef patties (Fernandez-Lopez et al. 2004; Sanchez-Escalante et al. 2001) and sausages fortified with folic acid have been produced at commercial level (Caceres et al. 2008).

Carotenoids are also being used in foods because of their anti-cancer, anti-oxidant and anti-inflammatory properties. For example, lycopene from tomato (Calvo et al. 2008; Sanchez-Escalante et al. 2003) and lutein (Csapo et al. 2006; Daly et al. 2010; Granado-Lorencio et al. 2010), carrot and sweet potato rich in provitamin A (Saleh and Ahmed 1998) and spinach rich in lutein and zeaxanthin (Pizzocaro et al. 1998). Extracts of fruits (grape, liquorice root, horsetail, arbutus berries, etc.) (Ganhao et al. 2010; Nissen et al. 2004), spices (cardamom, clove, nutmeg etc) and herbs for example oregano, rosemary, melissa, green tea, etc. (Haak et al. 2009; Sanchez-Escalante et al. 2001; Zhang et al. 2010) have also been used as a source of natural antioxidants. Frankfurters enriched with n-3 PUFA contain hydroxytyrosol which has antioxidant property (Cofrades et al. 2011a, b).

7.2.8 Other Added Compounds/Ingredients

A number of other compounds are added to meat products during formulation. They may be incorporated intentionally as separate compounds (sterols, taurine, etc.) or as components of one of the ingredients, including whole foods (walnut, soybean, seaweeds, carrot, herbs, etc.), employed in the formulation. In this regard, one tentative approach to the transformation of meat into a potential functional meat has consisted in the incorporation into the meat-matrix foods that have been recognized as healthy foods.

7.2.9 *Removal of Unhealthy Exogenous Compounds*

Different techniques and methods have been employed to reduce the content of different unhealthy exogenous compounds like sodium, phosphate, nitrite or allergens including gluten, lactose, etc. During meat processing these compounds are added as ingredients and additives (Jiménez-Colmenero 2007). Sodium intake can be controlled by adopting strategies to lessen salt contents. These strategies involve first to select raw materials of meat, varying manufacturing methodologies, replacement of salt substances, enriching flavor and modification of physical form of salt (Desmond 2006; Ruusunen and Puolanne 2005).

Control of phosphate content in meat products is also very important because dietary inorganic phosphate has been found to increase risk of lung cancer (Jin et al. 2009). Also excessive phosphate in diet can disturb the calcium, iron and magnesium balance in body which is a threat of bone diseases (Shahidi and Synowiecki 1997). Meat products processed without phosphate additives have been sold in market (Jiménez-Colmenero et al. 2010). Chemicals like methemoglobinemia and nitrosamines are formed from dietary nitrite. These chemicals are considered to be carcinogenic and teratogenic activities.

Technological methods to reduce the use of nitrite in meat processing have been devised. Also, strategies to prevent damage due to nitrosylation reactions have been proposed (Demeyer et al. 2008). Use of ingredients with high nitrite content in meat processing has been discouraged (Weiss et al. 2010). Allergen free meat products have also been produced by utilizing allergen free non-meat ingredients during meat product formulation (Arihara 2006).

7.3 Functional Evaluation of Meat Based Functional Foods

Many studies are available regarding strategies for optimization of bioactive ingredients in the meat based functional foods but still there is lake of in vitro studies to confirm that how much these supplementary compounds are bioavailable (Garci Martin et al. 2012; Vossen et al. 2012) especially even there is scarce information about assessment of the functionalization of these products in the humans who are ultimate end users. Evaluation of the efficacy of such functional foods is proved only from evidence base data of human studies, use of validated methods, assessment of the characteristic of food ingredients and the biological end points (Aggett et al. 2005, 2010). Meat based functional foods are mainly assessed for their probable positive role in lowering the risk of cardiovascular diseases as mostly meat is considered as one of the major causes of this health issue (Pan et al. 2012). Moreover, LDL-cholesterol and plasma are also used as biological markers for routine analysis to monitor cardiovascular diseases. In addition these products are also analysed for the any interventions made by the use of biomarkers.

The ultimate purpose of the foods obtained using any of the strategies mentioned above is to obtain a given benefit, to be functional, for the target user, the consumer. In the case of certain claims of health-promoting properties, based on generally accepted scientific proof, it is possible to admit a cause-effect relationship between a food category, a food, or one of its constituents, and the claimed effect. This has enabled the European Union (EU) to establish a list of authorized claims of health-promoting properties of foods other than those relative to the reduction of disease risk and the development and health of children (Regulation EU 432/2012).

The functional effects, in terms of improvements in health status and wellbeing or the reduction of disease risk, must be demonstrated by providing evidence to justify them and only then can the foods be claimed to have health-promoting properties in those target populations in which the effect produced by their consumption has been proved. When this cause-effect relationship is yet to be convincingly established, or substantiation of it is sought in a specific food matrix, it is necessary to evaluate the functional effect, which can be carried out using in vitro and in vivo models that provide information on mechanisms of action, the effect on function-response in vivo, the dose–response relationship and acute and chronic effects (Aggett et al. 2005; Howlett 2008). Studies in animals and in vitro should be considered parts of the entire range of trials to be performed, but human intervention studies (with the proper design, methodology and markers) are the most important tests required to confirm the claims of the health-promoting properties of a food or component (Regulation EU 1924/2006; Reglamento 353/2008).

In accordance with the indications established by different expert committees on the evidence-based process, the key points in the design and implementation of studies of this type are as follows: the research question (clearly identified), subjects (inclusion/exclusion criteria, characteristics of subjects described, representative sample of the relevant population), comparable groups (randomized method identified and distribution of disease status, prognosis, etc., and other factors that are similar across study groups at baseline, control group), methods of handling dropouts (follow-up described, number and characteristics of dropouts), blinding used to prevent introduction of bias, description of the intervention/exposure factors or procedures (protocol, amount of exposure, duration of intervention), clearly defined outcomes and valid and reliable measurements and finally, appropriate statistical analysis for the study design and type of outcome indicators (USDA/HHS 2013; Aggett et al. 2010).

When a clinical study is designed for the purpose of substantiating a health claim (i.e. cross-over vs parallel), access to the predefined claim is highly useful, since it helps in the formulation of the null hypothesis and in the selection of the markers of choice, the power of the study, etc. (Aggett et al. 2010). The design that provides evidence of a higher grade for demonstrating the relationship between a food or food component and a health objective is the randomized controlled trial (USDA/HHS 2013). Other types of designs to be considered in the assessment of effects are intervention studies (randomized controlled studies have a greater relevance than randomized studies that do not include a control group), observational studies (cohort, case–control, cross-sectional studies, etc.) and those designed to study possible mechanisms of action, which include bioavailability studies (Reglamento CE 353/2008).

7.4 Conclusion

Promoting health through nutrition is an increasingly important objective of nutrition and public health programs in a number of countries when it comes to making intake recommendations. Because of the amounts and frequency with which they are consumed, their contribution to dietary intakes of different nutrients, the diversity of presentations, the possibility of modifying their composition, their high consumer acceptability, etc., meat and meat products are excellent foods for delivering bioactive compounds without changing dietary habits. Based on different strategies used to improve the presence of bioactive compounds, a wide number of potential meat-based functional foods have been designed and developed. Although novel perspectives can be considered in meat production and new candidate bioactive components may be found for the modulation of different target functions, one of the most important challenges in this context would be to take serious steps toward substantiating the functional effects of healthy meat products in humans. It is true that certain claims of health-promoting properties can be established based on generally accepted scientific proof. However, while the cause-effect relationship is yet to be convincingly established, and the substantiation of this relationship in a meat matrix is still being sought, it is necessary to evaluate the functional effect.

References

Aggett PJ, Antoine JM, Asp NG, Bellisle F, Contor L, Cummings JH, Rechkemmer G (2005) Process for the assessment of scientific support for claims on foods. Eur J Nutr 44:1–31

Aggett P, Antoine JM, De Vries J, Gallagher A, Hendriks H, Kozianowski G, Meijer G, Richarson D, Rondeau V, Tweedie G, Welch R, Wittwer J (2010) Beyond Passclaim, guidance to substantiate health claims on foods. In: Yates K (ed) ILSI Europe Report Series, pp 7–19. (ISBN: 9789078637219)

Allan S, Bank P, de Ellis M (2019) Bioprocess design considerations for cultured meat production with a focus on the expansion bioreactor. Front Sustain Food Syst 3:44

Alvares TS, Conte-Junior CA, Pierucci AP, De Oliveira GV, Cordeiro EM (2018) Acute effect of fish protein hydrolysate supplementation on vascular function in healthy individuals. J Funct Foods 46:250–255

Ares G, Gambaro A (2007) Influence of gender, age and motives underlying food choice on perceived healthiness and willingness to try functional foods. Appetite 49:148–158

Arihara K (2006) Strategies for designing novel functional meat products. Meat Sci 74:219–229

Armero E, Navarro JL, Nadal MI, Baselga M, Toldra F (2002) Lipid composition of pork muscle as affected by sire genetic type. J Food Biochem 26:91–102

Arshad MS, Javed M, Sohaib M, Saeed F, Imran A, Amjad Z (2017) Tissue engineering approaches to develop cultured meat from cells: a mini review. Cogent Food Agric 3:1

Bass JJ, Butler-Hogg BW, Kirton AH (1990) Practical methods of controlling fatness in farm animals. In: Wood JD, Fisher AV (eds) Reducing fat in meat animals. Elsevier Applied Science, London, pp 145–200

Baublits RT, Pohlman FW, Brown AH, Johnson ZB, Proctor A, Sawyer J, Galloway DL (2007) Injection of conjugated linoleic acid into beef strip loins. Meat Sci 75:84–93

Bessa DP, Teixeira CE, Franco RM (2016) Functional sausage made from mechanically separated tilapia meat. Ital J Food Saf 28:426–439

Boada LD, Henriquez-Hernandez LA, Luzardo OP (2016) The impact of red and processed meat consumption on cancer and other health outcomes: epidemiological evidences. Food Chem Toxicol 92:236–244

Bouvard V, Loomis D, Guyton KZ (2015) International agency for research on cancer monograph working group. Carcinogenicity of consumption of red and processed meat. Lancet Oncol 16: 1599–1600

Burnett DD, Legako JF, Phelps KJ, Gonzalez JM (2020) Board invited review: biology, strategies, and fresh meat consequences of manipulating the fatty acid composition of meat. J Anim Sci 98: 1–12

Caceres E, Garcia ML, Selgas MD (2006) Design of a new cooked meat sausage enriched with calcium. Meat Sci 73:368–377

Caceres E, Selgas MD, Garcia ML (2008) Research & development: conventional and fat-reduced cooked meat sausages enriched with folic acid. Fleischwirtschaft Int 23:58–60

Calvo MM, Garcia ML, Selgas MD (2008) Dry fermented sausages enriched with lycopene from tomato peel. Meat Sci 80:167–172

Canales A, Bastida S, Librelotto J, Nus M, Benedí J, Sánchez-Muniz FJ (2009) Platelet aggregation, eicosanoid production and thrombogenic ratio in individuals at high risk consuming walnut-enriched meat. A crossover, placebo-controlled study. Br J Nutr 102:134–141

Canales A, Benedi J, Bastida S, Corella D, Guillen M, Librelotto J, Nus M, Sánchez-Muniz FJ (2010) The effect of consuming meat enriched in walnut paste on platelet aggregation and thrombogenesis varies in volunteers with different apolipoprotein A4 genotype. Nutr Hosp 25: 746–754

Canales A, Sánchez-Muniz FJ, Bastida S, Librelotto J, Nus M, Corella MD, Guillén M, Benedí J (2011) Effect of walnut-enriched meat on the relationship between VCAM, ICAM, and LTB4 levels and PON-1 activity in ApoA4 360 and PON-1 allele carriers at increased cardiovascular risk. Eur J Clin Nutr 65:703–710

Canto ACVCS, Lima BRCC, Suman SP (2014) Physico-chemical and sensory attributes of low-sodium restructured caiman steaks containing microbial transglutaminase and salt replacers. Meat Sci 96:623–632

Chen L, Guttieres D, Koenigsberg A, Barone PW, Sinskey AJ, Springs SL (2022) Large-scale cultured meat production: trends, challenges and promising biomanufacturing technologies. Biomaterials 280:121274

Choi KH, Lee DK, Kim SW, Woo SH, Kim DY, Lee CK (2019) Chemically defined media can maintain pig pluripotency network in vitro. Stem Cell Rep 13:221–234

Choi KH, Lee DK, Oh JN, Kim SH, Lee M, Woo SH, Kim DY, Lee CK (2020) Pluripotent pig embryonic stem cell lines originating from in vitro-fertilized and parthenogenetic embryos. Stem Cell Res 49:102093

Chriki S, Hocquette JF (2020) The myth of cultured meat: a review. Front Nutr 7:7

Clarke AD (1997) Reducing cholesterol levels in meat, poultry and fish products. In: Pearson AM, Dutson TR (eds) Production and processing of healthy meat, poultry and fish products. Blackie Academic & Professional, London, pp 101–117

Coates AM, Sioutis S, Buckley JD, Howe PRC (2009) Regular consumption of n−3 fatty acid-enriched pork modifies cardiovascular risk factors. Br J Nutr 101:592–597

Cofrades S, López-López I, Jiménez-Colmenero F (2011a) Applications of seaweed in meat based functional foods. In: Kim SSK (ed) Handbook of marine macroalgae: biotechnology and applied phycology. Wiley, Oxford, pp 491–499

Cofrades S, Sandoval LS, Delgado-Pando G, Lopez-Lopez I, Ruiz-Capillas C, Jiménez-Colmenero F (2011b) Antioxidant activity of hydroxytyrosol in frankfurters enriched with n-3 polyunsaturated fatty acids. Food Chem 129:429–436

Csapo I, Incze K, Kovacs A, Zelenak L, Zsigo J (2006) Development of meat products with lutein for eye health. In: Troy D, Pearce R, Byrne B, Kerry J (eds) 52nd international congress of meat science and technology. Wageningen Academic Publishers, The Netherlands, pp 687–688

Cunha LCM, Monteiro MLG, Lorenzo JM (2018) Natural antioxidants in processing and storage stability of sheep and goat meat products. Food Res Int 111:379–390

Daly T, Ryan E, Aherne SA, O'Grady MN, Hayes J, Allen P, O'Brien NM (2010) Bioactivity of ellagic acid-, lutein-or sesamol-enriched meat patties assessed using an in vitro digestion and Caco-2 cell model system. Food Res Int 43:753–760

Datar I, Betti M (2010) Possibilities for an in vitro meat production system. Innov Food Sci Emerg Technol 11:13–22

De Smet S, Claeys E, Demeyer D (2002) Muscle enzymes in relation to meat quality and muscularity. In: Toldra F (ed) Research advances in the quality of meat and meat products. Research Signpost, Trivandrum, pp 123–148

Delft CE. (2021) LCA of cultivated meat: future projections for different scenarios. https://policycommons.net/artifacts/2009833/lca-of-cultivated-meat-future-projections-for-different-scenarios/2762276/fragments/. Accessed 3 Feb 2023

Delgado-Pando G, Celada P, Sánchez-Muniz FJ, Jiménez-Colmenero F, Olmedilla AB (2013) Effects of improved fat content of frankfurters and pâtés on lipid and lipoprotein profile of volunteers at increased cardiovascular risk. A placebo controlled study. Eur J Nutr. https://doi.org/10.1007/s00394-013-0502-1

Demeyer D, Honikel K, De Smet S (2008) The world cancer research fund report 2007: a challenge for the meat processing industry. Meat Sci 80:953–959

Desmond E (2006) Reducing salt: a challenge for the meat industry. Meat Sci 74:188–196

DeZengotita VM, Kimura R, Miller WM (1998) Effects of CO_2 and osmolality on hybridoma cells: growth, metabolism and monoclonal antibody production. Cell Cult Eng 6:213–227

Ding S, Swennen GNM, Messmer T, Gagliardi M, Molin DGM, Li C, Zhou G, Post MJ (2018) Maintaining bovine satellite cells stemness through p38 pathway. Sci Rep 8:10808

Dubey AK, Lavanya L, Sadananda D, Gouthami K, Elfansu K, Singh A (2021) Inferences of carbon dioxide in present-day cell culture systems: an unacknowledged problem and perspectives. Austin Therap 6(1):1033

Fernández-Ginés JM, Fernández-López J, Sayas-Barberá E, Sendra E, Pérez-Álvarez JA (2004) Lemon albedo as a new source of dietary fiber: application to bologna sausages. Meat Sci 67:7–13

Fernandez-Lopez J, Fernandez-Gines JM, Aleson-Carbonell L, Sendra E, Sayas-Barbera E, Perez-Alvarez JA (2004) Application of functional citrus by-products to meat products. Trends Food Sci Technol 15:176–185

Fontana RJ, Lown KS, Paine MF, Fortlage L, Santella RM, Felton JS, Knize MG, Greenberg A, Watkins PB (1999) Effects of a chargrilled meat diet on expression of CYP3A, CYP1A, and P-glycoprotein levels in healthy volunteers. Gastroenterology 117:89–98

Furuhashi M, Morimoto Y, Shima A, Nakamura F, Ishikawa H, Takeuchi S (2021) Formation of contractile 3D bovine muscle tissue for construction of millimetre-thick cultured steak. NPJ Sci Food 5:6

Gamage D, Thompson M, Sutherland M, Hirotsu N, Makino A, Seneweera S (2018) New insights into the cellular mechanisms of plant growth at elevated atmospheric carbon dioxide concentrations. Plant Cell Environ 41(6):1233–1246

Ganhao R, Morcuende D, Estevez M (2010) Protein oxidation in emulsified cooked burger patties with added fruit extracts: influence on colour and texture deterioration during chill storage. Meat Sci 85:402–409

Garcia-Iniguez C, Larequi E, Rehecho S, Calvo MI, Cavero RY, Navarro-Blasco I, Astiasaran I, Ansorena D (2010) Selenium, iodine, ω-3 PUFA and natural antioxidant form *Melissa officialis* L: a combination of components for healthier dry fermented sausage formulation. Meat Sci 85:274–279

Garci-Martin A, Fonseca CS, Bastida S, Vazquez-Velasco M, Gonzalez-Torres L, Benedi J, Sanchez-Muniz FJ. (2012) Restructured meat containing Himanthalia elongata acts as a competitive α glucosidase inhibitor. Potential utilities 11 Encontro de Química dos alimentos. Bragança (Portugal): Qualidade dos alimentos: Novosdesafios

Garrison GL, Biermacher JT, Brorsen BW (2022) How much will large-scale production of cell-cultured meat cost? J Agric Food Res 10:100358

Genovese NJ, Domeier TL, Telugu BP, Roberts RM (2017) Enhanced development of skeletal myotubes from porcine induced pluripotent stem cells. Sci Rep 7:41833

Gilmore LA, Walzem RL, Crouse SF, Smith DR, Adams TH, Vaidyanathan V, Cao X, Smith SB (2011) Consumption of high-oleic acid ground beef increases HDL-cholesterol concentration but both high- and low-oleic acid ground beef decrease HDL particle diameter in normocholesterolemic men. J Nutr 141:1188–1194

Gomes VS, Mano SB, Freitas MQ (2016) Meat characteristics of cattle fed diets containing whole cottonseed. Arq Bras Med Vet Zootec (Online) 68:1069–1076

Goutefongea R, Dumont JP (1990) Developments in low-fat meat and meat products. In: Wood JD, Fisher AV (eds) Reducing fat in meat animals. Elsevier Applied Science, London, pp 398–436

Granado-Lorencio F, Lopez-Lopez I, Herrero-Barbudo C, Blanco-Navarro I, Cofrades S, Perez-Sacristan B, Jiménez-Colmenero F (2010) Lutein-enriched frankfurter-type products: physicochemical characteristics and lutein in vitro bioaccessibility. Food Chem 120:741–748

Grasso S, Brunton NP, Lyng JG, Lalor F, Monahan FJ (2014) Healthy processed meat products – regulatory, reformulation and consumer challenges. Trends Food Sci Technol 39:4–17

Guan X, Lei Q, Yan Q, Li X, Zhou J, Du G, Chen J (2021) Trends and ideas in technology, regulation and public acceptance of cultured meat. Future Foods 3:100032

Guedes-Oliveira JM, Conte-Junior CA, Salgado RL (2016) Alternativas tecnológicas para substituição de ingredientes em produtos carneos. Anim Bus Brasil 6:10–13

Haak L, Raes K, De Smet S (2009) Effect of plant phenolics, tocopherol and ascorbic acid on oxidative stability of pork patties. J Sci Food Agric 89:1360–1365

Haug A, Nyquist NF, Mosti TJ, Andersen M, Hostmark AT (2012) Increased EPA levels in serum phospholipids of humans after four weeks daily ingestion of one portion chicken fed linseed and rapeseed oil. Lipids Health Dis 11:104

Hays VW, Preston RL (1994) Nutrition and feeding management to alter carcass composition of pigs and cattle. In: Hafs HD, Zimbelman RG (eds) Low-fat meats: design strategies and human implications. Academic, London, pp 13–34

Higgs JD (2000) The changing nature of red meat: 20 years of improving nutritional quality. Trends Food Sci Technol 11:85–95

Howlett J (2008) Functional foods: from science to health and claims: ILSI Europe concise monographs series. ILSI Europe

Hygreeva D, Pandey MC, Radhakrishna K (2014) Potential applications of plant based derivatives as fat replacers, antioxidants and antimicrobials in fresh and processed meat products. Meat Sci 98:47–57

Jahreis G, Vogelsang H, Kiessling G, Schubert R, Bunte C, Hammes WP (2002) Influence of probiotic sausage (lactobacillus paracasei) on blood lipids and immunological parameters of healthy volunteers. Food Res Int 35:133–138

Jiménez Colmenero F, Herrero A, Cofrades S, Ruiz-Capillas C (2012) Meat and functional foods. In: Hui YH (ed) Handbook of meat and meat processing, 2nd edn. CRC Press. Taylor & Francis Group, Boca Raton, pp 225–248

Jiménez-Colmenero F (1996) Technologies for developing low-fat meat products. Trends Food Sci Technol 7:41–48

Jiménez-Colmenero F (2007) Meat based functional foods. In: Hui YH (ed) Handbook of food products manufacturing. Wiley, New Jersey, pp 989–1015

Jiménez-Colmenero F, Delgado-Pando G (2013) Fibre-enrichedmeat products. In: Delcour JA, Poutanen K (eds) Fibre-rich and wholegrain foods. Woodhead Publishing Limited, Oxford, pp 329–347

Jiménez-Colmenero F, Carballo J, Cofrades S (2001) Healthier meat and meat products: their role as functional foods. Meat Sci 59:5–13

Jiménez-Colmenero F, Sanchez-Muniz F, Olmedilla-Alonso B (2010) Design and development of meat-based functional foods with walnut: technological, nutritional and health impact. Food Chem 123:959–967

Jiménez-Colmenero F, Cofrades S, Herrero AM, Ruiz-Capilla C (2017) Implications of domestic food practices for the presence of bioactive components in meats with special reference to meat based functional foods. Crit Rev Food Sci Nutr 58:2334–2345

Jin H, Xu CX, Lim HT, Park SJ, Shin JY, Chung YS, Lee YS (2009) High dietary inorganic phosphate increases lung tumorigenesis and alters Akt signaling. Am J Respir Crit Care Med 179:59–68

Johnson MH, Calkins CR, Huffman RD, Johnson D, Hargrove DD (1990) Differences in cathepsin B+L and calcoium dependent protease activities among breed type and their relationship to beef tenderness. J Anim Sci 68:2371–2378

Joosen AM, Kuhnle GG, Aspinall SM, Barrow TM, Lecommandeur E, Azqueta A, Collins AR, Bingham SA (2009) Effect of processed and red meat on endogenous nitrosation and DNA damage. Carcinogenesis 30:1402–1407

Juarez M, Marco A, Brunton N, Lynch B, Troy DJ, Mullen AM (2009) Cooking effect on fatty acid profile of pork breakfast sausages enriched in conjugated linoleic acid by dietary supplementation or direct addition. Food Chem 117:393–397

Keeton JT (1994) Low-fat meat products – technological problems with processing. Meat Sci 36: 261–276

Kerry JF, Kerry JP (2006) Producing low-fat meat products. In: Williams C, Buttriss J (eds) Improving the fat content of foods. Woodhead Publishing Limited and CRC Press LLC, Cambridge, pp 336–379

Khalaf AT, Wei Y, Alneamah SJA, Al-Shawi SG, Kadir SYA, Zainol J, Liu X (2021) What is new in the preventive and therapeutic role of dairy products as nutraceuticals and functional foods? Biomed Res Int. https://doi.org/10.1155/2021/8823222. PMID: 33681381; PMCID: PMC7925044

Kumar P, Chatli MK, Mehta N, Singh P, Malav OP, Verma AK (2017) Meat analogues: health promising sustainable meat substitutes. Crit Rev Food Sci Nutr 57:923–932

Kumar P, Mehta N, Malav OP, Verma AK, Umraw P, Kanth MK (2019) The structure of meat analogs. Encycl Food Chem 105–109. https://doi.org/10.1016/B978-0-08-100596-5.21705-8

Kumar P, Sharma N, Sharma S, Mehta N, Verma AK, Chemmalar S, Sazili AQ (2021a) In-vitro meat: a promising solution for sustainability of meat sector. J Anim Sci Technol 63: 693–724

Kumar P, Verma AK, Umaraw P, Mehta N, Ranjan R (2021b) Food utilization to the fullest: the prospects of 3D-printing in the meat industry. Fleischwirtschaft Int 4:44–47

Kumar P, Abubakar AA, Verma AK, Umaraw P, Nizam MH, Mehta N, Ahmed MA, Kaka U, Sazili AQ (2022a) New insights in improving sustainability in meat production: opportunities and challenges. Crit Rev Food Sci Nutr 63(33):11830–11858. https://doi.org/10.1080/10408398.2022.2096562

Kumar P, Mehta N, Abubakar AA, Verma AK, Kaka U, Sharma N, Sazili AQ, Pateiro M, Kumar M, Lorenzo JM (2022b) Potential alternatives of animal proteins for sustainability in the food sector. Food Rev Int 39(8):5703–5728. https://doi.org/10.1080/87559129.2022.2094403

Kumar P, Sharma N, Ahmed MA, Verma AK, Umaraw P, Mehta N, Abubakar AA, Hayat MN, Kaka U, Lee SJ, Sazili AQ (2022c) Technological interventions in improving the functionality of proteins during processing of meat analogs. Front Nutr 9:1044024

Lai LX, Kang JX, Li RF, Wang JD, Witt WT, Yong HY, Hao YH, Wax DM, Murphy CN, Rieke A, Samuel M, Linville ML, Korte SW, Evans RW, Starzl TE, Prather RS, Dai YF (2006) Generation of cloned transgenic pigs rich in omega-3 fatty acids. Nat Biotechnol 24:435–436

Lynch PB, Kerry J (2000) UtiIizing diet to incorporate bioactive compounds and improve the nutritional QuaIity of MuscIe foods. In: Decker E, Faustman C, Lopez-Bote C (eds) Antioxidants in muscle foods: nutritional strategies to improve quality. Wiley, New York, pp 455–480
Lynch J, Pierrehumber R (2019) Climate impacts of cultured meat and beef cattle. Front Sustainable Food Syst 3:5
Madden UA, Osweiler GD, Knipe BGW, Beitz DG (1999) Effects of Eubacterium coprostanoligenes and lactobacillus on pH, lipid content, and cholesterol of fermented pork and mutton sausage-type mixes. J Food Sci 64:903–908
Martín D, Ruiz J, Kivikari R, Puolanne E (2008) Partial replacement of pork fat by conjugated linoleic acid and/or olive oil in liver pâtés: effect on physicochemical characteristics and oxidative stability. Meat Sci 80:496–504
Matvienko OA, Lewis DS, Swanson M, Arndt B, Rainwater DL, Stewart J, Alekel DL (2002) A single daily dose of soy bean phytosterols in ground beef decreases serum total cholesterol and LDL cholesterol in young, mildly hypercholesterolemic men. Am J Clin Nutr 76:57–64
Melzener L, Verzijden KE, Buijs AJ, Post MJ, Flack JE (2021) Cultured beef: from small biopsy to substantial quantity. J Food Sci Agric 101:7–14
Metcalf RG, James MJ, Mantzioris E, Cleland LG (2003) A practical approach to increasing intakes of n-3 polyunsaturated fatty acids: use of novel foods enriched with n-3 fats. Eur J Clin Nutr 57: 1605–1612
Moloney AP (2006) Reducing fats in raw meat. In: Williams C, Buttriss J (eds) Improving the fat content of foods. CRC Press and Woodhead Publishing Limited, Boca Raton, pp 313–335
Monteiro MLG, Marsico ET, Canto ACV (2015) Partial sodium replacement in tilapia steak without loss of acceptability. Food Sci Technol Int 21:295–305
Navas-Carretero S, Pérez-Granados AM, Sarriá B, Vaquero MP (2009) Iron absorption from meat pâté fortified with ferric pyrophosphate in iron-deficient women. Nutrition 25:20–24
Navas-Carretero S, Cuervo M, Abete I, Zulet MA, Martinez JA (2011) Frequent consumption of selenium-enriched chicken meat by adults causes weight loss and maintains their antioxidant status. Biol Trace Elem Res 143:8–19
Nissen LR, Byrne DV, Bertelsen G, Skibsted LH (2004) The antioxidative activity of plant extracts in cooked pork patties as evaluated by descriptive sensory profiling and chemical analysis. Meat Sci 68:485–495
Nus M, Frances F, Librelotto J, Canales A, Corella D, Sánchez-Montero JM, Sánchez-Muniz FJ (2007) Arylesterase activity and antioxidant status depend on PON1-Q192R and PON1-L55M polymorphisms in subjects at increased cardiovascular disease consuming a walnut-enriched meat. J Nutr 137:1783–1788
Olmedilla-Alonso B, Granado-Lorencio F, Herrero-Barbudo C, Blanco-Navarro I, Blázquez-García S, Pérez-Sacristán B (2008) Consumption of restructured products with added walnuts has a cholesterol-lowering effect in subjects at high cardiovascular risk: a randomised, cross-over, placebo-controlled study. J Am Coll Nutr 27:342–348
Olmedilla-Alonso B, Jiménez-Colmenero F, Sanchez-Muniz FJ (2013) Development and assessment of healthy properties of meat and meat products designed as functional foods. Meat Sci 95: 919–930
Pakseresht A, Ahmadi KS, Canavari M (2022) Review of factors affecting consumer acceptance of cultured meat. Appetite 170:105829
Pan A, Sun Q, Bernstein AM, Schulze MB, Manson JE, Stampfer MJ, Hu FB (2012) Red meat consumption and mortality: results from 2 prospective cohort studies. Arch Intern Med 172: 555–563
Pizzocaro F, Senesi E, Veronese P, Gasparoli A. (1998) Mechanically deboned poultry meat hamburgers. 2: protective and antioxidant effect of the carrot and spinach tissues during frozen storage [chicken-Turkey-Veneto]. IndustrieAlimentari (Italy)
Post MJ, Levenberg D, Kaplan L (2020) Scientific, sustainability and regulatory challenges of cultured meat. Nat Food 1:403–415

Raes K, De Smet S, Demeyer D (2004) Effect of dietary fatty acids on incorporation of long chain polyunsaturated fatty acids and conjugated linoleic acid in lamb, beef and pork meat: a review. Anim Feed Sci Technol 113:199–221
Reglamento (CE) no. 353/2008 de la Comisión, de 18 de abril de 2008, por el que se establecen normas de desarrollo para las solicitudes de autorización de declaraciones de propiedades saludables con arreglo al artículo 15 del Reglamento (CE) no. 1924/2006 del Parlamento Europeo y del Consejo
Regulation EU 432/2012. Commission regulation (EU) No 432/2012 of 16 May 2012 Establishing a list of permitted health claims made on foods, other than those referring to the reduction of disease risk and to children's development and health. Off J Eur Union, L136, 1–40
Regulation EU 1924/2006 (2006) No 1924/2006 of the European parliament and of the council of 20 December 2006 on nutrition and health claims made on foods. Off J Eur Union L 404:9–25
Reiss J, Robertson S, Suzuki M (2021) Cell sources for cultivated meat: applications and considerations throughout the production workflow. Int J Mol Sci 22:7513
Rosser J, Thomas-Vazquez D (2018) Bioreactor processes for maturation of 3D bioprinted tissue. In: Whitaker TJ (ed) 3D bioprinting for reconstructive surgery, 1st edn. Elsevier Wordmark, Amsterdam
Rubio NR, Datar I, Stachura D, Kaplan D, Krueger K (2019) Cell-based fish: a novel approach to seafood production and an opportunity for cellular agriculture. Front Sustain Food Syst 3:43
Ruusunen M, Puolanne E (2005) Reducing sodium intake from meat products. Meat Sci 70:531–541
Saeki K, Matsumoto K, Kinoshita M, Suzuki I, Tasaka Y, Kano K, Taguchi Y, Mikami K, Hirabayashi M, Kashiwazaki N, Hosoi Y, Murata N, Iritani A (2004) Functional expression of a 12 fatty acid desaturase gene from spinach in transgenic pigs. Proc Natl Acad Sci USA 101: 6361–6366
Saleh NT, Ahmed ZS (1998) Impact of natural sources rich in provitamin a on cooking characteristics, color, texture and sensory attributes of beef patties. Meat Sci 50:285–293
Salim APAA, Canto ACVCS, Costa-Lima BRC (2018) Inhibitory effect of acid concentration, aging, and different packaging on Escherichia coli O157:H7 and on color stability of beef. J Food Process Preserv 42:13402
Sanchez-Escalante A, Djenane D, Torrescano G, Beltran JA, Roncales P (2001) The effects of ascorbic acid, taurine, carnosine and rosemary powder on colour and lipid stability of beef patties packaged in modified atmosphere. Meat Sci 58:421–429
Sanchez-Escalante A, Torrescano G, Djenane D, Beltran JA, Roncales P (2003) Combined effect of modified atmosphere packaging and addition of lycopene rich tomato pulp, oregano and ascorbic acid and their mixtures on the stability of beef patties. Food Sci Technol Int 9:77–84
Sánchez-Muniz FJ, Canales A, Nus M, Bastida S, Guillen M, Corella D, Olmedilla-Alonso B, Granado-Lorencio F, Benedi J (2012) The antioxidant status response to low fat- and walnut paste enriched-meat differs in volunteers at high cardiovascular risk carrying different PON-1 polymorphisms. J Am Coll Nutr 31:194–205
Schmid A, Collomb M, Sieber R, Bee G (2006) Conjugated linoleic acid in meat and meat products: a review. Meat Sci 73:29–41
Scollan N, Hocquette JF, Nuernberg K, Dannenberger D, Richardson I, Moloney A (2006) Innovations in beef production systems that enhance the nutritional and health value of beef lipids and their relationship with meat quality. Meat Sci 74:17–33
Shahidi F, Synowiecki J (1997) Protein hydrolyzates from seal meat as phosphate alternatives in food processing applications. Food Chem 60:29–32
Sharma M, Kaur S, Kumar P, Mehta N, Umaraw P, Ghosh S (2022) Development, prospects, and challenges of meat analogs with plant-based alternatives. In: Recent advances in food biotechnology. Springer Nature Singapore, Singapore, pp 275–299
Solomon MB, Pursel VG, Mitchell AD (2002) Biotechnology for meat quality enhancement. In: Toldra F (ed) Research advances in the quality of meat and meat products. Trivandrum, Research Signpost, pp 17–31

Solomon MB, Eastridge JS, Paroczay EW (2008) Transgenic farm animal. In: Toldrá F (ed) Meat biotechnology 3–20. Springer Science + Business Media LLC, New York

Stephens N, Di Silvio L, Dunsford I, Ellis M, Glencross A, Sexton A (2018) Bringing cultured meat to market: technical, socio-political, and regulatory challenges in cellular agriculture. Trends Food Sci Technol 78:155–166

Tapola NS, Lyyra ML, Karvonen HM, Uusitupa MI, Sarkkinen ES (2004) The effect of meat products enriched with plant sterols and minerals on serum lipids and blood pressure. Int J Food Sci Nutr 55(3):389–397

Taub D (2010) Effects of rising atmospheric concentrations of carbon dioxide on plants. J Coll Nat Sci 3(10):21

Tikkanen MJ, Högström P, Tuomilehto J, Keinänen K, Sundvall J, Karpannen H (2001) Effect of a diet based on low-fat foods enriched with nonesterified plant sterols and mineral nutrients on serum cholesterol. Am J Cardiol 88:1157–1162

Tuomisto HL, de Mattos MJ (2011) Environmental impacts of cultured meat production. Environ Sci Technol 45:6117–6123

USDA/HHS commitment to evidence-based process. http://www.nel.gov/default.cfm?library=DGAC. Accessed 13 Jan 2013

Verma AK, Banerjee R (2010) Dietary fibre as functional ingredient in meat products: a novel approach for healthy living, a review. J Food Sci Technol 47:247–257

Vossen E, Raes K, Maertens L, Vandenberge V, Haak L, Chiers K, De Smet S (2012) Diets containing n-3 fatty acids-enriched pork: effect on blood lipids, oxidative status and atherosclerosis in rabbits. J Food Biochem 36:359–368

Weiss J, Gibis M, Schuh V, Salminen H (2010) Advances in ingredient and processing systems for meat and meat products. Meat Sci 86:196–213

Wheeler TL, Savell JW, Cross HR, Lunt DK, Smith SB (1990) Mechanisms associated with the variations in the tenderness of meat from Brahman and Hereford cattle. J Anim Sci 68:4206–4220

Wood JD, Enser M, Richardson RI, Whittington FM (2008) Fatty acids in meat and meat products. In: Chow CK (ed) Fatty acids in foods and their health implications. CRC Press, Taylor & Francis Group, Boca Raton, pp 87–107

Zhang S, Wang H (2019) Current progress in 3D bioprinting of tissue analogs. SLAS Technol 24: 70–78

Zhang W, Xiao S, Samaraweera H, Lee EJ, Ahn DU (2010) Improving functional value of meat products. Meat Sci 86:15–31

Chapter 8
Reduced and Low Fat Meat Products

Sajad A. Rather and F. A. Masoodi

8.1 Introduction

Consumption of meat by early humans began more than two million years ago and it is believed that meat contributed for cerebral evolution due to its high content in energy and nutrients (Rodrigues et al. 2023). Meat and meat products consumption is beneficial because meat contains many essential nutrients such as high quality proteins (i.e. source of all essential amino acids) with high biological value, iron, zinc, selenium, vitamins, etc. (Kumar 2019; Zhang et al. 2022). These substances are crucial for the proper functioning and development of the human body, serving as essential components within a varied diet. Thus, it is an essential part of a mixed diet.

Nevertheless, despite its status as a fundamental and exceptional source of nutrients nowadays, consumers are becoming more and more aware of the link between diet and health and are also concerned about the fat content of processed meat products. Consumption of high-fat processed meat products is correlated with diseases like cardiovascular diseases (CVD), diabetes, hypertension, etc. (WHO 2009; ADA 2012; Henry et al. 2013). Consumer's interest in low fat processed meat products has increased in recent past in view of the risk posed by food products containing high animal fat. On the other hand fat plays an important role in many quality attributes (stabilizing meat emulsions, reducing cooking loss, improving water holding capacity, providing juiciness and hardness) of processed meat products in addition to being a source of energy, essential fatty acids and liposoluble vitamins (Sarteshnizi et al. 2015; Grasso et al. 2014; Brewer 2012). However, high fat contents in particular animal fat provides higher amounts of saturated fatty acids and cholesterol (Huang et al. 2021; Ozvural and Vural 2008). The cholesterol oxidation products formation is also enhanced with processing and storage time in meat and meat products. Generally heat, pH, light, oxygen, water activity and the

S. A. Rather (✉) · F. A. Masoodi (✉)
Department of Food Science and Technology, University of Kashmir, Srinagar, India

S. A. Rather, F. A. Masoodi (eds.), *Hand Book of Processed Functional Meat Products*, https://doi.org/10.1007/978-3-031-69868-2_8

presence of unsaturated fatty acids are the major factors that influence cholesterol oxidation products formation during food processing or storage (Larkeson et al. 2000). Cholesterol oxidation products have been shown to have atherogenic, cytotoxic, mutagenic and carcinogenic affects (Dobarganes and Marquez-Ruiz 2003; Bauer et al. 2014). In addition to serious health risks of high animal fat diets, oxidation of lipids is major threat to meat quality. The onset of oxidative reactions in muscle foods during handling, processing and storage leads to undesirable sensory changes and deterioration of nutritive value (Ganhão et al. 2010; Hęś et al. 2012). A deterioration of nutritive value may be a consequence of interactions between lipid oxidation products and proteins (Hęś et al. 2012). In view of such implications, various international institutions, like FAO (2010) and WHO (2003) have drawn up the following nutritional recommendations for adults: fat should provide between 20% and 35% of dietary energy. This should include 10% of energy from saturated fatty acids, 6–11% of energy from polyunsaturated fatty acids and the remainder coming from monounsaturated fatty acids, but with 1% or less of energy from trans fatty acids (TFAs). For polyunsaturated fatty acids, dietary energy should include 2.5–9% n-6 fatty acids and 0.5–2% n-3 fatty acids and the n-3 fatty acids should supply 0.25–2 g of combined eicosapentaenoic acid (EPA) and docosahexanoic acid (DHA). Thus, fat reduction has generally been seen as an important strategy to improve the nutritional value of foods and produce healthier products. This aspect is especially relevant to the meat industry since many processed meat products contain high proportions of animal fat.

Low fat meat products may represent an opportunity for the meat industry and are in great demand as they have been perceived as more healthy by consumers than the traditional high fat counterparts. Simply lowering the fat content or using leaner raw materials can lead to problems concerning the acceptance of these products, for example when fat levels are reduced the meat products become firmer, more rubbery, less juicy, darker in colour, more costly and less acceptable in terms of skin formation, mouth feel and processing yield (Grasso et al. 2014; Youssef and Barbut 2011; Weiss et al. 2010; Sandrou and Arvanitoyyanis 2000; Jimenez-Colmenero 2000; Keeton 1994). For these reasons, manufacturers have introduced several modifications to mitigate the detrimental effects of reducing the fat level in meat products. These modifications include the use of non-meat ingredients as fat replacers could help to convey desirable texture, palatability and more important in enhancing water-holding capacity (Akoh 1998; Keeton 1994). Fat replacers or substitutes are ingredients that contribute minimum of calories to formulated meat products and do not dramatically alter flavour, juiciness, mouth feel, viscosity or other organoleptic and processing properties. Fat replacers may be used for several purposes, one of which is to facilitate a reduction in total and saturated fat consumption and decrease the incidence of obesity and certain chronic diseases through the development and promotion of low fat products. A second purpose is to decrease total energy consumption to improve health and a third is to maintain the palatability and appeal of foods reduced in fat content (Akoh 1998). Thus, a variety of nonmeat ingredients including carbohydrate based substances such as gums (alginate, carrageenan, xanthan), fibers (pea fiber, oat fiber), starches (tapioca starch, modified food starches) and other carbohydrate compounds (pectin, micro-crystalline cellulose), protein based compounds (whey, collagen, legume proteins) and fat based compounds (soy lecithin which serves as an emulsifier, healthier oils and fats) have been assessed in development of low fat processed meat products (Desmond et al. 1998a, b).

8.2 Dietary Animal Fat and Human Health

The consumption of red meat and particularly processed meat, has been linked to an elevated susceptibility to chronic ailments and pathological conditions (Giromini and Givens 2022), as cardiovascular disease due to its saturated fat and cholesterol content, or cancer (Corpet 2011; Lippi et al. 2016; Huang et al. 2021). There is a wide understanding among consumers of the need to reduce their intake of fat of animal origin. Fat is one of the major food constituents, which influences the sensory characteristics of meat products. Fat in red meat generally consists of 40% saturated fatty acids, 50% monounsaturated fatty acids, 5% trans fatty acids and 4% polyunsaturated fatty acids (Frida 2020). The most ubiquitous fatty acids are oleic (C18:1), palmitic (C16:0) and stearic (C18:0) acids. Poultry and pork meat contain somewhat more unsaturated fatty acids (10–15% of total fatty acids) than beef and lamb and also a notable amount of polyunsaturated fatty acids (PUFAs). Linoleic acid (C18:2) is the predominant polyunsaturated fatty acid (0.5–7%), followed by alpha-linolenic acid (up to 0.5%) (Livsmedelsverket 2004). Trans-fatty acids comprise about 1–2% of total fatty acids across all types of meat and in ruminant meats they represent 2–4%. Conjugated linoleic acid (CLA), a group of polyunsaturated fatty acids that appear in dairy products and are thought to have beneficial effects on health, are found at low milligram levels in meat, especially in beef and lamb (Belury 2002). In addition to fatty acids, cholesterol is a nutritionally important component of meat and meat products. The cholesterol content of meat varies between 30 and 120 mg/100 g, being even higher in organ meats (Valsta et al. 2005; National Public Health Institute 2001; Ministry of Agriculture, Fisheries and Food 1998). In most developed countries, dietary guidelines produced since the early 1980's have proposed a reduction in total fat and saturated fatty acid intake to reduce the risk of coronary heart disease (Volek and Marounek 2011). Kennedy et al. (2009) have reported that an excessive consumption of saturated fatty acids induce white adipose tissue expansion and hypertrophy leading to apoptosis. These mechanisms could promote the release of proteins like cytokines and chemokines (inflammatory proteins) inducing inflammation and insulin resistance, thus increasing the risk of cardiovascular disease and metabolic syndrome (Haffner 2006).

Several epidemiological early case control and cohort studies suggested that consumption of a food rich in animal fat has a positive correlation with incidences of prostate-, breast- and colon cancer (Lin et al. 2004; Schottenfeld and Fraumeni 1996). It is suggested that animal fat rich in saturated fatty acids, trans fatty acids and cholesterol is more likely related to carcinogenesis (Slattery et al. 1999; Biesalski

2005). The saturated fatty acids, specifically myristic acid (14:0) and palmitic (16:0) are typically among the most concentrated fatty acids in meat and have been associated with cardio vascular diseases, colourectal cancer and type 2 diabetes (McAfee et al. 2010; Micha et al. 2010). Trans fatty acid consumption increases the risk of cardiovascular diseases and has been reported to the result of several mechanisms including increased levels of seric cholesterol, LDL and decreased levels of HDL cholesterol, increased plasma levels of triglycerides, inhibition of the Δ6-desaturase enzyme which participates in the metabolism of essential fatty acids and prostaglandins that may lead to thrombogenesis (Ospina-E et al. 2012; Riserus et al. 2009). Dietary recommendations to reduce the incidence of cardio vascular diseases are to lower the contribution to daily nutrient intakes of total fat, saturated fatty acids (SFA) and trans fatty acids, in order to avoid their cholesterol raising effects (Gidding et al. 2005). De Vries (2004) reported that, excessive consumption of fat contributes to obesity, arteriosclerosis and colon cancer. Obesity poses a public health problem with its growing incidence due to acculturation. Osteoporosis and obesity, two disorders of body composition, are growing in high proportion in the USA, as well as worldwide (Rosen and Bouxsein 2006). Obesity specifically refers to having a high amount of body fat, which is usually accompanied by abnormalities in leptin and insulin secretion and their action, together with defects in lipid and carbohydrate metabolism (Batsis et al. 2007; Lorenzo et al. 2007). The Dietary Reference Intake (DRI) report on macronutrients recommended that saturated fatty acids and trans fatty acids be as low as possible, while consuming a diet that provides an adequate intake of all essential nutrients (DRI-US 2002).

8.3 Role of Fat in Quality of Meat Products

The functional and organoleptic properties of processed meat products are affected by the type, quality, proportion and contents of fat. Fat affect the physicochemical and sensory characteristics of food and participate in eating pleasure (Fig. 8.1). Fat composition complexity and its properties greatly affect food structure and properties e.g., high fat content is reflected by high viscosity, high lubrication, high elastic properties, high sensations of creaminess, low sensations of roughness and low density (Kupirovic et al. 2016; Prinz and Lucas 2000). Creaminess is a complicated sensory perception, generally perceived and explained in terms of the perception of fat, smoothness, thickness, flavour and mouth-coating (Jervis et al. 2014). The fat molecules influence creaminess of food products through several mechanisms: firstly, coalescence or destabilization of fat molecules, leading to the formation of a fatty mouth coating; secondly, the existence of colloidal-sized fat molecules that can influence system viscosity and enhance the perception of thickness; and finally, lipophilic aroma molecules are entrapped by fat globules, which might affect the rate at which volatile flavour components are released (Dickinson 2018; Van Aken et al. 2011). The creaminess was better correlated at lower sliding speed, corresponding to the boundary lubrication regime due to the coalescence of the fat molecules on the

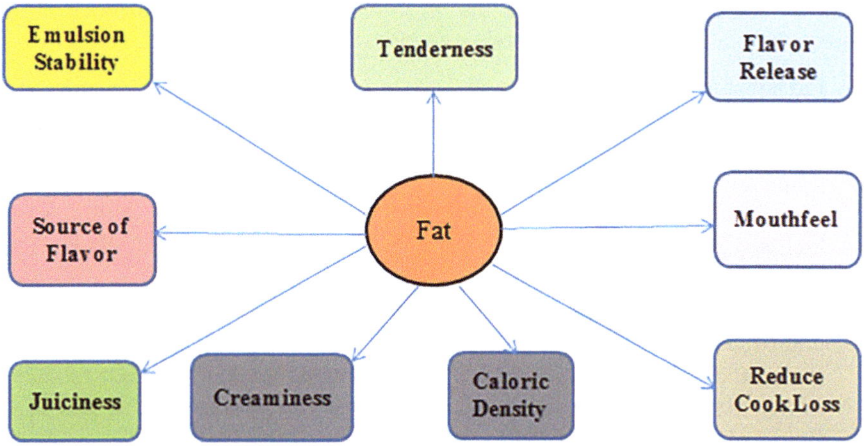

Fig. 8.1 Physico-chemical and sensory characteristics of fat in processed emulsion type meat products

tongue's surface. At higher speeds, a reversion of coalescence occurs due to the breakdown of fat globules. Fats also influence flavour perception (smell, taste and trigeminal components). They serve as a carrier for these flavours because most aroma compounds found in food are soluble in fats. Fat also influences the gastro-nervous hormonal pathway which is responsible for flavour perception and lubrication effect (Kumar 2019). In processed meat products fat plays important role in stabilizing meat emulsions, reducing cooking loss, improving water holding capacity and providing juiciness and hardness (Yoo et al. 2007; Carballo et al. 1995). Hughes et al. (1997) have also found that fats have considerable effects on the binding, rheological and structural properties of meat products. Thus, fat removal or reduction in foods causes numerous defects in the final product quality.

8.4 Fat Replacers

Reduction of fat in processed emulsion type meat products is extremely challenging and poses difficulties in terms of texture and sensorial properties compared to those of full-fat ones (Keeton 1994; Youssef and Barbut 2011). Hence, the use of fat replacers that imitate not only the functional role of fat but also its sensory features is essential to improve the quality attributes of low fat foods (Susan Brewer 2012; Nourmohammadi et al. 2023). For this reason, a number of substances with high water-binding capacity, able to promote the texture, mouthfeel and form gels have been examined for their ability to replace fat and are called fat replacers. Fat replacers are ingredients which structurally may be considered fats, proteins or carbohydrates, that is, designed to replace all or part of the fat in a product, with minimum impact on the organoleptic quality of the food product, yielding fewer calories than fat. Depending on the functional properties and manufacturing approaches, fat replacers fall into two categories:

(a) **Fat substitutes:** typically are ingredients that simulate the chemical and physical properties of fats and oils and can directly replace fat on a one-to-one or weight to weight basis. Because they are fat based, they are often stable at cooking and frying temperatures and provide all the functions of fat while yielding 9 kcal/g, which could be zero calories if none is absorbed (ADA 2005). Fat substitutes can be classified into two categories: liquids and plastics (Tan et al. 2006). The former is represented by vegetal oils and have been seen to have a positive impact on nutritional aspects as represented by reduced cholesterol content and improved PUFA/SFA and n-6/n-3 ratios. Sunflower, olive, avocado, corn, soybean, canola, grape seed, linseed, palm, cotton, hazelnut and their mixtures as well as fish oil have been evaluated (Rodríguez-Carpena et al. 2012; Delgado-Pando et al. 2011; Yildiz-Turp and Serdaroglu 2008; Hsu and Chung 2000; Ozvural and Vural 2008; Vural et al. 2004). Chemical and enzymatic alteration of some oils produce plastic fats; they appear to be solid and are resistant to light forces but when these forces pass a certain value plastic fats flow like liquids (Ospina-E et al. 2012). The chemically altered fats include triglycerides that become less energy dense i.e., to provide fewer than 9 Kcal/g, examples are medium chain triglycerides (MCT), caprenin and salatrim. Medium chain triglycerides are triglycerides made up of 8–12 carbon fatty acids that provide 7–8 Kcal/g.
(b) **Fat mimetics:** typically are ingredients that partially mimic the organoleptic and functional properties of animal fat in the food belonging to other macronutrient categories i.e., carbohydrates or proteins, but cannot replace fat on a one-to-one, gram-for-gram basis. Examples include carbohydrate-based substances, e.g., gums (alginate, carrageenan, xanthan and guar gum), fibers (pea fiber, oat fiber), starches (tapioca starch, modified food starches) and other carbohydrate compounds (pectin, micro-crystalline cellulose) and protein-based compounds e.g., whey, collagen, legume proteins, zein, microparticulated egg, protein aggregates, protein–polysaccharides composites and emulsion gels (Nourmohammadi et al. 2023). They provide lubricity, mouth feel and other characteristics of fat by holding water. The caloric value of fat mimetics ranges from 0 to 4 kcal/g. They cannot be used for frying because they bind excessive water (David Mela 1992; ADA 2012).

8.4.1 Carbohydrate Based Fat Replacers

Carbohydrate based fat replacers such as cellulose, dextrins, maltodextrins, polydextrose, gums, fiber, pectins and starches are food ingredients that are commonly used in low-fat or reduced-fat food products to mimic the functional properties of fat. They are commonly used in food preparation for controlling texture, mouth feel and other sensory characteristics of low fat food products (Lai and Lii 2004). Carbohydrate based fat replacers function mainly as thickeners and stabilizers and are typically used in a variety of foods, including dairy-type products, frozen desserts, sauces, salad dressings, emulsified meat products, baked goods, spreads, chewing gums and sweets. Dextrins and modified starches are used which absorb water and form gels that impart a texture and mouth feel similar to fat. Gums are used in restructured meat products, baked goods and salad dressings in which they act as stabilizers and thickeners and retain moisture in the reduced fat product. Pectins can also be used as fat replacers because of their gelling properties. Indigestible fiber such as cellulose is ground into microparticles that can form gels for use as fat replacer. When hydrated, polydextrose forms a gel that mimics some of the functional characteristics of fat. Replacing part or all of the fat content with carbohydrate based fat replacers in a food product can reduce the overall calorie and fat content of product thus, making them a healthier option for consumers. Overall, carbohydrate based fat replacers are alternative to animal fat in restructured meat products that can help to improve the nutritional profile of products with minimum or negligible loss of organoleptic and textural attributes (Colla et al. 2018; Kumar 2019).

8.4.1.1 Gums

Gum is the common term for hydrophilic colloids (polysaccharides) with key characteristics such as stabilizing potential, viscosity enhancement, emulsifying and surface-active property, extensive adaptability and affordability leading to their application in the formulation of food products (Eghbaljoo et al. 2022). They provide thickening effect and creamy texture in meat products and also have various health benefits such as protective agent against CVD, diverticulitis, constipation, irritable colon, colon cancer and diabetes (Chavan et al. 2016; Raibagkar and Patange 2023). Water soluble gums are used for various applications as dietary fiber, texture modifiers, gelling agents, thickeners, stabilizers and emulsifiers, coating agents and packaging films (Mirhosseini and Tan 2010a, b; Mirhosseini et al. 2010; Koocheki et al. 2009; Williams and Phillips 2000). The natural gums are categorized based on their origins, behaviours and chemical structures. Gums are known as complex polysaccharides from various sources e.g. endosperm of plant seeds (guar gum), plant exudates (tragacanth), tree or shrub exudates (gum arabic, karaya gum), sea weed extracts (carrageenan, agar), bacteria (xanthan gum) and animal sources (chitin) (Laaman 2011; Williams and Phillips 2000; Luruena-Martinez et al. 2004). When hydrocolloids are used in low fat meat formulations, a certain amount of water (around 10–20% depending on product type) needs to be added. Extra water addition provides an environment for a desirable texture by interacting with hydrocolloids (Candogan and Kolsarici 2003a, b).

8.4.1.2 Carrageenan

One of the most interesting hydrocolloid gums, which could be used in meat industries, is carrageenan. Carrageenan is a water soluble polysaccharide produced from red seaweeds (*Rhodophyceae*). In the emulsified meat products carrageenan could be used as a gelling agent, stabilizer and emulsifier due to its hydrophilic nature so it could bind the water and stabilize the emulsion (Anggraini and Lo 2023; Ramasari et al. 2012). It is hydrocolloid consisting of potassium, sodium, magnesium and calcium sulphate esters of galactose and 3, 6-anhydrogalactose (AG) copolymers. There are different kinds of carrageenan: kappa (I and II), iota and lambda, with varying numbers and positions of the sulphate groups on the galactose dimmer (Bloukas et al. 1997; Cofrades et al. 2000a). Carrageenan kappa I forms gels of high strength in water which in food systems contribute to moderate viscosity. This type of carrageenan has the highest gelling power due to its chemical composition: 24–25% sulphate esters and 35–40% AG. Because of its high gel strength it can be used in low levels to obtain desired textures. Complete dissolution of kappa I carrageenan is achieved at 75 °C and gels are formed at room temperature without refrigeration. Kappa II carrageenan is characterized by high reactivity with dairy proteins. It contains 25–28% sulphate ester and 32–34% of 3, 6 AG. The temperature at which kappa II carrageenan dissolves completely is about 70 °C. Similar, to kappa I carrageenan, gelling of carrageenan kappa II solutions occurs at room temperature. Gels formed by this type of carrageenan are thermo-reversible, firm and elastic. In food systems they contribute to higher viscosity compared to kappa I carrageenan's, due to their higher molecular weight. Iota carrageenan forms weak and thermally reversible elastic gels. It contains 30–32% sulphate ester and 28–32% AG. It shows good stability to freezing and thawing and thus is recommended for products subjected to these operations. For complete dissolution it requires a lower temperature than kappa-carrageenan (65 °C) and contributes to the higher viscosity of products (www.gelymar.com). Lambda carrageenan is capable of retaining added water, but does not form a gel (Cernikova et al. 2008). This carrageenan fraction contains about 35% sulphate ester and does not contain AG. It is soluble in cold water and provides high viscosity of food products (www.gelymar.com). Carrageenan creates homogenous structures whose shape and size is typical of fat droplets in full fat products (Huffman et al. 1991). Many researchers have noted that kappa and iota carrageenans improve the texture of low fat meat products, since these compounds create complexes with water and proteins (Cofrades et al. 2000a; Ulu 2006; Barbut and Mittal 1992). Meat pH is an important factor which affects the water holding capacity of meat. Even slight changes in pH can affect product quality (Hamm 1986). Textural properties of processed meat products are affected by the structure of the matrix formed by proteins gel, solutes and particles entrapped in the gel, which in turn depends on factors such as pH, protein, water binding ability, salt, fat content and the addition of non-meat ingredients (Flores et al. 2007). Changes of meat pH could reduce or increase water holding capacity of meat so that the effect of carrageenan might differ in meat batters of different pH and create diverse mechanical properties in sausages. It is widely used in the food industry for a broad range of applications because of its water binding, thickening and gelling properties. In the meat industry, carrageenan is used as a gelling agent in canned meats and pet foods and it allows reduction in fat content in comminuted meat products like frankfurters (Candogan and Kolsarici 2003a, b). In cooked sliced meat products carrageenan is used to improve moisture retention, cooking yields, slicing properties, mouth feel and juiciness (Imeson 2000). The effect of carrageenan addition on the functional properties of formulated meat products has been the subject of numerous studies. Bater et al. (1992) found that carrageenan caused an increase in yield, sliceability and rigidity and a decrease in expressible juice in roasted turkey breasts. In breakfast sausages, carrageenan was also found to increase the hardness of meat batters when replacing fat by water gum solution, whereas carrageenan importantly improved the water holding ability (Barbut and Mittal 1992). Emulsion stability and cooking yield of low fat chevon patties increased with the increase in levels of carrageenan (0.3, 0.6 and 0.9%) (Nayak and Pathak 2016). De Freitas et al. (1997) reported increased gel strength and water retention when adding carrageenan to salt soluble meat protein gels. Xiong et al. (1999) reported that carrageenan increased the cooking yield, hardness and bind strength of low fat sausages. Pietrasik (2003) studied the binding and textural properties of beef gels processed with carrageenan, egg albumin and microbial transglutaminase. Hsu and Chung (2000) observed an increase in cooking yield, hardness and other textural profile analysis parameters by adding up to 2% carrageenan to low fat emulsified meat balls. Influence of carrageenan addition on the properties of turkey meat sausages was studied. The results obtained show that carrageenan causes a decrease in emulsion stability and an increase in water holding capacity, hardness and cohesiveness of the formulated sausage samples. In turkey meat sausages carrageenan addition at low levels (0.2% and 0.5%) increases gel elasticity. However, a higher carrageenan concentration causes a reduction in sausages elasticity. Sensorial analysis shows that carrageenan presence has no significant effect on sausages taste. However, it improves sausage appearance and texture (Ayadi et al. 2009). Cierach et al. (2009) found that, frankfurters produced with carrageenan, irrespective of fat content, were characterized by significantly higher sensory evaluation scores for texture, colour and taste in comparison to control sausages. Candogan and Kolsarici (2003a) reported that better process yield and emulsion stability and less purge were observed with increasing carrageenan concentration (0.3, 0.5, or 0.7%) and carrageenan (0.3, 0.5, or 0.7%) with a pectin gel (20%) on some quality characteristics of low fat beef frankfurters in comparison to a high fat control (17%) and with increasing carrageenan concentration, WHC increased and penetrometer value decreased in low fat frankfurters. Koutsopoulos et al. (2008) found that low fat (10%) fermented sausages produced with two types of carrageenan (i- and *k*) in four levels (0%, 1%, 2% and 3%), i-Carrageenan had a better effect than *k*-carrageenan on pH, weight loss and lipid oxidation of the sausages, as well as, on sensory attributes. But the carrageenan level of 3% negatively affected the firmness of the sausages. Addition of salt, polyphosphates and *k*-carrageenan at around 2.7%, 0.17% and 2%, respectively, in of low fat (<10%)

emulsified meatballs or Kung-wans produced products that are more acceptable (Hsu and Chung 2000).

8.4.1.3 Locust Bean Gum (LBG)

Locust bean gum (LBG) also known as carob bean gum (CBG) obtained from the seed endosperm of the carob tree (*C. siliqua* L.) is a hetero polysaccharide galactomannan (Dakia et al. 2008). The final powder has a yellowish to whitish colour. The galactomannan is a linear polysaccharide consisting of β-(1–4)-mannose backbone with single D-galactopyranosyl units attached via α-(1–6) linkages as side branch and side branches not distributed uniformly in the main backbone chain (Da Silva and Gonçalves 1990; Daas Piet et al. 2000; Mudgil et al. 2014). The average M/G ratio of LBG is approximately 3.5 which is higher than guar gum (GG) with the average ratio of ~1.8 (Dea and Morrison 1975). As reported by Samil (2007), M/G ratio of cLBG and rLBG varied from 3.1 to 3.9. The molecular size and structure (M/G ratio, galactose distribution in the mannose linear chain) of galactomannan influence the viscosity, solubility mechanism and ability to self-associate (intra-chain and inter-chain interactions). A higher M/G ratio leads to higher thickening properties (Lazaridou et al. 2000). Dakia et al. (2008) also demonstrated that the degree of galactose substitution affected the water solubility and rheological properties of LBG. LBG shows the low solubility at the ambient temperature; therefore, the heat is required to achieve the maximum solubilisation and reach the highest water holding capacity (WHC) (Gainsford et al. 1986; Maier et al. 1993). In food applications LBG is mainly used as a thickener and stabilizer. It can be also used in other industrial products such as paper, textile, pharmaceutical and cosmetic products (Dakia et al. 2008). LBG do not form gels on its own only increase viscosity. However, in combination with other gums like xanthan gum produces, after heating and cooling a very elastic gel (Pedersen 1979; Symes 1980). Substitution of pork fat with olive oil and addition of locust bean/xanthan gum (0.5% and 0.6%) produced a significant increase in hydration/binding properties of low fat frankfurters characterised by lower cook losses, increasing yield, better emulsion stability and lower jelly and fat separation (Luruena-Martinez et al. 2004). LBG and κ-carrageenan exhibited synergistic effect in low fat sodium reduced sausages improved cooking yield, texture and water retention with only minor effects on sausage colour (García-García and Totosaus 2008).

8.4.1.4 Xanthan Gum

Xanthan gum is a non-linear anionic polysaccharide obtained by aerobic fermentation from *Xanthomonas campestris*. The primary structure of xanthan gum is a linear (1 → 4) linked β-D-glucose backbone (as in cellulose) with a trisaccharide side chain on every other glucose at C-3, containing a glucuronic acid residue linked (1 → 4) to a terminal mannose unit and (1 → 2) to a second mannose that connects to the backbone (Palaniraj and Jayaraman 2011; Jansson et al. 1975; Melton et al. 1976). It is soluble in hot and cold water. Even when this gum is used in low concentration, it forms a pseudoplastic and viscous solution (Pettit 1979). Viscosity is increased when salt is added to the formulation. This solution is stable although a variation in pH and temperature occurs. Pseudoplasticity is important for flavour, mouth feel and outward appearance of the product (Luruena-Martinez et al. 2004; Glicksman 1982). Addition of xanthan gum by itself is not suitable for using in emulsified or restructured meat product development, but use of Xanthan gum in combination of other hydrocolloids gave better results (Sarteshnizi et al. 2015). Barbut and Mittal (1992) reported that addition of 0.5% xanthan gum to low fat breakfast sausage increased moisture and fat, but it negatively affected the textural and organoleptic properties (Barbut and Mittal 1992). Solheim and Ellekjær (1993) who mentioned that xanthan gum by itself decreased firmness, elasticity and made off odour in low fat sausage. Partial replacement of fat with guar/xanthan gum resulted in higher emulsion stability and cooking yield but lower penetration force, MetMb% and lipid oxidation in meat emulsions (Rather et al. 2016a). In another study mentioned that xanthan gum (XG) as a fat replacer in traditional meat product (*Goshtaba*) of India resulted in lower cholesterol oxidation products during canning and storage up to 18 months (Rather et al. 2021). The use of xanthan gum in reduced fat and sodium content pork sausage resulted in a decrease of 27.27% in fat content, 19.54% in the energy value and an increase in the percentage of moisture, thus giving a higher yield (Maia Junior et al. 2013). Barbut and Mittal (1992) analysed the effect of xanthan addition (0.5%) to pork sausage and found that xanthan gum improved water and fat holding capacity at the expense of a gummy and chewy texture.

8.4.1.5 Guar Gum

Guar gum, a galactomannan rich flour extracted from the seeds of the leguminous plant *Cyamopsis tetragonoloba* (L.) Taub, is widely used as a thickening and stabilizing agents, in a range of industrial applications (Goldstein et al. 1973). This polysaccharide consists of galactomannan at a ratio of 2:1 mannose to galactose (M/G) (Kulicke et al. 1996). Generally, galactomannans are composed of a mannose backbone or main chain, with galactose residues or side chains. Therefore, with respect to guar gum, for every two mannose units there is one galactose side chain. Guar gum is mainly composed of high molecular weight polysaccharides with a molecular weight range of 50,000–8,000,000 Da (Kawamura 2008). Guar gum has a plethora of uses, mainly as an additive in the food industry. Typically guar gum is incorporated into foods at 1 g/100 g to facilitate gelling, thickening, firming and emulsification of food products (Flammang et al. 2006). In meat products, guar gum imparts a slick creamy mouth feel which mimics fat and is being used in low fat restructured meat products (Pearson and Gillett 1997, Rather et al. 2016b). Moreover, guar gum serves as a source of dietary fiber with associated health benefits, including reduction in cholesterol level and a significant decrease in the risk of heart disease (Theocharidou et al. 2022; Mudgil et al. 2014; Flammang et al. 2006;

Grabitske and Slavin 2009; Kawamura 2008; Slavin and Greenberg 2003). According to Butt et al. (2007) a number of clinical studies (both long term and short term) have shown a reduction in postprandial glycemia and insulinemia on consumption of guar gum, with the mechanism being attributed to an increased transit time in the stomach and small intestine, which may have been due to the viscosity of the meal hindering the access of glucose to the epithelium. Partial replacement of fat with guar gum resulted in higher emulsion stability and cooking yield but lower penetration force, MetMb% and lipid oxidation in meat emulsions (Rather et al. 2016a). Another study by Rather et al. (2016c) revealed that the use of guar gum as a fat replacer influenced the quality including textural and sensory characteristics of low fat mutton Goshtaba, an Indian meat product. In another study mentioned that guar gum (GG) as a fat replacer in Goshtaba resulted in lower cholesterol oxidation products during canning and storage up to 18 months (Rather et al. 2021b). Hasret Ulu (2006), found that low fat meat balls (10% fat), formulated with 0.5–1% carrageenan or guar gum exhibited better cooking yield and fat retention with respect to high fat meat balls (25%).

8.4.1.6 Pectins

Pectins are mainly industrially obtained by aqueous extraction of citrus peel and apple by products of juice manufacturing and consisting of partial methyl esters of polygalacturonic acid. Commercial pectins are classified into high-methoxyl pectin (HMP) and low-methoxyl pectin (LMP) according to the degree of estrification (DE) (Assoi et al. 2017). In the food industry, pectins are used in many food products as thickening and gelling agents. The gel formed by LMP primarily depends on ion regulation such as Ca^{2+} as the ions combines with the carboxyl group on the molecular chain of LMP to form an "egg box" structure. The gel formed by HMP at pH < 3.5 is primarily affected by hydrogen bonding and hydrophobic interaction (XU et al. 2022; Peng et al. 2020; Attallah and Mamdouh 2021). Hydration properties of pectins are crucial for the food application (Panchev et al. 2010). Pectin has a good interaction with meat proteins based on the H-bond formation and free binding energy (Ahmad et al. 2020). They interact with water by binding it to macromolecules and in capillaries (swelling and thickening) as well as by including it in network voids, formed by junction zones (gelling) (Ulrike Einhorn-Stoll et al. 2012). Pectin has become highly valued since it is a dietary fiber. Accumulating evidence suggests that it can reduce cholesterol (Brown et al. 1999), delay gastric emptying (Schwartz et al. 1988) and induce apoptosis of colon cancer cells (Olano-Martin et al. 2003). When used in food products, pectin functions as a gelling and thickening agent to modify texture and rheology. Water soluble pectin enriched materials (PEMs) from apple pomace, were evaluated as a fat replacer in a model food system (Min et al. 2010). Incorporation of apple pomace has potential in improving gel formation during heating in restructured low fat meat products (Raija et al. 2006). Méndez-Zamora et al. (2015) studied the effect of substitution of animal fat with different formulations of pectin and inulin on chemical composition, textural and sensory properties of emulsion type sausages and they recorded combined using of 15% pectin and 15% inulin could be used as animal fat replacer. Effects of carrageenan and/or pectin gel (20%) were evaluated in low fat beef frankfurters, frankfurters formulated with either carrageenan or carrageenan/ pectin gel had acceptable sensory scores (Candogan and Kolsarıcı 2003a, b). Han and Bertram (2017) evaluated using pectin in fat reduced pork meat model system; pectin added samples showed similar properties with the normal fat controls. Apple pomace was used as fat replacer in different meat products such as chicken sausages (Yadav et al. 2016), buffalo meat sausage (Younis and Ahmad 2015); mutton Goshtaba (Rather et al. 2015) and reduced-fat chicken sausages (Choi et al. 2016). All these studies showed enhanced cooking yield, WHC, emulsion stability and textural and sensory attributes of meat products.

8.4.1.7 Gellan Gum

Gellan gum is a linear anionic microbial heteropolysaccharide produced from the bacteria *Sphingomonas elodea* (Zia et al. 2018; Sanderson 1990). It consists of tetrasaccharide repeated units including one β-D-glucuronate, one α-L-rhamnose and two β-D-glucoses, which exhibits good gelling properties to produce a strong gel at low concentrations (Cassanelli et al. 2018; Grasdalen and Smidsrød 1987). It requires both heating cycle and cations for gelation (Morris 1990). Divalent cations have a more pronounced effect on gellan gum rheological properties than monovalent cations due to differences between the gelation mechanisms of these cation types. Divalent cations directly cross-link double helices while monovalent cations do so indirectly (Huang et al. 2004). Gellan gum could be used in the jams, fabricated foods, water-based gels, dough, confectionery products, meat systems as a stabilizer, emulsifier, binder, gelling agent and thickening agent (CassanelliIan and Mills 2017; Pérez-Campos et al. 2012; Oliveira et al. 2009; Bajaj et al. 2007). Gellan gum suspension improved gel strength of bigeye snapper surimi and increased overall acceptability (Petcharat and Benjaku 2018). Totosaus and Perez-Chabela (2009) found that addition of magnesium chloride in tandem with gellan gum could effectively compensated for the structural degradation induced by the reduction of fat and sodium in beef and pork meat batters. Addition of gellan gum and water has significant effects on the physico-chemical and rheological properties of low fat chicken meat batters (Li et al. 2019). Duda et al. (1995) reported that incorporation of gellan gum into comminuted scalded sausages increased production yield and reduced thermal drip. Gellan gum has also been used in conjunction with konjac flour to improve the quality of reduced-fat frankfurters (Lin and Huang 2003). Use of magnesium chloride in tandem with gellan gum in low fat, reduced sodium meat batters effectively compensated for the structural differences caused by fat and sodium reduction (Totosaus and Perez-Chabela 2009).

8.4.1.8 Starches and Flours

Starch based fat replacers are used either as native starch or modified starch or maltodextrins. These are made from starchy foods such as maize, rice, wheat, potato, tapioca, etc. Starch based fat replacers have been used widely in various food products, such as dairy products, baked goods, salad dressings, meat systems and mayonnaise. Native or modified starches are considered effective fat mimetics because some starch granules have similar sizes and shapes as fat globules. More importantly, native or modified starch has unique physiochemical properties, including moisture absorption and holding, thickening, viscous flow, bulking and gel formation (Chavan et al. 2016). Therefore, incorporation of starch may be beneficial in low fat, high added water processed meat products (Khalil 2000; Shewry and Tatham 2000). Native starch granules can be dispersed into food matrix and have similar sensory attributes as fat globules (Peng and Yao 2017). However, modified starch is more widely used in the food industry because native starch may not be suitable for all processing conditions, such as low pH, high temperature and freezing. Modified starches are produced either by chemical modification such as crosslinking by introducing mono- or difunctional chemical blocking groups to the starch polymer backbone or by substitution with chemical groups like sodium octenyl succinate or enzyme- or acid-hydrolysing methods (Peng and Yao 2017; Chen et al. 2020). Maltodextrins are produced by enzymatic or chemical hydrolysis of native starch. Low-dextrose-equivalence (DE < 20) maltodextrins are ideal fat replacers (Chavan et al. 2016). Maltodextrins can be produced from various raw materials, such as corn starch, wheat starch, tapioca starch and rice starch (Chavan et al. 2016). However, the functionality of maltodextrins as fat replacers is slightly different than that of native or modified starch because maltodextrins are hydrolyzed products and therefore do not have globular structure. However, the strong water holding capacity of maltodextrins gives it the ability to form gels with water in food systems. Maltodextrin gel has a similar size (1–3 μm) as fat globules, making it an ideal fat replacer and providing fat like texture to foods (Peng and Yao 2017). Interestingly, the maltodextrin gel only contains 1 part maltodextrin and 3 parts water molecules, which can result in a greater energy reduction of the food products (9 cal/g → 1 cal/g). Low DE maltodextrins are believed to be potential fat replacers and have the ability to improve physiochemical properties of food products such as viscosity, lubricity and softness (Grzelak et al. 2016). There are several commercial maltodextrin based fat replacers available on the market, such as Paselli SA2 from potato starch, Maltrin M040 from corn starch, N-oil from tapioca starch and Oatrim from oat starch (Chavan et al. 2016). The USDA has also approved Oatrim-5, which is a combination of oat beta-glucan and oat maltodextrins and the unique combination could deliver additional health benefits to the human body, such as cholesterol lowering effects. However, disadvantages associated with using maltodextrin based fat replacers should also be noted, including low freeze thaw stability and low acid and heat stability (Chavan et al. 2016). Nisar et al. (2009) reported that addition of tapioca starch (2%, 3%, 4%) to low fat (<10% fat) and medium fat

(>18% fat) ground buffalo patties increases cooking yield linearly and shrinkage decreased as tapioca level increases. Chevance et al. (2000) reported that tapioca starch is found to delay the release of volatile maillard products from beef burger thereby affecting flavour (Chevance et al. 2000). Legumes, such as peas, chickpeas and lentils, contain considerable proportion of starch and are characterized by high amylose content (24–65%), responsible for higher resistant starch content that may account for its lower digestibility (Ratnayake et al. 2002). Resistant starches contribute to low glycemic index in starch based food products, a feature beneficial in the management of diabetes and hyperlipidemia (Bjorck and Asp 1994). This important nutritional characteristic may enable the use of legume starches in dietetic foods, to reduce the risk of obesity, cardiovascular diseases and colon cancer. Addition of pea starch and fiber to low fat bolognas, lowered cooking and purge losses and increased water holding capacity (Pietrasik and Janz 2010). Consumer acceptance of low fat bolognas extended with pea starch and fibre fractions found equivalent to high fat. Serdaroglu et al. (2005) reported that addition of blackeye bean flour, chickpea flour, lentil flour and rusk (10%) to ground beef meatballs increase cooking yield, fat and moisture retention. Meat balls containing blackeye bean flour and chickpea flour had higher water holding capacity than other treatments. However, all meat balls containing legume flours were tougher than controls. Rice starch mesona blume gum (MBG) mixtures have been studied as potential fat replacer and binder for meat products because of their ability to display strong gel like behaviour (Feng et al. 2010). Addition of mesona bumes gum (MBG)/rice starch mixed gels as the fat substitute of Chinese Cantonese-style sausage showed superior total acceptability compared with low fat one after a 2 week refrigerated storage, but was slightly less than that of high fat one at the same condition (Tao Feng et al. 2013). There were no significant differences among pH, yield, residual nitrite and TBARS of high fat (28%), low fat (18%) and fat substituted (18%) sausages. In addition, the emulsifying stability and water holding capacity of fat substituted sausages were better than those of the other groups. Modified corn starch added to low fat ground beef patties increases juiciness and tenderness and had higher moisture content and cooking yield, without affecting flavour scores (Khalil 2000). Serdaroglu and Ozlem Degırmencioglu (2004) reported that addition of corn flour (0%, 2% and 4%) to Turkish type meat balls (koefte) at fat levels (5%, 10% and 20%) increased cooking yield and fat retention and at each fat level, incorporation of corn flour significantly increased protein content. Luo and Zhongyue Xu (2011) used enzyme-modified carboxymethyl starch (ECMS), prepared by etherification of starch and enzymatic hydrolysis as fat replacer in 10 g/100 g and 20 g/100 g target fat sausages. The addition of ECMS in sausages increased water holding capacity and emulsion stability and sensory analysis showed that the 10 g/100 g fat sausage with added ECMS was similar to the 20 g/100 g fat sausage control for juiciness, oiliness, texture, flavour and overall acceptability at storage days of 0, 7, 14 and 21. Thus, the new multiplex modified starch ECMS could be used as a suitable fat replacer since it offset some of the changes led by fat reduction. Heat-stable amylase-modified potato starch (2%, 4%) added as a fat replacer in low fat (5%, 15% fat) beef sausage emulsion reduced total energy (15–49%) compared to control (30% fat) sausage and increased sensory tenderness scores (Liu et al. 2008). Sausages containing 15% fat and 2% potato starch had similar hardness scores to controls (30% fat). Ichiro et al. (1987) and Nippon (1984) reported that, addition of cross-linked hydroxypropyl starch and/or cross-linked starch acetic acid ester, cross-linked starch phosphoric acid ester, cross-linked starch octenyl succinic acid markedly increases elasticity and organoleptic test scores of kamaboko and other fish paste products. Konjac glucomannan is a neutral polysaccharide obtained from the tuber of *Amorphophallus konjac* plant. Konjac flour consists of D-glucose and D-mannose joined by β-1, 4 glycosidic linkages with ~1 acetyl group in every 19 sugar units (Lin and Huang 2003; Williams and Phillips 2000). The molecular weight of the glucomannan in this product is approximately 300,000 daltons. Konjac flour is considered a low calorie ingredient which, given its content in non-digestible fiber, presents numerous physiological effects and therapeutic applications (Ruiz-Capillas et al. 2012; Al-Ghazzewi et al. 2007; González Canga et al. 2004; Tye 1991; Zhang et al. 2001). Konjac has been reported to lower serum cholesterol and LDL cholesterol and triacylglycerol (Lin and Huang 2003). It presents notable functional effects and exceptional technological properties which offer great potential for use in food processing applications. Konjac does not form gels in pure water but rather yields highly viscous pseudoplastic solutions that can be used as thickening agents. However, gelation may take place in the presence of alkaline coagulants (e.g., calcium hydroxide) after deacetylation, yielding thermo irreversible gels and gels are thermally stable to temperature greater than 20 °C (Thomas 1997). Moreover, because of its technological properties, when combined with other ingredients (starch, carrageenates, gellan gum) konjac can be used as a 'fat analogue' in the formulation of reduced/low fat meat products. Its use as a food additive is authorized in Europe (E-425) and it is classified as GRAS by the FDA. Konjac (added in different concentrations) has been used to reduce fat in products such as dry fermented sausages (Ruiz-Capillas et al. 2012), frankfurters (Lin and Huang 2003), bologna (Chin et al. 1998), fresh sausages (Osburn and Keeton 1994) and pork nuggets (Berry and Bigner 1996), pâté (Delgado-Pando et al. 2011) and emulsified meat balls (Hsu and Chung 2000). The konjac gel (KG) showed high stability under chilling storage conditions. Water binding capacity (WBC) of KG ranged between 87% and 90%, with no changes during chilling storage. Texture profile analysis parameters were not affected by chilling storage. The rheological behaviour and microstructure confirm the excellent chilling storage-stability of KGs. KG presented excellent thermal water binding capacity. Rheological analyses indicate that the behaviour of KG, which is thermally stable, is predominantly elastic during heating and exhibits rheological thermal behaviour (at over 40 °C) similar to that of pork back fat (Jiménez-Colmenero et al. 2012). The effect of freezing/thawing on fat analogue texture seems to be largely overcome by heating (Jiménez-Colmenero et al. 2013). Huang et al. (1998) found acceptable quality of low fat Chinese style sausage containing 10 or 20% konjac gels. The sensory characteristics of low-fat dry fermented sausage were acceptable by reducing fat content from 29.96% to 13.79%, replaced by an equal proportion of konjac gel (Ruiz-Capillas et al. 2012). Park (1996) reported that adding konjac flour enhanced the shear stress of surimi gel and produced heat-tolerant and stable freeze thaw fish gels. Chin et al. (1998) reported that low fat (<3%) bologna containing varying amounts of konjac blends had similar textural properties to regular bologna (~30% fat) control. One can find konjac listed in the ingredients statement of fat free meat products along with other additives that are combined in order to produce products acceptable to consumers as replacement products for the traditional fatter ones.

8.4.1.9 Fiber

Total fiber is defined as the sum of functional and dietary fiber, that is the sum of non-starch polysaccharides (NSP) and non-digestible oligosaccharides. The Health Canada, defined dietary fiber as non-digestible carbohydrates, either natural, synthetic or isolated with a degree of polymerization of three or more that are not digested and absorbed by the small intestine. In addition dietary fiber should have at least one physiological effect which includes: improving laxation or regularity by increasing stool bulk, reducing blood total or LDL cholesterol levels, reducing post-prandial blood glucose or insulin levels and providing energy yielding metabolites through colonic fermentation (Health Canada 2022). Addition of fibre in the daily diet is recommended because of its well known impact on reducing the risk of colon cancer, obesity, cardiovascular diseases and other disorders (Pietrasik and Janz 2010). The current recommendation for dietary fiber intake varies according to the regulations of each country for example, WHO recommends 25 g/day, Canada recommends 38 g/day for men and 25 g/day for women; the United States urges 28 g/day, México 25 g/day and ICMR recommends about 40 g/2000 kcal (ICMR 2020). Fiber is derived from various plants, fruits and nuts and has different functional characteristics depending on processing conditions such as grinding and drying (Choe et al. 2013). Depending on the type of dietary fiber used, energy value varies from 0 to 2 calories per gram. Different fiber-rich ingredients such as amorphous cellulose, glucomannan (KGM), β-glucan, inulin (IN), psyllium husk (PH), chia seed mucilage (CM), etc. have recently been used as functional additives to food products to support and ensure binding. This has resulted in the production of more stable low fat meat products with better textural properties (Pietrasik and Janz 2010; Bloukas and Paneras 1996; Claus and Hunt 1991; Cofrades et al. 2000a; Desmond and Troy 2002; Grigelmo-Miguel et al. 1999). Dietary fiber can lead to a compact gel formation due to the fiber's ability to retain fat and water (Fernández-Ginés et al. 2005), which can improve structural integrity, yield and adhesiveness in reduced fat meat products (Tokusoglu and Kemal 2003).

Inulin is obtained from different sources such as Jerusalem artichoke and chicory root (Zahn et al. 2010). It consists of up to sixty monomers of fructose bound by β-2-1 glycosidic linkages and is currently used in the food industry as prebiotic and food additive (Shoaib et al. 2016). Inulin is considered a functional food ingredient and is used in different food products as a fat replacer, energy reducing agent, improving water-holding and emulsion stability, as well as to modify the texture and viscosity (Boeckner et al. 2001). In a variety of processed meat products, such as scalded sausages (García et al. 2006; Tröger et al. 2005), canned meat products (Florowski and Adamczak 2010), meat balls (Flaczyk et al. 2009), liver pâté (Florowski et al. 2008) and fermented sausages, several researchers have used rehydrated inulin (inulin gel) (Mendoza et al. 2001). It has been used as a fat replacer in low fat sausages to improve quality of the final product (Nitsch 2006). Franck (2002) found that inulin can be used to give meat products creamier and juicier mouth feel without compromising taste and texture. Among its main health benefits, it has been reported to reduce the risk of arteriosclerosis, prevent osteoporosis (Franck 2000), control the level of glucose in blood, maintain low levels of triglycerides and cholesterol in serum and stimulate the immune system (Desmedt and Jacobs 2001). Tomaschunas et al. (2013), reported that addition of inulin and citrus fibre in reduced fat Lyon-style sausages (3–17% fat) and liver sausages (3–20% fat) improved colour intensity, spiciness and firmness. Berizi et al. (2017) reported that replacing fat with inulin in emulsion type sausages led to energy content reduction of up to 64% in addition, colour, sensory and textural attributes were comparable to that of traditional product. Inulin as a fat replacer in meat burger did not cause a significant effect on sensory or texture properties and at the same time reduced fat content in the product (Kamil et al. 2024).

Amorphous cellulose is an insoluble fiber obtained from cereals, is used as a fat replacer in foods. Amorphous cellulose can be directly incorporated in the form of a powder, or it can be added in the form of a gel (Chung and Min 2004). It increases viscosity, which confers a sensory property of succulence and a texture similar to that of fat, due to its high water retention capacity. Amorphous cellulose has already been successfully used to reduce fat levels in cheeses, hamburgers and baking products (Cho and Prosky 1999; Warner and Inglett 1997). Campagnol et al. (2012) reported that fermented sausages formulated with substitution of 25%, 50%, 75% or 100% of their pork back fat by amorphous cellulose gel. The substitution of up to 50% of the pork back fat content by amorphous cellulose gel can be accomplished without a loss of product quality, enabling the production of fermented sausages with the levels of fat and cholesterol decreased by approximately 45% and 15%, respectively.

The addition of Laminaria to various foods has been shown to improve the quality and increase the functionality of products. A brown seaweed, *Laminaria japonica*, is a common seafood in Korea and many other Asian countries (Bang et al. 2006; Kim et al. 2008) and is a major source of dietary fiber such as alginic acid, fucoidan and laminaran, all of which are beneficial to human health (Lee et al. 1998). Reduced fat pork patties produced with the addition of *Laminaria japonica* powder were evaluated for the chemical composition, cooking characteristics and sensory properties. Choi et al. (2012), reported that reducing fat content in pork patties from 20% to 10% and supplemented with 1 or 3% *L. japonica* powder had greatest overall acceptability than control patties containing a fat content of 20%. Reduced fat pork patties containing *L. japonica* powder had significantly higher moisture, ash, carbohydrate content, yellowness and springiness than the control sample. Pork patties with fat contents reduced from 20% to 10% and supplemented with 1 or 3% *L. japonica* had improved quality characteristics that were similar to the control patties containing a fat content of 20%.

The use of oat fibre source of β-glucan is useful for replacing saturated fats in foods products (Pinero et al. 2008). Oat incorporated products have also achieved a very positive consumer image due to the potential health benefits that have been associated with the consumption of fibre products (Aleson et al. 2005; Steenblock et al. 2001; Yilmaz and Daglioglu 2003). Different health benefits of β-glucan in isolated form or as a constituent of oat and barley products, are related to its viscosity: attenuation of post-prandial plasma glucose and insulin responses, higher transport of bile acids towards lower parts of the intestinal tract and higher excretions of bile acids (Dongowski et al. 2003) or lowering of serum cholesterol levels (Braaten et al. 1994; Hecker et al. 1998), giving a highly positive perception to the diet/health oriented consumers (Food and Drug Administration 1997; Lyly et al. 2007). Due to the potential health benefits offered by glucan it become interesting to use as an ingredient in common meat products, such a beef patties. Moreover, various researchers have reported the use of oat fibre in the elaboration of several meat products such as patties, hamburgers, sausages and meat balls (Pinero et al. 2008; Aleson et al. 2005; Steenblock et al. 2001; Troy et al. 1999; Warner and Inglett 1997; Yang et al. 2007; Yilmaz and Daglioglu 2003). Adding oat fiber as a source of beta-glucan (13.45%) to low fat (<10%) beef patties improves cooking yield and fat and moisture retention compared to 20% fat control Patties, due to water binding ability of beta glucan (Pinero et al. 2008). Addition of oat's soluble fibre had no effect on appearance, tenderness and colour. Desmond et al. (1998a, b) reported that adding oat fiber (0–20 g/kg) had limited effects on yield and water holding capacity of low fat beef burgers.

Barley β-glucan is a non-starch polysaccharide that shows potential to be used as a fat replacer in low fat sausage products (Temelli 1997). It is a water soluble hydrocolloid that functions to provide water control by thickening and/or gelling. As a soluble fibre component, β-glucan has the health benefits of reducing blood serum cholesterol (Hecker et al. 1998) and regulating blood glucose levels. Morin et al. (2004), reported that barley β-glucan reduces cooking loss in cooked meat products more than carboxymethyl cellulose, due to the physical entrapment of water. Mansour and Khalil (1997) reported that hydrated wheat fiber improved the physicochemical properties of low fat beef burgers.

Pea fiber is another alternative as fat replacer in ground meat products. Anderson and Berry (2000) reported that incorporating pea fiber to lower fat content in ground beef patties resulted in improved tenderness and cooking yields in addition to reducing change in thickness during cooking compared to all beef controls. When added to 10% and 14% in beef patties, pea fiber had no effect on beef flavour or juiciness. Fat retention in high fat ground beef (40% and 50%) increased from 33% to between 85% and 98% when pea fiber was added (Anderson and Berry 2001). Cooking yield was substantially greater in ground beef containing pea fiber (87–94%) compared to all beef controls (52%).

8.4.2 Protein Based Fat Replacers

Protein is a crucial macronutrient and its deficiency is one of the biggest public health issues, particularly in underdeveloped countries (Gomes et al. 2009). The regulation of body health including bone, gastrointestinal function and bacterial ecology, glucose homeostasis, cell communication and satiety are all areas where amino acids play important role (Millward et al. 2008). In food system proteins have various functional characteristics including solubility, viscosity, water binding, emulsification, gelation, binding of fat and taste, foaming and textural properties (Kinsella and Melachouris 1976; Syan and Bawa 2022). Protein based fat replacement in meat products benefits with lower fat content. Direct replacement of fat with non-meat proteins is an attractive avenue in approaching fat reduction due to the excellent functional and nutritional properties these non-meat proteins can impart. Proteins have been used as fat substitutes in a variety of meat products to improve yields (water and fat binding), reduce formulation costs, maintain nutritional value, enhance functional properties (water holding capacity) and decrease fat and cholesterol content (Abiola and Adegbaju 2001; Carballo et al. 1995; Chizzolini et al. 1999; Cofrades et al. 2000b; Hughes et al. 1998; Hsu and Chung 2000; Keeton 1994; Pietrasik and Duda 2000). Plant proteins such as soy bean (Gujral et al. 2002; Pietrasik and Duda 2000; Serdaroglu and Sapancı-Ozsumer 2003; Trout et al. 1992), sunflower protein (Wills and Rabirullah 1981), buck wheat protein (Bejesano and Corke 1998), wheat germ proteins (Gnanasambandam and Zayas 1992) and corn germ protein (Linn and Zayas 1987; Zayas and Lin 1988) have been used as binders and extenders in comminuted meat products. Lanier (1991) reported that non-meat proteins can interact directly with meat proteins, occupy the interstitial spaces in a gel matrix or gel as discrete pockets in the matrix structure. The potential of microbial transglutaminase in catalysing cross-linking of food proteins and modifying the functional properties of these proteins has been demonstrated recently (Pietrasik et al. 2007). Dickinson (1997); Motoki and Seguro (1998); Sakamoto et al. (1994) have reported that transglutaminase induces polymerization of food proteins such as milk proteins, soybean proteins, egg proteins, beef, pork, chicken, fish gelatin and myosin. Askin and Kilic (2009) studied the effects of microbial transglutaminase (MTGase) and its combinations with sodium caseinate (SC) or non-fat dry milk (NFDM) on quality of salt free, low fat turkey doner kebab. Use of microbial transglutaminase (MTGase) can minimize textural quality loss caused by a lack of salt in turkey doner kebab. When the enzyme was combined with SC, or NFDM, its effect on texture was enhanced and improved cooking yield, moisture and sensory properties compared to the corresponding control manufactured without the enzyme, salt, SC and NFDM. Transglutaminase in combination with other proteins therefore has been proposed as a useful tool to improve quality in salt free and low fat meat products.

8.4.2.1 Animal Protein Based Fat Replacers

Fat replacers based on animal proteins are derived from poultry, eggs, fish and milk and contain all essential amino acids. Animal proteins are considered superior than plant proteins due to their easy digestion and higher biological value (Millward et al. 2008). Animal proteins are an excellent source of fat substitute due to their various functional qualities including emulsion stability, ability to retain fat and flavour, foaming capacity, etc. (Yashini et al. 2019).

Egg White Proteins

Egg white contains many functionally important proteins like ovalbumin (54%), ovotransferrin (12%), ovomucoid (11%), ovomucin (3.5%) and lysozyme (3.5%) have a lot of industrial applications (Abeyrathne et al. 2013). The proteins from egg white are extracted using various techniques including salting out, aqueous biphasic systems, etc. The extracted egg white proteins can be employed in the food industry either unmodified or after being modified by enzymes. Lysozyme is one of the major bacteriolytic protein found in egg white used as a food preservative (Radziejewska et al. 2008). The WHO and many countries allow the use of lysozyme in food as a preservative (Mine et al. 2004). Ovomucin good inhibitory agent against *E. coli*, *Bacillus sp.*, and *Pseudomonas sp.* used in food industry as a food preservative (Omana et al. 2010). In addition used as nutritional supplement and has good emulsifying and forming characteristics in food systems (Abeyrathne et al. 2013; Stadelman and Cotterill 2001). Ovotransferrin is known to have a strong antimicrobial activity and thus, can be used to improve the safety of foods. Wu and Acero-Lopez (2012) reported that ovotransferrin has an antioxidant effect on poultry meat by establishing the cellular redox environment. Ovalbumin major protein in egg white is a spherical monomeric phosphoglyco protein with strong gelation ability (Xiong et al. 2019). During gelation, ovalbumin unfolds in a molten spherical structure and then aggregates to form a three-dimensional gel network structure (Campo-Deaño and Tovar 2009). Ovalbumin contains hydrophilic groups that absorb water and swell improve the gel strength; non-muscle protein can directly interact with myogenic fibrin to fill the gel network and enhance the texture of surimi or other muscle based comminuted products (Li et al. 2020). Rovers et al. (2016) compared the rheological and tribological characteristics of model food systems to examine the potential of microbubble dispersions in egg white protein as fat substitute and texture modifier. When introduced into a model food system, it was demonstrated to be an effective fat replacer (liquid with and without thickeners). Enzyme-modified soy protein isolate, sea fish collagen, and ovalbumin were used to prepare the traditional surimi-based low fat product (Nemipterus virgatus fish sausage). The sensory score of fish sausage was increased by 21.79%, maintaining its good flavour and sensory quality (Yan et al. 2023). Microparticulated protein (marketed under the brand names Simplesse and Trailblazer) is made from microparticulated milk or egg-white proteins, sugar, pectin and citric acid provide creamy mouth feel of fat and fat type favour (Raibagkar and Patange 2023).

Whey Proteins

Milk proteins are widely used as protein based fat replacers and are divided into two major fractions, casein (80%) and whey protein (20%) (Akalin et al. 2008). Whey protein, a by-product of dairy industry recognized as a valuable food ingredient with important nutritional and functional properties is gaining acceptance as functional food ingredient. Commercial whey proteins are considered a GRAS substance for food product applications. Much attention is focused on whey proteins. Whey protein has an exceptional biological value (BV) that exceeds that of egg protein by about 15% (Geoffrey Smithers 2008). In this regard, whey protein excels and is the protein of choice for body builders, elite athletes, and those whose health is compromised (Buckley et al. 1998; Carey et al. 2006; Coombes et al. 2000; Ha and Zemel 2003; Playford et al. 1999; Rankin and Darragh 2006). Whey protein is a rich source of the essential amino acids when compared with other typical food proteins and is also rich in the branched chain amino acids (leucine, isoleucine, and valine) (420%, w/w). These latter amino acids are thought to play a role as metabolic regulators in protein and glucose homoeostasis, in lipid metabolism and as such may play a role in weight control (Geoffrey Smithers 2008). Whey protein is also a rich and balanced source of the sulphur amino acids (methionine and cysteine). These amino acids serve a critical role as antioxidants, as precursors to the potent intracellular antioxidant glutathione and in one-carbon metabolism. In addition to these components whey contains a multitude of biologically active proteins and peptides and these and other components have formed the basis for the neutraceutical potential of whey. Apart from the major whey proteins-β-lactoglobulin, α-lactalbumin, and glycomacropeptide- whey contains a number of other proteins like lactoferrin, lactoperoxidase, immunoglobulins, insulin-like growth factor (IGF), platelet-derived growth factor (PDGF), transforming growth factor (TGF) and fibroblast growth factor (FGF) with potent bioactivity (Geoffrey Smithers 2008). Due to these bioactive components whey proteins have implicated a variety of nutritional and physiological effects, including (i) physical performance, recovery after exercise and prevention of muscular atrophy (Carey et al. 2006; Farnfield et al. 2005; Ha and Zemel 2003; Ohr 2004; Rankin and Darragh 2006; Tipton et al. 2004), (ii) satiety and weight management (Ohr 2004; Tome 2006; Zemel 2004), (iii) cardiovascular health (Ohr 2004; Pins et al. 2006), (iv) anti-cancer effects (Bounous et al. 1991; Gill and Cross 2000; McIntosh et al. 1995), (v) wound care and repair (Regester et al. 2003; Smithers 2004), (vi) management of infections (Bounous et al. 1993; Playford et al. 1999; Regester and Belford 1999), (vii) infant nutrition (Chatterton 2006) and (viii) healthy aging (Smilowitz et al. 2005). Evidence is growing that specific whey proteins, such as lactoferrin and lactoperoxidase, serum albumin and peptides have potential anti-cancer effects against certain tumours (Bounous et al. 1991; Gill and Cross 2000).

The application of whey protein concentrates (WPCs) in the manufacture of reformed and restructured meat proteins has received much attention (Giese 1994). Dairy proteins have been incorporated as water and fat binders and have the potential to modify the textural characteristics of low fat comminuted meat products (Comer et al. 1986; Ellekjaer et al. 1996; Keeton 1997). One of the key functional properties of whey protein concentrates (WPCs) is their ability to form heat induced three dimensional gel structures with increased water-holding capacity and potential texture modifying properties (Lyons et al. 1999). Dairy ingredients have been used as fillers and binders in comminuted meat products to improve texture and sensory properties and minimize cooking losses (Hung and Zayas 1992). Extensive research has been done on the using whey protein concentrate in emulsion type meat products such as frankfurters and bologna sausages (Hung and Zayas 1992; Lyons et al. 1999; Serdaroglu and Sapancı-Ozsumer 2003; Yetim et al. 2001). Meltem Serdaroğlu (2006), reported that addition of whey powder (WP) at levels of 0%, 2% and 4% in beef meat balls formulated with 5%, 10% and 20% fat levels, significantly increased cooking yield, fat and moisture retention of meat balls regardless of fat levels and WP had no detrimental effect on sensory properties of meat balls. Addition of whey protein concentrate (1–4%) to low fat (<11% fat) ground beef patties increases chewiness, juiciness and overall acceptability of low fat ground beef patties containing 4% whey protein were scored more highly than those containing less whey (El-magoli et al. 1995). The interactions between proteins (muscle and non-muscle) and polysaccharides are also important in determining meat products binding and textural properties and have been the subject of many investigations (Bernal et al. 1987; DeFreitas et al. 1997; Montero et al. 2000; Trius and Sebranek 1996). Pietrasik and Jarmoluk (2003), reported that sodium caseinate and k-carrageenan content favoured hydration properties and thermal stability, yielding lower cooking loss and higher water holding capacity of pork muscle gels.

Gelatin

Gelatin is another animal protein produced from the partial hydrolysis of collagen obtained from animal sources such as skin, bones and connective tissue of pigs, cows, fish, birds and insects (Syan and Bawa 2022). Gelatin has various technological qualities including gels formation, texture improvement and water-holding capacity and is protein-based fat replacer due to its ability of mimicking the smooth and lubrication characteristic of fat (Niua et al. 2020; Wu and McClements 2015). The utilization of gelatin as a fat replacer is linked to its capacity to replicate the sensory attributes, smoothness and lubrication characteristics of fat. It has been employed to replace fat in various products because it can form thermally reversible gels with a melting point close to body temperature, which can be utilized to imitate the creamy and spreadable properties of fat (Wu and McClements 2015). Niua et al. (2020) reported that gelatin soluble dietary fiber composite gels can be added in the processed meat as a fat replacer with good texture while formulating the meat balls.

Bovine Globin and Plasma

Protein from bovine blood plasma is inexpensive. In the slaughterhouses bovine blood is mixed with anticoagulant and centrifuged at 1000 g for 15 min to separate the bovine plasma (Yashini et al. 2019). Bovine globin and plasma showed important functional and nutritional properties (Silva et al. 2003; Silva and Silvestre 2003) and indicated as potential fat replacers in meat products (Jimenez-Colmenero 1996). Plasma is a good source of soluble proteins and could be useful in meat industries to replace fat (Cofrades et al. 2000). Viana et al. (2005) examined how the quality of ham paté would be affected by the addition of globin (10%), plasma (10%) or both (5% each) as fat replacers. Following the fat replacement and 25–35% fat reduction, the results indicated an increase in moisture and protein contents. The consistency, flavour and aroma remained unchanged in the low fat products.

8.4.2.2 Plant Protein Based Fat Replacers

In order to replace protein ingredients made from animal sources, the food industry is constantly looking for less expensive alternatives that are also healthy. Plant proteins make up the majority of the world's available protein per capita and are often regarded as incomplete proteins. Compared to animal proteins, plant proteins lack lysine, cysteine and methionine; however, this nutritional deficiency could be overcome by mixing different types of plant proteins (Nourmohammadi et al. 2023; Young and Pellett 1994). However, due to larger consumption of non-essential amino acids, plant protein may be able to reduce the risk of developing chronic degenerative illnesses (Nishinari et al. 2014). Furthermore, the incorporation of plant based protein as fat replacers levels up the protein content in foods, which enhances protein nutrition. Plant based proteins can be obtained from various sources like pulses, cereals, fruits and vegetables. Some plant based proteins used as fat replacers are pea protein, soya protein and wheat gluten. Plant proteins are being used by food makers because of their ability to replicate the qualities of fat in low fat diets (Massey 2003).

Soya Protein

Soya beans contain roughly 40% protein, 20% oil, 15% mono- and oligosaccharides, and 15% dietary fibre (Nishinari et al. 2014). Soya protein being an economical source of food protein can be added to meat as soya flour (50% protein), soya protein concentrate (70% protein) or soya protein isolate (90%). The four fractions of soya proteins are 2S (15%), 7S (34%), 11S (41.9%) and 15S (9.1%), categorized by their sedimentation coefficient. The b-conglycinin (7S) and glycinin (11S) fractions are the principal components represent 65–80% of the total seed proteins (Jambrak et al. 2009). Among protein additives, soya bean proteins are the most widely employed proteins in meat commodities due to various advantages which include (i) very little off-flavour; (ii) low cost; (iii) high nutritional value (contain all the essential amino acids required by humans and its digestibility is comparable to that of meat, fish, milk or egg proteins); (iv) interesting functional properties (easily associate with water and fat showing good hydration, gelling and emulsifying properties); (v) their presence in meat products improves the appearance and organoleptic characteristics of these products (Ospina-E et al. 2012). Lecomte et al. 1993; Rentfrow et al. (2004), reported that soya protein can be used as binder for improving yields, as a gelling agent to enhance emulsion stability and as a meat replacement to reduce costs. Decker et al. (1986) reported that frankfurters formulated to a 15% fat endpoint had sensory characteristics similar to the 30% fat control (CTL), when 20–30% of meat was replaced with hydrated soya protein isolate. Goat meat patties incorporated with full fat soya protein found more tender and juicy and highly acceptable (Das et al. 2006). Dignam et al. (1979); Kilic et al. (2010), found that, 20% textured soya protein incorporated into ground beef decreased cooking loss and evaporative loss however, it affects sensory characteristics. The texturized soya alone increases moisture binding and protein level, while it lowers flavour score of low fat ((14%) beef burgers compared to controls, but a combination of carrageenan (0.5%), texturized soya (1.5%) and trisodium phosphate (0.5%) improves all sensory attributes (Angor and Al-Abdullah 2010). Soya proteins were successfully used to reduce the fat content of beef patties and pork sausages (Decker et al. 1986). Soya protein together with potato flour decreased the cooking loss and a firm sausage was processed (Skrede 1989).

Pea Protein

Pea protein is known to be highly digestible, gluten free, nutritive and has low allergenicity; it offers potential for using in meat as fat replacer (Lam et al. 2018). In addition it has good emulsification and other functional properties (Baugreet et al. 2016). Pea proteins have an exemplary hydrophilic oil structure and during foaming, proteins can adsorb to these interfaces and reduce interfacial tension and can also form network structures through interactions and impart some stability (Ali et al. 2011).The reduction of fat and its replacement with texturized pea protein was feasible, with no detrimental impact on the acceptability of dry fermented sausages (Colomer Sellas et al. 2021) and low fat frankfurters (Revilla et al. 2022). Duck meat batters were prepared with different concentrations of pea protein isolate (0%, 3%, 6% or 9%, w/w) and evaluated for cooking yield, water-holding capacity, water distribution and migration, rheological properties and texture profile (Zhu et al. 2022). The study found that addition of pea protein improved the forming ability of duck breast mince and significantly enhances the properties of duck meat batters, increasing the water retention, cooking yield, rheological properties and textural properties of the matrix.

Wheat Gluten

Gluten is a wheat protein that on a dry basis, is made up of 75% protein and 25% carbohydrate and lipids. The hardest of all wheat varieties, durum wheat has a protein content that ranges from 9% to 18% (Delcour et al. 2012). In addition to the inherent attributes of wheat, which are anchored in genetic as well as environmental factors, gluten functionality and its contribution to the quality of the finished product depend on the processing of wheat flour. In the industrial setting, gluten is created using wet processing. The wheat gluten (WG) is the primary storage protein in wheat (*Triticum spp.*) grains. Wheat gluten (WG) is comprised of two sub-units; gliadin and glutenin. Gliadins are low or medium molecular weight monomeric proteins with intramolecular disulphide bonds, while glutenins are much larger molecules containing different polypeptides connected by intermolecular disulphide bonds (Wieser 2007). It is one of several vegetable proteins used as meat extenders or fillers in comminuted meat products to enhance functional characteristics and reduce cost (Serdaroglu and Sapancı-Ozsumer 2003; Jimenez-Colmenero 1996; Keeton 1994).

8.4.3 Fat Based Fat Replacers

Regarding lipid based fat substitutes animal fat is replaced either by vegetable and/or marine sources, whose characteristics are more in line with health recommendations: i.e., contain smaller proportions of saturated fatty acids (SFA) and larger proportions of monounsaturated (MUFA) or polyunsaturated (PUFA) fatty acids, especially long chain n-3 PUFA, better n-6/n-3 PUFA and PUFA/SFA ratios and if possible cholesterol free (Gonzalo Delgado-Pando et al. 2011; Jiménez-Colmenero 2007; Herrero et al. 2012). As a strategy to improve fatty acid profile (decrease SFA and increase MUFA and PUFA levels) of meat products vegetable oils such as sunflower oil (Ferro et al. 2021; da Silva et al. 2019), linseed oil (Gómez-Estaca et al. 2020), soyabean oil (Adili et al. 2020), sesame oil (Moghtadaei et al. 2018; de Souza et al. 2019), olive oil (Gómez-Estaca et al. 2020), canola oil (Oh et al. 2019; Alejandre et al. 2019), and chia oil (Vargas-Ramella et al. 2020) and marine oils such as fish oil (Marchetti et al. 2014) in general, are used to replace animal fat in meat products by means of direct addition, or organogelation, interesterification and structured emulsions (Jimenez-Colmenero et al. 2015), among which organogelation is the most commonly employed strategy.

The pre-emulsification of fats and or oils can improve fat binding ability, enhance physical stability and is usually easier to disperse into a water based system such as a meat batter (Jiménez-Colmenero 2007). Pre-emulsification is a process of preparing an emulsion (oil in water) stabilized with an emulsifier, which is typically a protein of a non-meat origin (Cáceres et al. 2008). Hoogenkamp (1987) reported that, when oil/fat is pre-emulsified with a non-meat protein, more meat proteins can become available for gel formation of the matrix.

The addition of individual lipids (of plant or marine origin) does improve the fatty acid profile of meat products, but a better approximation to an optimal lipid profile from a health standpoint can be achieved using healthier oil combinations as animal fat replacers (Delgado-Pando et al. 2010; García-Iníguez de Ciriano et al. 2010; López-López et al. 2009; Paneras et al. 1998). The healthier lipid combination was formed by vegetable oils (olive and linseed) and fish oils in suitable amounts and proportions to provide a fatty acid profile adjusted to healthier intake goals. This oil combination was designed to produce a lipid material with a small proportion of SFA, large proportions of MUFA and PUFA (including long chain n-3 PUFA) and balanced n-6/n-3 PUFA and PUFA/SFA ratios (Delgado-Pando et al. 2011).

Pork liver pâtés with partial or total replacement of pork back fat with healthier oil combination (olive, linseed and fish oils) and konjac gel (0–15%), had lower levels of saturated fatty acids (SFA) (27.4% and 21.3%) and monounsaturated fatty acids (MUFA) (49.8% and 42.5%) and higher levels of polyunsaturated fatty acids (PUFA) (22.4% and 35.6%) compared with control pâtés (32.2%, 58.2% and 9.04% respectively). The fat reduction produced a softer and more spreadable pâté, with good fat and water binding properties and acceptable sensory characteristics, similar to the control sample (Delgado-Pando et al. 2011). Konjac gel addition (15%) produced stiffer structures (as compared with 0 and 7%) which are very close to those of the control samples. López-López et al. (2009), reported that low salt (0.5%) and low fat (<10%) beef patties with partial or total replacement of pork back fat with olive oil-in-water emulsion and wakame seaweed (3%) were judged acceptable by the sensory panel and however softening effect ($P < 0.05$) induced by olive oil emulsion was observed only in seaweed-free patties. Herrero et al. (2012) reported that frankfurter formulated with olive oil-in-water emulsion stabilized with soy protein isolate (SPI) as pork back fat replacer had increase of hardness, springiness, cohesiveness and chewiness but a reduction of adhesiveness. Yun-Sang Choi et al. (2009), used vegetable oils prepared from olive, corn, soybean, canola or grape seed and rice bran fiber to replace the pork fat to 10% meat batter system. The vegetable oil and rice bran fiber addition increased moisture, protein, ash content, uncooked and cooked pH values, yellowness (*b**) value, hardness, cohesiveness, gumminess, chewiness and viscosity of meat batters than the control containing 30% pork fat. In addition, batters supplemented with vegetable oil and rice bran fiber had lower cooking loss and better emulsion stability. Replacing animal fat with ground poppy seed paste or oil is effective and may pave the way for innovative meat products. *Papaver somniferum* (poppy) seeds are mainly used to produce edible oil (Bozan and Temelli 2003; Nergiz and Ötles 1994) and the oil content of poppy was reported to range from 45% to 50% (Nergiz and Ötles 1994). Poppy oil has approximately 73% linoleic, 10% palmitic, and 13% oleic acid as the major fatty acids (Nergiz and Ötles 1994; Sengupta and Mazumder 1976). The health benefits of oils rich in linoleic acid, like corn or safflower oils, in lowering serum cholesterol levels were reported (Gotterbos 1988; Imaijumi et al. 2000). Turkish style meatballs and burger (koefte) containing poppy seed paste (5–20%) as animal fat replacer had reduced cholesterol content and saturated fatty acid contents (Gök et al. 2011). Samples having 20% ground poppy seed had significantly better texture and juiciness score than any other sample which could be explained by increased moisture retention. Products containing 20% ground poppy seed had significantly better texture and juiciness score than any other sample which could be explained by increased moisture retention.

Structured lipids (SLs) can also be used for fat replacement. These are triacylglycerols restructured or modified to change the fatty acid composition and/or their positional distribution in glycerol molecules by chemical or enzymatic processes (Zam 2020). Other fat replacers are based on the fact that level of 9 kcal/g does not apply when short chain fatty acids are present and also long chain stearic acid are incompletely metabolized and yield calorie value less than 9 kcal/g. By combining these two types of fatty acids into glycerides fats are obtained that have energy values of 5 kcal/g. Structured lipids are tailor-made lipids with desired physical characteristics, chemical properties and/or nutritional benefits (Raibagkar and Patange 2023).

8.5 Safety of Fat Replacers

The use of fat replacers to reduce the fat content in processed food products also raises the concern of consumers safety. An ideal fat replacer would be a substance that has no negative health implications or health risks and would provide fewer calories than fat. The safety of the currently used carbohydrate, protein and fat based fat replacers is ensured by the GRAS status by the FDA. The current research shows that carbohydrate, protein and fat based fat replacers don't have health implications except for polydextrose and olestra. In the EU, polydextrose (E 1200) has been evaluated by the Scientific Committee for Food (SCF) in 1990 and reported that large doses of polydextrose exerted a laxative effect with a mean laxative threshold of 90 g/day or 50 g as a single dose. Olestra, a non-caloric fat replacer, may cause leaky and fatty stools and interferes with the absorption of fat-soluble substances, including the fat-soluble vitamins (A, E, D, and K) and carotenoids (Raibagkar and Patange 2023). As a part of healthy dietary pattern fat replacers when used judiciously, may be safe for use in food products. Although long term clinical trials are needed to study the safe usage levels of different fat replacers in food products especially for children and immune compromised individuals to fully determine the longer-term health effects and the potential interaction with food ingredients. In addition no single fat replacer can provide all quality attributes and physical functions of fat. Further, there is a need to develop novel fat replacers, which can be used in a number of food products with no adverse effect on sensory quality and health (Chavan et al. 2016; Gopika and Yatnatti 2021).

8.6 Conclusion

Despite nutritional benefits of meat products, the consumption of high fat processed meat products has been associated with the occurrence of certain degenerative diseases due to considerable level of saturated fatty acids and cholesterol. Therefore, the link between human health and high fat processed meat products has prompted the meat industry to develop low fat and low calorie meat products with certain nutritional claims. Various physical modification strategies can be applied by use of carbohydrate or protein or lipid based fat replacers in meat products to reduce total or partial fat and improve the fatty acid profile and quality of processed meat products, in addition to serving as a nutritional alternative to prevent chronic diseases. However, more efforts are needed to find novel functional fat replacers, to achieve a balance between the fat replacement ratio and meat product quality, scaling of production to reduce cost for manufacturers and promoting the development of reformulated low fat meat products at commercial scale.

References

Abeyrathne EDNS, Lee HY, Ahn DU (2013) Egg white proteins and their potential use in food processing or as nutraceutical and pharmaceutical agents – a review. Poult Sci 92:3292–3299

Abiola SS, Adegbaju SW (2001) Effect of substituting pork back fat with rind on quality characteristics of pork sausage. Meat Sci 58(4):409–412

ADA (2012) American Diabetes Association. Standards of medical care in diabetes. Diabetes Care 35(Supp 1):S11–S63

Adili L, Roufegarinejad L, Tabibiazar M, Hamishehkar H, Alizadeh A (2020) Development and characterization of reinforced ethyl cellulose based oleogel with adipic acid: its application in cake and beef burger. LWT – Food Sci Technol 126(March):article 109277. https://doi.org/10.1016/j.lwt.2020.109277

Ahmad SS, Khalid M, Younis K (2020) Interaction study of dietary fibers (pectin and cellulose) with meat proteins using bioinformatics analysis: an *in-silico* study. LWT – Food Sci Technol 119:108889

Akalin AS, Karagozlü C, Ünal G (2008) Rheological properties of reduced-fat and low-fat ice cream containing whey protein isolate and inulin. Eur Food Res Technol 227:889–895

Aken VGA, Vingerhoeds MH, de Wijk RA (2011) Textural perception of liquid emulsions: role of oil content, oil viscosity and emulsion viscosity. Food Hydrocoll 25:789–796

Akoh CC (1998) Fat replacers. Food Tech 52:47–53

Alejandre M, Astiasarán I, Ansorena D, Barbut S (2019) Using canola oil hydro-gels and organogels to reduce saturated animal fat in meat batters. Food Res Int 122:129–136

Aleson L, Fernández J, Pérez J, Kuri V (2005) Functional and sensory effects of fibre-rich ingredients on breakfast fresh sausages manufacture. Food Sci Technol Int 11:89–97

Al-Ghazzewi FH, Khanna S, Tester RF, Piggott J (2007) The potential use of hydrolysed konjac glucomannan as a prebiotic. J Sci Food Agric 87(9):1758–1766

Ali MS, Kim GD, Seo HW, Jung EY, Kim BW, Yang HS, Joo ST (2011) Possibility of making low-fat sausages from duck meat with addition of Rice flour. Asian Australas J Anim Sci 24: 421–428

Anderson ET, Berry BW (2000) Sensory, shear, and cooking properties of lowerfat beef patties made with inner pea fiber. J Food Sci 65(5):805–810

Anderson ET, Berry BW (2001) Effects of inner pea fiber on fat retention and cooking yield in high fat ground beef. Food Res Int 34(8):689–694
Anggraini J, Lo D (2023) Health impact of carrageenan and its application in food industry: a review. Earth Environ Sci 1169:012098
Angor MM, Al-Abdullah BM (2010) Attributes of low-fat beef burgers made from formulations aired at enhancing product quality. J Muscle Foods 21(2):317–326
Askin OO, Kilic B (2009) Effect of microbial transglutaminase, sodium caseinate and non-fat dry milk on quality of salt-free, low fat Turkey doner kebab. LWT – Food Sci Technol 42:1590–1596
Assoi A, Konan K, Agbo GN, Dodo H, Holser R, Wicker L (2017) Palmyra palm (*Borassus aethiopum Mart.*) fruits: novel raw materials for the pectin industry. J Sci Food Agric 97(7): 2057–2067
Attallah OA, Mamdouh W (2021) Development and optimization of pectin/chitosan magnetic sponge for efficient cationic dyes removal using Box–Behnken design. Int J Environ Sci Technol 18:131–140
Ayadi MA, Kechaou A, Makni I, Attia H (2009) Influence of carrageenan addition on Turkey meat sausages properties. J Food Eng 93:278–283
Bajaj IB, Survase SA, Saudagar PS, Singhal RS (2007) Gellan gum: fermentative production, downstream processing and applications. Food Technol Biotechnol 45(4):341–354
Bang SJ, Shin IS, Chung DH, Kim SM (2006) Quality characteristics of surimibased product with sea tangle cell detritus. Korean J Food Sci Technol 38(3):337–341
Barbut S, Mittal GS (1992) Use of carrageenans and xanthan gum in reduced fat breakfast sausages. LWT – Food Sci Technol 25(6):509–513
Bater B, Descamps O, Maurer AJ (1992) Quality characteristics of hydrocolloid added oven-roasted Turkey breasts. J Food Sci 57(5):1068–1070
Batsis JA, Nieto-Martinez RE, Lopez-Jimenez F (2007) Metabolic syndrome: from global epidemiology to individualized medicine. Clin Pharmacol Ther 82:509–524
Bauer LC, de Andrade Santana D, Macedo M, Torres AG, de Souzad NE, Simionato JI (2014) Method validation for simultaneous determination of cholesterol and cholesterol oxides in milk by RP-HPLC-DAD. J Braz Chem Soc 25(1):161–168
Baugreet S, Kerry JP, Botine Stean C, Allen P, Hamill RM (2016) Development of novel fortified beef patties with added functional protein ingredients for the elderly. Meat Sci 122:40–47
Bejesano FP, Corke H (1998) Amarantus and buck wheat protein concentrate effects on emulsion-type meat products. Meat Sci 50(3):343–353
Belury MA (2002) Dietary conjugated linoleic acid in health: physiological effects and mechanisms of action. Annu Rev Nutr 22:505–531
Berizi E, Shekarforoush SS, Mohammadinezhad S, Hosseinzadeh S, Farahnaki A (2017) The use of inulin as fat replacer and its effect on texture and sensory properties of emulsion type sausages. Iran J Vet Res 18:253–257
Bernal VM, Smajda CH, Smith JL, Stanley DW (1987) Interactions in protein/polysaccharide/calcium gels. J Food Sci 52:1121–1125
Berry BW, Bigner ME (1996) Use of carrageenan and konjac flour gel in low-fat restructured pork nuggets. Food Res Int. https://doi.org/10.1016/0963-9969(96)00019-1
Biesalski HK (2005) Meat as a component of a healthy diet – are there any risks or benefits if meat is avoided in the diet. Meat Sci 70:509–524
Bjorck I, Asp NG (1994) Controlling the nutritional properties of starch in foods – a challenge to the food industry. Trends Food Sci Technol 5:213–218
Bloukas JG, Paneras ED (1996) Quality characteristics of low-fat frankfurters manufactured with potato starch, finely ground toasted bread and rice bran. J Muscle Foods 7:109–129
Bloukas JG, Paneras ED, Papadima S (1997) Effect of carrageenan on processing and quality characteristics of low-fat frankfurters. J Muscle Foods 8(1):63–83
Boeckner LS, Schnepf MI, Tungland BC (2001) Inulin: a review of nutritional and health implications. Adv Food Nutr Res 43:1–63

Bounous G, Batist G, Gold P (1991) Whey proteins in cancer prevention. Cancer Lett 57:91–94
Bounous G, Baruchel S, Falutz J, Gold P (1993) Whey proteins as a food supplement in HIV-seropositive individuals. Clin Invest Med 16:204–209
Bozan B, Temelli F (2003) Extraction of poppy seed oil using supercritical CO_2. J Food Sci 68: 422–426
Braaten T, Wood P, Scott F, Wolynetz M, Lowe M, Bradley P (1994) Oat b-glucan reduces blood cholesterol concentration in hypercholesterolemic subjects. Eur J Clin Nutr 48:465–474
Brewer MS (2012) Reducing the fat content in ground beef without sacrificing quality: a review. Meat Sci 91:385–395
Brown L, Rosner B, Willett WW, Sacks FM (1999) Cholesterol lowering effects of dietary fiber: a meta-analysis. Am J Clin Nutr 69:30–42
Buckley J, Abbott M, Martin S, Brinkworth G, Whyte P (1998) Effects of an oral bovine colostrum supplement (intacts) on running performance. Australian Conference of Science and Medicine in Sport, Adelaide
Bureau of Nutritional Sciences Food Directorate, Health Products & Food Branch, Health Canada: Policy for labelling and advertising of dietary fibre-containing food. (May 2022). Retrieved April 12 2022 from https://www.canada.ca/en/health-canada/services/publications/food-nutrition/labellingadvertising-dietary-fibre-foodproducts.html#:~:text=In%202012%2C%20the%20Food%20Directorate%20released %20the%20Pol
Butt MS, Shahzadi N, Sharif MK, Nasir M (2007) Guar gum: a miracle therapy for hypercholesterolemia, hyperglycemia and obesity. Crit Rev Food Sci Nutr 47:389–396
Cáceres E, García ML, Selgas MD (2008) Effect of pre-emulsifiedfish oil – as source of PUFA n-3 – on microstructure and sensory properties ofmortadella, a Spanish bologna-type sausage. Meat Sci 80:183–193
Campagnol PCB, dos Santos BA, Wagner R, Terra NN, Pollonio MAR (2012) Amorphous cellulose gel as a fat substitute in fermented sausages. Meat Sci 90:36–42
Campo-Deaño L, Tovar C (2009) The effect of egg albumen on the viscoelasticity of crab sticks made from Alaska Pollock and Pacific whiting surimi. Food Hydrocoll 23:1641–1646
Candogan K, Kolsarici N (2003a) Storage stability of low-fat beef frankfurters formulated with carrageenan or carrageenan with pectin. Meat Sci 64:207–214
Candogan K, Kolsarici N (2003b) The effects of carrageenan and pectin on some quality characteristics of low-fat beef frankfurters. Meat Sci 64(2):199–206
Carballo J, Barreto G, Jimenez-Colmenero F (1995) Starch and egg white influence on properties of bologna sausage as related to fat content. J Food Sci 60:673–677
Carey K, Larsen A, Rowney M, Cameron-Smith D (2006) Application of whey proteins to enhance the molecular adaptations and strength gains following resistance exercise training. Proceedings of the fourth international whey. Dairy J 18(2008):695–704
CassanelliIan M, Mills NT (2017) Effect of alcohols on gellan gum gel structure: bridging the molecular level and the three-dimensional network. Food Struct 14:112–120
Cassanelli M, Norton I, Mills T (2018) Role of gellan gum microstructure in freeze drying and rehydration mechanisms. Food Hydrocoll 75:51–61
Cernikova M, Bunka F, Pavlinek V, Brezina P, Hrabe J, Valašek P (2008) Effect of carrageenan type on viscoelastic properties of processed cheese. Food Hydrocoll 22(6):1054–1061
Chatterton DEW (2006) Digestion of novel milk protein ingredients for use in infant formulas: relationship to bioactivity. In: Proceedings of the fourth international whey conference, Chicago, USA. American Dairy Products Institute, Elmhurst, pp 170–184
Chavan RS, Khedkar CD, Bhatt S (2016) Fat replacer. In: Caballero B, Finglas P, Toldrá F (eds) The encyclopedia of food and health, vol 2. Academic, Oxford, pp 589–595
Chen Y, She Y, Zhang R, Wang J, Zhang X, Gou X (2020) Use of starch-based fat replacers in foods as a strategy to reduce dietary intake of fat and risk of metabolic diseases. Food Sci Nutr 8: 16–22

Chevance FFV, Farmer LJ, Desmond EM, Novelli E, Troy DJ, Chizzolini R (2000) Effect of some fat replacers on the release of volatile aroma compounds from low-fat meat products. J Agric Food Chem 48(8):3476–3484
Chin KB, Keeton JT, Longnecker MT, Lamkey JW (1998) Low-fat bologna in a model system with varying types and levels of konjac blends. J Food Sci 63(5):808–813
Chizzolini R, Zanardi E, Dorigoni V, Ghidini S (1999) Calorific value and cholesterol content of normal and low-fat meat and meat products. Trends Food Sci Technol 10(4–5):119–128
Cho SS, Prosky L (1999) Application of complex carbohydrates to food product fat mimetics. In: Cho SS, Prosky L, Dreher M (eds) Complex carbohydrates in foods. Marcel Dekker Inc, New York, pp 422–429
Choe JH, Kim HY, Lee JM, Kim YJ, Kim CJ (2013) Quality of frankfurter-type sausages with added pig skin and wheat fiber mixture as fat replacers. Meat Sci 93:849–854
Choi YS, Choi JH, Han DJ, Kim HY, Lee MA, Kim HW, Jeong JY, Kim CJ (2009) Characteristics of low-fat meat emulsion systems with pork fat replaced by vegetable oils and rice bran fiber. Meat Sci 82(2):266–271
Choi YS, Choi JH, Han DJ, Kim HY, Kim HW, Lee MA, Chung HJ, Kim CJ (2012) Effects of *Laminaria japonica* on the physico-chemical and sensory characteristics of reduced-fat pork patties. Meat Sci 91:1–7
Choi YS, Kim YB, Hwang KE, Song DH, Ham YK, Kim HW, Sung JM, Kim CJ (2016) Effect of apple pomace fiber and pork fat levels on quality characteristics of uncured, reduced-fat chicken sausages. Poult Sci 95(6):1465–1471
Chung HJ, Min DB (2004) Developments in fat replacers. CRC Press, Inc., Boca Raton
Cierach M, Modzelewska-Kapituła M, Szaciło K (2009) The influence of carrageenan on the properties of low-fat frankfurters. Meat Sci 82:295–299
Claus JR, Hunt MC (1991) Low-fat, high-added bologna formulated with texture-modifying ingredients. J Food Sci 56:643–647
Cofrades S, Hughes E, Troy DJ (2000a) Effects of oat fiber and carrageenan of the texture of frankfurters formulated with low and high fat. Eur Food Res Technol 211(1):19–26
Cofrades S, Guerra MA, Carballo J, Fernandez-Martin F, Jimenez Colmenero F (2000b) Plasma protein and soy fiber content effect on bologna sausage properties as influenced by fat level. J Food Sci 65(2):281–287
Colla K, Costanzo A, Gamlath S (2018) Fat replacers in baked food products. Food Secur 7(12):192
Colomer Sellas M, de Souza DL, Vila-Marti A, Torres-Moreno M (2021) Effect of pork back-fat reduction and substitution with texturized pea protein on acceptability and sensory characteristics of dry fermented sausages. CyTA – J Food 19:429–439
Comer FW, Chew N, Lovelock L, Allan-Wojtas P (1986) Comminuted meat products: functional and microstructural effects of fillers and meat ingredients. Can Inst Food Technol J 19:68–74
Coombes JS, Conacher M, Austen SK, Marshall PA (2000) Dose effects of oral bovine colostrum supplement (intacts) on physical work capacity in cyclists. In: Proceedings of sports medicine and physical education international congress on sport science pre-olympic congress, Brisbane
Corpet DE (2011) Red meat and colon cancer: should we become vegetarians, or can we make meat safer? Meat Sci 89:310–316
Da Silva JAL, Gonçalves MP (1990) Studies on a purification method for locust bean gum by precipitation with isopropanol. Food Hydrocoll 4:277–287
da Silva SL, Amaral JT, Ribeiro M, Sebastião EE, Vargas C, de Lima FF, Campagnol PCB (2019) Fat replacement by oleogel rich in oleic acid and its impact on the technological, nutritional, oxidative, and sensory properties of Bologna-type sausages. Meat Sci 149:141–148
Daas Piet JH, Schols Henk A, De Jongh Harmen HJ (2000) On the galactosyl distribution of commercial galactomannans. Carbohydr Res 329(3):609–619
Dakia PA, Blecker C, Roberta C, Watheleta B, Paquota M (2008) Composition and physicochemical properties of locust bean gum extracted from whole seeds by acid or water dehulling pre-treatment. Food Hydrocoll 22:807–818

Das AK, Anjaneyulu ASR, Kondaiah N (2006) Development ofreduced beany flavour full-fat soy paste for comminuted meat products. J Food Sci 71:S395–S400
De Freitas Z, Sebranek J, Olson D, Carr J (1997) Carrageenan effects on thermal stability of meat proteins. J Food Sci 62(3):544–547
de Souza PC, de Figueiredo FG, Honório AR, Mokarzel L, da Silva Vidal VA, Ribeiro APB, Pollonio MAR (2019) Functional emulsion gels as pork back fat replacers in Bologna sausage. Food Struct 20:100105
de Vries J (2004) Hydrocolloid gelling agents and their applications. In: Philips GO, Williams PA (eds) Gums and stabilizers for the food industry, vol 12. Oxford: RSC Publ; pp. 22–30
Dea ICM, Morrison A (1975) Chemistry and interactions of seed galactomannans. Adv Carbohydr Chem Biochem 31:241–242
Decker CD, Conley CC, Richert SH (1986) Use of isolated soy protein in the development of frankfurters with reduced level of fat, calories, and cholesterol. Proc Eur Meeting Meat Res Workers 7(1):333–349
DeFreitas Z, Sebranek JG, Olson DG, Carr JM (1997) Carrageenan effects on salt soluble meat proteins in model systems. J Food Sci 62:539–543
Delcour JA, Joye IJ, Pareyt B, Wilderjans E, Brijs K, Lagrain B (2012) Wheat gluten functionality as a quality determinant in cereal-based food products. Annu Rev Food Sci Technol 1:469–492
Delgado-Pando G, Cofrades S, Ruiz-Capillas C, Solas MT, Jiménez-Colmenero F (2010) Healthier lipid combination oil-in-water emulsions prepared with various protein systems. An approach for the development of functional meat products. Eur J Lipid Sci Technol 112:791–801
Delgado-Pando G, Cofrades S, Ruiz-Capillas C, Solas MT, Trikia M, Jiménez-Colmenero F (2011) Low-fat frankfurters formulated with a healthier lipid combination as functional ingredient: microstructure, lipid oxidation, nitrite content, microbiological changes and biogenic amine formation. Meat Sci 89:65–71
Desmedt A, Jacobs H (2001) Soluble fibre. In: Young J (ed) Guide to functional food ingredients, Illustrated edn. Leatherhead Publishers, Surrey, pp 112–140
Desmond EM, Troy DJ (2002) The effect of dietary fibres on the properties of pork sausages. In: Proceedings of the 48th international congress on meat science technology, vol 2. ICoMST, Rome, pp 840–841
Desmond E, Troy D, Buckley D (1998a) The effects of tapioca starch, oat fibre and whey protein on the physical and sensory properties of low-fat beef burgers. Lebensm Wiss Technol 31:653–657
Desmond EM, Troy DJ, Buckley DJ (1998b) Comparative studies of nonmeat adjuncts used in the manufacture of low-fat beef burgers. J Muscle Foods 9(3):221–241
Dickinson E (1997) Enzymic crosslinking as a tool for food colloid rheology control and interfacial stabilization. Trends Food Sci Technol 8:334–339
Dickinson E (2018) On the road to understanding and control of creaminess perception in food colloids. Food Hydrocoll 77:372–385
Dietary Reference Intakes for Energy (2002) Carbohydrate, fiber, fat, fatty acids, cholesterol, protein and amino acids. National Academies Press, Washington, DC
Dignam MM, Tseng RY, Smith-Nury E (1979) Lipid content and oxidation in soy protein and ground beef mixture. Home Econ Res J 7(6):356–360
Dobarganes C, Marquez-Ruiz G (2003) Oxidize fat in foods. Curr Opin Clin Nutr Metab Care 6: 157–162
Dongowski G, Drzikova B, Senge B, Blochwitz R, Gebhardt E, Habel A (2003) Rheological behaviour of b-glucan preparations from oat products. Food Chem 93:279–291
Duda Z, Pietrasik Z, Cieslik D, Tubaj M (1995) Gellan gum and carrageenan used as recipe component of comminuted scalded sausages. In: 41st international congress on meat science and technology proceedings, vol II, San Antonio, pp 431–432
Eghbaljoo H, Sani IK, Sani MA, Rahati S, Mansouri E, Molaee-Aghaee E, Fatourchchi N, Kadi A, Arab A, Sarabandi K, Samborska K, Jafari SM (2022) Advances in plant gum polysaccharides; sources, techno-functional properties, and applications in the food industry – a review. Int J Biol Macromol 1:2327–2340

Einhorn-Stoll U, Hatakeyama H, Hatakeyama T (2012) Influence of pectin modification on water binding properties. Food Hydrocoll 27:494–502

Ellekjaer MR, Naes T, Baardseth P (1996) Milk proteins a€ect yield and sensory quality of cooked sausages. J Food Sci 61:660–666

El-magoli SB, Larola S, Hansen PMT (1995) Ultrastructure of low-fat ground beef patties with added whey protein concentrate. Food Hydrocoll 9(4):291–306

FAO (2010) Fats and fatty acids in human nutrition. FAO report of an expert consultation. In: FAO Food and Nutrition paper, vol 91. Food and Agriculture Organization of the United Nations, Rome

Farnfield MM, Carey KA, Cameron-Smith D (2005) Whey protein supplementation and resistance training to enhance muscle growth in young and older adults. Asia Pac J Clin Nutr 14(Suppl): S69

Feng T, Gu ZB, Jin ZY, Zhuang HN (2010) Rheological properties of rice starch–Mesona Blumesgum mixtures. J Texture Stud 41:685–702

Feng T, Ye R, Zhuang H, Rong Z, Fang Z, Wang Y, Jin ZGZ (2013) Physicochemical properties and sensory evaluation ofMesona Blumesgum/rice starch mixed gels as fat-substitutes in Chinese Cantonese-style sausage. Food Res Int 50:85–93

Fernández-Ginés JM, Fernández-López J, Sayas-Barberá E, Pérez-Álvarez JA (2005) Meat products as functional foods: a review. J Food Sci 70:37–43

Ferro AC, de Souza Paglarini C, Rodrigues Pollonio MA, Lopes Cunha R (2021) Glyceryl monostearate-based oleogels as a new fat substitute in meat emulsion. Meat Sci 174:Article 108424

Flaczyk E, Górecka D, Kobus J, Szymander-Buszka K (2009) The influence of inulin addition as fat substitute on reducing energy value and consumer acceptance of model pork meatballs. Żywność Nauka Technologia Jakość 16(4):41–46

Flammang AM, Kendall DM, Baumgartner CJ, Slagle TD, Choe YS (2006) Effect of a viscous fibre bar on postprandial glycemia in subjects with type 2 diabetes. J Am Coll Nutr 25:409–414

Flores M, Giner E, Fiszman SM, Salvador A, Flores J (2007) Effect of a new emulsifier containing sodium stearoyl-2-lactylate and carrageenan on the functionality of meat emulsion systems. Meat Sci 1:9–18

Florowski T, Adamczak L (2010) Influence of fat replacement with inulin on selected quality traits of model canned meat products. Zesz Probl Postep Nauk Rol 553:91–98

Florowski T, Adamczak L, Fuertez-Hernández I, Moreno-Franco MB, Tyburcy A (2008) Effect of fat replacement by inulin on quality of baked poultry liver pâté. Roczniki Instytutu Przemysłu Mięsnego i Tłuszczowego 46(2):119–126

Food and Drug Administration (FDA) (1997) Food labeling: health claims; soluble fibre from whole oats and risk of coronary heart disease. Fed Regist 62:15343–15344

Franck AM (2000) Inulin and oligofructose. In: Gibson G, Angus F (eds) LFRA ingredient handbook: prebiotics and probiotics. Leatherhead Publishing, Surrey, pp 1–18

Franck A (2002) Technological functionality of inulin and oligofructose. Br J Nutr 87:S287–S291

Frida. (2020) Food data. Food. Available online: https://frida.fooddata.dk/

Gainsford SE, Harding SE, Mitchell JR, Bradley TD (1986) A comparison between the hot and cold water-soluble fractions of two locust bean gum samples. Carbohydr Polym 6:423–442

Ganhão R, Morcuende D, Estévez M (2010) Protein oxidation in emulsified cooked burger patties with added fruit extracts: influence on colour and texture deterioration during chill storage. Meat Sci 85:402–409

García ML, Cáceres E, Selgas MD (2006) Effect of inulin on the textural and sensory properties of mortadella, a Spanish cooked meat product. Int J Food Sci Technol 41:1207–1215

García-García E, Totosaus A (2008) Low-fat sodium-reduced sausages: effect of the interaction between locust bean gum, potato starch and κ-carrageenan by a mixture design approach. Meat Sci 78(4):406–413

García-Iníguez de Ciriano M, Larequi E, Rehecho S, Calvo MI, Cavero RY, Navarro-Blasco Í, Astiasarán I, Ansorena D (2010) Selenium, iodine, omega-3 PUFA and natural antioxidant from

Melissa officinalis L.: a combination of components from healthier dry fermented sausages formulation. Meat Sci 85:274–279

Gidding SS, Dennison BA, Birch LL, Daniels SR, Gillman MW, Lichtenstein AH (2005) Dietary recommendations for children and adolescents: a guide for practitioners: consensus statement from the American Heart Association. Circulation 112(13):2061–2075

Giese J (1994) Proteins as ingredients: types, functions, applications. Food Technol 48(10):49–60

Gill HS, Cross ML (2000) Anticancer properties of bovine milk. Br J Nutr 84:S161–S166

Giromini C, Givens DI (2022) Benefits and risks associated with meat consumption during key life processes and in relation to the risk of chronic diseases. Food Secur 11:2063

Glicksman M (1982) Food hydrocolloids, vol 1. CRC Press, Boca Raton

Gnanasambandam R, Zayas JF (1992) Functionality of wheat germ protein in comminuted meat products as compared with corn germ and soy proteins. J Food Sci 57:829–833

Gök V, Akkaya L, Obuz E, Bulut S (2011) Effect of ground poppy seed as a fat replacer on meat burgers. Meat Sci 89(4):400–404

Goldstein AM, Alter EN, Seaman JK (1973) Guar gum. In: Whistler RL, BeMiller JN (eds) Industrial gums. Academic, London, pp 303–321

Gomes SP, Nyengaard JR, Misawa R, Girotti PA, Castelucci P, Blazquez FHJ et al (2009) Atrophy and neuron loss: effects of a protein-deficient diet on sympathetic neurons. J Neurosci Res 87(16):3568–3575

Gómez-estaca J, Pintado T, Jiménez-colmenero F, Cofrades S (2020) The effect of household storage and cooking practices on quality attributes of pork burgers formulated with PUFA- and curcumin-loaded oleogels as healthy fat substi-tutes. LWT – Food Sci Technol 119:108909. https://doi.org/10.1016/j.lwt.2019.108909

González Canga A, Fernández Martínez N, Sahagún AM, García Vieitez JJ, Díez Liébana MJ, Calle Pardo AP et al (2004) Glucomanano: propiedades y aplicaciones terapéuticas. Nutr Hosp 19(1): 45–50

Gopika CM, Yatnatti S (2021) Fat replacers and their applications in food products. J Food Process Technol 3(10):110

Gotterbos JJ (1988) Nutritional evaluation of n-6 and n-3 polyunsaturated fatty acids. In: Rogers B (ed) Dietary fat requirements in health and development. AOLS Press, Chanpaing, p 107

Grabitske HA, Slavin JL (2009) Gastrointestinal effects of low digestible carbohydrates. Crit Rev Food Sci Nutr 49:327–360

Grasdalen H, Smidsrød O (1987) Gelation of gellan gum. Carbohydr Polym 7(5):371–393

Grasso S, Brunton NP, Lyng JG, Lalor F, Monahan FJ (2014) Healthy processed meat products – regulatory, reformulation and consumer challenges. Trends Food Sci Technol 39:4–17

Grigelmo-Miguel N, Abadía-Serós MI, Martín-Belloso OA (1999) Characterisation of low-fat high-dietary fiber frankfurters. Meat Sci 52:247–256

Grzelak T, Grupinska J, Pelczynska M, Sperling M, Czyzewska K (2016) The merits of fat replacers in low-calorie food. Acta Scientiarum Polonorum Hortorum Cultus 16(2):87–94

Gujral HS, Kaur A, Singh N, Sodhi SN (2002) Effect of liquid whole egg, fat and textured soy protein on the textural and cooking properties of raw and baked patties from goat meat. J Food Eng 53:377–385

Ha E, Zemel MB (2003) Functional properties of whey, whey components, and essential amino acids: mechanisms underlying health benefits for active people. J Nutr Biochem 14:251–258

Haffner SM (2006) The metabolic syndrome: inflammation, diabetes mellitus, and cardiovascular disease. Am J Cardiol 97(2A):3A–11A

Hamm R (1986) Functional properties of the myofibrillar system and their measurements. In: Bechtel PJ (ed) Muscle as food. Academic, Orlando, pp 135–199

Han M, Bertram HC (2017) Designing healthier com-minuted meat products: effect of dietary fibers on water dis-tribution and texture of a fat-reduced meat model system. Meat Sci 133:159–165

Hecker K, Marg M, Newman R, Newman W (1998) Barley b-glucan is effective as a hypocholesterolaemic ingredient in foods. J Sci Food Agric 77:179–183

Henry RR, Chilton R, Garvey WT (2013) New options for the treatment of obesity and type 2 diabetes mellitus (narrative review). J Diabetes Complicat 27:508–518

Herrero AM, Carmona P, Pintado T, Jiménez-Colmenero F, Ruiz-Capillas C (2012) Lipid and protein structure analysis of frankfurters formulated with olive oil-in-water emulsion as animal fat replacer. Food Chem 135:133–139

Hęś M, Waszkowiak K, Szymandera-Buszka K (2012) The effect of iodine salts on lipid oxidation and changes in nutritive value of protein in stored processed meats. Meat Sci 92:139–143. https://www.canada.ca/en/health-canada/services/nutrients/fibre.html

Hoogenkamp HW (1987) Low fat sausage: making it work. Meat Process 26:28–34

Hsu SY, Chung HY (2000) Interactions of konjac, agar, curdlan gum, *k*-carrageenan and reheating treatment in emulsified meatballs. J Food Eng 44:199–204

Huang CC, Su HP, Lin CW (1998) Utilization of konjac in low-fat Chinese-style sausage. Food Sci (Taiwan) 25(4):437–445

Huang Y, Singh PP, Tang J, Swanson BG (2004) Gelling temperatures of high acyl gellan as affected by monovalent and divalent cations with rheological analysis. Carbohydr Polym 56:27–33

Huang Y, Cao D, Chen Z, Chen B, Li J, Guo J, Dong Q, Liu L, Wei Q (2021) Red and processed meat consumption and cancer outcomes: umbrella review. Food Chem 356:129697

Huffman D, Egbert W, Chen C, Dylewski D (1991) Technology for low-fat ground beef. In: Proceedings of 44th annual reciprocal meat conference. June 9–12 1991. Kansas City University, Publication National Live Stock and Meat Board, Manhattan/Chicago, p 73

Hughes E, Cofrades S, Troy DJ (1997) Effects of fat level, oat fiber and carrageenan on frankfurters formulated with 5, 12 and 30% fat. Meat Sci 45(3):273–281

Hughes E, Mullen AM, Troy DJ (1998) Effects of fat level, tapioca starch and whey protein on frankfurters formulated with 5% and 12% fat. Meat Sci 48(1–2):169–180

Hung SC, Zayas JF (1992) Functionality of milk proteins and corn germ protein flour in comminuted meat products. J Food Qual 15:139–152

Ichiro J, Naoyuki I, Norishige O, Yasuo E, Toru N (1987) Manufacture of meat pastes containing starch esters. Japan Kokai Tokkyo Koho, JP 62:248–470

ICMR-NIN, RDA and EAR – A Short Report, ICMR-NIN, Secunderabad, India. (2020). https://www.nin.res.in/nutrition2020/RDA_short_report.pdf

Imaijumi K, Adan Y, Shibata K (2000) Role of dietary lipids in arteriosclerosis in experimental animals. Biofactors 13(1–4):25–28

Imeson AP (2000) Carrageenan. In: Phillips GO, Williams PA (eds) Handbook of hydrocolloids. Woodhead Publishing Ltd., Cambridge, pp 87–102

Jambrak AR, Lelas V, Mason TJ, Krešic G, Badanjak M (2009) Physical properties of ultrasound treated soy proteins. J Food Eng 93(4):386–393

Jansson PE, Kenne L, Lindberg B (1975) Structure of the exocellular polysaccharide from *Xanthomonas campestris*. Carbohydr Res 45:275–282

Jervis SM, Gerard P, Drake S, Lopetcharat K, Drake MA (2014) The perception of creaminess in sour cream. J Sens Stud 29(4):248–257

Jimenez-Colmenero F (1996) Technologies for developing low-fat meat products. Trends Food Sci Technol 7:41–48

Jimenez-Colmenero F (2000) Relevant factors in strategies for fat reduction in meat products. Trends Food Sci Technol 11:56–66

Jiménez-Colmenero F (2007) Healthier lipid formulation approaches in meat-based functional foods. Technological options for replacement of meat fats by non-meat fats. Trends Food Sci Technol 18:567–578

Jiménez-Colmenero F, Cofrades S, Herrero AM, Fernández-Martín L, Rodríguez-Salas F, Ruiz-Capillas C (2012) Konjac gel fat analogue for use in meat products: comparison with pork fats. Food Hydrocoll 26(1):63–72

Jiménez-Colmenero F, Triki M, Herrero AM, Rodríguez-Salas L, Ruiz-Capillas C (2013) Healthy oil combination stabilized in a konjac matrix as pork fat replacement in low-fat, PUFA-enriched, dry fermented sausages. LWT – Food Sci Technol 51(1):158–163

Jimenez-Colmenero F, Salcedo-Sandoval L, Bou R, Cofrades S, Herrero AM, Ruiz-Capillas C (2015) Novel applications of oil-structuring methods as a strategy to improve the fat content of meat products. Trends Food Sci Technol 44(2):177–188

Kamil MM, Mohamed GF, Ali HS, Ragb GH, Zaky AA (2024) Evaluation of inulin as a fat replacer in meat burger. Egypt J Chem 67(3):13–23

Kawamura Y (2008) Guar gum chemical and technical assessment. Prepared for the 69th joint FAO. WHO Expert Committee on Food Additives

Keeton JT (1994) Low-fat meat products-technological problems with processing. Meat Sci 36(1–2):261–276

Keeton JT (1997) Non-meat ingredients for low/non-fat processed meats. Recip Conf Proc 49:23–31

Kennedy A, Martinez K, Chuang C, Lapoint K, Mcintosh M (2009) Saturated fatty acid mediated inflammation and insulin resistance in adipose tissue: mechanisms of action and implications. J Nutr 1:1–4

Khalil AH (2000) Quality characteristics of low-fat beef patties formulated with modified corn starch and water. Food Chem 68:61–68

Kilic B, Kankaya T, Ekici YK, Orhan H (2010) Effect of textured soy protein on quality characteristics of low fat cooked kofte (Turkish meatball). J Anim Vet Adv 9(2):3048–3054

Kim JH, Kim JH, Yoo SS (2008) Impacts of the proportion of sea-tangle on quality characteristics of muffin. Korean J Food Cook Sci 24(5):565–572

Kinsella JE, Melachouris N (1976) Functional properties of proteins in foods: a survey. Crit Rev Food Sci Nutr 7(3):219–280

Koocheki A, Kadkhodaee R, Mortazavi SA, Shahidi F, Taherian AR (2009) Influence of alyssum homolocarpum seed gum on the stability and flow properties of O/W emulsion prepared by high intensity ultrasound. Food Hydrocoll 23:2416–2424

Koutsopoulos DA, Koutsimanis GE, Bloukas JG (2008) Effect of carrageenan level and packaging during ripening on processing and quality characteristics of low-fat fermented sausages produced with olive oil. Meat Sci 79:188–197

Kulicke WM, Eidam D, Kath F, Kix M, Kull AH (1996) Hydrocolloids and rheology: regulation of visco-elastic characteristics of waxy rice starch in mixtures with galactomannans. Starch-Starke 48:105–114

Kumar Y (2019) Development of low-fat/reduced-fat processed meat products using fat replacers and analogues. Food Rev Intl 37:296–312

Kupirovic UP, Elmadfa I, Juillerat MA, Raspor P (2016) Effect of saliva on physical food properties in fat texture perception. Crit Rev Food Sci Nutr 57(6):1061–1077

Laaman TR (2011) Hydrocolloids: fifteen practical tips. In: Hydrocolloids in food processing. Blackwell Publishing Ltd. and Institute of Food Technologists, Ames, Iowa, USA

Lai VMF, Lii CY (2004) Role of saccharides in texturization and functional properties of foodstuffs. In: Tomasik P (ed) Chemical and functional properties of food saccharides. CRC Press, Boca Raton, pp 159–179

Lam ACY, Can Karaca A, Tyler RT, Nickerson MT (2018) Pea protein isolates: structure, extraction, and functionality. Food Rev Intl 34(2):126–147

Lanier TC (1991) Interactions of muscle and non-muscle protein affecting heat set gel rheology. In: Parris N, Barford R (eds) Interactions of food proteins. ASC Series, Washington, DC, p 454

Larkeson B, Dutta PC, Hansson I (2000) Effects of frying and storage on cholesterol oxidation in minced meat products. J Am Oil Chem Soc 77:675–680

Lazaridou A, Biliaderis G, Izydorczyk MS (2000) Structural characterization and rheological properties of locust bean galactomannans: a comparison of samples from different carob tree populations. J Sci Food Agric 81:68–75

Trout ES, Hunt MC, Johnson DE, Claus JR, Kastner CL, Kropf DH (1992) Chemical, physical and sensory characteristics ground beef containing 5–30% fat. J Food Sci 57:19–24

Troy D, Desmond E, Buckley D (1999) Eating quality of low-fat beef burgers containing fat-replacing functional blends. J Sci Food Agric 79:507–516

Tye RJ (1991) Konjac flour: properties and applications. Food Technol 45:82–92

Ulu H (2006) Effects of carrageenan and guar gum on the cooking and textural properties of low fat meatballs. Food Chem 95(4):600–605

Valsta LM, Tapanainen H, Mannisto S (2005) Meat fats in nutrition. Meat Sci 70:525–530

Vargas-ramella M, Pateiro M, Barba FJ, Franco D, Campagnol PCB, Munekata PES, Tomasevic I, Domínguez R, Lorenzo JM (2020) Microen-capsulation of healthier oils to enhance the physicochemical and nutritional proper-ties of deer pâté. LWT – Food Sci Technol 125:109223

Viana FR, Silva VDM, Delvivo FM, Bizzotto CS, Silvestre MPC (2005) Silvestre quality of ham pate containing bovine globin and plasma as fat replacers. Meat Sci 70:153–160

Volek Z, Marounek M (2011) Effect of feeding growing–fattening rabbits a diet supplemented with whole white lupin (Lupinus albuscv. Amiga) seeds on fatty acid composition and indexes related to human health in hind leg meat and perirenal fat. Meat Sci 87:40–45

Vural H, Javidipour I, Ozbas OO (2004) Effects of interesterified vegetable oils and sugar beet fiber on the quality of frankfurters. Meat Sci 67:65–72

Warner K, Inglett G (1997) Flavor and texture characteristics of foods containing Z-trim corn and oat fibers as fat and flour replace. Cereal Food World 42(10):821–825

Weiss J, Gibis M, Schuh V, Salminen H (2010) Advances in ingredient and processing systems for meat and meat products. Meat Sci 86:196–213

WHO (2003) Diet, nutrition and the prevention of chronic diseases. In: Report of a joint WHO/FAO expert consultation, WHO Technical Report Series, vol 916. Geneva, Switzerland

WHO (2009) World health organization. Preventing chronic diseases: a vital investment. W.H. Organization, Geneva

Wieser H (2007) Chemistry of gluten proteins. Food Microbiol 24(2):115–119

Williams PA, Phillips GO (2000) Introduction to food hydrocolloids. In: Phillips GO, Williams PA (eds) Handbook of hydrocolloids. CRC Press, New York, pp 1–19

Wills RBH, Rabirullah M (1981) Use of sunflower protein in sausages. J Food Sci 46:1657–1658

Wu J, Acero-Lopez A (2012) Ovotransferrin: structure, bioactivities and preparation. Food Res Int 46:480–487

Wu BC, McClements DJ (2015) Microgels formed by electrostatic complexation of gelatin and OSA starch: potential fat or starch mimetics. Food Hydrocoll 47:87–93

Xiong YL, Noel DC, Moody WG (1999) Textural and sensory properties of low-fat beef sausages with added water and polysaccharides as affected by pH and salt. J Food Sci 64(3):550–554

Xiong WF, Deng QC, Li J, Zhong QX (2019) Ovalbumin-carboxymethyl cellulose complex coacervates stabilized high internal phase emulsions: comparison of the effects of pH and polysaccharide charge density. Food Hydrocoll 98:105282

Xu X, Zhang H, Li L, Sun L, Jia B, Yang H, Feng Zuo F (2022) Preparation of fat substitute based on the high-methoxyl pectin of citrus and application in moon-cake skin. Food Sci Technol Campinas 42:92121

Yadav S, Malik A, Pathera A, Islam RU, Sharma D (2016) Development of dietary fibre enriched chicken sausages by incorporating corn bran, dried apple pomace and dried tomato pomace. Nutr Food Sci 46(1):16–29

Yan G, Yu L, Chen X, Liu Z, Chen H (2023) Optimization of the application of protein matrix fat substitution methods. Gels 9:724

Yang H, Choi S, Jeon J, Park G, Joo S (2007) Textural and sensory properties of low fat pork sausages with added hydrated oat meal and tofu as texture modifying agents. Meat Sci 75(2): 283–289

Yashini M, Sunil CK, Sahana S, Hemanth SD, Chidanand DV, Ashish R (2019) Protein-based fat replacers – a review of recent advances. Food Rev Int:1–27

Yetim H, Muller WD, Eber M (2001) Using fluid whey in comminuted meat products: effects of technological, chemical and sensory properties of frankfurter type sausages. Food Res Int 34: 97–101

Yildiz-Turp G, Serdaroglu M (2008) Effect of replacing beef fat with hazelnut oil on quality characteristics of sucuk: a Turkish fermented sausage. Meat Sci 78:447–454

Yilmaz I, Daglioglu O (2003) The effect of replacing fat with oat bran on fatty acid composition and physicochemical properties of meatballs. Meat Sci 65(2):819–823

Yoo SS, Kook SH, Park SY, Shim JH, Chin KB (2007) Physicochemical characteristics, textural properties and volatile compounds in comminuted sausages as affected by various fat levels and fat replacers. Inter J Food Sci Tech 42:1114–1122

Young VR, Pellett PL (1994) Plant proteins in relation to human protein and amino acid nutrition. Am J Clin Nutr 59(5):1203S–1212S

Younis K, Ahmad S (2015) Waste utilization of apple pomace as a source of functional ingredient in buffalo meat sausage functional ingredient in buffalo meat sausage. Cogent Food Agric 4(1): 1–10

Youssef MK, Barbut S (2011) Fat reduction in comminuted meat products-effects of beef fat, regular and pre-emulsified canola oil. Meat Sci 87(4):356–360

Zahn S, Pepke F, Rohm H (2010) Effect of inulin as a fat replacer on texture and sensory properties of muffins: inulin as a fat replacer in muffins. Int J Food Sci Technol 45(12):2531–2537

Zam W (2020) Structured lipids: synthesis, health effects and nutraceutical applications. In: Lipids and edible oils: properties, processing, and applications, pp 289–327. ISBN: 9780128173725

Zayas FJ, Lin CS (1988) Quality characteristics of frankfurters containing corn germ protein. J Food Sci 53:1587–1591

Zemel MB (2004) Role of calcium and dairy products in energy partitioning and weight management. Am J Clin Nutr 79(Suppl.):907S–912S

Zhang H, Yoshimura M, Nishinari K, Williams MAK, Foster TJ, Norton IT (2001) Gelation behaviour of konjac glucomannan with different molecular weights. Biopolymers 59(1):38–50

Zhang H, Zhang W, Zeng X, Zhao X, Xu X (2022) Recent progress of fat reduction strategies for emulsion type meat products. Food Mater Res 2:10

Zhu X, Zhang J, Liu S, Gu Y, Yu X, Gao F, Wang R (2022) Relationship between molecular structure and heat-induced gel properties of duck myofibrillar proteins affected by the addition of pea protein isolate. Food Secur 11:1040

Zia MK, Tabasum S, FarisKhana M, Akram N, Akhterb N, Noreen A, Zuber M (2018) Recent trends on gellan gum blends with natural and synthetic polymers: a review. Int J Biol Macromol 109:1068–1087 term probiotics as "live microbial feed supplement which has beneficial effects on host animals by improving the microbial balance of the intestine". The microbial components and metabolites are not included within the ambit of probiotics. Further, human intestines have a wide range of microbial populations, and the intestinal balance of these microbials in the gut has vital role in maintaining the health of the host.

The concept of probiotics has a rich historical context, dating back to 1907 when Russian scientist Elie Metchnikoff attributed the robust health and longevity of Bulgarian and Asian farmers to their consumption of fermented milk and milk products, particularly yoghurt. (Metchnikoff 2004). Yoghurt, a rich source of lactic acid bacteria (LAB), plays a crucial role in probiotics consumption. These LAB replace harmful bacteria in the gut through competitive exclusion. The WHO/FAO (World Health Organization/Food and Agricultural Organization 2001) provides a comprehensive definition of probiotics as live microorganisms that provide health benefits to the host when ingested in adequate levels. However, due to the challenges posed by pH, enzymes, chemicals, and temperatures, these microorganisms need to be consumed in sufficient quantities (10^6–10^8 cells per gram of food) to ensure their presence in the intestine in live status.

9.2 Desired Properties and Mode of Action

A good probiotic should be non-toxic and safe for human consumption. The isolates of the human intestine are preferred for probiotic uses due to safety concerns. The probiotics should be a good fermenter and should adhere well and colonize the intestinal wall. It should have acid and bile resistance and should have high longevity and stability. It should have immune modulator properties and have wide availability. Probiotics should be homogenously distributed in the food matrix, and the bacterial culture should be properly tested for health claims, quality, safety, efficacy, and effectiveness both *in-vivo* (laboratory trials) and *in-vitro* (live animal) (Granato et al. 2020). Probiotics should be able to survive the conditions of human digestive systems (high acidity and gastric enzymes and low acidity and bile salt), should colonize the intestinal walls, and should not have virulence factor and production of enterotoxins (Munekata et al. 2020). These should have desirable effects on the physicochemical, textural, and sensory changes in meat products.

At present, lactic acid bacteria and *Bifidobacterium* dominate the list of most used probiotics in the food industry. This is attributed to the presence of desirable properties such as rapid growth and multiplication, high tolerance to gastric acid and bile, high acidifying properties and fermentation capabilities, proper adhesion to the intestine wall, and improving the nutritive value of food due to fermentation, and positive effect on the health of the host. LAB are facultative anaerobes and easily colonize in the gut and replace the harmful pathogenic bacteria of the gut. In addition to the LAB, a non-pathogenic strain of *Escherichia coli* was isolated from the feces of a First World War soldiers who did not develop enterocolitis during a severe outbreak of shigellosis. This isolate was also developed as the first non-LAB probiotic by Alfred Nissle in 1907. *Later, Bifidobacterium* was isolated from infants by Henry Tissier and developed as probiotics.

Mostly, the intestinal microbes that are usually associated with the beneficial effect on consumers' health and maintain the intestinal microbial balance are used as probiotics. The most common species used as probiotics are *Lactobacillus spps.* (*L. acidophilus, L. johnsonni, L. gasseri, L. casie, L. rhamnosus, L. plantarum*), and *Bifidobacteriaum spps (B. lomgum, B. breve, B. bifidum, B. inantis*). Some microbes that do not usually inhabit the intestine but are used as probiotics in dairy products, such as *L. bulgaricus*, *Streptococcus thermophilus*, *Leuconostoc*, and *Lactococcus* spps. These bacteria did not effectively colonize the intestine, thus causing minor effects on the intestinal microbial balance (Alm 1991). Among LAB, *L. casei Shirota*, *L. johnsonni* La1, *L. plantarum* 299 V, *L. rhamnosus* LB 21, *L. reuteri* SD 2112, *L. casei* Imunitass, *L. casei* F19 and *L. rhamnosus* GG strains are commonly used as probiotics. *B. animalis* subsp. *lactis* Bb 12, *B. animalis* subsp. *Bifidus Actiregularis*, *B. beve* Yakult, VSL # 3 (8 strains) and *B. longum* BB 536 are commonly used *Bifidobacterium* strains used in probiotics. *E. coli* Nissle 1917, *Saccharomyces boulardii, Lactococcus*, *Enterococcus*, *Saccharomyces* and *Propionibacterium* are also available probiotics for use in food products.

Lactobacillus (mostly *L. gasseri*) was identified as the most prevalent genus isolated from infant feces making upto 48% of total microflora followed by *Enterococcus* (*E. faecalis* 21%) (Rubio et al. 2014a). *Lactobacilli* have the ability to grow in vitro in processing conditions, survive the human gastrointestinal tract conditions, and have antimicrobial properties against foodborne pathogen (Rubio et al. 2014a). For selection as probiotic starter culture in fermented sausage, the Lactobacilli strains of *L. casei/paracasei*, *L. rhamnosus*, *L. gasseri* and *L. fermentum* could be used as starter cultures in fermented sausage and dominate the endogenous lactic acid bacteria (Rubio et al. 2014b).

Enterococci are present abundantly in certain types of fermented cheese and fermented sausages. *E. faecium* and *E. faecalis* due to their proteolytic and enzymatic activity (Ben Braïe and Smaoui 2019). These are used as probiotics in pharmaceutical and food preparations to treat diarrhea, and irritable bowel syndrome, lowering blood cholesterol levels, and improve host immunity (Franz et al. 2011). Enterococci application as food probiotics is debated due to associated health risks due to expression of virulence factors (such as enterococcal surface protein, invasion, proteases, capsular polysaccharide, hemolysin, cytolysin, pili, aggregation substance, extracellular superoxide), biofilm formation, translocating in the intestinal mucosa, thereby resisting innate immune response, acquisition of DNA or antibiotic resistance due to lack of CAS loci, and possible association with the colorectal cancers (Wang et al. 2020). Thus, these are not approved to be used for the cure and treatment of human diseases by the USDA and the European Food Safety Authority (EFSA) put enterococci under "Qualified Presumption of safety" (Wang et al. 2020) (Tables 9.1 and 9.2).

Table 9.1 Commonly used probiotics and their effect

Microbial culture	Brand name	Applications
Lactobacilli spps.		
L. casei Imunitass	Actimel	Immunomodulatory responses
L. casei F19	Cultura	Improving digestive health and immune responses
L. rhamnosus GG	Giefilus. vifit	
L casei Shirota	Yakult	Improving gut health and natural defense
L. johnsonni Lal	LC1	
L. plantarum 299V	Proviva	Maintaining health of gastro-intestinal health and improving immunity
L. rhamnosus LB 21	Verum	
L. reuteri SD 2112	Bio Gaia	Colonizing in the gastrointestinal tract and controlling diarrhoea
L. reuteri SD 2112	Bio Gaia	
Bifidobacterium spps		
B. animalis subsp lactis Bb 12	Various names	Maintaining gut microbial balance and improving immune responses
B. beve Yakult	Befiene	
B. longum BB 536	Various names	
B. animalis subsp Bifidus Actiregularis	Activia	Improving transit of ingesta from through gut
VSL # 3 (8 strains)	VSL # 3 powder	As a biotherapeutic agent, controlling irritable bowel syndrome
Miscellaneous		
E. coli Nissle 1917	Mutaflor	As a biotherapeutic agent, controlling irritable bowel syndrome
Saccharomyces boulardii	Enterol	
Lactococcus lactis	–	Improving immune function and maintaining gut microbial balance
Saccharomyces boulardii	Enterol	Treating gastrointestinal disorders and controlling travellers and antibiotic associated diarrhoea
Propionibacterium/ Cutibacterium freudenreichii	–	Anticancer, anti-inflammatory, anti-pathogenic and immunomodulatory properties

Adopted and modified from (Vuyst et al. 2008)

9.3 Fermented Meat Products

Fermentation of food products is an ancient method for improving the nutritive and functional value of these products. Fermentation of meat is a very old and economical process applied for improving nutritional value by increasing the availability of vitamins and essential amino acids, increasing the solubility of nutrients, and lowering anti-nutritional factors. Further, fermentation also causes organoleptic properties such as taste, flavor, color, juiciness, tenderness, and sliceability. The release of antimicrobial peptides and the production of acids improve the keeping quality and extend the storage stability of meat and meat products (Kumar et al.

Table 9.2 Mode of action of probiotics

Effect	Pathways	Probiotics	References
Blocking enteropathogens entry	Improving mucin secretion from goblet cells	*Lactobacillus plantarum 229* v, *Lctobacillus rhamnosus*	Hardy et al. (2013)
	Inhibiting water and chloride secretion	*Streptococcus thermophilus* and *Lactobacillus acidophilus*	Brown (2011)
	Higher production of junction protein by T84 epithelial cells due to genetic modulation	–	Syngai et al. (2016)
	Repairing intestinal damage by stimulating production of tight junction protein	*Escherichia coli* Nissle 1917	Goudarzi et al. (2014)
Gut immunity	Cytokines and chemokines release	*Lactobacillus plantarum* 299v	Hemaiswarya et al. (2013)
Competitive exclusion and growth inhibition	Organic acids production	–	Goudarzi et al. (2014)
	Blocking available sites for attachment of microbes	–	
	Competing for essential nutrient and energy	–	Brown (2011)
	Release of substances as arginine, histidine, CLA exerting gut protective and antimicrobial properties	–	Hemaiswarya et al. (2013)
Antimicrobial peptides	Lantibiotics (class I), bacteriocins (class II) viz. lactacin B (*L. acidophilus*), plantaricin (*L. plantarum*), nisin (*Lactococcus lactis*), and bacteriolysins (class III)		Saulnier et al. (2009)
	Reuterin exerts activity against bacteria, fungi, protozoa and viruses	*Lactobacillus reuteri*	
	Defensins-small cysteine-rich antimicrobial peptides especially at mucosal sites capable of destroying cytoplasmic membrane	–	Hardy et al. (2013)
	Inducing expression of human beta-defensin 2 in Caco-2 intestinal epithelial cells	*E. coli* Nissle 1917	Ng et al. (2009)
Immune modulation	Inducing macrophages for phagocytosis	*L. acidophilus* La1, *Bifidobacterium lactis* Bb12	Delcenserie et al. (2008)
	Interleukin-12 (IL-12) production	–	Yaqoob (2014)
		–	

(continued)

Table 9.2 (continued)

Effect	Pathways	Probiotics	References
	Stimulate B cells for production of IgA in mesenteric lymphnodes and Peyers patch		Delcenserie et al. (2008)
	Induce IL-12 and increase NK cell activity and TH1 pathways	–	Yaqoob (2014)
	Inducing IL-10	–	
	Degrading auto-inducers	*Lactobacillus*, *bifidobacterium* and *Bacillus cereus*	Goudarzi et al. (2014)

2017). These products have unique organoleptic properties and cultural heritage but also have improved stability and convenience. These products had high salt and nitrite content due to use in the curing ingredients such as salt, nitrite or nitrates, sugar, and fat (Flores et al. 2021). The application of nitrite and nitrates in fermented meat products contributed to the safety and the development of characteristics of dry-cured aroma.

Raw-cured fermented meat products have traditionally been used in the preparation of fermented meat products by native or added lactic acid bacteria culture and carbohydrates. The commercial application of probiotics is still not practiced on a wide scale (Kołożyn-Krajewska and Dolatowski 2012). Fermented meat and meat products are very popular in Europe and as per one estimate, these products especially sausages make up nearly 3–5% of total meat consumed (Hutkins 2006), which comes up to 20–40% of the processed meat products (Hamm et al. 2008). The sausages are very popular meat products among fermented meat products and a wide variety (more than 350 types) of fermented sausages are available in the market. The fermentation of meat is done by a mixture of native microflora. However, this method requires a longer fermentation period/ripening period and resulting in inconsistent quality of the fermented products due to lack of proper control on fermentation cultures. This can be overcome by the introduction of commercial starter cultures consisting of the defined microflora resulting in shortening the ripening period and maintaining the quality of the end products. At present, homofermentative *Lactobacilli species* and *Pediococcus acidilacti*, *P. pentosaceus*, Gram positive catalase positive cocci, non-pathogenic coagulase negative *Staphylococcus xylosus* and *S. carosus* are widely used for fermentation (Lücke 1994; Kumar et al. 2017).

Fermentation of meat is performed by adding starter culture or microbial culture from previous batch (backslopping). However, there are instances when accidental inoculation of meat leads to the fermentation of meat products. These microbes cause the necessary biochemical changes and oxidize the organic substrate incompletely, especially carbohydrates, into organic acids, gases, and alcohol in anaerobic conditions. On the basis of acidity developed due to the production of acids during the ripening process, fermented meat products can be categorized under the following two categories as low acid fermented meat products and high acid fermented

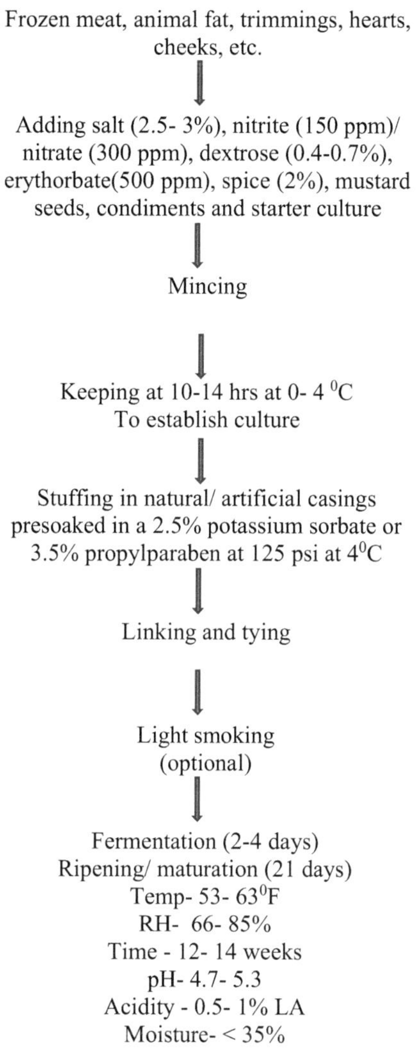

Fig. 9.1 Flowchart for preparation of dry-fermented sausage

meat products. The fermentation is also affected by the type of culture, inoculation size, temperature, presence of antioxidants, salt levels, etc. (Raccach 1992) (Fig. 9.1).

9.3.1 Fermented Sausage

Sausages are one of the oldest and most popular ready-to-eat processed meat products. These are available in different varieties, characteristic flavour, high nutritive value, convenience etc. The term 'sausage' is originated from latin word '*Salsus*' meaning salt. The process of sausage making, and its consumption was mentioned in the Babylonian and Chinese civilization around 1500 BC and had been described in classical Odyssey (Pearson and Gillett 1996). The fermented sausage containing probiotic strains are comes under functional meat products. Sausage is prepared by chopping the emulsified frozen meat, animal fat, curing salts, spices, seasoning ingredients and sugar. Sugar is added for providing substrate for the microbial growth, hence obtaining the desired level of acidity. Starter culture or backslopping or probiotics is incorporated into sausage mix and kept at refrigeration temperature for establishing the cultures. After that, this raw mix is stuffed into casing and put for ripening chamber under controlled conditions (time, temperature and relative humidity) to get the desired degree of fermentation.

The microbial safety of the dry fermented sausages is ensured by lowering the pH to 4.5–5.0 and fall in water activity below 0.90, so most of the pathogenic microbes do not grow (Barbuti and Parolari 2002). However, this process of microbial fermentation (ripening or maturation) also leads to growth of microbes that could have effect on the organoleptic and nutritive value of the end products. For example, the Iberian dry-cured sausages produced in the Central-West Spain had about 400 different strains of Micrococcaceae, most of strains were of Staphylococcus spps (*S. aureus*, *S. lugdunensis*, *S. saprophyticus*, *S. sciuri*, *S. chromogenes*, and *S. capitis*) and Kocuria spps (Martín et al. 2007). These microbes were able to survive and grow and at lower water activity ranging from 0.5 to 0.85 (Martín et al. 2007). These strains were reported to exert proteolytic and lipolytic activities in fat and sarcoplasmic pork proteins with lower production of biogenic amines without producing histamine, spermine, tyramine, and phenylethylamine (Martín et al. 2007).

In-game meat (wild boar and deer) fermented sausages, *L. sakei* and *L. curvatus* form about 20.55% of the initial microbiota, which further jumps to 70.48% at the end of ripening. At the same time, the microbial load of spoilage microorganisms was reduced to 13.98% at the end of the ripening period from the initial 44.15% (Mrkonjic Fuka et al. 2020).

Combinations of various probiotic strains have been used to improve the ripening and quality of meat products. Inoculation of the combination of *L. plantarum* TN8 and *P. acidilactici* MA 18/5 M in reformulated beef sausage at refrigeration temperature for 12 days was observed to have bioprotective function by reducing the Enterobacteriaceae counts to below 2 log CFU/g (Ben Slima et al. 2018). The inoculation of probiotic cultures resulted in improving the textural and sensory attributes along with significantly affecting the free fatty acid profile of the reformulated beef sausage (Ben Slima et al. 2018).

9.3.2 Fermented Camel Sausage

The demand for young camel meat is increasing due to lower fat and cholesterol and higher glycogen, protein, and PUFA content in camel meat as compared to other red meat (Kadim et al. 2006). Camel meat has lower sarcoplasmic proteins and is more prone to proteolytic degradation. It is considered tougher than the beef (Kadim et al. 2008). The nutritive and technological quality attributes of camel meat could be significantly enhanced by applying suitable processing interventions such as sausage preparations and fermentation. In addition to having a positive effect on the product's quality, the consumption of probiotic meat products also has a positive effect on the health of consumers. The health-promoting benefits of probiotic culture isolated from camel milk *L. plantarum* KX881772 and inoculated in the semi-dry fermented camel sausage were reported to have a higher degree of hydrolysis and higher ACE-inhibitions (50% higher than in fermented beef sausage) (Ayyash et al. 2019). After the end of 7 days of storage, the fermented camel sausage recorded 70% higher cytotoxicity activity against Caco-2 cell lines (Ayyash et al. 2019).

Camel hump fat is used in the preparation of prime-quality sausages and in frying of food products (Sbihi et al. 2013). The reformulated fermented camel meat sausage prepared by using camel hump fat was observed to have differences in chemical, textural, and sensorial properties as compared to sausage prepared with beef meat and beef sausage (Kargozari et al. 2014). Sausages prepared from camel meat had lower pH, and water activity but higher redness and lightness value with harder and higher PUFA/SFA ratio than beef sausages (Kargozari et al. 2014). Similarly, fermented camel meat sausages by using minced camel meat and hump fat (80:20 ratio) were prepared by inoculating *L. plantarum* (Kalalou et al. 2004). The meat mixture was stuffed into natural casings of 50 g each piece and dried at 15–18 °C and 70–80% relative humidity. On storage during the ripening phase, there was a decreasing count of proteolytic microorganisms (6 CFU/g to 1 CFU/g) recorded at the end of the process after 21 days. The pH was decreased to 4.5 and water activity reached 0.70 in the ripened sausage (Kalalou et al. 2004).

Various bioactive compounds are generated during the fermentation of meat that contribute to flavor and have positive effects on health. During the ripening of dry fermented camel sausage inoculated with starter culture was observed to have a high content of bioactive peptides with below 3 kDa molecular weight exerting antioxidant and antihypertensive roles as compared to pork or beef fermented sausages (Mejri et al. 2017a, b). Further, the application of probiotic mixture cultures of Staphylococcus and *L. sakei* were used in the preparation of traditional dry fermented camel meat sausage prepared by using camel meat and hump fat (Mejri et al. 2017a). After 7 days of ripening, the LAB counts in the sausages were recorded significantly higher with a lower pH. The inoculated product had good texture and low hardness value as compared to control samples (Mejri et al. 2017b).

9.3.3 Dry Sausages as Carrier of Probiotics

Dry sausages acclaimed as 'Ne-plus Ultra of industry' such as salami, mortadella, geneo, pepperoni, cervel etc., need not undergo cooking process. These are typical Mediterranean meat products, dried at low temperatures for about 90 days (Pearson and Gillett 1996). Dry sausages are sometimes smoked mildly. Dry sausages are not cooked and thus have a higher chance of survival of probiotics. The presence of fat in meat products also forms a protective coating around probiotics, further increasing the survival (Muguerza et al. 2004). However, low water activity, lower pH and presence of salts create hostile conditions for probiotics. Probiotic cultures that are resistance to these conditions in addition to bile and acid resistance such as *Lactobacillus sakei* and *Pediococcus acidilactici* are preferred for incorporation in dry sausages (Klingberg and Budde 2006). To compensate it, higher numbers of probiotics are used in the development of fermented meat products. The lactobacilli strains (*L. plantarum* MF1291, MF1298, DC13, *L. pentosus* MF1300 and *L. salivarius* DC5) were demonstrated to have higher chances of survival through human gastric system and could be reisolated from human feces (Klingberg and Budde 2006). The sausage matrix protected these probiotic strains, thereby resulting in higher survival through the gastrointestinal tract.

Several fermented meat products have been developed with the addition of probiotics with a starter culture. In general, in the beginning, the probiotic culture dominates due to high initial inoculation and as the ripening/maturation/fermentation progresses under suitable conditions such as temperature, humidity, salt concentration etc., the starter culture dominates and makes the most of the dominant flora of fermented probiotic meat products (Papamanoli et al. 2003). After the ripening phase, a reduction in LAB count was also observed. Burdychová and Šulcerová (2014) also documented the initial dominance of probiotic culture followed by high LAB count after ripening (at 11–13 °C, 75% relative humidity for 28 days) in fermented sausage with the added probiotic culture of *L. casei* with a starter culture of *Staphylococcus carnosus* and *L. curvatus; or Pediococcus acidilactici*. Erkkilä et al. (2001) also noted similar trends in decreasing *L. rhamnosus* probiotic strains (GG, E-97800 or LC-705) after the initial increase after the completion of 28 days ripening phase in fermented probiotic sausages. However, compensatory high inoculation of probiotics resulted in the availability of sufficient numbers (7 log CFU/g) in Iberian dry fermented sausages even at the end of 4 months of ripening (Benito et al. 2007).

Overall preparation of fermented probiotics meat products is a little bit difficult as the viability of probiotics bacteria is largely affected by high salt concentration, acidic environment, and lower water activity due to drying of dry sausages (Vuyst et al. 2008). The drying step in these types of sausages is very crucial as such a long period of drying makes these sausages more vulnerable to microbial and chemical degradation. The drying is done at 15–35 °C with air velocity at 15–20 cycles/h. During ripening and drying, up to 25–30% weight loss occurs with 0.5–1% lactic acid acidity, 4.8–5.3 pH, moisture: protein ratio (M: P ratio) less than 2.3: 1, and moisture ranging from 25% to 50% moisture. The preparation of semi-dry sausages, such as Thuringer, summer sausage, and cervelat, is completed within 7–30 days of heat drying and maturation. These can be packaged before or after the drying process. As compared to dry sausages, these sausages have a sharp, tangy taste attributed to the high acidity (0.5–1.3% lactic acid acidity with 4.7–5.3 pH). The salt, moisture, and moisture-to-protein ratio of these sausages ranges from 30% to 50%, 3.5%, and 2.3–3.7 respectively. This high salt, low moisture, and acidity prolong the shelf life of these sausages (Pearson and Gillett 1996).

In the Spanish dry-cured sausages, *L. sakei* strain is most sensitive and *E. faecium* is most resistant to antibiotics (Landeta et al. 2013). The various strains isolated from Spanish dry-cured sausage (*L. plantarum*, *L. paracsei* and *Enterococcus faecium*) exerted nitrate reductase activity. E. faecium and *L. sakei* AI-142 strains produced tyramine biogenic amines. The *L. paracasei* AI-128 and *L. casei* AI-143 strains possessed good quality and could be used in the preparation of fermented meat products (Landeta et al. 2013).

Sidira et al. (2014) developed dry-fermented probiotic sausage by applying the immobilized *L. casei* ATCC 393 on wheat for increasing survival under heat treatment. In the final product, a 7 log CFU/g increase in lactobacilli counts was accompanied by a decrease in pseudomonas, Enterobacter, and staphylococci count during the ripening stage. Even after 66 days of ripening, the *L. casei* ATCC 393 counts in the sausage were recorded more than 6 log CFU/g, thus exerting a probiotic effect.

9.4 Probiotic Meat Products

Probiotic meat products are gaining popularity nowadays. These are very common in dairy products. Probiotic meat products are meat products that have beneficial live bacteria in sufficient amounts to get the desired health benefits. The incorporation of probiotics in meat products is considered one of the main strategies in development and production of fermented meat products with improved safety and nutritive value (Munekata et al. 2022; Ojha et al. 2015). These products could play an important role in meeting the Sustainable Development Goals of the 2030 Agenda set out by the United Nations for good health and well-being by reducing malnutrition, preventing spoilage, and improving the health of consumers (Daliri and Lee 2015).

These products were developed in Japan and Germany by adding the lactic acid bacteria isolates of human intestinal origin. Probiotic salami was prepared in Germany in 1998 by using three human intestinal LAB isolates exerting probiotic properties viz. *Lactobacillus acidophilus*, *Lactobacillus casei*, *Bifidobacterium spp*. Similarly, *fermented probiotic meat spread was* developed by using *Lactobacillus rhamnosus FERM P-15120* culture. These spreads contained 200 ppm nitrite and 3.3% sodium chloride and were kept for ripening below 20 °C (Sameshima et al. 1998). The inoculation of the probiotic cultures of *L. rhamnosus* FERM P-15120 and *L. paracasei* subsp. *paracasei* FERM P-15121 was observed to inhibit the growth and production of Staphylococcus aureus enterotoxin production in the fermented sausage, thereby making them microbiologically safe (Sameshima et al. 1998).

Further, fermented moist type sausages by inoculating probiotic cultures of *L. paracasei* (2 × 107 CFU/g) were developed. The highly resistant strain of *L. paracasei* LTH 2579 towards low pH and bile was isolated from fruit mash, with this hardiness attributed to 235 base pairs DNA fragments of the LTH2579 strain (Bunte et al. 2000). These sausages exert beneficial effects on consumers' health and are made up of 5–12% of total lactobacilli in the fecal material. The sausage consumption caused an increase in CD4, T helper cells, and phagocytosis index and lowered the expression of CD54 in the three volunteers in the experiment. The consumption of these sausages is attributed to lower LDL (low-density lipoproteins) and increased CLA (conjugated linoleic acid) (Bunte et al. 2000).

The lactic acid bacteria produce bacteriocins and other peptides during the fermentation process. These bioactive compounds play an important role in improving nutritive value and microbial safety by inhibiting the growth of pathogenic and spoilage microorganisms and functional attributes of fermented meat products. The pediocin PA-1 produced by the *Pediococcus acidilactici* MCH14 pediocin-producing strain during the ripening stage of Spanish dry fermented sausages and in frankfurters was known to inhibit the growth of L. monocytogenes and *Clostridium perfringens* by 0.6–2 log cycle during refrigerated storage for 2 months (Nieto-Lozano et al. 2010).

Fermented sausages were developed by applying a mixed microbial culture of LAB and *Staphylococcus carnosus* to decrease ripening time and desired levels of acidity (Hugas et al. 1995; Hugas and Monfort 1997). The presence of these cultures prevented the growth of *Listeria* spps. in fermented sausages, mainly attributed to the presence of bacteriocinogenic LAB such as *Lactobacillus sakei*, *Lactobacillus curvatus*, *L. plantarum*. *L. casei* produces bacteriocins Lactocin 705, which prevent the growth of *L. plantarum, L. monocytogenes, S. aureus*, and Gram-negative bacteria. Further, these cultures also increase the competitiveness of the starter cultures. The *L. sake* CTC494 isolated from the fermented sausage was observed to produce proteolytic bacteriocin (sakacin K), which inhibited the growth of *L. monocytogenes* (by 1.25 log) and *L. innocua* in the model sausage system as well in dry sausages (Hugas et al. 1995). Further, the fermentation efficiency of *L. sake* CTC494 is good, and its inoculation in meat products resulted in improved nutritive and organoleptic attributes.

In traditional Greek fermented sausage, *L. plantarum* L125 stain was isolated (Pavli et al. 2016) and was inoculated with commercial starter culture during the preparation of dry fermented pork sausage. These sausages were ripened at 4 and 12 °C and at the end of ripening, the sausages fermented with potential probiotic strain exhibited pH values of 4.42–4.46 and water activity values ranging from 0.85 to 0.87, which were similar to values for these parameters recorded in sausage prepared by using commercial starter culture (Pavli et al. 2020).

Song et al. (2018) isolated *Bifidobacterium longum* from healthy Korean infants and encapsulated with glycerol was used in the preparation of functional fermented sausages. The encapsulation protects the *B. longum* bacterium even after 4 days of fermentation, and about 50% of these bacteria survived in the product even after 22 days of ripening. The sausage containing encapsulated bacteria recorded comparable sensory attributes, a marked decrease in lipid oxidation, no production of biogenic amines, had higher unsaturated fatty acids with desirable n-6/n-3 fatty acids profile as compared to the control (Song et al. 2018).

9.5 Health Benefits of Probiotic Meat Products

The consumption of fermented sausage containing probiotic microbes had a positive effect on the gut intestinal health and blood lipid profile of humans. A regular consumption of 50 g fermented probiotic sausage containing *L. paracasei* LTH 2579 for 5 weeks was recorded with improving immunity and phagocytosis index in 10 women and 10 men (Jahreis et al. 2002). However, in the same study, the probiotic sausage consumption did not have significant effect on various serum cholesterol fractions and triglycerides, but with increased antibodies against oxidized cholesterol. Further the probiotic sausage consumption increased the numbers of cells expressing CD4 and decreased the cells expressing CD54 (ICAM-1) (Jahreis et al. 2002).

The fermented probiotics sausage was also developed by inoculating *Bifidobacterium* culture (Lücke 2017). The poor survival of *Bifidobacterium* is countered by and can be taken care of by using high inoculation, so to maintain minimum prescribed number of these probiotics bacteria (6 log cfu/g) in fermented sausage. In traditional Scandinavian type fermented dry sausages, *L. plantarum* and *L. pentosus* have been isolated and these bacteria possess probiotic characteristics by good bile and acid resistance as well as competitive exclusion of harmful intestinal pathogens. Various strains of LAB such as *L. casei, L. paracasei, L. rhamnosus and L. sakei* have been isolated from traditional Italian dry fermented sausage. These strains are capable to inhibit commonly occurring gut pathogens as *E. coli* and *Salmonella enterica* (Klingberg and Budde 2006).

In Iberian dry fermented sausages, out of 1000 strains of LAB and bifidobacterium isolated only 51 strains were found to have probiotic potential by their ability to survive for 1.5 h under high acid (pH 2.5) and salt conditions (Ruiz-Moyano et al. 2008). The acid-resistant strains from human feces (*L. casei* and *L. fermentum*) and pig feces (*L. reuteri, L. animalis, L. murinus,* and *L. vaginalis*) could be used as potential probiotic strains in fermented meat products (Ruiz-Moyano et al. 2008). In two types of Iberian dry fermented sausages, out of 192 LAB, *P. pentosaceus* (31.2%), *P. acidilactici* (26.9%), *Lactococcus lactis* (18.6%), and *L. brevis* (17.6%) are the predominant LAB with sporadic presence of *Leuconostoc mesenteroides*, *L. plantarum*, and *L. curvatus* strains (Benito et al. 2007). Further, some of these probiotic strains had exerted antimicrobial effects against *L. monocytogenes*. In these sausages, the desired level of probiotic counts (7 log CFU/g) was reached within 2 months of the ripening phase and maintained that level up to 4 months of the ripening period (Benito et al. 2007).

In the development of Spanish fermented sausages, the inoculation of probiotic *Lactobacillus plantarum* 299 V and *Lactobacillus rhamnosus* GG strains (at a very high level of inoculation of 10^7 CFU/g) prevented the growth of Enterobacteriaceae throughout the maturation period (Rubio et al. 2013). These inoculations at high doses resulted in a rapid fall in the pH of the product and inhibited the growth of Gram-positive catalase + cocci. Furthermore, *L. plantarum* 299 V inoculation at 10^5 CFU/g resulted in increasing their counts during the ripening stage and touching at 10^8 CFU/g and maintaining the good organoleptic quality of the end product (Rubio et al. 2013).

9.6 Survival of Probiotics During Processing

Microencapsulation of probiotic cultures by enclosing live microbial strains within a semi-permeable protective coating for physical isolation from harsh conditions to ensure better survival rate, controlled release, and production of microbial metabolites. The size of these microencapsulated beads ranges from 1 to 100 μm, and the internal environment inside these beads remains conducive to the growth and survival of bacteria. Depending upon the type of shell/protective layer to be digested by specific enzymes or affected by the specific pH of our gut, controlled release of these probiotics at the desirable part of the gut is ensured. Another aspect of microencapsulation is that microencapsulated microbial cultures do not affect the organoleptic properties of meat products. However, most of the probiotics do not alter the sensory attributes of meat products. Pidcock et al. (2002) did not found any deterioration in sensory attributes of various fermented meat products upon incorporating probiotic cultures from human intestinal isolates viz. *L. paracasei* L26 and *Bifidobacterium lactis* B94. Muthukumarasamy and Holley (2007) added microencapsulated *L. reuteri* ATCC 55730 in fermented meat products and did not report any deterioration in the organoleptic properties of these products.

The potential probiotic strain *L. rhamnosus* CTC1679 isolated from human source was used for the production of reduced fat, sodium, and low-acid fermented sausages (fuets) containing 8 log CFU/g counts of *L. rhamnosus* CTC1679 (Rubio et al. 2014b). These strains were observed to survive gastrointestinal conditions and recovered in the feces of human volunteers who had consumed the product. Further, the strains could be isolated from the feces of these volunteers even after 3 days of non-consumption (Rubio et al. 2014b). These strains inhibited the growth of *L. monocytogenes* and eliminated Salmonella during the ripening phase. The application of other technological interventions, such as high hydrostatic pressure application of 600 MPa for 5 min on 14 days of ripening, further reduced the *L. monocytogenes* count to below 1 log CFU/g, and no pathogen was detected after 35d days of storage under refrigeration conditions (Rubio et al. 2014b).

9.7 Challenges

The selection of a proper probiotic culture to produce functional meat products is a complex task as it comprises a joint effort of microbiology, food science, and medicine (Granato et al. 2020). The probiotic strains that fulfill all the criteria of a desirable probiotic are the best ones to use in the production of probiotic meat products, but this also has resulted the drastically reducing the number of options of potential probiotic strains to be used for the development of probiotic meat products. For example, out of 169 Lactobacillus strains isolated from the fermented pork sausages, only 2 strains viz., *L. fermentum* 3007 and 3010 have exerted the probiotic potential (Klayraung 2008). Similarly, from traditional Serbian sausage (Sokobanja), *E. faecium* sk 7-5, sk-7-8 and sk 7-15 isolates possessed probiotic potential and survived under the simulated gastrointestinal conditions (Žugić Petrović et al. 2020). Similarly, 42 LAB cultures isolated from Ciauscolo salami from Central Italy possessed good resistance towards gastric juice and moderately tolerated the bile juice (Federici et al. 2014). The main probiotic strains isolated were *P. pentosaceus* (62781-3, 46035-1 and 46035-4) and *Leucosostoc mesenteroides* (14324-8) (Federici et al. 2014).

The characterization of microbial strain is of great importance while selecting microbial cultures for fermentation in functional meat products. These should not have genes involving in antibiotic resistance, biogenic amines, enterotoxins and toxic compounds (Ben Braïe and Smaoui 2019). The application of advanced technologies such as 16 s RNA sequencing, RAPD-PCR (Random amplified polymorphic DNA polymerase chain reaction) as compared to traditional methods (Yadav and Shukla 2017). These technologies have been used to identify the potential probiotic strains from fermented meat products such as *L. plantarum* L125 strains isolated from Greek fermented sausages (Cao et al. 2019), *P. pentosaceus* R1 from Harbin sausages (Han et al. 2017) and *L. sakei* 23 k strains isolated from Tunisian dry-fermented sausages (Najjari et al. 2020).

Although meat and meat products are vital sources of animal protein and other essential nutrients, buts its higher consumption is associated with colorectal cancer and cardiovascular diseases, mostly attributed to the high content of salt, nitrite, fat, low dietary fiber, and presence of carcinogenic compounds in cured and smoked meat products (Bouvard et al. 2015; Domingo and Nadal 2017; Han et al. 2019). In meat and meat product processing, salt is commonly added to solubilize myofibrillar proteins, improving technological properties and reducing microbial load. The reduction of salt in sausage production is quite challenging which on average have on an average 230–3300 mg sodium per 100 g serving (Allemandi et al. 2015). Similarly, dietary fat in meat products is very crucial for flavor, juiciness, texture, and nutritive value and plays an important role in determining the shelf life of meat products, especially by protecting them from oxidative rancidity (Asyrul-Izhar et al. 2022; Kumar 2021). The sausages are rich in fat, with an average fat content ranging from 10% to 20%, out of which about half of the total fat is saturated fat (Cunningham et al. 2015). Nitrates/nitrites are also present in sausages as food additives for their important role in the development of stable pink color and characteristic sensory attributes, prevention of lipid oxidation, and food safety, especially against Clostridium botulinum (Munekata et al. 2020). The nitrates/ nitrites also cause the generation of carcinogenic compounds, and WHO put processed meat products into the category of tobacco and asbestos (Crowe et al. 2019). Thus there is a need for reformulating these products by using suitable alternatives and processing technologies followed by a proper marketing approach. This would lead to increased consumer acceptance and marketability of these products, consequently leading to positive health effects on human health (Shan et al. 2017a, b).

The health claims pertaining to the consumption of probiotic meat products are, to date, mainly claimed based on the clinical trials performed with fermented sausages. The microbial population present in fermented sausage is associated with the fermentation process, thereby affecting the nutritive, sensory, and microbial safety of such products. The industry is in the process of selecting microbial cultures that have both fermentation and probiotic qualities. The probiotic culture is thermal sensitive (more than 45 °C), hence prone to a significant decrease in their numbers if the consumer prefers to cook or thermal process these products at home prior to consumption (Terpou et al. 2019). The probiotic dosage in the fermented sausage and consumption frequency has a significant effect on the overall health benefits associated with their consumption. In addition, there is a lack of prescribed dosage of probiotics in fermented sausage, and clarity on this issue will further boost this sector.

9.8 Conclusion

The fermented meat products are very popular since time immemorial and are found to be the carrier of probiotics. Probiotic meat products are not as popular as dairy probiotic products due to a lack of research on probiotics in meat and the adverse processing conditions and composition of meat and meat products. The dry sausages are best suited for the preparation of probiotic meat products due to the absence of heat treatment in their preparation. The probiotics, besides increasing the functionality of meat products without compromising sensory attributes, also improve the health benefits of the consumers. The demand for healthy foods such as probiotics, prebiotics, and symbiotic meat products provides an open arena full of opportunities for the meat industry. The need of the hour is to have improved probiotics with good viability and health-promoting benefits.

References

Allemandi L, Tiscornia MV, Ponce M, Castronuovo L, Dunford E, Schoj V (2015) Sodium content in processed foods in Argentina: compliance with the national law. Cardiovasc Diagn Ther 5(3): 197–206. https://doi.org/10.3978/j.issn.2223-3652.2015.04.01

Alm L (1991) The therapeutic effects of various cultures – an overview. In: Therapeutic properties of fermented milks. Elsevier Applied Science, London, pp 45–64

Asyrul-Izhar AB, Bakar J, Sazili AQ, Meng GY, Ismail-Fitry MR (2022) Incorporation of different physical forms of fat replacers in the production of low-fat/reduced-fat meat products: which is more practical? Food Rev Intl 39(9):6387–6419. https://doi.org/10.1080/87559129.2022.2108439

Ayyash M, Liu S-Q, Al Mheiri A, Aldhaheri M, Raeisi B, Al-Nabulsi A, Osaili T, Olaimat A (2019) In vitro investigation of health-promoting benefits of fermented camel sausage by novel probiotic *lactobacillus plantarum*: a comparative study with beef sausages. LWT – Food Sci Technol 99:346–354. https://doi.org/10.1016/j.lwt.2018.09.084

Barbuti S, Parolari G (2002) Validation of manufacturing process to control pathogenic bacteria in typical dry fermented products. Meat Sci 62(3):323–329. https://doi.org/10.1016/S0309-1740(02)00124-9

Ben Braïe O, Smaoui S (2019) Enterococci: between emerging pathogens and potential probiotics. Biomed Res Int 2019:1–13. https://doi.org/10.1155/2019/5938210

Ben Slima S, Ktari N, Triki M, Trabelsi I, Abdeslam A, Moussa H, Makni I, Herrero AM, Jiménez-Colmenero F, Ruiz-Capillas C, Ben Salah R (2018) Effects of probiotic strains, lactobacillus plantarum TN8 and Pediococcus acidilactici, on microbiological and physico-chemical characteristics of beef sausages. LWT – Food Sci Technol 92:195–203. https://doi.org/10.1016/j.lwt.2018.02.038

Benito MJ, Martín A, Aranda E, Pérez-Nevado F, Ruiz-Moyano S, Córdoba MG (2007) Characterization and selection of autochthonous lactic acid bacteria isolated from traditional Iberian dry-fermented Salchichón and chorizo sausages. J Food Sci 72(6):419–M201. https://doi.org/10.1111/j.1750-3841.2007.00419.x

Bouvard V, Loomis D, Guyton KZ, Grosse Y, Ghissassi FEL, Benbrahim-Tallaa L, Guha N, Mattock H, Straif K (2015) Carcinogenicity of consumption of red and processed meat. Lancet Oncol 16(16):1599–1600. https://doi.org/10.1016/S1470-2045(15)00444-1

Brown M (2011) Modes of action of probiotics: recent developments. J Anim Vet Adv 10(14): 1895–1900. https://doi.org/10.3923/javaa.2011.1895.1900

Bunte C, Hertel C, Hammes WP (2000) Monitoring and survival of lactobacillus paracasei LTH 2579 in food and the human intestinal tract. Syst Appl Microbiol 23(2):260–266. https://doi.org/10.1016/S0723-2020(00)80013-2

Burdychová R, Šulcerová H (2014) Senzorické hodnocení fermentovaných salámů herkules s přídavkem probiotického kmene Lactobacillus casei 01. Acta Univ Agric Silvic Mendel Brun 57(5):57–64

Cao C-C, Feng M-Q, Sun J, Xu X-L, Zhou G-H (2019) Screening of lactic acid bacteria with high protease activity from fermented sausages and antioxidant activity assessment of its fermented sausages. CyTA – J Food 17(1):347–354. https://doi.org/10.1080/19476337.2019.1583687

Crowe W, Elliott CT, Green BD (2019) A review of the in vivo evidence investigating the role of nitrite exposure from processed meat consumption in the development of colorectal cancer. Nutrients 11(11):2673. https://doi.org/10.3390/nu11112673

Cunningham J, Nguyen V, Adorno P, Droulez V (2015) Nutrient composition of retail samples of Australian beef sausages. Nutrients 7(11):9602–9617. https://doi.org/10.3390/nu7115491

Daliri EB-M, Lee BH (2015) New perspectives on probiotics in health and disease. Food Sci Human Wellness 4(2):56–65. https://doi.org/10.1016/j.fshw.2015.06.002

Delcenserie V, Martel D, Lamoureux M, Amiot J, Boutin Y, Roy D (2008) Immunomodulatory effects of probiotics in the intestinal tract. Curr Issues Mol Biol. 10.21775/cimb.010.037 10: 37–54

Domingo JL, Nadal M (2017) Carcinogenicity of consumption of red meat and processed meat: a review of scientific news since the IARC decision. Food Chem Toxicol 105:256–261. https://doi.org/10.1016/J.FCT.2017.04.028

Erkkilä S, Suihko M-L, Eerola S, Petäjä E, Mattila-Sandholm T (2001) Dry sausage fermented by *lactobacillus rhamnosus* strains. Int J Food Microbiol 64(1–2):205–210. https://doi.org/10.1016/S0168-1605(00)00457-8

Federici S, Ciarrocchi F, Campana R, Ciandrini E, Blasi G, Baffone W (2014) Identification and functional traits of lactic acid bacteria isolated from Ciauscolo salami produced in Central Italy. Meat Sci 98(4):575–584. https://doi.org/10.1016/j.meatsci.2014.05.019

Flores M, Perea-Sanz L, Belloch C (2021) Nitrite reduction in fermented meat products and its impact on aroma. In: Advances in food and nutrition research, pp 131–181. https://doi.org/10.1016/bs.afnr.2020.10.002

Food and Agricultural Organization of the United Nations and World Health Organization (2001) Health and nutritional properties of probiotics in food including powder milk with live lactic acid bacteria. Report of a Joint FAO/WHO Expert Consultation on Evaluation of Health and Nutritional Properties of Probiotics in Food Including Powder Milk with Live Lactic Acid Bacteria. https://openknowledge.fao.org/server/api/core/bitstreams/382476b3-4d54-4175-803f-2f26f3526256/content (accessed on 21-08-2024)

Franz CMAP, Huch M, Abriouel H, Holzapfel W, Gálvez A (2011) Enterococci as probiotics and their implications in food safety. Int J Food Microbiol 151(2):125–140. https://doi.org/10.1016/j.ijfoodmicro.2011.08.014

Fuller R (1989) Probiotics in man and animals. J Appl Bacteriol 66(5):365–378

Fuller R (1992) History and development of probiotics. In: Probiotics. Springer, Edinburgh, UK, pp 1–8. https://doi.org/10.1007/978-94-011-2364-8_1

Goudarzi M, Goudarzi H, Rashidan M (2014) Probiotics: an update on mechanisms of action and clinical applications. Novelty Biomed 2(1):22–30

Granato D, Barba FJ, Bursać Kovačević D, Lorenzo JM, Cruz AG, Putnik P (2020) Functional foods: product development, technological trends, efficacy testing, and safety. Annu Rev Food Sci Technol 11(1):93–118. https://doi.org/10.1146/annurev-food-032519-051708

Hamm WP, D Haller, MG Ganzle (2008) Fermented meat products. In: Edward, R. Farnworth (Ed.), Handbook of fermented functional foods, 2nd Ed. CRPress, USA. https://doi.org/10.1201/9781420053289

Han Q, Kong B, Chen Q, Sun F, Zhang H (2017) In vitro comparison of probiotic properties of lactic acid bacteria isolated from Harbin dry sausages and selected probiotics. J Funct Foods 32: 391–400. https://doi.org/10.1016/j.jff.2017.03.020

Han MA, Zeraatkar D, Guyatt GH, Vernooij RWM, El DR, Zhang Y, Algarni A, Leung G, Storman D, Valli C, Rabassa M, Rehman N, Parvizian MK, Zworth M, Bartoszko JJ, Lopes LC, Sit D, Bala MM, Alonso-Coello P, Johnston BC (2019) Reduction of red and processed meat intake and cancer mortality and incidence. Ann Intern Med 171(10):711–720. https://doi.org/10.7326/M19-0699

Hardy H, Harris J, Lyon E, Beal J, Foey A (2013) Probiotics, prebiotics and immunomodulation of gut mucosal defences: homeostasis and immunopathology. Nutrients 5(6):1869–1912. https://doi.org/10.3390/nu5061869

Hemaiswarya S, Raja R, Ravikumar R, Carvalho IS (2013) Mechanism of action of probiotics. Braz Arch Biol Technol 56(1):113–119. https://doi.org/10.1590/S1516-89132013000100015

Hugas M, Monfort JM (1997) Bacterial starter cultures for meat fermentation. Food Chem 59(4): 547–554. https://doi.org/10.1016/S0308-8146(97)00005-8

Hugas M, Garriga M, Aymerich MT, Monfort JM (1995) Inhibition of *Listeria* in dry fermented sausages by the bacteriocinogenic *Lactobacillus sake* CTC494. J Appl Bacteriol 79(3):322–330. https://doi.org/10.1111/j.1365-2672.1995.tb03144.x

Hutkins RW (2006) Starter cultures. In: Microbiology and technology of fermented foods. Blackwell Publishing, Oxford, UK, pp 67–106. ISBN: 9788126547975, 9788126547975

Jahreis G, Vogelsang H, Kiessling G, Schubert R, Bunte C, Hammes WP (2002) Influence of probiotic sausage (*Lactobacillus paracasei*) on blood lipids and immunological parameters of healthy volunteers. Food Res Int 35(2–3):133–138. https://doi.org/10.1016/S0963-9969(01)00174-0

Kadim IT, Mahgoub O, Al-Marzooqi W, Al-Zadjali S, Annamalai K, Mansour MH (2006) Effects of age on composition and quality of muscle longissimus thoracis of the Omani Arabian camel (*Camelus dromedaries*). Meat Sci 73(4):619–625. https://doi.org/10.1016/j.meatsci.2006.03.002

Kadim IT, Mahgoub O, Purchas RW (2008) A review of the growth, and of the carcass and meat quality characteristics of the one-humped camel (*Camelus dromedaries*). Meat Sci 80(3):555–569. https://doi.org/10.1016/j.meatsci.2008.02.010

Kalalou I, Faid M, Ahami TA (2004) Improving the quality of fermented camel sausage by controlling undesirable microorganisms with selected lactic acid bacteria. Int J Agric Biol 6:447–451

Kargozari M, Moini S, Akhondzadeh Basti A, Emam-Djomeh Z, Ghasemlou M, Revilla Martin I, Gandomi H, Carbonell-Barrachina ÁA, Szumny A (2014) Development of Turkish dry-fermented sausage (sucuk) reformulated with camel meat and hump fat and evaluation of physicochemical, textural, fatty acid and volatile compound profiles during ripening. LWT Food Sci Technol 59(2):849–858. https://doi.org/10.1016/j.lwt.2014.05.033

Klayraung S (2008) Probiotic properties of lactobacilli isolated from Thai traditional food. Sci Pharm 76(3):485–503. https://doi.org/10.3797/scipharm.0806-11

Klingberg TD, Budde BB (2006) The survival and persistence in the human gastrointestinal tract of five potential probiotic lactobacilli consumed as freeze-dried cultures or as probiotic sausage. Int J Food Microbiol 109(1–2):157–159. https://doi.org/10.1016/j.ijfoodmicro.2006.01.014

Kołożyn-Krajewska D, Dolatowski ZJ (2012) Probiotic meat products and human nutrition. Process Biochem 47(12):1761–1772. https://doi.org/10.1016/j.procbio.2012.09.017

Kumar Y (2021) Development of low-fat/reduced-fat processed meat products using fat replacers and analogues. Food Rev Intl 37(3):296–312. https://doi.org/10.1080/87559129.2019.1704001

Kumar P, Chatli MK, Verma AK, Mehta N, Malav OP, Kumar D, Sharma N (2017) Quality, functionality, and shelf life of fermented meat and meat products: a review. Crit Rev Food Sci Nutr 57(13):2844–2856. https://doi.org/10.1080/10408398.2015.1074533

Landeta G, Curiel JA, Carrascosa AV, Muñoz R, de las Rivas B (2013) Technological and safety properties of lactic acid bacteria isolated from Spanish dry-cured sausages. Meat Sci 95(2):272–280. https://doi.org/10.1016/j.meatsci.2013.05.019

Lücke FK (1994) Fermented meat products. Food Res Int 27(3):299–307. https://doi.org/10.1016/0963-9969(94)90098-1

Lücke F-K (2017) Fermented meat products – an overview. In: Fermented meat products – health aspects, Food Biology Series, CRC Press Taylor & Francis, Boca Raton Florida, USA, pp 1–14

Martín A, Colín B, Aranda E, Benito MJ, Córdoba MG (2007) Characterization of Micrococcaceae isolated from Iberian dry-cured sausages. Meat Sci 75(4):696–708. https://doi.org/10.1016/j.meatsci.2006.10.001

Mejri L, Vásquez-Villanueva R, Hassouna M, Marina ML, García MC (2017a) Identification of peptides with antioxidant and antihypertensive capacities by RP-HPLC-Q-TOF-MS in dry fermented camel sausages inoculated with different starter cultures and ripening times. Food Res Int 100:708–716. https://doi.org/10.1016/j.foodres.2017.07.072

Mejri L, Ziadi A, El Adab S, Boulares M, Essid I, Hassouna M (2017b) Effect of commercial starter cultures on physicochemical, microbiological and textural characteristics of a traditional dry fermented sausage reformulated with camel meat and hump fat. J Food Meas Charact 11(2):758–767. https://doi.org/10.1007/s11694-016-9445-6

Metchnikoff II (2004) The prolongation of life: optimistic studies. Springer Publishing Company, New York

Mrkonjic Fuka M, Tanuwidjaja I, Zgomba Maksimovic A, Zunabovic-Pichler M, Kublik S, Hulak N, Domig KJ, Schloter M (2020) Bacterial diversity of naturally fermented game meat sausages: sources of new starter cultures. LWT – Food Sci Technol 118:108782. https://doi.org/10.1016/j.lwt.2019.108782

Muguerza E, Gimeno O, Ansorena D, Astiasarán I (2004) New formulations for healthier dry fermented sausages: a review. Trends Food Sci Technol 15(9):452–457. https://doi.org/10.1016/j.tifs.2003.12.010

Munekata PES, Pateiro M, Zhang W, Domínguez R, Xing L, Fierro EM, Lorenzo JM (2020) Autochthonous probiotics in meat products: selection, identification, and their use as starter culture. Microorganisms 8(11):1833. https://doi.org/10.3390/microorganisms8111833

Munekata PES, Pateiro M, Tomasevic I, Domínguez R, da Silva Barretto AC, Santos EM, Lorenzo JM (2022) Functional fermented meat products with probiotics—a review. J Appl Microbiol 133(1):91–103. https://doi.org/10.1111/jam.15337

Muthukumarasamy P, Holley R (2007) Survival of Escherichia coli O157:H7 in dry fermented sausages containing micro-encapsulated probiotic lactic acid bacteria. Food Microbiol 24(1):82–88. https://doi.org/10.1016/j.fm.2006.03.004

Najjari A, Boumaiza M, Jaballah S, Boudabous A, Ouzari H (2020) Application of isolated *Lactobacillus sakei* and *Staphylococcus xylosus* strains as a probiotic starter culture during the industrial manufacture of Tunisian dry-fermented sausages. Food Sci Nutr 8(8):4172–4184. https://doi.org/10.1002/fsn3.1711

Nieto-Lozano JC, Reguera-Useros JI, Peláez-Martínez, M del C, Sacristán-Pérez-Minayo G, Gutiérrez-Fernández ÁJ, la Torre AH de (2010) The effect of the pediocin PA-1 produced by *Pediococcus acidilactici* against *Listeria monocytogenes* and *Clostridium perfringens* in Spanish dry-fermented sausages and frankfurters. Food Control 21(5):679–685. https://doi.org/10.1016/j.foodcont.2009.10.007

Ng SC, AL Hart, MA Kamm, AJ Stagg, SC Knight (2009) Mechanisms of action of probiotics: recent advances. Inflamm Bowel Dis 15(2):300–310

Ojha KS, Kerry JP, Duffy G, Beresford T, Tiwari BK (2015) Technological advances for enhancing quality and safety of fermented meat products. Trends Food Sci Technol 44(1):105–116. https://doi.org/10.1016/j.tifs.2015.03.010

Papamanoli E, Tzanetakis N, Litopoulou-Tzanetaki E, Kotzekidou P (2003) Characterization of lactic acid bacteria isolated from a Greek dry-fermented sausage in respect of their technological and probiotic properties. Meat Sci 65(2):859–867. https://doi.org/10.1016/S0309-1740(02)00292-9

Parker RB (1974) Probiotics, the other half of the antibiotics story. Anim Nutr Health 29:4–8

Pavli FG, Argyri AA, Papadopoulou OS, Nychas GJE, Chorianopoulos NG, Tassou CC (2016) Probiotic potential of lactic acid bacteria from traditional fermented dairy and meat products: assessment by in vitro test and characterization. J Probiotics Health 4:157

Pavli FG, Argyri AA, Chorianopoulos NG, Nychas G-JE, Tassou CC (2020) Effect of lactobacillus plantarum L125 strain with probiotic potential on physicochemical, microbiological and sensorial characteristics of dry-fermented sausages. LWT – Food Sci Technol 118:108810. https://doi.org/10.1016/j.lwt.2019.108810

Pearson AM, Gillett TA (1996) Processed meats. Springer Science & Business Media, New York

Pidcock K, Heard GM, Henriksson A (2002) Application of nontraditional meat starter cultures in production of Hungarian salami. Int J Food Microbiol 76(1–2):75–81. https://doi.org/10.1016/S0168-1605(02)00002-8

Raccach M (1992) Some aspects of meat fermentation. Food Microbiol 9(1):55–65. https://doi.org/10.1016/0740-0020(92)80062-9

Rubio R, Aymerich T, Bover-Cid S, Guàrdia MD, Arnau J, Garriga M (2013) Probiotic strains *Lactobacillus plantarum* 299V and *Lactobacillus rhamnosus* GG as starter cultures for fermented sausages. LWT Food Sci Technol 54(1):51–56. https://doi.org/10.1016/j.lwt.2013.05.014

Rubio R, Jofré A, Martín B, Aymerich T, Garriga M (2014a) Characterization of lactic acid bacteria isolated from infant faeces as potential probiotic starter cultures for fermented sausages. Food Microbiol 38:303–311. https://doi.org/10.1016/j.fm.2013.07.015

Rubio R, Martín B, Aymerich T, Garriga M (2014b) The potential probiotic lactobacillus rhamnosus CTC1679 survives the passage through the gastrointestinal tract and its use as starter culture results in safe nutritionally enhanced fermented sausages. Int J Food Microbiol 186:55–60. https://doi.org/10.1016/j.ijfoodmicro.2014.06.013

Ruiz-Moyano S, Martín A, Benito MJ, Nevado FP, de Guía CM (2008) Screening of lactic acid bacteria and bifidobacteria for potential probiotic use in Iberian dry fermented sausages. Meat Sci 80(3):715–721. https://doi.org/10.1016/j.meatsci.2008.03.011

Sameshima T, Magome C, Takeshita K, Arihara K, Itoh M, Kondo Y (1998) Effect of intestinal Lactobacillus starter cultures on the behaviour of Staphylococcus aureus in fermented sausage. Int J Food Microbiol 41(1):1–7. https://doi.org/10.1016/S0168-1605(98)00038-5

Saulnier DM, Spinler JK, Gibson GR, Versalovic J (2009) Mechanisms of probiosis and prebiosis: considerations for enhanced functional foods. COBIOT 20(2):135-141. https://doi.org/10.1016/j.copbio.2009.01.002

Sbihi HM, Nehdi IA, Al-Resayes SI (2013) Characterization of Hachi (Camelus dromedarius) fat extracted from the hump. Food Chem 139(1–4):649–654. https://doi.org/10.1016/j.foodchem.2013.01.118

Shan LC, De Brún A, Henchion M, Li C, Murrin C, Wall PG, Monahan FJ (2017a) Consumer evaluations of processed meat products reformulated to be healthier – a conjoint analysis study. Meat Sci 131:82–89. https://doi.org/10.1016/j.meatsci.2017.04.239

Shan LC, Regan Á, Monahan FJ, Li C, Lalor F, Murrin C, Wall PG, McConnon Á (2017b) Consumer preferences towards healthier reformulation of a range of processed meat products. Br Food J 119(9):2013–2026. https://doi.org/10.1108/BFJ-11-2016-0557

Sidira M, Karapetsas A, Galanis A, Kanellaki M, Kourkoutas Y (2014) Effective survival of immobilized *Lactobacillus casei* during ripening and heat treatment of probiotic dry-fermented sausages and investigation of the microbial dynamics. Meat Sci 96(2): 948–955. https://doi.org/10.1016/j.meatsci.2013.09.013

Song M-Y, Van-Ba H, Park W-S, Yoo J-Y, Kang H-B, Kim J-H, Kang S-M, Kim B-M, Oh M-H, Ham J-S (2018) Quality characteristics of functional fermented sausages added with encapsulated probiotic *Bifidobacterium longum* KACC 91563. Korean J Food Sci Anim Resour 38(5): 981–994. https://doi.org/10.5851/kosfa.2018.e30

Sperti GS (1971) Probiotics. AVI Pub. Co. https://lccn.loc.gov/71151608

Syngai GG, Gopi R, Bharali R, Dey S, Lakshmanan GMA, Ahmed G (2016) Probiotics – the versatile functional food ingredients. J Food Sci Technol 53(2):921–933. https://doi.org/10.1007/s13197-015-2011-0

Terpou A, Papadaki A, Lappa I, Kachrimanidou V, Bosnea L, Kopsahelis N (2019) Probiotics in food systems: significance and emerging strategies towards improved viability and delivery of enhanced beneficial value. Nutrients 11(7):1591. https://doi.org/10.3390/nu11071591

Vuyst LD, Falony G, Leroy F (2008) Probiotics in fermented sausages. Meat Sci 80(1):75–78. https://doi.org/10.1016/j.meatsci.2008.05.038

Wang X, Yang Y, Huycke MM (2020) Risks associated with enterococci as probiotics. Food Res Int 129:108788. https://doi.org/10.1016/j.foodres.2019.108788

Yadav R, Shukla P (2017) An overview of advanced technologies for selection of probiotics and their expediency: a review. Crit Rev Food Sci Nutr 57(15):3233–3242. https://doi.org/10.1080/10408398.2015.1108957

Yaqoob P (2014) Ageing, immunity and influenza: a role for probiotics? Proc Nutr Soc 73(2): 309–317. https://doi.org/10.1017/S0029665113003777

Žugić Petrović TD, Ilić PD, Grujović MŽ, Mladenović KG, Kocić-Tanackov SD, Čomić LR (2020) Assessment of safety aspect and probiotic potential of autochthonous enterococcus faecium strains isolated from spontaneous fermented sausage. Biotechnol Lett 42(8):1513–1525. https://doi.org/10.1007/s10529-020-02874-5

Chapter 10
Low-Salt Meat Products

Awis Qurni Sazili and Pavan Kumar

10.1 Introduction

Meat processing is very vital for improving the shelf life and increasing the variety and convenience of the developed products. Salt has been traditionally used in the preservation of meat especially due to dehydration and antimicrobial effects. Primitive humans used salt to preserve the meat. Besides improving the shelf life, salt plays a major role in the development of the flavor and taste of meat products. Ancient Egyptians were documented to preserve meat by adding salting and sun-drying. With the introduction of refrigeration systems/cold chains, the salt content is still present at high levels in processed meat products. This has led to higher dietary salt intake by the majority of consumers more than the recommended levels (Desmond 2006).

As per one estimate in 2011, the dietary intake was estimated to be 8.5 g for Irish women and 11.1 g for Irish men, and need to reduce this to 6 g per day as per recommendation by the Irish Food Safety Authority (IFSA) (Trieu et al. 2015). Furthermore, the World Health Organization (WHO) in 2012 recommended reducing the daily salt intake to less than 5 g per day and less than 2 g of sodium in adults. The average daily intake of sodium in Ireland, the UK, and India was reported in between 8.3 and 10 g per day per person (Kapil et al. 2003). High dietary sodium

A. Q. Sazili (✉)
Department of Animal Science, Faculty of Agriculture, Universiti Putra Malaysia, Serdang, Selangor, Malaysia

Halal Products Research Institute, Universiti Putra Malaysia, Serdang, Selangor, Malaysia
e-mail: awis@upm.edu.my

P. Kumar
Department of Livestock Products Technology, College of Veterinary Science, Guru Angad Dev Veterinary and Animal Sciences University, Ludhiana, Punjab, India
e-mail: pavankumar@gadvasu.in

S. A. Rather, F. A. Masoodi (eds.), *Hand Book of Processed Functional Meat Products*, https://doi.org/10.1007/978-3-031-69868-2_10

intake has been associated with the increased risk of high blood pressure, hypertension, and cardiovascular diseases (Aburto et al. 2013) and is widely associated with the top two dietary factors for causing global burden diseases among 67 risk factors studies in 21 countries from 2001 to 2010 (Lim et al. 2012). Excessive intake of salt is linked with several chronic diseases such as high blood pressure, kidney problems (Stock et al. 2022), and cardiovascular diseases (Kapoor et al. 2019).

Meat has a low quantity of sodium (approx. 50–90 mg of Na/100 g meat), but the addition of common salt (Sodium chloride/NaCl, containing 40% sodium) during the preparation of processed meat products makes a significant contributor to the overall dietary sodium intake (Verma and Banerjee 2012). On average, consumption of meat and meat products contributed to 20–30% of the total salt consumption in Western countries (Wirth 1991). In a study in North America, Europe, and Australia, the consumption of processed food was observed to account for about 70% of total salt consumed out of 20% of salt consumption was attributed to the consumption of processed meat products (Ruusunen and Puolanne 2005). As such, meat has low salt content, but the incorporation of salt during processing resulted in significantly/exponentially increasing the salt level in processed meat products. For example, fresh pork contains about 70 mg sodium per 200 g of meat, but the salt concentration increased to 21 folds (about 1480 mg per 100 g) in bacon (Henney et al. 2010). Further, about 77% of total sodium intake is met by packaged and restaurant foods. Twelve percent of total sodium intake is catered naturally from food and 11% by adding salt during cooking or consumption (Inguglia et al. 2017).

The increased awareness about the adverse effects of higher dietary intake of sodium or salt intake has led to the designing of various strategies for reducing the salt content in processed meat products. Various salt replacers or reformulations are used to reduce the salt content in processed meat products (Doyle and Glass 2010; Pietrasik and Gaudette 2014). Several countries in the 1980s have formulated dietary guidelines and undertaken various interventions, such as increasing public awareness to reduce the salt content in food (Inguglia et al. 2017). The replacement of sodium chloride with other chloride salts, such as potassium, calcium, magnesium, and lithium, but this strategy resulted in negatively affecting the functional and organoleptic attributes of meat products such as texture, flavor, taste, moisture content, juiciness and storage stability of these products. This has resulted in shifting the focus from using chloride salts to fat replacers, flavor enhancers, incorporation of various plant hydrolysates, and naturally salty plant ingredients such as seaweeds and yeast extracts (Verma et al. 2018).

10.2 Role of Salt in Meat Processing

Salt plays a vital role in the development of processed meat products and its reduction poses several challenges for the meat industry. The salt effect has an effect on the shelf-life of meat products, improves the color and flavor, and solubilizes the myofibrillar proteins, thus improving the emulsion stability, adhesion, and cohesion properties in processed meat products (Pietrasik and Gaudette 2014).

10.2.1 Antimicrobial Role of Salt

Salt exerts antimicrobial effects by reducing the free water available for the growth of microorganisms (water activity) by binding with water available in meat products. Overall, this antimicrobial effect due to salt in meat depends upon the availability of salt in the aqueous phase of the meat (Verma and Banerjee 2012). The sodium ions of salt efflux the water from the semipermeable membrane of the bacterial cells, consequently causing serious injury or death of the bacterial cells, thereby preventing bacterial growth. In addition to this, in some microbial cells, salt reduces oxygen solubility, leading to impaired energy production by interfering with the functioning of cellular enzymes in microbial cells (Shelef and Seiter 2005).

The salt reduction did not alter the microbial load but affected the growth and survival of the microorganisms over time. The meat products with reduced salt content were recorded with higher microbial growth as compared to their control counterparts having normal salt levels, such as in ham, chicken rolls, bacons, and beef burgers (Stringer and Pin 2005). However, it should also be mentioned that microbes have different survival and adaptation strategies to counter the effect of salt. Some microbes accumulate sugar, amino acids, and potassium to prevent the efflux of water from cells and help in survival. In addition to this, some microbes have the ability to change the cell morphology and alter the fatty acid profile of cell membranes or produce specific proteins, thus affecting the permeability of cells and countering the negative effect of salt on their growth and survival.

10.2.2 Effect on Sensory Quality

Salt is used in various food preparations to improve the organoleptic and functional attributes of the developed products. Salt improves the flavor and palatability of food products (Ruusunen and Puolanne 2005; Inguglia et al. 2017). In addition to the flavor and palatability, salt has the following effect on the function and quality of meat products such as:

(a) Salt solubilizes the salt-soluble myofibrillar and sarcoplasmic proteins and improves the emulsifying and gelling properties of meat products.
(b) Salt improves the water-holding capacity (WHC) of meat by extracting proteins and increasing hydration. This also resulted in increasing cooking yield and juiciness of meat products.

(c) Salt improves the viscosity of meat batter, consequently forming heat-stable emulsions. It results in improving the texture and tenderness of the meat products.
(d) Salt decreases fluid and nutrient loss.

Thus, reducing the salt content in meat products poses a challenge to maintaining the functional and quality attributes of meat products by applying suitable technological interventions. Saltiness and taste are the most common quality attributes of meat products affected by salt reduction. The texture could be improved by increasing the pH in low-salt meat products, but the low-salt and high pH could favor the microbial growth in the meat products, thereby affecting the food safety and shelf-life of such products (Stringer and Pin 2005). The salt reduction also alters the after-taste of the meat products (Paulsen et al. 2014).

The effect of salt on the quality attributes of meat products also depends upon the fat content of the products due to the interaction between salt and fat. The reduced salt content in high-fat meat products was not noticed to have adverse effects on the structure of low-salt and high-fat products (Tobin et al. 2012a, b, 2013). In beef patties reducing salt to 1% with 40% fat was reported to have good sensory attributes and was liked by the consumer panel of 20 members (Tobin et al. 2012a). Furthermore, authors noted that these patties have a 20% lower fat content and 50% lower salt content as compared to these levels in the commercial patties available in Ireland and UK (Tobin et al. 2012a). In pork breakfast sausages, the salt reduction from 2.4% to 0.8% led to tender and paler products, whereas the reduction in fat (37.5–22.5%) led to less dark, tougher, and less juicy products (Tobin et al. 2013). The pork sausage containing 1–1.4% salt was reported to have higher consumer acceptability as compared to sausages prepared with other salt levels (Tobin et al. 2013).

Salt improves the flavor and palatability of food products. It masks the bitterness and sweetness of the food products, thus having a major role in determining the overall acceptability (Barcenilla et al. 2022). It facilitates the volatilization of some molecules, thereby improving the aroma of the food. Even some meat products such as dry-cured ham are quite popular owing to the characteristic sensory attributes in which salt plays a vital role (Zhang et al. 2022).

10.3 Strategies for Salt Reduction in Meat Products

The salt reduction strategies in the meat products are centered around the following strategies such as-

(a) Reduction in the dietary salt content by partial or complete replacement of sodium chloride.
(b) Replacement of sodium chloride with low-sodium mixtures.
(c) Application of flavor enhancers such as yeast extracts, hydrolyzed vegetable proteins, and monosodium glutamate.

(d) Changing the physical form/state of the salt
(e) Application of novel technologies to improve the diffusion of salt such as high hydrostatic pressure and ultrasound.

The industry has followed the approach to reduce the salt content in meat products over a long period of time. The total reduction in salt level in food products in this approach happens at a very slow rate that consumers could perceive the changes in the saltiness and other organoleptic attributes (Kilcast and den Ridder 2007; Liem et al. 2011). In the UK adaptation of this strategy resulted in a 20–30% reduction in the salt levels in processed food products over a period of 3 years (He and MacGregor 2009). In a study conducted on 110 volunteers were used to detect the effect of gradually decreasing the salt content over a period of 6 weeks by using a single-blind test and the panelist did not notice a significant change in organoleptic attributes of sliced white bread while reducing 5% salt for over 6 weeks period resulting 25% reduction in salt levels (Girgis et al. 2003).

There are some limitations in using this strategy of salt reduction, the main concern being the time-consuming process. Moreover, to get the desirable impact on the consumer levels and improve the economics of the production process, this method needs to be applied to industry levels (Inguglia et al. 2017). There is a limitation on the overall salt reduction by using this strategy as reduction in salt to higher levels could lead to the making product unpalatable and poor in taste. As salt has an important role in the preservation of meat products, and this aspect should be considered while designing any approach to salt reduction; for example, a 2.3% w/w reduction in salt content in bacon was reported to have a shelf-life of 28 days as compared to 56 days shelf-life of bacon with 3.5% w/w salt content (Stringer and Pin 2005).

10.4 Salt Alternatives/Replacers and Mixtures

Various salt alternatives or replacers are available in the market that are used to reduce the salt content in processed meat products. These compounds have a similar role as salt in improving the functional and sensory qualities such as flavour or saltiness of meat products. The type of salt replacer or alternative depends upon several factors such as the type of the end product prepared (such as patties, balls, sausages), formulation of the product, desired physiological and sensory properties of the products (Fellendorf et al. 2016). A range of restructured meat products, such as sausages, nuggets, and burger patties, have been prepared by using lower-sodium ingredients. In these products, the functional and structural function of salt-soluble proteins was replaced by using milk or soy protein, starches, and gums (Fellendorf et al. 2016).

Potassium chloride is the most commonly used salt substitute. This has an antimicrobial effect equivalent to sodium chloride, and on a molar basis, KCl had an equivalent antimicrobial effect on *Yersinia enterocolitica*, *Aeromonas hydrophila*, *Shigella flexneri*, *Staphylococcus aureus* and *Enterobacter sakazakii* in laboratory media (Bidlas and Lambert 2008). Furthermore, KCl has an adverse effect on the texture and flavor of meat products. A 40% substitution of KCL of potassium lactate to NaCl in fermented sausages and dry-cured pork loins resulted in marked flavor defects in these compounds (Gou et al. 1996). Similarly, a 20% replacement of salt by glycerine in dry-cured loin had flavor defects. However, a 30% salt substitution by potassium lactate and more than 50% glycerine incorporation was observed to affect the cohesiveness of the fermented sausages (Gou et al. 1996). Similar to this study, a partial replacement of salt over 40% by using a mixture of potassium chloride + glycerine or potassium lactate + glycerine was observed to cause flavor and textural defects in fermented sausages (Gelabert et al. 2003). The calcium, potassium, and sodium lactate salts exert an antimicrobial effect and prevent the growth of bacteria during storage. A combination of lactate (sodium diacetate and potassium lactate) was successfully replaced the 40% salt and maintained the organoleptic and microbial quality of packaged cooked meat (Devlieghere et al. 2009).

Sodium lactate is abundantly used in processed meat products to inhibit Listeria and extend the shelf-life of meat products. It ranks the second most common source of sodium in meat products after sodium chloride (Gou et al. 1996). Sodium lactate is increasingly replaced by potassium lactate in the meat industry due to its similarity in antimicrobial and structural functions (Gou et al. 1996). Furthermore, the lactate solution (2–3% solution containing 4% sodium diacetate and 56% potassium lactate) exerted antimicrobial activity against *L. monocytogenes*, *C. botulinum,* and *Staphylococcus aureus* in addition to positively improving the water-holding capacity of meat, resulting in improving shelf life (up to 75–125%), cooking yield, texture, and juiciness of the cooked frankfurter type sausages (Stekelenburg 2003). The combination of lactates exerted a synergistic effect in inhibiting the growth of *L. monocytogenes* (Stekelenburg 2003).

The incorporation of citric and lactic acids in the meat products enhances the perceived saltiness of the sodium chloride, hence used in reduced salt meat products (Dötsch et al. 2009). The incorporation of salt replacers could enhance the spiciness, saltiness and fatness perceptions, but this may vary with the type of salt replacer used and formulations. Further, some flavor enhancers, such as arginine, are also used to improve the intensity and perceived saltiness of low-salt meat products. In the preparations of reduced salt and fat black puddings, products prepared with 5% fat and 0.6% sodium along with pectin and various salts of potassium (citrate, chlorides, and phosphates), as well as samples with 10% fat and 0.4% sodium with waxy maize starch was preferred by the consumers (Fellendorf et al. 2016).

To reduce the overall salt content in meat products, various solutions with low sodium content were developed. These are commercially available such as a commercial low salt mixture containing iodine, L-lysine hydrochloride, potassium chloride, and magnesium sulphate is available with the trade name Pansalt®. It imparts the same flavor and savory taste with a 43% reduction in sodium content. The application of Pansalt® in ground beef patties did not have any adverse effect on the sensory properties as compared to control ground beef patties with sodium chloride (Ketenoğlu and Candoğan 2011). Similarly, another commercial preparation known as Sub4salt® is available for the development of low-sodium meat products. This is made up of sodium chloride, potassium chloride, and sodium gluconate and reported to reduce the sodium content by upto 30% in ham and emulsified sausages without having any perceivable effect on the taste of these products (Wilson et al. 2012). A mixture of 70% sodium chloride and 30% potassium chloride or 70% sodium chloride and 30% magnesium chloride was used in the preparation of ham (Collins 1997). The developed product was recorded with comparable sensory and structural attributes as with the hams prepared with 100% sodium chloride (Collins 1997).

Salt enhancers are compounds that improve the salty flavor. These compounds such as hydrolyzed vegetable protein, yeast extracts, nucleotides, monosodium glutamate, seaweeds are commonly used to improve the taste and flavor of low salt meat products (Brandsma 2006). Monosodium glutamate (MSG) improves the flavor and taste perception of food products by improving the salivation in the mouth cavity. In processed meat products, MSG is commonly used in the preparation of low-salt sausages to compensate for the replacement of sodium chloride with potassium chloride. The low-salt fermented cooked sausage was prepared by replacing 50–75% sodium chloride with potassium chloride and adding flavor enhancers such as MSG, taurine, disodium guanylate, lysine and disodium inosinate (dos Santos et al. 2014). The incorporation of the various salt enhancers masked the undesirable organoleptic attributes of the salt replacers and the developed product was noted with good overall acceptance and upto 68% sodium reduction (dos Santos et al. 2014).

The potassium chloride imparts the metallic flavor to the products and masking of this metallic flavor was reported by the application of various commercially available yeast autolysates such as Provesta and Aromild (Desmond 2006). However, the yeast autolyzates imparts a strong broth flavor to the products, which is not liked by the consumers. This could be taken care of by optimizing the formulation of meat products and optimizing the umami effect (Lilić and Matekalo-Sverak 2011).

An increase ($p < 0.001$) in the redness and decrease in yellowness ($p < 0.001$) score of cooked emulsified pork sausage was noted on replacing salt with a mixture of calcium chloride, magnesium chloride, and potassium chloride (Jin et al. 2018). The magnesium chloride at 15% replacement of salt improved texture but increased the lipid oxidation during 5 weeks of refrigerated storage of sausages. Calcium chloride at 5% replacement of salt did not affect the quality of sausages despite showing the highest ionic strength (Jin et al. 2018). Sausages prepared by a combination of calcium chloride, magnesium chloride, and potassium chloride replacing 70% of the salt had poor sensory attributes (Jin et al. 2018).

Salt (sodium chloride) plays an important role in decreasing the formation of non-polar heterocyclic amines during thermal processing which are potentially carcinogenic compounds (Jiayun et al. 2021; Zhao et al. 2023). However, it should also be noted that sodium chloride promotes the formation of polar (Li et al. 2020). Basic amino acids have the potential to reduce the formation of Has by changing the gelation of myosin gel and improving the solubility of myosin at a low-salt concentration (1 mM). This results in increasing the WHC and gel strength (Zhang et al. 2017, 2022). Furthermore, the basic amino acids could exert an antioxidant role in meat products by scavenging free radicals and inhibiting phospholipase activity (Xu et al. 2018).

Overall, there are high chances of the formation of HAs in low-salt meat products. The incorporation of basic amino acids such as histidine, lysine, and arginine in low-fat meat products helps in improving the sensory and functional quality attributes. In addition, the incorporation of these amino acids was noted to decrease the formation of heterocyclic amines (HA) (Deng et al. 2024). Lysine addition at 1% effectively reduced (70.88% inhibition, followed by 1% histidine at 67.03% and 1% arginine at 59.51% inhibition) the HA formation in beef patties as compared control fried beef patties prepared with 3% salt. Furthermore, the low-salt fried beef patties with 1% sodium chloride had significantly ($p < 0.05$) improved quality characteristics (texture, frying loss, moisture, and instrumental color) as compared to control patties with 3% sodium chloride (Deng et al. 2024).

Dry-cured lacon is a famous traditional pork product of Spain. The 55% and 70% sodium chloride replacement by potassium chloride, calcium chloride, and magnesium chloride in these sausages had good sensory attributes as compared to products with 50% salt replacement and 100% salt (Lorenzo et al. 2015). Low-salt frankfurter with good sensory, technological, and microbial quality can be prepared by adding 0.85% potassium chloride and 0.25% phosphates which would have a 25% reduction in sodium chloride and an 18% reduction in sodium content (Yotsuyanagi et al. 2016).

In jerked beef, 50% sodium chloride replacement with potassium chloride and calcium chloride ($KCl + CaCl_2$ in equal concentration) caused a significant reduction in sodium content in the reformulated products (Vidal et al. 2019). Calcium chloride incorporation resulted in a bitter taste and rancid aroma due to higher lipid oxidation in the dried product. Furthermore, the authors reported no effect of salt substitution on the microbiological quality of the restructured beef jerky (Vidal et al. 2019).

10.5 Limitation of Salt Replacers

The metallic flavour of potassium chloride is the main hindrance in the application of potassium chloride as a salt replacer. Thus, an optimal level of a combination of NaCl and KCl is commonly used in the meat industry for reducing salt in processed meat products. However, the higher consumption of KCl results in a high potassium load in the body that could also adversely impact type 1 diabetes, adrenal insufficiency, and renal diseases (Khaw and Barrett-Connor 1984).

The use of salt enhancers itself contains high levels of salt, such as hydrolyzed vegetable protein containing 40% salt; their application in meat processing has its limitations. Monosodium glutamate (MSG) acts on glutamate receptors and facilitates the release of neurotransmitters, which play an important role in physiological and pathological functions (Abdallah et al. 2014). MSG consumption, in the long run, decreases taste sensation and is potentially associated with gastric disturbances, hyperacidity, headache, obesity, metabolic disorders, neurotoxic, adverse effects on reproduction, and sickness (Fernstrom 2007; Niaz et al. 2018). Furthermore, this is classified under food additives, hence not preferred by the consumers in general (Shim et al. 2011; Niaz et al. 2018). The application of various salt replacers in mortadella was noted to decrease the flavor scores (Horita et al. 2011).

10.6 Modifying the Physical State of Salt

The size of salt particles plays an important role in food preparation by affecting the saltiness of the food products and altering the physiological changes in the food matrix (Inguglia et al. 2017). A smaller salt particle size will increase the overall effect of salt, have a higher dissolution rate leading to higher salt perception, and be homogenously mixed in the food matrix (Desmond 2006). The manipulation of the size of salt particles could be a useful tool in achieving the reduction of salt in the product. In fried potato crisps, smaller crystals of salt were noted to enhance the perception of saltiness by faster delivery of sodium ions into the saliva (Rama et al. 2013).

The application of sea salt with a small crystal size (20 mm) was reported to improve the salt diffusion into the food, thereby increasing the salt perceptions by the consumers (Johnson et al. 2008). Furthermore, changing the particle's physical form, such as flakes, also dissolves faster in saliva due to an increase in surface area and lowered bulk density, and could be helpful in salt reduction. The flaked salt crystals are also commercially available now, such as Alberger flake salt and Star flake Salt manufactured by Cargill.

10.7 Natural Salt Replacers

The application of natural compounds (plants, herbs and spices) owing to salty taste in place of synthetic minerals could be a suitable strategy for producing low-salt meat products. These compounds are more efficient and effective and do not have adverse effects on the health of consumers. In addition, these compounds cost less, improve the nutritive and functional quality of meat products, and are readily available (Aprilia and Kim 2022).

Halophyte plants that grow in marshy and high-saline environments have been utilized as salt replacers by the meat industry. Glasswort (*Salicornia herbacea*), a halophyte, is widely used as a salt replacer in the food industry. It is a rich source of various bioactive compounds exerting antioxidant, antimicrobial, and anticancer properties (Kim and Lee 2009), salt (3.5–20%) and dietary fiber (73–74%) (Seung-Kwan 2004). A 50% replacement of salt by glasswort powder (20 g NaCl and 40 g glasswort per kg of meat cuts) in various healthier low-sodium dry-cured pork cuts (biceps femoris, semimembranous, quadriceps, semitendinous, adduct, gluteus, and gracilis) did not adversely affect the organoleptic and textural properties (Seong et al. 2017). The incorporation of glasswort in jerky inhibited microbial growth and lipid oxidation (Lim et al. 2013). Similarly, the incorporation of glasswort powder in low-salt frankfurters (1.5% glasswort and 0.75% sodium chloride) improved protein solubility, cooking yield, emulsion stability, and textural qualities mainly attributed to the high fiber and salt present in the glasswort powder (Kim et al. 2014).

Various hydrated states of glasswort powder exerted different functions in the meat products. The glasswort hydrates having non-meat ingredients such as sodium caseinate, carboxymethyl cellulose and isolated soy protein have significant effects on the colour and pH of the low-salt low-fat frankfurters (Lim et al. 2015). In raw sausage batter, the glasswort hydrates addition resulted in lower redness and higher yellowness value, increasing cooking loss and decreasing hardness, juiciness, and tenderness of the cooked products (Lim et al. 2015). However, the glasswort hydrates with non-meat ingredients were noted to improve the water-holding capacity, viscosity, hardness, and emulsion stability of low-salt low-fat frankfurters (Lim et al. 2015).

Various plant extracts are a rich source of phenolic compounds and minerals. These have been used in the development of functional meat products with improved antioxidant and antimicrobial potential (Kumar et al. 2020b; Awad et al. 2021, 2022). In low-salt meat products, the incorporation of these compounds could improve the microbial quality and shelf-life of these products. The application of various plant powders, spices, and herbs in the development of low-salt meat products is presented in Table 10.1.

10.8 Application of Novel Processing Technologies

The thermal processing of meat could have adverse effects on the flavour, texture, and nutritive value of the developed products. Various non-thermal technologies are increasingly used in meat processing to extend the shelf-life of the products and improve the structural and functional quality attributes. Some of the novel non-thermal meat processing technologies are high-pressure processing, ultrasound application, ultrafiltration, pulsed electric field, pulsed light and pulsed electric field (Kumar et al. 2020a; Mehta et al. 2020, 2022). Out of these, ultrasound and high-pressure processing have been used in the development of low-fat meat products.

High-pressure processing of food products is undertaken by applying a pressure of 300–600 MPa to food products without significantly increasing the temperature (less than 45 °C). Thus, the food products are preserved due to the inactivation of spoilage and pathogenic microorganisms without having adverse effects on the nutritive value, flavor, color, and texture of the products (Bolumar et al. 2021). Application HPP in the food processing was observed to inactivate food pathogens (such as *E. coli. L. monocytogenes* and *Salmonella*) and vegetative spoilage

Table 10.1 Application of various plant powders, spices and herbs in the development of low-salt meat products

Product	Processing protocol	Salient findings	References
Chicken nuggets	Winter mushroom powder (0.5–1%), Sodium pyrophosphate: 0.3%	No significant effect on the sensory attributes and soften textural attributes of the developed nuggets Increase in pH and jelly and melted fat Inhibition of lipid oxidation	Jo et al. (2018)
Stir-fried pork	Brewed soy sauce	17–50% salt replacement by brewed soy sauce in stir-fried pork without affecting the flavor and taste	Kremer et al. (2009)
Beef patties	RWPS at 2%,	Increased fiber, calcium, potassium and phenolic compounds Reduced growth of aerobic mesophilic and LAB Reduced salt up to 1 g/100 g of patties	García-Lomillo et al. (2017)
Low salt ready meal of chicken with other dishes	Herbs, spices, and whey proteins	Addition of garlic, rosemary, oregano and sage at 0.1–0.5% improved the flavor and overall acceptability Clove and pimento adversely affected the acceptability	Mitchell et al. (2013)
Brazilian franfurters	Fresh garlic, garlic oil, garlic extract (0.2%) obtained by PLE, 50% NaCl replaced by KCl and $CaCl_2$	Lowered sodium content and inhibited lipid oxidation Improved microbial safety by inhibiting the growth of spoilage bacteria Poor sensory score with the use of garlic oil	Horita et al. (2016)
Rabbit burger	Garlic powder (0.25%) and salt (1%)	Significant effect on pH (lower value), WHC and color attributes (lower redness) Reduced drip loss and cooking losses in burgers added with salt and garlic powder	Mancini et al. (2020)
Smoked hilsa fish	Salt, garlic and coriander 10% each	No detection of Salmonella and other indicator organisms Highest sensory and textural attributes were recorded for product having 10% salt and 105 garlic	Hossain et al. (2012)
Pork sausages	CEO: 0.12 μL/g and Sodium nitrite at 60 ppm	Higher redness of the cooked product Inhibited microbial growth and improved the antioxidant potential	Šojić et al. (2019)

(continued)

Table 10.1 (continued)

Product	Processing protocol	Salient findings	References
Bologna sausage	KCl, mixture of coriander, cardamon, pepper, and onion	Production of healthier products by 50% salt replacement by KCl Improved microbial safety without negatively affecting palatability	Carraro et al. (2012)

RWPS red wine pomace seasoning, *LAB* lactic acid bacteria, *PLE* pressurized liquid extraction, *WHC* water-holding capacity, *CEO* Coriander essential oil, *KCl* potassium chloride, $CaCl_2$ calcium chloride, *NaCl* sodium chloride

organisms (such as lactic acid bacteria, yeast and Pseudomonas) (Hugas et al. 2002; Kamenik et al. 2015). The pressurization of meat resulted in the interaction between salt and proteins, thereby increasing the perception of saltiness by the consumers.

Ultrasound technology applies sound waves with frequencies more than 20 kHz (20–100 kHz) and intensities mostly at 1 w/cm^2. This application of sound wave energy led to several changes in the meat products, such as the inactivation of enzymes, improving mass transfer rate, inducing redox reactions, extraction of proteins, and inactivation of microorganisms (Awad et al. 2012; Mehta et al. 2022). The microbial inactivation is attributed to the compression and rarefaction of particles observed while passing the sound waves through the liquid solutions. This results in the formation of cavitations or bubbles, and these become unstable and collapse with the passing of each sound wave cycle. It leads to localized increases in the temperature and pressures, thereby disrupting the microbial cell membranes and functional ingredients (Dolatowski et al. 2007). The imposition of bubbles also leads to the microjet effect, thereby improving the mass transfer (liquid) in meat. This feature is used in the development of low-salt meat products by accelerating the curing process and better and uniform salt distribution. This results in the increased perception of saltiness by the consumers even when the salt content is lower (Dolatowski et al. 2007; Alarcon-Rojo et al. 2015).

Pulse electric field (PEF) is a non-thermal technology that decontaminates and improves food quality by using short high-voltage pulses (Singh et al. 2012; Mehta et al. 2020). This technology could improve the diffusion and absorption of salt by cellular tissues, thereby improving the perception of saltiness. Although this technology improves the functional properties of meat proteins, but there are chances of spoiling the salty taste if the technology is not used properly (Pinton et al. 2021). Overall PEF increases the salt uptake and reduces the curing time and salt concentration (Galindo et al. 2009). The application of novel processing technologies in the development of low-salt meat products is presented in Table 10.2.

Table 10.2 Application of novel processing technologies in the development of low-salt meat products

Meat products	Processing parameters	Salient findings	References
High pressure processing			
Raw pork meat	High pressure (350 MPa at 20 °C for 6 min), NaCl: 1.5%, sodium nitrite (0–100 ppm)	Salt and HPP have synergistic action due to pressure sensitization of the microbial cells HPP delayed the spoilage but the combination of HPP with salt decreased microbial count to less than 2 log CFU/g	Duranton et al. (2012)
Smoked dry-cured ham	600 MPa for 5 min; NaCl replaced by KCl and potassium lactate	No elimination of *Salmonella* and *L. monocytogenes* in non-pressurized products HPP resulted in the elimination of both pathogens in smoked ham after 14 days NaCl free products had lower stability and both pathogens were identified on 28 and 56 days	Stollewerk et al. (2012)
Pork breakfast sausages	HPP at 150 MPa, 0.5–2.5% salt	Salt level below 1.5% had effect on the fat loss, color, sensory and texture of the product 2% salt level necessary to maintain desirable saltiness of the sausages HPP improved the emulsion stability and cooking yield HPP reduced the juiciness and cohesiveness of the sausages	O'Flynn et al. (2014)
Pork sausage emulsion and frankfurters	HPP at 100–400 MPa at 10 °C for 2 min, salt: 1–2.5%,	HPP at 200 MPa for 2 min at 10 °C had desirable effect Improved cooking yield and emulsion stability at 1% salt in emulsion and 2.5% salt in frankfurters Increased lightness and textural properties in sausage with 20% added fat and 1% salt	Yang et al. (2015)
Beef and pork frankfurters	HPP at 150–300 MPa at 20 °C for 5 min, salt: 1.5–2.5%	Improved textural attributes such as chewiness, cohesiveness and gumminess Lower cooking loss in product with 1% salt and 150 MPa due to increased emulsion stability	Crehan et al. (2000)
Dry-cured ham	HPP at 600 MPa, sliced and vacuum packaged	HPP increased lightness and physiological, microbiological and sensory attributes No effect on oxidative stability	Clariana et al. (2011)

(continued)

Table 10.2 (continued)

Meat products	Processing parameters	Salient findings	References
		Increased saltiness of the product upon pressurization	
Ham	HPP at 100 MPa for 5 min, NaCl: 1.1%, KCl: 0.2%	HPP at 100 MPa after tumbling significantly reduced salt content in ham HPP on raw meat or during injecting brine deteriorate structure and poor water retention	Tamm et al. (2016)
Sausage (chicken meat and pork fat)	HPP at 400 MPa for 10 min, cinnamon and clove extracts, 1.75% salt, 0.75% KCl and 0.5% phosphate and sugar each	HPP treated sausage exhibited higher pH throughout the storage period The HPP and spice extract combinations improved the microbial quality and extended the storage life	Xiao et al. (2021)
Pork sausage	HPP at 400–800 for 5 min MPa, carrot fiber, potato starch, 1.2–1.8% salt	Increased whiteness with the incorporation of starch, fiber and pressurization HPP, fiber and starch addition improved the water binding properties of low-fat sausages without affecting the sensorial attributes	Grossi et al. (2012)
Restructured chicken nuggets	HPP at 200–600 MPa for 5 min at 15 °C, KCl	50% NaCl replacement with KCl did not affect the sensory, textural and color attributes Higher cooking yield with increased HPP	Luckose et al. (2015)
Ultrasound application			
Pork loin	US—at 40 kHz, 37.5 W/dm^3 Salt: 50–280 NaCl/m^3	Increased mass transfer rate of salt in meat Ultrasound application improved the texture attributes	Ozuna et al. (2013)
Cured-ham	US at 4.2–19 W/cm$_2$ for 10–40 min	Salting could be a surface phenomenon US accelerated the curing process and improved mass transfer and protein extraction The desired salt level (2.5% NaCl) achieved within 2 h with US and 4 h without US	McDonnell et al. (2014a, b)
Pork	US 54.9 W/cm^2 for 2 h Salt/salt replacer: 5%	No improvement in salt diffusion rate for salt replacers Lowered cooking loss with salt replacers Only higher US resulted in improving curing rate	Ojha et al. (2016)

(continued)

Table 10.2 (continued)

Meat products	Processing parameters	Salient findings	References
Spiced beef	US 800 W, 20 kHz for 120 min	US increased the permeability of muscle tissues resulting in improved salt diffusion and stable product Marked reduction in cooking time Significantly ($p < 0.05$) increase in salt content in beef Improved flavor and taste of product but high US increased lipid oxidation	Zou et al. (2018)
Restructured cooked ham	US 600 W/cm^2 for 10 min, KCl: 0.5%, Na: 324 mg/100 g	US increased flavor, acceptability and fluid release KCl more effective than US in improving sensory and textural attributes of ham	Barretto et al. (2020)
	US 600 W/cm^2 for 10 min, salt: 0.75–1.5%	US increased total fluid loss, redness, and hardness 0.75% salt level resulted in 30% reduction in sodium content in the product US improved the acceptability of product with 0.75% salt	Barretto et al. (2018)
Basic Electrolyzed water			
Meat emulsion	BEW replacing 0–50% salt, US at 175 W, 25 kHz for 20 min	BEW-treated meat emulsion had high fat and water retention properties BEW and US could replace 30% salt in the emulsion with good stability and cooking yeild	Sena Vaz Leães et al. (2020)

HPP high pressure processing, *US* ultrasound, *WHC* water-holding capacity, *KCl* potassium chloride, $CaCl_2$ calcium chloride, *NaCl* sodium chloride, *Na* sodium, *BEW* basic electrolyzed water

10.9 Microbial Quality of Low-Salt Meat Products

Sodium chloride is very effective in controlling microbial growth thus in the case of low-salt meat products, proper strategies should be taken by using appropriate preservatives to control the growth of microorganisms (Perea-Sanz et al. 2018). The growth of microorganisms was reported to be higher in low-salt meat products as compared to their control products containing conventional salt levels. (Stringer and Pin 2005) noted the higher growth rate of *Bacillus cereus*, *Clostridium botulinum*, *Aeromonas hydrophilia*, *Listeria monocytogenes* and *Yersinia enterocolitica* in meat products with low-salt levels (salt levels—chicken rolls: 1.61%, ham: 2.81% and bacon: 2.85%) as compared to their control counterparts having higher salt levels (3.04% in chicken rolls, 5.57% in ham and 5.5% in bacon). Furthermore,

Clostridium botulinum is sensitive to salt concentration and did not grow at a salt concentration of 5.5% or above (Stringer and Pin 2005).

It should be noted that the overall effect of salt depends upon several factors such as pH, redox potential, temperature, availability of nutrients, sensitivity of microorganisms to salt, and presence of other antimicrobial compounds and processing conditions (Shelef and Seiter 2005). Further, in most meat products, salt is added at a 1.5–2.5% level, and at this level, it has a limited role in inhibiting microbial growth. The overall effect of sodium chloride in inhibiting spoilage and pathogens depends upon the pathogen of interest (as clostridium botulinum is sensitive to salt), formulations, and processing conditions (Taormina 2010). Thus, there is a need to undertake further in-depth research on the optimum and desirable levels of sodium chloride and its replacers in developing meat products.

Salt at high concentrations may interfere with the proper functioning of cellular enzymes and initiate the efflux of sodium ions from the microbial cells, resulting in microbial inhibition. Sat also increases lipid oxidation. Salt at 0.5–2.0% acts as a pro-oxidant on ground beef and at 1.5% salt level in pork, attributed to the disrupting the cell membrane by releasing ions and inhibiting antioxidant enzymes (Mariutti and Bragagnolo 2017). The alteration in the water activity upon salt incorporation could also have an impact on the enzymatic activity of some enzymes such as neutral lipases, cathepsins, and aminopeptidases inhibition by low water activity (0.85) caused due to dry curing of ham. However, on the other hand enzymatic activity of some enzymes such as acid phospholipase, calpain, and acid lipase, do not affected by the water activity (Toldrá 2006). In curing ingredients, sodium nitrites are also added to inhibit the growth of some pathogens in meat products and stabilization of typical cured meat product (Armenteros et al. 2012).

However, it should be noted that microorganisms are more resistant to thermal or pressurization at low water activity conditions/low moisture foods (Finn et al. 2013). The effect of salt depends upon the type of microorganisms such as some halophiles even grow at high salt content in food products due to the presence of enzymes that are active at very high ionic strength (Ortega et al. 2011). The *L. monocytogenes* has high resistance to salt, whereas *Clostridium botulinum* is salt sensitive. *L. monocytogenes* can survive very high salt concentrations (upto 10%) (García-Díez and Patarata 2017).

In cured meat products, lowering salt and nitrite levels results in higher growth of lactic acid bacteria and proteolytic bacteria, thereby reducing the shelf-life of meat products (Stringer and Pin 2005). The dry fermented sausages prepared by 58% salt replacement with potassium chloride and calcium chloride was noted to have higher growth of lactobacilli as compared to the growth of these organisms in the control samples having 2.4% sodium chloride (Beriain et al. 2011). The authors (Beriain et al. 2011) noted a decrease in the lactobacilli counts during the ripening phase in control samples, whereas in the treated samples, these levels were noted to be stable. The suppression of bacterial growth in the presence of salt concentration could lead to the higher growth of yeasts. (Aaslyng et al. 2014) observed higher yeast count (4.7–5.4 log CFU/g) in high-salt bacon (salt content: 10%) as compared to lower yeast count of 1.3–3.9 log CFU/g in low-salt bacon having 1.4% salt.

10.10 Microbial Safety of Low Salt Fermented Meat Products

Salt replacers such as potassium chloride, magnesium sulfate, calcium chloride, and magnesium chloride had poor inhibitory activity against microorganisms as compared to common salt (sodium chloride) and could not exert the desirable inhibition of microorganisms (Samapundo et al. 2010; Tamm et al. 2016). In traditional sour meat products, 4–6% salt is added to extend the shelf-life, but due to the fermentation process involving the efflux of water, the salt concentration of the end product could be higher (Wang et al. 2022). The low-salt fermented meat had higher levels of branched-chain amino acids (isoleucine, leucine, and valine) and volatile compounds (hexanal, butanol, heptanal, etc.) due to fermentation by the *Lactobacillus*, *Aspergillus*, *Staphylococcus*, *Mucor*, *Pediococcus*, and Debaryomyces causing typical sour meat flavor (Wang et al. 2022). The high water activity of Portuguese low-salt dry-fermented sausages resulted in an increase in the lactic acid bacteria growth and masked the negative effect of salt reduction in flavor and overall acceptability of these products (Laranjo et al. 2015). Salt reduction in fermented Iberain ham resulted in improving flavor and a higher generation of volatile flavor compounds, but salt reduction had a negative effect on the texture of the products (Lorido et al. 2015).

The fermentation temperature had an important role in affecting the flavor and texture of meat products. The optimum quality of sour meat in terms of microbial safety, flavor compounds, hard texture, and sensory attributes was recorded by fermenting at 20–25 °C (Lv et al. 2019a, b). At this temperature, the high fermentation rate resulted in the inhibition of Pseudomonas and Enterobacteriaceae. Furthermore, at the initial stage of fermentation, *L. sakei* dominated the microflora, which changed to *L. curvatus* in the middle stage of fermentation and *L. plantarum* in the later stage of fermentation (Lv et al. 2019a). In addition to bacteria, fungi (*Debaryomyces*, *Pichia*, and *Wickerhamomyces*) also play a role in the development of flavor in sour meat by producing lipase and proteases hydrolyzing lipids and proteins, resulting in the generation of volatile compounds (Zhong et al. 2021, 2022).

10.11 Conclusion

There is a need to use a multi-pronged strategy for the development of low-salt meat products to compensate for the various functions of salt, such as taste and flavor enhancers, antimicrobial compounds and proper processing technology. The salt substitute should be safe, available at low cost and should not have any adverse effect on the functional and sensory qualities of the meat. Various novel processing technologies such as ultrasonication and high-pressure application have promising results in the development of low-salt meat products by improving the mass transfer rate, accelerating curing time and inhibiting microorganisms.

References

Aaslyng MD, Vestergaard C, Koch AG (2014) The effect of salt reduction on sensory quality and microbial growth in hotdog sausages, bacon, ham and salami. Meat Sci 96:47–55. https://doi.org/10.1016/j.meatsci.2013.06.004

Abdallah CG, Jiang L, De Feyter HM, Fasula M, Krystal JH, Rothman DL, Mason GF, Sanacora G (2014) Glutamate metabolism in major depressive disorder. Am J Psychiatry 171:1320–1327. https://doi.org/10.1176/appi.ajp.2014.14010067

Aburto NJ, Ziolkovska A, Hooper L, Elliott P, Cappuccio FP, Meerpohl JJ (2013) Effect of lower sodium intake on health: systematic review and meta-analyses. BMJ 346:f1326–f1326. https://doi.org/10.1136/bmj.f1326

Alarcon-Rojo AD, Janacua H, Rodriguez JC, Paniwnyk L, Mason TJ (2015) Power ultrasound in meat processing. Meat Sci 107:86–93. https://doi.org/10.1016/j.meatsci.2015.04.015

Aprilia GHS, Kim HS (2022) Development of strategies to manufacture low-salt meat products – a review. J Anim Sci Technol 64:218–234. https://doi.org/10.5187/jast.2022.e16

Armenteros M, Aristoy M-C, Toldrá F (2012) Evolution of nitrate and nitrite during the processing of dry-cured ham with partial replacement of NaCl by other chloride salts. Meat Sci 91:378–381. https://doi.org/10.1016/j.meatsci.2012.02.017

Awad TS, Moharram HA, Shaltout OE et al (2012) Applications of ultrasound in analysis, processing and quality control of food: a review. Food Res Int 48:410–427. https://doi.org/10.1016/j.foodres.2012.05.004

Awad AM, Kumar P, Ismail-Fitry MR, Jusoh S, Aziz MFA, Sazili AQ (2021) Green extraction of bioactive compounds from plant biomass and their application in meat as natural antioxidant. Antioxidants 10:1465. https://doi.org/10.3390/antiox10091465

Awad AM, Kumar P, Ismail-Fitry MR, Jusoh S, Aziz MFA, Sazili AQ (2022) Overview of plant extracts as natural preservatives in meat. J Food Process Preserv. https://doi.org/10.1111/jfpp.16796

Barcenilla C, Álvarez-Ordóñez A, López M, Alvseike O, Prieto M (2022) Microbiological safety and shelf-life of low-salt meat products – a review. Food Secur 11:2331. https://doi.org/10.3390/foods11152331

Barretto TL, Pollonio MAR, Telis-Romero J, da Silva Barretto AC (2018) Improving sensory acceptance and physicochemical properties by ultrasound application to restructured cooked ham with salt (NaCl) reduction. Meat Sci 145:55–62. https://doi.org/10.1016/j.meatsci.2018.05.023

Barretto TL, Bellucci ERB, Barbosa RD, Pollonio MAR, Romero JT, da Silva Barretto AC (2020) Impact of ultrasound and potassium chloride on the physicochemical and sensory properties in low sodium restructured cooked ham. Meat Sci 165:108130. https://doi.org/10.1016/j.meatsci.2020.108130

Beriain MJ, Gómez I, Petri E et al (2011) The effects of olive oil emulsified alginate on the physico-chemical, sensory, microbial, and fatty acid profiles of low-salt, inulin-enriched sausages. Meat Sci 88:189–197. https://doi.org/10.1016/j.meatsci.2010.12.024

Bidlas E, Lambert RJW (2008) Comparing the antimicrobial effectiveness of NaCl and KCl with a view to salt/sodium replacement. Int J Food Microbiol 124:98–102. https://doi.org/10.1016/j.ijfoodmicro.2008.02.031

Bolumar T, Orlien V, Sikes A, Aganovic K, Bak KH, Guyon C, Stübler A, de Lamballerie M, Hertel C, Brüggemann DA (2021) High-pressure processing of meat: molecular impacts and industrial applications. Compr Rev Food Sci Food Saf 20:332–368. https://doi.org/10.1111/1541-4337.12670

Brandsma I (2006) Reducing sodium – a European perspective. Food Technol 60:24–29

Carraro CI, Machado R, Espindola V, Campagnol PCB, Pollonio MAR (2012) O efeito da redução de sódio e o uso de ervas e especiarias sobre a qualidade e segurança de mortadela. Food Sci Technol 32:289–297

Clariana M, Guerrero L, Sárraga C, Díaz I, Valero Á, García-Regueiro JA (2011) Influence of high pressure application on the nutritional, sensory and microbiological characteristics of sliced skin vacuum packed dry-cured ham. Effects along the storage period. Innov Food Sci Emerg Technol 12:456–465. https://doi.org/10.1016/j.ifset.2010.12.008

Collins JE (1997) Reducing salt (sodium) levels in processed meat, poultry and fish products. In: Production and processing of healthy meat, poultry and fish products. Springer, Boston, pp 282–297

Crehan CM, Troy DJ, Buckley DJ (2000) Effects of salt level and high hydrostatic pressure processing on frankfurters formulated with 1.5 and 2.5% salt. Meat Sci 55:123–130. https://doi.org/10.1016/S0309-1740(99)00134-5

Deng P, Teng S, Zhou Y, Liu Y, Liao B, Ren X, Zhang Y (2024) Effects of basic amino acids on heterocyclic amines and quality characteristics of fried beef patties at low NaCl level. Meat Sci 215:109541. https://doi.org/10.1016/j.meatsci.2024.109541

Desmond E (2006) Reducing salt: a challenge for the meat industry. Meat Sci 74:188–196. https://doi.org/10.1016/j.meatsci.2006.04.014

Devlieghere F, Vermeiren L, Bontenbal E, Lamers P, Debevere J (2009) Reducing salt intake from meat products by combined use of lactate and diacetate salts without affecting microbial stability. Int J Food Sci Andence Technol 44:337–341. https://doi.org/10.1111/j.1365-2621.2008.01724.x

Dolatowski ZJ, Stadnik J, Stasiak D (2007) Applications of ultrasound in food technology. Acta Sci Pol Technol Aliment 6:88–99

dos Santos BA, Campagnol PCB, Morgano MA, Pollonio MAR (2014) Monosodium glutamate, disodium inosinate, disodium guanylate, lysine and taurine improve the sensory quality of fermented cooked sausages with 50% and 75% replacement of NaCl with KCl. Meat Sci 96: 509–513. https://doi.org/10.1016/j.meatsci.2013.08.024

Dötsch M, Busch J, Batenburg M, Liem G, Tareilus E, Mueller R, Meijer G (2009) Strategies to reduce sodium consumption: a food industry perspective. Crit Rev Food Sci Nutr 49:841–851. https://doi.org/10.1080/10408390903044297

Doyle ME, Glass KA (2010) Sodium reduction and its effect on food safety, food quality, and human health. Compr Rev Food Sci Food Saf 9:44–56. https://doi.org/10.1111/j.1541-4337.2009.00096.x

Duranton F, Guillou S, Simonin H, Chéret R, de Lamballerie M (2012) Combined use of high pressure and salt or sodium nitrite to control the growth of endogenous microflora in raw pork meat. Innov Food Sci Emerg Technol 16:373–380. https://doi.org/10.1016/j.ifset.2012.08.004

Fellendorf S, O'Sullivan MG, Kerry JP (2016) Impact of ingredient replacers on the physicochemical properties and sensory quality of reduced salt and fat black puddings. Meat Sci 113:17–25. https://doi.org/10.1016/j.meatsci.2015.11.006

Fernstrom JD (2007) Health issues relating to monosodium glutamate use in the diet. In: Reducing salt in foods. CRC Press, Boca Raton Boston, New York, Washington DC, pp 55–76

Finn S, Condell O, McClure P, Amézquita A, Fanning S (2013) Mechanisms of survival, responses and sources of Salmonella in low-moisture environments. Front Microbiol 4. https://doi.org/10.3389/fmicb.2013.00331

Galindo FG, Dejmek P, Lundgren K, Rasmusson AG, Vicente A, Moritz T (2009) Metabolomic evaluation of pulsed electric field-induced stress on potato tissue. Planta 230:469–479. https://doi.org/10.1007/s00425-009-0950-2

García-Díez J, Patarata L (2017) Influence of salt level, starter culture, fermentable carbohydrates, and temperature on the behaviour of *L. monocytogenes* in sliced chouriço during storage. Acta Aliment 46:206–213

García-Lomillo J, MAL G-SI, Del Pino-García R, MAD R-P, Muñiz-Rodríguez P (2017) Alternative natural seasoning to improve the microbial stability of low-salt beef patties. Food Chem 227:122–128. https://doi.org/10.1016/j.foodchem.2017.01.070

Gelabert J, Gou P, Guerrero L, Arnau J (2003) Effect of sodium chloride replacement on some characteristics of fermented sausages. Meat Sci 65:833–839. https://doi.org/10.1016/S0309-1740(02)00288-7

Girgis S, Neal B, Prescott J, Prendergast J, Dumbrell S, Turner C, Woodward M (2003) A one-quarter reduction in the salt content of bread can be made without detection. Eur J Clin Nutr 57:616–620. https://doi.org/10.1038/sj.ejcn.1601583

Gou P, Guerrero L, Gelabert J, Arnau J (1996) Potassium chloride, potassium lactate and glycine as sodium chloride substitutes in fermented sausages and in dry-cured pork loin. Meat Sci 42:37–48. https://doi.org/10.1016/0309-1740(95)00017-8

Grossi A, Søltoft-Jensen J, Knudsen JC, Christensen M, Orlien V (2012) Reduction of salt in pork sausages by the addition of carrot fibre or potato starch and high pressure treatment. Meat Sci 92: 481–489. https://doi.org/10.1016/j.meatsci.2012.05.015

He FJ, MacGregor GA (2009) A comprehensive review on salt and health and current experience of worldwide salt reduction programmes. J Hum Hypertens 23:363–384. https://doi.org/10.1038/jhh.2008.144

Henney JE, Taylor CL, Boon CS, Institute of Medicine; Committee on Strategies to Reduce Sodium Intake (2010) Preservation and physical property roles of sodium in foods. In: Strategies to reduce sodium intake in the United States. National Academies Press, Washington

Horita CN, Morgano MA, Celeghini RMS, Pollonio MAR (2011) Physico-chemical and sensory properties of reduced-fat mortadella prepared with blends of calcium, magnesium and potassium chloride as partial substitutes for sodium chloride. Meat Sci 89:426–433. https://doi.org/10.1016/j.meatsci.2011.05.010

Horita CN, Farías-Campomanes AM, Barbosa TS, Esmerino EA, da Cruz AG, Bolini HMA, Meireles MAA, Pollonio MAR (2016) The antimicrobial, antioxidant and sensory properties of garlic and its derivatives in Brazilian low-sodium frankfurters along shelf-life. Food Res Int 84:1–8. https://doi.org/10.1016/j.foodres.2016.02.006

Hossain M, Adhikary RK, Mahbub KR, Begum M, Ul Islam MR (2012) Effect of 10% concentrations of salt, garlic and coriander on the quality of smoked hilsa fish (Tenualosa ilisha). Am J Food Technol 7(8):501–512

Hugas M, Garriga M, Monfort JM (2002) New mild technologies in meat processing: high pressure as a model technology. Meat Sci 62:359–371. https://doi.org/10.1016/S0309-1740(02)00122-5

Inguglia ES, Zhang Z, Tiwari BK, Kerry JP, Burgess CM (2017) Salt reduction strategies in processed meat products – a review. Trends Food Sci Technol 59:70–78. https://doi.org/10.1016/j.tifs.2016.10.016

Jiayun K, Haibin L, Yu H, Siyu S, Ting L, Wenhui L, Meixia P, Jinghua Q, Fang W (2021) Factors affecting the formation of heterocyclic amines in sauced pork. J Phys Conf Ser 2009:012018. https://doi.org/10.1088/1742-6596/2009/1/012018

Jin S-K, Hwang J, Hur SJ, Kim G-D (2018) Quality changes in fat-reduced sausages by partial replacing sodium chloride with other chloride salts during five weeks of refrigeration. LWT – Food Sci Technol 97:818–824. https://doi.org/10.1016/j.lwt.2018.08.004

Jo K, Lee J, Jung S (2018) Quality characteristics of low-salt chicken sausage supplemented with a winter mushroom powder. Korean J Food Sci Anim Resour 38:768–779. https://doi.org/10.5851/kosfa.2018.e15

Johnson C, Jensen M, Schilmoeller L, Smith G (2008) Seasoning and method for seasoning a food product utilizing small particle sea salt. US7923047B2. https://patents.google.com/patent/US7923047B2/en

Kameník J, Saláková A, Hulánková R, Borilova G (2015) The effect of high pressure on the microbiological quality and other characteristics of cooked sausages packed in a modified atmosphere or vacuum. Food Control 57:232–237. https://doi.org/10.1016/j.foodcont.2015.04.010

Kapil U, Pathak P, Singh P (2003) Benefits and safety of dietary iodine intake in India. Pak J Nutr 2: 43–45

Kapoor K, Fashanu O, Post WS, Lutsey PL, Michos ED, deFilippi CR, McEvoy JW (2019) Relation of dietary sodium intake with subclinical markers of cardiovascular disease (from MESA). Am J Cardiol 124:636–643. https://doi.org/10.1016/j.amjcard.2019.05.014

Ketenoğlu O, Candoğan K (2011) Effect of low-sodium salt utilization on some characteristics of ground beef patties. Blood Press 4:5

Khaw KT, Barrett-Connor E (1984) Dietary potassium and blood pressure in a population. Am J Clin Nutr 39:963–968. https://doi.org/10.1093/ajcn/39.6.963

Kilcast D, den Ridder C (2007) Sensory issues in reducing salt in food products. In: Reducing salt in foods. CRC Press, Boca Raton Boston, New York, Washington DC, pp 201–220

Kim HJ, Lee JH (2009) Physicochemical properties of Salicornia herbacea powder as influenced by drying methods. Food Eng Prog 13:105–109

Kim H-W, Hwang K-E, Song D-H, Kim Y-J, Lim Y-B, Ham Y-K, Yeo E-J, Chang S-J, Choi Y-S, Kim C-J (2014) Effect of glasswort (*Salicornia herbacea* L.) on the texture of frankfurters. Meat Sci 97:513–517. https://doi.org/10.1016/j.meatsci.2014.03.019

Kremer S, Mojet J, Shimojo R (2009) Salt reduction in foods using naturally brewed soy sauce. J Food Sci 74:S255–S262. https://doi.org/10.1111/j.1750-3841.2009.01232.x

Kumar P, Verma AK, Umaraw P, Kumar D, Mehta N (2020a) Superchilling preserves freshness: the novel technology has emerged to improve long term storage life. Fleischwirtschaft Int 2:45–49

Kumar P, Verma AK, Umaraw P, Mehta N, Malav OP (2020b) Plant phenolics as natural preservatives in food system. In: Plant phenolics in sustainable agriculture. Springer Singapore, Singapore, pp 367–406

Laranjo M, Agulheiro-Santos AC, Potes ME, Cabrita MJ, Garcia R, Fraqueza MJ, Elias M (2015) Effects of genotype, salt content and calibre on quality of traditional dry-fermented sausages. Food Control 56:119–127. https://doi.org/10.1016/j.foodcont.2015.03.018

Li Y, He J, Quan W, He Z, Qin F, Tao G, Wang Z, Zeng M, Chen J (2020) Effects of polyphosphates and sodium chloride on heterocyclic amines in roasted beef patties as revealed by UPLC-MS/MS. Food Chem 326:127016. https://doi.org/10.1016/j.foodchem.2020.127016

Liem DG, Miremadi F, Keast R (2011) Reducing sodium in foods: the effect on flavor. Nutrients 3: 694–711. https://doi.org/10.3390/nu3060694

Lilić S, Matekalo-Sverak V (2011) Salt reduction in meat products – challenge for meat industry. Sci J Meat Technol 52:22–30

Lim SS, Vos T, Flaxman AD, Danaei G, Shibuya K, Adair-Rohani H, AlMazroa MA, Amann M, Anderson HR, Andrews KG, Aryee M, Atkinson C, Bacchus LJ, Bahalim AN, Balakrishnan K, Balmes J, Barker-Collo S, Baxter A, Bell ML, Blore JD, Blyth F, Bonner C, Borges G, Bourne R, Boussinesq M, Brauer M, Brooks P, Bruce NG, Brunekreef B, Bryan-Hancock C, Bucello C, Buchbinder R, Bull F, Burnett RT, Byers TE, Calabria B, Carapetis J, Carnahan E, Chafe Z, Charlson F, Chen H, Chen JS, Cheng AT-A, Child JC, Cohen A, Colson KE, Cowie BC, Darby S, Darling S, Davis A, Degenhardt L, Dentener F, Des Jarlais DC, Devries K, Dherani M, Ding EL, Dorsey ER, Driscoll T, Edmond K, Ali SE, Engell RE, Erwin PJ, Fahimi S, Falder G, Farzadfar F, Ferrari A, Finucane MM, Flaxman S, Fowkes FGR, Freedman G, Freeman MK, Gakidou E, Ghosh S, Giovannucci E, Gmel G, Graham K, Grainger R, Grant B, Gunnell D, Gutierrez HR, Hall W, Hoek HW, Hogan A, Hosgood HD, Hoy D, Hu H, Hubbell BJ, Hutchings SJ, Ibeanusi SE, Jacklyn GL, Jasrasaria R, Jonas JB, Kan H, Kanis JA, Kassebaum N, Kawakami N, Khang Y-H, Khatibzadeh S, Khoo J-P, Kok C, Laden F, Lalloo R, Lan Q, Lathlean T, Leasher JL, Leigh J, Li Y, Lin JK, Lipshultz SE, London S, Lozano R, Lu Y, Mak J, Malekzadeh R, Mallinger L, Marcenes W, March L, Marks R, Martin R, McGale P, McGrath J, Mehta S, Memish ZA, Mensah GA, Merriman TR, Micha R, Michaud C, Mishra V, Hanafiah KM, Mokdad AA, Morawska L, Mozaffarian D, Murphy T, Naghavi M, Neal B, Nelson PK, Nolla JM, Norman R, Olives C, Omer SB, Orchard J, Osborne R, Ostro B, Page A, Pandey KD, Parry CD, Passmore E, Patra J, Pearce N, Pelizzari PM, Petzold M, Phillips MR, Pope D, Pope CA, Powles J, Rao M, Razavi H, Rehfuess EA, Rehm JT, Ritz B, Rivara FP, Roberts T, Robinson C, Rodriguez-

Portales JA, Romieu I, Room R, Rosenfeld LC, Roy A, Rushton L, Salomon JA, Sampson U, Sanchez-Riera L, Sanman E, Sapkota A, Seedat S, Shi P, Shield K, Shivakoti R, Singh GM, Sleet DA, Smith E, Smith KR, Stapelberg NJ, Steenland K, Stöckl H, Stovner LJ, Straif K, Straney L, Thurston GD, Tran JH, Van Dingenen R, van Donkelaar A, Veerman JL, Vijayakumar L, Weintraub R, Weissman MM, White RA, Whiteford H, Wiersma ST, Wilkinson JD, Williams HC, Williams W, Wilson N, Woolf AD, Yip P, Zielinski JM, Lopez AD, Murray CJ, Ezzati M (2012) A comparative risk assessment of burden of disease and injury attributable to 67 risk factors and risk factor clusters in 21 regions, 1990–2010: a systematic analysis for the global burden of disease study 2010. Lancet 380:2224–2260. https://doi.org/10.1016/S0140-6736(12)61766-8

Lim D-G, Choi K-S, Kim J-J, Nam K-C (2013) Effects of Salicornia herbacea powder on quality traits of sun-dried Hanwoo beef jerky during storage. Food Sci Anim Res 33:205–213

Lim Y-B, Kim H-W, Hwang K-E, Song D-H, Kim Y-J, Ham Y-K, Jang S-J, Lee C-H, He F-Y, Choi Y-S, Kim C-J (2015) Effects of glasswort (*Salicornia herbacea* L.) hydrates on quality characteristics of reduced-salt, reduced-fat frankfurters. Korean J Food Sci Anim Resour 35: 783–792. https://doi.org/10.5851/kosfa.2015.35.6.783

Lorenzo JM, Bermúdez R, Domínguez R, Guiotto A, Franco D, Purriños L (2015) Physicochemical and microbial changes during the manufacturing process of dry-cured lacón salted with potassium, calcium and magnesium chloride as a partial replacement for sodium chloride. Food Control 50:763–769. https://doi.org/10.1016/j.foodcont.2014.10.019

Lorido L, Estévez M, Ventanas J, Ventanas S (2015) Salt and intramuscular fat modulate dynamic perception of flavour and texture in dry-cured hams. Meat Sci 107:39–48. https://doi.org/10.1016/j.meatsci.2015.03.025

Luckose F, Pandey MC, Chauhan OP, Sultana K, Abhishek V (2015) Effect of high pressure processing on the quality characteristics and shelf life of low-sodium re-structured chicken nuggets. J Food Nutr Res 54(4):334

Lv J, Li C, Li S, Liang H, Ji C, Zhu B, Lin X (2019a) Effects of temperature on microbial succession and quality of sour meat during fermentation. LWT – Food Sci Technol 114:108391. https://doi.org/10.1016/J.LWT.2019.108391

Lv J, Yang Z, Xu W, Li S, Liang H, Ji C, Yu C, Zhu B, Lin X (2019b) Relationships between bacterial community and metabolites of sour meat at different temperature during the fermentation. Int J Food Microbiol 307:108286. https://doi.org/10.1016/j.ijfoodmicro.2019.108286

Mancini S, Mattioli S, Nuvoloni R, Pedonese F, Dal Bosco A, Paci G (2020) Effects of garlic powder and salt on meat quality and microbial loads of rabbit burgers. Food Secur 9:1022. https://doi.org/10.3390/foods9081022

Mariutti LRB, Bragagnolo N (2017) Influence of salt on lipid oxidation in meat and seafood products: A review. Food Res Int 94:90–100. https://doi.org/10.1016/j.foodres.2017.02.003

McDonnell CK, Allen P, Morin C, Lyng JG (2014a) The effect of ultrasonic salting on protein and water–protein interactions in meat. Food Chem 147:245–251. https://doi.org/10.1016/j.foodchem.2013.09.125

McDonnell CK, Lyng JG, Arimi JM, Allen P (2014b) The acceleration of pork curing by power ultrasound: a pilot-scale production. Innovative Food Sci Emerg Technol 26:191–198. https://doi.org/10.1016/j.ifset.2014.05.004

Mehta P, Kumar P, Verma AK, Umaraw P (2020) Prospects in the meat industry: the new technology could be used for accelerated drying, curing, and other production steps. Fleischwirtschaft Int 1:22–27

Mehta N, Jeyapriya S, Kumar P, Verma AK, Umaraw P, Khatkar SK, Khatkar AB, Pathak D, Kaka U, Sazili AQ (2022) Ultrasound-assisted extraction and the encapsulation of bioactive components for food applications. Food Secur 11:2973. https://doi.org/10.3390/foods11192973

Mitchell M, Brutnon NP, Fitzgerald RJ, Wilkinson MG (2013) The use of herbs, spices, and whey proteins as natural flavor enhancers and their effect on the sensory acceptability of reduced-salt chilled ready-meals. J Culin Sci Technol 11:222–240. https://doi.org/10.1080/15428052.2013.769869

Niaz K, Zaplatic E, Spoor J (2018) Extensive use of monosodium glutamate: a threat to public health? EXCLI J 17:273–278. https://doi.org/10.17179/excli2018-1092

O'Flynn CC, Cruz-Romero MC, Troy D, Mullen AM, Kerry JP (2014) The application of high-pressure treatment in the reduction of salt levels in reduced-phosphate breakfast sausages. Meat Sci 96:1266–1274. https://doi.org/10.1016/j.meatsci.2013.11.010

Ojha KS, Keenan DF, Bright A, Kerry JP, Tiwari BK (2016) Ultrasound assisted diffusion of sodium salt replacer and effect on physicochemical properties of pork meat. Int J Food Sci Technol 51:37–45. https://doi.org/10.1111/ijfs.13001

Ortega G, Laín A, Tadeo X, López-Méndez B, Castaño D, Millet O (2011) Halophilic enzyme activation induced by salts. Sci Rep 1:6. https://doi.org/10.1038/srep00006

Ozuna C, Puig A, García-Pérez JV, Mulet A, Cárcel JA (2013) Influence of high intensity ultrasound application on mass transport, microstructure and textural properties of pork meat (*Longissimus dorsi*) brined at different NaCl concentrations. J Food Eng 119:84–93. https://doi.org/10.1016/j.jfoodeng.2013.05.016

Paulsen MT, Nys A, Kvarberg R, Hersleth M (2014) Effects of NaCl substitution on the sensory properties of sausages: temporal aspects. Meat Sci 98:164–170. https://doi.org/10.1016/j.meatsci.2014.05.020

Perea-Sanz L, Montero R, Belloch C, Flores M (2018) Nitrate reduction in the fermentation process of salt reduced dry sausages: impact on microbial and physicochemical parameters and aroma profile. Int J Food Microbiol 282:84–91. https://doi.org/10.1016/j.ijfoodmicro.2018.06.004

Pietrasik Z, Gaudette NJ (2014) The impact of salt replacers and flavor enhancer on the processing characteristics and consumer acceptance of restructured cooked hams. Meat Sci 96:1165–1170. https://doi.org/10.1016/j.meatsci.2013.11.005

Pinton MB, dos Santos BA, Lorenzo JM, Cichoski AJ, Boeira CP, Campagnol PCB (2021) Green technologies as a strategy to reduce NaCl and phosphate in meat products: an overview. Curr Opin Food Sci 40:1–5. https://doi.org/10.1016/j.cofs.2020.03.011

Rama R, Chiu N, Carvalho Da Silva M, Hewson L, Hort J, Fisk ID (2013) Impact of salt crystal size on in-mouth delivery of sodium and saltiness perception from snack foods. J Texture Stud 44: 338–345. https://doi.org/10.1111/jtxs.12017

Ruusunen M, Puolanne E (2005) Reducing sodium intake from meat products. Meat Sci 70:531–541. https://doi.org/10.1016/j.meatsci.2004.07.016

Samapundo S, Ampofo-Asiama J, Anthierens T, Xhaferi R, Van Bree I, Szczepaniak S, Goemaere O, Steen L, Dhooge M, Paelinck H (2010) Influence of NaCl reduction and replacement on the growth of Lactobacillus sakei in broth, cooked ham and white sauce. Int J Food Microbiol 143:9–16. https://doi.org/10.1016/j.ijfoodmicro.2010.07.014

Sena Vaz Leães Y, Basso Pinton M, de Aguiar T, Rosa C, Sasso Robalo S, Wagner R, Ragagnin de Menezes C, Smanioto Barin J, Cezar Bastianello Campagnol P, José Cichoski A (2020) Ultrasound and basic electrolyzed water: a green approach to reduce the technological defects caused by NaCl reduction in meat emulsions. Ultrason Sonochem 61:104830. https://doi.org/10.1016/j.ultsonch.2019.104830

Seong P-N, Seo H-W, Cho S-H, Kim Y-S, Kang S-M, Kim J-H, Kang G-H, Park B-Y, Moon S-S, Hoa V-B (2017) Potential use of glasswort powder as a salt replacer for production of healthier dry-cured ham products. Czech J Food Sci 35:149–159. https://doi.org/10.17221/152/2016-CJFS

Seung-Kwan H (2004) Antioxidant effect of fermented *Salicornia herbacea* L. liquid with EM (effective microorganism) on pork. Food Science of Animal Resource 24:298–302

Shelef LA, Seiter J (2005) Indirect and miscellaneous antimicrobials. In: Food science and technology, vol 145. Marcel Dekker, New York, p 573

Shim S-M, Seo SH, Lee Y, Moon G-I, Kim M-S, Park J-H (2011) Consumers' knowledge and safety perceptions of food additives: evaluation on the effectiveness of transmitting information on preservatives. Food Control 22:1054–1060. https://doi.org/10.1016/j.foodcont.2011.01.001

Singh PK, Kumar S, Kumar P, Bhat ZF (2012) Pulsed light and pulsed electric field-emerging non thermal decontamination of meat. Am J Food Technol 7. https://doi.org/10.3923/ajft.2012.506.516

Šojić B, Pavlić B, Ikonić P, Tomović V, Ikonić B, Zeković Z, Kocić-Tanackov S, Jokanović M, Škaljac S, Ivić M (2019) Coriander essential oil as natural food additive improves quality and safety of cooked pork sausages with different nitrite levels. Meat Sci 157:107879. https://doi.org/10.1016/j.meatsci.2019.107879

Stekelenburg FK (2003) Enhanced inhibition of *Listeria monocytogenes* in Frankfurter sausage by the addition of potassium lactate and sodium diacetate mixtures. Food Microbiol 20:133–137. https://doi.org/10.1016/S0740-0020(02)00098-9

Stock JM, Chelimsky G, Edwards DG, Farquhar WB (2022) Dietary sodium and health: How much is too much for those with orthostatic disorders? Auton Neurosci 238:102947. https://doi.org/10.1016/j.autneu.2022.102947

Stollewerk K, Jofré A, Comaposada J, Arnau J, Garriga M (2012) The effect of NaCl-free processing and high pressure on the fate of *Listeria monocytogenes* and Salmonella on sliced smoked dry-cured ham. Meat Sci 90:472–477. https://doi.org/10.1016/j.meatsci.2011.09.009

Stringer SC, Pin C (2005) Microbial risks associated with salt reduction in certain foods and alternative options for preservation. Institute of Food Research, Norwich. https://acmsf.food.gov.uk/sites/default/files/mnt/drupal_data/sources/files/multimedia/pdfs/acm740a.pdf. Accessed 20 June 2017

Tamm A, Bolumar T, Bajovic B, Toepfl S (2016) Salt (NaCl) reduction in cooked ham by a combined approach of high pressure treatment and the salt replacer KCl. Innovative Food Sci Emerg Technol 36:294–302. https://doi.org/10.1016/j.ifset.2016.07.010

Taormina PJ (2010) Implications of salt and sodium reduction on microbial food safety. Crit Rev Food Sci Nutr 50:209–227. https://doi.org/10.1080/10408391003626207

Tobin BD, O'Sullivan MG, Hamill RM, Kerry JP (2012a) Effect of varying salt and fat levels on the sensory and physiochemical quality of frankfurters. Meat Sci 92:659–666. https://doi.org/10.1016/j.meatsci.2012.06.017

Tobin BD, O'Sullivan MG, Hamill RM, Kerry JP (2012b) Effect of varying salt and fat levels on the sensory quality of beef patties. Meat Sci 91:460–465. https://doi.org/10.1016/j.meatsci.2012.02.032

Tobin BD, O'Sullivan MG, Hamill RM, Kerry JP (2013) The impact of salt and fat level variation on the physiochemical properties and sensory quality of pork breakfast sausages. Meat Sci 93:145–152. https://doi.org/10.1016/j.meatsci.2012.08.008

Toldrá F (2006) The role of muscle enzymes in dry-cured meat products with different drying conditions. Trends Food Sci Technol 17:164–168. https://doi.org/10.1016/j.tifs.2005.08.007

Trieu K, Neal B, Hawkes C, Dunford E, Campbell N, Rodriguez-Fernandez R, Legetic B, McLaren L, Barberio A, Webster J (2015) Salt reduction initiatives around the world – a systematic review of progress towards the global target. PLoS One 10:e0130247. https://doi.org/10.1371/journal.pone.0130247

Verma AK, Banerjee R (2012) Low-sodium meat products: retaining salty taste for sweet health. Crit Rev Food Sci Nutr 52:72–84. https://doi.org/10.1080/10408398.2010.498064

Verma AK, Umaraw P, Kumar P, Singh VP, Kumar D (2018) Prospect and challenges of functional meat products: a review of functional ingredients in the global meat industry. Fleischwirtschaft 9:95–106

Vidal VAS, Biachi JP, Paglarini CS, Pinton MB, Campagnol PCB, Esmerino EA, da Cruz AG, Morgano MA, Pollonio MAR (2019) Reducing 50% sodium chloride in healthier jerked beef: An efficient design to ensure suitable stability, technological and sensory properties. Meat Sci 152:49–57. https://doi.org/10.1016/j.meatsci.2019.02.005

Wang Q, Li X, Xue B, Wu Y, Song H, Luo Z, Shang P, Liu Z, Huang Q (2022) Low-salt fermentation improves flavor and quality of sour meat: microbiology and metabolomics. LWT – Food Sci Technol 171:114157. https://doi.org/10.1016/j.lwt.2022.114157

Wilson R, Komitopoulou E, Incles M (2012) https://bfff.co.uk/wp-content/uploads/2013/06/Frozen-Food-Report-2-Nov-10.pdf

Wirth F (1991) Reducing the fat and sodium content of meat products. What possibilities are there? Fleischwirtsch Int 71:294–297

Xiao Q, Xu M, Xu B, Chen C, Deng J, Li P (2021) Combined effect of high-pressure processing with spice extracts on quality of low-salt sausage during refrigerated storage. Food Secur 10: 2610. https://doi.org/10.3390/foods10112610

Xu P, Zheng Y, Zhu X, Li S, Zhou C (2018) L-lysine and L-arginine inhibit the oxidation of lipids and proteins of emulsion sausage by chelating iron ion and scavenging radical. Asian-Austral J Anim Sci 31:905–913. https://doi.org/10.5713/ajas.17.0617

Yang H, Han M, Wang X, Han Y, Wu J, Xu X, Zhou G (2015) Effect of high pressure on cooking losses and functional properties of reduced-fat and reduced-salt pork sausage emulsions. Innovative Food Sci Emerg Technol 29:125–133. https://doi.org/10.1016/j.ifset.2015.02.013

Yotsuyanagi SE, Contreras-Castillo CJ, Haguiwara MMH, Cipolli KMVAB, Lemos ALSC, Morgano MA, Yamada EA (2016) Technological, sensory and microbiological impacts of sodium reduction in frankfurters. Meat Sci 115:50–59. https://doi.org/10.1016/j.meatsci.2015.12.016

Zhang Y, Wu J, Jamali MA, Guo X, Peng Z (2017) Heat-induced gel properties of porcine myosin in a sodium chloride solution containing L-lysine and L-histidine. LWT – Food Sci Technol 85: 16–21. https://doi.org/10.1016/j.lwt.2017.06.059

Zhang Y, Guo X, Peng Z, Jamali MA (2022) A review of recent progress in reducing NaCl content in meat and fish products using basic amino acids. Trends Food Sci Technol 119:215–226. https://doi.org/10.1016/j.tifs.2021.12.009

Zhao T, Zhang C, Ma Y, Wang X (2023) Effect of simultaneous application of antioxidants in oil and in chicken on heterocyclic aromatic amine formation in deep-fat fried chicken breast. LWT – Food Sci Technol 187:115287. https://doi.org/10.1016/j.lwt.2023.115287

Zhong A, Chen W, Duan Y, Li K, Tang X, Tian X, Wu Z, Li Z, Wang Y, Wang C (2021) The potential correlation between microbial communities and flavors in traditional fermented sour meat. LWT – Food Sci Technol 149:111873. https://doi.org/10.1016/j.lwt.2021.111873

Zhong A, Chen W, Hu L, Wu Z, Xiao Y, Li K, Li Z, Wang Y, Wang C (2022) Characterisation of key volatile compounds in fermented sour meat after fungi growth inhibition. LWT – Food Sci Technol 165:113662. https://doi.org/10.1016/j.lwt.2022.113662

Zou Y, Kang D, Liu R (2018) Effects of ultrasonic assisted cooking on the chemical profiles of taste and flavor of spiced beef. Ultrason Sonochem 46:36–45. https://doi.org/10.1016/j.ultsonch.2018.04.005

Chapter 11
Nitrite Reduction/Replacement in Processed Meat Products

Humaira Fayaz, Sheikh Rafeh Ahmad, Syed Arshad Hussain, Asif H. Sofi, and Tahir Nazir

11.1 Introduction

Meat and meat products are important protein sources for human nutrition. The important position of meat and meat products in human foods is shown by the long history of their consumption, which can be traced back to about three million years ago (Chris 2020; Guo et al. 2005; McKenna 2019). Fresh meat is highly perishable and is not stable to be stored for a long term. It will deteriorate within several days if not stored under feasible environmental conditions. Roasting and smoking thus became the earliest preservation methods for meat and meat products. Possibly because of the effectiveness of smoking for preserving meat and the habit of eating meat with a smoky flavor, roasted or smoked meat products are still popular and widely consumed globally. Meat curing is an ancient method of food preservation that is still widely used today (Alvarado and McKee 2007). It refers to adding nitrite/nitrate salt, common salt and spices to fresh meat in varying degrees of comminution and at various processing phases (Patarata et al. 2020). Prior to the invention of refrigeration, meat was preserved using methods discovered to be efficient in controlling deterioration after slaughter and extending the food supply during times of shortage. Despite being lost in time, the curing process is thought to be evolved from salt preservation methods as early as 3000 B.C. (Pegg and Honikel 2014). Among the various additives used in meat curing, nitrite salt is mostly used. Nitrite is a major intermediary throughout the biological nitrogen-cycle present in soil and water surface (Pourreza et al. 2012). It is a versatile chemical with a wide range of uses, including dye manufacturing and food preservation (Afkhami et al. 2004). Nitrites in various meat products are significant preservatives and impede the

H. Fayaz (✉) · S. R. Ahmad (✉) · S. A. Hussain · A. H. Sofi · T. Nazir
Division of Livestock Products Technology, FVSc & AH, SKUAST-K, Srinagar, Jammu and Kashmir, India
e-mail: sh.rafeh@skuastkashmir.ac.in

S. Rather, F. A. Masoodi (eds.), *Hand Book of Processed Functional Meat Products*,
https://doi.org/10.1007/978-3-031-69868-2_11

growth of several unwanted microorganisms (Lopez et al. 2021). Nitrite is added to cured meat at levels less than 150 ppm to prevent the development of microbiological organisms like *Clostridium botulinum*, which causes food poisoning (Badea et al. 2004). The main reasons for using nitrite as a preservative in meat are: (1) To inhibit the *Clostridium botulinum* from spreading and secreting toxins that cause food toxicity (Patarata et al. 2022) (2) to provide the necessary bright red color in meat products like sausage, ham, salami, etc. (Candan and Bağdatlı 2018). (3) To give cured meats their characteristic texture and aroma (Altunay and Elik 2020). In addition, nitrite also inhibits the oxidation of lipids in meat products and thus prevents rancidity (Aminzare et al. 2019).

The real reason for the commercial use of nitrate and nitrite salts in cured meat products is related to the ancient salting practices for meat preservation. Nitrate and nitrite were contaminant in salts used in salted meats. The advances in science allowed the discovery of nitrate (NO_3-) first, and then nitrite (NO_2-) as actual ingredients involved in cured meat products preservation (Bedale et al. 2016). NO_2 salts and NO_3 salts, such as those of sodium and potassium, are typical food preservatives used in meat processing. Both of them can inhibit the growth of microorganisms, delay the onset of rancidity, produce cured meat flavor or smell and stabilize the meat's red color (Ferysiuk and Wojciak 2020). The efficiency of nitrite and nitrate makes them indispensable additives for meat and meat products, especially for cured meat products (Ferysiuk and Wojciak 2020). However, the discovery of N-nitroso compounds (NOCs) in the 1950s raised concerns about the safety of using nitrite or nitrate in meat processing (Bedale et al. 2016), which seriously threatened the healthy development of the global meat industry. In addition, with widely acceptance of the clean label movement, having fewer ingredients becomes mainstream in food processing (Asioli et al. 2017).

Since the middle of the 1980s, research has shown that nitrite is a major chemical with substantial impacts on human health. Vegetables are a great source of dietary nitrates and they have been proven to be an important source of endogenous nitrite as well as nitric oxide (NO) in the human body (Dos Santos et al. 2021). Nitric oxide (NO), produced through enzymatic synthesis, regulates blood pressure, wound healing, immunological response and neurological processes in the human body (Hunault et al. 2009). New research has demonstrated that NO (nitric oxide) regulates blood circulation in cardiac tissues and perhaps in other body tissues (Milkowski et al. 2010). Furthermore, regular nitric oxide and nitrite production may help to prevent cardiovascular diseases like hypertension, atherosclerosis, and stroke (Bryan and van Grinsven 2013).

11.2 Health Concerns Associated with Nitrite in Meat

Regardless of all of sodium nitrite's benefits, its use in meat has been a bone of contention and concern. Due to nitrite's high chemical reactivity, it can combine with a variety of components in meat systems. The heat used throughout the thermal treatment of cured meat products increases its reactivity. Particularly, nitrite ions are highly reactive when the pH is lower than 7. It may react with a variety of meat components, including amino acids, sulfhydryl, amines, phenolic compounds, ascorbic acid and myoglobin. Nitrite can play a role as a nitrosating agent and form various nitroso compounds (Zanardi et al. 2002). Other nitrosating agents include nitrous acid and nitric oxide which are also derived from nitrite. Nitrous acid participates in the processes that result in the formation of endogenous N-nitroso compounds (NOCs). NO, on the other hand, may be a generator of nitrates and nitrites, which circulate in the body of human (Bedale et al. 2016). High nitrite concentrations, on the other hand, are extremely dangerous for infants since they can develop an infant's methemoglobinemia (Brender 2020). Furthermore, cancer-causing nitrosamines are formed when nitrite reacts with secondary or tertiary amines (Oliveira et al. 2004). Recently, the International Agency for Research on Cancer (IARC) stated processed meat as carcinogenic by evaluating sufficient epidemiological data (Bouvard et al. 2015). IARC also reported that ingested nitrite from processed meat can lead to colorectal cancer in human. Because of these harmful effects, many countries have severely restricted their use on processed food products (Wang et al. 2017). The toxicity of nitrite is ten times that of nitrate. For humans, the fatal oral dosage is 80 to 800 mg nitrate per Kg body weight and just 33 to 250 mg nitrite per Kg body weight. Long-term intake of increasing amounts of red meat, especially processed meat, is linked to a higher rate of mortality, colorectal cancer, type-2 diabetes and heart diseases in both male and female, according to large prospective United States, E.U. cohort studies as well as meta-analyses of epidemiology (Richi et al. 2015). Concerning these issues, an acceptable daily intake (ADI) of 0.07 mg nitrite per kg of body weight was set by the Joint Expert Committee of the Food and Agriculture Organization (JECFA) and the World Health Organization (WHO) that appears to be safe for healthy newborns, children, and adults (Casoni et al. 2019).

The World Cancer Research Fund (WCRF) and the American Institute for Cancer Research (AICR) published research in 2007 that found a moderate but significant link between increasing consumption of processed meat and a higher risk of colorectal cancer (Hung et al. 2015). As a result, specialists advised limiting red and processed meat consumption. Anywise, eating less meat alone may not result in a significant reduction in carcinogenic effects and it may be associated with several disadvantages, including the loss of nutritive value, especially iron (Fe). So, it will be wise to invest proactively in the processing of healthier meat products rather than anticipating processed meat consumption to fall.

Because of growing concerns regarding sodium nitrite's long-term adverse effects, their use in cured meat products is strictly regulated among most developed countries. Parallel to this, the new ideas of all-natural and clean label have raised a demand for healthy and high-quality meat products. Due to health hazards, consumers choose natural additives over chemicals in processed meat. As a result, research on substituting natural ingredients for the chemical additive nitrite has grown over the years. To reduce the risk of nitrosamine formation and mitigate potential human health hazards, researchers are trying to find effective ways in meat curing. One such method is the replacement of nitrite salt with alternative ingredients that have similar properties while posing no health risk (Sindelar and Milkowski 2011). However, no single alternative that provides the multi-functions of nitrite in meat products has yet been found. The employment of "hurdle technology" in meat curing is one proposed solution to this issue where low amounts of nitrite are mixed with other ingredients (Alahakoon et al. 2015).

Generally, N-nitroso compounds are classified into six types: non-volatile N-nitrosamines, volatile N-nitrosamines, N-nitrosated heterocyclic carboxylic products, N-nitrosamides, Amadori compounds and N-nitrosated glycosylamines (Pourazrang et al. 2002). The majority of volatile nitrosamines are categorized in group 2B, which means they are potentially carcinogenic to the human body (Van den Brand et al. 2020). The number of nitrosamines in processed meat products varies depending on the type of meat product. The quantity of N-nitrosamines in processed meat might be less than the detection limit (one microgram per kilogram) (Herrmann et al. 2014). As sodium nitrite can be a predecessor of nitrosamines, its usage in meat curing has gathered public concern. It is currently considered that the amount of nitrite added and the production of N-nitrosamines has a positive relationship but the relationship is not linear (Demeyer et al. 2016). Furthermore, they exhibit teratogenic effects too. There are about 300 varieties of nitrosamines and almost all of them (97%) have been demonstrated to be teratogenic in experimental animals (Redondo-Solano and Echandi 2011).

The development of nitrosamines in meat products is a complicated process and it may be influenced by a wide range of factors. Nitrite, nitrate, primary and secondary amines, amides, peptides, proteins and various amino acids are the initial compounds for nitrosamine synthesis in meats and these are converted into nitrosamines precursors by microbial activity. Microorganisms may contribute to the formation of nitrosamines by converting nitrates to nitrites and degrading proteins to amino acids and amines (Yurchenko and Mölder 2007). N-nitrosamines can develop in meat throughout the production processes, during home cooking and in the digestive tract after ingestion (Herrmann et al. 2015). They are mostly generated from secondary amines, nitrite and other nitrosating agents. In cured meats, residual nitrite may combine with amines and free amino acids and yield nitrosamines under specific conditions, such as the existence of secondary amines, low pH, product temperature >130 °C and the NO_2 availability to react (Bedale et al. 2016). During the grilling or frying of cured meats, nitrosamines may occur in little amounts and are expected to cause cancer in the human body (even with the little exposure over prolonged time) (Stoica 2019).

The chemical reactions that result in the developments of nitrosamines in cured meat systems are noted below:

$$NaNO_2 + H^+ \rightarrow Na^+ + HNO_2$$

$$HNO_2 + H^+ \rightarrow NO^+ + H_2O$$

$$2HNO_2 \rightarrow N_2O_3 + H_2O$$

$$N_2O_3 \rightarrow NO + NO_2$$

$$NO + M^+ \rightarrow NO^+ + M$$

$$RNH_2 \text{ (Primary amine)} + NO^+ \rightarrow RNH - N = O + H^+ \rightarrow ROH + N_2$$

$$R_2NH \text{ (Secondary amine)} + NO^+ \rightarrow R_2N - N = O + H^+$$

$$R_3N \text{ (Tertiary amine)} + NO^+ \rightarrow \text{no nitrosamine formation}$$

These chemical reactions exhibit the same process leading to the formation of nitric oxide and nitrous acid. As a result, the same consequences can lead to the nitrite reduction and favors the production of nitrosamine. Among the primary, secondary and tertiary amines, the secondary amines generate more persistent nitrosamines. In addition, the mixture of secondary amines and nitrite cause lung adenomas in mice. An investigation into mice treated with 0.5% sodium nitrite and 0.85% butyl urea showed the elevated occurrence of malignant lymphomas. Numerous epidemiological studies have found a link between nitrosamines (N.A.s) and various type of cancer risk (Iqbal 2017; Stepien et al. 2016; Keszei et al. 2013). In 2006, a working group of IARC (International Agency for Research on Cancer) stated that "ingested nitrite under certain conditions resulting in endogenous nitrosation is presumably carcinogenic to human body" (Chan 2011). An epidemiological study conducted in 2008 showed that there is an increased risk of colorectal cancer related to high processed meat intake (Santarelli et al. 2008). Excessive nitrite intake can also result in tissue poisoning, respiratory center paralysis and other hypoxia-related symptoms. In extreme cases, it can cause suffocation as well as death by decreasing the O_2 carrying capability of hemoglobin in human blood (Huang et al. 2020a, b). High nitrite consumption can impair iodine metabolism and decrease iodine absorption by the thyroid, which can result in the enlargement of the thyroid gland.

Methemoglobinemia, also known as "blue baby syndrome", is another health concern of high nitrite intake. It develops when nitrate is converted to reactive nitrite by reducing bacteria in the saliva or digestive system of humans. The blue baby syndrome is named after the blue color of a newborn's skin when their blood nitrite levels are high. As a result, methemoglobinemia is often known as "blue baby syndrome," and it is a life-threatening disease. When nitrite enters the bloodstream, it causes the hemoglobin (the protein that transports oxygen in the bloodstream to the body's tissues) to be oxidized to methemoglobin. The Fe^{2+} of the hemoglobin group is oxidized to Fe^{3+} as shown in the following reaction (Fakhre and Qader 2013; Kroupova et al. 2005).

$$4Hb\left(Fe^{2+}\right)O_2 + 4NO_2{}^- + 4H^+ \rightarrow 4Hb\left(Fe^{3+}\right) + 4NO_3{}^- + O_2 + 2H_2$$

This reaction produces methemoglobin which is responsible for the reduced oxygen supply to body tissues, causing the skin to become blue and possibly causing asphyxia. In the initial stages of methemoglobinemia, the blue color can be observed in the nose, lips and ears and in extreme cases it can affect the peripheral tissues (Gassara et al. 2016). Infants under 6 months of age are the most sensitive to methemoglobinemia. Meanwhile, this disease has been reported in both school-going children and adults (Rehman et al. 2018).

Furthermore, decreased tissue oxygenation can have a variety of negative consequences for the children, involving coma and eventually death. The toxic amounts of nitrites responsible for methemoglobinemia range from 0.4 mg to more than 200 mg per kg of body weight. The nitrite ion limit for newborns is up to 3 ppm (Gassara et al. 2016). The U.S. Environmental Protection Agency reported contradictory evidence over the relationship between higher nitrite intake and the elevated incidence of cancer in children and adults. In certain studies, it has been found that a high intake of nitrite can lead to the elevated occurrence of leukemia, nasopharyngeal and brain tumors in some children (Gassara et al. 2016).

11.3 Sources of Nitrite

Nitrites play a significant role in the biogeochemical cycle of nitrogen in natural water. They can be found in soils (Shariati-Rad et al. 2015), waterways (Lefevre et al. 2011) foodstuffs (Bijad et al. 2017), plants (Boora et al. 2014), air (as nitrogen dioxide) (Tsikas et al. 2010) and biological samples (Yang et al. 2003).

In the biological nitrogen cycle, nitrogen is converted to nitrate by bacteria, which is taken by plants and incorporated into tissues (Fig. 11.1). Animals that consume plants utilize nitrate to produce proteins. Animal excrement and microbial breakdown of animals and plants after death return nitrate to the environment. Nitrate or the ammonium ion can be converted to nitrite by micro-organisms; this reaction occurs in the environment, digestive tracts of humans and other animals. Once bacteria in the environment convert nitrate to nitrite and subsequently convert the nitrite to nitrogen, the cycle is completed.

In the whole nitrogen cycle, nitrite is the intermediary compound derived from nitrification. The microbiological process, carried out successively by multiple species of bacteria (*Nitrosomonas* and *Nitrobacter*), which converts ammonium to nitrates via the intermediate production of nitrite, is called nitrification (Li et al. 2017). The following are the two steps in the nitrification process, along with their equations:

1. Nitrosomonas bacteria converts ammonia to nitrite.

$$2NH_4^+ + 3O_2 \rightarrow 2NO_2^- + 4H^+ + 2H_2O$$

2. The nitrite is converted to nitrate by Nitrobacter bacteria (Ward et al. 2011).

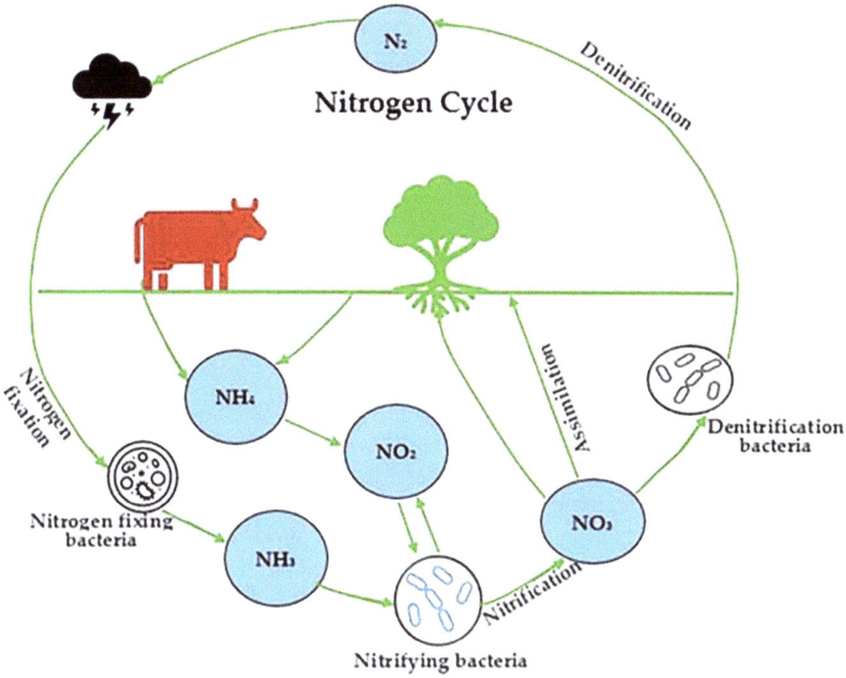

Fig. 11.1 Nitrogen cycle in the environment including nitrogen assimilation

$$2NO_2^- + O_2 \rightarrow 2NO_3^-$$

11.4 Role of Nitrites in Processed Meat

11.4.1 Cured Color Development

The primary influence of nitrite is to provide the cured meat color. Meat color is highly variable and is influenced by a variety of factors. When nitrite is introduced to meat, it is converted to nitric oxide (NO) via the reactions listed below (Honikel 2008).

$$NO_2^- + H^+ \leftrightarrow HNO_2 \tag{11.1}$$

$$2HNO_2 \leftrightarrow N_2O_3 + H_2O \tag{11.2}$$

$$N_2O_3 \leftrightarrow NO + NO_2 \tag{11.3}$$

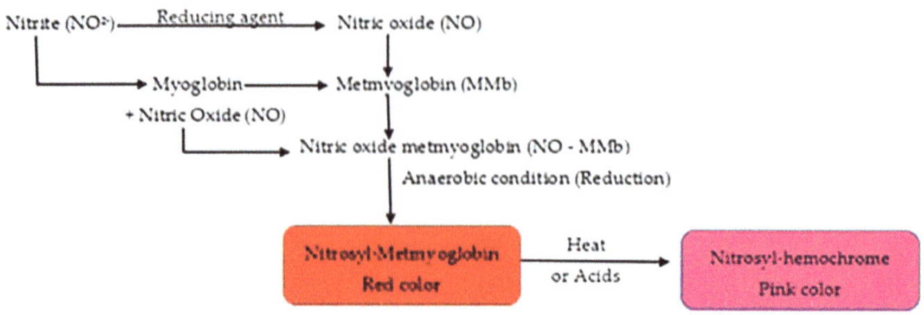

Fig. 11.2 Mechanism of color development in cured meats

Here, nitrite reacts with hydrogen ions (H^+) of water to produce nitrous acid. After that, nitrous acid progressively decomposes into water molecules (H_2O) and dinitrogen-trioxide [Eqs. (11.1) and (11.2)]. Then, nitric oxide and nitrogen dioxide are generated from dinitrogen trioxide (N_2O_3). The major component responsible for nitrite's apparent function in cured meat products is nitric oxide.

Nitric oxide combines with the iron of both myoglobin (Fe^{2+}) and metmyoglobin (Fe^{3+}) to produce a cured pink color in meat (Alahakoon et al. 2015). Myoglobin is the sarcoplasmic protein responsible for the red color in meat, and metmyoglobin (brown in color) is the oxidized form of myoglobin (Mb). NO-myoglobin is formed when nitric oxide (NO) reacts with myoglobin (Fe^{2+}). The bright red nitrosyl-myoglobin complex provides the foundation for the distinct color of cured meat (Parthasarathy and Bryan 2012). This complex is extremely unstable, and it turns into a stable, eye-catching reddish-pink pigment (nitroso-hemochrome) during heat treatment (Fig. 11.2).

Furthermore, myoglobin combines with nitrous acid and forms metmyoglobin (Fe^{3+}) by oxidation. Metmyoglobin (Fe^{3+}) then reacts with NO to produce NO-metmyoglobin. NO-metmyoglobin is also produced from the reduction of metmyoglobin (Suman and Joseph 2013) As a result, the meat becomes brown in color. NO-metmyoglobin can be converted to NO-myoglobin by a reductant, causing the formation of the cured color (pink) again when heated.

The presence of other additives in cured meats also affects the color development. Antioxidants including erythorbate, ascorbic acid and polyphenols stimulate the production of NO by allowing the N_2O_3 reduction (Weitzberg and Lundberg 2013). Ascorbic acid reduces Fe^{3+} to Fe^{2+} effectively and enhances the reduction process of NO-metmyoglobin. Thus, antioxidants with reducing activity aids in the cured meat color development by raising NO production and lowering NO-metmyoglobin levels. NaCl, generally added to meat for curing, reacts with HNO_2 to generate nitrosyl chloride, which is more sensitive than N_2O_3 in terms of generating nitric oxide (NO) and initiating the formation of NO-myoglobin (Sindelar and Milkowski 2012).

The rate of nitrosyl myoglobin production has been found to increase with increased salt concentration (Ahmed 2017). The sensory panelists rated the bacon with a high sodium chloride level as having more redness (Delgado-Pando et al. 2018). The pH also controls nitric oxide formation from nitrite. Nitrous acid (HNO_2) and nitrite reactivity increase as pH decreases (Skibsted 2011). The rate of nitric oxide (NO) formation is doubled when the pH is slightly reduced by 0.2–0.3 units (Gassara et al. 2016).

Basically, a very small quantity of nitrite is required for the development of the cured color in meats, usually approximately 2–14 ppm (Sindelar and Milkowski 2011). However, the level of residual nitrite in cured meats gradually decreases owing to oxidation during storage time. As a result, the meat starts to lose its cured color and become faded. Color loss also occurs when meat is exposed to air and light, while the presence of adequate residual nitrite as well as reducing chemicals delay this process (Sullivan 2011). Usually, 10–15 ppm of residual nitrite is recommended, which can act like a reservoir for the cured meat color regeneration (Sindelar and Milkowski 2011) On the other side, higher levels of sodium nitrite (>600 ppm/kg of meat) and low pH value may lead to nitrite burn (discoloration) where meat shows a green color due to the formation of nitrihemin, a green-brown pigment (Feiner 2016).

Furthermore, color formation effect Nitrite or nitrate can produce a pinkish-red color via interaction with the myoglobin in muscle. In addressing the food safety concerns, most researchers have, in recent years, focused on finding replacement color formation ingredients containing natural nitrite or nitrate, such as vegetable extracts and bacteria. Jarulertwattana et al. (2021) found that 4 × 10–6 g/g of nitrite was suitable for chicken meat marination when it was used with ginger paste, whereas more than 4 × 10–6 g/g of nitrite will produce pink color defect in the drumsticks. Zhu et al. (2020) used *Lactobacillus plantarum* to partially replace standard nitrite in pork sausages, finding that 50 mg/kg $NaNO_2$ combined with 7 log CFU/g *L. plantarum* produced similar effects to 100 mg/kg NaNO2 alone. Ozaki et al. (2020) added radish powder (0.5% or 1.0%) and oregano essential oil (100 mg/kg) to fermented cooked pork and beef sausage, and found that both treatments could improve the red color of these products. Higuero et al. (2020) found that 75 mg/kg nitrate or 37.5 mg/kg nitrite produced a similar color to that seen following 150 mg/kg nitrate or 150 mg/kg nitrite treatment. This research suggests that, in terms of red color formation, there is an optimal amount of nitrite or nitrate. Therefore, appropriate process optimization may be an advisable method for controlling the amount of nitrite or nitrate added. Typical process optimization methods are orthogonal experiment design and response surface experimental design (Zhang et al. 2012). The latter, in particular, has been successfully used in the optimization of processes involving meat or other food products (Zhang et al. 2014, 2015, 2019, 2021; Zhao et al. 2017a). Recently, Kim and Chin (2021) evaluated the effects of paprika oleoresin, sunflower seed oil, and sodium nitrite on the storage quality of emulsified pork sausage, and the results indicated that this combination, given as 7.5 × 10–5 NaNO2+ 0.1% paprika oleoresin solution (1% paprika oleoresin +99% sunflower seed oil or 5% paprika oleoresin +95% sunflower seed oil) could increase redness.

sodium nitrite on emulsified sausage; the results indicated that the thiobarbituric acid values of the PNHb group and NaNO2 group were 0.62 mg/kg and 0.67 mg/kg, respectively, suggesting that PNHb has a stronger antioxidant effect than NaNO2. Despite the results, nitrite is still widely accepted as a color formation and antiseptic agent in most countries.

11.4.4 Shelf Life Extension

The improvement in shelf life is actually a combination of many factors delaying the loss of quality, including the hygiene of raw meat, processing treatments, storage temperature and relative humidity, packaging method, etc. (Singh et al. 2011). For nitrite and nitrate, they can extend the shelf-life of meat and meat products by inhibiting the outgrowth of pathogenic and spoilage bacteria (Rivera et al. 2019). Recently, Chatkitanan and Harnkarnsujarit (2020) combined thermoplastic starch, sodium nitrite, and low-density polyethylene to develop a starch-based, nitrite-containing film. They used it to vacuum-package pork, and the results indicated that the film not only increased the redness of pork by more than 30% during chilled storage, but also inhibited microbial growth and lipid oxidation.

Nitrite acts as an antioxidant by protecting the lipid molecules of meat from oxidation. In cured meats, nitrite works as an antioxidant through different mechanisms. Nitrite serves as a chelating agent of metallic ions (main prooxidants in meats) and it also stabilizes the heme Fe (Sebranek 2009). Furthermore, nitric oxide, produced from nitrite, may be readily converted to NO_2 by reacting with oxygen (Honikel 2008). Nitric oxide also reacts with radicals of lipid to break the oxidation chain reactions (Skibsted 2011). Lipid oxidation may be started in a variety of ways and once initiated, grows exponentially due to free radical interactions. Once they are generated in the starting phase, lipid radicals are continually oxidized through radical chain reactions. Nitrite can inhibit lipid oxidation initiation by reacting with ROS (reactive oxygen species), such as hydroxyl radicals. Nitric oxide (NO) can also inhibit lipid oxidation by combining with lipid peroxyl radicals and produce non-radical molecules. Nitrite has been shown to have an antioxidant property at concentrations as low as 40 mg per kg (Al-Shuibi and l-Abdullah 2002). A reduction of about 65% in lipid oxidation has been reported when 50 ppm sodium nitrite was added to the meat products (Girouard et al. 2011).

Proteins, in addition to lipids, are oxidized during the preparation of meat. Moreover, the antioxidant action of nitrite in the inhibition of protein oxidation is yet unknown. As the protein oxidation mechanism is similar to the mechanism of lipid oxidation, it is believed that nitrite might hinder protein oxidation. The quantity of peroxide value, sulfhydryl, carbonyl groups and thiobarbituric acid-reactive compounds (TBARS) produced during meat processing are commonly used to assess meat oxidation (Falowo et al. 2014). The application of sodium nitrite to meat products results in a considerably lower TBARS value than that of controls (without sodium nitrite), but no influence on the carbonyl compound concentration, used to evaluate protein oxidation (Rysman et al. 2016). Sodium nitrite has been shown to have both antioxidant and pro-oxidant properties in meat products. As evidenced by the decreased generation of carbonyl compounds, sodium nitrite exhibits an antioxidant property towards protein oxidation. However, nitrite was also discovered to possess a pro-oxidant effect on protein oxidation by lowering the total sulfhydryl concentration and increasing disulfide bond formation in cooked sausage proteins. By absorbing oxygen from sensitive molecules or producing reactive nitrogen species, nitrite can serve as a pro-oxidant (Feng et al. 2016). Protein oxidation causes a variety of physicochemical as well as nutritional changes in meat proteins along with a reduction in amino acid bioavailability, difference in composition of amino acids, decline in protein solubility, reduction in protein digestibility and lack of proteolytic activity (Lund et al. 2011). All these changes can be minimized by the antioxidant activity of nitrite. Therefore, it can be said that nitrite plays a great role as an antioxidant by inhibiting lipid and protein oxidation and thus it can prevent meat quality deterioration.

11.4.5 Antimicrobial Effect

Nitrite has been found to be very effective as a bacteriostatic and bactericidal agent in inhibiting or regulating the development of bacteria to various degrees in meat products. Nitrite has been shown to impede the reproduction of *Clostridium botulinum*. The application of nitrite has been shown to inhibit the formation of botulinal toxins from inoculated *Clostridium botulinum* in wiener sausages during storage. There are two effects of nitrite found in controlling the growth of *Clostridium botulinum*. The first effect is inhibiting vegetative cells developing from surviving spores. The second effect is the prevention of vegetative cell division (Pierson et al. 1983). During meat preservation, nitrite lowers the amount of *Clostridium sporogenes*, which have comparable characteristics to *Clostridium botulinum*. In addition, numerous studies have found that nitrite inhibits the development of *Listeria monocytogenes*, *Bacillus cereus*, *Clostridium perfringens* and *Staphylococcus aureus* in various meat products (Myers et al. 2013a, b). The impact of nitrite and inhibitory mechanisms varies with several bacterial species (Cui et al. 2010). The effectiveness of antimicrobial activity is dependent on various factors like pH, residual nitrite level, salt concentration, Fe content, reductants presence, storage temperature, etc. (Tompkin 2005) At acidic pH, nitrite hinders the growth of unwanted microorganisms more effectively (Allaker et al. 2001). Most of the nitrite applied to cured meat products is used to suppress *C. botulinum*, with only a little amount (about 25 ppm) required for color development. Suppression of *C. botulinum* development and toxin generation rises when nitrite levels rise. The level of additional nitrite is thought to have a greater influence on inhibiting *C. botulinum* than that of the residual nitrite during storage, implying that the production of antimicrobial compounds as a consequence of nitrite-related reactions might be noteworthy (Lim et al. 2015). The growth of starter cultures and bacteriocin production has been shown to be inhibited when the nitrite concentration was 100 ppm in sausage (fermented using *Lactococcus lactis*). An estimation predicts that when the nitrite content in sausage fermented with *Lactococcus lactis* reached 100 ppm, the development of starter cultures and bacteriocin synthesis were suppressed (Scannell et al. 2001).

Nitrite attacks bacteria at numerous sites by blocking metabolic enzymes, restricting oxygen absorption, and breaking the gradient of protons. Furthermore, nitric oxide binds to iron and reduces the availability of iron which is required for enzyme activity as well as bacterial metabolic activity and development (Tompkin et al. 2020). Because of the strong reactivity of Fe and nitrite, heme ion centers of enzymes and Fe-sulfur complexes are the major target of nitrite. The antibacterial activity of nitrite may be due to the peroxynitrite (ONOO) formation and nitric oxide formation from nitrite (Majou and Christieans 2018) Acid catalysis may cause oxymyoglobin to be autoxidized, generating superoxide radicals. The interaction of nitric oxide with superoxide radicals as well as the reaction of nitrite with hydrogen peroxide can produce peroxynitrite. Under physiological environments, peroxynitrite and peroxynitrous acid (ONOOH) stay in equilibrium. These two compounds are strong oxidants as well as nitrating agents (Prolo et al. 2014). They penetrate the bacterial cells by passive anionic diffusion and disrupt the microorganisms by causing protein and lipid oxidation or by damaging DNA (McLean et al. 2010). Nitric oxide (NO) can also inhibit microbial growth by forming protein-bound dinitrosyl iron complexes when it reacts with iron-sulfur proteins, which are engaged in critical physiological activities including energy metabolism & DNA synthesis (Landry et al. 2011).

11.5 Potential Alternatives to Nitrite in Processed Meat

As nitrite is involved in the production of nitrosamines, meat industries are recently focusing on new strategies to substitute traditional $NaNO_2$ in cured meat with the aim of minimizing nitrite intake. Consumer's interest is also growing in the development of natural alternatives and other preservation methods that are comparatively healthier. Although nitrite's broad-spectrum action makes it hard to replace it with a sole antimicrobial agent, a mixture of nitrite and other antimicrobial agents might become effective (Sindelar and Milkowski 2011). Nevertheless, any improvements in terms of consumption safety should be made without compromising the distinctive features of the organic and natural processed meats and this must be linked to the consumer's desire to purchase such foods. It is possible that a replacement for nitrite might be found and new products may be developed, but it is questionable if this might be good enough to entice people to buy. Consumers prefer meat products which contain lower nitrite levels and the decision of buying new meat products depend on the function of nitrite, their application reasons and their outcome (Hung et al. 2016a, b). Therefore, a successful nitrite reduction in meat products, along with

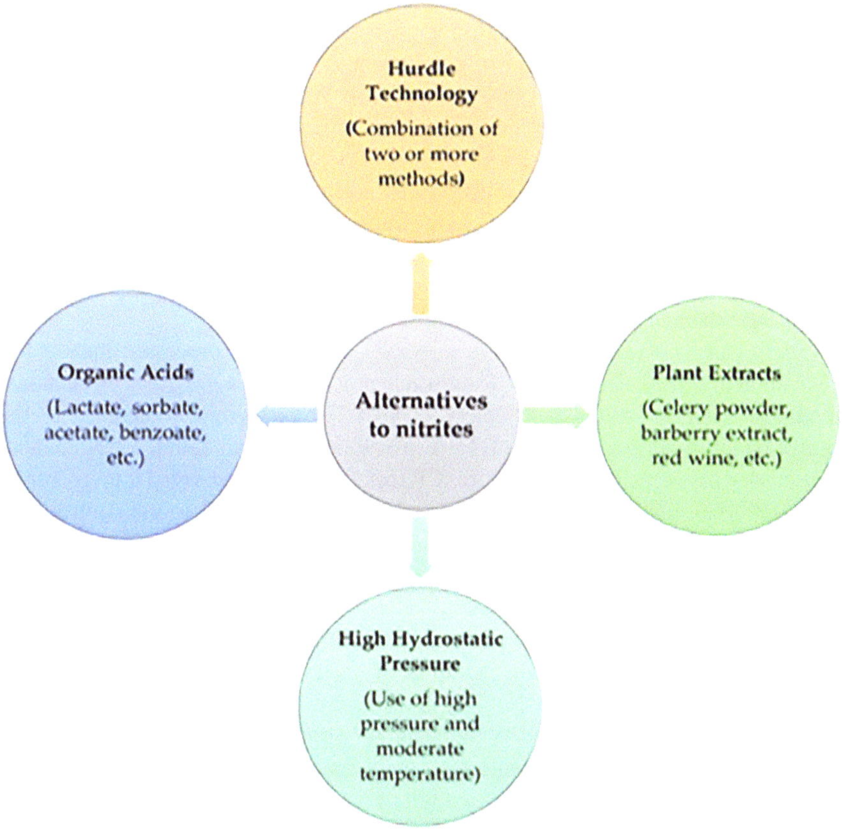

Fig. 11.3 Alternatives to nitrite reduction in processed meat products

the addition of several alternatives (Fig. 11.3), would provide a variety of benefits for the consumers, including a reduction in cancerogenic substances.

11.5.1 Plant Extracts

Nitrate is abundant in leafy green vegetables. Vegetables such as celery, lettuce, cress, spinach, rucola, etc., have been found to contain more than 2500 mg nitrate/kg (Colla et al. 2018). As nitrate can be reduced to nitrite by several microorganisms, these vegetables can be utilized as a partial or whole alternative to chemical nitrite in meat curing.

Parsley (*Petroselinum crispum*) has a high nitrate content (1000–2500 mg/kg on average) and it can be used as an alternative to nitrite in processed meat (Riel et al. 2017). Mortadella-type sausage produced with nitrite (greater than 60 ppm) from parsley extract powder appears to be equivalent to traditionally cured sausages in terms of *L. monocytogenes* inhibition and microbiological deterioration during storage. The microbial cell count of *L. monocytogenes* reduced as the percentage of added extract increased. By using parsley extract powder, reduced residual nitrite levels can be attained, and by marketing these meat products, customers' nitrite intake can be reduced. Application of 4.29 g PEP/kg sausage meat (120 ppm nitrate) appeared to be equivalent to the conventional use of nitrite salt in mortadella-type sausages in terms of the red color during most of the storage period. Consumer acceptability for products made with greater quantities of parsley extract was found in a recent study, suggesting that these items might be commercial. Parsley is also less allergic than celery extract (Riel et al. 2017).

Celery can be a viable alternative to $NaNO_2$ in cured or processed meats. The application of 0.8% celery powder for sausage production can meet the conventional standards. Sausages made with celery powder showed similar outcomes to the control (sodium nitrite) in terms of TBA, pH, total microbial count, VBN content, and sensory assessments. Furthermore, CP (celery powder) provided a more attractive meat color and reduced the residual nitrite content similar to the control as it naturally contains both pigment (betalains) and nitrate. The use of celery (as a concentrate or as a concentrate fortified with citric acid) seemed to have no effect on the product's redness; the a* (redness) value was reported identical to the value of the samples made with $NaNO_2$ (Horsch et al. 2014). The sensory qualities of sausages were unaffected by the addition of celery powder. As a result, celery powder improved the physicochemical characteristics of sausages during refrigerated conditions. A 75.6 g celery concentrate suppressed the multiplication of *L. monocytogenes* in a comparable way as $NaNO_2$ at 100 mg/kg. However, the most effective outcome of the application of celery powder was observed when combined with 10% citric acid (Horsch et al. 2014).

Spray-dried Swiss chard powder has recently been utilized as a natural nitrate source. This item is comparable to celery powder and it contains approximately 3.0–3.5% nitrate (Sebranek et al. 2012). The color consistency of cooked pork patty was enhanced by adding Swiss chard powder (PS). The shelf life of the product was also extended due to the inhibitory actions of PS on coliform bacteria. PS also enhanced the redness value (a*) of the samples due to greater nitrosoheme pigment concentration. The production of nitrosoheme pigments was favorably influenced by the addition of pre-converted nitrite (2%) from Swiss chard powder. The acidic pH of PS also showed a reduced residual nitrite concentration in the cooked pork patties (Shin et al. 2017). Moreover, PS prevented lipid oxidation in the cooked pork patties. This is possibly due to the presence of phenolic acids (antioxidant) and flavonoids, such as kaempferol and syringic acid (Pyo et al. 2004). As a consequence, cooked pork patties containing PS gained improved flavor and high acceptability ratings. The major benefit of Swiss chard powder is that it is free from allergens (Sebranek et al. 2012).

Barberry extract can be used as an alternative to nitrite in processed meat. In cooked sausages, the application of a high concentration of barberry extract (90 ppm) combined with a low concentration of nitrite (30 or 60 ppm) showed an increased shelf life under cold storage (4 °C). The antioxidant properties of the sample were improved by barberry extract addition, indicating that partial substitution of sodium nitrite with this plant extract might produce a healthier meat product (Khalehgi et al. 2016). Due to the antioxidant properties of barberry extract, it can be used in meat products instead of nitrite to prevent carcinogenic nitrosamine formation. In the sensory analysis, samples prepared with 90 ppm barberry extract and 30 or 60 ppm nitrite (lowest) gained the highest rank (Khalehgi et al. 2016).

Red wine or red wine mixed with garlic, has been demonstrated to be a viable option for controlling biological hazard in the production of chouriço (dry-cured sausage). In a recent study, it was revealed that the addition of wine or wine and garlic increased *Salmonella* destruction during meat processing (Patarata et al. 2020). The color of chouriço prepared with wine or wine mixed with garlic was better preferred by the consumers. On the contrary, the samples produced with traditional nitrite salt was rated as artificial and over-cured. The use of 7.5% red wine to chouriço showed a high redness value (a*) and the same value was observed with the addition of 150 ppm sodium nitrite. Furthermore, 7.5% red wine mixed with 1% garlic enhanced the yellowness (b*) of the sausage due to the reaction between these two items (Patarata et al. 2020). The combined effect of these two ingredients also showed a slight brownish color in the product. Moreover, a strong cured flavor was exhibited in samples produced with red wine or a mixture of wine and other ingredients. Furthermore, chouriço prepared with red wine was better rated than samples made with only nitrite salt (Patarata et al. 2020).

In a recent study, beetroot powder (*Beta vulgaris*) was used as a nitrite alternative in Turkish fermented sausage (Sucu and Turp 2018). The beetroot powder exhibited a change in the quality traits of this new product. Depending on the quantity of the powder used, the redness value (a*) of the sausage changed. The researchers also reported that beetroot powder reduced the yellowness value and increased the lightness value during the early period of storage. The highest redness value was obtained from the sample formulated with 0.35% of the beetroot extract. The redness value was well conserved during the storage period when a higher quantity of beetroot extract was added. However, samples containing an increased amount of beetroot powder and a low amount nitrite/no nitrite exhibited a higher oxidation. Although having polyphenolic compounds, *B. vulgaris* did not reduce the TBARS values during the storage period. However, sensory analysis of the sausage formulated with different amounts of beetroot extract (0.35%, 0.24%, 0.12%) reported no change when compared to the control (150 ppm $NaNO_2$) (Sucu and Turp 2018).

The effect of grape seed extract with olive pomace hydroxytyrosol and chestnut extract with olive pomace hydroxytyrosol, on the color of cinta senese dry-fermented sausage was observed by some researchers. The addition of grape seed extract resulted in high redness values which was probably due to the effect of Zn-protoporphyrin formation (Aquilani et al. 2018). The microbial counts of *L. monocytogenes*, *E. coli*, *Clostridium* spp., *Staphylococcus* spp. and *Salmonella* sp. were found in low numbers for up to 3 weeks of ripening. The researchers also found that these extracts can inhibit lipid oxidation as they contain polyphenols (Aquilani et al. 2018).

Tomato pulp powder (TPP) can be applied successfully to pork luncheon rolls at a concentration of 1.5% without affecting the physicochemical characteristics. The impact of tomato pulp powder on luncheon roll products was examined by some researchers (Hayes et al. 2013) who discovered that the redness (a*) values and the yellowness (b*) values increased while the lightness (L*) values dropped. However, the preparation of luncheon rolls supplemented with 3% TPP (tomato pulp powder) would be unfeasible due to the negative physicochemical and sensory effects. The decrease in nitrite concentration from 100 to 50 mg/kg showed no adverse effects on texture or microbiological stability (TVC) and boosted the overall product acceptance. According to ANOVA and chemometric techniques, the pork luncheon roll made with 1.5% TPP and 50 mg sodium nitrite exhibited the same ($p > 0.05$) or improved sensory characteristics in comparison to the luncheon roll with no TPP and 100 mg/kg nitrite. In another study, cooked sliced pork luncheon roll formulated with 1.5% TPP mixed with lower nitrite showed better technological and sensory qualities. The interaction between nitrite and plant extract may lead to a pro-oxidant state. In luncheon rolls, a mixture of tomato pulp powder and nitrite has shown a pro-oxidant impact (Hayes et al. 2013). Moreover, tomato pulp powder gives the product a savory taste. Furthermore, tomato processing wastes, such as peels and seeds, contain various bio-active components such as carotenoids (b-carotene, lycopene, phytofluene, phytoene and lutein) (Tang et al. 2021). Moreover, tomato contains important natural pigments which are used as coloring agents in various meat products (Francis et al. 2000). In addition, tomato powder can retard the oxidation process and improve the consumer acceptability when used as a natural alternative to nitrite in frankfurters. It has been reported to lower the residual nitrite concentration in frankfurters (Eyiler and Oztan 2011).

Grape pomace can be used as an alternative to nitrite in meat processing. Several physicochemical, microbiological, and sensory characteristics of beef sausages were affected by the addition of grape pomace. These findings show that adding large quantities of grape pomace (about 2%, *w/w*) as a partial replacement of nitrite might alter the color characteristics of sausages, resulting in a darker product. Nitrite-reduced sausage systems such as beef sausages prepared with 1 or 2% grape pomace and nitrite (30 mg/kg) showed an elevated yellowness of the products. On the other hand, grape pomace reduced the redness (a*) and lightness (L*) values of beef sausages in comparison to the control (120 mg/kg nitrite). On the first day of storage, sausages made with 2% grape pomace had higher oxidation values ($p < 0.05$). This was supported by a decrease in phenolic compounds and an elevation in total microbial counts after 30 days of storage at 4 °C (Riazi et al. 2016). Although there were changes in the sensory characteristics of beef sausages made with grape pomace, sensory panelists found these items to be acceptable. The addition of dry red grape pomace to sausages prevented bacterial growth during 1 month of cold storage. It also reduced residual nitrite levels, pH and extended the sausage's shelf life. Furthermore, adding DRGP to the beef product increased the health-promoting aspects as well as stabilized the samples generated (Riazi et al. 2016). According to these findings, grape pomace powder can be used as a healthy food ingredient in diet formulation.

Pomegranate peel extract can be used as a natural curing agent in meat processing as it contains the highest quantity of polyphenols among other fruits. Many researchers have investigated pomegranate peel as it has natural antioxidants including flavonoids and phenolic acids that have anti-cancer, anti-mutagenic, anti-cardiovascular disorder and anti-inflammatory properties (Negi et al. 2003; Li et al. 2006). Beef sausages prepared the highest quantity of PPE, pistachio green hulls (1250 ppm) and no added nitrite showed higher yellowness (b*) values than the control (where 120 ppm sodium nitrite was added) (Aliyari et al. 2020). Another study reported that the use of PPE and pistachio green hulls decreased the MDA formation. In compared to essential oils (EO) or synthetic antioxidants, PPE and PGHE are affordable, plentiful, and economical sources of phenolic content. They have a high TPC content and excellent antioxidant activity. The TBARS and PV values of samples containing less nitrite and more plant extracts were comparable or better than the control. Microbial analysis revealed that both extracts provided complimentary effects in keeping total viable count and harmful bacterial growth below the threshold level. Therefore, these plant extracts can be used with decreased levels of nitrite (up to 50%) in cooked sausages to improve functional characteristics and reduce the production of carcinogenic nitrosamines (Al-Gubory et al. 2016).

The seed of the annatto plant (*Bixa orellana* L.) has been used as a natural colorant in Asian cuisine. Among all natural colorants, annatto took second place in terms of economic value (Yolmeh et al. 2014). Annatto seed powder may be an effective alternative to nitrite in pork sausage production as it improves the color and textural properties of sausages. In a recent study, pork sausage made with various concentration of annatto seed powder (0.2%, 0.1%, 0.05%, 0.025%) reported an increased value of a* and b* parameters with an increasing amount of *Bixa orellana* (L.) seed powder. Furthermore, annatto added to sausages with 37.5 ppm nitrite was more effective in the lipid oxidation retardation process as well as in the inhibition the microorganism growth than the control sample (150 ppm nitrite). The greatest inhibition of TBARS formation was observed when the highest amount (0.2%) of the extract was used. The reduction of lipid oxidation products (primary and secondary) was observed to be comparable to that observed in the control sample (150 ppm nitrite) (Van Cuong and Thoa 2017). Therefore, annatto seed powder can be utilized for the substitution of nitrite in the preparation of meat products as a source of natural anti-microbial and antioxidant agents for extending the shelf life of the product.

11.5.2 Organic Acids and Salts

In meat industries, organic acids are used to prevent microbial development, decrease the pH of meat products, and increase the curing performance of processed meats (Drosinos et al. 2006). The use of organic acid to cured meat enhances the color development process while inhibiting microbiological growth. Lactate, sorbate, acetate, and benzoate are some important organic acids that have been widely used as food additives for many years. The rationale for employing organic acids is that they have the potential to lower pH to a level that prevents bacteria from proliferating (Mcclure 2009).

Lactate (lactic acid salt) can be applied in the manufacturing of processed meats and it has been proven to improve color stability, flavor and shelf life. Although various organic acids might be employed as antimicrobial agent, lactate is the most effective among all since it has the ability to enhance the meat quality by imparting a salty flavor, while also maintaining color and contributing to a greater ability to hold water (Doores 2005; Vijayakumar et al. 2008). Adding a little quantity of sodium/potassium lactate, salts, and water to meat can enhance the flavor, color and tenderness while also delivering antibacterial and antioxidant properties (McClure et al. 2011; Kim et al. 2009a, b). When introduced to raw meat, sodium lactate prevents the formation of foul and rancid flavors and appears to be an effective flavor enhancer without any negative effects (Serdengecti et al. 2006). One possible mechanism behind the inhibition of off-flavors by sodium lactate treatment is the radical scavenging activity. It binds to the free radicals present in meat and prevents lipid oxidation. However, the mixture of sodium lactate and NaCl significantly decreases the oxidative reactions caused by NaCl, observed by a decrease in the TBA value (0.463 to 0.384). Ground pork treated with a combination of NaCl and sodium lactate exhibited a significant ($p < 0.05$) reduction in lipid oxidation when stored at 2 °C (Tan and Shelef 2002).

Lactate promotes the color stability of fresh meat through increased LDH (lactate dehydrogenase) activity. LDH converts lactate to pyruvate while reduced NADH is also replenished. This reproduced NADH then reduces metmyoglobin (Fe^{3+}) to deoxymyoglobin (Fe^{2+}). As deoxymyoglobin can reduce NO_2 to NO (McClure et al. 2011) increased deoxymyoglobin generation is thought to result in increased nitric oxide production from nitrite and low residual nitrite remains in the samples. The NO produced from nitrite can then be combined with the residual deoxymyoglobin in the form of iron-nitrosylated myoglobin (Hendgen-Cotta et al. 2008). Through this mechanism fresh meat can maintain a pinkish-red color over an extended period of storage. In a previous study, researchers have observed that application of lactate in fresh meat products preserved the myoglobin pigments from being oxidized under oxidizing environments, allowing more red oxymyoglobin formation and retained the oxymyoglobin pigment for a longer period (Kim et al. 2009a, b). In another study, it was noticed that sodium lactate (1%) can retain the red cured color of the cured pork sausages while improving the salty flavor of the product. It was also shown that sodium lactate lessens taste loss and improves palatability during storage for every 1% increase in sodium lactate (Sedghi et al. 2022).

Lactate's antimicrobial impact on meat has been researched, and its effectiveness against a variety of pathogenic microbes and diseases has been demonstrated. It is worth noting that as the addition of sodium lactate rises, the number of aerobic as well as anaerobic microbes reduces. The conversion of pyruvate to lactate may be altered by high amounts of lactate ions, which is similar to thermodynamic equilibrium (Vasavada et al. 2003). Furthermore, sodium lactate salt works as a fragmented acid that penetrates through a bacteria's membranes, creating an acidic intracellular state (Carpenter and Broadbent 2009). Experiments have shown that bacterial intracellular pH as well as cellular metabolism is rapidly reduced due to cell body denaturation and eventual cell death of the bacteria (Bradley et al. 2011). For this reason, sodium lactate addition is said to cause reduced water activity and thus prevent microbial development. The inhibition of *C. perfringens* growth by added Na or K lactate depends on the quantity added, type of product, moisture content, p^H, adjunct salt and spice levels, as well as storage temperature (Juneja and Thippareddi 2004a, b; Sabah et al. 2004). Sodium lactate added to the sausages was demonstrated to improve the microbiological quality, provide a longer shelf life and enhance antimicrobial action compared to sausages treated with sodium nitrite (Bingöl and Bostan 2007) The effect of sodium lactate and NaCl together (20 and 20 g/kg) or NaCl alone on the chemical and microbial quality of uncooked ground beef (at 2 °C storage conditions) were studied; sodium lactate in combination with NaCl or NaCl alone slowed the growth of APC (aerobic plate count), lactic acid bacteria, psychotropic count, Enterobacteriaceae, as well as prolonged the shelf life of the product up to 21 days and 15 days, respectively (Sallam and Samejima 2004). Sodium and potassium lactate also inhibited *Clostridium perfringens* development in processed beef goulash while stored under isothermal conditions (Aran 2001). The use of 2% sodium lactate in poultry sausages prevented aerobic psychotropic and lactic acid bacteria from growing during cold storage (Cegielska-Radziejewska and Pikul 2004). *Lactobacillus alimentarius* and other LAB (lactic acid bacteria) did not grow in cooked or acidified chicken meat for 14 days after being exposed to 3.3% sodium lactate. The total viable counts of frankfurters (low-fat) made with 2% sodium lactate was lower in comparison to the control (NaCl) after 28 days of storage under refrigerated conditions (Bingöl and Bostan 2007).

The most important impact of sodium lactate in meats is its capability to extend the shelf life. It has been observed that sodium lactate extends the shelf life of meat products such as ham and cured sausages by around 1 to 2 weeks. After 2 months of refrigerated storage, sausages treated with 1.8% sodium lactate revealed no signs of deterioration, but the control group, which did not employ a preservative, showed some signs of deterioration after 1 month. The addition of 1% or 2% sodium lactate to sliced poultry sausages prolonged the shelf life by 3 or 4 times, respectively when stored at 5–7 °C (Cegielska-Radziejewska and Pikul 2004). The use of sodium-lactate salt in fresh pork-sausages was shown to be quite effective in reducing bacterial count and increasing the shelf life for around 14 days longer than the control experiment.

The application of potassium lactate in beef slices can enhance or retain the similar characteristic color, flavor, sensory taste, lipid oxidation, shear and cooking characteristics to conventionally processed beef patties (Quilo et al. 2009). The addition of 3% potassium lactate on the outer surface of frankfurters exhibited a slightly reduced water activity (a_w); the sensory panelists, on the other hand, only saw a minor effect or no change in sensory characteristics (De Gonzalez et al. 2004).

Buffered sodium citrate or the mixture of buffered sodium citrate and sodium diacetate may be successfully utilized to prevent the proliferation of *C. perfringens* during the chilling of injected pork and roast beef. The addition of sodium citrate during roast beef preparation led to a reduction in the growth of *C. perfringens* by 2.47, 1.87, 0.98 log CFU/g at 2.0%, 1.0% and 0.5% concentrations, respectively (Thippareddi et al. 2003). Moreover, a similar experiment was conducted on ground turkey where two extra organic acids were used (sodium lactate and sodium acetate); 1% concentration of sodium lactate and sodium acetate impeded *C. Perfringens* germination as well as proliferation (less than 1.0 log CFU/g) (Juneja and Thippareddi 2003).

Sorbate, propionate, and benzoate are commonly recorded as "safe additives" and these additives have been conventionally used to prevent the growth of mold in numerous foods (Food and Drug Administration 2003). These novel antimicrobial additives suppress the development of various Gram-positive bacteria such as *Staphylococcus aureus*, *Listeria monocytogenes* and *Clostridium botulinum* in meat systems (Islam et al. 2002; Samelis et al. 2001). Various research data in this regard has shown that 0.1% benzoate–sorbate and benzoate–propionate inhibited *L. monocytogenes* growth in some ready-to-eat meat products (Glass et al. 2007). It is difficult to completely replace sodium nitrite with sodium sorbate as it cannot perform all of nitrite's functions, such as color development. A possible alternative to sodium nitrite (120 ppm) alone is the combination of sodium–sorbate (2600 ppm) and sodium nitrite (80 ppm). When compared to nitrite–sorbate mixtures or nitrite alone, sorbate increased product rancidity. A high concentration (2600 mg/kg) of sorbate seems to affect (although not considerably) the sensorial quality of mortadella when compared to relatively low concentrations (1000 mg/kg) and nitrite alone. (Glass et al. 2007). The development of *Listeria monocytogenes* on the exterior of frankfurters was suppressed by a benzoate mixture at 25% (*w/v*) (Islam et al. 2002). Furthermore, in broth tests with no salt, sodium benzoate (2000 ppm) showed an inhibiting effect on the growth of non-proteolytic *Clostridium botulinum* (Mulvey et al. 2010).

11.5.3 Microbial Degradation

Beyond the addition of vegetable juice or extracts to reduce the residual nitrite in meat or meat products, some microbes also showed nitrite reduction and conversion effects. Microbial enzymes can convert metmyoglobin to NO-Mb, thereby making the cured meat pink. Huang et al. (2019) added *Lactobacillus fermentum* RC4 and *L. plantarum* B6, without any nitrite, during preparation of cured meat, producing a bright color and low nitrite content. Huang et al. (2020b) adopted three staphylococcal species (*S. vitulinus, S. carnosus, and S. equorum*) to induce the production of NO-Mb through nitric oxide synthase catalysis, which, in all three species, increased the a* value of the dry sausage model. Zhang et al. (2020b) inoculated Lactobacillus curvatus LAB26 and *Pediococcus pentosaceus* SWU73571, isolated from the sour meat of the Dong minority in China, into sour meat. Both bacteria decreased the nitrite content of sour meat and increased the a* value significantly. Luo et al. (2020)

added *Lactobacillus fermentum* AS1.1880 into pork batters and confirmed that the species increased their redness by producing nitric oxide synthase. Tian et al. (2020a) found that *Lactobacillus fermentum* RC4 significantly reduced the nitrite content of salted meat. They combined Lactobacillus fermentum RC4 and *L. plantarum* B6 with beet red, Monascus red, and nisin to produce Chinese bacon, and found that the combined agents not only improved the cured meat quality, but also decreased the nitrite content. Liu et al. (2021) used the commercial starter PRO-MIX5 (*Staphylococcus xylose, Lactobacillus sakei, Lactobacillus plantarum*) to prepare bacon and found that the starter resulted in significantly lower NDMA and total N-nitrosamine levels.

11.5.4 Cooking Method

Nitrite is easily oxidized in high-temperature aerobic environments, such as cooking. Therefore, cooking can decrease the nitrite content of meat or meat products. Nitrate reduction and nitrite accumulation were observed in the uncooked sausage models, but Waga et al. (2017) found that nitrate inhibits nitrite auto-decomposition in pork ham cooked in a water bath at 75 °C. Asioli et al. (2017) investigated the effects of dry-frying temperatures (100, 150, 200, and 250 °C) on the residual nitrite and N-nitrosamine levels in smoked bacon, and found that the residual nitrite content initially increased (from unheated control to 150 °C) and then sharply decreased between 150 °C and 250 °C. Sallan et al. (2020) investigated the effects of cooking extent on volatile nitrosamine formation in the heat-treated semi-dry fermented sausage sucuk, and found that higher temperature and higher heat intensity induced more nitrite controverted into nitrosamines, the content of nitrite and cooking levels influenced the content of N-nitrosopiperidine (NPIP) more than other nitrosamines, well done and very well done cooking levels combining with higher nitrite contents (100 and 150 mg/kg) resulted in significant increases in NPIP content.

11.5.5 Irradiation

Food irradiation includes treatment with gamma rays, X-rays, or electron beams. Among these methods, gamma irradiation has the highest potential. However, its development and commercialization have been hampered by unfavorable public perceptions. The endorsement of food irradiation by many international food and health organizations has counteracted this perception to some degree; thereby increasing consumers' confidence and raising food industry interest. Gamma irradiation is effective in reducing the residual nitrite and N-nitrosamines in meat products. A model sausage that was irradiated with γ-irradiation found that high dose irradiation (>10 kGy) could decrease the residual nitrite levels significantly (Ahn et al. 2002). A dose of 5 kGy γ-irradiation significantly decreased residual nitrite in

Chinese Rugao ham before ripening (Wei et al. 2009). In recent years, limited investigations on effect of irradiation on reduction of the residual nitrite were reported.

11.5.6 Plasma-Treated Water

The interaction of plasma with water can result in the generation of nitrate and nitrite, as well as reactive oxygen species (Chen et al. 2021; Moutiq et al. 2020). Therefore, plasma-treated water can be used as a substitute for nitrite or nitrate. Moutiq et al. (2020) used atmospheric cold plasma to treat chicken breast meat and found that levels of natural microflora were approximately 2 log CFU/g lower following treatment with 100 kV atmospheric cold plasma for 5 min, and that the shelf life of the chicken breast was extended by sterilization of the mesophiles, psychrotrophs, and Enterobacteriaceae in the meat. Dong and Tu (2006) used atmospheric nonthermal plasma to treat roasted lamb, and found that 45 min plasma treatment resulted in 30% lower residual nitrite compared with the lamb treated with added nitrite, while the overall sensory scores were similar in both groups.

11.5.7 High Hydrostatic Pressure

Treatment with high pressure (100–800 MPa) is used uniformly to meat products at moderate temperature (less than 45 °C) as an anti-microbial process with the purpose of extending the shelf life of that product. HHP increases the meat product's shelf life by reducing the growth of pathogenic microorganisms (San Martin et al. 2002; Patterson et al. 2007). The use of HHP also aids in the inactivation of enzymes for a greater duration of time without the use of synthetic additives. However, in order to ensure food safety and to increase the shelf life, proper application of pressure and temperature has to be set in accordance with the product's characteristics (Bolumar et al. 2021). Meat processors can now satisfy the growing demand of consumers for natural and "preservative-free" meat products while retaining the stable sensory qualities over a longer storage period and ensuring product safety by processing meats using HHP (Considine et al. 2008; Yordanov and Angelova 2010).

The pressure used in HHP treatment is isostatic, which means that the pressure is distributed uniformly and instantaneously to the product. Moreover, it is an adiabatic process which implies that there is a little temperature difference with the rising pressure (the temperature rises by nearly 3 °C for every 100 MPa rise, based on the food properties) irrespective of the product's size or shape (Wilson et al. 2008). The product which is to be treated is loaded into the pressure vessel, and the vessel is filled and sealed after covering the product with a fluid. This fluid, often water or a glycol solution, is used to transmit the pressure throughout the system. A pressure pump with extra fluid injection is used to support the system. The treatment efficacy is largely determined by the amount of pressure used and the time frame held (Rendueles et al. 2011). The pressure level ranges from 100 to 800 MPa and the retention time varies from a few seconds to 20 min. As per the isostatic theory, consistent pressure is applied to food uniformly in every direction. After applying and releasing pressure, the product will revert to its original form. The process of microorganism inhibition by HHP is affected by microorganism species, time, pressure level, temperature, water activity, and food composition (Chotyakul and Boonnoon 2016) HHP seeks to preserve food in a mild manner while eliminating harmful and pathogenic microorganisms. The cell membrane of microorganisms is most vulnerable to pressure damage. Cells are damaged and lose their integrity because they are unable to regulate the passage of fluids and electrolytes through their membranes under pressure. As a result, they eventually lose their reproductive ability (Hugas et al. 2002). Generally, Gram-negative bacteria are more pressure resistant than Gram-positive bacteria (Chotyakul and Boonnoon 2016).

The application of HHP treatments in salted white chicken meat develops the cooked meat texture and the color of ground beef. This data may be utilized to create HHP treatment as a processing option to minimize NaCl and nitrite application while retaining the physical properties, meat quality and functionality (Ros-Polski et al. 2015). HHP is a strong strategy for minimizing *Salmonella* spp., *Campylobacter* spp., *Listeria monocytogenes* in sliced cured ham (Ros-Polski et al. 2015). *Salmonella* spp. and *Campylobacter* spp. was not found in sliced cured ham when conducted with HHP at 600 MPa (31 °C) for 6 min, but *L. monocytogenes* was found in untreated samples (25 g) at the initial time. However, *Listeria monocytogenes* was not found in HHP-treated samples studied for 4 months at 4 ° C (Chotyakul and Boonnoon 2016). A reduction (10 logs) in the most resistant strain of *Listeria monocytogenes* was found in pork sausages when treated with HHP at 400 MPa and 50 °C for 6 min (Campus 2010).

HHP at 400 MPa, 17 °C for 10 min substantially decreased the levels of *Enterococcus* and *Enterobacteriaceae* in cooked sausages (Marcos et al. 2008). HHP has also been shown to retard lipid oxidation in pork meat exposed to 800 MPa and therefore extend the shelf life (Simonin et al. 2012; Rubio et al. 2007). HHP treatment as a non-thermal processing approach can be an effective strategy for reducing *L. monocytogenes* in sliced vacuum-packaged chorizo. The use of HHP at 400 MPa for 2.5–16 min enhanced the sensory characteristics of Portuguese chorizo as well as the overall brightness, firmness, and cohesiveness (Alfaia et al. 2015). In accordance with this investigation, HHP at 500 MPa for 5 min and 600 MPa for 3 min had no negative impact on the sensory characteristics of salchichón and salami, respectively (Gill and Ramaswamy 2008).

HHP has a substantial impact on the microbiological qualities of vacuum-packed wieners under refrigerated storage. HPP at 600 MPa, 8 °C for 3 min had no detrimental influence on the processing parameters and helped to increase the shelf life of all wieners of up to 84 days in storage. At 56 days of storage, HPP-treated wieners were evaluated and were substantially higher in overall acceptability than non-HPP-treated wieners. Regarding flavor acceptability, the magnitude of variation between the mean rated like values of HPP- and non-HPP-treated was the greatest at

6.1 vs. 4.7, respectively (Pietrasik et al. 2017). As a result, HPP can significantly improve the sensory quality of products, such as sodium-reduced formulations containing organic forms of nitrite. Moreover, an APC and LAB count in HHP-treated wiener was found below the detection limit of 1.0 log CFU/g for all preparations during the 12-weeks of storage period. Much prior research on RTE products found similar reductions in microbial population after HHP, indicating that HHP can significantly increase the shelf life of vacuum-packaged meat products (Vercammen et al. 2011; Myers et al. 2012). According to LAB populations, the use of HHP technique to vacuum-packaged sliced ham enhanced the product's shelf life. In this investigation, HHP at 400 MPa for 15 min extended the product's shelf life from 19 days to over 85 days. This result demonstrates unequivocally that HHP is a highly useful method for extending the shelf life of vacuum-packaged sliced ham (Slongo et al. 2009).

In sliced cured meats, HHP treatment is a highly promising method of preservation (Rubio et al. 2007). HHP at 400 MPa, 17 °C for 10 min after fermentation also enhanced the microbiological quality of the fermented sausages, without compromising the product's quality (Marcos et al. 2007). HHP treatment may be recommended as a finishing step in the preparation of low-acid fermented sausages with suitable starter cultures (Marcos et al. 2008). HHP reduces the *L. monocytogenes* population in RTE processed beef products. The viable *L. monocytogenes* populations remained below the detectable limit of the sample after a 600 MPa HHP treatment for 4 min (Lavieri et al. 2014). Some researchers looked at the possibility of utilizing HHP to extend the shelf life and enhance the food-safety of refrigerated ready-to-eat meat products when prepared with sodium nitrite (Hayman et al. 2004). They discovered that treating products with HHP at 600 MPa, 20 °C for 3 min was enough to keep the numbers of aerobic, anaerobic bacteria, and lactobacilli below the detection level for 95 days when stored at 4 °C. Combination of nitrite (75 ppm) and 450 MPa HHP produced satisfactory frankfurters in terms of functional characteristics (color, texture, and water holding capacity) (Jonas et al. 2017). As a result, *L. monocytogenes* inhibition programs can include a combination of HHP and antimicrobial chemicals, such as nitrite.

Combining high pressure technology with other hurdle methods can create a synergistic effect from the two hurdles and can improve the inhibitory impact. The combined impact of HHP (200 MPa for around 10 min) and enterocin LM-2 (2560 AU/g) on the shelf life of sliced cooked ham during refrigerated storage has been studied. Application of HHP (400 MPa) and enterocin (256 or 2560 AU/g) showed a suppression in the development of *Salmonella enteritidis* and *Listeria monocytogenes* below the detection level in sliced cooked ham. Furthermore, an increased shelf life of sliced cured ham of up to 70 to 90 days was observed under refrigerated storage (Aymerich et al. 2005).

11.5.8 Other Technologies

As the food industry develops, clean labels are widely preferred and demanded by consumers. Therefore, natural food additives, nondestructive technologies, and minimally harmful processing technologies will be developed to reduce or replace the addition of nitrite or nitrate (Ferysiuk and Wojciak 2020; Fraqueza et al. 2018; Ranasinghe 2018; Zhang et al. 2016; Zhang et al. 2020a). In terms of naturally sourced additives, there are more than 2000 edible vegetables, extracts of most of which remain to be evaluated for nitrite reduction. Among non-destructive technologies, ohmic heating and its combination with other non-destructive technologies have not yet been investigated.

11.6 Regulation of Nitrite and Nitrate Use

The multiple functions of nitrite and nitrate in meat and meat products cannot be fully replaced by other food additives at present (Karwowska and Kononiuk 2020). Complete restriction of their use in this context may not be acceptable to producers and consumers. In order to regulate the application of nitrite and nitrate in meat or meat products, many countries have established directives and regulations (Gassara et al. 2016). Germany is the fourth largest meat producer in the world and cured meat products are popular there. It was the first to propose limits to the addition of nitrite to meat and meat products, restricting the nitrite content in curing salt to between 0.5% and 0.6% (Ferysiuk and Wojciak 2020). The United Nations only limits residual nitrite levels according to the category of meat product. The United States is the second largest meat producer globally and its cured meat and meat products are frequently consumed as hotdogs or other convenience foods. The maximum nitrite level for pumped and/or massaged bacon (the ingoing level of sodium nitrite $\leq$120 (or 148 ppm potassium nitrite) ppm) and other meat products were restrictedly limited. More specific restrictions have been imposed by the European Union and European countries. They have restricted the levels of nitrite and nitrate in treatments and the residue levels in different meat products. China is the world's largest meat producer of cured meat products being widely-consumed traditional foods. To regulate the application of nitrite and nitrate in meat or meat products, similar limits have been proposed within Chinese additive standard (National Health and Family Planning Commission 2015). Canada, Korea, Japan and other countries are all restricting input and residue levels of nitrite and nitrate in meat or meat products. Restricting input and residue levels is beneficial, in that it prevents the overuse of nitrite and nitrate in meat or meat products, but it is also essential to ensure that directives or regulations are implemented appropriately. Although most countries have established food inspection agencies to supervise the use of nitrite and nitrate in meat or meat products, there will be some omissions. Abd Hamid et al. (2020) investigated the nitrite intake of school children in Brunei Darussalam, and found that more than 20% of children exceeded the Acceptable Daily Intake (ADI) of nitrite (0.06 mg/kg bw/day). Therefore, stricter application of regulatory restrictions and more frequent sampling surveys are required in countries where the ADIs for nitrite and nitrate are greater than 0.06 and 3.7 mg/kg bw/day, respectively. To avoid indiscriminate use of nitrite and nitrate in meat or meat products, the scope for their addition is rigorously restricted in most countries. As ingredient or hazardous compound levels are not monitored by regulation in raw meat materials in some countries, and considering that nitrite overuse in processed meat products has occurred (Abd Hamid et al. 2020; Neri and Patalinghug 2021), the application of nitrite or nitrate in raw meat materials may be a supervisory blind spot that increases the risk of overuse. Therefore, it is vital to strengthen the sampling regime for meat and meat products, especially raw meat where it is not currently included. Fried meat products comprise one of the most widely consumed meat categories, but there are not many reports on their residual levels of nitrite or nitrate. This may be because fried meat products are prepared meat products, which are readily consumed mixed with other foods, and usually eaten immediately after cooking. These features make them popular among adults or children and they are widely consumed in snack bars and homes. An investigation of nitrite intake in school children in Brunei indicated that more than 20% exceeded their nitrite ADI (0.06 mg/kg bw/day) by eating cured meat products (Abd Hamid et al. 2020). Therefore, in order to evaluate the safety hazards of eating fried meat products, more surveys of the residual nitrite or nitrate in fried meat products should be conducted, especially in products consumed by infants or children.

11.7 Conclusions

Nitrite is used as a versatile additive in the meat industry. It is liable for the pinkish-red color and unique flavor of cured meat products. It also acts as an antioxidant that prevents the development of a warmed-over flavor as well as a bacteriostatic effect that prevents the formation of botulinum toxins from *Clostridium botulinum*. Despite its many advantages in meat curing, sodium nitrite has been the subject of debate due to its probable carcinogenic impact on humans, according to various research. Ingesting too much nitrite can induce methemoglobinemia in children and raise the risk of developing colorectal cancer in adults. On the other hand, consumers' desire for organic or nitrite-reduced meat keeps growing. As a result, the meat industry is now focusing on finding efficient ways for minimizing residual nitrite content in meat products and safer nitrite alternatives for the preparation of organic meat products. As nitrite replacements, various plant extracts, organic acids (lactate, sorbate, etc.) and HHP can be employed efficiently in processed meats. Unfortunately, still no sole alternative for nitrite has been found that can fulfil all of nitrite's functions simultaneously. Hurdle technology using reduced levels of nitrite combined with other additives or processing techniques might have potential in producing the antimicrobial effects against the most prevalent microbial pathogens while also improving sensory characteristics. However, additional research is required to find a single alternative to nitrite that can be used to perform the nitrite broad-spectrum activities in a cost-effective way.

References

Abd Hamid NFH, Khan MM, Lim LH (2020) Assessment of nitrate, nitrite and chloride in selected cured meat products and their exposure to school children in Brunei Darussalam. Food Compos Anal 91(8):1–10. https://doi.org/10.1016/j.jfca.2020.103520

Afkhami A, Masahi S, Bahram M (2004) Spectrophotometric determination of nitrite based on its reaction with p-nitroaniline in the presence of diphenylamine in micellar media. Bull Korean Chem Soc 25:1009–1011

Ahmed MMAA (2017) Spectrophotometric determination of nitrite content in cured meats. Master's Thesis, Sudan University of Science and Technology, Khartoum, Sudan

Ahn H, Kim J, Jo C, Lee C, Byun M (2002) Reduction of carcinogenic N-nitrosamines and residual nitrite in model system sausage by irradiation. J Food Sci 67(4):1370–1373

Alahakoon AU, Jayasena DD, Ramachandra S, Jo C (2015) Alternatives to nitrite in processed meat: up to date. Trends Food Sci Technol 45(1):37–49. https://doi.org/10.1016/j.tifs.2015.05.008

Alfaia A, Alfaia CM, Patarata L, Fernandes MJ, Fernandes MH, Elias M, Ribeiro MH, Fraqueza MJ (2015) Binomial effects of high isostatic pressure and time on the microbiological, sensory characteristics and lipid composition stability of vacuum packed dry fermented sausages "chouriço". Innov Food Sci Emerg Technol 32:37–44

Al-Gubory KH, Blachier F, Faure P, Garrel C (2016) Pomegranate peel extract decreases small intestine lipid peroxidation by enhancing activities of major antioxidant enzymes. J Sci Food Agric 96:3462–3468

Aliyari P, Kazaj B, Barzegar M, Gavlighi HA (2020) Production of functional sausage using pomegranate peel and pistachio green hull extracts as natural preservatives. J Agric Sci Technol 22:159–172

Allaker RP, Mendez LS, Hardie JM, Benjamin N (2001) Antimicrobial effect of acidified nitrite on periodontal bacteria. Oral Microbiol Immunol 16:253–256

Al-Shuibi AA, l-Abdullah B (2002) Substitution of nitrite by sorbate and the effect on properties of mortadella. J Meat Sci 62:473–478

Altunay N, Elik AA (2020) Green and efficient vortex-assisted liquid-phase microextraction based on supramolecular solvent for UV–VIS determination of nitrite in processed meat and chicken products. J Food Chem 332:127395

Alvarado C, McKee S (2007) Marination to improve functional properties and safety of poultry meat. J Appl Poultry Res 16:113–120

Aminzare M, Hashemi M, Ansarian E, Bimikar M, Azar HH, Mehrasbi MR, Daneshamooz S, Raesi M, Jannat B, Afshari A (2019) Using natural antioxidants in meat and meat products as preservatives: a review. Adv Anim Vet Sci 7:417–426

Aquilani C, Sirtori F, Flores M, Bozzi R, Lebert B, Pugliese C (2018) Effect of natural antioxidants from grape seed and chestnut in combination with hydroxytyrosol, as sodium nitrite substitutes in Cinta Senese dry-fermented sausages. J Meat Sci 145:389–398

Aran N (2001) The effect of calcium and sodium lactates on growth from spores of Bacillus cereus and Clostridium perfringens in a 'sous vide' beef goulash under temperature abuse. Int J Food Microbiol 63:117–123

Asioli D, Aschemann-Witzel J, Caputo V, Vecchio R, Annunziata A, Naes T, Varela P (2017) Making sense of the "clean label" trends: a review of consumer food choice behavior and discussion of industry implications. Food Res Int 99(1):58–71

Aymerich T, Jofre A, Garriga M, Hugas M (2005) Inhibition of Listeria monocytogenes and *Salmonella* by natural antimicrobials and high hydrostatic pressure in sliced cooked ham. J Food Prot 68:173–177

Badea M, Amine A, Benzine M, Curulli A, Moscone D, Lupu A, Volpe G, Palleshi G (2004) Rapid and selective electrochemical determination of nitrite in cured meat in the presence of ascorbic acid. Microchim Acta 147:51–58

Bedale W, Sindelar JJ, Milkowski AL (2016) Dietary nitrate and nitrite: benefits, risks, and evolving perceptions. J Meat Sci 120(10):85–92. https://doi.org/10.1016/j.meatsci.2016.03.009

Bijad M, Maleh HK, Farsi M, Shahidi SA (2017) Simultaneous determination of amaranth and nitrite in foodstuffs via electrochemical sensor based on carbon paste electrode modified with CuO/SWCNTs and room temperature ionic liquid. Food Anal Methods 10:3773–3780

Bingöl EB, Bostan K (2007) Effect of sodium lactate on the microbiological quality and shelf life of sausages. Turk J Vet Anim Sci 31:333–339

Bolumar T, Orlien V, Sikes A, Aganovic K, Bak KH, Guyon C, Stubler AS, de Lamballerie M, Hertel C, Bruggemann DA (2021) High-pressure processing of meat: molecular impacts and industrial applications. Compr Rev Food Sci Food Saf 20:332–368

Boora F, Chirisa E, Mukanganyama S (2014) Evaluation of nitrite radical scavenging properties of selected Zimbabwean plant extracts and their phytoconstituents. J Food Process 2014:918018

Bouvard V, Loomis D, Guyton KZ, Grosse Y, Ghissassi FE, Talla LB, Guha N, Mattock H, Straif K (2015) Carcinogenicity of consumption of red and processed meat. Lancet Oncol 16:1599–1600

Bradley EM, Williams JB, Schilling MW, Coggins PC, Crist C, Yodre S, Campano SG (2011) Effects of sodium lactate and acetic acid derivatives on the quality and sensory characteristics of hot-boned pork sausage patties. J Meat Sci 88:145–150

Brender JD (2020) Human health effects of exposure to nitrate, nitrite, and nitrogen dioxide. In: Sutton MA (ed) Just enough nitrogen. Springer Nature, Cham, pp 283–294

Bryan NS, van Grinsven H (2013) The role of nitrate in human health. Adv Agron 119:153–182

Campus M (2010) High pressure processing of meat, meat products and seafood. Food Eng Rev 2: 256–273

Candan T, Bağdatlı A (2018) Natural applications for nitrite/nitrate reduction in meat products. Pamukkale Univ J Eng Sci 24:1382–1387

Carpenter C, Broadbent JR (2009) External concentration of organic acid anions and pH: key independent variables for studying how organic acids inhibit growth of bacteria in mildly acidic foods. J Food Sci 74:R12–R15

Casoni D, Badiu RR, Frențiu T (2019) Spectrophotometric determination and assessment of potential health risk of nitrite from meat and processed meat products. Stud U Babes-Bol Chem 2:265–277

Cegielska-Radziejewska R, Pikul J (2004) Sodium lactate addition on the quality and shelf life of refrigerated sliced poultry sausage packaged in air or nitrogen atmosphere. J Food Prot 67:601–606

Chan TYK (2011) Vegetable-borne nitrate and nitrite and the risk of methaemoglobinaemia. Toxicol Lett 200:107–108

Chatkitanan T, Harnkarnsujarit N (2020) Development of nitrite compounded starchbased films to improve color and quality of vacuum-packaged pork. J Food Packag Shelf Life 25:1–11

Chen L, Wang ZL, Ji LL, Zhang JM, Zhao ZP, Zhang R, Wang W (2021) Flavor composition and microbial community structure of mianning ham. Front Microbiol 11(6):1–9. https://doi.org/10.3389/fmicb.2020.623775

Chotyakul N, Boonnoon N (2016) High pressure food processing: an alternative technology to reduce food additives used in processed meat products. Panyapiwat J 8:327–341

Chris D (2020) Red meat republic: a hoof-to-table history of how beef changed America. Histories of economic life. By joshua specht. West Hist Q 2:1–2

Colla G, Kim HJ, Kyriacou MC, Rouphael Y (2018) Nitrate in fruits and vegetables. J Sci Hortic 237(7):221–238. https://doi.org/10.1016/j.scienta.2018.04.016

Considine KM, Kelly AL, Fitzgerald GF, Hi C, Sleator RD (2008) High-pressure processing–effects on microbial food safety and food quality. FEMS Microbiol Lett 281:1–9
Cui H, Gabriel AA, Nakano H (2010) Antimicrobial efficacies of plant extracts and sodium nitrite against *Clostridium botulinum*. Food Control 21:1030–1036
De Gonzalez MN, Keeton J, Ringer L (2004) Sensory and physicochemical characteristics of frankfurters containing lactate with antimicrobial surface treatments. J Food Sci 69:S221–S228
Delgado-Pando G, Fischer E, Allen P, Kerry JP, Sullivan MGO, Hamill RM (2018) Salt content and minimum acceptable levels in whole-muscle cured meat products. J Meat Sci 139:179–186
Demeyer D, Mertens B, Smet SD, Ulens M (2016) Mechanisms linking colorectal cancer to the consumption of (processed) red meat: a review. Crit Rev Food Sci Nutr 56:2747–2766
Domínguez R, Gomez M, Fonseca S, Lorenzo JM (2014) Effect of different cooking methods on lipid oxidation and formation of volatile compounds in foal meat. J Meat Sci 97:223–230
Domínguez R, Pateiro M, Gagaoua M, Barba FJ, Zhang W, Lorenzo JMA (2019) Comprehensive review on lipid oxidation in meat and meat products. Antioxidants 8:429
Dong Q, Tu K (2006) Research progress on bacteriostatic mechanism of nitrite in preserved meat. J Progr Mod Biomed 2(3):48–52
Doores S (2005) Organic acids. In: Davidson PM (ed) Antimicrobials in food, 3rd edn. CRC Press, Boca Raton, pp 91–140
Dos Santos BD, da Silva DVT, Paschoalin VMF (2021) A narrative review on dietary strategies to provide nitric oxide as a non-drug cardiovascular disease therapy: beetroot formulations – a smart nutritional intervention. Food Secur 10:859
Drosinos EH, Mataragas M, Kampani A, Kritikos D, Mataxopoulos I (2006) Inhibitory effect of organic acid salts on spoilage flora in culture medium and cured cooked meat products under commercial manufacturing conditions. J Meat Sci 73:75–81
Eyiler E, Oztan A (2011) Production of frankfurters with tomato powder as a natural additive. LWT-Food Sci Technol 44:307–311
Fakhre NA, Qader HA (2013) Flow-injection spectrophotometric determination of nitarate in wastewater samples using diazotization coupling reaction. Appl Sci Rep 3:125–131
Falowo AB, Fayemi PO, Muchenje V (2014) Natural antioxidants against lipid–protein oxidative deterioration in meat and meat products: a review. Food Res Int 64:171–181
Feiner GS (2016) Practical science and processing technology. Academic, Cambridge, MA, p 230
Feng X, Li C, Jia X, Guo Y, Lei N, Hackman RM, Chen L, Zhou G (2016) Influence of sodium nitrite on protein oxidation and nitrosation of sausages subjected to processing and storage. J Meat Sci 116:260–267
Ferysiuk K, Wojciak KM (2020) Reduction of nitrite in meat products through the application of various plant-based ingredients. Antioxidants 9(8):1–28. https://doi.org/10.3390/antiox9080711
Food and Drug Administration (2003) Substances generally recognized as safe. Code of Federal Regulations, Office of the Federal Register
Francis DM, Barringer SA, Whitmoyer RE (2000) Ultrastructural characterization of yellow shoulder disorder in a uniform ripening tomato genotype. Hortic Sci 35:1114–1117
Fraqueza MJ, Borges A, Patarata L (2018) Strategies to reduce the formation of carcinogenic chemicals in dry cured meat products. In: Handbook of food bioengineering series, Food control and biosecurity, vol 16. Elsevier
Gassara F, Kouassi AP, Brar SK, Belkacemi K (2016) Green alternatives to nitrates and nitrites in meat-based products-A review. Crit Rev Food Sci Nutr 56(13):2133–2148. https://doi.org/10.1080/10408398.2013.812610
Gill AO, Ramaswamy HS (2008) Application of high pressure processing to kill *Escherichia coli* O157 in ready-to-eat meats. J Food Prot 71:2182–2189
Girouard J, Frenette G, Sullivan R (2011) Comparative proteome and lipid profiles of bovine epididymosomes collected in the intraluminal compartment of the caput and cauda epididymidis. Int J Androl 34:e475–e486

Glass K, Preston D, Veesenmeyer J (2007) Inhibition of *Listeria monocytogenes* in Turkey and pork-beef bologna by combinations of sorbate, benzoate, and propionate. J Food Prot 70:214–217

Guo H, Huang D, Xue Y (2005) The history of meat processing in China. Meat Res 6:19–24

Hayes J, Canonico I, Allen P (2013) Effects of organic tomato pulp powder and nitrite level on the physicochemical, textural and sensory properties of pork luncheon roll. J Meat Sci 95:755–762

Hayman MM, Baxter I, Riordan PJ, Stewart CM (2004) Effects of high-pressure processing on the safety, quality, and shelf life of ready-to-eat meats. J Food Prot 67:1709–1718

Hendgen-Cotta UB, Merx MW, Sjiva S, Schmitz J, Becher S, Klare JP, Steinhoff HJ, Goedecke A, Schrader J, Gladwin MT et al (2008) Nitrite reductase activity of myoglobin regulates respiration and cellular viability in myocardial ischemia-reperfusion injury. Proc Nat Acad Sci USA 105:10256–10261

Herrmann SS, Granby K, Duedahl-Olesen L (2014) Formation and mitigation of N-nitrosamines in nitrite preserved cooked sausages. Food Chem 74:516–526

Herrmann SS, Duedahl-Olesen L, Christensen T, Olesen PT, Granby K (2015) Dietary exposure to volatile and non-volatile N-nitrosamines from processed meat products in Denmark. J Food Chem Toxicol 80:137–143

Higuero N, Moreno I, Lavado G, Vidal-Aragon MC, Cava R (2020) Reduction of nitrate and nitrite in Iberian dry cured loins and its effects during drying process. J Meat Sci 163(5): 108062.108061–108062.108068

Honikel K-O (2008) The use and control of nitrate and nitrite for the processing of meat products. J Meat Sci 78:68–76

Horsch A, Sebranek JG, Dickson JS, Niebuhr SE, Larson EM, Lavieri NA, Ruther BL, Wilson LA (2014) The effect of pH and nitrite concentration on the antimicrobial impact of celery juice concentrate compared with conventional sodium nitrite on *Listeria monocytogenes*. J Meat Sci 96:400–407

Huang L, Sun Z, Wu A, Sun J, Zeng Z, Zeng X, Pan D (2019) Optimization of processing technology for no-nitrite-added cured meat. J Ningbo Univ (NSEE) 32:21–27, 03

Huang L, Zeng XQ, Sun Z, Wu AJ, He J, Dang YL, Pan DD (2020a) Production of a safe cured meat with low residual nitrite using nitrite substitutes. J Meat Sci 162(4):1–9. https://doi.org/10.1016/j.meatsci.2019.108027

Huang P, Xu BC, Shao XF, Chen CG, Wang W, Li PJ (2020b) Theoretical basis of nitrosomyoglobin formation in a dry sausage model by coagulase-negative *staphylococci:* behavior and expression of nitric oxide synthase. J Meat Sci 161(3):1–9. https://doi.org/10.1016/j.meatsci.2019.108022

Huang L, Sun Z, Zeng X, Pan D, He J, Shen J (2021) Effects of multi-ingredients for nitrite on the volatile flavor compounds in cured meat. J Chin Inst Food Sci Technol 21(3):324–333

Hugas M, Garriga M, Monfort J (2002) New mild technologies in meat processing: high pressure as a model technology. J Meat Sci 62:359–371

Hunault CC, van Velzen AG, Sips AJAM, Schothorst RC, Meulenbelt J (2009) Bioavailability of sodium nitrite from an aqueous solution in healthy adults. Toxicol Lett 190:48–53

Hung Y, Verbeke W, de Kok TM (2015) Stakeholder and consumer reactions towards innovative processed meat products: Insights from a qualitative study about nitrite reduction and phytochemical addition. Food Control 60:690–698

Hung Y, de Kok TM, Verbeke W (2016a) Consumer attitude and purchase intention towards processed meat products with natural compounds and a reduced level of nitrite. J Meat Sci 121: 119–126

Hung Y, Verbeke W, de Kok TM (2016b) Stakeholder and consumer reactions towards innovative processed meat products: insights from a qualitative study about nitrite reduction and phytochemical addition. Food Control 60:690–698

Iqbal A (2017) Effect of food on causation and prevention of gastric cancer. J Cancer Prev Curr Res 8:327–330

Islam M, Chen J, Doyle MP, Chinnan M (2002) Control of *Listeria monocytogenes* on Turkey frankfurters by generally-recognized-as-safe preservatives. J Food Prot 65:1411–1416

Jarulertwattana PP, Wanaloh T, Charurungsipong P, Pittarate C, Asavasanti S (2021) Influence of nitrate-nitrite contamination on pink color defect in ginger marinated steamed chicken drumsticks. Appl Sci Eng Progress 14(3):417–424

Ji X, Ma G, Chen Y, Jin J, Yu Q (2020) Effect of sodium nitrite on fatty acid composition and lipid oxidation of mutton in marinating process. J Gansu Agric Univ 55:195–202, 05

Jo K, Lee S, Yong HI, Choi YS, Jung S (2020) Nitrite sources for cured meat products. LWT-Food Sci Technol 129:109583. https://doi.org/10.1016/j.lwt.2020.109583

Jonas G, Csehi B, Palotas P, Toth A, Kenesei G, Huszar KP, Friedrich L (2017) Combined effects of high hydrostatic pressure and sodium nitrite on color, water holding capacity and texture of frankfurter. J Phys Conf Ser 950:042006

Juneja V, Thippareddi H (2003) Control of Clostridium perfringens in a model roast beef by salts of organic acids during chilling. J Food Saf 24:95–108

Juneja V, Thippareddi H (2004a) Control of Clostridium perfringens in a model roast beef by salts of organic acids during chilling. J Food Saf 24:95–108

Juneja V, Thippareddi H (2004b) Inhibitory effects of organic acid salts on growth of *Clostridium perfringens* from spore inocula during chilling of marinated ground Turkey breast. Int J Food Microbiol 93:155–163

Karwowska M, Kononiuk A (2020) Nitrates/nitrites in food-risk for nitrosative stress and benefits. Antioxidants 9(3):1–17. https://doi.org/10.3390/antiox9030241

Keszei AP, Goldbohm RA, Schouten LJ, Jakszyn P, van den Brandt PA (2013) Dietary N-nitroso compounds, endogenous nitrosation, and the risk of esophageal and gastric cancer subtypes in the Netherlands Cohort Study. Am J Clin Nutr 97:135–146

Khalehgi A, Kasaai R, Darani KK, Rezaei K (2016) Combined use of black barberry (*Berberis crataegina* L.) extract and nitrite in cooked beef sausages during the refrigerated storage. J Agric Sci Technol 18:601–614

Khatib N, Varidi MJ, Mohebbi M, Varidi M, Hosseini S (2020) Replacement of nitrite with lupulon–xanthohumol loaded nanoliposome in cooked beef-sausage: experimental and model based study. J Food Sci Technol 57(7):2629–2639

Kim GH, Chin KB (2021) Evaluation of quality characteristics of low-nitrite pork sausages with paprika oleoresin solution during refrigerated storage. Food Sci Anim Resour 41(3):428–439

Kim Y, Keeton JT, Smith SB, Berghman LR, Savell JW (2009a) Role of lactate dehydrogenase in metmyoglobin reduction and color stability of different bovine muscles. J Meat Sci 83:376–382

Kim Y, Keeton JT, Smith SB, Maxim JE, Yang HS, Savell JW (2009b) Evaluation of antioxidant capacity and colour stability of calcium lactate enhancement on fresh beef under highly oxidising conditions. J Food Chem 115:272–278

Kroupova H, Machova J, Svobodova Z (2005) Nitrite influence on fish: a review. Vet Med Praha 50:461

Landry AP, Duan X, Huang H, Dimg H (2011) Iron–sulfur proteins are the major source of protein-bound dinitrosyl iron complexes formed in *Escherichia coli* cells under nitric oxide stress. Free Radic Biol Med 50:1582–1590

Lavieri NA, Sebranek JP, Cordray JS, Dickson JS, Horsch AM, Jung S, Manu DK, Strecher BFB, Mendonca AF (2014) Effects of different nitrite concentrations from a vegetable source with and without high hydrostatic pressure on the recovery of *Listeria monocytogenes* on ready-to-eat restructured ham. J Food Prot 77:781–787

Lee SH, Cho KI, Kim JY, Ahn YK, Rha SW, Kim YJ, Choi YS, Choi SW, Jeon DW, Min PK et al (2012) Non-lipid effects of rosuvastatin–fenofibrate combination therapy in high-risk Asian patients with mixed hyperlipidemia. Atherosclerosis 221:169–175

Lefevre S, Jensen FB, Huong DTT, Wang T, Phuong NT, Bayley M (2011) Effects of nitrite exposure on functional haemoglobin levels, bimodal respiration, and swimming performance in the facultative air-breathing fish Pangasianodon hypophthalmus. Aquat Toxicol 104:86–93

Li Y, Chapman SJ, Nicol GW, Yao H (2017) Nitrification and nitrifiers in acidic soils. Soil Biol Biochem 116:290–301

Li Y, Guo C, Yang J, Wei J, Xu J, Chneg S (2006) Evaluation of antioxidant properties of pomegranate peel extract in comparison with pomegranate pulp extract. J Food Chem 96: 254–260

Lim SC, Foster NF, Riley TV (2015) Susceptibility of Clostridium difficile to the food preservatives sodium nitrite, sodium nitrate and sodium metabisulphite. Anaerobe 37:67–71

Liu L, Li X, Chen Y, Ma L, Yang H (2021) Effects of microbial nitrosation inhibitors on the quality of new bacon. Meat Res 35(2):1–8

Lopez CM, Dallolio G, Bonilauri P, Rebecchi A (2021) Strategies for nitrite replacement in fermented sausages and effect of high pressure processing against *Salmonella* spp. and *Listeria innocua*. Food Secur 10:2617

Lund MN, Heiononen M, Baron CVP, Estevez M (2011) Protein oxidation in muscle foods: a review. Mol Nutr Food Res 55:83–95

Luo HT, Li PJ, Zhang HW, Diao XP, Kong BH (2020) Nitrosylmyoglobin formation in meat by *Lactobacillus fermentum* AS1.1880 is due to its nitric oxide synthase activity. J Meat Sci 166(8):1–7. https://doi.org/10.1016/j.meatsci.2020.108122

Ma X, Liu Y, Sun Y, Pan D, Cao J (2021) Effects of phosphorylated nitroso porcine hemoglobin as partial nitrite substitute on the quality of emulsified sausage. J Food Sci (NY). https://doi.org/10.7506/spkx1002-6630-20201201-018

Majou D, Christieans S (2018) Mechanisms of the bactericidal effects of nitrate and nitrite in cured meats. J Meat Sci 145:273–284

Marcos B, Aymerich T, Guardia MD, Garriga M (2007) Assessment of high hydrostatic pressure and starter culture on the quality properties of low-acid fermented sausages. J Meat Sci 76:46–53

Marcos B, Aymerich T, Monfort JM, Garriga M (2008) High-pressure processing and antimicrobial biodegradable packaging to control *Listeria monocytogenes* during storage of cooked ham. J Food Microbiol 25:177–182

Mcclure BN (2009) The effect of lactate on nitrite in a cured meat system. Master's thesis, Iowa State University, Ames, IA, USA, p 68

McClure BN, Sebranek JG, Kim YH, Sullivan GA (2011) The effects of lactate on nitrosylmyoglobin formation from nitrite and metmyoglobin in a cured meat system. J Food Chem 129:1072–1079

McKenna E (2019) Meat makes people powerful: a global history of the modern era. Agric Hist 93(1):187–189

McLean S, Bowman LAH, Sanguinetti G, Read RC, Poole RK (2010) Peroxynitrite toxicity in *Escherichia coli* K12 elicits expression of oxidative stress responses and protein nitration and nitrosylation. J Biol Chem 285:20724–20731

Milkowski A, Garg HK, Coughlin JR, Bryan NS (2010) Nutritional epidemiology in the context of nitric oxide biology: a risk–benefit evaluation for dietary nitrite and nitrate. Nitric Oxide 22: 110–119

Moutiq R, Misra N, Mendonca A, Keener K (2020) In-package decontamination of chicken breast using cold plasma technology: microbial, quality and storage studies. J Meat Sci 159(1):1–9. https://doi.org/10.1016/j.meatsci.2019.107942

Mulvey L, Everis L, Leeks D, Hughes H, Wood A (2010) Alternatives to nitrates and nitrites in organic meat products. Campden BRI Group, London, p 91

Myers K, Montoya D, Cannon J, Dickson J, Sebranek J (2012) The effect of high hydrostatic pressure, sodium nitrite and salt concentration on the growth of Listeria monocytogenes on RTE ham and turkey. Meat Sci 93:263–268

Myers K, Cannon J, Montoya D, Dickson J, Lonergan S, Sebranek J (2013a) Effects of high hydrostatic pressure and varying concentrations of sodium nitrite from traditional and vegetable-based sources on the growth of *Listeria monocytogenes* on ready-to-eat (RTE) sliced ham. J Meat Sci 94:69–76

Myers K, Montoya D, Cannon J, Dickson J, Sebranek J (2013b) The effect of high hydrostatic pressure, sodium nitrite and salt concentration on the growth of *Listeria monocytogenes* on RTE ham and Turkey. J Meat Sci 93:263–268

National Health and Family Planning Commission, P. (2015) Standard for use of food additives, GB vols 2760–2014

Negi P, Jayaprakasha G, Jena B (2003) Antioxidant and antimutagenic activities of pomegranate peel extracts. J Food Chem 80:393–397

Neri E, Patalinghug ME (2021) Nitrate concentration analysis using spectrophotometry on locally processed meat products in a province in The Philippines. Int Multidiscip Res J 3(1):42–47

Oliveira S, Lopes T, Rangel AO (2004) Spectrophotometric determination of nitrite and nitrate in cured meat by sequential injection analysis. J Food Sci 69:C690–C695

Ozaki MM, Santos M, Ribeiro WO, Ferreira N, Pollonio M (2020) Radish powder and oregano essential oil as nitrite substitutes in fermented cooked sausages. Food Res Int 140(3):109855. https://doi.org/10.1016/j.foodres.2020.109855

Parthasarathy DK, Bryan NS (2012) Sodium nitrite: the "cure" for nitric oxide insufficiency. J Meat Sci 92:274–279

Patarata L, Martins S, Silva J, Fraqueza MJ (2020) Red wine and garlic as a possible alternative to minimize the use of nitrite for controlling *Clostridium sporogenes* and *Salmonella* in a cured sausage: safety and sensory implications. Food Secur 9(2):1–18. https://doi.org/10.3390/foods9020206

Patarata L, Carvalho F, Fraqueza MJ (2022) Nitrite-free implications on consumer acceptance and the behavior of pathogens in cured pork loins. Food Secur 11:796

Patterson MF, Linton M, Doona CJ (2007) Introduction to high pressure processing of foods. In: Doona CJ (ed) High pressure processing of foods. Blackwell Publishing, Hoboken, pp 1–14

Pegg RB, Honikel KO (2014) Principles of curing. In: Fidel T (ed) Handbook of fermented meat and poultry. Wiley, Hoboken, pp 19–30

Perea-Sanz L, Lopez-Díez J, Belloch C, Flores M (2020) Counteracting the effect of reducing nitrate/nitrite levels on dry fermented sausage aroma by *Debaryomyces hansenii* inoculation. J Meat Sci 164(6):108103. https://doi.org/10.1016/j.meatsci.2020.108103

Pierson MD, Smoot LA, Robach MC (1983) Nitrite, nitrite alternatives, and the control of *Clostridium botulinum* in cured meats. Crit Rev Food Sci Nutr 17:141–187

Pietrasik Z, Gaudette N, Johnston S (2017) The impact of high hydrostatic pressure on the functionality and consumer acceptability of reduced sodium naturally cured wieners. J Meat Sci 129:127–134

Pourazrang H, Moazzami A, Bazzaz BF (2002) Inhibition of mutagenic N-nitroso compound formation in sausage samples by using L-ascorbic acid and α-tocopherol. J Meat Sci 62:479–483

Pourreza N, Fat'hi MR, Hatami A (2012) Indirect cloud point extraction and spectrophotometric determination of nitrite in water and meat products. Microchem J 104:22–25

Prolo C, Álvarez MN, Radi R (2014) Peroxynitrite, a potent macrophage-derived oxidizing cytotoxin to combat invading pathogens. Biofactors 40:215–225

Pyo YH, Lee TC, Logendra L, Rosen RT (2004) Antioxidant activity and phenolic compounds of Swiss chard (*Beta vulgaris* subspecies cycla) extracts. J Food Chem 85:19–26

Quilo S, Pohlman FW, Brown AH, Crandall PG, Dias-Morse PN, Baublits RT, Aparico JL (2009) Effects of potassium lactate, sodium metasilicate, peroxyacetic acid, and acidified sodium chlorite on physical, chemical, and sensory properties of ground beef patties. J Meat Sci 82:44–52

Ramarathnam N, Rubin L (2012) The flavour of cured meat. In: Shahidi F (ed) Flavor of meat and meat products. Springer, Berlin/Heidelberg, pp 174–198

Ranasinghe S (2018) Nitrate and nitrite content of vegetables: a review. J Pharmacogn Phytochem 7(4):322–328

Redondo-Solano M, Echandi MLA (2011) Comparación de métodos para el análisis de coliformes totales y fecales en muestras de agua mediante la técnica de Número Más Probable (NMP). UNED Res J 3:41–43

Rehman A, Shehadeh M, Khirfan D, Jones A (2018) Severe acute haemolytic anaemia associated with severe methaemoglobinaemia in a G6PD-deficient man. BMJ Case Rep:bcr2017223369. https://doi.org/10.1136/bcr-2017-223369

Rendueles E, Omer MK, Alvseike O, Alonso-Calleja C, Capita R, Prieto M (2011) Microbiological food safety assessment of high hydrostatic pressure processing: a review. LWT-Food Sci Technol 44:1251–1260

Riazi F, Zeynali F, Hoseini E, Behmadi H, Savadkoohi S (2016) Oxidation phenomena and color properties of grape pomace on nitrite-reduced meat emulsion systems. J Meat Sci 121:350–358

Richi EB, Baumer B, Conrad B, Darioli R, Schmid A, Keller U (2015) Health risks associated with meat consumption: a review of epidemiological studies. Int J Vitam Nutr Res 85:70–78

Riel G, Boulaaba A, Popp J, Klein G (2017) Effects of parsley extract powder as an alternative for the direct addition of sodium nitrite in the production of mortadella-type sausages–impact on microbiological, physicochemical and sensory aspects. J Meat Sci 131:166–175

Rivera N, Bunning M, Martin J (2019) Uncured-labeled meat products produced using plant-derived nitrates and nitrites: chemistry, safety, and regulatory considerations. J Agric Food Chem 67(29):8074–8084. https://doi.org/10.1021/acs.jafc.9b01826

Ros-Polski V, Koutchma T, Xue J, Defelice C, Balamurugan S (2015) Effects of high hydrostatic pressure processing parameters and NaCl concentration on the physical properties, texture and quality of white chicken meat. Innov Food Sci Emerg Technol 30:31–42

Rubio B, Martinez B, Garcia-Cachan D, Rovira J, Jaime I (2007) The effects of high pressure treatment and of storage periods on the quality of vacuum-packed "salchichón" made of raw material enriched in monounsaturated and polyunsaturated fatty acids. Innov Food Sci Emerg Technol 8:180–187

Rysman T, Hecke TV, Poucke CV, Smet SD, Royen GV (2016) Protein oxidation and proteolysis during storage and in vitro digestion of pork and beef patties. J Food Chem 209:177–184

Sabah J, Juneja V, Fung D (2004) Effect of spices and organic acids on the growth of *Clostridium perfringens* during cooling of cooked ground beef. J Food Prot 67:1840–1847

Sallam KI, Samejima K (2004) Microbiological and chemical quality of ground beef treated with sodium lactate and sodium chloride during refrigerated storage. LWT-Food Sci Technol 37: 865–871

Sallan S, Kaban G, Ogras SS, Celik M, Kaya M (2020) Nitrosamine formation in a semi-dry fermented sausage: effects of nitrite, ascorbate and starter culture and role of cooking. J Meat Sci 159(1):1–7. https://doi.org/10.1016/j.meatsci.2019.107917

Samelis J, Sofos JN, Kain ML, Scanga JA, Belk KE, Smith GC (2001) Organic acids and their salts as dipping solutions to control *Listeria monocytogenes* inoculated following processing of sliced pork bologna stored at 4 °C in vacuum packages. J Food Prot 64:1722–1729

San Martin M, Barbosa-Cánovas G, Swanson B (2002) Food processing by high hydrostatic pressure. Crit Rev Food Sci Nutr 42:627–645

Santarelli RL, Pierre F, Corpet DE (2008) Processed meat and colorectal cancer: A review of epidemiologic and experimental evidence. Nutr Cancer 60:131–144

Scannell AG, Schwarz G, Hill C, Ross RP, Arendt EK (2001) Pre-inoculation enrichment procedure enhances the performance of bacteriocinogenic *Lactococcus lactis* meat starter culture. Int J Food Microbiol 64:151–159

Sebranek JG (2009) Basic curing ingredients. In: Tarte R (ed) Ingredients in meat products. Springer, Berlin/Heidelberg, pp 1–23

Sebranek JG, Jackson-Daavis AL, Myer KL, Lavieri NA (2012) Beyond celery and starter culture: advances in natural/organic curing processes in the United States. J Meat Sci 92:267–273

Sedghi H, Mohammadi T, Shariati MA, Atarod S, Branch Q (2022) The effect of substituting sodium nitrite, with sodium acetate on microbial, physiochemical and sensory characteristics of meat products. Available online: https://www.semanticscholar.org/paper/The-effect-of-

substituting-sodium-nitrite%2C-with-on-Sedghi-Mohammadi/a689e9e16ff3660a1b90adc57a3d1cf1fc865db5 (accessed on 26 June 2022)
Serdengecti N, Yildirim I, Gokoglu N (2006) Effects of sodium lactate, sodium acetate and sodium diacetate on microbiological quality of vacuum-packed beef during refrigerated storage. J Food Saf 26:62–71
Shahidi F (2002) Lipid-derived flavors in meat products. In: Kerry J (ed) Meat processing: improving quality. Woodhead Publishing, Sawston, Cambridge, pp 105–121
Shariati-Rad M, Irandoust M, Mohammadi S (2015) Spectrophotometric determination of nitrite in soil and water using cefixime and central composite design. Spectrochim Acta A Mol Biomol Spectrosc 149:190–195
Shin DM, Hwang KE, Lee CW, Kim TK, Park YS, Han SG (2017) Effect of Swiss chard (*Beta vulgaris* var. cicla) as nitrite replacement on color stability and shelf-life of cooked pork patties during refrigerated storage. Korean J Food Sci Anim Resour 37:418
Simonin H, Duranton F, de Lamballerie M (2012) New insights into the high-pressure processing of meat and meat products. Compr Rev Food Sci Food Saf 11:285–306
Sindelar JJ, Milkowski AL (2011) Sodium nitrite in processed meat and poultry meats: a review of curing and examining the risk/benefit of its use. Am J Meat Sci Assoc White Pap Ser 3:1–14
Sindelar JJ, Milkowski AL (2012) Human safety controversies surrounding nitrate and nitrite in the diet. Nitric Oxide 26:259–266
Singh P, Wani AA, Saengerlaub S, Langowski HC (2011) Understanding critical factors for the quality and shelf-life of MAP fresh meat: a review. Crit Rev Food Sci Nutr 51(2):146–177
Skibsted LH (2011) Nitric oxide and quality and safety of muscle based foods. Nitric Oxide 24: 176–183
Slongo AP, Rosenthai A, Camargo LMQ, Deliza R, Mathias SP, de Aragio GMF (2009) Modeling the growth of lactic acid bacteria in sliced ham processed by high hydrostatic pressure. LWT-Food Sci Technol 42:303–306
Stepien M, Chajes V, Romieu I (2016) The role of diet in cancer: the epidemiologic link. Salud Publica Mex 58:261–273
Stoica M (2019) Overview of sodium nitrite as a multifunctional meat-curing ingredient. Ann Univ Dunarea Jos Galati Fasc VI Food Technol 43:155–167
Sucu C, Turp GY (2018) The investigation of the use of beetroot powder in Turkish fermented beef sausage (sucuk) as nitrite alternative. J Meat Sci 140:158–166
Sullivan GA (2011) Naturally cured meats: quality, safety, and chemistry. Ph.D. thesis, Iowa State University, Ames, IA, USA
Suman SP, Joseph P (2013) Myoglobin chemistry and meat color. Ann Rev Food Sci Technol 4:79–99
Tan W, Shelef L (2002) Effects of sodium chloride and lactates on chemical and microbiological changes in refrigerated and frozen fresh ground pork. J Meat Sci 62:27–32
Tang R, Peng J, Chen L, Liu D, Wang W, Guo X (2021) Combination of Flos Sophorae and chili pepper as a nitrite alternative improves the antioxidant, microbial communities and quality traits in Chinese sausages. Food Res Int 141(3):1–9
Thippareddi H, Juneja VK, Phebus RK, Marsden JL, Kastner CL (2003) Control of *Clostridium perfringens* germination and outgrowth by buffered sodium citrate during chilling of roast beef and injected pork. J Food Prot 66:376–381
Tian Q, Wang X, Ye C, Pan Q, Zeng X, Pan D, Shen J (2020a) Application of *Lactobacillus fermentum* RC4 on salted meat for nitrite degradation and quality improvement. J Ningbo Univ 33(1):38–44
Tian Q, Wang X, Ye C, Pan Q, Zeng X, Pan D, Shen J (2020b) Application of *Lactobacillus fermentum* RC4 on salted meat for nitrite degradation and quality improvement. J Ningbo Univ Sci Technol Ed 33(1):38–44
Tompkin RB (2005) Nitrite. In: Davidson PM (ed) Antimicrobials in food, vol 2005, 3rd edn. CRC Press, Boca Raton, pp 169–236
Tompkin RB, Bedale W, Milkowski A, Glass K, Sindelar JJ (2020) Nitrite. In: Davidson PM (ed) Antimicrobials in food, 4th edn. CRC Press, Boca Raton, pp 219–308

Tsikas D, Mitscke A, Gutzki FM, Engeli S, Jordan J (2010) Evidence by gas chromatography–mass spectrometry of ex vivo nitrite and nitrate formation from air nitrogen oxides in human plasma, serum, and urine samples. Anal Biochem 397:126–128

Ursachi CŞ, Perţa-Crişan S, Munteanu FD (2020) Strategies to improve meat products' quality. Food Secur 9:1883

Van Cuong T, Thoa NT (2017) Effects of partial replacement of nitrite by annatto (*Bixa orellana*, L.) seed powder on the properties of pork sausages. J Sci Technol 55:178–187

Van den Brand AD, Beukers M, Niekerk M, Van Donkersgoed G, Van Der Aa M, Van De Ven B, Bulder A, Van Der Voet H, Sprong CR (2020) Assessment of the combined nitrate and nitrite exposure from food and drinking water: application of uncertainty around the nitrate to nitrite conversion factor. Food Addit Contam: Part A 37:568–582

Vasavada M, Carpenter CE, Cornforth DP, Ghorpade V (2003) Sodium levulinate and sodium lactate effects on microbial growth and stability of fresh pork and Turkey sausages. J Muscle Foods 14:119–129

Vercammen A, Vanoirbeek KGA, Lurquin I, Steen L, Goemaere O, Szczepaniak S, Paelinck H, Hendrickx MEG, Michiels CW (2011) Shelf-life extension of cooked ham model product by high hydrostatic pressure and natural preservatives. Innov Food Sci Emerg Technol 12:407–415

Vijayakumar J, Aravindan R, Viruthagiri T (2008) Recent trends in the production, purification and application of lactic acid. Chem Biochem Eng Q 22:245–264

Villaverde A, Ventanas J, Estévez M (2014) Nitrite promotes protein carbonylation and Strecker aldehyde formation in experimental fermented sausages: are both events connected? J Meat Sci 98:665–672

Waga M, Takeda S, Sakata R (2017) Effect of nitrate on residual nitrite decomposition rate in cooked cured pork. J Meat Sci 129(7):135–139. https://doi.org/10.1016/j.meatsci.2017.03.002

Wang QH, Yu LJ, Liu Y, Lin L, Lu RG, Zhu JP, He L, Lu ZL (2017) Methods for the detection and determination of nitrite and nitrate: a review. Talanta 165:709–720

Wang ZL, Wang ZX, Ji LL, Zhang JM, Zhao ZP, Zhang R, Chen L (2021) A review: microbial diversity and function of fermented meat products in China. Front Microbiol 12(6):1–8. https://doi.org/10.3389/fmicb.2021.645435

Ward BB, Arp DJ, Klotz MG (2011) Nitrification. ASM Press, Washington, DC

Wei F, Xu X, Zhou G, Zhao G, Li C, Zhang Y, Qi J (2009) Irradiated Chinese Rugao ham: changes in volatile N-nitrosamine, biogenic amine and residual nitrite during ripening and post-ripening. J Meat Sci 81(3):451–455

Weitzberg E, Lundberg JO (2013) Novel aspects of dietary nitrate and human health. Ann Rev Nutr 33:129–159

Wilson DR, Dabrowski L, Stringer S, Moezelaar R, Brocklehurst TF (2008) High pressure in combination with elevated temperature as a method for the sterilisation of food. Trends Food Sci Technol 19:289–299

Yang BK, Vivas EX, Reiter CD, Gladwin MT (2003) Methodologies for the sensitive and specific measurement of S-nitrosothiols, iron-nitrosyls, and nitrite in biological samples. Free Radic Res 37:1–10

Yolmeh M, Najafi MBH, Farhoosh R (2014) Optimisation of ultrasound-assisted extraction of natural pigment from annatto seeds by response surface methodology (RSM). J Food Chem 155: 319–324

Yordanov D, Angelova G (2010) High pressure processing for foods preserving. Biotechnol Biotechnol Equip 24:1940–1945

Yurchenko S, Mölder U (2007) The occurrence of volatile N-nitrosamines in Estonian meat products. J Food Chem 100:1713–1721

Zanardi E, Dazzi G, Madarena G, Chizzolini R (2002) Comparative study on nitrite and nitrate ions determination. Ann Fac Med Vet Univ Parma 2002(22):70–86

Zhang Y, Wang W, Zhang J, Pu D (2012) Comparison of the optimized hydrolysis processing for bone extract obtain by orthogonal and response surface design experiments. J Food Res Dev 33(7):53–56

Zhang Y, Wang W, Wang X, Zhang J (2014) Bone soup: protein nutrition and enzymatic hydrolysis process optimized by response surface method. J Food Nutr Res 53(1):1–12

Zhang Y, Pan Z, Venkitasamy C, Ma H, Li Y (2015) Umami taste amino acids produced by hydrolyzing extracted protein from tomato seed meal. LWT-Food Sci Technol (Lebensmittel-Wissenschaft-Technol) 62(2):1154–1161

Zhang Y, Chandrasekar V, Pan Z, Liu W, Zhao L (2016) Novel umami ingredients: umami peptides and their taste. J Food Sci 82(1):16–23

Zhang Y, Zhang L, Venkitasamy C, Guo S, Pan Z, Ke H, Zhao L (2019) Improving the flavor of microbone meal with Flavourzyme by response surface method. J Food Process Eng 42(4): e13040.13041–e13040.13011

Zhang Y, Zhang L, Venkitasamy C, Pan Z, Ke H, Guo S, Zhao L (2020a) Potential effects of umami ingredients on human health: pros and cons. Crit Rev Food Sci Nutr 60(13):2294–2302

Zhang YL, Hu P, Xie YY, Wang XY (2020b) Co-fermentation with *Lactobacillus curvatus* LAB26 and *Pediococcus pentosaceus* SWU73571 for improving quality and safety of sour meat. J Meat Sci 170(12):1–10. https://doi.org/10.1016/j.meatsci.2020.108240

Zhang Y, Ke H, Bai T, Chen C, Guo T, Mu Y, Zhao L (2021) Characterization of umami compounds in bone meal hydrolysate. J Food Sci 6(86):1–12

Zhao Y (2020) Study on the antioxidative function of lactic acid bacteria and its application in mutton fermented sausage (Master Master of Agriculture). Inner Mongolia Agricultural University, Inner Mongolia

Zhao L, Yin Z, Venkitasamy C, Pan Z, Gou X (2017a) Preparation of umami octopeptide with recombined *Escherichia coli:* feasibility and challenges. Bioengineered 9(1):166–169

Zhu Y, Guo L, Yang Q (2020) Partial replacement of nitrite with a novel probiotic *Lactobacillus plantarum* on nitrate, color, biogenic amines and gel properties of Chinese fermented sausages. Food Res Int 137:1–9. https://doi.org/10.1016/j.foodres.2020.109351

Chapter 12
Natural Antioxidants and Antimicrobials in Processed Meat Products

Sajad A. Rather, F. A. Masoodi, Syed Muzaffar, and Tariq Ahmad Ganie

12.1 Introduction

Meat and meat products are an important source of high-quality proteins, essential fatty acids, vitamins, and minerals. However, the high nutritional value, moisture content and neutral pH makes meat and meat products susceptible to spoilage because of microbial growth and oxidative rancidity together with other biochemical and enzymatic deterioration (Olvera-Aguirre et al. 2023; Tůmova et al. 2021; Echegaray et al. 2021). The microbial, enzymatic and oxidation degradation are the limiting factors in the quality and acceptability of meat products that deteriorate the organoleptic properties, texture, nutritional value and alters the pH (Das et al. 2020). The processes of auto-oxidation and microbial deterioration of meat and meat products lead to the formation of undesirable flavours, odours and toxic compounds, such as reactive oxygen species (ROS) and nitrogen (RNS) that can cause mutagenesis, thus posing a risk to human health (Estevez 2021; Embuscado 2015).

Oxidation of lipids and proteins in meat products is the second most important cause of spoilage (following microbial spoilage) (Falowo et al. 2014; Domínguez et al. 2019). It is associated with intrinsic and extrinsic quality degradation, results in the formation of reactive chemical compounds with cytotoxic, mutagenic and oxidative effects on body tissues, thereby promoting cancer, atherosclerosis, inflammation and aging processes (Domínguez et al. 2019). Like lipid peroxides associated with oxidative damage to colon cells and inflammation and cancer in the digestive

S. A. Rather (✉) · F. A. Masoodi (✉)
Department of Food Science and Technology, University of Kashmir, Srinagar, India

S. Muzaffar
Department of Chemistry AAAM Degree College Bemina, Srinagar, Jammu & Kashmir, India

T. A. Ganie
Department of Food Technology, Islamic University of Science and Technology, Pulwama, Jammu & Kashmir, India

S. A. Rather, F. A. Masoodi (eds.), *Hand Book of Processed Functional Meat Products*, https://doi.org/10.1007/978-3-031-69868-2_12

tract (Angeli et al. 2011), cholesterol oxidation products (COPs) disrupt the junctions between arterial endothelial cells and are related to atherosclerosis and coronary diseases (Broncano et al. 2009) and aldehydes and oxysterols exhibit cytotoxic and mutagenic action in different types of cells, contributing to the manifestation of chronic diseases and cancer (Sottero et al. 2019).

The application of preservatives i.e., antioxidants and antimicrobials is one approach to protect the meat and meat products from auto-oxidation and microbial spoilage (Olvera-Aguirre et al. 2023). To avert the effects of oxidation in meat and meat products, synthetic antioxidants such as butyl-hydroxy-anisole (BHA), sulphur dioxide, butyl-hydroxy-toluene (BHT), tert-butyl-hydroquinone (TBHQ), propyl-gallate (PG), nitrites, nitrates, ascorbates and liquid smoke are added in order to lengthen the shelf-life of the products and prevent spoilage (Wang et al. 2021; Amarachukwu Uzombah 2022; Shahidi and Ambigaipalan 2015). Despite their effectiveness, in prevention of oxidation possible health implications like allergic reactions, disorders in pregnant women and children, potential carcinogenic action, etc. of the extended use of synthetic antioxidants has been confirmed (Kumariya et al. 2019). Because of the potential health risks associated with the use of synthetic antioxidants, the meat industry is looking for natural antioxidant compounds. Plant extracts and essential oils rich in phenolic acids, flavonoids, tannins, terpenes, catechins, alkaloids, saponins and coumarins are the main natural sources of bioactive compounds that are used in meat and meat products as antioxidants (Jiang et al. 2021; Olvera-Aguirre et al. 2022; Radünz et al. 2020). Antioxidants can prevent lipid oxidation through following mechanisms: (i) preventing chain inhibition by scavenging initiating radicals, (ii) breaking chain reaction, (iii) decomposing peroxides, (iv) decreasing localized oxygen concentrations and (v) binding chain initiating catalysts, such as metal ions (Rather et al. 2016; Dorman et al. 2003).

Meat and meat products are also exposed to foodborne microorganisms at different levels in the processing chain including slaughtering, dressing, retail handling, storage and distribution and during manufacturing into ready to eat products. Microorganisms associated with the spoilage of meat and meat products including bacteria such as *Pseudomonas spp.*, *Acinetobacter, Brochothrix thermosphacta, Lactobacillus spp., Enterobacter,* etc., and yeasts and molds cause quality defects such as off flavour, off-odour, etc. The pathogenic microorganisms associated with the meat and meat products if not preserved and handled properly are *Salmonella spp., Campylobacter jejuni, Escherichia coli O157:H7, Listeria monocytogenes, Clostridium spp., Aeromonas hydrophila,* etc. can result in foodborne illnesses to consumers (Jayasena and Jo 2013). As a result, the contamination of meat and meat products can end in significant morbidity, mortality, economic loss and environmental impact worldwide (Bodie et al. 2023). To extend the storage life and eliminate or reduce spoilage and pathogenic microorganisms many synthetic preservatives such as nitrites and nitrates are being used (Chen et al. 1992). However, because of the potential health risks like carcinogenic and toxic properties associated with synthetic preservatives, the meat industry is looking for safer alternatives (Jayasena and Jo 2013). One such alternative is use of natural antimicrobials like essential oils (EOs) or plant extracts (PE), organic acids and bacteriocins which have been found to have antimicrobial properties in meat and meat products (Al-Juhaimi et al. 2016; Gonçalves et al. 2011; Pateiro et al. 2021; Santiesteban-López et al. 2022). Natural antimicrobials were proposed to control the growth of spoilage as well as pathogenic microorganisms in foods.

12.2 Natural Antioxidants in Processed Meat Products

12.2.1 Mechanism of Oxidation in Meat Products

The lipid oxidation in meat products causes degradation of unsaturated fatty acids and leads to the formation of aldehydes, ketones and alcohols that confer the reheated flavour characterized by a rancid-odour/taste (Jayathilakan et al. 2007). Malondialdehyde (MDA) main aldehyde generated during secondary lipid oxidation in meat products and is used to measure the lipid oxidation process in such products (Wojtasik-Kalinowska et al. 2021). The mechanism of lipid auto-oxidation is presented in Fig. 12.1, which is a chain of three steps-initiation, propagation and termination reactions. Undesirable flavours and odours are attributed to the production of volatile compounds of radical fatty acids (such as superoxide anion, hydroxyl [LOH], peroxy [LOO], hydroperoxy [HO_2], lipid radical [RO], nitric oxide [LNO], nitrosyl cation [$LONH_2$] and hydroperoxides [LOOH]) (Jayathilakan et al. 2007; Guedes-Oliveira et al. 2018; Zeb 2021).

Proteins and amino acids in meat products are another source for oxidation reactions (Fig. 12.2). The oxidation of proteins and amino acids in meat products generate cytotoxic substances like ROS (peroxyl radicals [LOO], alkyl peroxide [LOOH], carboxyl [LCOOH], alkoxyl [LO], hydroxyl [LOH]), as well as biogenic amines and their precursors (histamine, cadaverine, tyramine, putrecine, tryptamine, histidine, lysine, tyrosine, ornithine and tryptophan) (Xiong and Guo 2021).

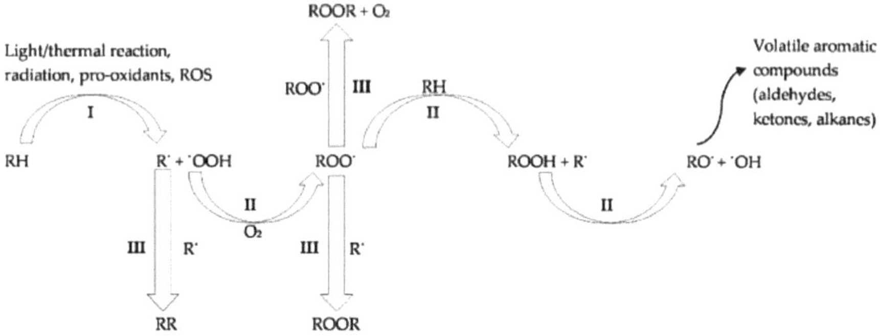

Fig. 12.1 Lipid oxidation mechanism. I: initiation, II: propagation, III: completion, RH: unsaturated fatty acid, ROS: reactive oxygen species, $R^\bullet$: alkyl radical, RR: non-radical product, $ROO^\bullet$: peroxide radical, ROOR: organic peroxide, ROOH: hydroperoxides, $^\bullet OOH$: hydroperoxyl radical, $RO^\bullet$: alkoxy radical, $^\bullet OH$: hydroxyl radical, O_2: oxygen. (Source: Manessis et al. 2020)

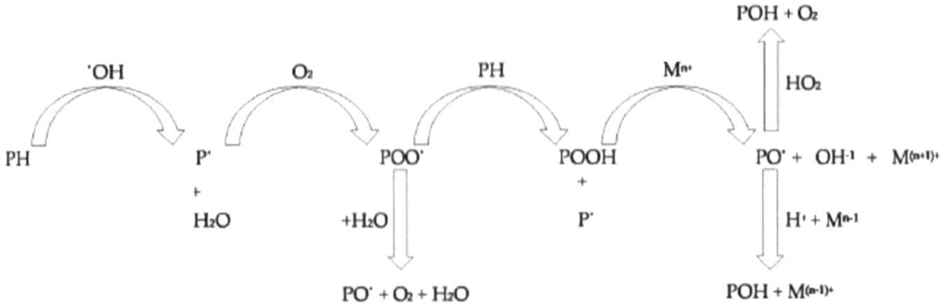

Fig. 12.2 Protein oxidation mechanism. PH: protein, •OH: hydroxyl radical, P•: protein carbon-centered radical, H_2O: water, O_2: oxygen, POO•: analkylproxyl radical, PO•: alkoxyl radical, POOH: alkylperoxide, Mn+: reduced forms of transition metals (Fe^{2+} or Cu^{1+}), OH^{-1}: hydroxide, POH: hydroxyl derivative, HO_2: hydroperoxyl, H+: hydron. (Source: Manessis et al. 2020)

The colour of meat is conferred by myoglobin and hemoglobin. Myoglobin the heme protein is the main pigment responsible for colour of meat and meat products. Oxidation of the central Fe atom within the heme group is responsible for discolouration, a change from red OxyMb to brownish MetMb. When ferrous heme iron oxidizes to its ferric form, oxygen is released and replaced by a water molecule. The rate of discolouration in meat and meat products is muscle-specific (Jeong et al. 2009). Muscles that contain higher proportions of red fibers and thus more lipid and higher oxygen consumption rates, appear to discolour more quickly. Many factors affect OxyMb oxidation including temperature, pH, MetMb reducing activity, partial oxygen pressure and lipid oxidation (Faustman and Cassens 1990; Renerre 2000). OxyMb oxidation is enhanced by higher temperatures, lower pH values and the presence of non-heme iron (Brown and Mebine 1969; Gotoh and Shikama 1974; Allen and Cornforth 2006). MetMb reducing activity can be enzymatic or non-enzymatic and favours maintenance of ferrous forms of myoglobin in meat (Bekhit et al. 2005). Partial oxygen pressures (pO_2) in which a complete vacuum exists or in which oxygen saturation is attained favour ferrous myoglobin forms. Low non-zero pO_2 favours MetMb formation (George and Stratman 1952; Ledward 1970; Neill and Hastings 1925). Lipid oxidation enhances OxyMb oxidation and meat discolouration. The mechanisms by which lipid oxidation could enhance myoglobin oxidation have been explained primarily on the reactivity of primary and secondary products derived from unsaturated fatty acids. Muscles with less oxygen consumption and less lipid oxidation exhibits greater colour stability. Muscle microsomes with greater degrees of unsaturated fatty acid promoted greater OxyMb oxidation in vitro (Yin and Faustman 1994). Overall oxidation of meat products damages the texture, smell, flavour and colour, thus impacts storage life. Therefore, it is important to protect oxidation of lipids, proteins and heme groups in meat and meat products.

12.2.2 Role of Natural Antioxidants in Prevention of Oxidation

Natural antioxidant refers to any naturally occurring substance (plant, animal or microbial origin) that, when present at a lower concentration compared to that of an oxidizable substrate, is able to either delay or inhibit the oxidation of the substrate (Rather et al. 2016; Pisoschi and Pop 2015). The mechanism of action of antioxidants include (i) the inhibition of chain-reaction initiation by scavenging oxidation-initiating radicals, (ii) the breaking of chain reactions that abstract hydrogen for prolonged times, (iii) the decreasing of localized oxygen concentrations, (iv) the decomposition of peroxides and the prevention of their conversion to initiation radicals, and (v) the chelation of chain-reaction initiating catalysts, such as metal ions (Pokorný 2007; Shah et al. 2014).

The different types of natural antioxidants exhibits separate mechanisms to retard or inhibit oxidation of meat products including (i) Primary antioxidants like phenolic compounds and tocopherols terminate the free-radical chain reactions by donating electrons or hydrogen, (ii) Oxygen scavengers like vitamin C react with oxygen and reduce it in closed systems, (iii) Enzymatic antioxidants like glucose oxidase or ROS (e.g., superoxide dismutase, catalase, etc.) remove oxygen and (iv) Chelating agents like citric acid or amino acids capture metallic ions such as copper and iron that act as catalysts of lipid oxidation (Manessis et al. 2020; Kumar et al. 2015).

12.2.3 Application of Natural Antioxidants in Meat Products

The processing treatments like thermal, irradiation, high hydrostatic pressures, salt incorporation and grinding methods promote the oxidation of lipids and proteins in meat and meat products via Strecker reactions, the release of metals such as iron, the exposure of fatty acids by cell membrane disruption and the increase in the redox potential of products. Antioxidants, on the other hand, exert their effects to retard or inhibit oxidation through the resonance of their conjugated double bonds and the ease with which they transfer electrons, create bonds, or donate hydrogen (Petcu et al. 2023). The natural antioxidants exploited in meat industry are classified based on their origin (plants, animals or bacteria) and their chemical structure (e.g., phenols, tannins, flavonoids and isoflavonoids, anthocyanins, lignans, stilbenes, tocopherols, carotenoids or vitamin C). Plant-derived natural antioxidants are derived from fruits, teas, herbs, seeds, spices, vegetables, cereals and trees (Manessis et al. 2020). Various plant organs such as leaves, flowers, fruits, stems or roots accumulate antioxidants in high concentrations, which vary according to the plant species and the antioxidant substance itself. The plant parts are used as antioxidants in meat and meat products or other food products either directly (e.g., fruit puree or juice) or after the extraction and purification of the antioxidant substances they contain (e.g., rosemary extract, mint extract, etc.) (Shah et al. 2014; Karre et al.

2013). The extracts, powders and pastes from various plant sources like fruits, herbs and spices have been added as natural antioxidants in meat and meat products by earlier researchers to reduce lipid oxidation (Table 12.1). The natural antioxidants are also available commercially and several studies have been carried out by different researchers applying commercially available natural antioxidants of plant origin to meat. The antioxidants most commonly found in spices and herbs are phenolic acids (caffeic, p-coumaric, ferulic and neochlorogenic acid) and flavonoids (quercetin, luteolin, apigenin, kaempferol and isorhamnetin). Additionally, herbs contain phenolic diterpenes (rosmanol and carnosol), phenylpropanoids (thymol, carvacrol and eugenol) and volatiles (1, 8 cineole and α, β-pinene) (Brewer 2011). Carlsen et al. (2010) reported that spices and herbs present the highest antioxidant contents among all the animal- and plant-based food categories. Most commonly studied spices and herbs in meat and meat products are oregano, rosemary, sage, thyme, tea and black pepper (Manessis et al. 2020). Fruits and vegetables and their products that are most commonly utilized by the meat industry for their antioxidant properties are grapes, plums, berries (bearberries, cranberries and strawberries), pomegranates, peaches, apricots, curry leaf, carrot, lotus, broccoli, etc. (Rather et al. 2016; Manessis et al. 2020).

12.3 Natural Antimicrobials in Processed Meat Products

Food spoilage microorganisms are threats to food quality and etiological agents for several pathologies—some debilitating and some fatal. Due to the detrimental effects of synthetic preservatives are coming forth, consumers are demanding safe and natural antimicrobial preservatives. Natural means of preservation, involving the use of plant derived compounds/extracts, natural organic acids and bacteriocins.

12.3.1 Plant Derived Natural Antimicrobials

12.3.1.1 Mechanism of Action of Plant Derived Natural Antimicrobials

Plant based antimicrobial compounds such as polyphenolic substances, essential oil compounds, isothiocyanates, alkaloids and thiols have multiple mechanisms of antimicrobial action (Fig. 12.3). The main mechanisms of action are the modification of the membrane permeability, the intracellular enzyme inactivation, modification of intracellular pH, the interference with the ATP-generating system, the inhibition of DNA synthesis, interact with the cell membrane intercalating the phospholipid layer, increase in osmotic pressure, inhibition of key enzymes necessary for the protein synthesis, damage lipids and proteins in the cell and inhibit bacterial enzymes. As regards the antifungal mechanism, these compounds produce oxidative stress and

Table 12.1 Literature review on source of natural antioxidants used in meat and meat products

Source	Main antioxidants	Product	References
Herbs and spices			
Oregano (oregano essential oil)	Carvacrol, thymol, p-cymene, rosmarinic acid, α-thujene, α,-β-pinene, p-coumaric acid and γ-terpinene	Chicken and pork meat, chicken liver meat, lamb burgers and sausages, foal steaks	Chouliara et al. (2007); Fasseas et al. (2008); Hasapidou, and Savvaidis (2011); Lorenzo et al. (2014); Fernandes et al. (2017, 2018)
Rosemary (rosemary extract)	Non-volatile phenolic substances (carnosic and rosmarinic acid, carnosol, rosmanol, rosmariquinone, rosmaridiphenol, rosmadial, 12-methoxycarnosic acid, epi- and iso-rosmanol, and caffeic acid).	Pork patties, ham, pork sausages, fresh beef meat, chicken frankfurters	Haak et al. (2009); Lara et al. (2011); Jiang et al. (2013); Armenteros et al. (2016); Jongberg et al. (2013); Barbosa-Pereira et al. (2014); Rižnar et al. (2006)
Sage (sage extracts)	Carnosol, carnosic acid, rosmanol, apigenin, luteolin methyl carnosate, rosmadial, 9-ethylrosmanol ether, epirosmanol, isorosmanol and galdosol	Liver pates, meatballs and sausages	Estévez et al. (2007); Zhang et al. (2013); Akcan et al. (2017); Šojić et al. (2018)
Thyme (thyme essential oils and extracts)	Thymol, p-cymene, carvacrol, γ-terpinene and linalool	Minced beef, chicken meat and chicken sausages	Fratianni et al. (2010); Zengin and Baysal (2015); Sharma et al. (2017)
Tea (green and black tea extracts)	Green tea-catechin, epicatechin, gallocatechin, epicatechin gallate, epigallocatechin and epigallocatechin gallate Black tea-theaflavins, thearubigenes and chlorogenic, caffeic, p-coumaric and quinic acids	Pork meat, patties and sausages, liver pate, bacon and frankfurters	Lorenzo et al. (2014); Haak et al. (2009); Jongberg et al. (2013); Bozkurt (2006); Qin et al. (2013); Pateiro et al. (2014); Wang et al. (2015); Alirezalu et al. (2017); Jayawardana et al. (2019);
Black pepper	Phenolic acids (caffeic acid and p-coumaric acid), other volatiles (α, β-caryophylline, limonene, α, β—Pinene, camphene and cymene), phenylpropanoids	Pork and beef meat and meat products like cooked ground beef	Jayasena and Jo (2014); Zhang et al. (2016b); Vasavada et al. (2006)

(continued)

Table 12.1 (continued)

Source	Main antioxidants	Product	References
	(cinnamaldehyde and eugenol) And sabinene		
Fruits and vegetables			
Grapes (grape skins and seed extracts)	Phenolic compounds including anthocyanins, resveratrol, quercetin, kaempferol, catechins, phenolic acids and procyanidins.	Cooked minced Turkey breast, raw pork patties, and raw and cooked minced chicken meat	Mielnik et al. (2006); Carpenter et al. (2007); Selani et al. (2011)
Plums (plum extract, dried plums and dried plum puree, plum juice)	Anthocyanins (cyanidin 3-rutinoside, cyanidin 3-glucoside and peonidin 3-rutinoside), phenolic compounds like hydroxycinnamic acid (chlorogenic acid and neochlorogenic acid) and quercetin derivatives, vitamins, carotenoids, flavonoids	Beef roasts, irradiated Turkey breast rolls, low-fat Beef patties, breast chicken patties.	Nuñez De Gonzalez et al. (2008); Lee et al. (2005); Yldz-Turp and Serdaroglu (2010); Basanta et al. (2018)
Bearberries (crude bearberry leaf and bearberry extracts)	Glycosides arbutin, methylarbutin and free aglycones, ursolic acid, tannic acid, gallic acid, p-coumaric acid, syringic acid, galloylarbutin, gallotannins and flavonoids	Raw pork patties	Carpenter et al. (2007)
Cranberries (cranberry powder and extract)	Phenolic compounds such as phenolic acids, flavonoids, anthocyanins, p-hydroxybenzoic acid and their derivatives, anthocyanin–anthocyanidin contents	Turkey (MST) and ground cooked pork	Lee et al. (2006); Kathirvel et al. (2009)
Strawberries (strawberry pulp and extract)	Vitamins, anthocyanins, flavonoids, and phenolic acids. Pelargonidin 3-glucoside, pelargonidin 3-rutinoside, pelargonidin 3-glucoside-succinate, cyanidin 3-glucoside and cyanidin 3-glucoside-succinate	Cooked chicken patties	Jo et al. (2006)
Pomegranate (pomegranates’	Anthocyanins, with 3-glucosides and	Cooked chicken patties, chicken breast	Chandralekha et al. (2012); Mielnik et al.

(continued)

Table 12.1 (continued)

Source	Main antioxidants	Product	References
peels, arils, seeds and juice)	3,5-diglucosides of delphinidin, cyanidin and pelargonidin, punicalagin (tannin), ellagic acid, ellagitannins, punic acid, flavanols, flavan-3-ols and flavones	meat, ground pork meat, beef meatballs	(2006); Andrés et al. (2017)
Avocado (avocado seed extract)	Phenolic components and procyanidins (catechin and epicatechin) vitamin C, vitamin E, and the carotenoid lutein	Porcine patties, cooked and chilled loins from treated Pigs	Rodríguez-Carpena et al. (2011); Hernández-López et al. (2016)
Banana (Banana inflorescences-male flowers extracts)	Catechins, epicatechins, proanthocyanidins (procyanidin B2), carotenoids (β-carotene, lutein), flavonoids (kampferol, quercetin, rutin), vitamin C, vitamin E	Sausages	Rodrigues et al. (2020)
Peach (peach skin powder)	Chlorogenic acid, neochlorogenic acid, caffeic Acid, gallic acid, ellagic acid, quercetin, β-carotene, lutein, vitamin C	Cooked Turkey meat	Zhang et al. (2016c)
Prunus mume (Japanese Apricot) methanolic extracts	Phenolic acids, flavonoids, anthocyanins, vitamin C	Cooked chicken Breast meat	Jo et al. (2006)
Ginkgo biloba leaf extract	Flavonoids and terpenoids, glycosides and ginkgolides	Meat ball	Kobus-Cisowska et al. (2014)
Lychee seed extract	Polyphenols, flavonoids, and proanthocyanidins, vitamin C.	Raw meat paste	Qi et al. (2015)
Lyophilized and powdered *Gentiana lutea* root	Secoiridoids, xanthones, and flavonoids.	Fresh beef patties	Azman et al. (2014)
Broccoli leaf extract	Chlorophylls and phenolic compounds, constituting mainly	Ground beef patties, goat meat nugget	Kim et al. (2013a), Banerjee et al. (2012)

(continued)

Table 12.1 (continued)

Source	Main antioxidants	Product	References
	kaempferol and quercetin glucosides.		
Butterbur leaf extract	Polyphenols, glycosides, and some flavonoids as patuletin.	Ground beef patties	Kim et al. (2013b)
Curry leaf extracts (*Murrayakoenigii* L.)	Phenolic acids including tannic, gallic, caffeic, cinnamic, chlorogenic, ferulic, and vanillic acids.	Pork meat	Biswas et al. (2012)
Lotus rhizome knot (LRK) and lotus leaf (LL) extracts	Flavonoids, including flavonoid *C*-glycosides like schaftoside, quercetin derivatives, and catechin derivatives	Raw and cooked porcine and bovine meat samples	Huang et al. (2011)
Carrot juice	β-Carotene, vitamin C and K	Irradiated beef sausage	Badr and Mahmoud (2011)
Olive leaf extracts	Oleuropein, hydroxytyrosol, and tyrosol; caffeic acid; and ligstroside	Minced beef patties	Hayes et al. (2010)

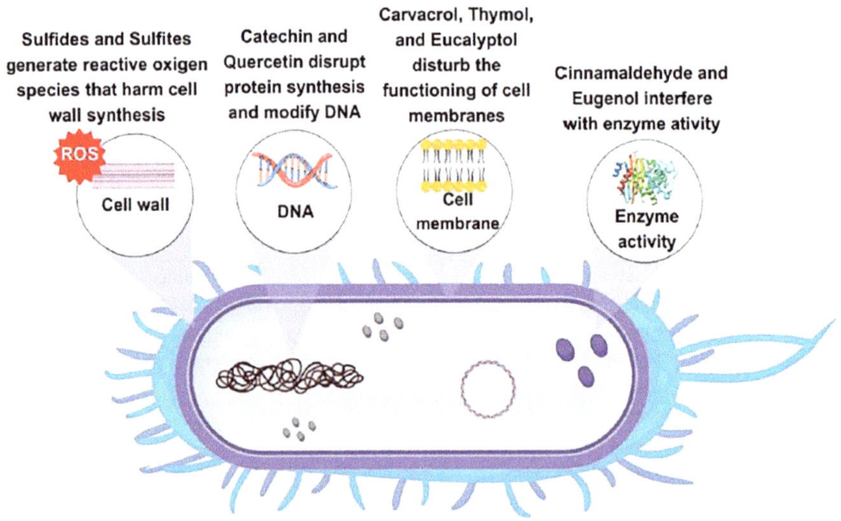

Fig. 12.3 Mechanisms of action of plant derived antimicrobials against foodborne bacteria

disorganization of the wall or membrane of the hyphae (Bouarab Chibane et al. 2019; Ansari et al. 2013; Morales et al. 2017; You et al. 2022; Luo et al. 2020).

12.3.1.2 Application of Plant Derived Natural Antimicrobials in Meat Products

Plant secondary metabolites were proposed as natural antimicrobials in the meat industry in order to improve the safety and storage life of meat and meat products. On the basis of structure plant antimicrobials are classified into two classes: antimicrobials with one or several nitrogen atoms in their structures, such as alkaloids, glucosinolates; and antimicrobials without nitrogen, such as terpenoids and phenolic substances. Alkaloids, glucosinolates, and phenolic substances are water-soluble compounds obtained by aqueous or hydroalcoholic extractions, whereas terpenoids are lipophilic obtained using distillation procedures or extractions with organic solvents (Wink 2022). Table 12.2 summarize the different classes of plant antimicrobials and their antimicrobial action against main food-related spoilage and pathogenic microorganisms. Meat products generally have a shorter storage life than whole meat due to their more porous structure resulting from processes such as grinding. The restructuring process creates more exposure to contamination and creates favorable environment for the proliferation of spoilage and pathogenic microorganisms, which can lead to potential consumer health threat and food safety concern (Olvera-Aguirre et al. 2023). The natural antimicrobial compounds or whole plant extracts from a huge number of plant sources have been studied in meat products (Table 12.3). In summary the use of natural antimicrobials offers a promising approach to improve the safety, quality, sustainability and shelf-life of meat products. The shelf life of meat and meat products is measured by colour coordinates, microbiology, sensory evaluation and nutrient oxidation. Thus the proper use of natural antimicrobials in meat and meat products has the effect of enhancing storage life, safety, quality, sustainability by preventing contamination by pathogenic and spoilage microorganisms.

12.3.2 Organic Acids

12.3.2.1 Mechanism of Action of Natural Organic Acids

The mechanism of action of organic acids against the spoilage and pathogenic microorganisms is the diffusion of the protonated acid through cell membranes, followed by intracellular dissociation, resulting in cytoplasm acidification and intracellular acid anion accumulation, disruption of the proton motive force, enzymes and protein modifications and DNA changes (Ricke 2003; Ben Braïek and Smaoui 2021; Lu et al. 2011; Dittoe et al. 2018). This inhibits the metabolism of the microbial cells and causes death. Factors that affect the antimicrobial activity of organic acids are

Table 12.2 Literature review on different classes of plant antimicrobial compounds and their antimicrobial action against food-related microorganisms

Plant antimicrobial	Minimum inhibitory concentration	Antimicrobial activity spectrum	References
Polyphenols			
(a) Phenolic acids	1000–2000 μg mL^{-1}	*Escherichia coli, Staphylococcus aureus, Listeria monocytogenes, Salmonella enteritidis, Alicyclobacillus acidoterrestris, Botrytis cinerea, Alternaria alternate*	Bae et al. (2022); Borges et al. (2013); Xu et al. (2022); Li et al. (2022); Shi et al. (2018); Shi et al. (2021); Patzke and Schieber (2018); Yuan et al. (2019); Li et al. (2018)
(b) Stilbenes,	100–200 μg mL^{-1}	*S. aureus, Enterococcus faecalis, E. coli, Sal. Enterica, Aspergillus carbonarius*	Vestergaard, and Ingmer (2019); Cai et al. (2022); Puljula et al. (2020)
(c) Tannins,	0.7–5 mg mL^{-1}	*S. aureus, E. coli*	Wang et al. (2020); Peng et al. (2022)
(d) Lignins	2 mg mL^{-1}	*Bacillus subtilis, S. aureus*	Wang et al. (2019)
(e) Flavonoids	0.5–8 μg mL^{-1} to 300–600 μg mL^{-1} (depending on type)	*L. monocytogenes, E. coli, E. coli O157:H7, Sal. Enteritidis, S. aureus, Penicillium expansum, Aspergillus flavus, Zygosaccharomyces parabailii*	Du et al. (2018); Xiong et al. (2017); Cetin-Karaca and Newman (2015); Amin et al. (2015); Zhang et al. (2018); Li et al. (2019); Cha et al. (2009); Kalli et al. (2022)
Terpenes and essential oils	0.16 μg mL^{-1} to 625 μg mL^{-1} (depending on type)	*E. coli, S. aureus, Salvia officinalis L., Lavandula dentata L., Laurus nobilis L., A. carbonarius, Enterococcus faecium, Sal. thyphimurium, Penicillium expansum, Aspergillus, Rhizopus oryzae, A. niger, B. cinerea, P. italicum, and Alt. alternata, L. monocytogenes, P. italicum.*	Sakkas et al. (2018); Valdivieso-Ugarte et al. (2021); Reyes-Jurado et al. (2019); Císarová et al. (2020); Marchese et al. (2017); Pinto et al. (2020); Cava-Roda et al. (2021); Yoplac et al. (2021); Zheng et al. (2015); Tang et al. (2018); Quintieri et al. (2022); Siddiqua et al. (2015); Sun et al. (2020); Xu et al. (2018)
Glucosinolate derivatives	4.2 μg mL^{-1} to 88 mg mL^{-1} (depending on type)	*Sal. thyphimurium, L. monocytogenes, A. flavus, P. verrucosum, E. coli, A. carbonarius, A. ochraceus, A. niger, S. aureus, Alt. alternate.*	Song, and Ku (2021); Olaimat et al. (2018); de Melo Nazareth et al. (2019, 2020); Yang et al. (2020); Wu et al. (2022); Yang et al. (2021); Nowicki et al. (2019); Zhang et al. (2020); Wu et al. (2022)

(continued)

Table 12.2 (continued)

Plant antimicrobial	Minimum inhibitory concentration	Antimicrobial activity spectrum	References
Alkaloids and thiols	30 μg mL^{-1} to 150 μg mL^{-1} (depending on type)	*E. coli, S. aureus, Aspergillus, P. italicum, Campylobacter jejuni, B. cereus.*	Othman et al. (2019); El-Zahar et al. (2022); Ma et al. (2022); He et al. (2022); Marchese et al. (2016); Jin et al. (2021); Tang et al. (2021)

pH, acid concentration, ionic strength, the acid's molecular weight, the ratio of protonated to unprotonated forms and the bacterial strains being attacked. Organic acids such as citric, acetic, lactic and tartaric acids have exhibited bactericidal properties (Mani-López et al. 2012).

12.3.2.2 Application of Natural Organic Acids in Meat Products

Natural organic acids are found in nature, usually in foods such as fruits, vegetables and fermented foods. Moreover, organic acids are produced through metabolism by lactic acid bacteria via carbohydrate fermentation. Organic acids as well as their salts have a potent antibacterial (Gram-positive and Gram-negative bacteria) and antifungal activity (Table 12.4). Organic acids such as lactic acid, acetic acid and citric acids are considered most interesting preservatives against spoilage and pathogenic microorganisms in addition improve the organoleptic quality of various foods and have techno-functional properties (Ben Braïek and Smaoui 2021). Organic acids and their salts like sodium lactate, sodium diacetate, acetic acid, etc. singly or in combination with other acids are generally applied in uncooked meats (ground beef), commercial RTE meat and poultry products, such as in packaged luncheon meat slices, wieners, smoked-cooked ham, light bologna, frankfurters and salami (Seman et al. 2002; Thompson et al. 2008; Glass et al. 2007; Porto et al. 2002; Samelis et al. 2002; Bodie et al. 2019; Bodie et al. 2022; Sallam, and Samejima 2004). They produce a significant antimicrobial effect against total plate count and total anaerobic count, *E. coli, Salmonella, Listeria monocytogenes, Enterobacteriaceae* and no additional changes in food product formulations are required. Therefore application of organic acids and their salts is a promising strategy to extend the storage life of meat and meat products as well as to ensure their safety due to their strong inhibitory effect against pathogenic and spoilage microorganisms.

Table 12.3 Literature review on natural plant antimicrobials used in meat products to reduce spoilage and pathogenic microorganisms in meat products

Plant material/ extract	Dosage in meat product	Meat product	Antimicrobial effect	References
Moringa leaf powder	100 g/kg	Chicken burgers	*Total coliforms, Escherichia coli, Staphylococcus aureus*	Elhadi et al. (2017)
Olive and apple extracts	1 and 3%	Beef patties	*E. coli O157:H7*	Rounds et al. (2013)
Extracts from green tea, stinging nettle, and olive leaves	500 ppm	Frankfurter-type sausage storage	*Total viable count, yeast, molds*	Alirezalu et al. (2017)
Lemongrass extract	1%	Fresh chicken sausages	*Total psychrotrophic, aerobic mesophilic count*	Boeira et al. (2018)
Lemongrass extract	1%	Cooked and shredded chicken breast	*Staphylococcus spp., Salmonella spp., Coliforms*	Kieling et al. (2019)
Roselle calyx extract	500 mg/L	Steaks and ground beef meat	*E. coli, Salmonella enterica, Salmonella typhimurium, S. aureus, L. monocytogenes, B. cereus*	Selene Márquez-Rodríguez et al. (2020); Chao and Yin (2009)
Beetroot extract	20 mg/mL	Cooked pork	*L. monocytogenes*	Gong et al. (2022)
Strawberry tree and pomegranate peel extracts	10%	Dry-cured ham-based model	*L. monocytogenes*	dos Santos et al. (2022)
Extracts from Jamaica	–	Hot dogs	*Listeria monocytogenes, Staphylococcus aureous*	Higginbotham et al. (2014)
Thyme essential oils	0.01–0.05% to 3%	Minced beef meat	*Escherichia coli, Salmonella typhimurium, Staphylococcus aureus, Pseudomonas aeruginosa*	Jayari et al. (2018)
Thyme and rosemary essential oils		Cook-chill beef	*Listeria monocytogenes*	Gouveia et al. (2016)
Black pepper essential oil		Fresh pork	*Pseudomonas spp.*	Zhang et al. (2016b)
Sage essential oil		Pork sausages	*Salmonella spp., E. coli, L. monocytogenes*	Šojić et al. (2018)
Coriander essential oil	0.075–0.150 μL/g	Pork sausages	*Total plate counts*	Šojic et al. (2019)

(continued)

Table 12.3 (continued)

Plant material/extract	Dosage in meat product	Meat product	Antimicrobial effect	References
Freeze-dried Myrtle (*Myrtus communis* L.) extract	5%	Grounded beef meat		Casaburi et al. (2015)
Extracts of rosemary and clove	1% v/w	Raw chicken meat	*Pseudomonas spp.*	Zhang et al. (2016a)
Thymol or carvacrol extract	0.4% w/w	Marinated beef	*Mesophilic total viable count, lactic acid bacteria, Broch. thermosphacta, Pseudomonas spp., total coliforms*	Karam et al. (2020)
Nettle extract	250–500 ppm	Ground beef	*Mesophilic, psychrotrophic, lactic acid bacteria, Pseudomonas, Enterobacteriaceae count, E. coli*	Cózar and Vergara (2018)

Table 12.4 Literature review on antimicrobial activity spectrum of organic acids and their salts

Organic acid or salt	Antimicrobial activity spectrum
Citric acid	*L. monocytogenes, S. typhimurium, E. coli O157:H7, A. flavus, P. purpurogenum, R. nigricans, F. oxysporum, S. cerevisiae, Z. bailii.*
Lactic acid	*L. monocytogenes, S. aureus, E. faecalis, B. cereus, Salmonella spp., E. coli, P. aeruginosa, Proteus spp., C. albicans, S. cerevisiae, P. nordicum, P. purpurogenum, A. flavus, R. nigricans, Rhodotorula spp.*
Propionic acid	*L. monocytogenes, E. coli, Salmonella spp., Cl. perfringens, A. flavus, Fusarium spp., Penicillium spp., R. nigricans.*
Tartaric acid	*S. typhimurium, A. flavus, Fusarium spp., Penicillium spp., R. nigricans.*
Formic acid	*E. coli, Salmonella spp., Cl. perfringens, A. flavus, Fusarium spp., Penicillium spp., R. nigricans.*
Benzoic acid	*E. coli, L. monocytogenes*
Succinic acid	*S. typhimurium, E. coli, B. subtilis, S. suis.*
Sorbic acid	*Fusarium spp., L. monocytogenes, E. coli*

Source: Ben Braïek and Smaoui (2021)

12.3.3 Bacteriocins

Bacteriocins are synthesized ribosomally and extracellularly released antimicrobial peptides of length 20–60 amino acids, cationic and hydrophobic, capable of inhibiting both pathogenic and food spoilage bacteria from Gram-negative and Gram-positive groups thus provide a new tool to combat food spoilage and food borne diseases (Walsh et al. 2015; Suganthi and Mohanasrinivasan 2015; Kumariya et al. 2019). They are considered safe for human consumption because they are digested rapidly by proteases in the human GIT therefore, have little influence on the gut microbiota (Darbandi et al. 2022).

12.3.3.1 Classification of Bacteriocins

Bacteriocins were initially classified into class I, II, III, and IV. However, class IV are not considered true bacteriocins due to their carbohydrate or lipid moieties (Kumariya et al. 2019). Therefore, majorly they are classified into three classes (Table 12.5). Classes I and II have low molecular weights (<10 kDa) and are thermostable, hydrophobic and cationic peptides, but class II has an amphiphilic helical structure, promoting membrane depolarization and cell death (Da Costa et al. 2019). Class III bacteriocins are large (>30 kDa) and heat-sensitive macromolecules (Santiesteban-López et al. 2023).

12.3.3.2 Mechanism of Action of Bacteriocins

The bacterial cell surface which is anionic made of phosphatidylethanolamine (PE), phosphatidylglycerol (PG), lipopolysaccharide (LPS), lipoteichoic acid (LTA) and cardiolipin (CL), is the common target exploited by the cationic bacteriocins (López-Lara and Geiger 2017). Bacteriocins assume configurations so as the positively charged groups interact electrostatically with the negatively charged bacterial cell surface (Kumariya et al. 2019). The hydrophobic surfaces of the bacteriocins are aligned toward the membrane and traverse the lipid bilayer and subsequently polymerize into complexes; permeabilization of the target bacteria cell membrane, probably by forming ion selective pores, which cause the dissipation of the proton motive force, the depletion of intracellular ATP, leakage of the intracellular substrates and eventual death of the bacterial cell (Bodie et al. 2023). Localized holes are created in the cell wall and membrane of *L. monocytogenes* by enterocin, results in collapse of proton motive force (Minahk et al. 2000). Docking molecule such as lipid II or mannose permease of the phosphotransferase system is required by bacteriocins to interact with the membranes (Héchard and Sahl 2002). However, some bacteriocins may not require target receptors for the docking. Garvicin ML, Lactococcus garvieae DCC43-derived circular bacteriocin, utilizes maltose ABC transporter and permease as receptors Gabrielsen et al. (2012).

Table 12.5 Classification of bacteriocins

Class		Properties	Example	Mode of action	Producer bacteria	Antibacterial spectrum	Refrences
I	Ia	Low molecular weight bacteriocins, heat stable, lantibiotics (<5 kDa peptides containing lanthionine and β-methyl lanthionine)	Nisin	Membrane permeabilization by pore formation	*L. lactis*	Gram-positive bacteria	Mazzotta et al. (1997); Szabo and Cahill (1999); Harris et al. (1991); Le Lay et al. (2016)
	Ib	Low molecular weight bacteriocins, heat stable, Carbacyclic lantibiotics containing labyrinthin and labionin	Labyrinthopeptin A1	–	*Actinomadura namibiensis*	HIV, HSV	Ferir et al. (2013)
	Ic	Low molecular weight bacteriocins, heat stable, Sactibiotics (Sulphur to alpha carbon-containing antibiotics)	Thuricin CD	–	*B. thuringiensis*	Gram-positive bacteria	Mathur et al. (2017)
II	IIa	Low molecular weight bacteriocins, heat stable, small heat-stable peptides, synthesized in a form of precursor which is processed after two glycine residues, active against Listeria, have a consensus sequence of YGNGV-C in the N-terminal	Pediocin PA-1, sakacins A and P, leucocin A.	Membrane permeabilization by pore formation	*P. pentosaceus, P. acidilactici, Lactobacillus sakei*	Gram-positive and gram-negative bacteria	Fimland et al. (1998); Kumariya et al. (2015); Trinetta et al. (2012); Balay et al. (2017)
	IIb	Low molecular weight bacteriocins, heat stable, two component systems: two different peptides required to form an active poration complex	Lactococcins G, plantaricin EF and plantaricin JK	Membrane permeabilization by pore formation	*L. lactis subsp. cremoris, Lb. plantarum*	Gram-positive bacteria	Oppegard et al. (2010); Ekblad et al. (2016, 2017)

(continued)

Table 12.5 (continued)

Class		Properties	Example	Mode of action	Producer bacteria	Antibacterial spectrum	Refrences
	IIc	Low molecular weight bacteriocins, heat stable, circular bacteriocins	Gassericin A, enterocin AS-48, garvicin ML	Membrane permeabilization by pore formation	*L. gasseri, E. faecalis, L. garvieae*	Gram-positive bacteria	Pandey et al. (2013); Grande Burgos et al. (2014); Borrero et al. (2011)
	IId	Low molecular weight bacteriocins, heat stable, unmodified, linear, leaderless, non-pediocin-like bacteriocins	Bactofencin A, LsbB	Membrane permeabilization by pore formation	*L. salivarius, L. lactis subsp. Lactis*	Gram-positive bacteria	O'Shea et al. (2013); O'Connor et al. (2018); Uzelac et al. (2015)
III		High molecular weight bacteriocins(>10 kDa), large molecules sensitive to heat	Helveticin M, helveticin J and enterolysin A	Membrane permeabilization by pore formation	*Lb. crispatus, L. helveticus, E. faecalis*	Gram-positive and gram-negative bacteria	Sun et al. (2018); Thompson et al. (1996); Nilsen et al. (2003)

12.3.3.3 Application of Bacteriocins in Meat Products

The antimicrobial potential of bacteriocins is dependent on the food constituents, such as salt content or curing salts, fat, pH, enzymes, distribution into the food, or the interaction with other additives. Thus, the application of bacteriocins should be studied in real foods and against target bacteria (Da Costa et al. 2019). Bacteriocins are used as inhibitory substances in different food products including meat, fish, dairy, cereal, fruit and vegetable beverage products and other foods usually stored under refrigeration. In meat products the most common bacteriocin used is nisin and is generally regarded as safe (GRAS). Nisin has been shown to reduce populations of Gram-positive bacteria, such as Listeria in RTE meat products (De Arauz et al. 2009; Leverentz et al. 2003). Leverentz et al. (2003) found that frankfurters treated with nisin resulted in a 2.35-log reduction in *L. monocytogenes*. RTE vacuum-packed diced turkey treated with nisin resulted in 4-log reduction in Listeria (Ruiz et al. 2010). In vacuum-packed sliced cooked ham addition of nisin (0.001–0.013%) produces a significant reduction in mesophilic aerobic (3–4 log reduction) and lactic acid bacteria (2–3 log reduction) (Kalschne et al. 2014). Bacteriocins pentocin 31–1 (40 and 80 AU/mL) and nisin (75 AU/mL) were tested in stored tray-packaged pork meat resulted in inhibition of the growth of total bacteria (about 2 log reduction after 6 storage days), Listeria and Pseudomonas (Zhang et al. 2010). Similarly, the bacteriocin BacFL31 (200 and 400 AU/g) was used as a preservative in ground turkey meat demonstrates an effective inhibition in the proliferation of *L. monocytogenes* (about 6 log reductions), *S. aureus* (>2.5 log reductions), and *Salmonella typhimurium* (about 1 log reductions), as well as a strong inhibition of spoilage bacteria, including aerobic plate (about 3 and 6 log reductions for 200 and 400 AU/g, respectively), psychrotrophic (about 4 and 5 log reductions for 200 and 400 AU/g, respectively) and Enterobacteriaceae counts (about 2 log reductions) (Chakchouk-mtibaa et al. 2017).

In addition to direct addition of bacteriocins to RTE meat products, other means exist for effective delivery. For example, edible films infused with bacteriocins can control *L. monocytogenes* in RTE meat products, as they gradually release bacteriocins onto the surface of the meat product during the entire storage life (Coma 2008). Aymerich et al. (2022) found that polyvinyl alcohol films infused with Enterocin A results inhibition of *L. monocytogenes* (1-log reduction) in vacuum-packed, dry-cured ham stored under refrigeration. Lungu and Johnson (2005) used zein coatings containing nisin on frankfurters and determined that nisin reduced *L. monocytogenes* counts by 6.6 CFU/g during 28 days of refrigerated storage.

In conclusion the promising results of bacteriocins in meat products preservation and safety could be increased when a bacteriocin mix is used. Bacteriocin mix is more efficient since it can inhibit bacteria resistance to another specific bacteriocin (Da Costa et al. 2019). Additionally, their stability despite changes in pH, the addition of salts, changes in temperature, and the addition of other additives makes them potential biopreservatives that ensure the safety of reformulated meat products, while reducing the use of synthetic or chemical additives.

12.4 Conclusion

During recent past demand for healthy, functional and wholesome meat products has increased as consumers are increasingly concerned about the harmful effects of synthetic antioxidants (BHT, BHA), fat, salt, or synthetic antimicrobials (nitrites) consumption. However, in general, the reduction of synthetic preservatives can lead to greater spoilage and risk to consumer health. Natural preservatives (antioxidants and antimicrobials) has gained considerable attention as promising alternatives to synthetic preservatives in the meat industry, offering numerous benefits that is very effective in increasing the safety, shelf life and consumer acceptance of meat products. The synergistic effects between plant derived extracts, essential oils, organic acids and bacteriocins and the successful integration of hurdle technologies, contributed to the scouting of novel food preservation methods. However, it is necessary to optimize the added dosage of natural preservatives to obtain the maximum preservation effect without affecting the sensory quality and to minimize their chemical degradation during the meat products manufacture process, such as during high temperature treatments, extrusion, or high-pressure treatments so that their activity can be retained.

Furthermore, additional studies are needed to assess the synergistic effect of different combinations of natural antioxidants and antimicrobials, to evaluate their optimum concentrations, and to assess their comprehensive toxicological studies. In addition, addressing the challenges and barriers to regulatory approval for natural antioxidants and antimicrobials as food additives at international level will accelerate their broader application in the food industry. In conclusion, further research efforts in the area will help to overcome the current challenges and pave the way towards a widespread adoption of plant antioxidants and antimicrobials in the food industry, contributing to safer, more sustainable, consumer-friendly and quality foods.

References

Akcan T, Estévez M, Serdaroglu M (2017) Antioxidant protection of cooked meatballs during frozen storage by whey protein edible films with phytochemicals from *Laurus nobilis* L. and Salvia offcinalis. LWT Food Sci Technol 77:323–331

Alirezalu K, Hesari J, Eskandari MH, Valizadeh H, Sirousazar M (2017) Effect of green tea, stinging nettle and olive leaves extracts on the quality and shelf life stability of frankfurter type sausage. J Food Process Preserv 41:1–11

Al-Juhaimi F, Ghafoor K, Hawashin MD, Alsawmahi ON, Babiker EE (2016) Effects of different levels of Moringa (*Moringa oleifera*) seed flour on quality attributes of beef burgers. CyTA – J Food 14:1–9

Allen KE, Cornforth DP (2006) Myoglobin oxidation in a model system as affected by nonheme iron and iron chelating agents. J Agric Food Chem 54:10134–10140

Amarachukwu Uzombah T (2022) The implications of replacing synthetic antioxidants with natural ones in the food systems. Tech Open, London. https://doi.org/10.5772/intechopen.103810

Amin MU, Khurram M, Khattak B, Khan J (2015) Antibiotic additive and synergistic action of rutin, morin and quercetin against methicillin resistant Staphylococcus aureus. BMC Complement Altern Med 15:59

Andrés AI, Petrón MJ, Adámez JD, López M, Timón ML (2017) Food by-products as potential antioxidant and antimicrobial additives in chill stored raw lamb patties. Meat Sci 129:62–70

Angeli JPF, Garcia CCM, Sena F, Freitas FP, Miyamoto S, Medeiros MHG, Di Mascio P (2011) Lipid hydroperoxide-induced and hemoglobin-enhanced oxidative damage to colon cancer cells. Free Radic Biol Med 51:503–515

Ansari MA, Anurag A, Fatima Z, Hameed S (2013) Natural phenolic compounds: a potential antifungal agent. In: Méndez-Vilas A (ed) Microbial pathogens and strategies for combating them: science, technology and education, 1st edn. Formatex Research Center, Badajoz, pp 1189–1195

Armenteros M, Morcuende D, Ventanas J, Estévez M (2016) The application of natural antioxidants via brine injection protects Iberian cooked hams against lipid and protein oxidation. Meat Sci 116:253–259

Aymerich T, Jofré A, Bover-Cid S (2022) Enterocin A-based antimicrobial film exerted strong antilisterial activity in sliced dry-cured ham immediately and after 6 months at 8 °C. Food Microbiol 105:104005

Azman NAM, Gordon MH, Monika SM, Segovia F, Almajano MP (2014) Use of lyophilised and powdered *Gentiana lutea* root in fresh beef patties stored under different atmospheres. J Sci Food Agric 95(9):1804–1811

Badr HM, Mahmoud KA (2011) Antioxidant activity of carrot juice in gamma irradiated beef sausage during refrigerated and frozen storage. Food Chem 127(3):1119–1130

Bae JY, Seo YH, Oh SW (2022) Antibacterial activities of polyphenols against foodborne pathogens and their application as antibacterial agents. Food Sci Biotechnol 31:985–997

Balay DR, Dangeti RV, Kaur K, McMullen LM (2017) Purification of leucocin a for use on wieners to inhibit *Listeria monocytogenes* in the presence of spoilage organisms. Int J Food Microbiol 255:25–31

Banerjee R, Verma AK, Das AK, Rajkumar V, Shewalkar AA, Narkhede HP (2012) Antioxidant effects of broccoli powder extract in goat meat nuggets. J Meat Sci 91(2):179–184

Barbosa-Pereira L, Aurrekoetxea GP, Angulo I, Paseiro-Losada P, Cruz JM (2014) Development of new active packaging fi lms coated with natural phenolic compounds to improve the oxidative stability of beef. Meat Sci 97:249–254

Basanta MF, Rizzo SA, Szerman N, Vaudagna SR, Descalzo AM, Gerschenson LN, Pérez CD, Rojas AM (2018) Plum (Prunus salicina) peel and pulp microparticles as natural antioxidant additives in breast chicken patties. Food Res Int 106:1086–1094

Bekhit AED, Simmons N, Faustman C (2005) Metmyoglobin reducing activity in fresh meat: a review. Meat Sci 71:407–439

Ben Braïek O, Smaoui S (2021) Chemistry, safety, and challenges of the use of organic acids and their derivative salts in meat preservation. J Food Quality. 1–20

Biswas AK, Chatli MK, Sahoo J (2012) Antioxidant potential of curry (*Murraya koenigii* L.) and mint (Mentha spicata) leaf extracts and their effect on colour and oxidative stability of raw ground pork meat during refrigeration storage. J Food Chem 133(2):467–472

Bodie AR, Dittoe DK, Feye KM, Knueven CJ, Ricke SC (2019) Application of an alternative inorganic acid antimicrobial for controlling *Listeria monocytogenes* in frankfurters. Front Sustain Food Syst 3:34

Bodie AR, Dittoe DK, Feye KM, Knueven CJ, Ovall C, Ricke SC (2022) Comparison of ready-to-eat "organic" antimicrobials, sodium bisulfate, and sodium lactate, on *Listeria monocytogenes* and the indigenous microbiome of organic uncured beef frankfurters stored under refrigeration for three weeks. PLoS One 17:0262167

Bodie AR, O'Bryan CA, Olson EG, Ricke SC (2023) Natural antimicrobials for *Listeria monocytogenes* in ready-to-eat meats: current challenges and future prospects. Microorganisms 11:1301

Boeira CP, Piovesan N, Soquetta MB, Flores DCB, Lucas BN, da Rosa CS, Terra NN (2018) Extraction of bioactive compounds of lemongrass, antioxidant activity and evaluation of antimicrobial activity in fresh chicken sausage. Ciênc Rural 48:20180477

Borges A, Ferreira C, Saavedra MJ, Simões M (2013) Antibacterial activity and mode of action of ferulic and gallic acids against pathogenic bacteria. Microb Drug Resist 19:256–265

Borrero J, Brede DA, Skaugen M, Diep DB, Herranz C, Nes IF, Cintas LM, Hernandez PE (2011) Characterization of garvicin ML, a novel circular bacteriocin produced by Lactococcus garvieae DCC43, isolated from mallard ducks (Anas platyrhynchos). Appl Environ Microbiol 77:369–373

Bouarab Chibane L, Degraeve P, Ferhout H, Bouajila J, Oulahal N (2019) Plant antimicrobial polyphenols as potential natural food preservatives. J Sci Food Agric 99:1457–1474

Bozkurt H (2006) Utilization of natural antioxidants: green tea extract and Thymbra spicata oil in Turkish dry-fermented sausage. Meat Sci 73:442–450

Brewer MS (2011) Natural antioxidants: sources, compounds, mechanisms of action, and potential applications. Compr Rev Food Sci Food Saf 10:221–247

Broncano JM, Petrón MJ, Parra V, Timón ML (2009) Effect of different cooking methods on lipid oxidation and formation of free cholesterol oxidation products (COPs) in latissimus dorsi muscle of Iberian pigs. Meat Sci 83:431–437

Brown WD, Mebine LB (1969) Autoxidation of oxymyoglobins. J Biol Chem 244:6696–6701

Cai X, Qi J, Xu Z, Huang L, Li Y, Ren X, Kong Q (2022) Three stilbenes make difference to the antifungal effects on ochratoxin A and its precursor production of *Aspergillus carbonarius*. Food Microbiol 103:103967

Carlsen MH, Halvorsen BL, Holte K, Bøhn SK, Dragland S, Sampson L, Willey C, Senoo H, Umezono Y, Sanada C et al (2010) The total antioxidant content of more than 3100 foods, beverages, spices, herbs and supplements used worldwide. Nutr J 9:1–11

Carpenter R, O'Grady MN, O'Callaghan YC, O'Brien NM, Kerry JP (2007) Evaluation of the antioxidant potential of grape seed and bearberry extracts in raw and cooked pork. Meat Sci 76: 604–610

Casaburi A, Di Martino V, Ercolini D, Parente E, Villani F (2015) Antimicrobial activity of *Myrtus communis* L. water-ethanol extract against meat spoilage strains of Brochothrix thermosphacta and Pseudomonas fragi in vitro and in meat. Ann Microbiol 65:841–850

Cava-Roda R, Taboada-Rodríguez A, López-Gómez A, Martínez-Hernández GB, Marín-Iniesta F (2021) Synergistic antimicrobial activities of combinations of vanillin and essential oils of cinnamon bark, cinnamon leaves, and cloves. Food Secur 10:1406

Cetin-Karaca H, Newman MC (2015) Antimicrobial efficacy of plant phenolic compounds against *Salmonella* and *Escherichia coli*. Food Biosci 11:8–16

Cha JD, Moon SE, Kim JY, Jung EK, Lee YS (2009) Antibacterial activity of sophoraflavanone G isolated from the roots of Sophora flavescens against methicillin-resistant Staphylococcus aureus. Phytother Res 23:1326–1331

Chakchouk-mtibaa A, Smaoui S, Ktari N, Sellem I, Najah S, Karray-rebai I, Mellouli L (2017) Biopreservative efficacy of Bacteriocin BacFL31 in raw ground Turkey meat in terms of microbiological, physicochemical, and sensory qualities. Biocontrol Sci 22:67–77

Chandralekha S, Jagadeesh Babu A, Moorthy PRS, Karthikeyan B (2012) Studies on the effect of pomegranate rind powder extract as natural antioxidant in chicken meat balls during refrigerated storage. J Adv Vet Res 2:107–112

Chao CY, Yin MC (2009) Antibacterial effects of Roselle calyx extracts and Protocatechuic acid in ground beef and apple juice. Foodborne Pathog Dis 6:201–206

Chen CH, Pearson AM, Gray JI (1992) Effects of synthetic antioxidants (BHA, BHT and PG) on the mutagenicity of IQ-like compounds. Food Chem 45:177–183

Chouliara E, Karatapanis A, Savvaidis IN, Kontominas MGÃ (2007) Combined effect of oregano essential oil and modified atmosphere packaging on shelf-life extension of fresh chicken breast meat, stored at 4°C. Food Microbiol 24:607–617

Císarová M, Hleba L, Medo J, Tančinová D, Mašková Z, Cubon J, Kováčik A, Foltinová D, Božike M, Kloucek P (2020) The *in vitro* and *in situ* effect of selected essential oils in vapour phase against bread spoilage toxicogenic aspergilli. Food Control 110:107007

Coma V (2008) Bioactive packaging technologies for extended shelf life of meat-based products. Meat Sci 78:90–103

Cózar A, Vergara H (2018) Lamb burgers made with low and high value cuts: effect of the spice added and the packaging method on shelf life. CyTA – J Food 16:1115–1124

Da Costa RJ, Voloski FLS, Mondadori RG, Duval EH, Fiorentini ÂM (2019) Preservation of meat products with Bacteriocins produced by lactic acid bacteria isolated from meat. J Food Quality. 1:1-12.

Darbandi A, Asadi A, Mahdizade A, Marzieh AM, Ohadi E, Talebi M, Halaj ZM, Darb EA, Ghanavati R, Kakanj M (2022) Bacteriocins: properties and potential use as antimicrobials. J Clin Lab Anal 36:e24093

Das AK, Nanda PK, Bandyopadhyay S, Banerjee R, Biswas S et al (2020) Application of nanoemulsion based approaches for improving the quality and safety of muscle foods: a comprehensive review. Compr Rev Food Sci Food Saf 19(5):2677–2700

De Arauz LJ, Jozala AF, Mazzola PG, Penna TCV (2009) Nisin biotechnological production and application: a review. Trends Food Sci Technol 20:146–154

de Melo Nazareth T, Quiles JM, Torrijos R, Luciano FB, Manes J, Meca G (2019) Antifungal and antimycotoxigenic activity of allyl isothiocyanate on barley under different storage conditions. LWT Food Sci Technol 112:108237

de Melo NT, Alonso-Garrido M, Stanciu O, Mañes J, Manyes L, Meca G (2020) Effect of allyl isothiocyanate on transcriptional profile, aflatoxin synthesis, and *Aspergillus flavus* growth. Food Res Int 128:108786

Dittoe DK, Ricke SC, Kiess AS (2018) Organic acids and potential for modifying the avian gastrointestinal tract and reducing pathogens and disease. Front Vet Sci 5:216

Domínguez R, Pateiro M, Gagaoua M, Barba FJ, Zhang W, Lorenzo JM (2019) A comprehensive review on lipid oxidation in meat and meat products. Antioxidants 8:429

dos Santos LR, Alía A, Martin I, Gottardo FM, Rodrigues LB, Borges KA, Furian TQ, Córdoba JJ (2022) Antimicrobial activity of essential oils and natural plant extracts against *Listeria monocytogenes* in a dry-cured ham-based model. J Sci Food Agric 102:1729–1735

Du W, Zhou M, Liu Z, Chen Y, Li R (2018) Inhibition effects of low concentrations of epigallocatechin gallate on the biofilm formation and hemolytic activity of *Listeria monocytogenes*. Food Control 85:119–126

Echegaray N, Pateiro M, Munekata PES, Lorenzo JM, Chabani Z, Farag MA, Domínguez R (2021) Measurement of antioxidant capacity of meat and meat products: methods and applications. Molecules 26

Ekblad B, Kyriakou PK, Oppegard C, Nissen-Meyer J, Kaznessis YN, Kristiansen PE (2016) Structure-function analysis of the two-peptide bacteriocin plantaricin EF. Biochemist 55:5106–5116

Ekblad B, Nissen-Meyer J, Kristiansen T (2017) Whole-genome sequencing of mutants with increased resistance against the two-peptide bacteriocin plantaricin JK reveals a putative receptor and potential docking site. PLoS One 12:e0185279

Elhadi DAE, Elgasim EA, Mohamed Ahmed IA (2017) Características microbianas y oxidativas de la hamburguesa de pollo refrigerada con moringa en polvo (Moringa oleifera). CyTA – J Food 15:234–240

El-Zahar KM, Al-Jamaan ME, Al-Mutairi FR, Al-Hudiab AM, Al-Einzi MS, Mohamed AAZ (2022) Antioxidant, antibacterial, and antifungal activities of the ethanolic extract obtained from Berberis vulgaris roots and leaves. Molecules 27:6114

Embuscado ME (2015) Spices and herbs: natural sources of antioxidants – a mini review. J Funct Foods 18:811–819

Estevez M (2021) Critical overview of the use of plant antioxidants in the meat industry: opportunities, innovative applications and future perspectives. Meat Sci 181:108610

Estévez M, Ramírez R, Ventanas S, Cava R (2007) Sage and rosemary essential oils versus BHT for the inhibition of lipid oxidative reactions in liver pâté. LWT Food Sci Technol 40:58–65
Falowo AB, Fayemi PO, Muchenje V (2014) Natural antioxidants against lipid–protein oxidative deterioration in meat and meat products: a review. Food Res Int 64:171–181
Fasseas MK, Mountzouris KC, Tarantilis PA, Polissiou M, Zervas G (2008) Antioxidant activity in meat treated with oregano and sage essential oils. Food Chem 106:1188–1194
Faustman C, Cassens RG (1990) The biochemical basis for discolouration in fresh meat: a review. J Muscle Foods 1:217–243
Ferir G, Petrova MI, Andrei G, Huskens D, Hoorelbeke B, Snoeck R, Vanderleyden J, Balzarini J, Bartoschek S, Bronstrup M, Sussmuth RD, Schols D (2013) The lantibiotic peptide labyrithopeptin A1 demonstrates broad anti-HIV and anti-HSV activity with potential for microbicidal applications. PLoS One 8:e64010
Fernandes RPP, Trindade MA, Toni FG, Pugine SMP, Lima CG, Lorenzo JM, Melo MP (2017) De evaluation of oxidative stability of lamb burger with Origanum vulgare extract. Food Chem 233: 101–109
Fernandes RPP, Trindade MA, Lorenzo JM, Melo MP (2018) De assessment of the stability of sheep sausages with the addition of different concentrations of Origanum vulgare extract during storage. Meat Sci 137:244–257
Fimland G, Jack R, Jung G, Nes IF, Nissen-Meyer J (1998) The bactericidal activity of pediocin PA-1 is specifically inhibited by a 15-mer fragment that spans the bacteriocin from the center toward the C terminus. Appl Environ Microbiol 64:5057–5060
Fratianni F, De Martino L, Melone A, De Feo V, Coppola R, Nazzaro F (2010) Preservation of chicken breast meat treated with thyme and balm essential oils. J Food Sci 75:M528–35
Gabrielsen C, Brede DA, Hernández PE, Nes IF, Diep DB (2012) The maltose ABC transporter in Lactococcus lactis facilitates high-level sensitivity to the circular bacteriocin garvicin ML. Antimicrob Agents Chemother 56:2908–2915
George P, Stratman CJ (1952) The oxidation of myoglobin to metmyoglobin by oxygen. 2. The relation between the first order rate constant and the partial pressure of oxygen. Biochem J 51: 418–425
Glass KA, McDonnell LM, Rassel RC, Zierke KL (2007) Controlling *Listeria monocytogenes* on sliced ham and Turkey products using benzoate, propionate, and sorbate. J Food Prot 70:2306–2312
Gonçalves LA, Lorenzo JM, Trindade MA (2011) Fruit and agro-industrial waste extracts as potential antimicrobials in meat products: a brief review. Food Secur 10:1–10
Gong S, Jiao C, Guo L (2022) Antibacterial mechanism of beetroot (Beta vulgaris) extract against *Listeria monocytogenes* through apoptosis-like death and its application in cooked pork. LWT Food Sci Technol 165:113711
Gotoh T, Shikama K (1974) Autoxidation of oxymyoglobin from bovine heart muscle. Arch Biochem Biophys 163:476–481
Gouveia AR, Alves M, Silva JA, Saraiva C (2016) The antimicrobial effect of rosemary and thyme essential oils against *Listeria monocytogenes* in sous vide cook-chill beef during storage. Procedia Food Sci 7:173–176
Grande Burgos MJ, Pulido RP, Del Carmen Lopez Aguayo M, Galvez A, Lucas R (2014) The cyclic antibacterial peptide enterocin AS-48: isolation, mode of action, and possible food applications. Int J Mol Sci 15:22706–22727
Guedes-Oliveira JM, da Costa-Lima BRC, Cunha LCM, de APA, Salim A, Baltar JD, Fortunato AR, Conte-Junior CA (2018) Impact of myrciaria dubia peel and seed extracts on oxidation process and colour stability of ground lamb. CyTA – J Food 16:931–937
Haak L, Raes K, De SS (2009) Effect of plant phenolics, tocopherol and ascorbic acid on oxidative stability of pork patties. J Sci Food Agric:1360–1365
Harris LJ, Fleming HP, Klaenhammer TR (1991) Sensitivity and resistance of *Listeria monocytogenes* ATCC 19115, Scott A, and UAL500 to nisin. J Food Prot 54:836–840

Hasapidou A, Savvaidis IN (2011) The effects of modified atmosphere packaging, EDTA and oregano oil on the quality of chicken liver meat. Food Res Int 44:2751–2756

Hayes JE, Stepanyan V, Allen P, O'Grady MN, Kerry JP (2010) Evaluation of the effects of selected phytochemicals on quality indices and sensorial properties of raw and cooked pork stored in different packaging systems. Meat Sci 85(2):289–296

He M, Wang Y, Hong M, Li T (2022) Berberine as a promising natural compound to control Penicillium italicum causing blue mold of citrus fruit. Sci Hortic 305:111370

Héchard Y, Sahl HG (2002) Mode of action of modified and unmodified bacteriocins from gram-positive bacteria. Biochimie 84:545–557

Hernández-López SH, Rodríguez-Carpena JG, Lemus-Flores C, Galindo-García J, Estévez M (2016) Antioxidant protection of proteins and lipids in processed pork loin chops through feed supplementation with avocado. J Food Sci Technol 53:2788–2796

Higginbotham KL, Burris KP, Zivanovic S, Davidson PM, Stewart CN (2014) Aqueous extracts of Hibiscus sabdariffa calyces as an antimicrobial rinse on hot dogs against *Listeria monocytogenes* and methicillin-resistant Staphylococcus aureus. Food Control 40:274–277

Huang B, He J, Ban X, Zeng H, Yao X, Wang Y (2011) Antioxidant activity of bovine and porcine meat treated with extracts from edible lotus (Nelumbo nucifera) rhizome knot and leaf. Meat Sci 87(1):46–53

Jayari A, El Abed N, Jouini A, Mohammed Saed Abdul-Wahab O, Maaroufi A, Ben Hadj Ahmed S (2018) Antibacterial activity of Thymus capitatus and Thymus algeriensis essential oils against four food-borne pathogens inoculated in minced beef meat. J Food Saf 38:e12409

Jayasena DD, Jo C (2013) Essential oils as potential antimicrobial agents in meat and meat products: a review. Trends Food Sci Technol 34(2):96–108

Jayasena DD, Jo C (2014) Potential application of essential oils as natural antioxidants in meat and meat products: a review. Food Rev Int 30:71–90

Jayathilakan K, Sharma GK, Radhakrishna K, Bawa AS (2007) Antioxidant potential of synthetic and natural antioxidants and its effect on warmed-over-flavour in different species of meat. Food Chem 105:908–916

Jayawardana BC, Warnasooriya VB, Thotawattage GH, Dharmasena VAKI, Liyanage R (2019) Black and green tea (*Camellia sinensis* L.) extracts as natural antioxidants in uncured pork sausages. J Food Process Preserv 43:1–8

Jeong JY, Hwang YH, Park GB, Joo ST, Hur SJ, Yang HS et al (2009) Discolouration characteristics of 3 major muscles from cattle during cold storage. J Food Sci 74:1–c5

Jiang J, Zhang X, True AD, Zhou L, Xiong YL (2013) Inhibition of lipid oxidation and rancidity in precooked pork patties by radical-scavenging licorice (*Glycyrrhiza glabra*) Extract J Food Sci 78:C1686-94

Jiang Y, Fang Z, Leonard W, Zhang P (2021) Phenolic compounds in Lycium berry: composition, health benefits and industrial applications. J Funct Foods 77:104340

Jin Z, Li L, Zheng Y, An P (2021) Diallyl disulfide, the antibacterial component of garlic essential oil, inhibits the toxicity of Bacillus cereus ATCC 14579 at sub-inhibitory concentrations. Food Control 126:108090

Jo SC, Nam KC, Min BR, Ahn DU, Cho SH, Park WP, Lee SC (2006) Antioxidant activity of Prunus Mume extract in cooked chicken breast meat. Int J Food Sci Technol 41:15–19

Jongberg S, Tørngren MA, Gunvig A, Skibsted LH, Lund MN (2013) Effect of green tea or rosemary extract on protein oxidation in Bologna type sausages prepared from oxidatively stressed pork. Meat Sci 93:538–546

Kalli S, Araya-Cloutier C, Chapman J, Sanders JW, Vincken JP (2022) Prenylated (iso) flavonoids as antifungal agents against the food spoiler *Zygosaccharomyces parabailii*. Food Control 132: 108434

Kalschne DL, Geitenes S, Veit MR, Sarmento CMP, Colla E (2014) Growth inhibition of lactic acid bacteria in ham by nisin: a model approach. Meat Sci 98:744–752

Karam L, Chehab R, Osaili TM, Savvaidis IN (2020) Antimicrobial effect of thymol and carvacrol added to a vinegar-based marinade for controlling spoilage of marinated beef (Shawarma) stored in air or vacuum packaging. Int J Food Microbiol 332:108769

Karre L, Lopez K, Getty KJK (2013) Natural antioxidants in meat and poultry products. Meat Sci 94:220–227

Kathirvel P, Gong Y, Richards MP (2009) Identification of the compound in a potent cranberry juice extract that inhibits lipid oxidation in comminuted muscle. Food Chem 115:924–932

Kieling DD, Delarco MF, Prudencio SH (2019) Lemongrass extract as a natural preservative of cooked and shredded chicken breast during storage. J Culin Sci Technol 19:55–66

Kim SJ, Cho AR, Han J (2013a) Antioxidant and antimicrobial activities of leafy green vegetable extracts and their applications to meat product preservation. Food Control 29(1):112–120

Kim SJ, Min SC, Shin HJ, Lee YJ, Cho AR, Kim SY, Han J (2013b) Evaluation of the antioxidant activities and nutritional properties of ten edible plant extracts and their application to fresh ground beef. Meat Sci 93(3):715–722

Kobus-Cisowska J, Flaczyk E, Rudzinska M, Kmiecik D (2014) Antioxidant properties of extracts from Ginkgo biloba leaves in meatballs. Meat Sci 97(2):174–180

Kumar Y, Yadav DN, Ahmad T, Narsaiah K (2015) Recent trends in the use of natural antioxidants for meat and meat products. Compr Rev Food Sci Food Saf. https://doi.org/10.1111/1541-4337.12156

Kumariya R, Sood SK, Rajput YS, Garsa AK (2015) Gradual pediocin PA-1 resistance in enterococcus faecalis confers cross-protection to diverse pore-forming cationic antimicrobial peptides displaying changes in cell wall and mannose PTS expression. Ann Microbiol 65:721–732

Kumariya R, Garsa AK, Rajputc YS, Soodc SK, Akhtard N, Patele S (2019) Bacteriocins: classification, synthesis, mechanism of action and resistance development in food spoilage causing bacteria. Microb Pathog 128:171–177

Lara MS, Gutierrez JI, Timón M, Andrés AI (2011) Evaluation of two natural extracts (*Rosmarinus offcinalis* L. and *Melissa offcinalis* L.) as antioxidants in cooked pork patties packed in MAP. Meat Sci 88:481–488

Le Lay C, Dridi L, Bergeron MG, Ouellette M, Fliss I (2016) Nisin is an effective inhibitor of Clostridium difficile vegetative cells and spore germination. J Med Microbiol 65:169–175

Ledward DA (1970) Metmyoglobin formation in beef stored in carbon dioxide enriched and oxygen depleted atmospheres. J Food Sci 35:33–37

Lee YS, Han CH, Kang SH, Lee SJ, Kim SW, Shin OR, Sim YC, Lee SJ, Cho YH (2005) Synergistic effect between catechin and ciprofloxacin on chronic bacterial prostatitis rat model. Int J Urol 12:383–389

Lee CH, Reed JD, Richards MP (2006) Ability of various polyphenolic classes from cranberry to inhibit lipid oxidation in mechanically separated Turkey and cooked ground pork. J Muscle Foods 17:248–266

Leverentz B, Conway WS, Camp MJ, Janisiewicz WJ, Abuladze T, Yang M, Saftner R, Sulakvelidze A (2003) Biocontrol of *Listeria monocytogenes* on fresh-cut produce by treatment with lytic bacteriophages and a bacteriocin. Appl Environ Microbiol 69:4519–4526

Li W, Yuan S, Sun J, Li Q, Jiang W, Cao J (2018) Ethyl p-coumarate exerts antifungal activity in vitro and in vivo against fruit Alternaria alternata via membrane-targeted mechanism. Int J Food Microbiol 278:26–35

Li XM, Li ZY, Wang YD, Wang JQ, Yang PL (2019) Quercetin inhibits the proliferation and aflatoxins biosynthesis of Aspergillus flavus. Toxins 11:154

Li J, Zhao N, Xu R, Li G, Dong H, Wang B, Li Z, Fan M, Wei X (2022) Deciphering the antibacterial activity and mechanism of p-coumaric acid against Alicyclobacillus acidoterrestris and its application in apple juice. Int J Food Microbiol 378:109822

López-Lara IM, Geiger O (2017) Bacterial lipid diversity. Biochim Biophys Acta Mol Cell Biol Lipids 1862:1287–1299

Lorenzo JM, Batlle R, Gómez M (2014) Extension of the shelf-life of foal meat with two antioxidant active packaging systems. LWT Food Sci Technol 59:181–188
Lu HJ, Breidt F, Perez-Diaz IM, Osborne JA (2011) Antimicrobial effects of weak acids on the survival of Escherichia coli O157:H7 under anaerobic conditions. J Food Prot 74:893–898
Lungu B, Johnson MG (2005) Fate of *Listeria monocytogenes* inoculated onto the surface of model Turkey frankfurter pieces treated with zein coatings containing nisin, sodium diacetate, and sodium lactate at 4 degrees C. J Food Prot 68:855–859
Luo J, Xu F, Zhang X, Shao X, Wei Y, Wang H (2020) Transcriptome analysis of Penicillium italicum in response to the flavonoids from Sedum aizoon L. World J Biochim Biophys Microbiol Biotechnol 36:62
Ma K, Zhe T, Li F, Zhang Y, Yu M, Li R, Wang L (2022) Sustainable films containing AIE-active berberine-based nanoparticles: a promising antibacterial food packaging. Food Hydrocoll 123: 107147
Manessis G, Kalogianni AI, Lazou T, Moschovas M, Bossis I, Gelasakis AI (2020) Plant-derived natural antioxidants in meat and meat products. Antioxidants 9:1215
Mani-López E, García HS, López-Malo A (2012) Organic acids as antimicrobials to control salmonella in meat and poultry products. Food Res Int 45:713–721
Marchese A, Barbieri R, Sanches-Silva A, Daglia M, Nabavi SF, Jafari NJ, Izadi M, Ajami M, Nabavi SM (2016) Antifungal and antibacterial activities of allicin: a review. Trends Food Sci Technol 52:49–56
Marchese A, Arciola CR, Barbieri R, Silva AS, Nabavi SF, Tsetegho Sokeng AJ, Izadi M, Jafari NJ, Suntar I, Daglia M et al (2017) Update on monoterpenes as antimicrobial agents: a particular focus on p-cymene. Materials 10:947
Mathur H, Fallico V, O'Connor PM, Rea MC, Cotter PD, Hill C, Ross RP (2017) Insights into the mode of action of the sanctibiotic thuricin CD. Front Microbiol 8:696
Mazzotta AS, Crandall AD, Montville TJ (1997) Nisin resistance in clostridium botulinum spores and vegetative cells. Appl Environ Microbiol 63:2654–2659
Mielnik MB, Olsen E, Vogt G, Adeline D, Skrede G (2006) Grape seed extract as antioxidant in cooked, cold stored Turkey meat. LWT Food Sci Technol 39:191–198
Minahk CJ, Farías ME, Sesma F, Morero RD (2000) Effect of Enterocin CRL35 on *Listeria monocytogenes* cell membrane. FEMS Microbiol Lett 192:79–83
Morales J, Mendoza L, Cotoras M (2017) Alteration of oxidative phosphorylation as a possible mechanism of the antifungal action of p-coumaric acid against Botrytis cinerea. J Appl Microbiol 123:969–976
Neill JM, Hastings AB (1925) The influence of the tension of molecular oxygen upon certain oxidations of hemoglobin. J Biol Chem 63:479–492
Nilsen T, Nes IF, Holo H (2003) Enterolysin A, a cell wall degrading bacteriocin from enterococcus faecalis LMG 2333. Appl Environ Microbiol 69:2975–2984
Nowicki D, Maciąg-Dorszynska M, Bogucka K, Szalewska-Pałasz A, Herman-Antosiewicz A (2019) Various modes of action of dietary phytochemicals, sulforaphane and phenethyl isothiocyanate, on pathogenic bacteria. Sci Rep 9:1–12
Nuñez De Gonzalez MT, Boleman RM, Miller RK, Keeton JT, Rhee KS (2008) Antioxidant properties of dried plum ingredients in raw and precooked pork sausage. J Food Sci 73:H63-71
O'Connor PM, O'Shea EF, Cotter PD, Hill C, Ross RP (2018) The potency of the broad spectrum bacteriocin, bactofencin A, against staphylococci is highly dependent on primary structure, N-terminal charge and disulphide formation. Sci Rep 8:11833
Olaimat AN, Al-Holy MA, Ghoush MA, Al-Nabulsi AA, Holley RA (2018) Control of salmonella enterica and *Listeria monocytogenes* in hummus using allyl isothiocyanate. Int J Food Microbiol 278:73–80
Olvera-Aguirre G, Mendoza-taco MM, Moo-huchin VM, Roque-jim A, Armando G (2022) Effect of extraction type on bioactive compounds and antioxidant activity of *Moringa oleifera Lam*, Leaves. Agriculture:1–9

Olvera-Aguirre G, Piñeiro-Vázquez ÁT, Sanginés-García JR, Z'arate AS, Ochoa-Flores AA, Segura-Campos MR, Vargas-Bello-P'erez E, Chay-Canul AJ (2023) Using plant-based compounds as preservatives for meat products: a review. Heliyon 9:e17071

Oppegard C, Rogne P, Kristiansen PE, Nissen-Meyer J (2010) Structure analysis of the two-peptide bacteriocin lactococcin G by introducing D-amino acid residues. Microbiology 156:1883–1889

O'Shea EF, O'Connor PM, O'Sullivan O, Cotter PD, Ross RP, Hill C (2013) Bactofencin A, a new type of cationic bacteriocin with unusual immunity, MBio 4:00498–e00513

Othman L, Sleiman A, Abdel-Massih RM (2019) Antimicrobial activity of polyphenols and alkaloids in middle eastern plants. Front Microbiol 10:911

Pandey N, Malik RK, Kaushik JK, Singroha G, Gassericin A (2013) A circular bacteriocin produced by lactic acid bacteria lactobacillus gasseri. World J Microbiol Biotechnol 29:1977–1987

Pateiro M, Lorenzo JM, Amado IR, Franco D (2014) Effect of addition of green tea, chestnut and grape extract on the shelf-life of pig liver pâté. Food Chem 147:386–394

Pateiro M, Gómez-Salazar JA, Jaime-Patlan M, Sosa-Morales ME, Lorenzo JM (2021) Plant extracts obtained with green solvents as natural antioxidants in fresh meat products. Antioxidants 10:1–21

Patzke H, Schieber A (2018) Growth-inhibitory activity of phenolic compounds applied in an emulsifiable concentrate-ferulic acid as a natural pesticide against Botrytis cinerea. Food Res Int 113:18–23

Peng M, Jiang C, Jing H, Du X, Fan X, Zhang Y, Wang H (2022) Comparison of different extraction methods on yield, purity, antioxidant, and antibacterial activities of proanthocyanidins from chokeberry (Aronia melanocarpa). J Food Meas Charact 16:2049–2059

Petcu CD, Mihai OD, Tăpăloagă D, Gheorghe-Irimia RA, Pogurschi EN, Militaru M, Borda C, Ghimpețeanu OM (2023) Effects of plant-based antioxidants in animal diets and meat products: a review. Food Secur 12:1334

Pinto L, Bonifacio MA, De Giglio E, Cometa S, Logrieco AF, Baruzzi F (2020) Unravelling the antifungal effect of red thyme oil (*Thymus vulgaris* L.) compounds in vapor phase. Molecules 25:4761

Pokorný J (2007) Are natural antioxidants better-and safer-than synthetic antioxidants? Eur J Lipid Sci Technol 109:629–642

Porto ACS, Franco BDGM, Santanna ES, Call JE, Piva A, Luchansky JB (2002) Viability of a five-strain mixture of *Listeria monocytogenes* in vacuum-sealed packages of frankfurters, commercially prepared with and without 2.0 or 3.0% added potassium lactate, during extended storage at 4 and 10 °C. J Food Prot 65:308–315

Puljula E, Walton G, Woodward MJ, Karonen M (2020) Antimicrobial activities of ellagitannins against *Clostridiales perfringens, Escherichia coli, Lactobacillus plantarum* and *Staphylococcus aureus*. Molecules 25:3714

Qi S, Huang H, Huang J, Wang Q, Wei Q (2015) Lychee (*Litchi chinensis Sonn.*) seed water extract as potential antioxidant and anti-obese natural additive in meat products. Food Control 50:195–201

Qin YY, Yang JY, Lu HB, Wang SS, Yang J, Yang XC, Chai M, Li L, Cao JX (2013) Effect of chitosan film incorporated with tea polyphenol on quality and shelf life of pork meat patties. Int J Biol Macromol 61:312–316

Quintieri L, Fancello F, Caputo L, Sorrentino A, Zara S, Lippolis V, Cervellieri S, Fanelli F, Corvino A, Pace B et al (2022) Effect of gaseous citral on table grapes contaminated by Rhizopus oryzae ITEM 18876. Food Secur 11:2478

Radünz M, dos Santos Hackbart HC, Camargo TM, Nunes CFP, de Barros FAP, Dal Magro J, Filho PJS, Gandra EA, Radünz AL, da Rosa ZE (2020) Antimicrobial potential of spray drying encapsulated thyme (Thymus vulgaris) essential oil on the conservation of hamburger-like meat products. Int J Food Microbiol 330:108696

Renerre M (2000) Chapter 5: Oxidative processes and myoglobin. In: Antioxidants in muscle foods. Nutritional strategies to improve quality. Wiley, New York

Reyes-Jurado F, Cervantes-Rincón T, Bach H, López-Malo A, Palou E (2019) Antimicrobial activity of Mexican oregano (Lippia berlandieri), thyme (Thymus vulgaris), and mustard (*Brassica nigra*) essential oils in gaseous phase. Ind Crop Prod 131:90–95
Ricke SC (2003) Perspectives on the use of organic acids and short chain fatty acids as antimicrobials. Poult Sci 82:632–639
Rižnar K, Celan Š, Knez Ž, Škerget M, Bauman D, Glaser R (2006) Antioxidant and antimicrobial activity of rosemary extract in chicken frankfurters. J Food Sci 71:425–429
Rodrigues AS, Kubota EH, da Silva CG, Dos Santos AJ, Hautrive TP, Rodrigues GS, Campagnol PCB (2020) Banana inflorescences: a cheap raw material with great potential to be used as a natural antioxidant in meat products. Meat Sci 161:107991
Rodríguez-Carpena JG, Morcuende D, Estévez M (2011) Avocado by-products as inhibitors of colour deterioration and lipid and protein oxidation in raw porcine patties subjected to chilled storage. Meat Sci 89(2):166–173
Rounds L, Havens CM, Feinstein Y, Friedman M, Ravishankar S (2013) Concentration-dependent inhibition of Escherichia coli O157: H7 and heterocyclic amines in heated ground beef patties by apple and olive extracts, onion powder and clove bud oil. Meat Sci 94:461–467
Ruiz A, Williams SK, Djeri N, Hinton A, Rodrick GE (2010) Nisin affects the growth of *Listeria monocytogenes* on ready-to-eat Turkey ham stored at four degrees Celsius for sixty-three days. Poult Sci 89:353–358
Sakkas H, Economou V, Gousia P, Bozidis P, Sakkas VA, Petsios S, Mpekoulis G, Ilia A, Papadopoulou C (2018) Antibacterial efficacy of commercially available essential oils tested against drug-resistant gram-positive pathogens. Appl Sci 8:2201
Sallam KI, Samejima K (2004) Microbiological and chemical quality of ground beef treated with sodium lactate and sodium chloride during refrigerated storage. Leb Technol 37:865
Samelis J, Bedie GK, Sofos JN, Belk KE, Scanga JA, Smith GC (2002) Control of *Listeria monocytogenes* with combined antimicrobials after post process contamination and extended storage of frankfurters at 4C in vacuum package. J Food Prot 65:299–307
Santiesteban-López NA, Gómez-Salazar JA, Santos EM, Campagnol PCB, Teixeira A, Lorenzo JM, Sosa-Morales ME, Domínguez R (2022) Natural antimicrobials: a clean label strategy to improve the shelf life and safety of reformulated meat products. Food Secur 11:2613
Selani MM, Contreras-Castillo CJ, Shirahigue LD, Gallo CR, Plata-Oviedo M, Montes-Villanueva ND (2011) Wine industry residues extracts as natural antioxidants in raw and cooked chicken meat during frozen storage. Meat Sci 88:397–403
Selene Márquez-Rodríguez A, Nevárez-Baca S, Lerma-Hernández JC, Hernández-Ochoa LR, Nevárez-Moorillon GV, Gutiérrez-Méndez N, Muñoz-Castellanos LN, Salas E (2020) In vitro antibacterial activity of *Hibiscus sabdariffa* L. Phenolic extract and its in situ application on shelf-life of beef meat. Foods 9:1080
Seman DL, Borger AC, Meyer JD, Hall PA, Milkowski AL (2002) Modeling the growth of *Listeria monocytogenes* in cured ready-to-eat processed meat products by manipulation of sodium chloride, sodium diacetate, potassium lactate, and product moisture content. J Food Prot 65: 651–658
Shah MA, Bosco SJD, Mir SA (2014) Plant extracts as natural antioxidants in meat and meat products. Meat Sci 98:21–33
Shahidi F, Ambigaipalan P (2015) Phenolics and polyphenolics in foods, beverages and spices: antioxidant activity and health effects – a review. J Funct Foods 18:820–897
Sharma H, Mendiratta SK, Agrawal RK, Gurunathan K, Kumar S, Singh TP (2017) Use of various essential oils as bio preservatives and their effect on the quality of vacuum packaged fresh chicken sausages under frozen conditions, vol 81. Elsevier Ltd, Amsterdam:9189305159
Shi YG, Zhu YJ, Shao SY, Zhang RR, Wu Y, Zhu CM, Liang XR, Cai WQ (2018) Alkyl ferulate esters as multifunctional food additives: antibacterial activity and mode of action against Escherichia coli in vitro. J Agric Food Chem 66:12088–12101

Shi YG, Zhang RR, Zhu CM, Liang XR, Ettelaie R, Jiang L, Lin S (2021) On the mechanism behind enhanced antibacterial activity of alkyl gallate esters against foodborne pathogens and its application in Chinese icefish preservation. Food Microbiol 99:103817

Siddiqua S, Anusha BA, Ashwini LS, Negi PS (2015) Antibacterial activity of cinnamaldehyde and clove oil: effect on selected foodborne pathogens in model food systems and watermelon juice. J Food Sci Technol 52:5834–5841

Šojić B, Pavlić B, Zeković Z, Tomović V, Ikonić P, Kocić-Tanackov S, Džinić N (2018) The effect of essential oil and extract from sage (*Salvia offcinalis* L.) herbal dust (food industry by-product) on the oxidative and microbiological stability of fresh pork sausages. LWT Food Sci Technol 89:749–755

Šojic B, Pavlic B, Ikonic P, Tomovic V, Ikonic B, Zekovic Z, Kocic-Tanackov S, Jokanovic M, Škaljac S, Ivic M (2019) Coriander essential oil as natural food additive improves quality and safety of cooked pork sausages with different nitrite levels. Meat Sci 157:107879

Song HJ, Ku KM (2021) Optimization of allyl isothiocyanate sanitizing concentration for inactivation of *Salmonella* typhimurium on lettuce based on its phenotypic and metabolome changes. Food Chem 364:130438

Sottero B, Leonarduzzi G, Testa G, Gargiulo S, Poli G, Biasi F (2019) Lipid oxidation derived aldehydes and oxysterols between health and disease. Eur J Lipid Sci Technol 121:1–16

Suganthi V, Mohanasrinivasan V (2015) Optimization studies for enhanced bacteriocin production by Pediococcus pentosaceus KC692718 using response surface methodology. J Food Sci Technol 52:3773–3783

Sun Z, Wang X, Zhang X, Wu H, Zou Y, Li P, Sun C, Xu W, Liu F, Wang D (2018) Class III bacteriocin helveticin-M causes sublethal damage on target cells through impairment of cell wall and membrane. J Ind Microbiol Biotechnol 45:213–227. https://doi.org/10.1007/s10295-018-2008-6

Sun Q, Li J, Sun Y, Chen Q, Zhang L, Le T (2020) The antifungal effects of cinnamaldehyde against aspergillus Niger and its application in bread preservation. Food Chem 317:126405

Szabo EA, Cahill ME (1999) Nisin and ALTA(TM) 2341 inhibit the growth of *Listeria monocytogenes* on smoked salmon packaged under vacuum or 100% CO_2. Lett Appl Microbiol 28:373–377

Tang X, Shao YL, Tang YJ, Zhou WW (2018) Antifungal activity of essential oil compounds (geraniol and citral) and inhibitory mechanisms on grain pathogens (*Aspergillus flavus* and *Aspergillus ochraceus*). Molecules 23:2108

Tang Y, Li F, Gu D, Wang W, Huang J, Jiao X (2021) Antimicrobial effect and the mechanism of diallyl trisulfide against *Campylobacter jejuni*. Antibiotics 10:246

Thompson JK, Collins MA, Mercer WD (1996) Characterization of a proteinaceous antimicrobial produced by Lactobacillus helveticus CNRZ450. J Appl Bacteriol 80:338–348

Thompson RL, Carpenter CE, Martini S, Broadbent JR (2008) Control of *Listeria monocytogenes* in ready-to-eat meats containing sodium levulinate, sodium lactate, or a combination of sodium lactate and sodium diacetate. J Food Sci 73:M239–M244

Trinetta V, Morleo A, Sessa F, Iametti S, Bonomi F, Ferranti P (2012) Purified sakacin a shows a dual mechanism of action against Listeria spp: proton motive force dissipation and cell wall breakdown, FEMS Microbiol. Lett 334:143–149

Tůmova E, Chodová D, Volek Z, Ketta M (2021) The effect of feed restriction, sex and age on the carcass composition and meat quality of nutrias (*Myocastor coypus*). Meat Sci 182:108625

Uzelac G, Miljkovic M, Lozo J, Radulovic Z, Tosic N, Kojic M (2015) Expression of bacteriocin LsbB is dependent on a transcription terminator. Microbiol Res 179:45–53

Valdivieso-Ugarte M, Plaza-Diaz J, Gomez-Llorente C, Gómez EL, Sabés-Alsina M, Gil Á (2021) In vitro examination of antibacterial and immunomodulatory activities of cinnamon, white thyme, and clove essential oils. J Funct Foods 81:104436

Vasavada MN, Dwivedi S, Cornforth D (2006) Evaluation of garam masala spices and phosphates as antioxidants in cooked ground beef. J Food Sci 71:292–297

Vestergaard M, Ingmer H (2019) Antibacterial and antifungal properties of resveratrol. Int J Antimicrob Agents 53:716–723
Walsh CJ, Guinane CM, Hill C, Ross RP, O'Toole PW, Cotter PD (2015) In silico identification of bacteriocin gene clusters in the gastrointestinal tract, based on the human microbiome Project's reference genome database. BMC Microbiol 15:183
Wang Y, Li F, Zhuang H, Li L, Chen X, Zhang J (2015) Effects of plant polyphenols and α-tocopherol on lipid oxidation, microbiological characteristics, and biogenic amines formation in dry-cured bacons. J Food Sci 80:547–555
Wang G, Pang T, Xia Y, Liu X, Li S, Parvez AM, Kong F, Si C (2019) Subdivision of bamboo kraft lignin by one-step ethanol fractionation to enhance its water-solubility and antibacterial performance. Int J Biol Macromol 133:156–164
Wang R, Zhang Y, Jia Y, Zhang M, Huang Y, Li C, Li K (2020) Persimmon oligomeric proanthocyanidins exert antibacterial activity through damaging the cell membrane and disrupting the energy metabolism of Staphylococcus aureus. ACS Food Sci Technol 1:35–44
Wang W, Xiong P, Zhang H, Zhu Q, Liao C, Jiang G (2021) Analysis, occurrence, toxicity and environmental health risks of synthetic phenolic antioxidants: a review. Environ Res 201: 111531
Wink M (2022) Current understanding of modes of action of multicomponent bioactive phytochemicals: potential for nutraceuticals and antimicrobials. Annu Rev Food Sci Technol 13:337–359
Wojtasik-Kalinowska I, Onopiuk A, Szpicer A, Wierzbicka A, Półtorak A (2021) Frozen storage quality and flavour evaluation of ready to eat steamed meat products treated with antioxidants. CyTA – J Food 19:152–162
Wu HY, Xu YH, Wei LN, Bi JR, Hou HM, Hao HS, Zhang GL (2022a) Inhibitory effects of 3-(methylthio) propyl isothiocyanate in comparison with benzyl isothiocyanate on *Listeria monocytogenes*. J Food Meas Charact 16:1768–1775
Wu TL, Hu YM, Sun Y, Zhang ZJ, Wu ZR, Zhao WB, Tang C, Du SS, He YH, Ma Y et al (2022b) Insights into the mode of action of 2-(4-methoxyphenyl) ethyl isothiocyanate on *Aspergillus Niger*. Food Control 136:108871
Xiong YL, Guo A (2021) Animal and plant protein oxidation: chemical and functional property significance. Food Secur 10:40. https://doi.org/10.3390/foods10010040
Xiong LG, Chen YJ, Tong JW, Huang JA, Li J, Gong YS, Liu ZH (2017) Tea polyphenol epigallocatechin gallate inhibits Escherichia coli by increasing endogenous oxidative stress. Food Chem 217:196–204
Xu L, Tao N, Yang W, Jing G (2018) Cinnamaldehyde damaged the cell membrane of Alternaria alternata and induced the degradation of mycotoxins in vivo. Ind Crop Prod 112:427–433
Xu JG, Hu HX, Chen JY, Xue YS, Kodirkhonov B, Han BZ (2022) Comparative study on inhibitory effects of ferulic acid and p-coumaric acid on Salmonella Enteritidis biofilm formation. World J Microbiol Biotechnol 38:136
Yang CX, Wu HT, Li XX, Wu HY, Niu TX, Wang XN, Lian R, Zhang GL, Hou HM (2020) Comparison of the inhibitory potential of benzyl isothiocyanate and phenethyl isothiocyanate on Shiga toxin-producing and enterotoxigenic Escherichia coli. LWT Food Sci Technol 118: 108806
Yang B, Li L, Geng H, Zhang C, Wang G, Yang S, Gao S, Zhao Y, Xing F (2021) Inhibitory effect of allyl and benzyl isothiocyanates on ochratoxin a producing fungi in grape and maize. Food Microbiol 100:103865
Yin MC, Faustman C (1994) The influence of microsomal and cytosolic components on the oxidation of myoglobin and lipid in vitro. Food Chem 51:159–164
Yldz-Turp G, Serdaroglu M (2010) Effects of using plum puree on some properties of low fat beef patties. Meat Sci 86:896–900
Yoplac I, Vargas L, Robert P, Hidalgo A (2021) Characterization and antimicrobial activity of microencapsulated citral with dextrin by spray drying. Heliyon 7:e06737

You S, Xie Y, Zhuang X, Chen H, Qin Y, Cao J, Lan T (2022) Effect of high antioxidant activity on bacteriostasis of lignin from sugarcane bagasse. Biochem Eng J 180:108335

Yuan S, Li W, Li Q, Wang L, Cao J, Jiang W (2019) Defense responses, induced by p-coumaric acid and methyl p-coumarate, of jujube (*Ziziphus jujuba* Mill.) fruit against black spot rot caused by *Alternaria alternata*. J Agric Food Chem 67:2801–2810

Zeb A (2021) Phenolic antioxidants in foods: chemistry, biochemistry and analysis, 1st edn. Springer Cham. https://doi.org/10.1007/978-3-030-74768-8

Zengin H, Baysal AH (2015) Antioxidant and antimicrobial activities of thyme and clove essential oils and application in minced beef. J Food Process Preserv 39:1261–1271

Zhang J, Liu G, Li P, Qu Y (2010) Pentocin 31-1, a novel meat-borne bacteriocin and its application as biopreservative in chill-stored tray-packaged pork meat. Food Control 21:198–202

Zhang L, Lin YH, Leng XJ, Huang M, Zhou GH (2013) Effect of sage (*Salvia offcinalis*) on the oxidative stability of Chinese-style sausage during refrigerated storage. Meat Sci 95:145–150

Zhang H, Wu J, Guo X (2016a) Effects of antimicrobial and antioxidant activities of spice extracts on raw chicken meat quality. Food Sci Human Wellness 5:39–48

Zhang J, Wang Y, Pan DD, Cao JX, Shao XF, Chen YJ, Sun YY, Ou CR (2016b) Effect of black pepper essential oil on the quality of fresh pork during storage. Meat Sci 117:130–136

Zhang Y, Han I, Bridges WC, Dawson PL (2016c) Peach skin powder inhibits oxidation in cooked Turkey meat. Poult Sci 95:2435–2440

Zhang M, Xu L, Zhang L, Guo Y, Qi X, He L (2018) Effects of quercetin on postharvest blue mold control in kiwifruit. Sci Hortic 228:18–25

Zhang M, Li Y, Bi Y, Wang T, Dong Y, Yang Q, Zhang T (2020) 2-Phenylethyl isothiocyanate exerts antifungal activity against Alternaria alternata by affecting membrane integrity and mycotoxin production. Toxins 12:124

Zheng S, Jing G, Wang X, Ouyang Q, Jia L, Tao N (2015) Citral exerts its antifungal activity against Penicillium digitatum by affecting the mitochondrial morphology and function. Food Chem 178:76–81

Chapter 13
Application of Enzymes in Processed Meat Products

Humaira Fayaz, Sheikh Rafeh Ahmad, Asma Irshad Qureshi, Syed Arshad Hussain, and Tahir Nazir

13.1 Introduction

The growing consumer demand for healthier products has stimulated the development of nutritionally enhanced meat foods. In order to achieve these nutritionally enhanced meat foods, changes such as the use of improved raw materials, reformulation of products and technological processes are necessary (Trespalacios and Pla 2007). These improvements, however, can bring undesirable consequences to the product, such as texture alterations in low-salt and low-phosphate meat foods (Zhu and Tramper 2008). Additionally, high costs have stimulated researchers to use all animal parts, including muscles of poorer technological quality, to maximize the yield of marketable products. This has required the development of methods to restructure low-valued cuts and trimmings, improving appearance and texture and increasing market value (Motoki and Seguro 1998). The interest of the food industry concerning enzymes stems from the constant search for foods with a long shelf life, in addition to the quest to reduce waste and increase the quality of products, transforming raw materials into the main product or acting as additives to modify desired properties such as flavor, texture, or machinability, among others (Singh et al. 2016; Yushkova et al. 2019; Bilal and Iqbal 2020). Thus, research has been carried out to explore enzymes which contribute to the quality and stability of food products. Moreover, several studies have explored the inactivation of enzymes that can accelerate undesirable processes, such as food degradation (Ray and Rosell 2017; Adesegun Kehinde et al. 2020). Enzymes are defined as organic molecules of protein origin, designed to catalyze biochemical reactions and do not effectively participate in the reactions as a reagent (Fennema et al. 2010). They are basically

H. Fayaz (✉) · S. R. Ahmad (✉) · A. I. Qureshi · S. A. Hussain · T. Nazir
Division of Livestock Products Technology, FVSc & AH, SKUAST-K, Srinagar, Jammu and Kashmir, India
e-mail: sh.rafeh@skuastkashmir.ac.in

S. A. Rather, F. A. Masoodi (eds.), *Hand Book of Processed Functional Meat Products*, https://doi.org/10.1007/978-3-031-69868-2_13

composed of two parts: the protein part called apoenzyme and the non-protein part known as the cofactor (maybe a metallic ion) or coenzyme (organic molecule), which together give rise to the complete functional enzyme called the holoenzyme (Ahmad and Sardar 2015). The enzymatic catalysis process results from its specificity, which is considered one of the main characteristics of the enzymes. In the absence of enzymes, some chemical reactions would probably not be possible. However, its presence alone is insufficient to catalyze reactions since reactions have variable rates (Ordoñez et al. 2005; Lajolo and Mercadante 2017). The application of enzymes covers many industries as food industries which basically use three sources: microorganisms, plants or from animal tissues (Ray et al. 2016; Ahmad et al. 2018), with enzymes mainly being used in dairy products, baking and beverages such as fruit juices, wine and beer (Guerrand 2018; Taheri-Kafrani et al. 2021).

13.2 Enzymes in Meat Processing

Tenderness of meat is considered as the most important quality distinguishing feature of meat in consumer evaluation (Koohmaraie and Geesink 2006; Destefanis et al. 2008; Zork et al. 2009). Tenderness in meat results from a combination of breakdown reactions within muscle fibres, primarily because of the activity of enzymes and loosening of connective tissue, particularly, collagen. Various pre-slaughter and post-slaughter factors and their mutual effect influence tenderness of meat (Destefanis et al. 2008). In meat industry and catering predominantly protein-degrading enzymes have been used. Of the protein cross-linking enzymes, transglutaminases (TGase) have been used as texture improvers for several years. Structure engineering by oxidative enzymes and flavour design by lipases, glutaminases, proteases and peptidases are examples of emerging enzyme technologies in the food sector (Whitehurst and van Oort 2010). One of the potential areas in meat processing is meat tenderizing using enzymes such as papain and bromelain derived from plant sources such as papaya and pine apple plant, respectively. Toughness of meat is generally due to the presence of collagen, elastin and actomyocin. Meat cuts that are considered as lower quality due to toughness are as nutritious as prime quality meat, which can be tenderized using enzymes to convert it to prime quality meat. Even the flavour of the meat depends on the peptides and the amino acids present in meat.

13.3 Conversion of Muscle into Meat

Meat is the result of a series of transformations that the muscle tissue of the animal undergoes after slaughtering. The conversion of muscle into meat entails structural transformations and biochemical reactions, which will produce changes affecting the technological and sensory quality of the meat. The process of converting muscle into meat comprises three stages: pre rigor mortis, rigor mortis and post rigor mortis (Braden 2013). The first stage (3–6 h) occurs immediately after the slaughter of the animal due to the interruption of blood circulation caused by bleeding. This process causes the arrival of oxygen and nutrients to be abruptly interrupted (Ouali et al. 2006). In the second stage (until 24 h), the depletion of energy components takes place, that is, adenosine triphosphate (ATP), phosphocreatinine and glucose. Finally, in the post rigormortis, endogenous proteolytic systems lead to the disintegration of the muscle structure of the myofibrils, which induces meat tenderization. The duration of the conversion of muscle into meat depends on three aspects: the animal species, the glycogen reserves at the time the animal is slaughtered and the storage temperature. Thus, the three stages of rigor mortis take at least 14 days in cattle, 7 to 10 days in sheep, 5 to 7 days in pigs and approximately 6 h in poultry (Nowak 2011). The three proteolytic systems involved in the three stages are calpains, cathepsins and ubiquitin–proteasomes (Salmerón et al. 2015).

In pre rigormortis, the meat must be kept at a temperature above 10 °C until reaching the rigor mortis phase (10–10 rule, temperature within meat on a carcass should not be below 10 °C within 10 h after slaughter). Electrical stimulation immediately after the sacrifice of the animal causes a muscular contraction that accelerates the consumption of glycogen, the drop in pH and the establishment of rigor mortis. In this way, it is possible to prevent meat toughening and the apparition of the "cold shortening" problem during rigor mortis (Savell et al. 2005). In pre rigor mortis, the metabolism of the animal muscle changes from aerobic to anaerobic and, therefore, undergoes a gradual decrease in energy intake. Muscle needs glycogen and phosphocreatine to synthesise ATP from glucose. Under these circumstances, the enzymes that lead the muscle metabolism begin to act, that is, those responsible for glycolysis. Among the enzymes that participate in anaerobic glycolysis, glucose 6-phosphate and phosphocreatine kinase are particularly relevant. These enzymes act until the glycogen and phosphocreatine reserves are depleted, after which ATP is reduced to form, first, adenosine diphosphate (ADP) and, subsequently, adenosine monophosphate (AMP), which can be deaminated by the enzymes responsible for the degradation of ATP. On the other hand, after the progressive reduction in ATP levels, inorganic phosphate is generated, which stimulates the degradation of glucose to pyruvate, and, subsequently, lactic acid is generated from the enzyme lactate dehydrogenase. Lactic acid causes the muscle pH to drop and the enzymes responsible for anaerobic metabolism (glycolysis) to be inactivated (Chauhan and England 2018). The decrease in muscle pH is one of the most significant post mortem changes, leading to the start of the rigor mortis. After the depletion of ATP, there is a depolarization of the membranes due to an ionic increase linked to the Ca^{2+}, Na^{+} and K^{+} pumps' standstill, which is dependent on the ATP content. That is why the Ca^{2+} ions react with troponin, which modifies the configuration of the active sites of actin; subsequently, myosin binds to actin, giving rise to the irreversible formation of actomyosin, which causes a reduction in the water retention capacity (WRC) and, therefore, a hardening of the muscle. The rigor mortis ends with the formation of actomyosin, which is characterised by muscle tension and stiffness.

In the post rigor mortis, the enzymatic reactions responsible for the tenderization of meat take place. First, the proteolytic system of calpain plays a central role in post mortem proteolysis and softening (Koohmaraie 1996). The calpain system is dependent on Ca^{2+} and its endogenous inhibitors, calpastatins and has been described as the main factor responsible for proteolysis in the early post mortem period (0–24 h) and meat tenderization since the calpain system acts at neutral pH and its activity declines when the pH drops. Caballero et al. (2007) postulated that the synergistic action of calpains with cathepsins leads to meat softening. Cathepsins, lysosomal enzymes, are activated at a lower pH than calpains; therefore, they become more important in the later phases of post mortem along with their endogenous inhibitors, cystatins. Finally, the proteasome is responsible for the degradation of most intracellular proteins, with the ubiquitin proteasome complex responsible for the intracellular turnover of damaged proteins (Ouali et al. 2013). However, to date, the role of the proteasome in tenderization has not been fully clarified, although it is known that its activation is one of the first cellular responses to oxidative stress.

The post-slaughter evolution of pH has a great effect on the technological properties of the meat, affecting the texture (tenderization), colour and aroma due to the generation of the volatile compounds resulting from the proteolytic and lipid degradation of the meat (Matarneh et al. 2017). In addition, the role played by the colour of the meat is important as it is indicative of the meat's freshness and is, therefore, a key factor for consumer acceptance. The colour depends on the concentration and degree of oxidation of the heme compounds, mainly myoglobin (Mb). Mb is a globular protein, which can be found in four chemical forms: deoxymyoglobin (DeoxyMb), carboxymyoglobin (CarboxyMb), metmyoglobin (MetMb) and oxymyoglobin (Mancini and Hunt 2005). Mb is a water-soluble protein, and, among the amino acid residues that it contains, histidine has received the most attention due to its key role in the structure and function of Mb. In addition, there are other heme proteins, such as haemoglobin and cytochrome C, that may also play a role in the colour of beef, lamb, pork and poultry. However, the mechanisms that control colour stability have not been completely elucidated (Suman and Nair 2017). The bright red colour of fresh meat depends on a triple balance of biochemical factors: the respiratory activities (O2 uptake rate), the auto-oxidation of Mb and the enzymatic reduction of MetMb, which in turn can be affected by time, temperature and muscle pH history (Ledward 1985). These enzymatic processes affecting texture, colour and aroma occur in all types of meat but are particularly important in the aging of beef cuts (Dashdorj et al. 2016). Finally, the treatments carried out, both before and after the slaughter of an animal, determine the final quality of the meat and can trigger two types of meat: dark, firm and dry (DFD) and pale, soft and exudative (PSE). The factors determining this type of meat are those related to the muscle glycogen content, which affects the pH of the meat and the temperature to which the meat is subjected after slaughtering the animal (Zhang and Barbut 2005).

13.4 Enzymatic Generation of Flavour in Meat Products

Flavour is defined as multimodal phenomenon in which several independent sensory modalities such as taste, aroma, trigeminal sensation and texture are involved. Flavour has a major role, along with tenderness, in acceptability of meat by consumers. Because of the relationship of flavour and consumer acceptability, it is important to understand the factors influencing meat flavour in order to produce good-quality meat products. The flavour of raw meat is quite bland. However, it contains non-volatile constituents that are essential flavour precursors which during processing and storage affect the taste of the meat product (Lawrie 1998). The flavour of processed meat is a result of either enzymatic action or chemical reactions such as pyrolysis of amino acids and peptides, sugar degradation, degradation of ribonucleotides, maillard reactions, thiamine degradation and degradation of lipids. The main enzymatic reactions affecting meat flavour or formation of flavour precursors are proteolysis and lipolysis. Both groups of reactions are due to the contribution of endogenous proteases and lipases, enzymes of microbial origin naturally present in the product or enzymes added during the manufacturing process.

Proteolysis and lipolysis play a role in meat flavour development such as in dry-cured meat products which are appreciated for their unique flavour. The compounds implicated in flavour generation arise from many sources, such as spices, sugar metabolism, lipolysis and lipid oxidation, proteolysis and amino acid degradation. Lipids in muscles and adipose tissues, changes during processing and affects the sensory properties of meat products. During ripening proteolysis takes place, yielding for example polypeptides, peptides and free amino acids, which are involved in taste and flavour development of meat products. Meat protein hydrolysis is mainly catalyzed by endogenous enzymes, such as cathepsin and trypsin-like peptidases as well as proteases produced by microorganisms are involved in the ripening process. These enzymes are mainly originating from *Micrococcaceae* but also from moulds and yeasts in those dry sausages in which they are present (Fernandez et al. 2000). Glutaminase addition to protease has an important role from the point of view of sausage production, especially regarding the deamidation of glutamine, since hydrolysis of the glutamine amide group produces ammonia, an acidity neutralizer and umami flavour. Umami can be described as savoury or broth-like taste with ability to enhance other flavours. Lipolysis constitutes another important group of enzymatic reactions which are related to aroma formation of fermented sausages. Phospholipases and lipases hydrolyze phospholipids and triacylglycerols forming free fatty acids. Unsaturated fatty acids are further oxidized to aroma volatile compounds. This oxidation may lead to the formation of aliphatic hydrocarbons, alcohols, aldehydes and ketones. Further alcohols react with free fatty acids forming some esters (Toldrá 1998).

13.5 Structure Engineering by Cross-Linking Enzymes-Restructured Meat Products

Apart from the hydrolytic enzymes affecting the tenderness of meat or generation of flavour, the functional properties of meat proteins can be modified by cross-linking enzymes. These enzymes are used to bind fresh meat pieces together and to tailor the structural properties of various processed meat products. The main target protein in meat for cross-linking enzymes is the myofibrillar protein myosin. Cross-linking enzymes are generally capable of positively affecting gelation and consequently the texture of meat gels. Potent TGase has been the main cross-linking enzyme studied and applied industrially in meat protein modification. The ability of TGase to cross-link meat proteins has been generally known for over two decades. Although the most obvious source of TGase for the meat industry would be blood, microbial enzymes have superseded the mammalian ones due to their Ca^{2+}-independency, favourable pH and temperature profiles and commercial availability for industrial use.

Restructured Meat Products

Meat industries are currently processing innovative products, such as restructured meat. By means of restructured meat, an attempt is undertaken to mimic the appearance of meat muscle. Thereby, restructured meat is considered an intermediate product between minced meat and a piece of meat with anatomical integrity. Its production begins with meat pieces of different sizes to achieve a consistent product by joining these pieces (Trindade et al. 2017). Using this process, it is possible to produce products of wider consumer acceptance from portions of meat of low commercial value, with poor texture and difficult commercialization. The restructuration technology is applied to all types of meat. Different blinders are used in the manufacturing of restructured meat products. One of the most important ingredients is the enzyme transglutaminase (Anzani et al. 2020).

Transglutaminase is an enzyme typically obtained from the microorganism *Streptoverticilium mobaraense* despite being naturally found in most tissues of living organisms. This enzyme has a range of action between 0 °C and 60 °C, the optimal being at 50 °C and pH 7. In addition, unlike endopeptidase enzymes such as calpain, it is independent of Ca^{2+} and has the ability to improve the functional characteristics of protein, including water retention and water solubility and the functional properties, such as food taste and toughness (Yang and Zhang 2019; Santhi et al. 2017). Transglutaminase is a transferase enzyme that is characterised by catalysing cross-breeding reactions between the residual γ-carboxyamide residual groups of glutamine and the residual ε-amino groups of lysine, resulting in a molecular cross of proteins by forming ε–γ glutamyl lysine bonds (Tarté 2009). The inter- and intramolecular bonds they form are highly resistant to proteolysis, which is why transglutaminase becomes an important ingredient that enhances the physical and functional properties of meat products (Kaufmann et al. 2012). Transglutaminase has been tested on a large number of meat products with different objectives, one of which is to develop new meat products or enriched meat products. There are two main types of transglutaminases: (1) calcium dependent, which are found in most animal tissues and bodily fluids (Aeschlimann and Paulsson 1994; Chen and Mehta 1999), and (2) calcium independent, which are produced through fermentation by some microorganisms such as *Streptoverticillium cinnamoneum* and *Streptoverticillium mobaraense*. Extracellular transglutaminase has also been reported by using *Corynebacterium glutamicum* (Yokoyama et al. 2004).

Transglutaminase forms cross-linked insoluble protein polymers. Based on this relevant property, glutaminase is used with surimi or other protein-rich foods to bind proteins together to give restructured fish or meat pieces a given form or shape resembling crab or steak. It is also used to join parts of deboned fish or hams, which are reconstituted to reproduce the intact original piece (Kuraishi et al. 1997; Carballo et al. 2006). It may take several hours at temperatures below 5 °C for the enzyme to cross-link the pieces of meat or fish. Transglutaminase may be used to bind a variety of protein-rich foods such as red meat, poultry and seafood pieces as well as casein, soybean globulins and egg proteins (Motoki and Seguro 1998; Yokoyama et al. 2004). Other applications include the enhanced retention of capsaicin in capsaicin-enriched layered noodles (Li et al. 2013a, b). Transglutaminase is also important for the gelation process and cross-linking of actomyosin during surimi manufacture (Nielsen and Nielsen 2012). Such bonds are quite resistant to proteolysis degradation and might affect its bio-availability. The dipeptide glutamine-lysine remains intact after ingestion and is absorbed through the intestinal brush-border where two enzymes, g-glutamylamine cyclotransferase and g-glutamyltransferase, are able to cleave such bonds (Yokoyama et al. 2004).

Baugreet et al. (2018) optimised a restructured beef fillet enriched with vegetable protein (lentil and rice) through the action of transglutaminase, from which meat with a high protein value (28 g protein/100 g meat) was obtained. Ahhmed et al. (2007) investigated the improvement in the physical properties of meat products, such as the texture of chicken and beef sausages. In addition, there are several studies that show how to achieve a good synergy between meat pieces using transglutaminase, without the need for the use of salt or phosphates, for the purposes of manufacturing cooked ham (Fulladosa et al. 2009) and sausages (Colmenero et al. 2005). This enzyme has also been used in restructured unsalted and low-fat kebab meat (Askin and Kilic 2009) improving its functional properties.

Thrombin is a serine peptidase (EC 3.4.21.5.) that is found in blood and converts soluble fibrinogen into insoluble strands of fibrin, as well as catalysing many other coagulation-related reactions. Thrombin has a highly specific proteolytic activity as this enzyme only cleaves certain arginine-glycine peptide bonds found in fibrinogen, which is its most abundant natural substrate. Optimal activity is at a pH of 7–8 in the presence of at least 0.1 M NaCl (Othrner and Kosow 1980; Barrett et al. 2004). The source of thrombin and fibrinogen is cattle and pig blood plasma obtained from slaughterhouses that are controlled and approved by veterinary inspection or from fish tissues (Manseth et al. 2003). Thrombin is then obtained from prothrombin, which is activated by thromboplastin in the presence of calcium ions (Manseth et al. 2003). Thrombin is used to convert fibrinogen into a fibrin gel that interacts with collagen and binds together pieces of meat or fish. The main applications are animal protein foods like meat, poultry and seafood to create restructured pieces. The binding strength of the reconstituted piece of meat or fish depends on the pH and temperature, as well as the moisture, sizes of the cuts and direction of the fibres. Binding is strongest at a pH near 7, with larger particle sizes and if the fibrin gel runs parallel to the available collagen in meat (Chen and Lin 2002).

13.6 Role of Enzymes on Ripening of Processed Meat Products

Aging of meat, also known as "ripening" or "conditioning", is a natural process and is useful in improving its palatability and overall flavor. On a commercial basis, aging is generally performed by storing the meat under controlled refrigerated conditions to improve several traits of beef that affect consumer satisfaction, such as its tenderness and flavor (McGee 2004; Piao et al. 2015). Meat aging is a combination of transformations that originate in the animal's muscle after slaughter, resulting in changes in colour, tenderness and aroma (Lonergan et al. 2018). The biochemical processes that occur during meat aging are mainly caused by endogenous enzymes, leading to glycolysis, proteolysis and lipolysis. In the glycolysis reactions, glucose is metabolised to produce lactic acid, which lowers muscle pH and depletes the energy reserves (ATP). The energy depletion leads to the degradation of myofibrillar proteins by the action of endopeptidases and the action of exopeptidases. Endogenous proteases (calpains, cathepsins and calpastatin) play a crucial role in the proteolysis of meat; however, exogenous proteases (peptidylpeptidases, aminopeptidases and carboxypeptidases) secreted from microorganisms involved in meat fermentation also contribute to increase the concentration of peptides and amino acids (Wang et al. 2022). Another reaction that takes place during the meat aging is the lipolysis in the muscle and the adipose tissue (Tatiyaborworntham et al. 2022).

The products resulting from the degradation of proteins and lipids are precursors of the characteristic flavour and aroma of the meat and meat products (Khan et al. 2016; Toldrá and Flores 1998; Fernandez et al. 2000). In addition, the fragmentation of the myofibrils also leads to changes in texture that lead to meat softening (Bekhit et al. 2014a, b). Therefore, post mortem changes in the muscle, which affect the organoleptic properties of meat, are mostly related to the action of enzymes. As the tenderness of the meat increases with an increase in the aging time, the time and temperature are two very important factors necessary to obtain a tender product (Mottram 1998; Nishimura et al. 1995). Lipolysis, proteolysis, and oxidation are some of the biochemical reactions that occur during postmortem aging. Proteolysis includes the breakdown of myofibrillar proteins that release peptides and amino acids, which act as water-soluble flavoring precursors (Spanier et al. 1997) responsible for the taste properties and flavoring characteristics of the meat (Koutsidis et al.

2008). These products react with the reducing sugars and contribute to the meat flavor (Imafidon and Spanier 1994). Further, the free fatty acids released due to the degradation of lipids result in the production of peroxides, which react with peptides to form the aroma compounds (Zhou and Zhao 2007). The quality of beef product depends on a number of pre and post-mortem factors. Pre-mortem factors that affect the quality of the final aged product include the breed, genetics, gender, energy level of the diet of the animal, and, most importantly, the type of muscle selected for the aging process (Monson et al. 2005). The tenderness of the aged meat varies with the type of muscles and amount of connective tissue present. This is because different muscles require variable aging periods to achieve appropriate tenderness according to the consumer's expectations (Bratcher et al. 2005; Monson et al. 2005). Aging improves the flavor, mouthfeel, and juiciness of beef (Irurueta et al. 2008). Aged beef possesses a peculiar salty or roasted flavor in comparison to the unaged beef (Bruce et al. 2005; Gorraiz et al. 2002). It has been reported that slaughtering results in an interruption of the circulatory system, thus, leading to accumulation of metabolic by-products in muscles and consequently decreasing the pH, which in turn causes favorable changes in the flavor of the beef (Yancey et al. 2005). The level of lipid oxidation increases as the aging time is increased, thus, resulting in the release of products that react with the protein degradation products and impart an intense flavor to the aged beef. The color of the meat surface and its stability are critical parameters governing the marketability of fresh beef. Furthermore, the internal color of the cooked meat is used as an indicator for the degree of doneness and point of consumption (Suman et al. 2014). Vitale et al. (2014) observed lower color stability in *longissimus thoracis* (LT) muscles of Friesian beef after 14–21 days of wet aging in comparison to 0–8 day-aged samples and attributed this difference to the myoglobin and lipid oxidation. The denaturation of proteins during aging results in a lighter color and reduced water activity in aged beef (Jayasooriya et al. 2007).

The two commercial types of aging for beef are dry aging and wet aging. In dry aging, the meat is placed in a refrigerated room without any packaging, while in wet aging, the meat is vacuum-packed and stored (Smith et al. 2008). The time for beef aging varied considerably with temperatures and an extended time has been reported to enhance the textural and flavor properties of the meat (Leroy et al. 2003). However, optimal time for dry aging is 14–21 days while for wet aging 7–10 days at specific temperature (0–1 °C). Dry-aged beef possesses a beefy and roasted flavor, which differs from the flavor of wet-aged beef, which is bloody and metallic in nature (Campbell et al. 2001; Warren and Kastner 1992). Various studies have been conducted on the palatability of aged beef, a number of which have shown that aging makes the meat tender (Parrish et al. 1991; Warren and Kastner 1992); however, a disagreement exists in these studies about the palatability aspect. Campbell et al. (2001) concluded that dry aging for a minimum of 14 days improves the palatability characteristics of meat compared to vacuum aging, and that the development of these attributes in the meat are sufficient to offset the increased expenses incurred due to dry aging. Dry-aged meat products have more characteristic flavor in comparison to wet-aged or un-aged products (Diles et al. 1994; Warren and Kastner 1992). However, no difference or decrease in the palatability attributes, other than tenderness, was observed (Savell et al. 1978). Hodges et al. (1974) observed an increase in the beef flavor intensity in the 15-days dry-aged USDA Choice short loin in comparison to the control, while USDA Standard short loin showed a lower beef flavor intensity compared to the control. A higher moisture loss during the dry aging process makes it more expensive and hence, makes the availability of dry-aged meat possible at specialty shops only rather than ordinary food stores (Stenstrom et al. 2014). A new bag technology that possesses a very high water vapor transmission rate has been introduced in dry aging in order to achieve beef with equal sensorial qualities compared to the dry aging process under less harder conditions. Ahnstrom et al. (2006) have conducted dry aging of beef for 14 and 21 d using the traditional dry aging and bag aging techniques and observed no differences in the flavor and meat physicochemical qualities. However, there was a significant decrease in the weight and trim loss for bag-aged sample. Xin et al. (2013) also produced dry aged beef using dry aging bag under controlled conditions without negative effects on sensory and quality attributes.

13.7 Physicochemical Changes During Postmortem Aging

The postmortem storage and processing conditions affect the flavor and tenderness of the meat (Perry 2012), and are important for improving the final quality of the product (Honkavaara et al. 2003). Animal slaughtering results in an interruption in the cell control systems and enzymes are produced that start attacking different cell molecules indiscriminately, thus, turning large flavorless molecules into smaller flavorsome fragments (McGee 2004). The proteins are broken down into savory amino acids and fats, into corresponding aromatic fatty acids that contribute to the characteristic nutty and meaty flavor of the dry-aged meat. Furthermore, these breakdown products react with other ingredients during cooking and enrich the aroma of the meat by producing new flavoring molecules. Aged beef possesses a strong, savory, and roasted odor compared to the un-aged beef, which has a weak and bland odor. Certain flavor notes like fatty, beefy, brothy, sweet, and caramelly are increased by aging the beef for up to 14 days, while some undesired flavors like painty, cardboard, bitter, and sour also get introduced during aging (Bruce et al. 2005; Gorraiz et al. 2002; Spanier and Miller 1993). The interruptions in the circulatory system upon slaughtering result in the accumulation of lactic acid and a drop in the pH, which facilitates these changes in the flavor. Presence of a lower pH (5.4) activates endogenous enzymes, like cathepsins B and L and thiol proteinases (which hydrolyze more peptide bonds in comparison to any other proteases), which are intracellularly redistributed during aging (Spanier et al. 1990, 1997). These enzyme activities are independent of the changes in the temperature, and cathepsins B and L retain their high activity even at cooking temperatures (70 °C). The production of flavoring compounds in the meat is influenced by the redistribution of enzymes and their activity due to the combined effects of postmortem aging and cooking.

The μ- and m-calpains responsible for postmortem textural changes increase the rancid, sour, and salty flavors of the meat by producing certain peptides. Bauer (1983) has stated in his study that glutamic acid content was observed to more than double during the 7 days of aging (9 mg/100 g on day 4 to 21 mg/100 g on day 7). Furthermore, the microbial breakdown of proteins also contributed to glutamic acid production. Production of carbonyls resulting from lipid oxidation was increased during aging and may contribute toward the development of considerably undesirable flavors. A longer aging period of 35 days may increase the metallic flavor of the meat, as aging for more than 21 days decreases the flavor identity (Yancey et al. 2005). Moreover, microbial degradation of lipids also has a significant role in increasing the concentration of flavor precursors during the aging process. The aging conditions influence the meat flavor; for instance, a high-oxygen environment results in the production of burnt and toasted off-flavors. A more ‘aged beef’ aroma is attributed to the presence of derivatives of 2-methyl-3-furanthiol (Rowe 2002). Dry aging results in a more intense beef flavor in comparison to vacuum or carbon dioxide packed aging (Jeremiah and Gibson 2003; Sitz et al. 2006). The inosine-5′-monophosphate (IMP) concentration is found to decrease during aging, with a simultaneous increase in inosine and hypoxanthine concentrations (Kato and Nishimura 1987). Further, higher concentrations of IMP and its degradation products, were found to be present at around 72 h of postmortem aging, thus, indicating that postmortem metabolism precedes by about 2 days post-slaughter. The proteolytic enzymes are responsible for the breakdown of muscle proteins into smaller fragments, resulting in rigor mortis during the process of conversion of muscles into meat (Miller et al. 1997).

13.8 Enzymes and Meat Tenderization

The tenderization process is the result of the proteolysis of muscle proteins by three large study systems, which are cathepsins, calpains or calcium-dependent peptidases and proteasomes after rigor mortis. On the other hand, meat aging is a process during which exogenous enzymes and microorganisms act upon the meat to break down the connective tissue, thereby tenderizing the meat and giving it a richer flavor (Koohmaraie and Geesink 2006). Tenderness has been rated by consumers as the most important organoleptic attribute of fresh meat. Regarding exogenous enzyme systems, they are mostly formed by enzymes of plant origin, papain, bromelain and ficin being the most widely studied. Other less commonly studied enzymes are zingibain, cucumisin and actinidin (Mazorra-Manzano et al. 2018).

13.8.1 Exogenous Enzymes Used in Meat Tenderization

13.8.1.1 Bromelain

Bromelain is a complex of proteolytic enzymes found in some fruits, specifically in the stem of pineapple (*Ananas comosus*), whose enzymatic activity is slightly more limited than that of papain. The profile of proteins subjected to enzymatic action indicates that papain degrades myosin and actin at similar rates, while bromelain mainly degrades myosin (Kim and Taub 1991). Bromelain is a sulfhydryl protease obtained either from the stems of pineapples (*Ananas* sp.) (EC 3.4.22.32) or from the fruit and leaves (EC 3.4.22.33) (Marques et al. 2010; Amid et al. 2011). This enzyme requires a free sulfhydryl group of a cysteine side-chain in the active site (Amid et al. 2011). Bromelain is a complex of proteolytic enzymes found in some fruits, specifically in the stem of pineapple (*Ananas comosus*), whose enzymatic activity is slightly more limited than that of papain. The profile of proteins subjected to enzymatic action indicates that papain degrades myosin and actin at similar rates, while bromelain mainly degrades myosin (Kim and Taub 1991). Ionescu et al. (2008) studied the effect of applying different concentrations of papain and bromelain (10, 15 and 20 mg enzyme/100 g of meat) to beef for 24 and 48 h at 4 °C; it is subsequently cooked, which leads to an improvement in the meat softening process that is more noticeable when it is papain that is applied rather than bromelain. The optimal concentration applied was 10 mg enzyme (papain or bromelain)/100 g meat for a tenderization time of 24 h at 4 °C in order to avoid an excessive structural degradation of the meat caused by the enzymatic treatment. It is added as a powder that is sprinkled on the surface of the uncooked meat (Weiss et al. 2010). The optimal temperature for bromelain is about 50 °C, which makes it good for meat tenderisation (Ketnawa et al. 2011; Amid et al. 2011), and this enzyme is inactivated at about 65 °C, during the cooking process. It is more active against collagen compared to papain (Sullivan and Calkins 2010).

Ionescu et al. (2008) studied the effect of applying different concentrations of papain and bromelain (10, 15 and 20 mg enzyme/100 g of meat) to beef for 24 and 48 h at 4 °C; it is subsequently cooked, which leads to an improvement in the meat softening process that is more noticeable when it is papain that is applied rather than bromelain. The optimal concentration applied was 10 mg enzyme (papain or bromelain)/100 g meat for a tenderization time of 24 h at 4 °C in order to avoid an excessive structural degradation of the meat caused by the enzymatic treatment.

13.8.1.2 Ficin

Ficin is a proteolytic substance obtained from the latex of trees of the genus Ficus. The cysteine proteases of *Ficus glabrata* and *Ficus carica* have demonstrated an ability to increase the solubilization of proteins and to improve the tenderness of meat products. Ficin is a cysteine peptidase (EC 3.4.22.3) that is obtained from the latex of ficus trees *(Ficus laurifolia* and *Ficus glabrata)*. It may be also used for meat tenderisation even though its use is not so extended as that of other peptidases. Its main proteolysis is observed against elastin, collagen and actomyosin complex (Maiti and Ahlawat 2008). However, the main application of ficin is in tropic regions as an antihelmintic drug to treat parasitic infections based on its ability to digest various living worms (e.g., roundworms and whipworms). Ramezani et al. (2003) observed how the quality of beef sausages could be improved by using (the enzyme) ficin together with chemically modified soy proteins. It was also observed how (the enzyme) ficin increased the solubility of meat proteins by degrading them into units of smaller molecular weight. In addition, the joint action of ficin and soy proteins boosted the softening effect of the meat compared to what may be achieved through their individual effects. This observation suggests that the addition of enzymes can be combined with other ingredients and processes to enhance the properties of the final products.

13.8.1.3 Zingibain

Zingibain is a vegetable proteolytic enzyme isolated from the ginger rhizome. The ginger rhizome is mainly used as a flavouring agent, but its application as a softening agent in meat products is of great interest. In this regard, although zingibain showed good potential for beef softening, the injection level was limited due to problems associated with taste (Naqvi et al. 2021). Consequently, if the enzyme could be purified, its application in meat would be more promising (Sullivan and Calkins 2010).

13.8.1.4 Papain

Papain is a cysteine peptidase (EC 3.4.22.2) present in papaya *(Carica papaya)* and mountain papaya (*Vasconcellea cundinamarcensis*) (Cheret et al. 2007; Marques et al. 2010). Papain is obtained by collecting the latex from the fruit of the papaya tree. Papain can be found in the latex of the papaya plant, *Carica papaya,* and is an enzyme that protects the papaya plant from insects (Konno et al. 2004). Papain is a very heat-stable enzyme, and, therefore, it is not easily deactivated, allowing a continuous change in product texture even after cooking (Dransfield and Etherington 1981a, b). The application of papain for beef tenderization purposes has been studied at different concentrations (0.003, 0.005, 0.007 and 0.01 mg/100 g meat), allowing the enzyme to act from 24 to 48 h at 4 °C with a subsequent heat treatment at 83 °C for 10 min. The results showed an increase in tenderness and juiciness as the concentration rose and the application time lengthened. Ashie et al. (2002) showed how, in beef, the application of papain to obtain a residual level of 0.002 to 0.05 units of proteolytic activity/100 g meat improved the tenderness of the meat by 25–30%. However, high doses of papain are not recommended because they can lead to meat pastiness (Istrati 2008). This enzyme is applied to tenderise meat but, due to its high proteolytic activity, tends to break down pro teins and create excessive tenderness and sometimes off flavours (Ashie et al. 2002; Sullivan and Calkins 2010). Papain shows good stability with an optimal temperature within 60–70 °C so that its action continues even during cooking (Ashie et al. 2002). Due to its stability, it takes intensive heat to inactivate this enzyme.

13.8.1.5 Actinidin or Actinidain

Actinidin or actinidain (EC 3.4.22.14), is obtained from the kiwi fruit (Actinidai deliciosa). The enzyme varies greatly among different cultivars of fruit and it ranges from non-detectable to 10.7 mg/mL juice (Nishiyama 2007). The actinidin was stable at a pH range of 7–10, however, it had an optimal activity at 58–62 °C (Yamaguchi et al. 1982). The best activity was found to be in the pH range of 7.3–7.6 (Yamaguchi et al. 1982), but a pH range of 5–7, reported by Boyes et al. (1997), reflects the variation in the cultivar and assay used in estimation of proteolytic activity. Actinidin hydrolyzes both myofibrillar proteins and connective tissue proteins (Christensen et al. 2009; Han et al. 2009) but appears to have higher proteolytic activity toward collagen (Wada et al. 2002). When porcine muscles were subjected to a brine solution containing actinidin, a decreased Warner-Blatzer shear force and improved sensory feature was observed. Restricted degradation of connective tissue proteins and myofibrils were observed when the enzyme was added to the raw meat and hence is considered a prime agent for improving meat tenderness.

13.8.1.6 Microbial Enzymes

In line with plant proteases, several enzymes have been isolated from microorganisms and bring a desired change in meat characteristics. With respect to meat tenderization, alkaline elastase from alkalophilic *Bacillus* sp. strain Ya-B demonstrated an optimum activity in the pH range of 5.5–6.0 and temperature range of 10–50 °C (Yeh et al. 2002). In another study, it was found that *Bacillus subtilis* and *Bacillus subtilis* var. amyloliquefaciens are the prime source of proteases like subtilisin (EC3.4.21.62) and neutral protease (EC3.4.24.28), each of which also have a GRAS status (FDA 1999). McConn et al. (1964) reported that neutral protease activity increases at 50 °C but drastically drops at temperatures above 65 °C. However, the enzyme is active at wide pH range of 5.0–9.0, with optimal activity at 7.0. Qihe et al. (2006) found that an elastase from Bacillus sp. (EL31410) on beef led to a 30% decrease in relative hardness. Similarly, it was reported that some microorganisms (e.g., Bacillus) play a very important role in protein degradation in fermented meat and fish products (Bekhit 2010). Collagenases from *Clostridium histolyticum* (Foegeding and Larick 1986) and Vibrio B-30 (Miller et al. 1989), and elastases from alkalophilic Bacillus spp. (Takagi et al. 1992) have been used in meat products. However, it has been found that collagenases and elastases found in fish and aquatic invertebrates have very different specificities compared to those in mammalian muscles (Shahidi and Kamil 2001).

Several fungal enzymes have also been found to have an appreciable effect on the meat system; *Aspergillus oryzae* produces an aspartic protease and has an effective proteolytic activity in meat (Ashie et al. 2002). The enzyme possesses minimal activity under refrigerated storage but increases to an optimum at 55 °C before drastically dropping at 60 °C. *Aspergillus oryzae* is a source of several proteolytic preparations available commercially which contain several GRAS (FDA 1995, 1999, 2009) proteases like neutral protease (EC 3.4.24.4), alkaline protease (EC 3.4.21.14), and aspartic protease (EC3.4.23.6), indicating a wide pH range for activity. Both collagen and elastin have good proteolytic action against connective tissue proteins (Ashie et al. 2002). Benito et al. (2003) tested the use of a fungal protease on pork, and observed the effect of EPg222, a protease from *Penicillium chrysogenum*, on products like dry-cured ham. Proteases having potential to act against both myofibrillar and connective tissue proteins from *Penicillium chrysogenum* Pg222 (Benito et al. 2003), *Aspergillus sojae* (Gerelt et al. 2000), *Aspergillus oryzae* (Ashie et al. 2002; Gerelt et al. 2000), have been explored for meat and other food items.

Other enzymes. A crude enzyme extract from *Calotropis procera* latex showed extensive proteolysis in pork and beef meats as protein breakdown and soluble TCA peptides were detected. Thus, it also constitutes an alternative source of proteolytic enzymes to effectively tenderise meat (Rawdkuen et al. 2013).

13.8.2 Endogenous Enzymes in Meat Tenderization

The proteolytic mechanisms present in a muscle: the system of calpain, lysosomal cathepsins, and proteasomes (MPC-multicatalytic proteinase complex) (Dransfield et al. 1992a, b; Kemp et al. 2010; Koohmaraie and Geesink 2006). However, the system of calpains is thought to be the prime factor for the postmortem-proteolysis of proteins resulting in meat tenderization (Bernard et al. 2007; Neath et al. 2007). Cathepsin-treated myofibrillar proteins have different degradation patterns in comparison to those occurring during postmortem storage of muscle. Additionally, it has been suggested that MPC does not play a significant role in tenderization, as myofibrils are very poor substrates for this type of protease (Koohmaraie 1992). Wheeler et al. (1992) reported that injection of calcium (calpain activator), in muscles accelerates postmortem proteolysis. and injection with calpain inhibitors prevents postmortem proteolysis and hence tenderization. Many studies confirm the importance of calpains in tenderization, and the same proteolytic patterns have been observed in postmortem muscle and the incubation of calpains with myofibrils (Geesink and Koohmaraie 1999).

References

Adesegun Kehinde B, Majid I, Hussain S, Nanda V (2020) Innovations and future trends in product development and packaging technologies. In: Prakash B (ed) Functional and preservative properties of phytochemicals, vol 13. Academic, Cambridge, pp 377–409. https://doi.org/10.1016/B978-0-12-818593-3.00013-0

Aeschlimann D, Paulsson M (1994) Transglutaminases: protein cross-linking enzymes in tissues and body fluids. J Thromb Haemost 71:402–415

Ahhmed AM, Kawahara S, Ohta K, Nakade K, Soeda T, Muguruma M (2007) Differentiation in improvements of gel strength in chicken and beef sausages induced by transglutaminase. Meat Sci J 76:455–462. https://doi.org/10.1016/j.meatsci.2007.01.002

Ahmad R, Sardar M (2015) Enzyme immobilization: an overview on nanoparticles as immobilization matrix. Biochem Anal Biochem 4(2):e1000178. https://doi.org/10.4172/2161-1009.1000178

Ahmad IZ, Tabassum H, Ahmad A, Kuddus M (2018) Food enzymes in pharmaceutical industry: perspectives and limitations. In: Kuddus M (ed) Enzymes in food technology. Springer, Singapore, pp 41–62. https://doi.org/10.1007/978-981-13-1933-4_3

Ahnstrom ML, Seyfert M, Hunt MC, Johnson DE (2006) Dry aging of beef in a bag highly permeable to water vapour. Meat Sci 73:674–679

Amid A, Ismail NA, Yusof F, Salleh HM (2011) Expression, purification, and characterization of a recombinant stem bromelain from *Ananas comosus*. Process Biochem 46:2232–2239

Anzani C, Boukid F, Drummond L, Mullen AM, Álvarez C (2020) Optimising the use of proteins from rich meat co-products and non-meat alternatives: nutritional, technological and allergenicity challenges. Food Res Int 137:109575. https://doi.org/10.1016/j.foodres.2020.109575

Ashie INA, Sorensen TL, Nielsen PM (2002) Effects of papain and a microbial enzyme on meat proteins and beef tenderness. J Food Sci 67:2138–2142. https://doi.org/10.1111/j.1365-2621

Askin OO, Kilic B (2009) Effect of microbial transglutaminase, sodium caseinate and non-fat dry milk on quality of salt-free, low fat Turkey döner kebab. LWT Food Sci Technol 42:1590–1596. https://doi.org/10.1016/j.lwt.2009.06.005

Barrett AJ, Rawlings ND, Woessner JF (2004) Thrombin. In: Handbook of proteolytic enzymes, vol 2, 2nd edn. Elsevier Academic Press, Amsterdam, pp 1667–1672

Bauer F (1983) Free glutamic acid in meat products. Ernahrung 7:688

Baugreet S, Kerry JP, Brodkorb A, Gomez C, Auty M, Allen P, Hamill RM (2018) Optimisation of plant protein and transglutaminase content in novel beef restructured steaks for older adults by central composite design. Meat Sci 142:65–77. https://doi.org/10.1016/j.meatsci.2018.03.024

Bekhit AED (2010) Fermentation of fish roe. In: Heldman DR, Hoover DG, Wheeler MB (eds) The encyclopedia of biotechnology in agriculture and food, vol 1. Taylor & Francis Group, Boca Raton, pp 251–256

Bekhit AA, Hopkins DL, Geesink G, Bekhit AA, Franks P (2014a) Exogenous proteases for meat tenderization. Crit Rev Food Sci Nutr 54:1012–1031. https://doi.org/10.1080/10408398.623247

Bekhit AEA, Carne A, Ha M, Franks P (2014b) Physical interventions to manipulate texture and tenderness of fresh meat: a review. Int J Food Prop 17:433–453

Benito MJ, Rodriguez M, Acosta R, Cordoba JJ (2003) Effect of the fungal extracellular protease EPg222 on texture of whole pieces of pork loin. J Meat Scie 65:877–884

Bernard C, Cassar-Malek I, Cunff ML, Dubroeucq H, Renand G, Hocquette JF (2007) New indicators of beef sensory quality revealed by expression of specific genes. J Agric Food Chem 55:5229–5237

Bilal M, Iqbal HMN (2020) State-of-the-art strategies and applied perspectives of enzyme biocatalysis in food sector: current status and future trends. Crit Rev Food Sci Nutr 60(12): 2052–2066. https://doi.org/10.1080/10408398.2019.1627284

Boyes S, Strubi P, Marsh H (1997) Actinidin levels in fruit of Actinidia species and some *Actinidia arguta* rootstock–scion combinations. LWT Food Sci Technol 30:379–389

Braden KW (2013) Converting muscle to meat: the physiology of rigor. In: The science of meat quality. Wiley, Hoboken, pp 79–97
Bratcher CL, Johnson DD, Littell RC, Gwartney BL (2005) The effects of quality grade, aging, and location within muscle on Warner-Bratzler shear force in beef muscles of locomotion. J Meat Sci 70:279–284
Bremner HA, Hallett IC (1985) Muscle fiber-connective tissue junctions in the fish blue grenadier (*Macruronus novaezelandiae*): a scanning electron microscope study. J Food Sci 50:975–980
Bruce HL, Beilken SL, Leppard P (2005) Variation in flavor textural descriptions of cooked steaks from bovine *M. longissimus thoracis et lumborum* from different production and aging regimes. J Food Sci 70:309–S316
Caballero B, Sierra V, Olivan M, Vega-Naredo I, Tomás-Zapico C, Alvarez-García Ó, Tolivia D, Hardeland R, Rodríguez-Colunga MJ, Coto-Montes A (2007) Activity of cathepsins during beef aging related to mutations in the myostatin gene. J Sci Food Agr 87:192–199. https://doi.org/10.1002/jsfa.2683
Campbell RE, Hunt MC, Levis P, Chambers E, (IV). (2001) Dry-aging effects on palatability of beef *longissimus* muscle. J Food Sci 66:196–199
Carballo J, Ayo J, Colmenero FJ (2006) Microbial transglutaminase and caseinate as cold set binders: influence of meat species and chilling storage. LWT Food Sci Technol 39:692–699
Casas E, White SN, Wheeler TL, Shackelford SD, Koohmaraie M, Riley DG, Chase CC, Johnson DD, Smith TPL (2006) Effects of calpastatin and μ-calpain markers in beef cattle on tenderness traits. J Anim Sci 84:520–525.
Chauhan SS, England EM (2018) Postmortem glycolysis and glycogenolysis: insights from species comparisons. J Meat Sci 144:118–126. https://doi.org/10.1016/j.meatsci.2018.06.021
Chen MJ, Lin CW (2002) Factors affecting the water-holding capacity of fibrinogen/plasmaprotein gels optimized by response surface methodology. J Food Sci 67:2579–2582
Chen JS, Mehta K (1999) Tissue transglutaminase: an enzyme with a split personality. Int J Biochem Cell Biol 31:817–836
Chen L, Feng XC, Lu F, Xu XL, Zhou GH, Li QY, Guo XY (2011) Effects of camptothecin, etoposide and Ca2+ on caspase-3 activity and myofibrillar disruption of chicken during post-mortem ageing. J Meat Sci 87:165–174
Cheret R, Delbarre-Ladrat C, Lamballerie-Anton M, Verrez-Bagnis V (2007) Effect of high pressure on the calpain–calpastatin system in fish muscle. J Food Sci 72:313–316
Christensen M, Torngren MA, Gunvig A, Rozlosnik N, Lametsch R, Karlsson AH, Ertbjerg P (2009) Injection of marinade with actinidin, increases tenderness of porcine *M. biceps* femoris and affects myofibrils and connective tissue. J Sci Food Agr 89:1607–1614
Colmenero FJ, Ayo MJ, Carballo J (2005) Physicochemical properties of low sodium frankfurter with added walnut: effect of transglutaminase combined with caseinate, KCl and dietary fibre as salt replacers. J Meat Sci 69:781–788. https://doi.org/10.1016/j.meatsci.2004.11.011
Dahlmann B, Ruppert T, Kloetzel PM, Kuehn L (2001) Subtypes of 20S proteasomes from skeletal muscle. Biochimie 83:295–299
Dashdorj D, Tripathi VK, Cho S, Kim Y, Hwang I (2016) Dry aging of beef; review. J Anim Sci Technol 58:1–11. https://doi.org/10.1186/s40781-016-0101-9
Destefanis G, Brugiapaglia A, Barge MT, Dal Molin E (2008) Relationship between beef consumer tenderness perception and Warner–Bratzler shear force. J Meat Sci 78:153–156
Diles JJB, Miller MF, Owen BL (1994) Calcium chloride concentration, injection time, and aging period effects on tenderness, sensory, and retail color attributes of loin steaks from mature cows. J Anim Sci 72:2017–2021
Dransfield E, Etherington D (1981a) Enzymes and food processing. Springer, Berlin/Heidelberg. Enzymes in the tenderisation of meat
Dransfield E, Etherington D (1981b) Enzymes in the tenderization of meat. In: Birch GG, Blakebrough N, Parker KJ (eds) Enzymes and food processing. Applied Science Publishers Ltd, London, pp 177–194

Dransfield E, Etherington DJ, Taylor MAJ (1992a) Modelling post-mortem tenderisation – II: enzyme changes during storage of electrically stimulated and non-stimulated beef. J Meat Sci 31:75–84

Dransfield E, Wakefield DK, Parkman ID (1992b) Modelling post-mortem tenderisation – I: texture of electrically stimulated and non-stimulated beef. J Meat Sci 31:57–73

FDA (Food and Drug Administration) (1995) Enzyme preparations from animal and plant sources; Affirmation of GRAS Status as Direct Food Ingredients 60 Fed. Reg. 32904–32912 (21 CFR PART 184 (Docket No. 84G-0257))

FDA (Food and Drug Administration) (1999) Carbohydrase and protease enzyme preparations derived from *Bacillus subtilis* or *Bacillus amyloliquefaciens*. Affirmation of GRAS Status as Direct Food Ingredients. 21 CFR Part 184. Federal Register, vol 64, no 78, 23 April 1999, pp 19887–19896

FDA (Food and Drug Administration) (2009) Database of Select Committee on GRAS Substances (SCOGS) reviews. CFR184.1585. http://www.accessdata.fda.gov/scripts/fcn/fcnNavigation.cfm?rpt=scogsListing&displayAll=true

Fennema OR, Damodaran S, Parkin KL (2010) Química de alimentos de Fennema, 4th edn. Artmed, Porto Alegre

Fernandez M, Ordóñez JA, Bruna JM, Herranz B, de la Hoz L (2000) Accelerated ripening of dry fermented sausages. Trends Food Sci Technol 11:201–209. https://doi.org/10.1016/S0924-2244(00)00077-7

Foegeding EA, Larick DK (1986) Tenderization of beef with bacterial collagenase. J Meat Sci 19: 201–214

Fulladosa E, Serra X, Gou P, Arnau J (2009) Effects of potassium lactate and high pressure on transglutaminase restructured dry-cured hams with reduced salt content. J Meat Sci 82:213–218. https://doi.org/10.1016/j.meatsci.2009.01.013

Geesink GH, Koohmaraie M (1999) Effect of calpastatin on degradation of myofibrillar proteins by l-calpain under postmortem conditions. J Anim Sci 77:2685–2692

Gerelt B, Ikeuchi Y, Suzuki A (2000) Meat tenderization by proteolytic enzymes after osmotic dehydration. J Meat Sci 56:311–318

Goll DE, Thompson VF, Li HQ, Wei W, Cong JY (2003) The calpain system. Physiol Rev 83:731–801

Gorraiz C, Beriain MJ, Chasco J, Insausti K (2002) Effect of aging time on volatile compounds, odor, and flavor of cooked beef from Pirenaica and Friesian bulls and heifers. J Food Sci 67: 916–922

Guerrand D (2018) Economics of food and feed enzymes: status and prospectives. In: Nunes CS, Kumar V (eds) Enzymes in human and animal nutrition, chap 26. Academic, Cambridge, pp 487–514. https://doi.org/10.1016/B978-0-12-805419-2.00026-5

Han J, Morton JD, Bekhit AED, Sedcole JR (2009) Pre-rigor infusion with kiwifruit juice improves lamb tenderness. J Meat Sci 82:324–330

Hernandez-Herrero MM, Duflos G, Malle P, Bouquelet S (2003) Collagenase activity and protein hydrolysis as related to spoilage of iced cod (*Gadus morhua*). Food Res Int 36(2):141–147

Hodges JH, Cahill VR, Ockerman HW (1974) Effect of vacuum packaging on weight loss, microbial growth and palatability of fresh beef wholesale cuts. J Food Sci 39:143–146

Honkavaara M, Rintasalo E, Ylonen J, Pudas T (2003) Meat quality and transport stress of cattle. Dtsch Tierarztl Wochenschr 110:125–128

Hopkins DL, Thompson JM (2001) Inhibition of protease activity 2. Degradation of myofibrillar proteins, myofibril examination of free calcium levels. J Meat Sci 59:199–209

Ilian MA, Bekhit AE, Bickerstaffe R (2004) The relationship between meat tenderization, myofibril fragmentation and autolysis of calpain 3 during post-mortem ageing. J Meat Sci 66:387–397

Imafidon GI, Spanier AM (1994) Unraveling the secret of meat flavour. Trends Food Sci Technol 5: 317–321

Ionescu A, Aprodu I, Pascaru G (2008) Effect of papain and bromelin on muscle and collagen proteins in beef meat. Ann Univ Dunarea Jos Galati Fasc VI Food Technol 32:9–16

Irurueta M, Cadoppi A, Langman L, Grigioni G, Carduza F (2008) Effect of aging on the characteristics of meat from water buffalo grown in the Delta del Paraná region of Argentina. J Meat Sci 79:529–533

Istrati D (2008) The influence of enzymatic tenderization with papain on functional properties of adult beef. J Agroaliment Process Technol 14:140–146

Jamdar SN, Harikumar P (2002) Sensitivity of catheptic enzymes in radurized chicken meat. J Food Sci Technol 39(1):72–73

Jayasooriya SD, Torley PJ, D'Arcy BR, Bhandari BR (2007) Effect of high power ultrasound and ageing on the physical properties of bovine Semitendinosus and *longissimus* muscles. J Meat Sci 75:628–639

Jeremiah LE, Gibson LL (2003) Cooking influences on the palatability of roasts from the beef hip. Food Res Int 36:1–9

Kato H, Nishimura T (1987) Taste components and conditioning of beef, pork, and chicken. In: Kawamura Y, Kare MR (eds) Umami: a basic taste. Marcel Dekker Inc, New York, pp 289–306

Kaufmann A, Köppel R, Widmer M (2012) Determination of microbial transglutaminase in meat and meat products. Food Addit Contam 29:1364–1373. https://doi.org/10.1080/19440049.2012.691557

Kemp CM, Bardsley RG, Parr T (2006) Changes in caspase activity during the post-mortem conditioning period and its relationship to shear force in porcine longissimus muscle. J Anim Sci 84:2841–2846

Kemp CM, Sensky PL, Bardsley RG, Buttery PJ, Parr T (2010) Tenderness – an enzymatic view. J Meat Sci 84:248–256. (Review)

Ketnawa S, Chaiwut P, Rawdkuen S (2011) Aqueous two-phase extraction of bromelain from pineapple peels ('Phu Lae' cultv.) and its biochemical properties. Food Sci Technol 20:1219–1226

Khan MI, Jung S, Nam KC, Jo C (2016) Postmortem aging of beef with a special reference to the dry aging. Korean J Food Sci Anim Resour 2016(36):159. https://doi.org/10.5851/kosfa.2016.36.2.159

Kim HJ, Taub IA (1991) Specific degradation of myosin in meat by bromelain. J Food Chem 40:337–343. https://doi.org/10.1016/0308-8146(91)90117-7

Kolczak T (2000) Influence of post-mortem factors on beef tenderness. Gosp Mięsna 5:28–31

Konno K, Hirayama C, Nakamura M, Tateishi K, Tamura Y, Hattori M, Kohno K (2004) Papain protects papaya trees from herbivorous insects: role of cysteine proteases in latex. Plant J 37: 370–378. https://doi.org/10.1046/j.1365-313X.2003.01968.x

Koohmaraie M (1992) The role of Ca2+-dependent proteases (calpains) in post mortem proteolysis and meat tenderness. Biochimie 74:239–245

Koohmaraie M (1996) Biochemical factors regulating the toughening and tenderization processes of meat. J Meat Sci 43:193–201. https://doi.org/10.1016/0309-1740(96)00065-4

Koohmaraie M, Geesink GH (2006) Contribution of postmortem muscle biochemistry to the delivery of consistent meat quality with particular focus on the calpain system. J Meat Sci 74: 34–43. https://doi.org/10.1016/j.meatsci.2006.04.025

Koutsidis G, Elmore JS, Oruna-Concha MJ, Campo MM, Woo JD, Mottram DS (2008) Water-soluble precursors of beef flavor. Part II: effect of post mortem conditioning. J Meat Sci 79:270–277

Kuraishi C, Sakamoto J, Yamazaki K, Susa Y (1997) Production of restructured meat using microbial transglutaminase without salt or cooking. J Food Sci 62:488–491

Kurebayashi N, Harkins AB, Baylor SM (1993) Use of fura red as an intracellular calcium indicator in frog skeletal muscle fibers. Biophys J 64:1934–1960

Lajolo F, Mercadante AZ (2017) Química e bioquímica de alimentos, 2nd edn. Atheneu, São Paulo

Lawrie RA (1998) The conversion of muscle to meat. Lawrie's meat science, 6th edn. Woodhead Publishing Ltd, Cambridge

Ledward DA (1985) Post-slaughter influences on the formation of metyyoglobin in beef muscles. J Meat Sci 15:149–171. https://doi.org/10.1016/0309-1740(85)90034-8

Leroy B, Lambotte S, Dotreppe O, Lecocq H, Istasse L, Clinquart A (2003) Prediction of technological and organoleptic properties of beef *Longissimus thoracis* from near-infrared reflectance and transmission spectra. J Meat Sci 66:45–54

Li LY, Easa AM, Liong MT, Tan TC, Foo WT (2013a) The use of microbial transglutaminase and soy protein isolate to enhance retention of capsaicin in capsaicin-enriched layered noodles. Food Hydrocoll 30:495–503

Li X, Babol J, Wallby A, Lundstron K (2013b) Meat quality microbiological status and consumer preference of beef gluteus medius aged in a dry ageing bag or vacuum. J Meat Sci 95:229–234

Lonergan SM, Topel DG, Marple DN (2018) The science of animal growth and meat technology. Academic, Cambridge, MA

Maiti AK, Ahlawat SS (2008) Application of natural tenderizers in meat a review. Agric Rev 29: 226–230

Mancini RA, Hunt M (2005) Current research in meat color. J Meat Sci 75:100–121. https://doi.org/10.1016/j.meatsci.2005.03.003

Manseth E, Skjervold PO, Fjear SO, Brosstad FR, Odegaard OR, Flensgrud R (2003) Developing a fish meat-binding agent: purification of salmon thrombin. J Food Sci 68:1648–1652

Marques AYC, Maróstica MR, Pastore GM (2010) Some nutritional, technological and environmental advances in the use of enzymes in meat products. Enzyme Res 48:9–23

Matarneh SK, England EM, Scheffler TL, Gerrard DE (2017) The conversion of muscle to meat. In: Lawrie's meat science. Woodhead Publishing, Sawston, pp 159–185

Mazorra-Manzano MA, Ramírez-Suarez JC, Yada RY (2018) Plant proteases for bioactive peptides release: a review. Crit Rev Food Sci Nutr 58:2147–2163. https://doi.org/10.1080/10408398.2017.1308312

McConn JD, Tsuru D, Yasunobu KT (1964) *Bacillus subtilis* neutral proteinase: I. A zinc enzyme of high specific activity. J Biol Chem 239:3706–3715

McGee H (2004) On food and cooking: the science and lore of the kitchen. Scribner, New York

Miller AJ, Strange ED, Whiting RC (1989) Improved tenderness of restructured beef steaks by microbial collagenase derived from Vibrio B-30. J Food Sci 54:855–857

Miller MF, Kerth CR, Wise JW, Lansdell JL, Stowell JE, Ramsey CB (1997) Slaughter plant location, USDA quality grade, external fat thickness, and aging time effects on sensory characteristics of beef loin strip steak. J Anim Sci 75:662–667

Monson F, Sanudo C, Sierra I (2005) Influence of breed and ageing time on the sensory meat quality and consumer acceptability in intensively reared beef. J Meat Sci 71:471–479

Motoki M, Seguro K (1998) Transglutaminase and its use for food processing. Trends Food Sci Technol 9:204–210

Mottram DS (1998) Flavour formation in meat and meat products: a review. J Food Chem 62:415–424

Moudilou EN, Mouterfi N, Exbrayat JM, Brun C (2010) Calpains expression during Xenopus laevis development. Tissue Cell 42:275–281

Naqvi ZB, Campbell MA, Latif S, Thomson PC, McGill DM, Warner RD, Friend MA (2021) Improving tenderness and quality of *M. biceps* femoris from older cows through concentrate feeding, zingibain protease and sous vide cooking. J Meat Sci 180:108563. https://doi.org/10.1016/j.meatsci.2021.108563

Neath KE, Del-Barrio AN, Lapitan RM, Herrera JRV, Cruz LC, Fujihara T, Muroya S, Chikuni K, Hirabayashi M, Kanai Y (2007) Protease activity higher in postmortem water buffalo meat than Brahman beef. J Meat Sci 77:389–396

Nielsen MK, Nielsen HH (2012) Seafood enzymes. In: Simpson BK, Nollet LML, Toldrá F, Benjakul S, Paliyath G, Hui YH (eds) Food biochemistry & food processing, 2nd edn. Blackwell Publishing, Ames, pp 247–262

Nishimura T, Hattori A, Takahashi K (1995) Structural weakening of intramuscular connective tissue during conditioning of beef. J Meat Sci 39:127–133

Nishiyama I (2007) Fruits of the actinidia genus. Adv Food Nutr Res 52:293–324

Nowak D (2011) Enzymes in tenderization of meat-the system of calpains and other systems – a review. Pol J Food Nutr Sci 2011:61,231–61,237. https://doi.org/10.2478/v10222-011-0025-5
Ordoñez JA, Rodríguez MIC, Álvarez LF, Sanz MLG, Minguillón G (2005) Tecnologia de alimentos: componentes dos alimentos e processos, 1st edn. Artmed, Porto Alegre
Othrner CL, Kosow DP (1980) Evidence that human a-thrombin is a monovalent cationactivated enzyme. Arch Biochem Biophys 202:63–75
Ouali A, Herrera-Mendez CH, Coulis G, Becila S, Boudjellal A, Aubry L, Sentandreu MA (2006) Revisiting the conversion of muscle into meat and the underlying mechanisms. J Meat Sci 74: 44–58. https://doi.org/10.1016/j.meatsci.2006.05.010
Ouali A, Gagaoua M, Boudida Y, Becila S, Boudjellal A, Herrera-Mendez CH, Sentandreu MA (2013) Biomarkers of meat tenderness: present knowledge and perspectives in regards to our current understanding of the mechanisms involved. J Meat Sci 3(95):854–870. https://doi.org/10.1016/j.meatsci.2013.05.010
Parrish FC, Boles JA, Rust RE, Olson DG (1991) Dry and wet aging effects on palatability attributes of beef loin and rib steaks from three quality grades. J Food Sci 56:601–603
Perry N (2012) Dry aging beef. Int J Gastron Food Sci 1:78–80
Piao MY, Jo C, Kim HJ, Lee HJ, Kim HJ, Ko JY, Baik M (2015) Comparison of carcass and sensory traits and free amino acid contents among quality grades in loin and rump of Korean cattle steer. Asian Australas J Anim Sci 28:1629–1640
Purslow PP, Ertbjerg P, Baron CP, Christensen M, Lawson MA (2001) Patterns of variation in enzyme activity and cytoskeletal proteolysis in muscle. In: 47th ICoMST, pp 38–43
Qihe C, Guoqing H, Yingchun J, Hui N (2006) Effects of elastase from a *Bacillus* strain on the tenderization of beef meat. J Food Chem:624–698, 629
Ramezani R, Aminlari M, Fallahi H (2003) Effect of chemically modified soy proteins and ficin-tenderized meat on the quality attributes of sausage. J Food Sci 68(1):85–88. https://doi.org/10.1111/j.1365-2621.2003.tb14119.x
Rawdkuen S, Jaimakreu M, Benjakul S (2013) Physicochemical properties and tenderness of meat samples using proteolytic extract from *Calotropis procera* latex. J Food Chem 136:909–916
Ray RC, Rosell CM (2017) Microbial enzyme technology in food applications, 1st edn. CRC Press, Boca Raton. https://doi.org/10.1201/9781315368405
Ray L, Pramanik S, Bera D (2016) Enzymes-an existing and promising tool of food processing industry. Recent Pat Biotechnol 10(1):58–71. https://doi.org/10.2174/1872208310666160727150153
Rees MP, Trout GR, Warner RD (2002) Effect of calcium infusion on tenderness and ageing rate of pork muscle longissimus thoracis et lumborum after accelerated boning. J Meat Sci 61:169–179
Rowe D (2002) High impact aroma chemicals. Part 2: the good, the bad, and the ugly. Perfum Flavor 27:24–29
Salmerón C, Navarro I, Johnston IA, Gutiérrez J, Capilla E (2015) Characterisation and expression analysis of cathepsins and ubiquitin-proteasome genes in gilthead sea bream (*Sparus aurata*) skeletal muscle. BMC Res Notes 8:1–15. https://doi.org/10.1186/s13104-015-1121-0
Santhi D, Kalaikannan A, Malairaj P, Arun PS (2017) Application of microbial transglutaminase in meat foods: a review. Crit Rev Food Sci Nutr 57:2071–2076. https://doi.org/10.1080/10408398.2014.945990
Savell JW, Smith GC, Carpenter ZL (1978) Beef quality and palatability as affected by electrical stimulation and cooler aging. J Food Sci 43:1666
Savell JW, Mueller SL, Baird BE (2005) The chilling of carcasses. J Meat Sci 70:449–459. https://doi.org/10.1016/j.meatsci.2004.06.027
Shahidi F, Kamil YVAJ (2001) Enzymes from fish and aquatic invertebrates and their application in the food industry. Trends Food Sci Technol 12:435–464
Singh R, Kumar M, Mittal A, Mehta PK (2016) Microbial enzymes: industrial progress in 21st century. 3. Biotech 6(2):174. https://doi.org/10.1007/s13205-016-0485-8
Sitz BM, Calkins CR, Feuz DM, Umberger WJ, Eskridge KM (2006) Consumer sensory acceptance and value of wet-aged and dry-aged beefsteaks. J Anim Sci 84:1221–1226

Smith RD, Nicholson KL, Nicholson JDW, Harris KB, Miller RK, Griffin DB, Savell JW (2008) Dry versus wet aging of beef: retail cutting yields and consumer palatability evaluations of steaks from US choice and US select short loins. J Meat Sci 79:631–639

Spanier AM, Miller JA (1993) Role of proteins and peptides in meat flavor. In: Spanier AM, Okai H, Tamura M (eds) Food flavor and safety, ACS Symposium series no 528. Am Chem Soc, Washington, DC, pp 78–97

Spanier AM, McMillin KW, Miller JA (1990) Enzyme activity levels in beef: effect of postmortem aging and endpoint cooking temperature. J Food Sci 55:318–326

Spanier AM, Flores M, Mcmillin KW, Bidner TD (1997) The effect of postmortem aging on meat flavor quality. Correlation of treatment, sensory, instrumental, and chemical descriptors. J Food Chem 59:531–538

Stenstrom H, Li X, Hunt MC, Lundström K (2014) Consumer preference and effect of correct or misleading information after ageing beef *longissimus* muscle using vacuum, dry ageing, or a dry ageing bag. J Meat Sci 96:661–666

Sullivan GA, Calkins CR (2010) Application of exogenous enzymes to beef muscle of high and low-connective tissue. J Meat Sci 85:730–734. https://doi.org/10.1016/j.meatsci.2010.03.033

Suman SP, Nair MN (2017) Current developments in fundamental and applied aspects of meat color. In: New aspects of meat quality. Woodhead Publishing, Sawston, pp 111–127

Suman SP, Hunt MC, Nair MN, Rentfrow G (2014) Improving beef color stability: practical strategies and underlying mechanisms. J Meat Sci 98:490–504

Taheri-Kafrani A, Kharazmi S, Nasrollahzadeh M, Soozanipour A, Ejeian F, Etedali P, Mansouri-Tehrani HA, Razmjou A, Yek SM, Varma RS (2021) Recent developments in enzyme immobilization technology for high-throughput processing in food industries. Crit Rev Food Sci Nutr 61(19):3160–3196. https://doi.org/10.1080/10408398.2020.1793726

Takagi H, Kondou M, Hisatsuka T, Nakamori S, Tsai YC, Yamasaki M (1992) Effects of an alkaline elastase from an alkalophilic *Bacillus* strain on the tenderization of beef meat. J Agric Food Chem 40:577–583

Tarté R (2009) Ingredients in meat products: properties, functionality and applications. Springer Link, New York

Tatiyaborworntham N, Oz F, Richards MP, Wu H (2022) Paradoxical effects of lipolysis on the lipid oxidation in meat and meat products. J Food Chem 14(100317). https://doi.org/10.1016/j.fochx.2022.100317

Taylor RG, Geesink GH, Thompson VF, Koohmaraie M, Goll DE (1995) Is z-disk degradation responsible for post-mortem tenderization. J Anim Sci 73:1351–1367

Toldrá F (1998) Proteolysis and lipolysis in flavour development of dry-cured meat products. J Meat Sci 49:s101–s110

Toldrá F, Flores M (1998) The role of muscle proteases and lipases in flavor development during the processing of dry-cured ham. Crit Rev Food Sci 38:331–352. https://doi.org/10.1080/10408699891274237

Trespalacios P, Pla R (2007) Synergistic action of transglutaminase and high pressure on chicken meat and egg gels in absence of phosphates. Food Chem 76(3):455–462.

Trindade MA, Rocha YJP, Bekhit AEDA (2017) Restructured Meat Products. In: Advances in meat processing technology. CRC Press, Boca Raton, pp 487–504

Vitale M, Pérez-Juan M, Lloret E, Arnau J, Realini CE (2014) Effect of aging time in vacuum on tenderness, and color and lipid stability of beef from mature cows during display in high oxygen atmosphere package. J Meat Sci 96:270–277

Wada M, Suzuki T, Yaguti Y, Hasegawa T (2002) The effects of pressure treatments with kiwi fruit protease on adult cattle semitendinosus muscle. J Food Chem 78:167–171

Wang D, Cheng F, Wang Y, Han J, Gao F, Tian J, Zhang K, Jin Y (2022) The changes occurring in proteins during processing and storage of fermented meat products and their regulation by lactic acid bacteria. Food Secur 11:2427. https://doi.org/10.3390/foods11162427

Warren KE, Kastner CL (1992) A comparison of dry-aged and vacuum-aged beef striploins. J Muscle Foods 3:151–157

Weiss J, Gibis M, Schuh V, Salminen H (2010) Advances in ingredient and processing systems for meat and meat products. J Meat Sci 86:196–213
Wheeler TL, Crouse JD, Koohmaraie M (1992) The effect of postmortem time of injection and freezing on the effectiveness of calcium chloride for improving beef tenderness. J Anim Sci 70: 3451–3457
Whitehurst RJ, Van Oort M (2010) Enzymes in meat processing. In: Enzymes in food technology (2nd edn). John Wiley & Sons, Ltd., UK
Xian-Xing X, Xue S, Zhi-Hang C, Cheng-Qi S, Yu-Nan H, Yuan-Guo C (2009) Development and application of a real-time PCR method for pharmacokinetic and biodistribution studies of recombinant adenovirus. J Mol Biotechnol 43:130–137
Yamaguchi Y, Yamashita Y, Takeda I, Kiso H (1982) Proteolytic enzymes in green asparagus, kiwifruit and miut:occurrence and partial characterisation. J Agric Biol Chem 46:1983–1986
Yamashita M, Konagaya S (1990) Participation of cathepsin L into extensive softening of the muscle of chum salmon caught during spawning migration. Nippon Suisan Gakkaishi 56(8): 1271–1277
Yancey EJ, Dikeman ME, Hachmeister KA, Chambers EIV, Milliken GA (2005) Flavor characterization of top blade, top sirloin, and tenderloin steaks as affected by pH, maturity, and marbling. J Anim Sci 83:2618–2623
Yang X, Zhang Y (2019) Expression of recombinant transglutaminase gene in *Pichia pastoris* and its uses in restructured meat products. J Food Chem 291:245–252. https://doi.org/10.1016/j.foodchem.2019.04.015
Yeh CM, Yang MC, Tsai YC (2002) Application potency of engineered G159 mutants on P1 substrate pocket of Subtilisin YaB as improved meat tenderizers. J Agric Food Chem 50:6199–6204
Yokoyama K, Nio N, Kikuchi Y (2004) Properties and applications of microbial transglutaminase. J Appl Microbiol Biotechnol 67:447–454
Yushkova ED, Nazarova EA, Matyuhina AV, Noskova AO, Shavronskaya DO, Vinogradov VV, Skvortsova NN, Krivoshapkina EF (2019) Application of immobilized enzymes in food industry. J Agric Food Chem 67(42):11553–11567. https://doi.org/10.1021/acs.jafc.9b04385
Zhang L, Barbut S (2005) Rheological characteristics of fresh and frozen PSE, normal and DFD chicken breast meat. Braz J Poult Sci 46:687–693. https://doi.org/10.1080/00071660500391516
Zhou G, Zhao G (2007) Biochemical changes during processing of traditional Jinhua ham. J Meat Sci 77:114–120
Zhu Y, Tramper J (2008) Novel applications for microbial transglutaminase beyond food processing. Trends Biotechnol 26(10):559–565.
Zork K, Ortiz R, Saatci E, Bardsley R, Parr T, Csöregi E, Nistor M (2009) Label free capacitive immunosensor for detecting calpastatin – a meat tenderness biomarker. Bioelectrochemistry 76: 93–99

Chapter 14
Cultured Meat

Shazia Akhter, Jahangir Ahmad Rather, Tawheed Amin, Zahida Naseem, Ruqaya Tariq, and Aaruba Maqbool Sofi

14.1 Introduction

As per the United Nations Department of Economics and Social Affairs, the global population is projected to reach 10 billion by the year 2050 (Guan et al. 2021). This increasing population demands 455 million metric tonnes of animals for consumption, representing a surge of about 40% compared to the levels in 2019 (Reddy et al. 2022). India, grappling with a growing population, grapples with recurring food security challenges stemming from limited arable land, the impacts of climate change, unsustainable food production practices, and governmental regulations affecting food marketing and distribution to the public. In 2021, the Food and Agriculture Organization of the United States (FAO) recorded an estimated meat production of 7361 thousand metric tonnes in India. FAO says this meat consumption fulfils 35% of the country's overall protein requirements.

Meat plays a crucial role in providing essential proteins, fats, vitamins, minerals, and micronutrients, making it a fundamental component of the human diet (Siddiqui et al. 2022). Livestock resources significantly contribute to India's GDP, constituting 4.11% of the overall GDP and 25.6% of the Agriculture sector's GDP. However, the per capita meat availability in India stands at approximately 15 grams per day, which is below the recommended daily intake of 30 grams per person suggested by the Indian Council of Medical Research (ICMR) (Islam et al. 2016). The majority of animals in our country are raised in extremely unnatural and disorganized situations, such as farms, which may lead to the development of various diseases, which have a high likelihood of spreading to people who consume this meat. In addition to

S. Akhter (✉) · T. Amin · Z. Naseem · R. Tariq · A. M. Sofi
Division of Food Science and Technology, SKUAST, Srinagar, Jammu and Kashmir, India

J. A. Rather
Department of Food Science and Technology, Islamic University of Science and Technology, Awantipora, Jammu and Kashmir, India

S. A. Rather, F. A. Masoodi (eds.), *Hand Book of Processed Functional Meat Products*, https://doi.org/10.1007/978-3-031-69868-2_14

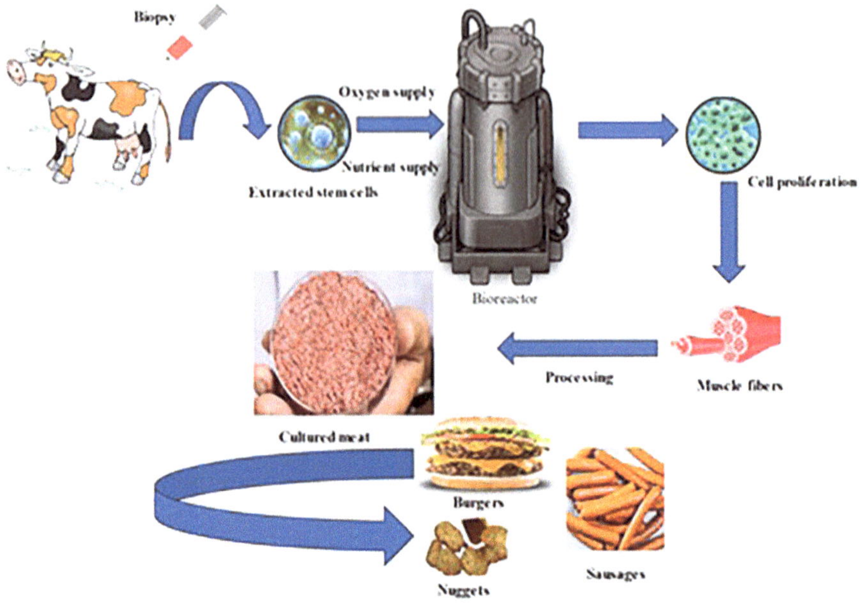

Fig. 14.1 Outlining the technology employed in cultured meat production

utilizing 70% of agricultural land, raising livestock also causes freshwater reservoirs to run dry, biodiversity to decline, soil erosion, habitat damage, etc. The production of meat, eggs, and milk contributes to roughly 15% of global greenhouse gas emissions, positioning it as the second-largest emitter. This surpasses the cumulative emissions from all modes of transportation worldwide.

According to projections, India's meat consumption will increase as consumer earnings improve. The sector that produces cultured meat is predicted to reduce the disparity between the supply and demand of proteins made from meat among the Indian population. In addition to animal husbandry, numerous effective methods of producing proteins are being developed to meet the rising food demand of the expanding human population. Considering the current obstacles confronting livestock, including environmental issues and concerns about animal welfare, cultured meat, alternatively known as in vitro meat, emerges as a favored choice among advocates as a sustainable option for individuals seeking ethical alternatives without altering their dietary preferences. Despite being a state-of-the-art technology, the production of animal-free or cultured meat employs fundamental methods. This process entails harvesting mature muscle cells that have undergone cultivation from loosely arranged myo-satellite cells on a substrate in a liquid medium, with mechanical stimulation, as illustrated in Fig. 14.1 In other words, in vitro meat is similar to the meat that comes from animals that we consume, but it avoids killing the animals to get the flesh. Due to the culturing methods used, the phrase "cultured meat" was more frequently used (Rosenfeld and Janet Tomiyama 2023). The term "clean meat," which has been promoted since 2015, then, seems more enticing and grabs more consumers' attention. According to (Hocquette 2016), he expressed disagreement by asserting that the term "meat" inherently involves the development within an animal and the subsequent need to slaughter the animal for tissue and organ extraction. Instead, he preferred to describe lab-produced meat as "artificial muscle proteins." Despite the various terms used for lab-grown meat, the core principles underlying each method generally remain consistent. The emergence of cultured meat, also known as cell-based or lab-grown meat, presents a promising alternative to traditional methods of meat production. Its potential to offer numerous benefits, including improved health attributes, environmental sustainability, and freedom from religious restrictions, makes it an attractive candidate for addressing the challenges posed by conventional livestock agriculture (Arshad et al. 2017). Furthermore, the potential to improve the nutritional content of cultured meat by including vital nutrients like vitamins, unsaturated fatty acids, trace elements, and amino acids offers additional benefits compared to traditional sources of meat (Bhat et al. 2020; Nobre 2022).

Quality and consumer acceptance are critical factors for the successful adoption of cultured meat. Research has shown that the aroma, texture, taste, and appearance of cultured meat should closely resemble those of conventional meat to gain consumer approval (Fraeye et al. 2020). Nevertheless, a significant hurdle in the production of cultured meat lies in the absence of myoglobin, the compound responsible for the characteristic red hue in traditional meat. In response to this challenge, scientists have explored alternatives, including the use of natural dyes like sugar beet or saffron, as well as haemoglobin extracted from animal blood or its derivatives, to replicate the red color in cultured meat (Mateti et al. 2022). Recent progress in this field indicates that incorporating serene muscle cells can enhance the functionality of muscle stem cells within scaffolds, resulting in improved sensory attributes and toughness in cultured meat (Ben-Arye and Levenberg 2019; Zheng et al. 2021).

14.2 Cultured Meat Production Techniques

Continuous improvements and dedicated research efforts are necessary to set up and inspire the industry to create cultured meat that is both highly structured unprocessed cultured meat. This remains fundamental to the various techniques involved in cultured meat production. Generally, there are three main classifications based on the techniques employed: scaffold-based technique, self-organizing technique, and 3-D printing meat.

14.2.1 Scaffold-Based Technique

Within scaffold-based methodologies, the proliferation of embryonic myoblasts or muscle cells (MCs) occurs as they adhere to scaffolds or carriers, such as collagen meshwork or microcarrier beads. Following this, the cultured cells are placed in a growth medium, and this cultivation process takes place in either a stationary or a rotating bioreactor (Bhat et al. 2014; Izhar Ariff Mohd Kashim et al. 2023). Different environmental cues have been employed, resulting in the merging of these cells into myotubes, followed by their progressive development into myofibers. Stem cells are cultivated and suspended alongside collagen spheres, serving as a substrate or scaffold to which myoblasts can attach and undergo the process of differentiation. On these collagen spheres, myoblasts evolve into myotubes. These myotubes then undergo additional differentiation, ultimately transforming into myofibers within the bioreactor. Subsequently, these myofibers can be harvested, processed, and utilized as consumable meat products (showen in Fig. 14.2) (Izhar Ariff Mohd Kashim et al. 2023). Scaffold-based techniques, while unable to generate intricately structured

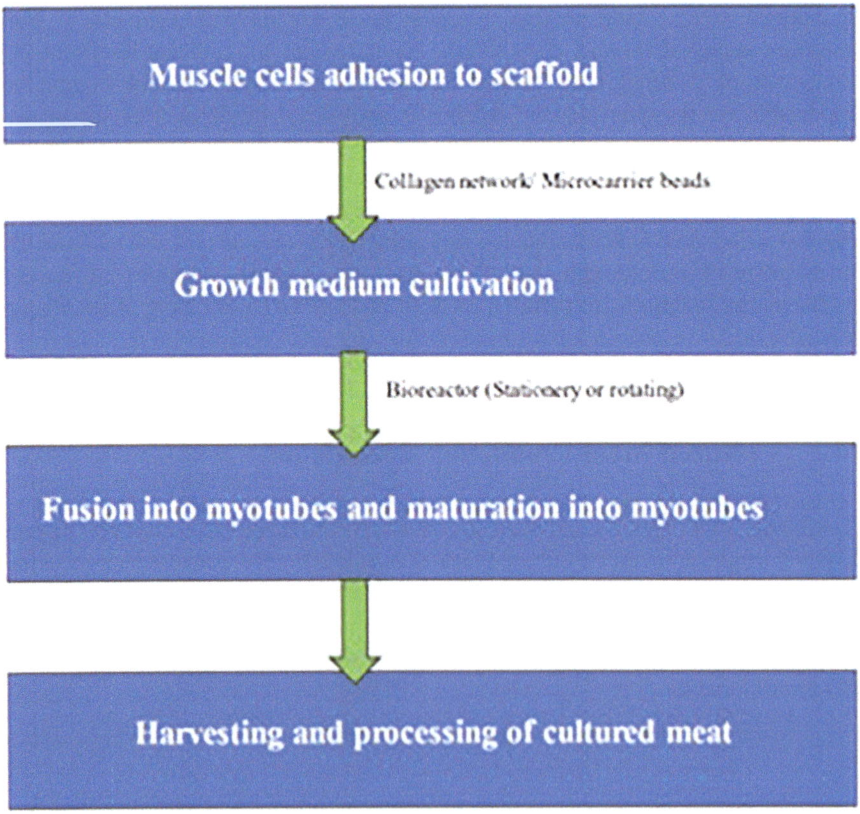

Fig. 14.2 Scaffold meat production

meats such as steaks, are suitable for the production of ground and boneless meats characterized by a tender consistency. Alternatively, cells can be cultured in substrates that facilitate the formation of "self-organizing constructs," leading to the development of more rigid structures (Bhat et al. 2015). Van Eelen, Van Kooten, and Westerhof are credited with securing a Dutch patent for the comprehensive methodology of cultivating meat, as outlined in their publication from 1999. Nonetheless, the groundbreaking accomplishment of actually producing meat using this technique appears to be attributed to Catts and Zurr (Singh et al. 2014).

Fresh meat is composed not solely of a single muscle tissue type but also includes connective tissues and fat. The synthesis of full-thickness complex tissue engineering necessitates the integration of various tissues cultivated independently and subsequently fused through a tissue engineering methodology (Post 2020; Izhar Ariff Mohd Kashim et al. 2023). Developing advanced meat products that incorporate muscle, connective tissue, fat and vascular cells using the scaffolding technique enhances the composition of cultured meat, aligning its flavor with that of conventional livestock meat. Crafting intricate cultured meat involves integrating muscle fibers, fibrous vascular cells, and adipose tissue. However, the simultaneous cultivation of diverse cells demands careful optimization of the growth medium and the application of specific protocols. The intricate development of muscle tissues relies on the characteristics of the scaffold biomaterial, customized to facilitate the growth of sturdy muscle and adipose tissue (Izhar Ariff Mohd Kashim et al. 2023).

14.2.2 Self-Organizing Technique

The production of well-structured meat, like steak, is possible by generating organized muscle tissues in the form of self-organizing constructs (Bhat et al. 2014; Izhar Ariff Mohd Kashim et al. 2023). Benjaminson, Gilchriest, and Lorenz employed this technique, involving the placement of skeletal muscle explants from goldfish (*Carassius auratus*) into various culture media for a seven-day duration. The results indicated an increase in surface area ranging from 5.2% to 13.8%. Particularly noteworthy was the substantial growth observed when these explants were placed in a culture containing dissociated Carassius skeletal muscle cells, resulting in a remarkable 79% expansion in explant surface area. The key advantage of explants lies in their ability to encompass all muscle cells in their inherent proportions, closely resembling an in vivo structure. Nevertheless, a notable constraint arises from the absence of blood circulation in these tissue samples, preventing substantial growth. This limitation is attributed to the tendency of cells to undergo necrosis when isolated for prolonged periods, surpassing a distance of more than 0.5 mm from a nutrient source (Singh et al. 2014, Bhat et al. 2014; Izhar Ariff Mohd Kashim et al. 2023).

14.2.3 3-D Printing Meat

Dr. Gabor Forgacs introduced this technique in 2011 for generating multiple tissue cultures of meat. This method entails combining billions of blood cells, fat cells, muscle cells, and other cellular components. The inventive method employs a 3-D printer to assemble cells by injecting ink onto a surface, enabling cell mobility. Subsequently, the cells that have been printed undergo cultivation in a medium similiar to the one employed in the scaffolding technique (Izhar Ariff Mohd Kashim et al. 2023). Figure 14.3 represents 3D printing of meat.

14.3 Potential Benefits of Cultured Meat

In recent years, advancements in cellular agriculture have enabled the customization and controlled alteration of the final composition of cultivated meat products, encompassing attributes such as taste, texture, and nutritional quality (Post 2012). This innovative approach holds the potential to address pressing concerns related to antibiotic contamination within the supply chain and the transmission of zoonotic diseases, ultimately ensuring the provision of safer, pathogen-free products to consumers (Reddy et al. 2022).

Notably, the cultivation of meat within laboratory settings eliminates the need for animal suffering and slaughter at any stage of the process, making it an appealing prospect for a wide range of stakeholders, including vegetarians and proponents of animal welfare (Zhang et al. 2020). Diverse consumer preferences can be accommodated through the culture of cells sourced from various animals, offering an

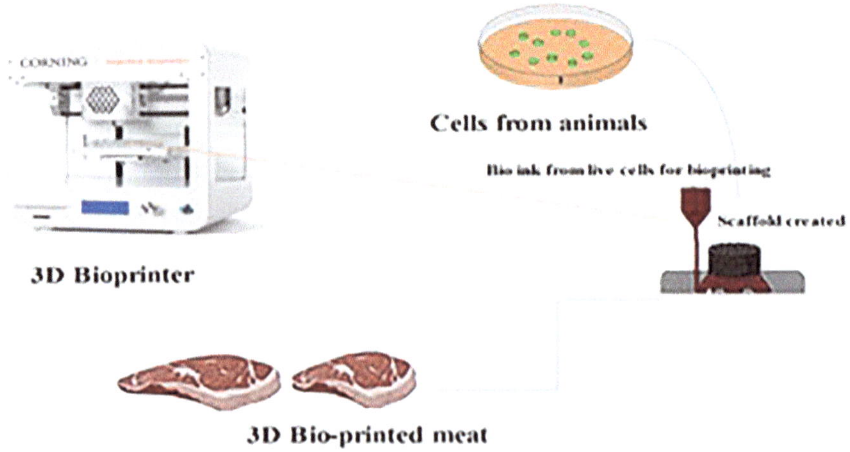

Fig. 14.3 3-D printing meat process

extensive selection of lab-grown meat products, including those from exotic species (Zaraska 2013).

The application of cellular agriculture also facilitates the retention of the traditional genetic pool, contributing to the preservation of genetic diversity within the cultivated meat sector. Furthermore, the potential for large-scale production translates to increased profitability and the reduction of food wastage. This is particularly significant given the lower environmental impact and decreased resource consumption associated with cellular agriculture, resulting in diminished emissions of greenhouse gases and fewer carbon footprints (Stephens et al. 2018).

A single cell line or tissue sample from a livestock source can serve as the foundation for an abundant and limitless supply of cultivated meat, circumventing the limitations of traditional livestock rearing (Rorheim et al. 2016; Gaydhane et al. 2018). This approach offers the added advantage of requiring reduced land and water usage, aligning with sustainable resource management practices.

Thus, cellular agriculture presents a transformative paradigm for meat production, offering tailored product characteristics, improved food safety, and environmental benefits. The integration of scientific advancements into the food industry has the potential to revolutionize the way meat is produced, meeting the demands of a growing population while minimizing ecological impact.

14.4 Technical Challenges of Production of Cultured Meat

Cultured meat, produced through cellular biomass cultivation, holds tremendous potential as a sustainable alternative to conventional meat production. However, the realization of its equivalence to traditional meat is contingent upon achieving similar texture, flavor, and nutritional content. Presently, the cultured meat industry is in its nascent stages, necessitating rigorous optimization of each production procedure and the resolution of several technological challenges before these products can become commonplace in the regular diet.

14.4.1 Bio-reactor Capacity and Environmental Replication

The challenges impeding the widespread adoption of large-scale production include difficulties associated with reactors and the upscaling process of cultured meat. Bioreactors play a crucial role in the expansion and differentiation of cells during the meat cultivation process. To illustrate, an estimated 8×10^{12} muscle cells necessitate a 5000-L stirred tank bioreactor. Although this volume is common in established bioprocessing industries, its application in tissue engineering and mesenchymal stem cell amplification is yet to be demonstrated. Recently, Aleph Farms declared the successful production of a batch of cell-based cultured steaks within just 3 weeks, achieved through extensive stem cell proliferation on a large scale (Ye et al.

texture, including marbling and 3D structural features similar to traditional cuts. In contrast to conventionally slaughtered meat, which contains bioactive substances like growth factors and immunoglobulins, cultured meat cannot entirely replicate these components. To enhance sensory qualities, cultured muscle tissue from bioreactors is supplemented with various food additives (Ye et al. 2022). Owing to the constraints of existing technology, the cultured meat products currently accessible primarily consist of burgers and other food items blended with minced meat.

14.6 Health and Safety

14.6.1 Contamination and Oversight

While proponents emphasize the controlled nature of cultured meat production, the potential for human errors and oversight cannot be dismissed. Industrial meat production provides a pertinent example, as incidents of contamination are not uncommon. A comprehensive evaluation of potential risks and the establishment of stringent quality control measures are imperative to mitigate the risk of contamination (Shapiro 2018).

14.6.2 Dysregulation and Health Implications

The proliferation of cell lines in cultured meat production raises concerns about unexpected biological mechanisms, analogous to dysregulation in cancer cells. The potential consequences for muscle structure, human metabolism, and consumer health necessitate further investigation (Hocquette 2016).

14.6.3 Antibiotic Resistance and Contamination

Advocates assert that controlled environments in cultured meat production can mitigate the issue of antibiotic resistance observed in conventional livestock. However, the use of antibiotics to prevent contamination could undermine this advantage, potentially leading to unintended consequences. The delicate balance between preventing contamination and safeguarding long-term health must be considered (Oliver Stephen et al. 2011; Chriki and Hocquette 2020).

14.6.4 Nutritional Considerations

14.6.4.1 Micronutrient Replication

Advocates maintain that the nutritional composition of cultured meat can be fine-tuned through meticulous control of production methods. Nonetheless, the challenge persists in replicating specific micronutrients present in animal products, such as iron and vitamin B12. Unique biological compounds and their interactions within cultured cells may impact micronutrient bioavailability and efficacy (Chriki and Hocquette 2020).

14.6.4.2 Bioavailability and Compound Interaction

Micronutrient uptake by cultured cells and their interactions within the cellular matrix are essential factors influencing the nutritional quality of cultured meat. The interplay between various compounds and their effects on human health necessitates thorough investigation (Scollan et al. 2014).

In vitro, cultured meat holds promise as a novel approach to sustainable and controlled food production. However, the safety and nutritional aspects of this technology demand careful consideration. While advocates highlight the controlled environment as a safeguard against contamination, the potential for errors and oversight remains. The occurrence of dysregulation in cell lines and the subsequent health implications requires further research. Additionally, challenges persist in replicating micronutrients unique to animal products and ensuring their bioavailability within cultured meat. A holistic approach, combining rigorous safety measures, comprehensive research, and transparent labelling, is essential to realize the full potential of in vitro cultured meat while safeguarding public health.

14.7 Market and Legislation of Cultured Meat

14.7.1 Market

In 2013, Professor Mark Post of Maastricht University successfully created the inaugural in vitro hamburger after 2 years of development. Valued at over $300,000, the elevated cost was attributed to Professor Post's utilization of products and compounds traditionally associated with medical science, including hormones and nutrients. After introducing this groundbreaking innovation, Professor Post attracted additional investments and formed a research team to advance the development of in vitro meat, culminating in the establishment of a new startup named Mosa Meat. As of now, cultured meat has not yet become available on store shelves, indicating the necessity for further research to decrease its production costs (Chriki et al. 2020). As per a recent survey, those demonstrating interest in cultured meat, which is currently in the developmental phase, are identified as a demographic comprising young, educated individuals who traditionally consume conventional meat. These potential consumers exhibit a modest level of awareness about in vitro meat but convey a readiness to reduce their intake of conventionally slaughtered meat (Mancini and Antonioli 2019; Chriki and Hocquette 2020). In the coming years, the meat industry is anticipated to evolve into a more intricate landscape, surpassing the current scenario. This transformation will involve a broader range of meat products or alternatives emerging from various sources and production processes (Hicks Talia et al. 2018). The success of novel products depends on their ability to provide economically viable alternatives to conventional meat production. The effectiveness of cultured meat, whether presented as an substitute, complement or alternative to traditional meat, is essential. This is due to consumers' inclination to favor products with similar market positioning (Van der Weele et al. 2019; Verbeke et al. 2015). Additionally, if alternative meat products can offer competitive pricing, consumers are more likely to consider adjusting their buying preferences. However, the implementation of the most technologically advanced alternatives to conventional meat also demands different levels of social and institutional transformation (Chriki et al. 2020; Chriki and Hocquette 2020).

14.7.2 Legislation

There exists a relatively small but notable collection of literature that explores the regulation of cultured meat. Schneider thoroughly examines the regulatory landscape in the United States, while Petetin concentrates on the considerations within the European Union (Stephens et al. 2018; Broucke et al. 2023). In 2023, (Broucke et al. 2023) provided a review on EU and USA regulatory aspects.

In the European context, the regulation of cultured meat falls under the purview of the Novel Foods regulation. Novel foods encompass those that haven't been significantly consumed by humans in the EU before May 15, 1997, and exhibit a new or intentionally altered molecular structure. This category includes foods derived from insects, microorganisms, fungi, or algae, as well as those produced through cell or tissue culture originating from animals, plants, fungi, or algae. To determine if a food product is subject to the Novel Food Regulation, a verification process takes place through consultation at the Member State level. Following this, the safety assessment of these novel foods is carried out by the European Food Safety Authority (EFSA). In cases where cultured meat products incorporate a genetically modified (GM) component, they would be regulated by Regulation (EC) No 1829/2003 on GM food and feed, as such foods are excluded from the EU Novel Food regulation. This scenario would apply if genetically engineered induced pluripotent stem cells (iPSC) are utilized in the production of cultured meat (Stephens et al. 2018; Broucke et al. 2023).

At the federal level in the United States, the Food and Drug Administration (FDA) holds the primary responsibility for overseeing food safety for all food types, excluding meat and poultry. The regulation of meat and poultry products falls under the jurisdiction of the US Department of Agriculture Food Safety and Inspection Service (USDA–FSIS) through the Federal Meat Inspection Act (FMIA). Regarding the regulatory framework for cell-based meat intended for human consumption, the USDA and the FDA jointly issued a statement in 2018 (Broucke et al. 2023). Specifically, this framework encompasses the supervision of human food produced using animal cell culture technology, derived from cell lines of USDA-amenable species, and mandated to bear a USDA mark of inspection. The FDA holds authority over the initial stages of cell-based meat production, including activities such as cell collection, establishment of cell banks, and cell growth and differentiation. This involves a thorough examination of the facility registry to ensure compliance with FDA requirements for food manufacturing, as well as to verify the safety and purity of cell cultures. During the harvest phase, the FDA oversees the transformation of harvested cells into meat or poultry products that bear the USDA mark of inspection. Following this, in the production and labeling stage, the USDA-FSIS takes the lead in ensuring the safety, purity, wholesomeness, and appropriate labeling of these products (Broucke et al. 2023).

14.8 Ethical Issues and Social Attitudes

The viewpoints of consumers and the ethical challenges they face are critical factors that can serve as a potential hindrance to the widespread adoption of lab-grown meat and its achievement of commercial success (Hocquette 2016). The objective is to create a product with sensory and organoleptic attributes closely mirroring those of natural meat, acknowledged and preferred by consumers. Yet, presently, attaining this goal proves challenging (Sergelidis 2019).

Cultured meat proponents advocate for its potential to mitigate the harsh treatment endured by animals in conventional livestock farming (Hopkins et al. 2008). The deplorable conditions of battery-farmed animals often result in disease, infections, behavioral issues, and suffering. However, the lack of a nervous system in cultured cells raises questions about the capacity for pain perception (Chauvet 2018). Although biopsies on animals for cell collection may raise concerns about animal welfare, cultured meat, devoid of a nervous system, is presumed to be devoid of pain (Chauvet 2018). Some authors even categorize cultured meat as a vegetarian product due to the absence of animal suffering (Hopkins 2015).

Cultured meat production is anticipated to reduce the need for conventional livestock farming, thereby potentially benefiting animal welfare. The claim is that fewer animals are required for muscle sampling, yet the necessity of animal involvement in cell harvesting remains (Alvaro 2019). This raises the question of whether lab-grown meat still involves animal exploitation, conflicting with the proponents' aim to eliminate animal harm (Alvaro 2019). The primary criteria for acceptance revolve around ethical considerations concerning technology, its application, and its alignment with the natural product, encompassing anticipated qualitative features and potential advantages or drawbacks. Presently, consumer awareness of lab-grown meat is limited. While existing surveys indicate hesitancy among most consumers when queried about their willingness to try cultured meat in the future, only a minority explicitly dismisses the concept (Sergelidis 2019).

Thus, the widespread acceptance of lab-grown meat hinges on addressing complex ethical dilemmas and aligning sensory characteristics with traditional meat. Advocates argue for improved animal welfare, yet questions about pain perception and the initial involvement of animals in cell harvesting persist. Limited consumer awareness adds further complexity. As the field advances, transparent communication and education will play a pivotal role in shaping attitudes toward lab-grown meat, emphasizing the delicate balance between technological innovation, ethical considerations, and consumer acceptance.

14.9 Conclusion

The advent of cultured meat represents a revolutionary stride towards addressing critical challenges in conventional meat production, offering a plethora of potential benefits. The ability to customize taste, texture, and nutritional composition provides a versatile solution to meet diverse consumer preferences. Notably, the elimination of animal suffering and slaughter aligns with ethical considerations, appealing to a broad spectrum of stakeholders, including vegetarians and advocates of animal welfare.

Moreover, the application of cellular agriculture holds promise in mitigating concerns related to antibiotic contamination and the transmission of zoonotic diseases, ensuring the production of safer, pathogen-free meat products. The potential to cultivate cells from various animal sources contributes to an extensive selection of lab-grown meat products, fostering culinary diversity and accommodating diverse cultural preferences.

From an environmental standpoint, the reduction in land and water usage, coupled with lower resource consumption, positions cultured meat as a sustainable alternative to traditional livestock rearing. The preservation of genetic diversity within the cultivated meat sector further underscores its potential long-term impact on the global food industry.

However, despite these promising aspects, the cultured meat industry faces formidable technical challenges that must be addressed for widespread adoption. Overcoming hurdles related to bioreactor capacity, environmental replication, cell line optimization, and scaffolding with direct perfusion is crucial for achieving production scalability and cost-effectiveness. Additionally, the quality of cultured meat products, including texture, color, and flavor replication, requires substantial advancements to meet consumer expectations.

In essence, while cultured meat holds immense promise as a sustainable, ethical, and customizable food source, its realization on a large scale hinges on overcoming the current technical challenges. Continued research, innovation, and collaboration across scientific, industrial, and regulatory domains will be pivotal in unlocking the full potential of cultured meat and reshaping the future of meat production for a growing global population.

References

Alvaro C (2019) Lab-grown meat and veganism: a virtue-oriented perspective. J Agric Environ Ethics 32(1):127–141. https://doi.org/10.1007/s10806-019-09759-2

Arshad MS et al (2017) Tissue engineering approaches to develop cultured meat from cells: a mini review. Cogent Food Agric 3(1). https://doi.org/10.1080/23311932.2017.1320814

Ben-Arye T, Levenberg S (2019) Tissue engineering for clean meat production. Front Sustain Food Syst 3(June):1–19

Bhat ZF, Bhat H, Pathak V (2014) Chapter 79 – prospects for in vitro cultured meat – a future harvest. In: Lanza R, Langer R, Vacanti JB (eds) Principles of tissue engineering, 4th edn. Academic Press, Boston, pp 1663–1683. https://www.sciencedirect.com/science/article/pii/B9780123983589000793

Bhat ZF, Kumar S, Fayaz H (2015) In vitro meat production: challenges and benefits over conventional meat production. J Integr Agric 14(2):241–248

Bhat ZF, Bhat H, Kumar S (2020) Cultured meat-a humane meat production system. In: Principles of tissue engineering. https://doi.org/10.1016/B978-0-12-818422-6.00075-7

Broucke K et al (2023) Cultured meat and challenges ahead: a review on nutritional, technofunctional and sensorial properties, safety and legislation. Meat Sci 195(September 2022):109006. https://doi.org/10.1016/j.meatsci.2022.109006

Chauvet DJ (2018) Should cultured meat be refused in the name of animal dignity? Ethical Theory Moral Pract 21(2):387–411. https://doi.org/10.1007/s10677-018-9888-4

Chriki S, Hocquette JF (2020) The myth of cultured meat: a review. Front Nutr 7(January)

Chriki S et al (2020) Analysis of scientific and press articles related to cultured meat for a better understanding of its perception. Front Psychol 11(August)

Fraeye I, Vandenburgh MKH, Thorrez L (2020) Sensorial and nutritional aspects of cultured meat in comparison to traditional meat: much to be inferred. Front Nutr 7(March)

Gaydhane M et al (2018) Cultured meat: state of the art and future. Biomanuf Rev 3(1). https://doi.org/10.1007/s40898-018-0005-1

Goodwin JN, Shoulders (2013) The future of meat: a qualitative analysis of cultured meat media coverage. Meat Sci 95(3):445–450. https://doi.org/10.1016/j.meatsci.2013.05.027

Guan X et al (2021) Trends and ideas in technology, regulation and public acceptance of cultured meat. Future Foods 3(February):100032. https://doi.org/10.1016/j.fufo.2021.100032

Hicks Talia M, Knowles SO, Farouk MM (2018) Global provisioning of red meat for flexitarian diets. Front Nutr 5(June):1–11

Hocquette JF (2016) Is in vitro meat the solution for the future? Meat Sci 120:167–176. https://doi.org/10.1016/j.meatsci.2016.04.036

Hopkins PD (2015) Cultured meat in Western media: the disproportionate coverage of vegetarian reactions, demographic realities, and implications for cultured meat marketing. J Integr Agric 14(2):264–272. https://doi.org/10.1016/S2095-3119(14)60883-2

Islam MM, Shabana Anjum RJ, Modi WKN (2016) Scenario of livestock and poultry in India and their contribution to national economy. Int J Sci Environ Technol 5(3):956–965

Izhar Ariff Mohd Kashim M et al (2023) Scientific and Islamic perspectives in relation to the halal status of cultured meat. Saudi J Biol Sci 30(1):103501. https://doi.org/10.1016/j.sjbs.2022.103501

Jairath G, Mal G, Gopinath D, Singh B (2021) An holistic approach to access the viability of cultured meat: a review. Trends Food Sci Technol 110(February):700–710. https://doi.org/10.1016/j.tifs.2021.02.024

Liu J, Ellies-oury MP, Stoyanchev T, Hocquette JF (2022) Consumer perception of beef quality and how to control, improve and predict it? Focus on eating quality. Food Secur 11(12):1–27

Mancini MC, Antonioli F (2019) Exploring consumers' attitude towards cultured meat in Italy. Meat Sci 150(December 2018):101–110. https://doi.org/10.1016/j.meatsci.2018.12.014

Mateti T, Laha A, Shenoy P (2022) Artificial meat industry: production methodology, challenges, and future. JOM 74(9):3428–3444

Nobre FS (2022) Cultured meat and the sustainable development goals. Trends Food Sci Technol 124(September 2021):140–153. https://doi.org/10.1016/j.tifs.2022.04.011

Oliver Stephen P, Murinda SE, Jayarao BM (2011) Impact of antibiotic use in adult dairy cows on antimicrobial resistance of veterinary and human pathogens: a comprehensive review. Foodborne Pathog Dis 8(3):337–355

Post MJ (2012) Cultured meat from stem cells: challenges and prospects. Meat Sci 92(3):297–301

Post MJ (2020) Scientific, sustainability and regulatory challenges of cultured meat. Nat Food 1(7): 403–415

Reddy MK et al (2022) Cultured meat – a review. Biol Forum Int J 14(4):363–367

Rorheim A, Mannino A, Baumann T, Caviola L (2016) Cultured meat: a pragmatic solution to the problems posed by industrial animal farming. Policy Pap by Sentience Politics, Basel

Rosenfeld DL, Janet Tomiyama A (2023) Toward consumer acceptance of cultured meat. Trends Cogn Sci 27(8):689–691. https://doi.org/10.1016/j.tics.2023.05.002

Scollan ND et al (2014) Enhancing the nutritional and health value of beef lipids and their relationship with meat quality. Meat Sci 97(3):384–394

Sergelidis D (2019) Lab grown meat: the future sustainable alternative to meat or a novel functional food? Biomed J Sci Tech Res 17(1):12440–12444

Shapiro P (2018) Clean meat: how growing meat without animals will revolutionize dinner and the world. Simon and Schuster

Siddiqui SA et al (2022) Cultured meat: processing, packaging, shelf life, and consumer acceptance. LWT 172(November):114192

Singh VP, Sachan N, Verma AK (2014) In vitro meat-the start of new era in meat production. Int J Livest Res 2(1):38–51

Specht EA, Welch DR, Rees Clayton EM, Lagally CD (2018) Opportunities for applying biomedical production and manufacturing methods to the development of the clean meat industry. Biochem Eng J 132:161–168

Stephens N et al (2018) Bringing cultured meat to market: technical, socio-political, and regulatory challenges in cellular agriculture. Trends Food Sci Technol 78(April):155–166. https://doi.org/10.1016/j.tifs.2018.04.010

Van der Weele C et al (2019) Meat alternatives: an integrative comparison. Trends Food Sci Technol 88:505–512

Verbeke W et al (2015) 'Would you eat cultured meat?' Consumers' reactions and attitude formation in Belgium, Portugal and the United Kingdom. Meat Sci 102:49–58

Ye Y, Zhou J, Guan X, Sun X (2022) Commercialization of cultured meat products: current status, challenges, and strategic prospects. Future Foods 6(August):100177. https://doi.org/10.1016/j.fufo.2022.100177

Zaraska M (2013) Lab-grown beef taste test: 'almost'like a burger'. The Washington Post

Zhang G et al (2020) Challenges and possibilities for bio-manufacturing cultured meat. Trends Food Sci Technol 97(January):443–450. https://doi.org/10.1016/j.tifs.2020.01.026

Zheng Y-Y et al (2021) Evaluation of the effect of smooth muscle cells on the quality of cultured meat in a model for cultured meat. Food Res Int 150:110786

Chapter 15
Reduction of Carcinogenic N-Nitroso Compounds (NOCs) and Polycylic Aromatic Hydrocarbons (PAHs) in Processed Meat Products

Syed Arshad Hussain, Sheikh Rafeh Ahmad, and Humaira Fayaz

15.1 Introduction

Meat stands out as a valuable reservoir of top-tier proteins, indispensable amino acids, B-vitamins, minerals and an array of other micronutrients (Marangoni et al. 2015). Incorporating red meat, particularly beef, pork, lamb and veal into the human diet is vital due to its rich nutrient profile. Recognized by the Food and Drug Administration as a prime source of protein (FDA 2013), meat serves as a significant contributor to meeting daily nutritional needs. The US Department of Agriculture has established the daily value (DV) for protein in beef and pork at 40%, in lamb at 37% and in veal and game at 36% (Cashman and Hayes 2017). In addition to its protein content, meat is instrumental in enhancing the body's utilization of vitamin D, especially in cases of limited sun exposure. A notable portion of the daily requirement for vitamin D, approximately 30–35% and 21% for iron, is fulfilled through meat consumption (Pogorzelska-Nowicka et al. 2022). These nutrients address critical health concerns outlined in the US Dietary Guidelines for Americans, which highlight vitamin D, calcium, potassium, iron and fiber as five shortfall nutrients in diets (Zhang et al. 2021; USDA 2015). Furthermore beef, pork, veal and lamb contain significant levels of potassium, another shortfall nutrient lacking in many diets. The levels present in these meats warrant the claim of being a "source of potassium" according to the EU Nutrient Reference Value, with at least 15% available in a 100 g serving (Pogorzelska-Nowicka et al. 2022). This underscores the importance of including meat in a balanced diet for overall nutritional well-being.

S. A. Hussain (✉) · S. R. Ahmad (✉) · H. Fayaz
Division of Livestock Products Technology, FVSc & AH, SKUAST-Kashmir, Srinagar, India
e-mail: sh.rafeh@skuastkashmir.ac.in

S. A. Rather, F. A. Masoodi (eds.), *Hand Book of Processed Functional Meat Products*, https://doi.org/10.1007/978-3-031-69868-2_15

Table 15.1 Primary pathways in the formation of carcinogenic compounds in processed meats

Compounds	Source of exposure	Conditions affecting their presence
N-nitroso compounds/Nitrosamines	Cured meats	Presence of Residual nitrite in cured meats along with secondary amines and catalyzed by temperature
Heterocyclic aromatic amines (HAAs)	High temperature Cooked meats	Type of cooking, time and temperature of cooking
Polycyclic aromatic hydrocarbons (PAHs)	Smoked meats	Smoking, high temperature long time traditional smoking
Biogenic amines	Fermented meats and/or non- hygienic meats	Type of microbial flora and/or Hygiene: decarboxylation of certain amino acids

Nevertheless, the intake of red meat is also associated with particular undesired outcomes. The International Agency for Research on Cancer (IARC) indicates that red meat consumption is likely carcinogenic to humans (IARC 2015). Research has shown that exceeding the recommended intake of processed meat by 50 grams raises the risk of cancer by approximately 18% (Bouvard et al. 2015). Numerous factors contribute to the adverse impact of meat consumption on human health, many of which are associated with its processing methods, including cooking, excessive salt usage, curing and smoking. Three categories of compounds have garnered significant scrutiny due to their potential carcinogenic properties. Even if not initially present in raw meat, they can arise during subsequent processing or cooking. These groups include N-nitroso-compounds (NOCs), heterocyclic aromatic amines (HAAs) and polycyclic aromatic hydrocarbons (PAHs).

Cooking meat enhances its digestibility and sensory appeal while crucially eliminating harmful microorganisms that pose health hazards to humans (Jonge 2019). Studies demonstrate that cooking meat to an internal temperature of 70 °C and holding it at this temperature for 2 min ensures microbiological safety (Ježek et al. 2019). However, exposure to high temperatures during cooking can result in the formation of heterocyclic aromatic amines (HAAs) and polycyclic aromatic hydrocarbons (PAHs), both potential and confirmed carcinogens (Pogorzelska-Nowicka et al. 2022). Furthermore, curing and smoking processes also contribute to the creation of carcinogenic substances such as nitrosamines and polycyclic aromatic hydrocarbons (PAHs) (Pogorzelska-Nowicka et al. 2022). Additionally, other compounds with potential detrimental effects on consumer health include biogenic amines, formed through microbial decarboxylation of particular amino acids, which possess potential vasoactive or psychoactive toxicological impacts. Amines also serve as a potential reservoir of secondary amines capable of reacting with nitrite to generate nitrosamines, which exhibit potential carcinogenic properties (Pogorzelska-Nowicka et al. 2022). Table 15.1 provides a concise overview of the primary pathways through which these compounds are found in meat.

15.2 Heterocyclic Aromatic Amines (HAAs)

Heterocyclic aromatic amines (HAAs) constitute a category of compounds generally characterized by a structure comprising 2–5 aromatic cycles, one or more nitrogen atoms and an exocyclic amino group, with certain exceptions (Alaejos and Afonso 2011). These compounds emerge during the heat treatment of protein-rich products surpassing temperatures of 120 °C. The principal substrates for the formation of heterocyclic aromatic amines (HAAs) are free amino acids abundant in muscle tissue, particularly creatine and creatinine. Their production during maillard reactions is contingent upon the pH of the meat. Over the past four decades, more than 25 HAAs have been isolated and identified in thermally processed foods (Pogorzelska-Nowicka et al. 2022). HAAs can be categorized into two main groups based on their polarity. Polar HAAs, such as amino-imidazo-azarenes (IQ group), are generated during food processing at temperatures ranging from 100 °C to 300 °C (e.g., smoking, frying, toasting and grilling). Non-polar HAAs like pyroindoles and pyroimidazoles form during heat treatment at temperatures exceeding 300 °C, earning them the designation of pyrolytic amines. Exposure to HAAs predominantly stems from compounds like 2-amino-1-methyl-6-phenylimidazo [4, 5-b] pyridine (PhIP) and 2-amino-3,8-dimethylimidazo[4,5-f] quinoxaline (MeIQx). Other minor compounds include 2-amino-9H-pyrido[2,3-b]indole (AC), 2-amino-3,4-dimethylimidazo[4,5-f]quinoline (IQ), and 2-amino-3,4,8-trimethylimidazo[4,5-f] quinoxiline (DiMeIQx) (reference: Jakszyn et al. 2004). These amines are closely associated with the onset of specific cancer types (Bogen 1994; Augustsson et al. 1999). Determining the intake of HAAs is challenging as their presence in meat varies depending on cooking methods, temperature and duration (Bjeldanes et al. 1983). Frying, broiling and grilling/barbecuing yield the highest levels of HAAs due to the extremely high temperatures employed (Sinha et al. 1998a, b). Conversely, oven roasting and baking at lower temperatures result in lower levels of HAAs. Generally, direct contact of meat with the heat source facilitates the formation of higher HAAs levels compared to when the meat has less intimate contact with the heating apparatus.

15.3 Polycyclic Aromatic Hydrocarbons (PAHs)

Polycyclic aromatic hydrocarbons (PAHs) represent pollutants characterized by the presence of two or more fused aromatic rings. Cooking techniques such as grilling, roasting, barbecuing and most importantly smoking, contribute to PAHs formation. For centuries, smoking has served as a method for preserving meat and more recently, it's been valued for its impact on the sensory attributes of meat. However, the smoke itself may harbor health-hazardous compounds such as PAHs, phenols and formaldehyde (Bem 1995). Several factors influence the formation of PAHs in cooked meat, including temperature, cooking duration, fat content, type of fuel utilized for cooking and the occurrence of melted fat dripping onto the heating surface. The gravest situations arise in heavily smoked meat products processed in old or inadequate smokehouses, where PAH levels may approach 100 μg/kg (Simko 2009). Numerous factors such as direct or indirect smoking methods, the type of generator utilized, the wood and herb types and compositions, the accessibility to oxygen and the temperature and duration of the smoking process influence the content of PAH formed in the smoked meat products. While PAHs can emerge from proteins and carbohydrates, their levels notably escalate during fat pyrolysis at temperatures surpassing 200 °C (Pogorzelska-Nowicka et al. 2022).

15.4 N-Nitroso Compounds (NOCs)

N-nitroso compounds (NOCs) exist in two forms: nitrosamines and nitrosamides arising from the reaction of amines and amides respectively with nitrosating agents derived from nitrite sources (Preston-Martin et al. 2006). Nitrosamides, which act as direct alkylating agents, are potent neuro-carcinogens, especially impactful during pregnancy or shortly after birth (Dietrich et al. 2005; Ohgaki 2009; Wrensch et al. 2002). Diets high in NOCs may heighten the risk of glioma.

Nitrite, the primary additive in preserving cured meats, inhibits the proliferation of spores from putrefactive and pathogenic bacteria like *Clostridium botulinum*. Besides its antimicrobial properties, nitrite contributes to the formation of nitrosylmyoglobin, enhancing the colour of cured meat and its lipid stability, while indirectly influencing its flavor (Ramarathnam 1998). However, concerns revolve around residual nitrite content in cured meat products, as it can yield nitrous acid and subsequently nitrosamines if secondary amines are present (Toldrá et al. 2009). Nitrous acid production escalates as the product's pH approaches the pKa of nitrous acid (pKa = 3.36). The presence of nitrosamines in meat products hinges on numerous factors, including nitrite levels, processing conditions, lean meat proportion, presence of catalysts or inhibitors, processing temperature, product pH, duration, storage conditions and inclusion of reducing agents like ascorbate or iso-ascorbate. Microorganisms capable of converting nitrate to nitrite or producing amines also contribute to nitrosamine formation. Nitrite levels generally decline rapidly during processing, resulting in low residual concentrations in the final product (Hill et al. 1973). Among nitrosamines, prevalent in cured foods, processed meats are a significant source, whereas the presence of chemically fragile nitrosamides is less common (Dietrich et al. 2005; IARC 2006). Notable nitrosamines identified in cured meats include N-nitrosodimethylamine, N-nitrosopirrolidine, N-nitrosopiperidine, N-nitrosodiethylamine, N-nitrosodi-n-propylamine, N-nitrosomorpholine, and N-nitrosoethylmethylamine (Reigand and Toldrá 2010).

Dietary habits primarily expose individuals to NOCs, with 45–75% of overall exposure attributed to endogenous formation (Tricker 1997). Processed meats, rich in nitrite, amines and amides, are crucial dietary reservoirs, providing precursors for endogenous NOC synthesis within the stomach (Dietrich et al. 2005; IARC 2006). Moreover, the consumption of nitrate, subsequently converted to nitrite, primarily by oral bacteria, can also trigger endogenous NOC production (Dietrich et al. 2005; Preston-Martin et al. 2006; IARC 2006). In addition to processed meats, dietary nitrite mainly originates from grain products and select vegetables like leafy greens which serve as the primary dietary nitrate sources (IARC 2006). Nutrients rich in fruits and vegetables, such as vitamins C and E and polyphenols, inhibit endogenous NOC formation (Preston-Martin et al. 2006; Ohgaki 2009). Heme, present in red meat, prompts the synthesis of endogenous NOCs within the gastrointestinal tract (Lunn et al. 2007), potentially allowing NOCs to reach the brain through the bloodstream (Dietrich et al. 2005). N-nitrosodimethylamine (NDMA) is recognized as the most carcinogenic NOC. Its presence is directly associated with the nitrogen concentration in meat, leading to significantly higher levels in cured foods (Ciemniak 2006). For example, the shoulder meat contains approximately 2.46 μg/kg NDMA, while the cured shoulder exceeds 9 μg/kg. Smoked products exhibit the highest NDMA levels, ranging between 10 and 17 μg/kg, whereas pasteurized canned products register the lowest values at around 1–2 μg/kg (Ciemniak 2006). NDMA is not confined to beef and pork; salted fish, for instance, exhibits very high levels ranging from 12.64 to 322.92 μg/kg, a significant disparity compared to fresh fish, which ranges from 0.04 to 3.5 μg/kg. Poultry demonstrates relatively low NDMA levels, with chicken containing about 0.52 μg/kg and duck meat containing 0.13 μg/kg (Park et al. 2015). The quantity of NOCs is directly influenced by the method of heat treatment utilized (e.g., smoking, steaming) and the curing process.

15.5 Effects of N-Nitroso Compounds (NOCs), Heterocyclic Aromatic Amines (HAAs) and Polycyclic Aromatic Hydrocarbons (PAHs) on Human Health

In processed meat products, the presence and hazard of heterocyclic aromatic amines (HAAs), polycyclic aromatic hydrocarbons (PAHs) and N-nitroso compounds (NOC's) raise significant concerns for both consumers and researchers. An elevated intake of meat products has been linked to an amplified cancer risk. This heightened risk could stem from various factors such as the elevated levels of fat and protein, or the creation of carcinogenic substances like heterocyclic aromatic amines and polycyclic aromatic hydrocarbons during meat processing under high temperatures (John et al. 2011). Furthermore, the abundance of heme iron, notably present in red meat, might play a role in the internal production of N-nitroso compounds from nitrates (Nogacka et al. 2019; John et al. 2011). These compounds have the potential to disrupt DNA synthesis, escalate cell proliferation, influence hormone metabolism, elevate insulin-like growth factors, and contribute to the formation of free radicals, all of which may foster the onset of cancer (Salamat et al. 2020). Epidemiological studies suggest a link between high meat intake and increased cancer risk, particularly due to the formation of HCAs during high-temperature cooking (Egeberg et al. 2013; Janoszka 2010; Liao et al. 2010; Oz and Kaya 2011a). Cohort studies have associated red and processed meat consumption with gastric non-cardiac and breast cancer (González et al. 2006; Stefani et al. 1997; Zheng et al. 1998). PAH exposure is associated with lung and skin cancer (USEPA 2008). Although processed meat is classified as carcinogenic to humans (IARC 1993), the level of carcinogens varies widely due to manufacturing processes and food materials (Rahman et al. 2014).

Among heat-treated food products, heterocyclic aromatic amines (HAAs) are considered among the most mutagenic and carcinogenic compounds. HAAs undergo metabolic activation through catalytic cytochrome P450 enzymes. In the initial phase of activation, the exocyclic amino group of these compounds undergoes N-hydroxylation to form hydroxylamine (-NHOH). Subsequently, enzymes from the N-acetyltransferases and sulfotransferases groups facilitate the formation of highly active metabolites, including esters and sulfates. These metabolites then bind with deoxyguanosine (dG) within DNA molecules or with protein molecules, resulting in the formation of stable adducts which have the potential to disrupt DNA replication, leading to mutations and ultimately, contributing to the development of cancer (Bulanda and Janoszka 2022). Research suggests that they are 2000 times more mutagenic than benzo[a]pyrene and 100 times stronger than aflatoxin (Oz and Kaya 2011b; Haskaraca et al. 2014). The International Agency for Research on Cancer (IARC), a branch of the World Health Organization, recognizes nine HAAs, including MeIQ, MeIQx, PhIP, AαC, MeAαC, Glu-P-1, Glu-P-2, Trp-P-1, and Trp-P-2, as potential human carcinogens (Group 2B), with one, 2-amino-imidazo [4,5-f]quinoline, classified as a probable carcinogen (Group 2A). Studies have demonstrated that HAAs can induce tumors in rodents and non-human primates at various sites, including the lungs, liver, mammary glands, prostate, and colorectum (Fu et al. 2014). In the United States, the National Toxicology Program identifies four HAAs (PhIP, IQ, MeIQ, 8-MeIQ) as carcinogenic to humans.

Polycyclic aromatic hydrocarbons (PAHs) mostly arise from the incomplete combustion of organic material like thermal decomposition of wood at temperatures exceeding 500 °C in oxygen-deprived conditions (Simko 2009) producing numerous harmful components of carcinogenic, mutagenic, and teratogenic properties, with benzo(a)pyrene being the most prevalent (Trafialek and Kolanowski 2014). They can accumulate in barbequed, grilled, fried, and smoked foods. Although approximately 100 compounds have been identified, humans primarily encounter PAHs through food, accounting for over 90% of exposure. Formaldehyde, identified as a promoter of cancerous tumors, is present in smoke. Some smoke-derived phenols have the potential to react and form highly toxic nitrosophenols, which can further undergo reactions to produce toxic byproducts like nitrophenols, polymeric nitrosic compounds, and other hazardous substances, or even catalyze the formation of nitrosamines (Bem 1995). The International Agency for Research on Cancer, categorizes three PAHs (benz[a]anthracene (BaA), benzo[a]pyrene (BaP), dibenz[a,h] anthracene) as potentially carcinogenic to humans (Group 2A) and three others (benzo[a]fluoranthene (BbF), benzo[k]fluoranthene, indeno[1,2,3-cd]pyrene) as possibly carcinogenic to humans (Group 2B). BaA and BaP, classified as probable carcinogens in humans (Group 2A), are present in smoked meats and have high Toxicity Equivalency Factors (TEFs) compared to other PAHs (IARC 2010). According to the European Food Safety Authority (EFSA), the acceptable level for benzo[a]pyrene is 2 μg/kg of the product, while the sum of four PAHs—benzo (a)pyrene (BaP), chrysene (CHR), benz(a)anthracene (BaA), and benzo(b)-fluoranthene (BbF)—must not exceed 12 μg/kg (Reinik et al. 2007). PAHs, being lipophilic compounds, are carried in the bloodstream bound to lipoproteins. Within the body, they undergo biotransformation into water-soluble compounds, resulting in the production of polar oxygen derivatives. Among these derivatives, epoxides are the primary products, along with other carcinogenic compounds. The metabolic activation of PAHs produces radical cations, diol-epoxides, and o-quinones, which have the potential to form adducts with DNA. This process can disrupt DNA replication, affect promoter methylation, and interfere with promoter binding, ultimately leading to DNA mutation and abnormal gene expression, which may contribute to tumor formation (Moorthy et al. 2015). Furthermore, metabolites of PAHs can form adducts with proteins within cells, potentially altering their normal function. Additionally, reactive oxygen species generated by PAH metabolites can initiate carcinogenesis by modifying DNA, lipids and proteins (Moorthy et al. 2015).

N-nitroso compounds (NOCs) have the ability to induce cancer in different animal species, including higher primates and have been found to be carcinogenic in various organs among animals (Bogovski and Bogovski 1981). In humans, there exists compelling evidence supporting the role of NOCs in the development of specific cancers such as gastric, esophageal, nasopharyngeal, and colon cancers (Tricker and Preussmann 1991; Mirvish 1995). Numerous NOCs are considered potential human carcinogens (Grosse et al. 2006; WCRF/AICR 2007), among them N-nitrosodimethylamine (NDMA) has been designated as "probably carcinogenic to humans" (Grosse et al. 2006; IARC 1987). Human exposure to preformed NOCs occurs through various sources including tobacco products, dietary intake, medications, occupational settings, as well as NOCs produced internally (Mirvish 1995). Internally, NOCs, which include nitrosamines and nitrosamides, can form when nitrite reacts with the breakdown products of amino acids within the stomach. It is estimated that endogenous synthesis accounts for 45–75% of total NOC exposure (Tricker 1997). Moreover, the abundance of heme iron present in red meat could potentially contribute to the internal generation of N-nitroso compounds derived from nitrates (Nogacka et al. 2019; John et al. 2011).

In a research study conducted in the Netherlands, it was observed that the intake of processed meat showed a positive correlation with overall head and neck cancer (HNC) and the specific type known as oral cavity cancer (OCC), but no such relationship was identified with other subtypes such as oropharyngeal cancer (OHPC) and laryngeal cancer (LC) (Perloy et al. 2017). Another investigation by Xu et al. (2014) indicated that the consumption of processed meat was linked to the development of cancers in the mouth and oropharynx; however, there was no discernible association between the consumption of red meat and oropharyngeal cancer. Furthermore, Li et al. (2016) corroborated previous findings suggesting that a diet high in red or processed meat is correlated with an elevated risk of nasopharyngeal carcinoma (NPC).

Another cancer which ranks among the most prevalent forms of cancer globally is Esophageal cancer (EC). There is speculation that a diet rich in protein might elevate the susceptibility to EC (Torre et al. 2015). Advanced case-control investigations have hinted at a heightened risk of esophageal (squamous cell) cancer among individuals with a history of consuming substantial quantities of processed meat (Rosato et al. 2019). These findings were substantiated in his meta-analysis by Choi et al. (2013), revealing a 30% increase in the risk of esophageal cancer. Salehi et al. (2013) asserted that esophageal cancer is linked to the intake of total red meat and processed meat, rather than higher consumption of poultry. In another study, Pournaghi et al. (2019) established a connection between the consumption of red meat, processed meat and poultry with the risk of EC. They observed a positive correlation between the frequency of consuming red meat, processed meats like sausages and chicken with skin, and the risk of EC, whereas the consumption of lamb and fish did not exhibit any association. Additionally, Cross et al. (2011) demonstrated in their study that elevated concentrations of two HAAs, MeIQx and PhIP, found in meat, were associated with an increased risk of esophageal adenocarcinoma. Moreover, for esophageal neoplasms, the dose-response relationship between PAH concentration and EC risk holds significance, indicating a potential causal role of PAH exposure in the pathogenesis of EC (Abedi-Ardekani et al. 2010).

According to some findings mutagenic HAAs have been shown to alter DNA structures within pancreatic cells, suggesting that the consumption of heat-treated meat could pose a risk factor for pancreatic cancer (Malfatti et al. 2016). In a separate case-control study the consumption of red and processed meat was linked to a heightened risk of pancreatic cancer. Interestingly, this association was observed to be more prevalent among men compared to women (Zhao et al. 2017).

Cross et al. (2011) conducted a study revealing a correlation between the intake of 2-amino-3,4,8-trimethylimidazo[4,5-f]quinoxaline (DiMeIQx) and gastric carcinoma. Individuals with the highest DiMeIQx consumption exhibited an elevated risk of gastric cardiac cancer, while benzo(a)pyrene displayed no significant association. This heightened risk was notably influenced by the consumption of smoked, grilled, and processed meat (Lazarevic et al. 2010). Moreover, the prevalence of dietary patterns, characterized by substantial meat consumption, was linked to an increased susceptibility to gastric adenocarcinoma (Castelló et al. 2018).

Exposure to heterocyclic aromatic amines (HAAs) and polycyclic aromatic hydrocarbons (PAHs) derived from food is recognized as a significant contributor to colorectal cancer (CRC) development. Studies conducted by Drew et al. (2013) have underscored the role of HAAs (particularly MeIQx and DiMeIQx) and PAHs present in red meat dishes in initiating the carcinogenesis of colon cancer. In a research study by Cross et al. (2010), both the consumption of red meat and processed meat exhibited positive associations with the incidence of colorectal cancer. These associations may stem from exposure not only to HAAs (MeIQx and DiMeIQx), which are elevated in meat cooked at high temperatures, but also to heme iron and nitrates inherent in meat. Steck et al. (2014) identified positive correlations between the intake of thermally processed meat, its HAAs compounds, and colon cancer. Furthermore, they explored how specific genotypes modulate the relationship between meat consumption, meat-derived carcinogens, and colon cancer. Various genotypes manifest distinct expressions of resultant HAA-DNA adducts. The presence of "at-risk" alleles, predisposed to CRC, is linked with high intakes of red meat, well-done red meat, pan-fried red meat, and meat carcinogens MeIQx and DiMeIQx.

A meta-analysis led by Fabiani et al. (2016) and a case-control investigation conducted in Argentina by Niclis et al. (2015) affirmed that dietary patterns known as "Western" and "carbohydrate," characterized by elevated consumption of red meat, processed meat, eggs, sweets, bread, pasta and rice, were significantly linked to an elevated risk of prostate cancer. In one research study by John et al. (2011), heightened prostate cancer risk was associated with frequent intake of grilled or well-done red meat. Interestingly, there wasn't an increased risk of advanced prostate cancer observed with animal fat consumption, suggesting that the link between diets rich in red meat and prostate cancer isn't attributable to fat content alone. The manner of cooking and the extent of meat roasting, influencing the concentration of carcinogenic HAAs and PAHs, have been shown to influence the risk of prostate cancer (John et al. 2011).

Numerous accounts suggest an elevated risk of lymphatic cancers linked to excessive intake of processed high-protein products. Dietary choices may influence lymphoma development by stimulating lymphoid tissue in the digestive tract via specific nutrients, thereby altering the immune system's response (Shivappa et al. 2017). An observable rise was noted correlating with the frequency of grilled or roasted chicken consumption, with daily intake associated with an 80% rise in NHL risk (Campagna et al. 2015). A study by Rohrmann et al. (2011) revealed a positive correlation between chronic lymphocytic leukemia (CLL) and increased consumption of processed meat. Daniel et al. (2011) conducted a case–control study involving US men and women, investigating meat consumption and meat-related mutagens concerning renal cell carcinoma (RCC). Exposure to PAH, commonly found in grilled meat, was linked to an elevated risk of kidney cancer. RCC risk surged alongside grilled meat consumption and exposure to PAHs like benzo(a)pyrene, with BaP consumption doubling the risk. The NIH (National Institutes of Health) and AARP (American Association of Retired Persons) Diet and Health study revealed that regular processed meat intake heightens bladder cancer risk. Findings indicated a positive correlation between frequent red meat consumption and dietary exposure to PhIP, which contributed to bladder carcinogenicity (Ferrucci et al. 2010).

15.6 Factors Affecting Formation of N-Nitroso Compounds (NOCs), Heterocyclic Aromatic Hydrocarbons (HAAs) and Polycyclic Aromatic Hydrocarbons (PAHs)

15.6.1 Formation of N-Nitroso Compounds (NOCs)

The formation of N-nitroso compounds (NOCs) is influenced by various factors, each contributing to the complex chemistry involved in their production. Understanding these factors is crucial for mitigating health risks associated with the consumption of NOCs. Some of the key factors affecting NOC formation are as follows:

Amine content-amines, including primary, secondary and tertiary amines, are naturally present in meat proteins. These amines as well as certain primary or secondary amino acids as well as quaternary ammonium salts act as precursors in the formation of N-nitrosocompounds (Fiddler et al. 1972a; Shephard et al. 1987). During processing, they can react with nitrite to form NOCs.

$$\text{NaN0}_2\ (\text{Nitrite}) + \text{HCl}\ (\text{Hydrochloric acid}) \rightarrow \text{HONO}\ (\text{Nitrous acid}) + \text{NaCl}\ (\text{Salt})$$

N-nitrosamines originate from the chemical interaction of secondary amines with nitrous acid derived from sodium nitrite, typically occurring under acidic conditions, with optimal activity around a pH of 3.5 (Fig. 15.1). Nonetheless, the speed of nitrosation concerning secondary amines is notably impacted by their basic nature (Walters 1985). Likewise, N-nitrosamides arise from the reaction between nitrous acid and the respective amides (Crosby 1983).

Nitrite content-nitrite (NO2-) is commonly used in the curing process of processed meats to prevent bacterial growth, enhance flavor and maintain the characteristic pink color. However, nitrite can react with amines present in meat to form NOCs through nitrosation reactions. Given that the rate of N-nitrosamine

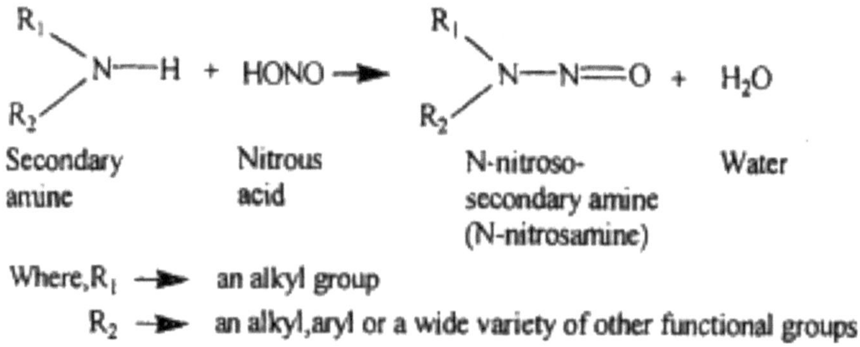

Fig. 15.1 Mechanism of nitrosamine formation in muscle foods

formation correlates with the square of nitrite concentration (Walters 1985), there exists a strong association between the initial levels of nitrite and the generation of these compounds. Sen et al. (1974) noted a strong correspondence between the quantity of NPYR (N- Nitrosopyrrolidine) and the initial concentrations of nitrite present in fried bacon.

pH level-the pH level of meat products plays a critical role in NOC formation. Lower pH levels, typically achieved through the addition of acidic ingredients like vinegar or citric acid, promote the nitrosation reaction between nitrite and amines, leading to increased NOC formation. For instance, in normal healthy individuals, gastric acidity itself serves to convert nitrate into nitrite primarily through oral microbial flora (NAS 1981). Consequently, factors such as gastric pH levels, nitrite concentration in the stomach, in vivo nitrate reductase activity, dietary nitrate intake, levels of ascorbic acid and the ingestion of amine/amide precursors can all affect the formation of NOCs and thus impact cancer susceptibility (Hotchkiss et al. 1992). Moreover, a notable positive association was identified between intragastric pH levels (ranging from pH 4 to 7.5) and the overall volatile N-nitrosamine content, although this relationship was absent within the lower pH range (pH 1–3) (Dallinga et al. 1998). More recent findings by Krul et al. (2004) indicated the formation of NDMA when nitrite was combined with dimethylamine or certain fish species, such as codfish, within an in vitro gastrointestinal model system.

Processing time, temperature and method of cooking-high-temperature cooking methods, such as grilling, frying and broiling, can promote the formation of NOCs. Sinha et al. (1998a, b) demonstrated that cooking methods like grilling or barbecuing, which expose meat to high temperatures, result in increased levels of NOCs compared to cooking methods like boiling or stewing. This finding underscores the role of heat and catalytic compounds in NOC formation during meat preparation. During cooking, nitrosation reactions are accelerated, particularly at temperatures above 120 °C. Additionally, the duration of cooking and the intensity of heat influence NOC formation. Osterdahl and Alriksson (1990) reported that the generation of nitrosamines showed significant alterations based on frying duration and temperature. Moreover, it was observed that microwavc-cookcd bacon contained notably lower levels ($p < 0.01$) of volatile N-nitrosamines compared to bacon fried in a pan. This observation was corroborated by Shahidi et al. (1995), who noted that frying led to the highest NDMA content of 0.79 pg/kg in chicken salami, whereas boiled or microwaved products contained up to 0.55 and 0.65 pg/kg of NDMA, respectively. Additionally, indirect heating methods resulted in lesser NDMA production in seafood products (Lee et al. 2003). Furthermore, both an increase in processing time and nitrite concentration amplified NDMA levels in frankfurters (Fiddler et al. 1972b). N-Nitrosopyrrolidine (NPYR) stands as another noteworthy NOC frequently detected in fried bacon. Pensabene et al. (1974) obscrvcd no NPYR formation in bacon fried at 99 °C, while a maximum level (17 pg/kg) emerged at 204 °C, with higher concentrations found on the bacon's exterior. Nonetheless, a direct correlation was established between slice thickness and frying duration in NPYR formation during bacon frying (Theiler et al. 1981). Fiddler et al. (1974) proposed potential pathways for NPYR formation in fried bacon, either through nitrite action on collagen itself or its pyrolytic breakdown during heating, with adipose tissue serving as a precursor for NPYR formation.

Genetic, environmental and ecological factors-nitrosamine concentrations in raw meat were influenced by diverse factors like environmental conditions, dietary habits, animal species, breeding practices and seasonal variations (Rywotycki 2003a, b). For instance, dimethylnitrosamine (DMNA) exhibited mean contents ranging from 6.02 to 9.11 pg/kg, while diethylnitrosamine-DENA ranged from 6.35 to 8.83 pg/kg across different wild animals (Rywotycki 2003a). Nitrosamine levels were lowest during winter and peaked in autumn. Additionally, non-organically bred pork and beef exhibited higher nitrosamine levels compared to organically bred ram, goat, and veal meats (Rywotycki 2003b). This observation aligns with the findings of Shahidi et al. (1995), who noted elevated NDMA levels (0.37–0.52 pg/kg) in chicken salami containing seal meat or seal protein hydrolysate compared to chicken-only meat salami (0.33 pg/kg).

Meat processing techniques-different processing techniques, such as smoking, curing, and fermentation, can impact NOC formation. For example, traditional smoking methods involving the incomplete combustion of wood can introduce PAHs, which may react with nitrite to form NOCs. While Kimoto et al. (1982) identified N-nitrosothiazolidine (NTHZ) in fried bacon too, however, the primary contributor to NTHZ formation in bacon was attributed to the smoking process, particularly wood smoke, as noted by Pensabene and Fiddler (1983). Cysteamine emerged as a probable precursor for NTHZ formation in bacon, as reported by Pensabene and Fiddler (1985). Moreover, Ikins et al. (1988) suggested that formaldehyde and acetaldehyde present in liquid smoke could generate N-nitrosothiazolidine-4-carboxylic acid (NTCA) and Methyl-nitrosothiazolidine-4-carboxylic acid (MNTCA) in vitro or in vivo, serving as precursors for NTHZ. Additionally, Sen et al. (1986) demonstrated a positive correlation between NTHZ formation from NTCA and frying temperature and time.

Processing and packaging materials-N-nitrosodi-n-butylamine, N-nitrosodibenzylamine (NDBzA), N-nitrosodiisobutylamine (NDIBA), and N-nitrosodibutylamine (NDBA) represent a significant group of NOCs formed when meat or its products interact with rubber or rubber-containing materials (Sen et al. 1987; Petersen 1993; Fiddler et al. 1998). The primary amine precursor in rubber, dithiocarbamate, reacts with nitrite in meats to yield NDEA and NDBA (Sen et al. 1987). A study analyzing boneless hams processed in elastic rubber nettings revealed the presence of over 100 ppb NDBzA on bacon surfaces (Fiddler et al. 1998). However, Danish cured meat products exhibited NDBA occurrences not exceeding 0.05 ppm (Petersen 1993). Netting contact time with meat emerged as another factor influencing NDBzA formation in ham products, suggesting the immediate removal of such netting post-processing (Fiddler et al. 1996). Moreover, the utilization of natural casing and dry smoking was associated with increased N-nitrosoproline (NPRO) formation, a precursor of NPYR, reaching levels as high as 353 mg/kg compared to synthetic casing and liquid smoking (Hamburg and Hamburg 1992).

Ingredients used-various types of food additives find application in muscle foods, serving either as flavor enhancers or preservatives. However, Sen et al. (1973) highlighted the presence of NPYR (up to 105 ppb) and NPIP (up to 60 ppb) in sausages, primarily arising from the reaction of curing mixtures containing amine components from black pepper and/or paprika with nitrites present in cured meats. This discovery was corroborated by Song and Hu (1988) in Chinese sausages. Similarly, Yamamoto et al. (1984) identified NPIP levels reaching up to 29.5 ng/g in Japanese salami, originating from spices and pepper utilized in the formulation.

Microbial activity-microorganisms naturally present in meat, as well as those introduced during processing, can influence NOC formation. Some bacteria possess nitrate reductase enzymes that convert nitrate (NO3-) to nitrite (NO2-), which can then react with amines to form NOCs. These enzymes facilitate the conversion of nitrate (NO3-) to nitrite (NO2-). For example, species like Staphylococcus, Micrococcus, and Lactobacillus have been identified as nitrate-reducing bacteria commonly found in meat products. Nitrite produced by these bacteria can serve as a precursor for NOC formation. Pegg and Shahidi (2000) highlighted the significance of bacterial nitrate reductase activity in the production of nitrite and subsequent formation of NOCs in meat products. Similarly, investigations by Corpet et al. (1997) provided insights into the influence of microbial activity on NOC formation during meat processing.

Presence of catalysts-certain compounds present in processed meats can act as catalysts in NOC formation, accelerating the conversion of nitrites and amines into carcinogenic NOCs during cooking and processing. Heme iron, has been identified as a potent catalyst in the formation of NOCs. When meat is cooked at high temperatures, heme iron catalyzes the reaction between nitrites and amines present in the meat, leading to the formation of nitrosamines and other NOCs. Research suggests that the interaction between heme iron, nitrites, and amines is a major pathway for NOC formation in processed meats (Santarelli et al. 2008). Metals like copper, present in food additives, cookware, or processing equipment, can also act as catalysts in NOC formation. These metals may interact with nitrites and amines during cooking or processing, facilitating the conversion of nitrites to NOCs. Studies have demonstrated the catalytic role of metals in NOC formation, highlighting the importance of controlling metal exposure during meat processing (Lijinsky 1999). Likewise, Sulfur-containing compounds, such as sulfides and thiols, present in processed meats can also serve as catalysts in NOC formation. These compounds may originate from additives or naturally occur in certain ingredients used in meat processing. During cooking or processing, sulfur compounds can react with nitrites and amines, promoting the formation of NOCs (Hedberg et al. 1983).

Storage conditions-improper storage conditions, such as prolonged exposure to light, heat, or oxygen, can promote the formation of NOCs in processed meats. Nitrite can undergo oxidative reactions under these conditions, leading to increased NOC levels (Puangsombat et al. 2012). Oxygen plays a critical role in facilitating oxidative reactions, which can convert nitrites into nitrosating agents capable of reacting with amines present in meat to form NOCs. These reactions are accelerated in the presence of heat and light, which further promote the formation of NOCs in processed meats (Scanlan 1983).

15.6.2 Formation of Heterocyclic Aromatic Hydrocarbons (HAHs) and Polycyclic Aromatic Hydrocarbons (PAHs)

The formation of heterocyclic aromatic hydrocarbons (HAHs) and polycyclic aromatic hydrocarbons (PAHs) are influenced by various factors such as thermal conditions, cooking methods, heating/cooking time, precursor type and amount, fat content and oxidative status, etc.

Thermal conditions-thermal conditions play a pivotal role in determining the levels of heterocyclic amines (HAAs) and polycyclic aromatic hydrocarbons (PAHs) in meat (Yang et al. 2019; Zhang et al. 2013). As temperatures increase, the formation rate of HAAs and PAHs accelerates, as noted in scientific literature (Keating et al. 1999). Cooking methods such as steaming, boiling, or microwave heating, which operate at temperatures below 100 °C, typically yield meat with minimal or undetectable HAA content (Solyakov and Skog 2002; Chiu et al. 1998). However, mutagenic substances begin forming at 125 °C, with a significant increase in HAA formation dynamics observed at 150 °C, peaking at 200 °C. For example, in pork fried at 125 °C, 58 ng/g of HAAs were detected, while at 150 °C, the concentration nearly doubled to 91 ng/g. Similarly, thermal processing of beef resulted in a significant increase in the generation of IQx, MeIQx, PhIP, and 4,8-DiMeIQx when the temperature rose from 170 °C to 200 °C (Kondjoyan et al. 2010). PAHs, on the other hand, commence formation at 200 °C, with the highest levels observed between 500 °C and 700 °C.

Cooking methods-the choice of cooking method significantly influences the mutagenic activity of heat-treated meat (Oz and Kotan 2016; Guo et al. 2014). Different cooking techniques yield varying concentrations of carcinogenic compounds. For instance, grilled drake meat exhibits the lowest HAA concentration compared to pan-fried, roasted, or deep-fried meat (Omojola et al. 2015). The energy source used also influences PAH levels; chicken grilled with charcoal contains higher PAHs compared to oven-grilled chicken (Farhadian et al. 2010). Generally, gentler cooking methods operating at lower temperatures, such as boiling and roasting, produce fewer mutagenic compounds, while higher temperatures associated with methods like pan frying and grilling lead to elevated mutagen levels (Alaejos and Afonso 2011). This trend is consistent across various studies, although temperature alone does not solely determine HAA content in meat. For instance, stone-barbecued beef steak was found to have a higher amount of HAAs compared to wire-barbecued beef steak, indicating the influence of factors like heat transfer mechanisms and surrounding media (Zorrilla and Singh 2003). Coefficient of heat transfer, better in methods with direct contact, significantly impacts mutagen levels;

for example, pan frying has a higher coefficient compared to air convection, resulting in higher mutagen levels (Persson et al. 2002). Indirect heating methods, like oven roasting and microwave cooking, generally yield lower levels of carcinogenic compounds compared to deep-frying or charcoal grilling (Jinap et al. 2013). Direct heating methods, such as pan frying, offer opportunities to mitigate HAA formation by frequently turning the meat and preventing melted fat from contacting the heat source, which leads to the formation of PAHs. PAH content is influenced by both the source of the product and the method of thermal processing, with the latter being more influential (EFSA 2008). For example, smoked sausage contained higher benzo[a]pyrene levels compared to grilled or cooked sausage, emphasizing the significance of processing methods in determining PAH content. For example, the benzo[a]pyrene content was found to be 0.7 μg/kg in smoked pork meat and 1.3 μg/kg in chicken meat. In the context of sausages processed through three different methods, smoked sausage contained 1.1 μg/kg of benzo[a]pyrene, grilled sausage contained 0.5 μg/kg, and cooked sausage contained 0.15 μg/kg (EFSA 2008). However, since these factors (temperature and cooking methods) take place simultaneously, verifying their additive effect is important in determining the final content of HAAs in meat products as presented in Table 15.2 below:

Heating/cooking time-heating/cooking time significantly influences the generation of carcinogenic compounds. On examining the aqueous model system of meat extracts, researchers observed that heating at 175 °C for 1 h resulted in a 1.2 ng/g content of 4,8-DiMeIQx. Doubling the time at the same temperature tripled the amount of 4,8-DiMeIQx (Bordas et al. 2004). Similar patterns emerged in a study investigating the levels of HAAs in ground beef cooked at temperatures ranging from 150 °C to 180 °C. Initially, low levels of these substances were detected after 3 min of cooking, but their concentrations increased over time (Turesky et al. 2005). Studies on beef have shown that HAAs do not form in less than 5 min of heat treatment, while certain mutagens (IQ, MeIQ, 4,8-DiMeIQx, AαC, and harman) were only detected in small quantities after at least 10 min. However, the influence of time on HAA formation is directly correlated with the cooking temperature. Conversely, the formation of PAHs is most prominent during the initial minutes of cooking. This phenomenon is likely due to the deposition of PAHs on the meat surface, which prevents further formation due to the isolation of fat (a precursor) from the heat source (Lee et al. 2016).

Precursor type and amount-the genesis of HAAs hinges upon the presence of specific precursors, including creatine, carbohydrates, amino acids, nitrogenous bases, and nucleosides. Upon exposure to heat, creatine transforms into the physiologically inert compound creatinine, which plays a pivotal role in the formation of the imidazole ring, crucial for the generation of certain HAAs like IQ and IQx, contingent upon the creatinine levels in meat (Skog and Jägerstad 1993). Research has demonstrated that beef with lower creatinine levels exhibits reduced mutagenicity compared to beef with higher creatinine concentrations (Jackson et al. 1994). However, the creation of non-polar HAAs remains unaffected by creatinine content, necessitating other amino acids as precursors. Various amino acids, including glycine, serine, threonine, lysine, and alanine, contribute to the synthesis of

Table 15.2 Combined effect of temperature and cooking method on the formation of HAAs in meat products

Meat product	Temperature/Method of cooking	Effect on formation of HAAs	Reference
Beef patties	Steam roasting at 100 °C, infrared grilling at temperatures of 180 °C, 200 °C, and 220 °C, charcoal grilling, and microwave cooking at power levels of 1000 W and 500 W	Charcoal-grilled beef samples had the highest content of HAAs	Wang et al. (2021)
Lamb patties	Frying at 200 °C; pan frying at 200 °C, roasting at 200 °C and stewing in seasonings at 100 °C	Patties stewed in seasonings had the higher content of HAAs in comparison to roasted, fried, or pan-fried	Guo et al. (2014)
Pork patties	Boiling to an internal temperature of 71 and 77 °C; oven-broiling at 177 and 225 °C; pan frying at 177 and 225 °C	HAAs formation was highest in pan-fried pork patties. HAA concentration increased in meat samples with the increase in internal temperature	Han-Seung (2005)
Beef chops	Sous-vide cooking at 75, 85, and 95 °C; pan frying at 75, 85, and 95 °C; boiling in pressure cooker	Pan-fried beef chops at the temperature of 95 °C had the highest total content of HAAs	Oz and Zikirov (2015)
Chicken breast and duck breast	Pan frying with no oil at 180 °C; deep fat frying at 180 °C; charcoal grilling at 200 °C and roasting (oven) at 200 °C	The highest amount of HAAs were noticed in charcoal-grilled chicken breast followed by pan-fried and charcoal-grilled duck breast	Liao et al. (2010)
Beef and chicken meatballs	Deep fat frying at 150 °C; pan-cooking at 180 °C; charcoal grilling (temperature of 280 °C on meatball surface); oven roasting at 180 °C	Charcoal grilling was responsible for the highest HAAs generation in beef meatballs while pan frying for chicken meatballs	Keskekoglu and Uren (2017)
Pork loin	Electric oven cooking at 180 °C; hot air frying at 180 °C and deep oil frying at maximum power	Meat samples subjected to cooking in an electric oven had the highest HAAs content	Yang et al. (2017)
Lamb patties	Charcoal grilling at 450–500 °C; infrared grilling at 240 °C and superheated steam roasting at 240 °C	The highest rate of HAAs formation was observed in charcoal grilling followed by infrared grilling and superheated steam roasting	Suleman et al. (2020)

mutagenic compounds such as MeIQx (Jägerstad and Skog 1991). Additionally, reducing sugars like glucose, fructose, lactose, and ribose serve as critical precursors for HAAs formation, with the ratio between amino acids and sugars playing a pivotal role. When the sugar content surpasses half the molar quantity of amino acids, HAAs generation diminishes (Skog and Jagerstad 1990). For instance, a notable 50% reduction in HAAs content was observed in pork from pigs with the RN allele, characterized by higher glycogen levels, compared to meat from normal pigs (Olsson et al. 2002). The HAA content significantly varies across meat types due to differences in precursor concentrations. Ground beef, for instance, contains 50 ng/g of PhIP and 2 ng/g of MeIQx (Zeng et al. 2018), with slightly lower PhIP values recorded at 200–230 °C frying temperatures, ranging from 10–25 ng/g. Poultry meat, on the other hand, exhibits markedly higher HAAs levels, with barbecued breast fillets containing up to 480 ng/g of PhIP, while oven-broiled meat records only 72 ng/g of PhIP (Sinha et al. 1995). Pork demonstrates the lowest HAA values, with grilled pork steaks containing approximately 28.62 ng/g of PhIP (Polak et al. 2020).

Fat content and oxidative status-the composition and oxidative condition of fat have been shown to influence the development of mutagenic substances. The superior heat transfer capability of fats allows for rapid attainment of higher temperatures, leading to increased production of HAAs and PAHs in meat. Additionally, higher fat content diminishes the relative presence of other crucial precursors necessary for HAAs formation, such as creatine, sugars, and amino acids. Studies have demonstrated that exceeding the optimal fat ratio decreases the HAAs content (Persson et al. 2008). Knize et al. (1985) observed that ground beef with 30% fat was less mutagenic (150,000 revertants per kg of fresh beef) compared to beef with 15% fat (230,000 revertants/kg). Furthermore, the oxidative state of fat is equally significant. Including oxidized soybean oil enhanced the production of PhIP, as did the incorporation of lipid oxidation byproducts like ω-6- and ω-3-derived lipid hydroperoxides, 4,5-epoxy-2-alkenals, 2,4-alkadienals, 2-alkenals, 4-oxo-2-alkenals, and 4-hydroxy-2-nonenal (Zamora et al. 2012). In contrast, the total PAH content in meat products primarily relies on the smoking technique and the type of wood employed during the process. For instance, the combined PAH (BaP, BaA, BbF, CHR, and BjF) content in pork smoked using plum wood chips measured 221 μg/kg, whereas pork smoked with apple wood chips registered 27.3 μg/kg (Stumpe-Vīksna et al. 2008).

15.7 Reduction of N-Nitroso Compounds (NOCs) and Polycyclic Aromatic Hydrocarbons (PAHs) in Processed Meat Products

The reduction of N-nitroso compounds (NOCs) and polycyclic aromatic hydrocarbons (PAHs) in processed meat products is essential for safeguarding consumer health, ensuring regulatory compliance and maintaining the overall quality of food products.

15.7.1 Mitigation Strategies for Reduction of Carcinogenic N-Nitroso Compounds (NOCs) in Processed Meat Products

Nitrosamine reduction can be approached from four angles-management of precursors, adjustment of processing conditions, controlling the external component and utilization of degradation techniques.

Precursor management-nitrites and biogenic amines play crucial roles as precursors in the formation of nitrosamines. Therefore, controlling the levels of these precursors is key to managing nitrosamine production. Alongside adhering strictly to nitrite and nitrate residue standards, finding suitable substitutes that can partially fulfill the functions of nitrite is imperative. Nitrite serves primarily to preserve color, inhibit bacterial growth, and prevent oxidation in processed meat items. Various alternatives have been identified for nitrite-based coloration, including Lactobacillus pentosus, Lactobacillus JCM1173, and Monascus pigment-17. As for antibacterial properties, substitutes like nisin, potassium sorbate, and hypophosphite are commonly utilized (Huhtanen and Feinberg 1980). Substances such as monascus pigment, streptococcus lactis, and potassium sorbate exhibit superior inhibitory effects against Clostridium botulinum compared to nitrous acid. Moreover, natural sources like celery powder and spinach powder offer viable alternatives to nitrite usage (Terns et al. 2011). These findings underscore the importance of exploring diverse substitutes to mitigate the reliance on nitrites and enhance food safety in processed meats.

In meat products, enzymes known as amino acid decarboxylases play a role in converting amino acids into biogenic amines. Hence, controlling the proliferation of enzyme-producing microorganisms is key to minimizing the presence of biogenic amines. This would involve rigorous control of raw material freshness and processing environment cleanliness. Optimizing factors like temperature, pH levels, osmotic pressure, and water activity can help manage the growth of microorganisms. Additionally, the inclusion of bacteria and additives that deter biogenic amine formation is important (Tasic et al. 2012). Moreover, techniques such as vacuum packaging, high-pressure treatment, irradiation, and freezing can further aid in reducing biogenic amine levels in meat products.

Adjustment of processing conditions-various processing conditions, including temperature, pH levels, cooking methods, and the fat-to-lean ratio of raw meat, play significant roles in determining the levels of nitrosamines. Studies have demonstrated a direct correlation between higher cooking temperatures and increased nitrosamine production (Yurchenko and Molder 2007; Skrypec et al. 1985). Regarding pH, the nitrosation reaction is theoretically most efficient within a range of 3.0–4.0. Excessively high pH levels hinder nitrogen oxide conversion, slowing the reaction rate, whereas excessively low pH levels lead to amine compound protonation, hindering interaction with nitrogen oxides. Different meat processing methods yield varying amounts of nitrosamines; for instance, fried and smoked barbecue products tend to contain higher levels, while those cooked via microwave typically have lower concentrations. These factors underscore the complexity involved in managing nitrosamine levels during meat processing. Deep frying and pan-frying methods tend to result in elevated levels of NDMA, NDEA, and NPYR in dry-cure sausages, while boiling and microwaving show a lack of N-nitrosamine production, indicating their suitability for cured meat preparation (Li et al. 2012). Direct heating techniques using coal, electric, or gas cookers generate more NDMA compared to indirect heating methods like electric ovens, steam pans, or microwave ovens (Lee et al. 2003). This discrepancy arises from nitrogen oxide production during the combustion of propane gas or coal, potentially leading to nitrosamine formation when interacting with amines in fish. Fat content also plays a crucial role in nitrosamine levels within meat products. Higher proportions of unsaturated fatty acids correlate with increased NPYR levels in sausages, possibly due to interactions between unsaturated fatty acids and nitric oxide, resulting in nitrosamine formation (Skrypec et al. 1985). In a study by Yurchenko and Molder (2007), fat separation conducted on cooked mutton revealed variations in nitrosamine content between lean meat and fat. Interestingly, while fat had higher NPYR levels compared to lean meat, the content was lower than in mutton without fat separation.

Controlling the external components-certain external substances have demonstrated effectiveness in inhibiting nitrosamine synthesis, including vitamins, phenols, sulfhydryls, quinones, spices, and various fruits and vegetables. Vitamins such as Vitamin C and Vitamin E employ a redox reaction with nitrite to impede nitrosamine formation. Vitamin C not only aids in nitrite reduction but also prevents NO oxidation, hindering imine and nitroso group combinations. Additionally, the hydroxyl group in the sixth position of the oxaman ring of Vitamin E's active group can capture free radical ammonia, thereby disrupting the free radical chain reaction and reducing nitrosamine synthesis. The incorporation of polyphenols into kimchi and meat products has also proven effective in blocking nitrosamine formation. Polyphenols exert their blocking effect primarily through the phenolic hydroxyl group. This group not only interacts with nitric oxide (NO) but also aids in the reduction of nitrous acid. The significant nitrite reduction activity observed in procyanidins and epicatechin gallate can be attributed to several factors. These include the ease of generating semiquinones, the presence of phenolic coupling, and the influence of acidic pH levels, which readily induce changes in molecular resonance (Bors et al. 2001; Ferreira et al. 2011). Proanthocyanidins and catechins stand out for their exceptional free radical scavenging abilities, attributed to specific structural features including the B ring, the 3-position hydroxyl group in the C ring, as well as the presence of 5-OH and 7-OH (Rocha et al. 2014). In a study by Liu et al. (2011), CO_2 supercritical fluid extraction technology was utilized to extract flavonoids from corn silk. Findings indicated that a concentration of 500 μg/ml of flavonoid extract achieved a maximum nitrite removal rate of 88.1% at pH 3.0. The capacity for nitrite removal positively correlated with concentration but inversely correlated with pH levels. Furthermore, the presence of quercetin and catechin in red wine has been shown to convert nitrite ions into NO, effectively reducing nitrite levels in the mouth and stomach (Takahama et al. 2010; Nadtochiy and Redman 2011). These discoveries highlight the potential of specific compounds in mitigating nitrite-related risks. Soybean phenolics exhibit the ability to disrupt nitric oxide (NO) metabolism particularly in acidic conditions by expediting the reduction of nitrite. Furthermore, the nitrite reduction activity of soybean phenolics shows a direct positive correlation with the phenolic content (Ferreira et al. 2011). Plant-derived polyphenols, including those from green tea and grape seeds, along with ascorbic acid, demonstrate remarkable efficacy in reducing residual nitrite levels. Among these, ascorbic acid stands out for its superior effectiveness (Li et al. 2013). Mercapto-based substances possess the capability to bind with nitrite, forming thionitrites. In a study by Choi et al. (2007), the impact of garlic juice on NDMA under simulated gastric juice conditions was investigated, revealing that garlic juice achieved a maximum blocking rate of 50.5% against NDMA. Quinones, owing to their redox properties, play significant roles in biological redox reactions, potentially facilitating or impeding various biochemical processes. Notably, certain spices such as Chinese pepper, star anise, cinnamon, and cumin have been identified for their inhibitory effects. For instance, a high level of black pepper (15 g/kg) was found to decrease NPIP content in both raw and well-done sucuk (Sallan et al. 2019). Furthermore, the inclusion of spices like onion and cumin during the processing of fermented sausages effectively inhibits the presence of NDMA, NDEA, NMOR, and NPIP, as well as the accumulation of biogenic amines (Sun et al. 2021). Moreover, natural plant extracts demonstrate efficacy in reducing nitrite levels. The addition of carambola dietary fiber concentrate to a pork and turkey mixture enhances the antioxidant capacity of Vienna sausages while significantly diminishing nitrite content (Vivar-Vera et al. 2018).

Since the identification of NDMA as the initial nitrosamine impurity in angiotensin II receptor antagonists in 2018, the presence of nitrosamine impurities in pharmaceuticals has sparked significant safety concerns within the industry. This has resulted in the recall and shortage of essential drugs. Over the past few years, nitrosamines have been detected in various sartan-containing medications, including losartan, irbesartan, and candesartan (Wichitnithad et al. 2023). Unlike sources of nitrosamine contamination in food, drugs are indispensable for patients requiring treatment for specific illnesses. Hence, there is a pressing need for inhibitors to minimize nitrosamine formation in pharmaceuticals. Recent studies have investigated five inhibitors, namely ascorbic acid, sodium ascorbate, α-tocopherol, caffeic acid, and ferulic acid, which demonstrated over 80% inhibition when spiked at approximately 1% level. These findings highlight the feasibility of inhibiting nitrosamine formation in oral solid dosage forms using appropriate inhibitors. Furthermore, several amino acids (including glycine, lysine, and histidine) have exhibited potential inhibitory effects on nitrosamine formation from 4-phenylpiperidine hydrochloride in solution (Nanda et al. 2021). These discoveries are not only crucial for the pharmaceutical industry to regulate nitrosamine formation in drug products but also offer insights for controlling nitrosamines in food processing.

Nitrosamine degradation techniques-technologies like irradiation and biodegradation offer potential for degrading nitrosamines. The efficacy of irradiation on Korean pork sausages was investigated by Ahn et al. (2004). The sausages underwent inflation and vacuum packaging respectively before exposure to irradiation doses of 0, 5, 10, and 20 kGy, followed by storage at 4 °C. Results revealed a reduction in NDMA content in vacuum-packaged sausages and NPYR in air-packaged sausages at doses of 10 and 20 kGy. Similarly, the NDMA and NPYR content in pepperoni and vegetable salami decreased significantly following γ-irradiation treatment (Byun et al. 2004). Furthermore, γ-rays demonstrated the ability to reduce NDMA, NDEA, and NPYR content in Chinese rugao ham (Wei et al. 2009). However, the precise mechanism behind radiation-induced nitrous gum degradation remains unclear, necessitating further investigation.

Research focusing on biodegradable nitrosamines is currently a prominent area of study. Rhodococcus RHA1, notably Rhodococcus jostii RHA1, demonstrated effective degradation of high concentrations (20–2000 μg/L) of nitrosamines, including NDMA, NDEA, NDPA, NPYR, and nitrosomorpholine (NMOR) (Homme and Sharp 2013). Additionally, Rhodococcus ENV425, or Rhodococcus ruber ENV425, exhibited significant degradation capability for NDMA in groundwater, reducing NDMA levels from 12 g/L to 100 ng/L (a degradation rate of 99.2%) (Wijekoon et al. 2013). Lactobacillus plantarum CP3 effectively inhibited the production of NDEA, NDMA, NMOR, and NMEA in fermented sausages. Moreover, co-inoculation with Lactobacillus saki M4 synergistically enhanced nitrosamine degradation. The variation in nitrosamine degradation rates by microorganisms may be linked to the presence of electron-withdrawing and electron-donating groups in nitrosamine molecules (Webster et al. 2013). Nitrosamines containing strong electron-donating groups, such as dimethyl, are readily degraded by organisms. Conversely, those with weak electron-donating groups tend to persist, resulting in lower degradation rates.

In conclusion, the reduction of N-nitroso compounds (NOCs) in processed meat products represents a critical endeavor in ensuring food safety and mitigating potential health risks associated with nitrosamines. Employing various strategies such as controlling processing conditions, managing precursors like nitrites and biogenic amines, and exploring alternative additives or technologies, manufacturers can effectively minimize the formation of NOCs. This concerted effort not only aligns with regulatory standards but also underscores the commitment to delivering high-quality, safe meat products to consumers. As research continues to unveil novel approaches and insights into NOC reduction, collaboration across industries and regulatory bodies remains essential in fostering continuous improvement and innovation in food safety practices.

15.7.2 Mitigation Strategies for Reduction of Carcinogenic Polycyclic Aromatic Hydrocarbons (PAHs) in Processed Meat Products

The strategies for reducing and controlling the formation of polycyclic aromatic hydrocarbons (PAHs) in processed meat products are multifaceted and can be broadly categorized into two main segments: pre- and post-thermal processing treatments. Pre-thermal processing involves actions like selecting lean meats, using appropriate fuels, filters, marinades and probiotics, as well as employing specific heating methods. Post-thermal processing treatments include techniques such as warm water rinsing and suitable packaging.

Use of suitable fuel sources and alternative cooking methods-raw materials such as meat, fuels, and cooking oils are susceptible to pre-contamination by Polycyclic Aromatic Hydrocarbons (PAHs) prior to processing (Zhu et al. 2022; Onopiuk et al. 2021). The majority of PAHs found in processed meat products arise during traditional and faulty cooking methods. Hence, taking preventive measures is crucial to minimize PAH formation while cooking. It is recommended to use hard woods with lower lignin content over soft woods, as the latter tend to burn quickly at high temperatures, resulting in increased PAH formation during smoking (Zhu et al. 2022; Onopiuk et al. 2021). Studies have shown that log fires, pine cones, and mesquite wood produce high levels of PAHs during smoking (Rose et al. 2015). Surprisingly, even coconut shells are utilized as "flameless and smokeless" charcoal to reduce the pyrolysis process (Viegas et al. 2012). Researchers have found that replacing woods with bagasse as a smoke generator effectively reduces PAH4 content in smoked fish (Essumang et al. 2013). Moreover, investigations with various wood chips revealed that poplar and hickory significantly reduce PAH contamination in meat products, while alder and beech tend to produce higher concentrations (Hitzel et al. 2014). Studies on smoked sausages using different types of wood showed that toxin concentrations remained below EU legislative limits when beech, neem, copper pod, ear leaf acacia, and eucalyptus wood were used (Malarut and Vangnai 2018). Researchers have also observed that white charcoal produces fewer PAHs compared to black and extruded charcoal during meat grilling (Kim et al. 2021). Therefore, traditional smoked meat producers need to opt for hard woods or white charcoal as preventive measures against PAH formation. Preparing wood or charcoal by pre-heating and charring to produce smoke at high temperatures, followed by extinguishing the flame before grilling or barbecuing meat, can be effective in altering the chemical composition of the smoke, particularly reducing PAHs. Additionally, reheating pre-burned charcoal to generate flames, prolonged exposure to high temperatures, dripping fat, and maintaining improper distances (< 25 cm) from heat sources should be avoided to minimize PAH concentration in meat products (Das et al. 2023). Chaemsai et al. (2016) conducted a study where mangrove charcoal was pre-heated to 650 °C for varying durations before grilling meat, emphasizing the importance of fully glowing charcoal before initiating grilling. It is evident from the discussion that various factors influence the pyrolysis and pyro-synthesis processes of fat and wood, necessitating remedial actions for each potential contributing factor.

Many commercial establishments opt for gas/electric ovens for roasting or barbecuing, employing indirect smoking methods with diverse filters such as zeolite, granular activated carbon and gravel filters to mitigate Polycyclic Aromatic Hydrocarbons (PAHs) in meat products. A study by Sampaio et al. (2021) demonstrated a substantial reduction in BaP content (90%) and PAH4 (85%) using zeolite and carbon filters. Innovative approaches, such as grilling chicken wrapped in banana leaf (34.7 μg/kg) and aluminum foil (45.4 μg/kg) to assess 3PAHs (Singh et al. 2020; Farhadian et al. 2012), yielded promising results and are commendable for addressing the issue, particularly for traditional food producers and roadside vendors. Eldaly et al. (2016) also applied the wrapping principle to meat grilling, introducing barriers between the product and flame, resulting in a significant reduction of BaP content in mutton and beef. Casings, whether natural or artificial, play a crucial role in preventing PAH contamination, particularly in sausage preparation, by establishing a barrier layer where smoke particles deposit, gradually forming clogging pores (Onopiuk et al. 2021). Compared to natural casings, cellulose or collagen casings prove more effective in blocking PAH penetration. This is attributed to the high porosity and uneven morphology of natural casings, increasing the risk of PAH contamination, while synthetic casings with smooth surfaces deposit minimal amounts of PAHs and exhibit reduced penetration ability (Onopiuk et al. 2021; Ledesma et al. 2015; Gomes et al. 2013).

When smoking or barbecuing meat, producers should opt for indirect heating or utilize electric or gas ovens instead of charcoal. Grilling leaner portions of meat and maintaining a proper distance from the heat source are also advisable practices to reduce the contamination of toxic carcinogens in food. Additionally, producers must take extra precautions to decontaminate raw materials from toxic environmental pollutants, which could potentially contribute to PAH contamination in food. Hence, extensive investigations are necessary to monitor raw materials throughout transportation, processing, and storage, considering possible environmental exposures such as air, soil, and surface water quality from a safety perspective. Moreover, traditional smoked meat manufacturers should adopt successful industrial strategies for reducing PAHs, regardless of consumer preferences and sensory attributes.

Washing of smoked meat products-the process of washing smoked meat products involves removing the skin or outer layer, rinsing with lukewarm water, and cleaning under running tap water (Sava et al. 2023; Vranešević et al. 2022). Some studies suggest that washing can effectively reduce the PAH content in smoked meat products. Mahugija and Njale (2018) found that washing smoked products with purified lukewarm water (60 °C, 2–3 min) significantly decreased the total PAH content. While this method is helpful in addressing the issue, it may sometimes not meet organoleptic standards. Additionally, washed products exhibited a shorter shelf life, indicating a need for further research on storage and preservation of washed smoked products. In another study, pork loins rinsed immediately after smoking showed reduced PAH content (BaP $= 3.43 \pm 0.21$ μg/kg; PAH4 $= 25.33 \pm 1.88$ μg/kg) on the surface compared to unrinsed samples (BaP $= 7.33 \pm 0.77$ μg/kg; PAH4 $= 61.14 \pm 1.72$ μg/kg) [97]. The PAH values (BaP and PAH4) obtained after rinsing smoked dry-cured pork loin are considered safe, as they fall within the permissible limits (BaP < 5 μg/kg; PAH4 < 30 μg/kg) set by the EU Regulation for traditional meat and meat products.

Use of marinades-marinating meat is a widespread culinary practice aimed at enhancing the texture and quality attributes such as tenderness, juiciness and flavor, particularly in heat-processed meat products. The selection of suitable marinade ingredients holds the potential to reduce the level of PAHs in the final product by influencing the physicochemical properties of the processed meat. This is because pretreating meat with a blend of spices, fruits or vegetables, or their extracts, containing unique active principles like citric acid, ascorbic acid, antioxidants, and phenolic components, before thermal processing can either inhibit the formation of carcinogenic components or inadvertently accelerate the faulty cooking process (Iko Afé et al. 2020).Various studies shed light on this phenomenon. For instance, the addition of onion or garlic during meat processing resulted in a significant decrease in PAH levels (Janoszka 2011). Preparing meat with a mixture of tomato juice, garlic paste, onion, salt and spices like cumin, coriander and black pepper significantly reduced PAH levels in chicken after thermal processing compared to using garlic alone, possibly due to the antioxidative properties of spice mixtures (Badry 2010). Marinating duck meat with pepper juice before charcoal grilling led to a notable reduction in BaP levels compared to the control group (Sinaga et al. 2016). In a recent study, different plant extracts, either individually or as a mixture, were incorporated as marinades for the pretreatment of pork neck before grilling. It was reported that the phenolic components, particularly jalapeno pepper, reduced almost 95% of 12PAHs, indicating the effectiveness of these extracts in reducing PAH formation in pork products (Onopiuk et al. 2021). The authors attributed this reduction to capsaicin, a biologically active substance present in jalapeno, which acted as a scavenger of free radicals, thereby enhancing the safety and shelf life of grilled meat products. Similarly, the use of different beers for marinating pork before heat processing resulted in significant reductions in BaP concentration (Viegas et al. 2014). Vinegar, a fermented product known for its active components like phenolic compounds, is frequently employed to enhance the microbiological quality, safety and shelf-life of various food items (Laranjo et al. 2019). In a study conducted by Cordeiro et al. (2020), different vinegars with varying levels of antioxidative properties demonstrated significant differences in their ability to inhibit the concentration of PAHs in smoked meat products. It was found that the use of acidic substances in marinades could notably decrease PAH levels during the thermal processing of meat, unlike alkaline ingredients or oils, which tend to increase the formation of heavy PAHs in meat. Lemon juice or tamarind juice, for example, showed promising results in inhibiting PAH formation in grilled meat products, in addition to their role in improving the texture and physicochemical properties of the meat (Sampaio et al. 2021; Farhadian et al. 2012; Wongmaneepratip and Vangnai 2017). Eldaly et al. (2016) investigated the impact of yogurt combined with various spice blends on marinating beef for kabab and kofta preparation. Marinating the beef before grilling resulted in reduced levels of PAHs, with grilled kebabs showing 57.93 μg/kg and grilled kofta exhibiting 30.2 μg/kg, compared to untreated samples which had 119.8 μg/kg for kabab and 59.2 μg/kg for kofta. Vinegar marinades also showed promise in reducing 4PAH content in charcoal-grilled pork loin, with Elderberry vinegar leading the way at 82%, followed by white wine vinegar (79%), red wine and cider vinegars (66%), and fruit vinegar with raspberry juice (55%) (Cordeiro et al. 2020). Natural food resources containing effective phenolic components, with their free radical scavenging capacity, hold potential as inhibitors for carcinogen formation in smoked meat products. However, further advanced research is necessary to unravel the exact mechanisms behind the application of acidic marinades and their inhibitory effects on PAH formation during intense heat treatment. Similarly, marinades containing various antioxidants extracted from green tea, such as epigallocatechin gallate, gallocatechin, catechin, epicatechin gallate, catechin gallate, eriodictyol, naringenin, and quinic acid, demonstrated significant reductions in PAH concentrations during grilling or roasting of chicken wings or pork meat (Wang et al. 2018, 2019; Park et al. 2017), possibly attributed to the presence of polyphenols as active components. Darwish et al. (2019) explored the antioxidative effects of micronutrients rosmaric and ascorbic acids against heat-treated meat, concluding that these molecules protected human colon (CaCo-2) cells from BaP-induced mutagenicity and oxidative stress. These findings suggest that marination with various phytochemicals containing polyphenols and flavonoids, spice blends, yogurt and associated precautions could effectively reduce PAH content in thermally processed meat products.

Use of probiotics and LAB-biological methods are emerging as promising alternatives to physical and chemical decontamination approaches for removing carcinogenic and mutagenic compounds from foods. This shift is due to their Generally Recognized as Safe (GRAS) status, cost-effectiveness and environmentally friendly characteristics. Many microorganisms, including bacteria, algae and fungi, can utilize PAHs as a carbon source necessary for their growth and development. In recent years, a novel and environmentally friendly strategy to reduce PAH contamination in food has gained attention through biological degradation using lactic acid bacteria (LAB) and probiotics (Yousef et al. 2022a, b). Literature also highlights the mechanisms involved in neutralizing the toxicity of food-borne carcinogens, particularly PAHs and acrylamides. The breakdown of PAHs involves activating antioxidative enzymes such as oxidase, manganese peroxidases and lipases. This process converts organic and stable carcinogens into less toxic, degradable hydrophilic metabolites. Moreover, the binding of heat-generated carcinogens to the cell wall and peptidoglycans of these bacteria may contribute to their biological detoxification (Yousef et al. 2022a; Khorshidian et al. 2020; Yousefi et al. 2021).

Using Fourier transform infrared spectroscopy (FTIR) and molecular dynamics (MD) techniques, researchers illustrated how organic toxins like Di-n-butyl phthalate (DBP) physically bind to the peptidoglycan layer of the cell wall of probiotics and lactic acid bacteria through C–O, OH, and/or NH functional groups, forming hydrogen bonds (Shoukat 2020; Lili et al. 2017). However, the efficiency of binding varies among different organisms due to factors like pH, temperature, incubation time, stability under acidic and heat conditions, nutrient availability and bacterial cell concentration. For example, in a study by Zhao et al. (2013) heat-treated inactive LAB exhibited a similar BaP binding ability to live viable pellets. Similarly, the binding rates of viable cells from strains like Lactobacillus pentosus CICC 23163 and Lactobacillus plantarum CICC 22135 were comparable to those of heat-treated pellets from the same strains (Lili et al. 2017).

Biological detoxification of PAHs presents a promising alternative to chemical methods, offering potentially high efficacy in the near future. Various strains of probiotics and lactic acid bacteria (LABs), including Lactobacillus delbrueckii subsp. lactis ATCC 4797, Lactobacillus plantarum ATCC 8014, Lactobacillus casei ATCC 11578, Lactobacillus sakei 23 K, Lactobacillus plantarum WCFS1, Bifidobacterium adoleascentis ATCC 15703, B. infantis, B. bifidum, B. adolescentis, B. longum, B. lactis, and B. breve, as well as Streptococcus thermophilus, have demonstrated the ability to detoxify various PAHs present in food systems, particularly in smoked meat products, through in-vivo or in-vitro methods (Shoukat 2020; Lili et al. 2017). Studies have shown that factors such as acidic pH, longer incubation times, higher temperatures and concentrations are directly related to the PAH binding ability of probiotics and LAB (Shoukat 2020). For instance, Yousefi et al. (2021) conducted an experiment using artificially contaminated PAH4 (BaA, BaP, BbF, and Chr) in phosphate buffer saline to assess the detoxification efficiency of Lactobacillus brevis TD4. They found that the maximum binding rate occurred at an initial concentration of 10 ppm, pH 5, a bacterial population of 10^9 CFU/mL and 24 h of incubation time. In another investigation, researchers explored the use of fermented potato juices containing metabolites like bacteriocin and pediocin from various strains of lactic acid bacteria. These strains, including Pediococcus acidilactici KTU05-7, Pediococcus pentosaceus KTU05-9 and Lactobacillus sakei KTU05-6, were isolated from spontaneous rye sour dough. The fermented potato juice was applied to the surface of pork sausages at 18 °C for 60 min, both before and after smoking (Bartkiene et al. 2017). Remarkably, the application of these juices not only reduced the levels of toxic carcinogens like BaP and chrysene on the sausage surface but also decreased the presence of various biogenic amines such as cadaverine, spermidine, and putrescine, found in either the surface or core of the sausages. Additionally, LAB and their metabolites have been found to effectively inhibit the growth of harmful pathogens and organisms responsible for food spoilage, including Pseudomonas aeruginosa and Escherichia coli, thereby extending the shelf life of food products. Furthermore, studies conducted using gastrointestinal digestion cell lines and animal models have provided evidence supporting the role of LAB in mitigating the presence of BaP and other dietary toxins (Slotkin et al. 2017).

While probiotics have demonstrated their capacity to detoxify diets contaminated with BaP and other PAHs, the potential introduction of genetically modified organisms (GMOs) could offer a new avenue for combatting PAH-induced cancer. Additionally, the synthesis of biosurfactants by safe and suitable microorganisms presents another potential method for bioremediation of toxins (Xu et al. 2022). Microbial biosurfactants have the unique ability to break down and disperse toxins and PAHs. In this context, probiotic bacteria adhere to PAHs present on the surface of meat. These bound PAHs, along with the probiotics, can subsequently be removed from the meat surface through additional processing methods such as slicing, washing, and marination. Given the limited number of studies conducted on the potential of probiotics to remove PAHs compared to other bioremediation strategies, further research is warranted.

Use of packaging systems and other applications-packaging serves the critical role of preserving quality, minimizing losses, and extending the shelf-life of processed food items until they reach consumers. With advancements in technology, packaging materials now present new opportunities for mitigating harmful compounds like PAHs found in various foods, including smoked meat products. Studies by Kuzmicz and Ciemniak (2018) have highlighted that packaging materials such as high density polyethylene (HDPE), low density polyethylene (LDPE), polypropylene, oxo-degradable and polyethylene terephthalate (PET) possess effective PAH adsorption properties, capable of trapping migrating PAHs from food. Independent research has shown that LDPE packaging systems can reduce BaP levels in smoked sausages and roasted meat products (Semanová et al. 2016; Chen and Chen 2005). For instance, Chen and Chen (Chen and Chen 2005) found that packaging roasted duck meat in LDPE at 25 °C led to a significant 73% reduction in BaP levels after 24 h of storage. Additionally, exposure of LDPE film to UV light for 2–3 h resulted in a specific reduction in BaP content, ranging from 13.5% to 29.2%.

In recent years, there has been significant interest in developing novel aerogels as effective alternatives to traditional adsorbents due to their favorable properties, such as low density, high porosity, and large surface area (Franco et al. 2021). In a study conducted by Kim et al. (2021), natural and environmentally friendly cellulosic aerogels (NaOH/urea, LiBr, and LiOH/urea) were utilized as absorbents to remove PAHs from smoked meat and meat products. The LiBr-functionalized absorbent exhibited the highest efficiency for adsorbing total PAHs due to its unique surface structure and pore size distribution. Moreover, chlorinated polyethylene (CPE), a microplastic polymer derived from modifying conventional polyethylene, has demonstrated the ability to adsorb various organic compounds, including PAHs and benzene derivatives found in freshwater (Gui et al. 2021). The adsorption behavior of CPE could be explored in different food matrices, including PAH-contaminated meat and meat products.

The use of irradiation as yet another alternative to diminish PAH formation has also been tried. In a recent study, gamma irradiation of smoked guinea fowl meat at a dose of 5 kilo gray (kGy) showed promise in reducing concentrations of PAHs and their carcinogenic derivatives compared to non-irradiated meat (Otoo et al. 2022). Hence, ongoing research into smart or active packaging systems and the application of ultraviolet rays hold potential as effective strategies for reducing PAHs and ensuring the safety of meat products.

15.8 Conclusion

The health risks posed by N-nitroso compounds (NOCs), heterocyclic aromatic amines (HAAs) and polycyclic aromatic hydrocarbons (PAHs), including their toxicity, carcinogenicity, teratogenicity and mutagenicity, has been significantly highlighted and documented by numerous epidemiological studies and governmental agencies. The link between dietary exposure to N-nitroso compounds, heterocyclic aromatic amines and polycyclic aromatic hydrocarbons (PAHs) and their potential role as carcinogens in humans has been firmly established. As a result, there is a pressing need to prioritize the prevention and management of their formation in processed meat products to address public health concerns and ensure safety. However, this task presents considerable challenges due to the wide variety of regional cuisines, food types, consumer preferences and cooking methods employed globally. The diversity in meat processing and preparation systems further complicates the issue. Given these complexities, implementing effective strategies to mitigate their formation requires careful consideration of various factors, including cultural practices, culinary traditions and cooking temperatures. The current strategies for mitigating nitrosamines in meat products encompass several key areas: addressing nitrosamine precursors, optimizing processing conditions, exploring alternative additives and employing degradation techniques (such as biodegradation and irradiation). One promising avenue involves finding substitutes for traditional nitrites, which can impede nitrosamine formation at the substrate level. Additionally, novel processing technologies require meticulous optimization of technical parameters to achieve optimal inhibition effects. Moreover, significant attention has been directed towards leveraging natural plant components to mitigate nitrosamine production. Agro-food by-products and waste materials offer abundant, environmentally friendly sources for potential natural inhibitors. Nitrosamine degradation technologies boast high efficiency and cost-effectiveness, yet each method has its own set of advantages and limitations. Therefore, a holistic approach that combines various strategies may be the most effective path forward. This entails maximizing inhibitory effects while minimizing any potential negative impacts on the nutritional and sensory qualities of food products. Furthermore, there is a need for further practical research to transition these methods into commercially viable and industrially relevant technologies. Furthermore, efforts to mitigate the presence of PAH carcinogens involve various strategies, including the use of marinades with antioxidant properties, the application of biological methods such as probiotics and lactic acid bacteria to detoxify the meat and the development of advanced packaging materials that can absorb or reduce the migration of harmful compounds. Moreover, innovative cooking techniques and equipment modifications are being explored to minimize the formation of N-nitroso compounds and PAHs during meat processing and preparation. Overall, addressing the presence of carcinogenic compounds in processed meats requires a comprehensive approach that encompasses research, regulation and consumer education. By implementing effective control measures, establishing stringent regulatory standards and raising awareness about the risks associated with these compounds, stakeholders can work together to ensure the safety and quality of processed meat products throughout the food supply chain.

References

Abedi-Ardekani B, Kamangar F, Hewitt S, Hainaut P, Sotoudeh M, Abnet C, Taylor P, Boffetta P, Malekzadeh R, Dawsey S (2010) Polycyclic aromatic hydrocarbon exposure in oesophageal tissue and risk of oesophageal squamous cell carcinoma in North-Eastern Iran. Gut 59:1178–1183

Ahn HJ, Kim JH, Jo C, Lee JW, Yook HS, Byun MW (2004) Effects of gamma irradiation on residual nitrite, residual ascorbate, color, and N-nitrosamines of cooked sausage during storage. Food Control 15(3):197–203

Alaejos MS, Afonso AM (2011) Factors that affect the content of heterocyclic aromatic amines in foods. Compr Rev Food Sci Food Saf 10:52–108

Augustsson K, Skog K, Jagerstad M, Dickman PW, Steineck G (1999) Dietary heterocyclic amines and cancer of the colon, rectum, bladder, and kidney: a population - based study. Lancet 353: 686–687

Badry N (2010) Effect of household cooking methods and some food additives on polycyclic aromatic hydrocarbons (PAHs) formation in chicken meat. World Appl Sci J 9(9):963–974

Bartkiene E, Bartkevics V, Mozuriene E, Krungleviciute V, Novoslavskij A, Santini A, Rozentale I, Juodeikiene G, Cizeikiene D (2017) The impact of lactic acid bacteria with antimicrobial properties on biodegradation of polycyclic aromatic hydrocarbons and biogenic amines in cold smoked pork sausages. Food Control 71:285–292

Bem Z (1995) Desirable and undesirable effects of smoking meat products. Die Fleischerei 3:3–8

Bjeldanes LF, Morris MM, Timourian H, Hatch FT (1983) Effects of meat composition and cooking conditions on mutagen formation in fried ground beef. J Agric Food Chem 31:18–21

Bogen KT (1994) Cancer potencies of heterocyclic amines found in cooked foods. Food Chem Toxicol 32:505–515

Bogovski P, Bogovski S (1981) Animal species in which N-nitroso compounds induce cancer. Int J Cancer 27:471–474

Bordas M, Moyano E, Puignou L, Galceran MT (2004) Formation and stability of heterocyclic amines in a meat flavour model system. J Chromatogr B 802:11–17

Bors W, Foo LY, Hertkorn N, Michel C, Stettmaier K (2001) Chemical studies of proanthocyanidins and hydrolyzable tannins. Antioxidants Redox Signal 3(6):995–1008

Bouvard V, Loomis D, Guyton KZ, Grosse Y, Ghissassi FE, Benbrahim-Tallaa L, Guha N, Mattock H, Straif K, International Agency for Research on Cancer Monograph Working Group (2015) Carcinogenicity of consumption of red and processed meat. Lancet Oncol 16(16):1599–1600

Bulanda S, Janoszka B (2022) Consumption of thermally processed meat containing carcinogenic compounds (polycyclic aromatic hydrocarbons and heterocyclic aromatic amines) versus a risk of some cancers in humans and the possibility of reducing their formation by natural food additives-a literature review. Int J Environ Res Public Health 19(8):4781

Byun MW, Ahn HJ, Kim JH, Lee JW, Yook HS, Han SB (2004) Determination of volatile N-nitrosamines in irradiated fermented sausage by gas chromatography coupled to a thermal energy analyzer. J Chromatogr A 1054(1–2):403–407

Campagna M, Cocco P, Zucca M, Angelucci E, Gabbas A, Latte G, Uras A, Rais M, Sanna S, Ennas G (2015) Risk of lymphoma subtypes and dietary habits in a mediterranean area. Cancer Epidemiol 39:1093–1098

Cashman KD, Hayes A (2017) Red meat's role in addressing 'nutrients of public health concern. Meat Sci 132:196–203

Castelló A, Fernández de Larrea N, Martín V, Dávila-Batista V, Boldo E, Guevara M, Moreno V, Castaño-Vinyals G, Gómez-Acebo I, Fernández-Tardón G, Peiró R, Olmedo-Requena R, Capelo R, Navarro C, Pacho-Valbuena S, Pérez-Gómez B, Kogevinas M, Pollán M, Aragonés N (2018) High adherence to the western, prudent, and mediterranean dietary patterns and risk of gastric adenocarcinoma: MCC-Spain study. Gastric Cancer 21:372–382

Chaemsai S, Kunanopparat T, Srichumpuang J, Nopharatana M, Tangduangdee C, Siriwattanayotin S (2016) Reduction of the polycyclic aromatic hydrocarbon (PAH) content of charcoal smoke during grilling by charcoal preparation using high carbonisation and a preheating step. Food Addit Contam Part A Chem Anal Control Expo Risk Assess 33(3):385–390

Chen J, Chen S (2005) Removal of polycyclic aromatic hydrocarbons by low density polyethylene from liquid model and roasted meat. Food Chem 90(3):461–469
Chiu CP, Yang DY, Chen BH (1998) Formation of heterocyclic amines in cooked chicken legs. J Food Prot 61:712–719
Choi SY, Chung MJ, Lee SJ, Shin JH, Sung NJ (2007) N-nitrosamine inhibition by strawberry, garlic, kale, and the effects of nitrite-scavenging and N-nitrosamine formation by functional compounds in strawberry and garlic. Food Control 18(5):485–491
Choi Y, Song S, Song Y, Lee JE (2013) Consumption of red and processed meat and esophageal cancer risk: meta-analysis. World J Gastroenterol 19:1020–1029
Ciemniak A (2006) A comparison of N-nitrosodimethylamine contents in selected meat products. Rocz Panstw Zakl Hig 57:341–346
Cordeiro T, Viegas O, Silva M, Martins ZE, Fernandes I, Ferreira IMLPVO, Pinho O, Mateus N, Calhau C (2020) Inhibitory efect of vinegars on the formation of polycyclic aromatic hydrocarbons in charcoal-grilled pork. Meat Sci 167:108083
Corpet DE, Yin Y, Archer MC (1997) Change in protein fermentation in the colon and an increase in specific bacterial enzyme activities are associated with the early stages of colorectal tumorigenesis in rats. J Nutr 127(3):487S–492S
Crosby NT (1983) Nitrosamines: a review of their chemistry, biological properties, and occurrence in the environment. In: Rechcigl M (ed) CRC handbook of naturally occurring food toxicants. CRC Press Inc., Boca Raton, Florida, pp 13–159
Cross A, Ferrucci L, Risch A, Graubard B, Ward M, Park Y, Hollenbeck A, Schatzkin A (2010) A large prospective study of meat consumption and colorectal cancer risk: an investigation of potential mechanisms underlying this association. Cancer Res 70:2406–2414
Cross A, Freedman N, Ren J, Ward M, Hollenbeck A, Schatzkin A, Sinha R, Abnet C (2011) Meat consumption and risk of esophageal and gastric cancer in a large prospective study. Am J Gastroenterol 106:432–442
Dallinga JW, Pachen DM, Lousberg AH, Van Geel JA, Houben GM, Stockburgger RW, Maanen JM, Kleinjans JC (1998) Volatile N-nitrosamines in gastric juice of patients with various conditions of the gastrointestinal tract determined by gas chromatography-mass spectrometry and related to Intragastic pH and nitrate and nitrite levels. Cancer Lett 124:119–125
Daniel CR, Schwartz KL, Colt JS, Dong LM, Ruterbusch JJ, Purdue MP, Cross AJ, Rothman N, Davis FG, Wacholder S, Graubard BI, Chow WH, Sinha R (2011) Meat-cooking mutagens and risk of renal cell carcinoma. Br J Cancer 105:1096–1104
Darwish WS, Chiba H, El-Ghareeb WR, Elhelaly AE, Hui SP (2019) Determination of polycyclic aromatic hydrocarbon content in heat-treated meat retailed in Egypt: health risk assessment, benzo[a]pyrene induced mutagenicity and oxidative stress in human colon (CaCo-2) cells and protection using rosmarinic and ascorbic acid. Food Chem 290:114–124
Das AK, Bhattacharya D, Das A, Nath S, Bandyopadhyay S, Nanda PK, Gagaoua M (2023) Current innovative approaches in reducing polycyclic aromatic hydrocarbons (PAHs) in processed meat and meat products. Chem Biol Technol Agric 10:109
Dietrich M, Block G, Pogoda JM, Buffler P, Hecht S, Preston-Martin S (2005) A review: dietary and endogenously formed N-nitroso compounds and risk of childhood brain tumors. Cancer Causes Control 16:619–635
Drew L, Helmus S, Thompson C, Zelenskiy S, Tucker T (2013) Red meat-derived heterocyclic amines increase risk of colon cancer: a population-based case-control study. Nutr Cancer 65: 1141–1150
EFSA (European Food Safety Authority) (2008) Polycyclic aromatic hydrocarbons in food-scientific opinion of the panel on contaminants in the food chain. EFSA J 6:724
Egeberg R, Olsen A, Christensen J, Halkjær J, Jakobsen MU, Overvad K, Tjønneland A (2013) Associations between red meat and risks for colon and Rectal cancer depend on the type of red meat consumed. J Nutr 143:464–472

Eldaly E, Hussein M, El-Gaml A, El-hefny D, Mishref M (2016) Polycyclic aromatic hydrocarbons (PAHs) in charcoal grilled meat (Kebab) and kofta and the efect of marinating on their existence. Zagazig Vet J 44(1):40–47
Essumang DK, Dodoo DK, Adjei JK (2013) Effect of smoke generation sources and smoke curing duration on the levels of polycyclic aromatic hydrocarbon (PAH) in different suites of fish. Food Chem Toxicol 58:86–94
Fabiani R, Minelli L, Bertarelli G, Bacci S (2016) A western dietary pattern increases prostate cancer risk: a systematic review and meta-analysis. Nutrients 8:626
Farhadian A, Jinap S, Abas F, Sakar ZI (2010) Determination of polycyclic aromatic hydrocarbons in grilled meat. Food Control 21:606–610
Farhadian A, Jinap S, Faridah A, Zaidul ISM (2012) Efects of marinating on the formation of polycyclic aromatic hydrocarbons (benzo[a]pyrene, benzo[b]fuoranthene and fuoranthene) in grilled beef meat. Food Control 28(2):420–425
FDA (2013) Guidance Document, Guidance for Industry: Food Labeling Guide
Ferreira PG, Lima MA, Bernedo-Navarro RA, Conceição RA, Linhares E, Sawaya AC, Yano T, Salgado I (2011) Stimulation of acidic reduction of nitrite to nitric oxide by soybean phenolics: possible relevance to gastrointestinal host defense. J Agric Food Chem 59(10):5609–5616
Ferrucci L, Sinha R, Ward M, Graubard B, Hollenbeck A, Kilfoy B, Schatzkin A, Michaud D, Cross A (2010) Meat and components of meat and the risk of bladder cancer in the NIH-AARP diet and health study. Cancer 116:4345–4353
Fiddler W, Pensabene JW, Doerr RC, Wasserman AE (1972a) Formation of N-nitrosodimethylamine from naturally occurring quaternary ammonium compounds and tertiary amines. Nature 236:307
Fiddler W, Piotrowski EG, Pensabene JW, Doerr RC, Wasserman AE (1972b) Effect of sodium nitrite concentration on N-nitrosodimethylamine formation in frankfurters. J Food Sci 37:668–670
Fiddler W, Pensabene JW, Fagan JC, Thorne EJ, Piotrowski EG, Wasserman AE (1974) The role of lean and adipose tissue on the formation of nitrosopyrrolidine in fried bacon. J Food Sci 39: 1070–1071
Fiddler W, Pensabene JW, Gates RA, Foster JM (1996) Investigations on nitrosamine reduction in boneless hams processed in elastic rubber nettings. J Muscle Foods 7(4):398–401
Fiddler W, Pensabene JW, Gates RA, Adam R (1998) Nitrosamine formation in processed hams as related to reformulated elastic rubber netting. J Food Sci 63(2):276–278
Franco P, Cardea S, Tabernero A, De Marco I (2021) Porous aerogels and adsorption of pollutants fromwater and air: a review. Molecules 26(15):4440
Fu Y, Zhao G, Wang S, Yu J, Xie F, Wang H, Xie J (2014) Simultaneous determination of fifteen heterocyclic aromatic amines in the urine of smokers and nonsmokers using ultra-high performance liquid chromatography–tandem mass spectrometry. J Chromatogr A 1333:45–53
Gomes A, Santos C, Almeida J, Elias M, Roseiro LC (2013) Effect of fat content, casing type and smoking procedures on PAHs contents of portuguese traditional dry fermented sausages. Food Chem Toxicol 58:369–374
González C, Jakszyn P, Pera G, Agudo A, Bingham S, Palli D, Ferrari P, Boeing H, del Giudice G, Plebani M, Carneiro F, Nesi G, Berrino F, Sacerdote C, Tumino R, Panico S, Berglund G, Simán H, Nyrén O, Hallmans G, Martinez C, Dorronsoro M, Barricarte A, Navarro C, Quirós J, Allen N, Key T, Day N, Linseisen J, Nagel G, Bergmann M, Overvad K, Jensen M, Tjonneland A, Olsen A, Bueno-de-Mesquita H, Ocke M, Peeters P, Numans M, Clavel-Chapelon F, Boutron-Ruault M, Trichopoulou A, Psaltopoulou T, Roukos D, Lund E, Hemon B, Kaaks R, Norat T, Riboli E (2006) Meat intake and risk of stomach and esophageal adenocarcinoma within the European prospective investigation into cancer and nutrition (EPIC). J Natl Cancer Inst 98(5):345–354
Grosse Y, Baan R, Straif K, Secretan B, El GF, Cogliano V (2006) Carcinogenicity of nitrate, nitrite, and cyanobacterial peptide toxins. Lancet Oncol 7:628–629

Gui B, Xu X, Zhang S, Wang Y, Li C, Zhang D, Su L, Zhao Y (2021) Prediction of organic compounds adsorbed by polyethylene and chlorinated polyethylene microplastics in freshwater using QSAR. Environ Res 197:111001

Guo H, Wang Z, Pan H, Li X, Chen L, Rao W, Gao Y, Zhang D (2014) Effects of traditional chinese cooking methods on formation of heterocyclic aromatic amines in lamb patties. Food Sci Biotechnol 23:747–753

Hamburg A, Hamburg A (1992) N-Nitrosoproline and N-nitrososarcosine In Estonian foodstuffs. J Food Saf 12(2):149–159

Han-Seung S (2005) Influence of food ingredients on the formation of heterocyclic aromatic amine in cooked pork patties. Food Sci Biotechnol 14:572–575

Haskaraca G, Demirok E, Kolsarıcı N, Öz F, Özsaraç N (2014) Effect of green tea extract and microwave pre-cooking on the formation of heterocyclic aromatic amines in fried chicken meat products. Food Res Int 63:373–381

Hedberg JJ, Turesky RJ, Österdahl BG (1983) Formation of volatile N-nitrosamines from aliphatic amines in heated soy protein isolate and in soy protein isolate–sucrose mixtures. J Agric Food Chem 31(6):1312–1318

Hill LH, Webb NB, Mongol LD, Adams AT (1973) Changes in residual nitrite in sausages and luncheon meat products during storage. J Milk Food Technol 36:515–519

Hitzel A, Pöhlmann M, Schwägele F, Speer K, Jira W (2014) Polycyclic aromatic hydrocarbons (PAH) and phenolic substances in meat products smoked with different types of wood and smoking spices. Food Chem 139:955–962

Homme CL, Sharp JO (2013) Differential microbial transformation of nitrosamines by an inducible propane monooxygenase. Environ Sci Technol 47(13):7388–7395

Hotchkiss JH, Helser MA, Maragos CM, Weng YM (1992) Nitrate, nitrite and N-Nitrosocompounds, food safety and biological implications. In: Finley JW, Robinson SF, Armstrong DJ (eds) Food safety assessment. American Chemical Society, Washington, DC, p 400

Huhtanen CN, Feinberg J (1980) Sorbic acid inhibition of clostridium-botulinum in nitrite-free poultry frankfurters. J Food Sci 45(3):453–457

IARC (1987) Overall evaluations of carcinogenicity: an updating of IARC monographs volumes 1 to 42. IARC monographs on the evaluation of carcinogenic risks to humans. Supplement 7. Lyon, France

IARC (1993) Some naturally occurring substances: food items and constituents, heterocyclic aromatic amines and mycotoxins. IARC Monogr Eval Carcinog Risks Hum 56:165–195

IARC (2006) Ingested nitrates and nitrites and cyanobacterial peptide toxins. IARC monographs on the evaluation of carcinogenic risks in humans, vol. 94

IARC (2010) Some non-heterocyclic polycyclic aromatic hydrocarbons and some related exposures. In: World Health Organization (ed) IARC monographs on the evaluation of carcinogenic risks to humans, vol 92. WHO Press, Lyon

IARC (2015) International Agency for Research on Cancer red meat and processed meat

Ikins WG, Gray JI, Mandagere AK, Booren AM, Pearson AM, Stachiw MA, Buckley DJ (1988) Contribution of wood smoke to in vivo formation of nitrosothiazolidine-4-carboxylic acid initial studies. Food Chem Toxicol 26(1):15–21

Iko Afé OH, Douny C, Kpoclou YE, Igout A, Mahillon J, Anihouvi V, Hounhouigan DJ, Scippo ML (2020) Insight about methods used for polycyclic aromatic hydrocarbons reduction in smoked or grilled fishery and meat products for future re-engineering: a systematic review. Food Chem Toxicol 141:111372

Jackson LS, Hargraves WA, Stroup WH, Diachenko GW (1994) Heterocyclic aromatic amine content of selected beef flavors. Mutat Res/Genet Toxicol 320:113–124

Jägerstad M, Skog K (1991) Formation of meat mutagens. Adv Exp Med Biol 289:83–105

Jakszyn P, Agudo A, Ibañez R, García-Closas R, Pera G, Amiano P, González CA (2004) Development of a food database of nitrosamines, heterocyclic amines, and polyccyclic aromatic hydrocarbons. J Nutr 134:2011–2014

Janoszka B (2010) Heterocyclic amines and azaarenes in pan-fried meat and its gravy fried without additives and in the presence of onion and garlic. Food Chem 120(2):463–473
Janoszka B (2011) HPLC-fuorescence analysis of polycyclic aromatic hydrocarbons (PAHs) in pork meat and its gravy fried without additives and in the presence of onion and garlic. Food Chem 126(3):1344–1353
Ježek F, Kameník J, Macharáčková B, Bogdanovičová K, Bednář J (2019) Cooking of meat: effect on texture, cooking loss and microbiological quality-a review. Acta VetBrno 88:487–496
Jinap S, Mohd-Mokhtar MS, Farhadian A, Hasnol NDS, Jaafar SN, Hajeb P (2013) Effects of varying degrees of doneness on the formation of heterocyclic aromatic amines in chicken and beef satay. Meat Sci 94:202–207
John E, Stern M, Sinha R, Koo J (2011) Meat consumption, cooking practices, meat mutagens and risk of prostate cancer. Nutr Cancer 63:525–537
Jonge RD (2019) Predictable and unpredictable survival of foodborne pathogens during non-isothermal heating. Int J Food Microbiol 291:151–160
Keating GA, Layton DW, Felton JS (1999) Factors determining dietary intakes of heterocyclic amines in cooked foods. Mutat Res 443:149–156
Keskekoglu H, Uren A (2017) Inhibitory effects of grape seed extract on the formation of heterocyclic aromatic amines in beef and chicken meatballs cooked by different techniques. Int J Food Prop 20:S722–S734
Khorshidian N, Yousef M, Shadnoush M, Siadat SD, Mohammadi M, Mortazavian AM (2020) Using probiotics for mitigation of acrylamide in food products: a mini review. Curr Opin Food Sci 32:67–75
Kim HJ, Cho J, Kim D, Park TS, Jin SK, Hur SJ, Lee SK, Jang A (2021) Effects of gochujang (Korean red pepper paste) marinade on polycyclic aromatic hydrocarbon formation in charcoal-grilled pork belly. Food Sci Anim Resour 41(3):481–496
Kimoto WI, Pensabene JW, Fiddler W (1982) Isolation and identification of N-nitrosothiozolodine in fried bacon. J Agric Food Chem 30:757–760
Knize MG, Andresen BD, Healy SK, Shen NH, Lewis PR, Bjeldanes LF, Hatch FT, Felton JS (1985) Effects of temperature, patty thickness and fat content on the production of mutagens in fried ground beef. Food Chem Toxicol 23:1035–1040
Kondjoyan A, Chevolleau S, Grève E, Gatellier P, Santé-Lhoutellier V, Bruel S, Touzet C, Portanguen S, Debrauwer L (2010) Formation of heterocyclic amines in slices of longissimus thoracis beef muscle subjected to jets of superheated steam. Food Chem 119:19–26
Krul CAM, Zeilmaker MJ, Schothorst RC, Havenaar R (2004) Intragastric formation and modulation of N-nitrosodimethylamine in a dynamic in vitro gastrointestinal model under hum an physiological conditions. Food Chem Toxicol 42:51–63
Kuzmicz K, Ciemniak A (2018) Assessing contamination of smoked sprats (Sprattus sprattus) with polycyclic aromatic hydrocarbons (pahs) and changes in its level during storage in various types of packaging. J Environ Sci Heal Part B 53(1):1–11
Laranjo M, Potes ME, Elias M (2019) Role of starter cultures on the safety of fermented meat products. Front Microbiol 10:843
Lazarevic I, Nagorni A, Rancic N, Milutinovic S, Stosic L (2010) Dietary factors and gastric cancer risk: hospital-based case control study. J BUON 15(1):89–93
Ledesma E, Rendueles M, Díaz M (2015) Spanish smoked meat products: Benzo(a)pyrene (BaP) contamination and moisture. J Food Compos Anal 37:87–94
Lee SJ, Shin JH, Sung NJ, Kim JG (2003) Effect of cooking on the formation of N – nitrosodimethylamine in Korean dried seafood products. Food Addit Contam 20(1):31–36
Lee JG, Kim SY, Moon JS, Kim SH, Kang DH, Yoon HJ (2016) Effects of grilling procedures on levels of polycyclic aromatic hydrocarbons in grilled meats. Food Chem 199:632–638
Li L, Wang P, Xu X, Zhou G (2012) Influence of various cooking methods on the concentrations of volatile N-nitrosamines and biogenic amines in dry-cured sausages. J Food Sci 77(5):C560–C565

Li L, Shao J, Zhu X, Zhou G, Xu X (2013) Effect of plant polyphenols and ascorbic acid on lipid oxidation, residual nitrite and N-nitrosamines formation in dry-cured sausage. Int J Food Sci Technol 48(6):1157–1164

Li F, Duan F, Zhao X, Song C, Cui S, Dai L (2016) Red meat and processed meat consumption and nasopharyngeal carcinoma risk: a dose-response meta-analysis of observational studies. Nutr Cancer 68:1034–1043

Liao GZ, Wang GY, Xu XL, Zhou GH (2010) Effect of cooking methods on the formation of heterocyclic aromatic amines in chicken and duck breast. Meat Sci 85(1):149–154

Lijinsky W (1999) The formation and occurrence of nitroso compounds in the environment. In: N-nitroso compounds: occurrence, biological effects and relevance to human cancer. IARC Scientific Publications, pp 87–100

Lili Z, Hongfei Z, Shoukat S, Xiaochen Z, Bolin Z (2017) Screening lactic acid bacteria strains with ability to bind di-n-butyl phthalate via Turbiscan technique. Antonie Van Leeuwenhoek 110(6): 759–769

Liu J, Lin S, Wang Z, Wang C, Wang E, Zhang Y, Liu J (2011) Supercritical fluid extraction of flavonoids from Maydis stigma and its nitrite-scavenging ability. Food Bioprod Process 89(C4): 333–339

Lunn JC, Kuhnle G, Mai V, Frankenfeld C, Shuker DE, Glen RC, Goodman JM, Pollock JR, Bingham SA (2007) The effect of haem in red and processed meat on the endogenous formation of N-nitroso compounds in the upper gastrointestinal tract. Carcinogenesis 28(3):685–690

Mahugija JAM, Njale E (2018) Effects of washing on the polycyclic aromatic hydrocarbons (PAHs) contents in smoked fish. Food Control 93:139–143

Malarut AJ, Vangnai K (2018) Influence of wood types on quality and carcinogenic polycyclic aromatic hydrocarbons (PAHs) of smoked sausages. Food Control 85:98–106

Malfatti M, Kuhn E, Turteltaub K, Vickers S, Jensen E, Strayer L, Anderson K (2016) Disposition of the dietary mutagen 2-Amino-3,8-dimethylimidazo[4,5-f]quinoxaline in healthy and pancreatic cancer compromised humans. Chem Res Toxicol 21:352–358

Marangoni F, Corsello G, Cricelli C, Ferrara N, Ghiselli A, Lucchin L, Poli A (2015) Role of poultry meat in a balanced diet aimed at maintaining health and wellbeing: An Italian consensus document. Food Nutr Res 59:1–11

Mirvish SS (1995) Role of N-nitroso compounds (NOC) and N-nitrosation in etiology of gastric, esophageal, nasopharyngeal and bladder cancer and contribution to cancer of known exposures to NOC. Cancer Lett 93:17–48

Moorthy B, Chu C, Carli D (2015) Polycyclic aromatic hydrocarbons: from metabolism to lung cancer. Toxicol Sci 145:5–15

Nadtochiy SM, Redman EK (2011) Mediterranean diet and cardioprotection: the role of nitrite, polyunsaturated fatty acids, and polyphenols. Nutrition 27(7–8):733–744

Nanda KK, Tignor S, Clancy J, Marota MJ, Allain LR, D'Addio SM (2021) Inhibition of N-nitrosamine formation in drug products: a model study. J Pharm Sci 110(12):3773–3775

NAS (National Academy of Science) (1981) The health effects of nitrate, nitrites, and Nitroso compounds. National Academy Press, Washington, DC

Niclis C, Román M, Osella A, Eynard A, Díaz M (2015) Traditional dietary pattern increases risk of prostate cancer in Argentina: results of a multilevel modeling and bias analysis from a case-control study. J Cancer Epidemiol 2015:179562

Nogacka A, Gómez-Martín M, Suárez A, González-Bernardo O, Reyes-Gavilán C, Gonzalez S (2019) Xenobiotics formed during food processing: their relation with the intestinal microbiota and colorectal cancer. Int J Mol Sci 25:2051

Ohgaki H (2009) Epidemiology of brain tumors. In: Verma M (ed) Methods of molecular biology, cancer epidemiology, vol 472. Humana Press, Totowa, pp 323–342

Olsson V, Solyakov A, Skog K, Lundström K, Jägerstad M (2002) Natural variations of precursors in pig meat affect the yield of heterocyclic amines-effects of RN genotype, feeding regime, and sex. J Agric Food Chem 50:2962–2969

Omojola AB, Ahmed SA, Attoh-Kotoku V, Wogar GI (2015) Effect of cooking methods on cholesterol, mineral composition and formation of total heterocyclic aromatic amines in Muscovy drake meat. J Sci Food Agric 95:98–102
Onopiuk A, Kołodziejczak K, Szpicer A, Wojtasik-Kalinowska I, Wierzbicka A, Półtorak A (2021) Analysis of factors that influence the PAH profile and amount in meat products subjected to thermal processing. Trends Food Sci Technol 115:366–379
Osterdahl BG, Alriksson E (1990) Volatile nitrosamines in microwave-cooked bacon. Food Addit Contam 7(l):51–54
Otoo EA, Ocloo FCK, Appiah V, Nuviadenu C, Asamoah A (2022) Reduction of polycyclic aromatic hydrocarbons concentrations in smoked Guinea fowl (Numida meleagris) meat using gamma irradiation. CYTA J Food 20(1):343–354
Oz F, Kaya M (2011a) Heterocyclic aromatic amines in meat. J Food Process Preserv 35:739–753
Oz F, Kaya M (2011b) The inhibitory effect of black pepper on formation of heterocyclic aromatic amines in high-fat meatball. Food Control 22:596–600
Oz F, Kotan G (2016) Effects of different cooking methods and fat levels on the formation of heterocyclic aromatic amines in various fishes. Food Control 67:216–224
Oz F, Zikirov E (2015) The effects of sous-vide cooking method on the formation of heterocyclic aromatic amines in beef chops. LWT-Food Sci Technol 64:120–125
Park J, Seo J, Lee J, Kwon H (2015) Distribution of seven N-nitrosamines in food. Toxicol Res 31(3):279–298
Park KC, Pyo HS, Kim WS, Yoon KS (2017) Efects of cooking methods and tea marinades on the formation of benzo[a]pyrene in grilled pork belly (Samgyeopsal). Meat Sci 129:1–8
Pegg RB, Shahidi F (2000) Nitrite curing of meat: the N-nitrosamine problem and nitrite alternatives. Trends Anal Chem 19(17):147–159
Pensabene JW, Fiddler W (1983) Factors affecting the N-nitrosothiazolidine content in Bacon. J Food Sci 48:1452–1454
Pensabene JW, Fiddler W (1985) Formation and inhibition of N-nitrosothiazolidine in bacon. Food Technol 39(l):91–94
Pensabene JW, Fiddler W, Gates RA (1974) Effects of frying and other cooking conditions on Nitrosopyrolidine formation in bacon. J Food Sci 39:316
Perloy A, Maasland D, van den Brandt P, Kremer B, Schouten L (2017) Intake of meat and fish and risk of head–neck cancer subtypes in The Netherlands cohort study. Cancer Caus Control 28: 647–656
Persson E, Sjöholm I, Skog K (2002) Heat and mass transfer in chicken breasts-effect on PhIP formation. Eur Food Res Technol 214:455–459
Persson E, Oroszvári BK, Tornberg E, Sjöholm I, Skog K (2008) Heterocyclic amine formation during frying of frozen beefburgers. Int J Food Sci Technol 43:62–68
Petersen A (1993) N-Nitrosodibutylamine and other volatile nitrosamines in cured meat packaged in rubber nettings. J Food Sci 58(1):47–48
Pogorzelska-Nowicka E, Kurek M, Hanula M, Wierzbicka A, Półtorak A (2022) Formation of carcinogens in processed meat and its measurement with the usage of artificial digestion-a review. Molecules 27:4665
Polak ML, Demšar L, Zahija I, Polak T (2020) Influence of temperature on the formation of heterocyclic aromatic amines in pork steaks. Czech J Food Sci 38:248–254
Pournaghi S, Noveyri F, Doust H, Ahmadi A, Hamedi H, Rahimi J, Ghasemi M, Hamidi H, Gholamalizadeh M, Doaei S, Lashkardoost H (2019) The association of consumption of animal proteins and the risk of esophageal cancer. Nutr Cancer 71(7):1094–1099
Preston-Martin S, Munir R, Chakrabarti I (2006) Nervous system. In: Schottenfeld D, Fraumeni Joseph F Jr (eds) Cancer epidemiology and prevention. Oxford University Press, New York, pp 1173–1195
Puangsombat K, Gadgil P, Houser TA, Hunt MC, Smith JS (2012) Occurrence of heterocyclic amines in cooked meat products. Meat Sci 90(3):739–746

Rahman U, Sahar A, Khan MI, Nadeem M (2014) Production of heterocyclic aromatic amines in meat: Chemistry, health risks and inhibition. A review. LWT-Food Sci Technol 59(1): 229–233

Ramarathnam N (1998) The flavour of cured meat. In: Shahidi F (ed) Flavor of meat, meat products and seafood. Blackie Academic & Professional, London

Reigand M, Toldrá F (2010) Detection of chemical hazards. In: Toldrá F (ed) Handbook of meat processing. Wiley-Blackwell Publishing, Iowa, pp 469–480

Reinik M, Tamme T, Roasto M, Juhkam K, Tenno T, Kiis A (2007) Polycyclic aromatic hydrocarbons (PAHs) in meat products and estimated PAH intake by children and the general population in Estonia. Food Addit Contam 24:429–437

Rocha BS, Nunes C, Pereira C, Barbosa RM, Laranjinha J (2014) A shortcut to wideranging biological actions of dietary polyphenols: modulation of the nitrate-nitritenitric oxide pathway in the gut. Food Funct 5(8):1646–1652

Rohrmann S, Linseisen J, Jakobsen M, Overvad K, Raaschou-Nielsen O, Tjonneland A, Boutron-Ruault M, Kaaks R, Becker N, Bergmann M, Boeing H, Khaw KT, Wareham NJ, Key TJ, Travis R, Benetou V, Naska A, Trichopoulou A, Pala V, Tumino R, Masala G, Mattiello A, Brustad M, Lund E, Skeie G, Bueno-de-Mesquita HB, Peeters PH, Vermeulen RC, Jakszyn P, Dorronsoro M, Barricarte A, Tormo MJ, Molina E, Argüelles M, Melin B, Ericson U, Manjer J, Rinaldi S, Slimani N, Boffetta P, Vergnaud AC, Khan A, Norat T, Vineis P (2011) Consumption of meat and dairy and lymphoma risk in the european prospective investigation into cancer and nutrition. Int J Cancer 128(3):623–634

Rosato V, Kawakita D, Negri E, Serraino D, Garavello W, Montella M, Decarli A, La Vecchia C, Ferraroni M (2019) Processed meat and risk of selected digestive tract and laryngeal cancers. Eur J Clin Nutr 73:141–149

Rose M, Holland J, Dowding A, Petch SRG, White S, Fernandes A, Mortimer D (2015) Investigation into the formation of PAHs in foods prepared in the home to determine the effects of frying, grilling, barbecuing, toasting and roasting. Food Chem Toxicol 78:1–9

Rywotycki R (2003a) The influence of environment, mode of nutrition and animal species on level of nitrosamine contamination in venison. Meat Sci 65:1045–1053

Rywotycki R (2003b) Meat nitrosamine contamination level depending on animal breeding factors. Meat Sci 65:669–676

Salamat F, Semnani S, Honarvar M, Fazel A, Roshandel G (2020) 10-year trends in dietary intakes in the high and low-risk areas for esophageal cancer: a population-based ecological study in northern Iran. Middle East J Dig Dis 12:89–98

Salehi M, Moradi-Lakeh M, Salehi MH, Nojomi M, Kolahdooz F (2013) Meat, fish, and esophageal cancer risk: a systematic review and dose-response meta-analysis. Nutr Rev 71:257–267

Sallan S, Kaban G, Kaya M (2019) Nitrosamines in sucuk: effects of black pepper, sodium ascorbate and cooking level. Food Chem 288:341–346

Sampaio GR, Guizellini GM, da Silva SA, de Almeida AP, Pinaffi-Langley ACC, Rogero MM, de Camargo AC, Torres EAFS (2021) Polycyclic aromatic hydrocarbons in foods: biological effects, legislation, occurrence, analytical methods, and strategies to reduce their formation. Int J Mol Sci 22(11):6010

Santarelli RL, Pierre F, Corpet DE (2008) Processed meat and colorectal cancer: a review of epidemiologic and experimental evidence. Nutr Cancer 60(2):131–144

Sava A, Uiuiu P, Lațiu C, Cocan D, Muntean GC, Papuc T, Ihut A, Răducu C, Becze A, Craioveanu C, Munteanu C, Constantinescu R, Mires V (2023) PAHs, physicochemical and microbiological analyses of trout processed by traditional smoking, in diferent types of packaging. Aust Fish 8(8):424

Scanlan RA (1983) Formation and occurrence of nitrosamines in food. Cancer Res 43(5):2435s–2440s

Semanová J, Sklářšová B, Šimon P, Šimko P (2016) Elimination of polycyclic aromatic hydrocarbons from smoked sausages by migration into polyethylene packaging. Food Chem 201:1–6

Sen NP, Miles WF, Donaldson B, Panalaks T, Iyengar JR (1973) Formation of nitrosamines in a meat curing mixture. Nature 245(5420):104–105
Sen NP, Iyengar JR, Donaldson BA, Panalaks T (1974) Effect of sodium nitrite concentration on the formation of Nitrosopyrrolidine and Dimethylnitrosamine in fried bacon. J Agric Food Chem 22(3):540–541
Sen NP, Baddoo PA, Seaman SW (1986) Nitrosothiazolidine and N-nitrosothiozolidine-4-carboxylic acid in smoked meats and fish. J Food Sci 51:821–825
Sen NP, Baddoo PA, Seaman SW (1987) Volatile nitrosamines in cured meats packaged in elastic rubber nettings. J Agric Food Chem 35:346–350
Shahidi F, Synowiecki J, Sen NP (1995) Nitrosamines in nitrite-cured chicken-seal salami. J Food Prot 58(4):446–448
Shephard SE, Schlatter CH, Lutz WK (1987) Assessment of the risk of formation of carcinogenic N-nitroso compounds from dietary precursors in the stomach. Food Chem Toxicol 25(1): 91–108
Shivappa N, Hébert JR, Taborelli M, Montella M, Libra M, Zucchetto A, Crispo A, Grimaldi M, La Vecchia C, Serraino D, Polesel J (2017) Dietary inflammatory index and non-Hodgkin lymphoma risk in an Italian case-control study. Cancer Causes Control 28(7):791–799
Shoukat S (2020) Potential anti-carcinogenic effect of probiotic and lactic acid bacteria in detoxifcation of benzo[a]pyrene: a review. Trends Food Sci Technol 99:450–459
Simko P (2009) Polycyclic aromatic hydrocarbons in smoked meats. In: Toldrá F (ed) Safety of meat and processed meat. Springer, New York
Sinaga K, Legowo AM, Supriatna E, Pramono YB (2016) Reduction of benzo(A)pyrene in charcoal grilled duck meat by marinating with andaliman (Zanthoxylum Acanthopodium, DC) fruit juice. J Indones Trop Anim Agric 41(4):204–208
Singh L, Agarwal T, Simal-Gandara J (2020) PAHs, diet and cancer prevention: cooking process driven-strategies. Trends Food Sci Technol 99:487–506
Sinha R, Rothman N, Brown ED, Salmon CP, Knize MG, Swanson CA, Rossi SC, Mark SD, Levander OA, Felton JS (1995) High concentrations of the carcinogen 2-amino-1-methyl-6-phenylimidazo-[4,5-b]pyridine (PhIP) occur in chicken but are dependent on the cooking method. Cancer Res 55:4516–4519
Sinha R, Rothman N, Brown ED, Mark SD, Hoover RN, Caporaso NE, Levander OA (1998a) High concentrations of the carcinogen 2-amino-1-methyl-6-phenylimidazo [4,5-b] pyridine (PhIP) occur in chicken but are dependent on the cooking method. Cancer Res 58(17):3749–3753
Sinha R, Rothman N, Salmon CP, Knize MG, Brown ED, Swanson CA, Rhodes D, Rossi S, Felton JS, Levander OA (1998b) Heterocyclic amine content in beef cooked by different methods to varying degrees of doneness and gravy made from beef drippings. Food Chem Toxicol 36:279–287
Skog K, Jagerstad M (1990) Effects of monosaccharides and disaccharides on the formation of food mutagens in model systems. Mutat Res 230:263–272
Skog K, Jägerstad M (1993) Incorporation of carbon atoms from glucose into the food mutagens MeIQx and 4,8-DiMeIQx using ^{14}C-labelled glucose in a model system. Carcinogenesis 14: 2027–2031
Skrypec DJ, Gray JI, Mandagere AK, Booren AM, Pearson AM, Cuppett SL (1985) Effect of bacon composition and processing on N-nitrosamine formation. Food Technol 39(1):74–79
Slotkin TA, Skavicus S, Card J, Giulio RTD, Seidler FJ (2017) In vitro models reveal diferences in the developmental neurotoxicity of an environmental polycylic aromatic hydrocarbon mixture compared to benzo[a] pyrene: neuronotypic PC12 cells and embryonic neural stem cells. Toxicology 377:49–56
Solyakov A, Skog K (2002) Screening for heterocyclic amines in chicken cooked in various ways. Food Chem Toxicol 40:1205–1211
Song PJ, Hu JF (1988) N-nitrosamines in Chinese foods. Food Chem Toxicol 26(3):205–208
Steck S, Butler L, Keku T, Antwi S, Galanko J, Sandler R (2014) Nucleotide excision repair gene polymorphisms, meat intake and colon cancer risk. Mutat Res 762:24–31

Stefani ED, Ronco A, Mendilaharsu M, Guidobono M, Deneo-Pellegrini H (1997) Meat intake, heterocyclic amines, and risk of breast cancer: a case-control study in Uruguay. Cancer Epidemiol Biomarkers Prev 6:573–581
Stumpe-Vīksna I, Bartkevičs V, Kukāre A, Morozovs A (2008) Polycyclic aromatic hydrocarbons in meat smoked with different types of wood. Food Chem 110:794–797
Suleman R, Hui T, Wang Z, Liu H, Zhang D (2020) Comparative analysis of charcoal grilling, infrared grilling and superheated steam roasting on the colour, textural quality and heterocyclic aromatic amines of lamb patties. Int J Food Sci Technol 55:1057–1068
Sun X, Xin X, Liu J, Jin Y, Zhao L (2021) Inhibition of compound fermentation agents and spices on N-nitrosamine formation in fermented sausage. J Chin Inst Food Sci Technol 21(5):194–202
Takahama U, Tanaka M, Hirota S (2010) Formation of nitric oxide, ethyl nitrite and an oxathiolone derivative of caffeic acid in a mixture of saliva and white wine. Free Radic Res 44(3):293–303
Tasic T, Ikonic P, Mandic A, Jokanovic M, Tomovic V, Savatic S, Petrovic L (2012) Biogenic amines content in traditional dry fermented sausage Petrovska klobasa as possible indicator of good manufacturing practice. Food Control 23(1):107–112
Terns MJ, Milkowski AL, Claus JR, Sindelar JJ (2011) Investigating the effect of incubation time and starter culture addition level on quality attributes of indirectly cured, emulsified cooked sausages. Meat Sci 88(3):454–461
Theiler RF, Aspelund TG, Sato K, Miller AF (1981) Model system studies on N-nitrosamine formation in cured meats: the effects of slice thickness. J Food Sci 46:691–693
Toldrá F, Aristoy MC, Flores M (2009) Relevance of nitrate and nitrite in dry-cured ham and their effects on aroma development. Grasas Aceites 60:291–296
Torre L, Bray F, Siegel R, Ferlay J, Lortet-Tieulent J, Jemal A (2015) Global cancer statistics, 2012. CA Cancer J Clin 65:87–108
Trafialek J, Kolanowski W (2014) Dietary exposure to meat-related carcinogenic substances: is there a way to estimate the risk? Int J Food Sci Nutr 65:774–780
Tricker AR (1997) N-nitroso compounds and man:sources of exposure, endogenous formation and occurrence in body fluids. Eur J Cancer Prev 6:226–268
Tricker AR, Preussmann R (1991) Carcinogenic N-nitrosamines in the diet:occurrence, formation, mechanisms and carcinogenic potential. Mutat Res 259:277–289
Turesky RJ, Taylor J, Schnackenberg L, Freeman JP, Holland RD (2005) Quantitation of carcinogenic heterocyclic aromatic amines and detection of novel heterocyclic aromatic amines in cooked meats and grill scrapings by HPLC/ESI-MS. J Agric Food Chem 53:3248–3258
US EPA (Environmental Protection Agency) (2008) Polycyclic aromatic hydrocarbons (PAHs)-EPA fact sheet. National Center for Environmental Assessment, Office of Research and Development, Washington, DC
USDA (2015) Department of Health and Human Services 2015–2020 dietary Guidlines for Americans, Washington, DC
Viegas O, Novo P, Pinto E, Pinho O, Ferreira IM (2012) Effect of charcoal types and grilling conditions on formation of heterocyclic aromatic amines (HAs) and polycyclic aromatic hydrocarbons (PAHs) in grilled muscle foods. Food Chem Toxicol 50(6):2128–2134
Viegas O, Yebra-Pimentel I, Martínez-Carballo E, Simal-Gandara J, Ferreira IM (2014) Effect of beer marinades on formation of polycyclic aromatic hydrocarbons in charcoal-grilled pork. J Agric Food Chem 62(12):2638–2643
Vivar-Vera MLA, Pérez-Silva A, Ruiz-López II, Hernández-Cázares AS, Solano-Barrera S, Ruiz-Espinosa H, Bernardino-Nicanor A, González-Cruz L (2018) Chemical, physical and sensory properties of Vienna sausages formulated with a starfruit dietary fiber concentrate. J Food Sci Technol 55(8):3303–3313
Vranešević J, Kartalović B, Vidaković Knežević S, Škaljac S, Jokanović M (2022) Reduction of polycyclic aromatic hydrocarbons to improve safety of traditional (Homemade) smoked dry-cured pork loin. Polycycl Aromat Compd 43(10):1–6

Walters CL (1985) Nitrosamines-can the hazard be controlled? In: Gibson GG, Walker R (eds) Food toxicology-real or imaginary problems? Taylor & Francis, London and Philadelphia, pp 342–354
Wang H, Pan L, Si L, Miao J (2018) The role of Nrf2-Keap1 signaling pathway in the antioxidant defense response induced by PAHs in the calm Ruditapes philippinarum. Fish Shellfsh Immunol 80:325–334
Wang L, Lee WW, Cui YR, Ahn G, Jeon YJ (2019) Protective effect of green tea catechin against urban fine dust particle-induced skin aging by regulation of NF-KB, AP-1, and MAPKs signaling pathways. Environ Pollut 252:1318–1324
Wang W, Dong L, Zhang Y, Yu H, Wang S (2021) Reduction of the heterocyclic amines in grilled beef patties through the combination of thermal food processing techniques without destroying the grilling quality characteristics. Food Secur 10:1490
WCRF (World Cancer Research Fund)/AICR (American Institute for Cancer Research) (2007) Food, nutrition, physical activity, and the prevention of cancer: a global perspective. AICR, Washington, DC
Webster TS, Condee C, Hatzinger PB (2013) Ex situ treatment of Nnitrosodimethylamine (NDMA) in groundwater using a fluidized bed reactor. Water Res 47(2):811–820
Wei F, Xu X, Zhou G, Zhao G, Li C, Zhang Y, Chen L, Qi J (2009) Irradiated Chinese Rugao ham: changes in volatile N-nitrosamine, biogenic amine and residual nitrite during ripening and post-ripening. Meat Sci 81(3):451–455
Wichitnithad W, Nantaphol S, Noppakhunsomboon K, Rojsitthisak P (2023) An update on the current status and prospects of nitrosation pathways and possible root causes of nitrosamine formation in various pharmaceuticals. Saudi Pharmaceut J 31(2):295–311
Wijekoon KC, Fujioka T, McDonald JA, Khan SJ, Hai FI, Price WE, Nghiem LD (2013) Removal of N-nitrosamines by an aerobic membrane bioreactor. Bioresour Technol 141:41–45
Wongmaneepratip W, Vangnai K (2017) Effects of oil types and pH on carcinogenic polycyclic aromatic hydrocarbons (PAHs) in grilled chicken. Food Control 79:119–125
Wrensch M, Minn Y, Chew T, Bondy M, Berger MS (2002) Epidemiology of primary brain tumors: current concepts and review of the literature. Neuro-Oncology 4:278–299
Xu J, Yang X, Wu Y, Li X, Bai B (2014) Meat consumption and risk of oral cavity and oropharynx cancer:A meta-analysis of observational studies. PLoS One 9:e95048
Xu M, Wu M, Zhang Y, Zhang H, Liu W, Chen G, Xiong G, Guo L (2022) Biodegradation of polycyclic aromatic hydrocarbons (PAHs) by bacterial mixture. Int J Environ Sci Technol 19(5): 3833–3844
Yamamoto M, Iwata R, Ishiwata H, Yamada T, Tanimura A (1984) Determination of volatile nitrosamine levels in foods and estimation of their daily intake in Japan. Food Chem Toxicol 22: 61–64
Yang Z, Lu R, Song H, Zhang Y, Tang J, Zhou N (2017) Effect of different cooking methods on the formation of aroma components and heterocyclic amines in pork loin. J Food Process Preserv 41:e12981
Yang D, He Z, Gao D, Qin F, Deng S, Wang P, Xu X, Chen J, Zeng M (2019) Effects of smoking or baking procedures during sausage processing on the formation of heterocyclic amines measured using UPLC-MS/MS. Food Chem 276:195–201
Yousef M, Khorshidian N, Hosseini H (2022a) In vitro PAH-binding ability of lactobacillus brevis TD4. Polycycl Aromat Compd 42(7):4343–4358
Yousef M, Hosseini H, Khorshidian N, Rastegar H, Shamloo E, Abdolshahi A (2022b) Effect of time and incubation temperature on ability of probiotics for removal of polycyclic aromatic hydrocarbon in phosphate bufer saline. J Microbiol Biotechnol Food Sci 12(1):e1918
Yousefi M, Khorshidian N, Mortazavian AM (2021) Detoxifcation properties of microorganisms in foods. In: Ray RC (ed) Microbial biotechnology in food and health. Elsevier, Amsterdam
Yurchenko S, Molder U (2007) The occurrence of volatile N-nitrosamines in Estonian meat products. Food Chem 100(4):1713–1721

Zamora R, Alcón E, Hidalgo FJ (2012) Effect of lipid oxidation products on the formation of 2-amino-1-methyl-6-phenylimidazo[4,5-b]pyridine (PhIP) in model systems. Food Chem 135: 2569–2574

Zeng M, Wang J, Zhang M, Chen J, He Z, Qin F, Xu Z, Cao D (2018) Chen J. Inhibitory effects of Sichuan pepper (*Zanthoxylum bungeanum*) and sanshoamide extract on heterocyclic amine formation in grilled ground beef patties. Food Chem 239:111–118

Zhang Y, Yu C, Mei J, Wang S (2013) Formation and mitigation of heterocyclic aromatic amines in fried pork. Food Addit Contam Part A 30:1501–1507

Zhang R, Fu J, Moore JB, Stoner L, Li R (2021) Processed and unprocessed red meat consumption and risk for type 2 diabetes mellitus: an updated meta-analysis of cohort studies. Int J Environ Res Public Health 18:10788

Zhao H, Zhou F, Qi Y, Dziugan P, Bai F, Walczak P, Zhang B (2013) Screening of lactobacillus strains for their ability to bind benzo(a)pyrene and the mechanism of the process. Food Chem Toxicol 59:67–71

Zhao Z, Yin Z, Pu Z, Zhao Q (2017) Association between consumption of red and processed meat and pancreatic cancer risk: a systematic review and meta-analysis. Clin Gastroenterol Hepatol 15:486–493

Zheng W, Gustafson DR, Moore D, Hong CP, Anderson KE, Kushi LH, Sellers TA, Folsom AR (1998) Well-done meat intake and the risk of breast cancer. J Natl Cancer Inst 90(22): 1724–1729

Zhu Z, Xu Y, Huang T, Yu Y, Bassey AP, Huang M (2022) The contamination, formation, determination and control of polycyclic aromatic hydrocarbons in meat products. Food Control 141:109194

Zorrilla SE, Singh RP (2003) Heat transfer in double-sided cooking of meat patties considering two-dimensional geometry and radial shrinkage. J Food Eng 57:57–65

Chapter 16
Nanotechnology in Meat Processing Industry

Sajad A. Rather, F. A. Masoodi, Tanveer Ahmad Mir, Tariq Ahmad Ganie, and Syed Muzaffar

16.1 Introduction

Currently due to the demand for sustainable production of meat products and emphasis on human health and wellness, meat industry is making great efforts to improve hygiene, extend shelf life of products, replace synthetic additives with natural and non-toxic, prevent food-borne illnesses and contamination by chemical and physical agents and to improve its detection and control if contamination has already occurred (Idamokoro and Hosu 2022). Thus, there is a constant search for safer, low-cost and innovative technologies that can help in achieving these objectives. Nanotechnology is one of the major innovative alternative that is increasingly applied in the different areas/fields of food science including meat production chain to ensure a longer shelf-life with improved product quality, safety and traceability (Fig. 16.1) (Ramachandraiah et al. 2015; Bayda et al. 2020). It is multidisciplinary field and can be defined as application of nano-sized materials of less than 100 nm (<100 nm) with unique and novel characteristics (Khan et al. 2016). Nanotechnology has its roots in a talk delivered in 1959 by physicist Richard Feynman to the American Physical Society (Ozimek et al. 2010). Nano-sized materials reach and act more efficiently on their target at very low concentrations than micro- and macro-sized materials due to their high surface area to volume ratio (Khan et al. 2019). According to Garcia et al. (2010), the word "nano" simply refers to something minuscule, minute and atomic in structure. The scientific use of "nano" gave rise

S. A. Rather (✉) · F. A. Masoodi (✉)
Department of Food Science and Technology, University of Kashmir, Srinagar, India

T. A. Mir · T. A. Ganie
Department of Food Technology, Islamic University of Science and Technology, Pulwama, Jammu & Kashmir, India

S. Muzaffar
Department of Chemistry AAAM Degree College Bemina, Srinagar, Jammu & Kashmir, India

S. A. Rather, F. A. Masoodi (eds.), *Hand Book of Processed Functional Meat Products*, https://doi.org/10.1007/978-3-031-69868-2_16

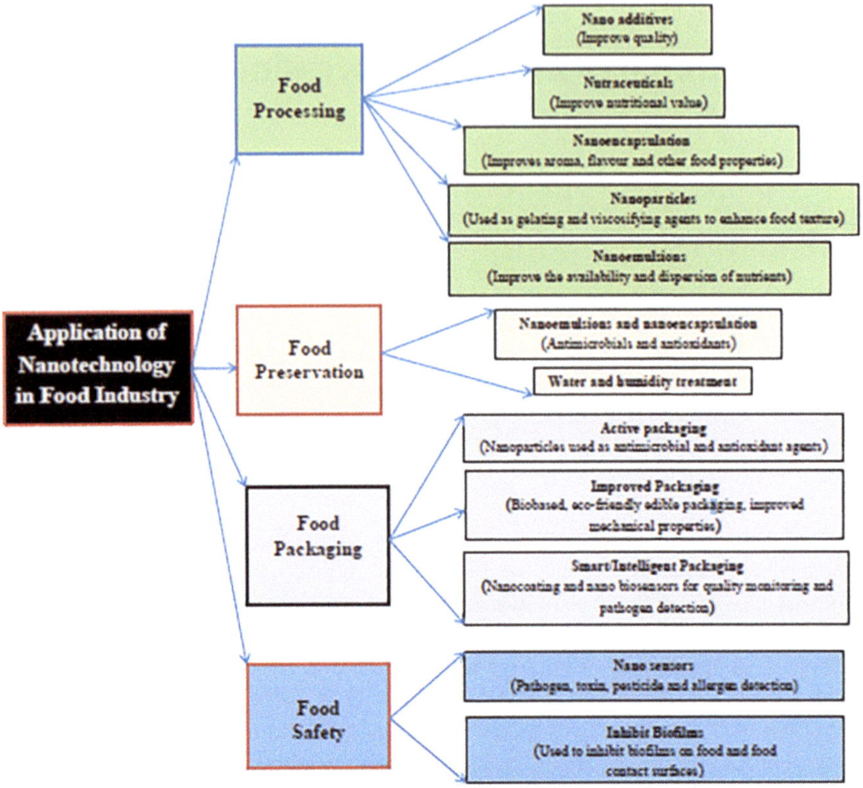

Fig. 16.1 Potential applications of nanotechnology in food industry

to the discipline of nanotechnology. Nanomaterials may include any of the following nano forms: nanoparticles, nanotubes, fullerenes, nanofibres, nanowhiskers, nanosheets (Cushen et al. 2012). A product can be classified with nanotechnique when following approaches have been used in the production system (Singh et al. 2015).

(a) Any of the food ingredients processed in such a manner to form nanostructures e.g. nano-sized nutrients to enhance their bioavailability and improve organoleptic properties of product.
(b) Nano-sized or nano-encapsulated additives and supplements have been incorporated in a product to enhance their efficacy.
(c) Packaging materials utilize nano-particles to improve functionality of packaging and product quality.
(d) Nanotechnology based device used in the processing or packaging of a product e.g. nano-sensors to monitor product quality and traceability.

The potential of a nanomaterial to enhance bioavailability and solubility while also preserving bioactive components during processing and storage is being used in a range of areas/field including agriculture, medicine/pharmaceutical, cosmetics, apparel, public health and food (Fu 2014; Hu et al. 2017; Pathakoti et al. 2017). In food industry the use of nanotechnology allows new applications including improved bioavailability, antimicrobial effects, targeted delivery of bioactive compounds, diversification in the food packaging system, the improved absorption of nutrients, better microorganism detection system and traceability (Ramachandraiah et al. 2015; Lamri et al. 2021). Even so, lower level of bulk functional ingredients can be achieved through nanotechnology such as higher levels of salt, sugar and synthetic preservatives can be minimized and colour, flavour, texture and sensory properties of food can be enhanced (Chaudhry and Castle 2011).

The potential applications of nanotechnology in meat industry was performed for several goals including: (i) better protection against pathogenic and spoilage microorganisms, (ii) improvement of sensory properties, (iii) enhancing the functional and nutritional aspects of the meat products, and (iv) to monitor the quality during storage and transportation (Ojha et al. 2017; Pinelli et al. 2021). There are many more positive impacts of nanotechnology in the meat industry and these are expected to grow in the near future.

16.2 Manufacturing of Nanomaterials

Nanomaterials are classified into different types based on their particle size, structure and characteristics including nanoparticles, nanocapsules, nanoclays, nanoemulsions, nanotubes, nanofibers and nanolaminates (Grumezescu and Holban 2018). Each class of these nanomaterials has applications in food industry and can be made through different methods. Nanomaterial manufacturing process involves 'top down' or 'bottom up' approaches. Top down approach involves breaking down larger particles of matter to nanometers dimension by physical or chemical means such as mechanical milling or grinding, homogenization, use of lasers and vaporization followed by cooling, etc. (Brody et al. 2008). While, bottom up approach involves self-assembly and self-organization such as vapour phase techniques, co-precipitation, hydrothermal, green/biological synthesis, liquid phase techniques, etc. (Altammar 2023). The self-organization on the nanometer scale can be achieved by setting a balance between the different non-covalent forces (Dickinson and Van Vliet 2003).

16.2.1 Top-Down Approaches

In top-down approaches, bulk materials are divided to produce nanostructured materials. Top-down methods include mechanical milling, laser ablation, etching, sputtering, and electro-explosion.

16.2.1.1 Mechanical Attrition (Milling/Mechanochemical Processing)

Mechanical milling method is used to prepare advanced materials with particular grain or interface-boundary design. Using mechanical milling two different techniques have been developed for the manufacturing of nanopowders. In first technique single phase powder is milled by controlling the balance point between fracturing and cold welding. Reduction of the average grain size from 50–100 μm up to 2–20 nm is obtained (Luther 2004). Plastic deformation occurs due to mechanical attrition, so a continuous refinement of the internal structure of the powder particles to nanometer scales is accomplished (Fect 1996). During the milling process temperature raises over 100–200 °C. As the mechanical milling is very sensitive to contamination, atmospheric control is used for chemical reactions between the environment and the milled powders. Due to this limitation a second method of mechanical attrition-mechanochemical processing (MCP) was developed. In mechanochemical processing a conventional ball mill can be used as a low temperature chemical reactor. The ball mill increases the reaction kinetics in the reacting powder mixture as a result of the intimate mixing and refinement of the grain structure to the nanometer scale. The reaction occurs during the actual milling and for this reason; a proper reactive gas, like O_2, N_2, atmospheric air, etc. and a suitable precursor are chosen (Luther 2004). For wet milling, an organic fluid can also be used. Various ball mills have been developed for mechanical milling, such as tumbler mills, attrition, shaker mills, vibratory mills, planetary mills, etc. (Altammar 2023). The material powder is placed in a sealed container with coated balls by hardened steel or tungsten carbide. The mass ratio for the ball to the powder shall be 5:10, with a typical particle diameter 50 μm for the powder. Kinetic energy of balls is a function of their mass and velocity, as a result, steel or tungsten, which are high density materials, are preferable as means of milling (Fect 1996).

16.2.1.2 Electrospinning

Electrospinning is one of the simplest top-down methods for the development of nanostructured materials. It is mainly used to produce nanofibers from various materials like polymers (Ostermann et al. 2011). The important breakthrough in electrospinning is coaxial electrospinning in which spinneret comprises two coaxial capillaries. In these capillaries, two viscous liquids, or a viscous liquid as the shell and a non-viscous liquid as the core, can be used to form core–shell nanoarchitectures in an electric field. It is an effective approach for achieving core–shell ultrathin fibers on a large scale. The lengths of these ultrathin nanomaterials can be extended to several centimeters. This method has been used for the development of core–shell and hollow polymer, inorganic, organic and hybrid materials (Kumar et al. 2014).

16.2.1.3 Lithography

Lithography is a useful technique for manufacturing nanoarchitectures. It uses a focused beam of light or electrons and is divided into two main types: masked lithography and maskless lithography (Pimpin and Srituravanich 2012). In masked nanolithography, nanopatterns are transferred over a large surface area using a specific mask or template. Masked lithography includes photolithography. Maskless lithography includes scanning probe lithography, focused ion beam lithography (Matsumoto et al. 2020) and electron beam lithography. In maskless lithography, arbitrary nanopattern writing is carried out without the involvement of a mask. 3D freeform micro-nano-fabrication can be achieved via ion implantation with a focused ion beam in combination with wet chemical etching (Brady et al. 2019).

16.2.1.4 Sputtering

Sputtering technique is used to produce nanomaterials via bombarding solid surfaces with high-energy particles such as intense plasma or gas particles and is an effective method for producing thin films of nanomaterials (Behrisch 1981). In the sputtering deposition technique, energetic gaseous ions bombard the target surface, causing the physical ejection of small atom clusters depending upon the incident gaseous-ion energy. Sputtering can be performed in various ways, such as utilizing magnetron, radio-frequency diode and DC diode sputtering. It is performed in an evacuated chamber, to which the sputtering gas is introduced. A high voltage is applied to the cathode target and free electrons collide with the gas to produce gas ions. The positively charged ions strongly accelerate in the electric field towards the cathode target, which these ions continuously hit, resulting in the ejection of atoms from the surface of the target (Muñoz-García et al. 2009). Magnetron sputtering is used to produce WSe_2-layered nanofilms on SiO_2 and carbon paper substrates. The sputtering technique is cost-effective compared with electron-beam lithography with fewer contaminants (Baig et al. 2021).

16.2.1.5 The Arc Discharge Method

The arc discharge method is useful for producing carbon-based materials, such as carbon nanotubes, amorphous spherical carbon nanoparticles, etc. (Shi et al. 2000). It uses a simple device to achieve the desired high temperature for carbon atom evaporation from the plasma. In this technique, a hollow graphite anode packed with a mixture of transition metals (Fe, Co, or Ni) and graphite powder is vaporized by the electric arc (Kumar et al. 2013). The flow of the inert gas is maintained at 100–1000 Torr, while other conditions are 100 amps and 20 V. The variations of the process parameters such as flow rate, metal concentration and gas pressure are significant to achieve the optimum yield of CNTs in this technique.

16.2.1.6 Laser Ablation

Laser ablation involves use of powerful laser beam that hits the target material for synthesis of nanoparticles. The source material or precursor vaporizes due to the high energy of the laser irradiation, resulting in nanoparticle formation. Laser ablation can be considered as a green technique, as there is no need for stabilizing agents or other chemicals (Amendola and Meneghetti 2009). Various nanomaterials can be synthesized through Laser ablation technique such as metal nanoparticles, carbon nanomaterials, oxide composites and ceramics (Su and Chang 2018). Another approach for production of monodisperse colloidal nanoparticle solutions without using surfactants or ligands is pulsed laser ablation in liquids. During this technique the nanoparticle characteristics such as average size and distribution, can be tuned via adjusting the fluence, wavelength and laser salt addition (Baig et al. 2021).

16.2.1.7 Sonication

Sonication is the most promising technique for creation of nanofluids. The fluid is magnetically stirred in a magnetic stirrer, the sonication process is carried in an ultrasonication path, ultrasonic vibrator and mechanical homogenizer. Sonicators have become the industry standard for probe sonication and are more powerful and effective when compared to ultrasonic cleaner baths for nanoparticle applications. Probe sonication is most effective for processing nanomaterials (carbon nanotubes, graphene, inks, metal oxides, etc.) (Zheng et al. 2010).

16.2.2 Bottom-up Approaches

Tiny atoms and molecules are self-assembled or self-organized or combined in bottom-up technique to create nano-structured particles (Baig et al. 2021). These include chemical and biological approaches:

16.2.2.1 Chemical Vapour Deposition (CVD)

Chemical vapor deposition technique is mainly used for the generation of carbon-based nanomaterials. Through a chemical reaction of vapor-phase precursors a thin film is formed on the substrate surface during CVD (Dikusar et al. 2009). The precursors shall exhibit is sufficient volatility, high chemical purity, good stability during evaporation, low cost, non-hazardous nature and a long shelf-life. Moreover, its breakdown should not result any contaminants or residual impurities. Vapor phase epitaxy, metal-organic CVD, atomic layer epitaxy and plasma-enhanced

CVD are only a few CVD variations. CVD is an excellent method for producing very pure and high-quality nanomaterials that are stiff, homogeneous, and strong (Ago 2015; Machac et al. 2020). It is used for production of two-dimensional nanoparticles (Baig et al. 2021).

16.2.2.2 Sol-Gel Process

The sol-gel method is a wet-chemical technique extensively utilized to create nanomaterials (Das and Srivasatava 2016; Baig et al. 2021). It is mostly used for the creation of various kinds of high-quality metal-oxide-based nanomaterials. In this method the liquid precursor is transformed to a sol and the sol is ultimately converted into a network structure that is called a gel. Metal alkoxides are the conventional precursors for the generation of nanomaterials using the sol-gel method. The steps involved in synthesis of nanoparticles via the sol-gel method are—(i) hydrolysis of metal oxide in presence of water or with the assistance of alcohol to form a sol; (ii) condensation takes place, results in an increase in the solvent viscosity to form porous structures that are left for ageing and during condensation or polycondensation process, hydroxo-(M–OH–M) or oxo-(M–O–M) bridges form, resulting in metal-hydroxo- or metal-oxo-polymer formation in solution; (iii) During the aging process, polycondensation continues, with changes to the properties, structure and porosity; (iv) after the aging process, drying takes place, in which water and organic solvents are removed from the gel; (v) lastly, calcination is carried to achieve nanoparticles (Baig et al. 2021). The generated nanoparticles are environmentally friendly and have many benefits (Patil et al. 2021).

16.2.2.3 Co-precipitation

Co-precipitation is a solvent displacement and wet chemical technique. Different solvents are used in this technique including ethanol, acetone, hexane, and non-solvent polymers. The polymer phases used can be synthetic or natural. Mixing the polymer solution, fast diffusion of the polymer-solvent into the non-solvent phase of the polymer results. The nanoparticles are formed due to the interfacial stress at two phases (Das and Srivasatava 2016).

16.2.2.4 Hydrothermal

Hydrothermal technique uses a wide temperature gradient from ambient to extremely high temperatures for the production of nanoparticles. This method offers several benefits comparing to physical and biological ones. However, at higher temperatures, the nanomaterials produced by hydrothermal synthesis could become unstable (Banerjee et al. 2008; Patil et al. 2021).

16.2.2.5 Green/Biological Synthesis

Diverse metal nanoparticles are produced utilizing natural materials such as plant materials, microbes and biowastes like vegetable waste, fruit peel waste, eggshell, agricultural waste, algae, etc. and is known as "green" or "biological" nanoparticle synthesis (Kumari et al. 2021). Sustainable green synthesis technologies are necessary to prevent the production of undesirable or dangerous by-products. This technique has several advantages, including being straightforward, affordable, producing nanoparticles with high stability, requiring little time, producing non-toxic by-products and being readily scaled up for large-scale synthesis (Malhotra and Alghuthaymi 2022).

16.2.2.6 Biological Synthesis Using Microorganisms

Several microorganisms are used in the synthesis of metal nanoparticles, via metal capture, enzymatic reduction and capping (Patil and Kim 2018). Use of microorganisms for synthesis of metalic nanoparticles is environmental friendly, fast and economical.

Biosynthesis of nanoparticles by bacteria: Bacteria may produce inorganic compounds either inside or outside of their cells (Hulkoti and Taranath 2014). *Desulforibrio caledoiensis* (Qi et al. 2013), *Enterococcus sp.* (Rajeshkumar et al. 2014), *Escherichia coli VM1* (Maharani et al. 2016) and *Ochrobactrum anhtropi* (Thomas et al. 2014) based metal nanoparticles are reported previously for their potential photocatalytic properties (Qi et al. 2013), antimicrobial activity (Rajeshkumar et al. 2014) and anticancer activity (Maharani et al. 2016).

Biosynthesis of Nanoparticles by Bacteria The extracellular synthesis of nanoparticles by bacteria involves bacterial extracellular reductase enzymes which shrink the silver ions to the nanoscale range. From the protein analysis of bacteria, the NADH-dependent reductase enzyme carries out the bioreduction of silver ions to AgNPs. The decline occurs within a few minutes in the quick extracellular creation of nanoparticles (Altammar 2023). The bacterium *R. capsulata* at pH 7 produced gold nanoparticles with sizes ranging from 10 to 20 nm. Numerous nanoplates and spherical gold nanoparticles were produced when the pH was changed to 4 (Sriram et al. 2012). The bacteria R. *capsulata's* release cofactor NADH and NADH-dependent enzymes may cause the bioreduction of Au (3+) to Au (0) and the generation of gold nanoparticles (Sriram et al. 2012). The intracellular synthesis of nanoparticles by bacteria involves three steps: trapping, bioreduction and capping.

Biosynthesis of Nanoparticles by Fungi Due to the presence of several enzymes in fungal cells and the ease of handling, they are thought to be potential candidates for producing metal and metal sulfide nanoparticles (Mohanpuria et al. 2008). The nanoparticles were synthesized on the surface of the mycelia.

It was determined that the Ag + ions are initially trapped on the surface of the fungal cells by an electrostatic interaction between gold ions and negatively charged carboxylate groups, which is facilitated by enzymes that are present in the mycelia's cell wall. Later, the enzymes in the cell wall reduce the silver ions, causing the production of silver nuclei. These nuclei then increase as more Ag ions are reduced and accumulate on them. Gold nanoparticles of 20–40 nm in size were reported to be synthesized by *Verticillium luteoalbum* (Erasmus et al. 2014). *Aspergillus terreus* and *Penicillium brevicompactum* KCCM 60390 based metal nanoparticles are reported for their antimicrobial (Li et al. 2011) activities.

Biological Synthesis Using Plant Extracts For creation of nanoparticles various plant extracts are mixed with a metal salt solution at ambient temperature. This technique has been used to create nanoparticles of silver, gold and various other metals (Li et al. 2011). Various parameters impact the production of nanoparticles such as kind of plant extract, its concentration, the concentration of the metal salt, pH, temperature and length of contact time (Mittal and Chisti 2013). A leaf extract from *Polyalthia longifolia* was used to create silver nanoparticles, the average particle size was around 58 nm (Kumar and Yadav 2009; Kumar et al. 2016). *Acacia auriculiformis* (Saini et al. 2016), *Anisomeles indica* (Govindarajan et al. 2016), *Azadirachta indica* (Velusamy et al. 2015), *Bergenia ciliate* (Phull et al. 2016), *Clitoria ternatea*, *Solanum nigrum* (Krithiga et al. 2013), *Coffea arabica* (Dhand et al. 2016), *Coleus forskohlii* (Naraginti et al. 2016), *Curculigo orchioides* (Kayalvizhi et al. 2016), *Digitaria radicosa* (Kalaiyarasu et al. 2016), *Dioscorea alata* (Pugazhendhi et al. 2016), *Diospyros paniculata* (Rao et al. 2016), *Elephantopus scaber* (Kharat and Mendhulkar 2016), *Emblica officinalis* (Ramesh et al. 2015), Euphorbia antiquorum L. (Rajkuberan et al. 2017), *Ficus benghalensis* (Nayak et al. 2016), *Lantana camara* (Ajitha et al. 2015), *Cinnamomum zeylanicum* (Soni and Sonam 2014) and *Parkia roxburghii* (Paul et al. 2016) are the few examples of plants which are reported for the green synthesis of metal nanoparticles (i.e., AgNPs). The nanoparticles synthesized by this technique were evaluated for their antifilaria activity (Saini et al. 2016), mosquitocidal activity (Govindarajan et al. 2016), antibacterial activity (Velusamy et al. 2015), catalytic activity (Edison et al. 2016), antioxidant activity (Phull et al. 2016) and Cytotoxicity (Patil et al. 2017).

Biological Synthesis Using Biomimetic "Biomimetic synthesis" refers to product synthesis through a chemical processes that resemble biological synthesis carried out by living things (Dahoumane et al. 2017). In this approach, proteins, enzymes, cells, viruses, pollen and waste biomass are used to synthesize nanoparticles. Biomimetic synthesis is classified into two categories i.e., functional biomimetic synthesis and process biomimetic synthesis. In functional biomimetic synthesis various materials and approaches are used to emulate particular characteristics of natural materials, structures and systems (Zan and Wu 2016).

However process biomimetic synthesis aims to create different desirable nanomaterials/structures by imitating the synthesis pathways, processes, or procedures of natural chemicals and materials/structures (Zan and Wu 2016).

16.2.2.7 Reverse Micelle Methods

Oil-in-water emulsion forms normal micelles, in which hydrophobic tails are aimed towards a core that has trapped oil droplets within it. However, reverse micelles are formed in a water-in-oil emulsion, in which the hydrophilic heads are pointing towards a core that contains water (Malik et al. 2012). The core of the reverse micelles acts as a nanoreactor for the creation of nanoparticles. The nanoreactors size can be controlled by varying the water-to-surfactant ratio, ultimately affecting the size of the nanoparticles synthesized through this method. If the water concentration is decreased, this result in smaller water droplets, resulting in the formation of smaller nanoparticles (Shiri et al. 2019). Thus, the reverse micelle method provides a facile route for synthesizing uniform nanoparticles with precisely controlled size.

16.3 Applications of Nanotechnology in Meat Industry

In meat industry, nanotechnology is an emerging approach to enhance the shelf-life, quality and safety of meat and meat products acting all along the production chain from processing, preservation to packaging (Table 16.1). In food processing, both inorganic and organic nanoparticles (e.g., nanoemulsion, nanofiber, nanostructured ingredients, etc.) are designed as colour/flavour additives, preservatives or neutraceutical food carriers (Lamri et al. 2021). The incorporation and use of bioactive and functional nanomaterials in meat products has generated significant amount of research (Delfanian and Sahari 2020). It has the potential to develop novel meat products with improved functionality, better preservation effects and reliable and smart packaging of meat application products. Nanotechnology can also play a significant role in meat authenticity to combat meat adulteration.

Table 16.1 Applications of nanotechnology in meat industry

Technology	Advantages	Description
Nanostructures	Improved sensory properties, texture; salt and sugar reduction; enhanced bioavailability	Nanosized food ingredients and additives
Nanoencapsulation	Aroma protection, flavour protection, bioactive delivery, nutrient delivery	Provide delivery systems for food supplements and sensitive ingredients
Nanocomposites	Improved packaging, active packaging and smart packaging	Active packaging with nano-sized antimicrobials and antioxidants, intelligent packaging with nanocoding for sensing.
Nanolayers and inks	Nanolithography depositions	Prevention from adulteration, authenticity, traceability
Nanosensors	Pesticide detection, toxin detection and pathogen detection	Nanomaterial-based sensors

16.3.1 Processing

During recent past meat industry is focusing on improving the healthiness of meat products through incorporating beneficial ingredients and reducing the use of unhealthy ingredients (Gullón et al. 2020). To improve the functional value of meat products use of plant-based materials such as oat fiber, soy fiber, citrus fiber, linseed, flaxseed, starches, gums and plant proteins as fat replacing agents, phytochemicals as antioxidants and antimicrobials as well as ensuring the reduction of saturated fatty acids and cholesterol with a possible reduction of salts and synthetic preservatives (Weiss et al. 2010). The use of nanotechnology can be a valid approach to replace unhealthy ingredients in a cost effective way with natural non-chemical ingredients. Thus, application of nanoparticles to processed meat products can be a strategy of the improvement of functional and nutritional motives. Nanoscale ingredients can also be added to meat products to mask the adverse effects of unpleasant odours and flavours and to improve the texture (Cushen et al. 2013).

Using nanoemulsions in food products can facilitate the use of less fat without a compromise in creaminess, thus offering the consumers a healthier option; examples include low fat nanostructured mayonnaise and spreads (Chaudhry et al. 2008). When the size of the droplets in an emulsion is reduced, there is less chances of the emulsion to break down and separate. Thus, nanoemulsification may lower the requirement for certain stabilizers in a product (Cushen et al. 2012). Functional food ingredients can be incorporated within the droplets, the interfacial region, or the continuous phase of the nano-emulsions (Jasinska et al. 2010). Hence the reduced fat meat products can be developed without compromising with the flavor of the products.

Use of paprika oleoresin nanoparticles in chicken breast fillets improved marination and organoleptic qualities of the fillets (Yusop et al. 2012). Application of bacterial nanocellulose (BNC) to low-lipid, low sodium meat emulsions formulated with pre-emulsified high-oleic sunflower oil stabilized meat sausages during 45 days under vacuum storage (Marchetti et al. 2017). Ultrasonic-assisted incorporation of nano-encapsulated omega-3 fatty acids was found to enhance the fatty acid profile (e.g., n3/n6 ratio values) of the generated product in terms of the nutritional modification of pork meat products (Ojha et al. 2017).

The combinations of essential oils (EOs) and their nanoemulsions can be used as an antioxidant and antimicrobial without affecting the technological characteristics of the meat products (Pinelli et al. 2021). Use of sodium caseinate containing a nanoemulsion of ginger EO on chicken breast fillets, exhibited a significant decrease in the total aerobic psychrophilic bacteria of the refrigerated chicken fillets (Noori et al. 2018). Liu and Liu (2020) developed thyme EO chitosan nanoemulsions and thymol chitosan nanoemulsions as natural novel antibacterial packaging materials resulting in effective meat preservation. Phenolics (e.g., benzoic and vanillic acids) when packaged in polylactic glycolic acid nanoparticles were effective at very low concentrations for inhibition of pathogenic microorganisms such as *Salmonella Typhimurium*, *Escherichia coli* O157:H7 and *Listeria monocytogenes*

(Ravichandran et al. 2011). Similarly zein nanofibers can be applied to improve the delivery of several antioxidants such as gallic acid (Neo et al. 2013). The nanostructuring can be used for production of low sodium meat products (Boskovic et al. 2013). Therefore, the use of nanotechnology is valid strategy for production of healthy meat products without compromising the quality of the products.

16.3.2 Preservation

Man has been trying to develop better methods of food preservation since ancient times. This predates the days when people used to store recently killed animals in caves. The wet climate of the caves helped to keep the kill alive for a while. The twenty-first century then saw the development of refrigeration technology. Other conventional techniques for food preservation included the use of cellars and cold streams (Alfadul and Elneshwy 2010). Early BC saw the usage of fermentation and sun drying and what is practised now is simply an improved and modified version of these ideas (Chellaram et al. 2014). The Romans were renowned for inventing and using pickling to prevent microbial infestation of the food, while the Egyptians were known for using sun-drying to keep their foods free from spoiling. There is evidence that these methods of food preservation were used in the past. Many techniques for preserving meats have evolved over time, from salt addition to sun drying, refrigeration, fermentation, canning, freezing, use of synthetic preservatives, etc. The fundamental principle of all of these preservation techniques revolved on two concepts: either killing or retarding the growth of microbiological organisms and inhibiting or retarding the biochemical reactions.

By the introduction of nanotechnology, food preservation and production have been revolutionised. In the myosystems nanoemulsions (micro or nano) are used mainly for the preservation of meat and fish products and therefore to extent their shelf life. In recent years, due to the consumers demand for healthy meat products meat industry is increasing focussing on natural antimicrobials and antioxidants such as essential oils for meat preservation. However, their application may be limited due to their low solubility in water, the high volatility of some of their constituents and their strong flavor and odour, which could affect the organoleptic properties of the meat products. One approach used to overcome these limitations is the encapsulation of such type of natural antimicrobials and antioxidants. Encapsulation of essential oils in nanoemulsions improves their bioactivity increases diffusion through biological membranes, owing to their nanoscale and lipophilic nature (Donsì et al. 2011). A nanoemulsion also protect essential oils against evaporation, oxidation, thermal or physical degradation and increases their stability, reduces the dose of essential oils required to ensure their bioactivity and diminishes their impact on the organoleptic properties of the product (Donsì et al. 2011). In this context, earlier researchers developed coatings with emulsions and nanoemulsions based on chitosan particles incorporated with essential oil from citrus and effect of these coatings on the shelf life of silver pomfret fillet stored under refrigeration was investigated. The nanoemulsions were more effective than conventional emulsions, extended the shelf life from 12 days to 16 days, while the control sample had a shelf life of only 8 days and were effective in inhibiting the growth of microorganisms and changes in the chemical properties (Wu et al. 2016). Alirezalu et al. (2021) used ε-polylysine along with ε-polylysine nanoparticles extracted from olive and green tea for nitrile-free sausages formulation. The nanoparticles were effective in extending the shelf-life and considerably improved the microbiological safety of sausages. Cellulose pads containing AgNPs have been shown to reduce the microbial counts of exudates from beef meat stored in modified atmosphere packaging (Fernandez et al. 2010). Similarly silver nanoparticles obtained from tea leave extracts were proven to have an antimicrobial activity and can be used in packaging of meat products against microbial growth such as Gram-negative pathogens (Loo et al. 2018). Ghaderi-Ghahfarokhi et al. (2016) investigated enrichment of beef burgers with thyme phenolics loaded chitosan nanoparticles and found desirable changes in antioxidant capacity, overall acceptability and sensory quality of beef burgers during storage under refrigerated conditions. Nanoemulsion based on sunflower oil applied in steaks made from Indo-Pacific king mackerel resulted in reducing the growth of microbes, thereby increasing the shelf-life (Joe et al. 2012). Moreover, nanoparticles exhibit promising antibacterial activity due to their increased specific surface area leading to improved surface reactivity (Kumar et al. 2020). Chitosan–gelatin based films incorporated with nanocapsules loaded with tarragon EOs significantly inhibit the deterioration of pork slice quality, as nanoencapsulation exhibited sustained release of tarragon EOs, resulting in improved antioxidant, antibacterial and sensory properties (Zhang et al. 2020). Nanoencapsulated garlic essential oil was successfully incorporated into edible films for extending shelf-life of vacuum-packed sausages (Esmaeili et al. 2020). Feridoni and Khademi Shurmasti (2020) investigated that nanoencapsulation of sour tea (*Hibiscus sabdariffa* L.) extracted with carboxymethyl cellulose significantly extended the storage life of chicken nuggets, therefore reducing the oxidative effect on the final meat product.

16.3.3 Packaging

Increased safety of meat and meat products is achieved by prolonging the shelf life and preventing spoilage or nutritional loss. Food packaging materials for muscle foods should possess proper mechanical, barrier and optical properties to ensure better shelf life. Moreover, the packaging materials currently used for the packaging of muscle foods are not biodegradable and cause environmental issues. On the other hand, environment friendly biodegradable packaging made of biopolymers may not have proper mechanical and barrier properties (Rhim et al. 2013). Thus application of nanotechnology in packaging of meat and meat products could result in longer shelf life and better product quality as well as the diversity in packaging technology. The key advantages of using nanotechnology for development of packaging are enhanced antimicrobial effects, improved mechanical, physical, heat-resistant and barrier properties, continuous monitoring during product shelf-life, improved biodegradability in comparison to the conventional packaging materials (Ramachandraiah et al. 2015; Lamri et al. 2021). In the food packaging industry nanotechnology can have following approaches (Chaudhry et al. 2008; Singh et al. 2015):

(a) Nanoparticle incorporation in the packaging materials to enhance flexibility, barrier and mechanical properties, etc.
(b) Active packaging: incorporation of nanoparticles having antimicrobial or antioxidant properties.
(c) Intelligent packaging: incorporation of nanoparticles for continuous monitoring during product shelf-life like nanosensors.
(d) Biodegradable polymer: nanomaterial composites.

16.3.3.1 Nanocomposites

The advanced materials, known as polymer nanocomposites (PNCs), are designed to combat mechanical strength issues as well as permeability and barrier-related issues in the packaging sector. In order to create polymer nanocomposites (PNCs), an inert, nanoscale filler is dispersed throughout a polymeric matrix. Possible filler materials include silica (SiO_2) nanoparticles, carbon nanotubes, graphene, starch nanocrystals, chitin or chitosan nanoparticles and clay and silicate nanoplatelets (Duncan 2011). An effective substitute for conventional packaging materials (paper, metal and glass) in food packaging is the use of polymers in nanocomposite packaging (Silvestre et al. 2011). With a minimum feature size of less than 100 nm, nanocomposites are polymer matrices reinforced with nanofillers (nanoclays, carbon nanotubes, nano-oxides, and cellulose microfibrils). Food packaging has made use of a variety of synthetic and natural polymers, but several of each kind have been used (Bastarrachea et al. 2011; Rhim et al. 2013). However, demand for biodegradable packaging made of biopolymers such as polyvinyl alcohol, polyglycolic acid and polylactide has increased due to environmental concerns.

Since polymers are not naturally impermeable to gases or water vapour, the use of nanoparticles can enhance the mechanical and heat resistance qualities of packaging materials while also extending their shelf life. Polymer silicate nanocomposites, for example, have better gas barrier, mechanical strength and heat resistance qualities (Schaefer 2005; Brody 2006). A novel and exciting substitute for traditional polymer composites is the use of nanoscale fillers in composite films (Alexandre and Dubois 2000; Sinha et al. 2002). Since the filler materials are essentially impermeable inorganic crystals, gas molecules must diffuse around them rather than following a straight path. This means that the dispersal of nano-sized fillers into the polymer matrix alters the barrier properties of a homogeneous polymer film by creating a tortuous path for gas diffusion. Due to its successful incorporation into composite materials and the beneficial features that the reinforced films exhibit, properties that are attributed to the high aspect ratio montmorillonite (MMT), has been extensively explored and is of particular interest (Rhim and Ng 2007). Other related clays, such as kaolinite, hectorite, and saponite, can also be employed in PNC applications; the natural structure of these different clays can have a significant impact on PNC's efficiencies. MMT is not used as only single layer silicate material in nanocomposite materials. (Duncan 2011). Picouet et al. (2014) applied nanoclays in their latest study to vacuum-pack beef loins. According to Brody et al. (2008), Bayer's meat-packaging product Durethan is made of a polyamide plastic film bonded with clay nanoparticles, which allows it to function as an oxygen, moisture and carbon dioxide barrier. According to Doulgeraki et al. (2011), storage conditions in addition to the use of suitable packaging can have a significant impact on the preservation of meat quality.

16.3.3.2 Active Packaging

Food-borne pathogens prevalent in meat, such as *Salmonella spp.*, *Campylobacter spp.*, *E. coli*, etc. continue to cause millions of episodes of disease each year worldwide despite efforts and advancements in slaughter cleanliness and food production procedures (Burt 2004; Sofos 2008; Newell et al. 2010; Duncan 2011; Linscott 2011; Boskovic et al. 2013). In light of the meat industry's demands, it is crucial to talk about the antibacterial qualities of nanoparticles and their applications. Metal or metal oxide nanoparticles have been added to polymer nanocomposites for strength, abrasion resistance, UV absorption or antibacterial bioactive packaging (Chaudhry et al. 2008). Duncan (2011), reported that silver (Ag), zinc (Zn), magnesium (Mg), copper (Cu) and titanium (Ti) are the metal nanoparticles that are frequently utilized for antibacterial activities. Although a variety of materials can be used to create antimicrobial films for food packaging, silver is the most frequently utilized due to its ability to combat both Gram-positive and Gram-negative bacteria, fungi, viruses and protozoa (Azeredo 2009; Azeredo 2013; Loncina et al. 2013). The use of silver as an antibacterial in food has a long history. Despite silver's long history of use as an antibacterial, nothing is known about the exact mechanism underlying this effect. The general explanation is that silver causes death by at least one of the following mechanisms: (a) binding to sulfhydryl or disulfide functional groups on the surfaces of membrane proteins and other enzymes, which interferes with essential cellular processes; (b) disrupting DNA replication; and (c) causing oxidative stress by catalysing the formation of reactive oxygen species (ROS) (Luoma 2008). The most commonly utilized metal oxide materials as inorganic nanomaterials are magnesium oxide (MgO), zinc oxide (ZnO) and titanium dioxide (TiO2). These materials also show antibacterial activity, which may be related to the production of reactive oxygen species (ROS) (Azeredo 2013). The antimicrobial activity of absorbent pads containing Ag nanoparticles was studied recently in the field of muscle foods by Fernnandez et al. (2009); the antimicrobial effect against *Staph. aureus* and *E. Coli* were confirmed during the modified atmosphere packaging preservation of poultry meat. In a different type of application, it has been demonstrated that cellulose pads containing AgNPs produced from silver ions in situ lower the microbial levels of exudates from beef meat kept in packaging under changed environment (Fernandez et al. 2010). According to a study on chicken breasts, this kind of packaging suppressed meat-borne pathogens such *Escherichia coli*, *Pseudomonas aeruginosa* and *Listeria monocytogenes* (Panea et al. 2013). Fedotova et al. (2010) demonstrated potent antibacterial and antifungal properties when they included silver nanoparticles into a cellulose and collagen film sausage casing.

16.3.3.3 Intelligent Packaging

Any deviation from the prescribed storage conditions may cause the product to deteriorate prematurely and may even pose a risk to consumer health due to the presence of toxins or infections. It has been documented that the use of nano biosensors can be used to identify infections in processing facilities or to notify manufacturers, distributors and customers about the product's safety (Cheng et al. 2006; Helmke and Minerick 2006). Also including in packaging, nanosensors can identify particular markers of pathogen metabolism or warn or educate the customer about a product's history of exposure to light, temperature, or oxygen (tampering) (Cushen et al. 2012). Various tools, including nanowires and antibodies, have been created to detect a wide range of toxins, infections and chemicals in food packaging (Dingman 2008). Additionally, it gives producers the ability to monitor any kind of product abuse occurring during the product's journey from the production facility to the end user. When employed as nanosensors, nanoparticles may identify chemical pollutants, gases, smells and changes in the surrounding environment (Azeredo 2009; Duncan 2011). Nanosensors make it simple to monitor gases in a headspace, such as oxygen, without destroying the package. Among these techniques is the use of photo-activated indicator ink for in-package oxygen monitoring, which is based on nanoscale TiO_2 or SnO_2 particles and methylene blue, whose colour changes in response to O_2 exposure (becoming blue when the film is exposed and bleached when it is not) (Milan et al. 2013). Other gases, such as gaseous amines, which are markers of fish and meat spoiling, may be detected by sensors in extremely small concentrations (Mills and Hazafy 2009; Duncan 2011). Like gas sensors, moisture sensors can indicate the moisture content without destroying a product's container. The humidity polymer-matrix of the packaging swells, resulting in larger degrees of inter-nanoparticle separation. These changes cause sensor strips to reflect or absorb different colours of light, which can be monitored. This sensor enables quick and accurate determination of package moisture levels without invasive sampling (Duncan 2011). In a different work, a xanthine amperometric sensor that measures the freshness of fish samples was created using calcium carbonate nanoparticles (Shan et al. 2009). In order to improve consumer knowledge about quality and safety, intelligent packaging can interact with the food inside the package, recognize any of its quality changes and use a variety of techniques to record and transmit information (Sharma et al. 2017). Indicators (gases, temperature, and freshness), data carriers (barcodes and radio-frequency identification systems) and sensors are among the developments in intelligent packaging; the latter group has the greatest promise for future advancements (Ghaani et al. 2016). According to Patel and Beveridge (2003), a sensor is an apparatus or system that possesses software, an interconnection network and control and processing electronics. The initial component of the system, the receptor, which is the detecting portion of the sensor and is represented by a sampling area where the surface chemistry takes place, is where nanotechnological applications are most significant. According to Bouwmeester et al. (2009), nanoscale particles are particularly interesting in this context because of their distinctive chemical and electro-optical characteristics. The primary advancements in nanotechnology for meat and fish are sensors and indicators for O_2 and CO_2, the ability to identify pathogenic bacteria and spoilage, freshness indications and temporal temperature indicators (Sharma et al. 2017).

Bacteria detection systems are a subset of intelligent packaging systems. These systems, in particular, use antibodies that have been attached to nanomaterials like quantum dots. The basic principle of this system is the interaction between antibodies and antigens, which is enhanced by the special optical and electrical qualities of nanoparticles as well as other features including large and easily functionalized surfaces (Sharma et al. 2017). Yang and Li (2006) produced an example of a quantum dot-based device with possible uses in meat packaging. Although *S. Typhimurium* and *E. Coli* O157:H7 were detected by this approach, further research is required before these methods may be used in food packing (Sharma et al. 2017). Freshness indicators give an indirect estimate of microbial contamination. In this instance, the intelligent gadget can interact with a microbial metabolite to provide visual data regarding the product's freshness. Food packaging materials made of plastic film or paper coated in a very thin coating of copper or silver were created by Smolander et al. (2004). This material reacts to the presence of sulfide volatiles produced by fresh meat undergoing deterioration, causing a change in colour. Systems that show the presence of volatile amines, which are markers of fish freshness, make up another significant group (Kuswandi et al. 2012; Ma et al. 2018; Maynor et al. 2007). Since temperature alterations during food distribution and storage significantly impacts the physical and microbiological decomposition of foods, time temperature indicators can also be thought of as "freshness indicators." (Angiolillo et al. 2017). Zhang et al. (2013) created an intelligent system for tracking time and temperature using plasmonic nanocrystals. This system can be customized to react within a particular temperature range, allowing for the tracking of perishable food items and providing quality information through vivid and multi-coloured changes. Smolander et al. (2004) developed a food packaging material made of paper or plastic film coated in a very thin coating of silver or copper. When fresh meat deteriorates, the material reacts by changing colour due to the presence of sulfide volatiles. According to Kuswandi et al. (2012), Ma et al. (2018), Maynor et al. (2007) and other researchers, a significant subset of packaging system developed, reported the existence of volatile amines, which are the markers of fish freshness.

16.3.3.4 Bionanocomposites

Bionanocomposites provide a way to create packaging materials that are carbon neutral and biodegradable, the use of nanocomposite with biopolymers is anticipated to increase. This will present opportunities for developing nations to utilize their agricultural and forestry resources, by products and wastes to create biopolymer nanocomposites (Chaudhry and Castle 2011). By adding nanoclays to a poly lactic acid biopolymer, food products have longer shelf life and better barrier qualities against oxygen and water vapour (Lagaron et al. 2005). To minimize the drawbacks of conventional packaging materials, bionanocomposites use biodegradable ingredients and nanomaterial fillers (Sorrentino et al. 2007). In addition to preserving the food, they surround and extending its shelf life with the help of nanofillers, the use of nanocomposite packaging materials minimizes the usage of plastics and requires less fossil fuel for production and biodegradation, making them more environmentally friendly (Sozer and Kokini 2009). Zein, a significant part of corn protein, has been used extensively in the food sector because of its special qualities and molecular makeup. Zein is appreciated to find new uses in the biodegradable plastics and specialty food industries; the credit goes to nanotechnology techniques. Zein has the ability to build a meshwork of tubular structures that, when treated with formaldehyde, can become inert and microbiologically resistant (Lawton 2002). For mechanical and tensile qualities, it is essential to regulate the homogeneity and structure of zein films at the nanoscale. Other significant natural polymers that could be enhanced even more by nanotechnology techniques are starch and its derivatives. When thermoplastic starch (TPS) hybrids with nanoclay were studied, the attributes that emerged showed that, in comparison to native TPS matrix, the strong contact between the TPS and the nanoclay enhanced tensile strength and decreased water-vapour permeability (Lopez-Rubio et al. 2006). Another class of biodegradable polymers made from biological monomers are called biopolyesters. Examples of these include polylactic acid (PLA), polyhydroxybutyrate (PHB), and polycaprolactone (PCL). Biopolyesters can be moulded into items or made into films since they are biocompatible and biodegradable (Tharanathan 2003). However, because of their brittleness and very weak gas-barrier qualities, biopolyesters have some significant drawbacks, when it comes to their efficiency in the food packaging industry. The use of nanoclays, namely kaolinite nanofillers into PLA films enhanced their mechanical and thermal stability while maintaining their barrier qualities, as demonstrated by Cabedo et al. (2006). The primary uses of nanocomposites are to extend the shelf life of food products and protect them from mechanical and high temperature shock (Honarvar et al. 2016). For instance, the food sector has paid a lot of attention to zinc oxide nanocomposite because of its antioxidant properties, which make it a popular choice for active food packaging (Kim et al. 2020). In most cases, polymers and nanoparticles combine to form nanocomposites, which enhance the polymer's inherent capabilities (Yotova et al. 2013). Because of their extremely diverse and chemical activity, nanocomposites often aid in preserving the freshness of food material and delaying microbial decomposition for longer periods of time. Consequently, their main application is in the creation of materials with strong barrier properties (Pandey et al. 2013). Carbon dioxide cannot escape from carbonated beverage bottles and cans because of the gas barrier properties of nanocomposites. Rather than utilizing bulky glass bottles and costly cans, the food industries, which mostly produce carbonated beverages, can apply a layer of nanocomposites at top of the beverage's cans or bottles to reduce leakage (Pandey et al. 2013). Usually, a nanoclay nanocomposite is used to create these gas barriers. Some of the commercially available nanoclay-based nanocomposites in the market are Durethan, Aegis, Imperium and Nanocar. Because of these nanoclays' transparency, low density, surface characteristics, biodegradability and excellent flow properties, they are highly appreciated (Flanagan and Singh 2006). By giving the clay better barrier qualities, the Aegis nanoclay, which is mostly utilized in the production of carbonated drinks, functions as an oxygen scavenger and aids in the retention of carbon dioxide (Flanagan and Singh 2006). Imperium is a nylon-based polymer based on nanoclay that also functions as an oxygen scavenger. In order to reduce carbon dioxide leakage from the beverage, nano-car is utilized during the production process of plastic beer bottles. Another polymer based on nanoclay is nano-car which serves as a gas barrier agent (He et al. 2010). These days, the market is filled with a wide range of marketed nanocomposite-based solutions that aid in the packaging and preservation of fresh produce as well as extending the shelf life of the product. Some of the available nanocomposites in the market among these are Nano-Ceram PAC, Top Screen DS13, and Guard IN Fresh (Burdo 2005). Based on elongation ability, film solubility and clarity, Bhattacharya Dipanwita and Kandeepan (2017) found that 2% carboxymethyl cellulose is the most suitable level for preparing edible film for packing luncheon chicken meat slices. According to Soni et al. (2016), 1.5% carrageenan is the ideal level for use as an edible film for chicken patties. According to Lim et al. (2014), meat patties that were packaged using biodegradable composite material managed to hold their moisture content without compromising their diameter during storage.

16.3.4 Nanotechnological Based Detection Methods

Nanosensors are potential tool in the meat industry for the quick monitoring and quantification of pathogenic microbes, spoilage causing agents, chemical residues and proteins responsible for allergic reactions (Silvestre et al. 2011). A nanosensor or bio-analytical device is an integrated system that consists of different nanostructured materials and biological receptors. They are easily integrated with analytes due to their specificity and high sensitivity. Different types of nanomaterials, such as nanoparticles (metallic, non-metallic, and metal oxide), nanorods, nanowires, carbon nanotubes and nanofibers are used in the development of nanosensors enhances its surface-to-volume ratios, optical and electrical properties (Ansari et al. 2010). Nanosensors detect colour changes in the food as well as gases such as ammonia, hydrogen, hydrogen sulfide, sulfur dioxide and nitrogen oxides that are being produced due to microbial or biochemical spoilage. Currently nanosensors are commonly used in the quick detection of food-borne pathogens, toxins and pesticides that are present in food products. The major components of nanosensors are electronic (data processing devices) and sensing devices that make them capable of detecting changes in heat, light, gas and chemicals by converting them into electrical signals (Lopez-Rubio et al. 2006). Commonly, gas sensors consist of metallic nanoparticles such as gold, platinum and palladium (Mannino and Scampicchio 2007).

Immunomagnetic separation (IMS) technique is used to isolate target organisms from food to avoid hindrances from food particles, prior to detection. Usually nanoscale magnetic particles used in this regard due to their extremely high surface-to-volume ratios, which facilitate large analyte capture efficiencies (Duncan 2011). Using universal protein G-liposomal nanovesicles and an immounomagnetic bead sandwich assay can simultaneously detect *E. coli* O157:H7, *Salmonella spp.* and *Listeria monocytogenes* (Cheng et al. 2006). A biosensor based on gold nanoparticle coated quartz crystal microbalance for detection of DNA has been used for monitoring of *E. coli* O157:H7 synthesized oligonucleotides (Rashidi and Kashravi-Darani 2011). Use of nanoparticles amplifies the signals and improves the efficacy of detection limit for pathogenic bacteria (Mao et al. 2006).

Another innovative class of biosensors is nanocantilevers. The principle of detection is based on their ability to detect biological-binding interactions, such as between antigen and antibody, enzyme and substrate or cofactor and receptor and ligand, through physical and/or electromechanical signalling (Hall 2002). Nanocantilevers consist of tiny pieces of silicon-based materials that have the capability of recognizing proteins and detecting pathogenic bacteria and viruses (Kumar 2006). Pathogen detection here is based on their ability to vibrate at different frequencies which in turn depends on the biomass of the pathogen under scrutiny.

Nanosensors using carbon nanotubes along with antibodies are used for identification of pathogens like salmonella on the surface of chicken fillets (Villamizar et al. 2008). These nanosensors can also be used for the detection of freshness in canned tuna with the help of several chemicals used as meat freshness indicator like xanthine and hypoxanthine (Çubukçua et al. 2007). Sharma et al. (2010) reported the potential application of functionalized-gold nanoparticles in aflatoxin B1 detection from food samples. Zhou et al. (2009), used gold-based nanoprobe immunochromatographic assays for the detection of brevetoxins in fishery products. Similarly, gold-based immunochromatographic assays have been used for the detection of botulinum neurotoxin type B from processed foods (Chiao et al. 2004). Overall nanotechnology based sensors are projected to have a brighter future in many areas in food industry although challenges in fabrication, integration and mass manufacture of such devices exist (Evans 2009).

16.4 Safety Concerns of Nanotechnology

As the technique of nanotechnology is new and the materials used have not fully explored perspective, hence there are many unanswered questions, regarding the safety of nanotechnology and its potential impact on human health as well as on environment (Chaudhry et al. 2017; Berekaa 2015; Gallocchio et al. 2015). Since, nanomaterials possess properties like large surface area, solubility, bioavailabilty, biopersistence, aggregation and adsorption, UV light activation, metal ion release from nanometals and nanometal oxides that are different from those of their bulk materials they could pose ill effects to the human body (Eleftheriadou et al. 2017). Although nanostructures used in foods have no direct effect on human health, however their nanoscale nature may result in some inevitable side effects. The nanomaterial toxicity has been widely investigated by He et al. (2014) and McShan et al. (2014). Overproduction of ROS by nanostructures results in oxidative stress and is likely to cause cell migration, cytotoxicity, apoptosis and the onset of cancer and cell damage (Fu 2014). Another concern is the ability of nanoparticles to penetrate cellular and membrane barriers due to their nanoscale size (Eleftheriadou et al. 2017). Therefore, the order of nanomaterial toxicity is as follows: NPs > nanotubes (nanowires) > nanostructures. Moreover, depending on the nature of nanoparticles, Chaudhry and Castle (2011) have classified them as follows:

(a) **Areas of least concern:** Those in which food products contain natural food nano-structures, which are either digested or solubilised in the gastrointestinal (GI) tract and are not biopersistent.
(b) **Areas of some concern:** Those in which food products contain encapsulated food additives in nano-sized carriers which may not be biopersistent but may carry the encapsulated substances across the GI tract. A greater uptake of food colours or preservatives could take the application outside of the conditions under which the ADI (acceptable daily intake value) was set for the additive.
(c) **Areas of major concern:** Those in which food products contain insoluble, indigestible, and potentially biopersistent nanoadditives (e.g. metals or metal oxides), or functionalized nanomaterials.
(d) ADME profile (adsorption, distribution, metabolism and elimination) and toxicological properties of which are not fully known at present. Some of the projected applications in the agricultural sector (e.g. nanopesticides) will also fall in this category.

Despite the growing concerns about the potential adverse effects on human health, nanotechnology applications are already being sold and implemented (Ravichandran 2010). Therefore, building a regulatory system is a key requirement to control and manage the potential risks and toxicity of nanoparticles for ensuring the safety of nanotechnology applications in the food industry. Furthermore, additional research is warranted to investigate the potential possible toxicity of nanoparticles on human health and the environment.

16.5 Conclusions and Future Prospects

By the introduction of nanotechnology food production, processing and preservation have been revolutionised. In meat production industry nanotechnology can play different roles from processing, preservation to packaging of fresh or processed meat products. They are instrumental for the delivery of bioactive ingredients to specific tissues. Remarkable outcomes in food processing and preservation have been achieved utilising nanomaterials, which may preserve food from moisture, lipids, gases, bad tastes, aromas, autoxidation reactions and inhibit or retard spoilage and pathogenic microorganisms. Promising applications of nanotechnology may be in meat packaging and could potentially benefit keeping quality of both fresh and processed meat if the nanomaterial is carefully chosen. Due to the more efficient delivery system development by incorporating nanocarriers, it is going to be an age of least use of chemicals like antibiotics and other veterinary drugs in meat production and consequently less carryover over of these residues in the meat produced. Other prospects of nanotechnology in meat processing include improvements in the nutrition and functional properties of meat products. Use of nanosensors a valuable weapon to detect food borne organisms and toxic chemicals. There will no longer be a threat in meat industry due to better detection system (nanosensors) having improved efficiency in terms of time and limit of detection. Overall the advancements in nanotechnology can supposedly bring an all-new digital future, which can be regarded as big data information, thereby leading to new research fields.

Despite the remarkable outcomes have been achieved utilising nanotechnology its concept has not touched the different aspects of food industry and it needs to be explored at all facets of the industry. Like, use of nanopowders is hindered due to lack of efficient equipments for processing, preparation methodologies which are difficult and lack of scalability and cost effectiveness.

The application of nanotechnology offers a number of disadvantages also. Currently the fate of nanoparticles and their potential toxicity is not fully known. Numerous health risks such as skin toxicity, acute toxicity, skin irritation and genotoxicity associated with nanomaterials' applications have been found and these concerns must be addressed (Yang and Li 2005). Therefore, prior to commercialization of nanofoods, robust studies of the potential risks to human health and environment is required. Standard regulation is indeed required to set guidelines for the use of nanotechnology in food along with its packaging and additives and provide guidelines for assessing the risks.

Due to the globalisation of food trade, responding to outbreaks associated with foodborne diseases is extremely challenging. The advancements in nanotechnology engineering may help to enhance global food security and safety (King et al. 2017). Further the innovative properties of the nanomaterials will urge the food industry to increase commercial applications of the technology.

Therefore the future of nanotechnology in meat products depends on the roles played by governments, regulatory agencies and manufacturers in addressing the challenges related to the application of nanomaterials in food.

References

Ago H (2015) CVD growth of high-quality single-layer graphene. In: Matsumoto K (ed) Frontiers of graphene and carbon nanotubes. Springer, Berlin, pp 3–20

Ajitha B, Reddy YAK, Reddy PS (2015) Green synthesis and characterization of silver nanoparticles using Lantana camara leaf extract. Mater Sci Eng C 49:373–381

Alexandre M, Dubois P (2000) Polymer-layered silicate nanocomposites: preparation, properties and uses of a new class of materials. Mater Sci Eng Rep 28(1–2):1–63

Alfadul SM, Elneshwy AA (2010) Use of nanotechnology in food processing, packaging and safety – review. Afr J Food Agric Nutr Dev 10:6

Alirezalu K, Hesari J, Yaghoubi M, Khaneghah AM, Alirezalu A, Pateiro M, Lorenzo JM (2021) Combined effects of ε-polylysine and ε-polylysine nanoparticles with plant extracts on the shelf life and quality characteristics of nitrite-free frankfurter-type sausages. Meat Sci 172:108318

Altammar KA (2023) A review on nanoparticles: characteristics, synthesis, applications, and challenges. Front Microbiol 14:1155622

Amendola V, Meneghetti M (2009) Laser ablation synthesis in solution and size manipulation of noble metal nanoparticles. Phys Chem Chem Phys 11:3805–3821

Angiolillo L, Spinelli S, Conte A, Alessandro Del Nobile M (2017) Recent advances in food packaging with a focus on nanotechnology. Recent Pat Eng 11(3):174–187

Ansari AA et al (2010) Prospects of nanotechnology in clinical immunodiagnostics. Sensors 10(7): 6535–6581

Azeredo MCH (2009) Nanocomposites for food packaging applications. Food Res Int 42:1240–1253

Azeredo MCH (2013) Antimicrobial nanostructures in food packaging. Trends Food Sci Technol 30:56–69

Baig N, Kammakakam I, Falath W (2021) Nanomaterials: a review of synthesis methods, properties, recent progress, and challenges. Mater Adv 2:1821–1871

Banerjee A, Krishna R, Das B (2008) Size controlled deposition of cu and Si nano-clusters by an ultra-high vacuum sputtering gas aggregation technique. Appl Phys A Mater Sci Process 90: 299–303

Bastarrachea L, Dhawan S, Sablani SS (2011) Engineering properties of polymeric-based antimicrobial films for food packaging: a review. Food Eng Rev 3:79–93

Bayda S, Adeel M, Tuccinardi T, Cordani M, Rizzolio F (2020) The history of nanoscience and nanotechnology: from chemical–physical applications to nanomedicine. Molecules 25:112

Behrisch R (1981) Sputtering by particle bombardment. Springer, Berlin-Heidelberg

Berekaa MM (2015) Nanotechnology in food industry; advances in food processing, packaging and food safety. Int J Curr Microbiol App Sci 4(5):345–357

Bhattacharya D, Kandeepan G (2017) Selection of biopolymers to develop a biodegradable and edible flm for packaging of luncheon chicken meat slices. Asian J Dairy Food Res 36(1):67–71

Boskovic M, Baltic ZM, Ivanovic J, Duric J, Loncina J, Dokmanovic M, Markovic R (2013) Use of essential oils in order to prevent food borne illness caused by pathogens in meat. Tehnologija Mesa 54(1):14–21

Bouwmeester H, Dekkers S, Noordam MY, Hagens WI, Bulder AS, de Heer C, ten Voorde SECG, Wijnhoven SWP, Marvin HJP, Sips AJAM (2009) Review of health safety aspects of nanotechnologies in food production. Regul Toxicol Pharmacol 53(1):52–62

Brady B, Wang PH, Steenhoff V, Brolo AG (2019) Nanostructuring solar cells using metallic nanoparticles. In: Kassab LRP, De Araujo CB (eds) Metal nanostructures for photonics. Elsevier, Amsterdam, pp 197–221

Brody AL (2006) Nano and food packaging technologies converge. Food Technol 60:92–94

Brody AL, Bugusu B, Han JH, Koelsch Sand C, McHugh TH (2008) Innovative food packaging. J Food Sci 73(8):107–117

Burdo OG (2005) Nanoscale effects in food-production technologies. J Eng Phys 78(1):90–96. https://doi.org/10.1007/s10891-005-0033-6

Burt S (2004) Essential oils: their antibacterial properties and potential applications in foods: a review. Int J Food Microbiol 94:223–253

Cabedo L, Feijoo JL, Villanueva MP, Lagaron JM, Gimenez E (2006) Optimization of biodegradable nanocomposites based on a PLA/PCL blends for food packaging applications. Macromol Symp 233:191–197

Chaudhry Q, Castle L (2011) Food applications of nanotechnologies: an overview of opportunities and challenges for developing countries. Trends Food Sci Technol 22:595–603

Chaudhry Q, Scotter M, Blackburn J, Ross B, Boxall A, Castle L (2008) Applications and implications of nanotechnologies for the food sector. Food Addit Contam Part A Chem Anal Control Expo Risk Assess 25(3):241–258

Chaudhry Q, Watkins R, Castle L (2017) Nanotechnologies in food: what, why and how? In: Nanotechnologies in food. Royal Society of Chemistry, Cambridge. https://doi.org/10.1039/9781782626879-00001

Chellaram C, Murugaboopathi G, John AA, Sivakumar R, Ganesan S, Krithika S, Priya G (2014) Significance of nanotechnology in the food industry. APCBEE Proc 8:109–113

Cheng MMC, Cuda G, Bunimovich YL, Gaspari M, Heath JR, Hill HD, Mirkin CA, Nijdam AJ, Terracciano R, Thundat T, Ferrari M (2006) Nanotechnologies for biomolecular detection and medical diagnostics. Curr Opin Chem Biol 10:11–19

Chiao DJ et al (2004) Colloidal gold-based immunochromatographic assay for detection of botulinum neurotoxin type B. J Chromatogr B 09(1):37–41

Çubukçua M, Timurb S, Anik U (2007) Examination of performance of glassy carbon paste electrode modified with gold nanoparticle and xanthine oxidase for xanthine and hypoxanthine detection. Talanta 74:434–439

Cushen M, Kerry J, Morris M, Cruz-Romero M, Cummins E (2012) Nanotechnologies in the food industry – recent developments, risks and regulation. Trends Food Sci Technol 24:30–46

Cushen M, Kerry J, Morris M, Cruz-Romer M, Cummins E (2013) Migration and exposure assessment of silver from a PVC nanocomposite. Food Chem 139:389–397

Dahoumane SA, Mechouet M, Wijesekera K, Filipe CD, Sicard C, Bazylinski DA et al (2017) Algae-mediated biosynthesis of inorganic nanomaterials as a promising route in nanobiotechnology – a review. Green Chem 19:552–587

Das S, Srivasatava VC (2016) Synthesis and characterization of ZnO–MgO nanocomposite by co-precipitation method. Smart Sci 4:190–195

Delfanian M, Sahari MA (2020) Improving functionality, bioavailability, nutraceutical and sensory attributes of fortified foods using phenolics-loaded nanocarriers as natural ingredients. Food Res Int 137:109555

Dhand V, Soumya L, Bharadwaj S, Chakra S, Bhatt D, Sreedhar B (2016) Green synthesis of silver nanoparticles using Coffea arabica seed extract and its antibacterial activity. Mater Sci Eng C 58:36–43

Dickinson E, Van Vliet T (2003) Food colloids biopolymers and materials. Royal Society of Chemistry, Cambridge

Dikusar A, Globa P, Belevskii S, Sidel'nikova S (2009) On limiting rate of dimensional electrodeposition at meso-and nanomaterial manufacturing by template synthesis. Surf Eng Appl Electrochem 45:171–179

Dingman J (2008) Nanotechnology: its impact on food safety. J Environ Health 70:47–50

Donsì F, Annunziata M, Sessa M, Ferrari G (2011) Nanoencapsulation of essential oils to enhance their antimicrobial activity in foods. LWT Food Sci Technol 44:1908–1914

Doulgeraki AI, Paramithiotis S, GJE Nychas (2011) Characterization of the Enterobacteriaceae community that developed during storage of minced beef under aerobic or modified atmosphere packaging conditions. Int J Food Microbiol 145(1):77–83

Duncan VT (2011) Applications of nanotechnology in food packaging and food safety: barrier materials, antimicrobials and sensors. J Colloid Interface Sci 363:1–24

Edison TNJI, Lee YR, Sethuraman MG (2016) Green synthesis of silver nanoparticles using Terminalia cuneata and its catalytic action in reduction of direct yellow-12 dye. Pectrochimica Acta Part A 161:122–129
Eleftheriadou M, Pyrgiotakis G, Demokritou P (2017) Nanotechnology to the rescue: using nano-enabled approaches in microbiological food safety and quality. Curr Opin 44:87–93
Erasmus M, Cason ED, VanMarwijk J, Botes E, Gericke M, Van Heerden E (2014) Gold nanoparticle synthesis using the thermophilic bacterium Thermus scotoductus SA-01 and the purification and characterization of its unusual gold reducing protein. Gold Bull 47:245–253
Esmaeili H, Cheraghi N, Khanjari A, Rezaeigolestani M, Basti AA, Kamkar A, Aghaee EM (2020) Incorporation of nanoencapsulated garlic essential oil into edible films: a novel approach for extending shelf life of vacuum-packed sausages. Meat Sci 166:108135
Evans HM (2009) Nanotechnology enabled sensing. Report of the national nanotechnology workshop. Nat Nanotechnol Initiat Workshop 9:27–34
Fect H (1996) Edelstein AS, Cammaratra RC (eds) Nanomaterials: synthesis, properties and applications. Taylor & Francis, Washington
Fedotova AV, AG SOA, Sdobnikova LG, Samoilova TA, Smurova AA, Revina Khailova EB (2010) Packaging materials manufactured from natural polymers modified with silver nanoparticles. Int Polym Sci Technol 37:59–64
Feridoni BS, Khademi Shurmasti D (2020) Effect of the nanoencapsulated sour tea (*Hibiscus sabdariffa* L.) extract with carboxymethylcellulose on quality and shelf life of chicken nugget. Food Sci Nutr 8:3704–3715
Fernandez A, Picouet P, Lloret E (2010) Reduction of the spoilage-related microflora in absorbent pads by silver nanotechnology during modified atmosphere packaging of beef meat. J Food Prot 73:2263–2269
Fernnandez A, Soriano E, Loopez-Carballo G, Picouet P, Lloret E, Gavara R (2009) Preservation of aseptic conditions in absorbent pads by using silver nanotechnology. Food Res Int 42(8): 1105–1112
Flanagan J, Singh H (2006) Microemulsions: a potential delivery system for bioactives in food. Crit Rev Food Sci 46(3):221–237. https://doi.org/10.1080/10408690590956710
Fu PP (2014) Introduction to the special issue: nanomaterials e toxicology and medical applications. J Food Drug Anal 22:1–2
Gallocchio F, Belluco S, Ricci A (2015) Nanotechnology and food: brief overview of the current scenario. Procedia Food Sci 5:85–88
Garcia M, Forbe T, Gonzalez E (2010) Potential applications of nanotechnology in the agro-food sector. Ciênc Tecnol Aliment 30(3):573–581
Ghaani M, Cozzolino CA, Castelli G, Farris S (2016) An overview of the intelligent packaging technologies in the food sector. Trends Food Sci Technol 51:1–11
Ghaderi-Ghahfarokhi M, Barzegar M, Sahari MA, Azizi MH (2016) Nanoencapsulation approach to improve antimicrobial and antioxidant activity of thyme essential oil in beef burgers during refrigerated storage. Food Bioprocess Technol 9:1187–1201
Govindarajan M, Rajeswary M, Veerakumar K, Muthukumaran U, Hoti S, Benelli G (2016) Green synthesis and characterization of silver nanoparticles fabricated using Anisomeles indica: mosquitocidal potential against malaria, dengue and Japanese encephalitis vectors. Exp Parasitol 161:40–47
Grumezescu AM, Holban AM (2018) Impact of nanoscience in the food industry, vol 12. Academic, London
Gullón B, Gagaoua M, Barba FJ, Gullón P, Zhang W, Lorenzo JM (2020) Seaweeds as promising resource of bioactive compounds: overview of novel extraction strategies and design of tailored meat products. Trends Food Sci Technol 100:1–18
Hall RH (2002) Biosensor technologies for detecting microbiological food borne hazards. Microbes Infect 4:425–432

He CX, He ZG, Gao JQ (2010) Microemulsions as drug delivery systems to improve the solubility and the bioavailability of poorly water-soluble drugs. Expert Opin Drug Deliv 7(4):445–460. https://doi.org/10.1517/17425241003596337

He X, Aker WG, Leszczynski J, Hwang HM (2014) Using a holistic approach to assess the impact of engineered nanomaterials inducing toxicity in aquatic systems. J Food Drug Anal 22:128–146

Helmke BP, Minerick AR (2006) Designing a nano-interface in a microfluidic chip to probe living cells: challenges and perspectives. Proc Natl Acad Sci 103:6419–6424

Honarvar Z, Hadian Z, Mashayekh M (2016) Nanocomposites in food packaging applications and their risk assessment for health. Electron Physician 8(6):2531. https://doi.org/10.19082/2531

Hu B, Liu X, Zhang C, Zeng X (2017) Food macromolecule based nanodelivery systems for enhancing the bioavailability of polyphenols. J Food Drug Anal 25:3–15

Hulkoti NI, Taranath T (2014) Biosynthesis of nanoparticles using microbes – a review. Colloids Surf B Biointerfaces 121:474–483

Idamokoro EM, Hosu YS (2022) Global research trends on the use of nanotechnology to boost meat production: a scientometric analysis. Front Res Metr Anal 6:793853

Jasinska M, Dmytrow I, Mituniewicz-Malek A, Wąsik K (2010) Cow feeding system versusmilk utility for yoghurt manufacture. Acta Sci Pol Technol Aliment 9(2):189–199

Joe MM, Chauhan PS, Bradeeba K, Shagol C, Sivakumaar PK, Sa T (2012) Influence of sunflower oil based nanoemulsion (AUSN-4) on the shelf life and quality of indo-Pacific king mackerel (*Scomberomorus guttatus*) steaks stored at 20 °C. Food Control 23:564–570

Kalaiyarasu T, Karthi N, Sharmila GV, Manju V (2016) In vitro assessment of antioxidant and antibacterial activity of green synthesized silver nanoparticles from *Digitaria radicosa* leaves. Asian J Pharm Clin Res 9:297–302

Kayalvizhi T, Ravikumar S, Venkatachalam P (2016) Green synthesis of metallic silver nanoparticles using Curculigo orchioides rhizome extracts and evaluation of its antibacterial, larvicidal, and anticancer activity. J Environ Eng 142:C4016002

Khan MI, Sahar A, Rahman UU (2016) Grumezescu AM (ed) Nanotechnology in healthier meat processing in novel approaches of nanotechnology in food. Academic, London, pp 313–345

Khan I, Saeed K, Khan I (2019) Nanoparticles: properties, applications and toxicities. Arab J Chem 12:908–931

Kharat SN, Mendhulkar VD (2016) Synthesis, characterization and studies on antioxidant activity of silver nanoparticles using Elephantopus scaber leaf extract. Mater Sci Eng C 62:719–724

Kim I et al (2020) ZnO nanostructures in active antibacterial food packaging: preparation methods, antimicrobial mechanisms, safety issues, future prospects, and challenges. Food Rev Int. https://doi.org/10.1080/87559129.2020.1737709

King T, Cole M, Farber JM, Eisenbrand G, Zabaras D, Fox EM, Hill JP (2017) Food safety for food security: relationship between global megatrends and developments in food safety. Trends Food Sci Technol 68:160–175

Krithiga N, Jayachitra A, Rajalakshmi A (2013) Synthesis, characterization and analysis of the effect of copper oxide nanoparticles in biological systems. Ind J Ns 1:6–15

Kumar CSSR (ed) (2006) Nanomaterials for biosensors. Wiley-VCH, Weinheim

Kumar V, Yadav SK (2009) Plant-mediated synthesis of silver and gold nanoparticles and their applications. J Chem Technol Biotechnol 84:151–157

Kumar R, Singh RK, Dubey PK, Kumar P, Tiwari RS, Oh IK (2013) Pressure-dependent synthesis of high-quality few-layer graphene by plasma-enhanced arc discharge and their thermal stability. J Nanopart Res 15:1–10

Kumar PS, Sundaramurthy J, Sundarrajan S, Babu VJ, Singh G, Allakhverdiev SI, Ramakrishna S (2014) Hierarchical electrospun nanofibers for energy harvesting, production and environmental remediation Energy Environ Sci 7:3192–3222

Kumar V, Bano D, Mohan S, Singh DK, Hasan SH (2016) Sunlight induced green synthesis of silver nanoparticles using aqueous leaf extract of Polyalthia longifolia and its antioxidant activity. Mater Lett 181:371–377

Kumar S, Mukherjee A, Dutta J (2020) Chitosan based nanocomposite films and coatings: emerging antimicrobial food packaging alternatives. Trends Food Sci Technol 97:196–209
Kumari SC, Dhand V, Padma PN (2021) Green synthesis of metallic nanoparticles: a review. Nano 2021:259–281
Kuswandi B, Jayus Restyana A, Abdullah A, Heng LY, Ahmad M (2012) A novel colorimetric food package label for fish spoilage based on polyaniline film. Food Control 25(1):18–189
Lagaron JM, Cabedo L, Cava D, Feijoo JL, Gavara R, Gimenez E (2005) Improving packaged food quality and safety. Part 2: nanocomposites. Food Addit Contam Part A 23(10):994–998
Lamri M, Bhattacharya T, Boukid F, Chentir I, Dib AL, Das D, Djenane D, Gagaoua M (2021) Nanotechnology as a processing and packaging tool to improve meat quality and safety. Food Secur 10:2633
Lawton JW (2002) Zein: a history of processing and use. Cereal Chem 79:1–18
Li X, Xu H, Chen ZS, Chen G (2011) Biosynthesis of nanoparticles by microorganisms and their applications. J Nanomater 2011:270974
Lim SL, Ishak W, Rosli W (2014) Nutritional composition and lipid oxidation stability of beef patties packed with biodegradable and non-biodegradable materials. Sains Malays 43:1197–1203
Linscott AJ (2011) Food-borne illnesses. Clin Microbiol News 33:41–45
Liu T, Liu L (2020) Fabrication and characterization of chitosan nanoemulsions loading thymol or thyme essential oil for the preservation of refrigerated pork. Int J Biol Macromol 162:1509–1515
Loncina J, Ivanovic J, Baltic T, Dokmanovic M, Duric J, Boskovic M, Baltic ZM (2013) Active system packaging of meat and meat products. Vet J Republic Srpska 13(1):5–16
Loo YY, Rukayadi Y, Nor-Khaizura MAR, Kuan CH, Chieng BW, Nishibuchi M, Radu S (2018) In vitro antimicrobial activity of green synthesized silver nanoparticles against selected gram-negative foodborne pathogens. Front Microbiol 9:1555
Lopez-Rubio A, Gavara R, Lagaron JM (2006) Bioactive packaging: turning foods into healthier foods through biomaterials. Trends Food Sci Technol 17(10):567–575
Luoma SN (2008) Silver nanotechnologies and the environment: old problems or new challenges? Woodrow Wilson International Center for Scholars: Project of Emerging Nanotechnologies, Washington. www.nanotechproject.org
Luther W (ed) (2004) Industrial application of nanomaterials – chances and risks. Future Technologies Division of VDI Technologiezentrum GmbH GrafRecke-Str, Dusseldorf
Ma X, Xu X, Xia Y, Wang Z (2018) SERS aptasensor for salmonella typhimurium detection based on spiny gold nanoparticles. Food Control 84:232–237
Machac P, Cichon S, Lapcak L, Fekete L (2020) Graphene prepared by chemical vapour deposition process. Graph Technol 5:9–17
Maharani V, Sundaramanickam A, Balasubramanian TJE (2016) In vitro anticancer activity of silver nanoparticle synthesized by Escherichia coli VM1 isolated from marine sediments of Ennore southeast coast of India. Enzym Microb Technol 95:146–154
Malhotra SPK, Alghuthaymi MA (2022) Biomolecule-assisted biogenic synthesis of metallic nanoparticles. In: Agri-waste and microbes for production of sustainable nanomaterials, pp 139–163
Malik MA, Wani MY, Hashim MA (2012) Microemulsion method: a novel route to synthesize organic and inorganic nanomaterials: 1st Nano update. Arab J Chem 5:397–417
Mannino S, Scampicchio M (2007) Nanotechnology and food quality control. Vet Res Commun 31(1):149–151
Mao X, Yang L, Su XL, Li Y (2006) A nanoparticle amplification based quartz crystal microbalance DNA sensor for detection of Escherichia coli O157:H7. Biosens Bioelectron 21:1178–1185
Marchetti L, Muzzio B, Cerrutti P, Andrés SC, Califano AN (2017) Bacterial nanocellulose as novel additive in low-lipid low-sodium meat sausages. Effect on quality and stability. Food Struct 14:52–59

Matsumoto R, Adachi S, Sadki EHS, Yamamoto S, Tanaka H, Takeya H, Takano Y (2020) Maskless patterning of gallium-irradiated superconducting silicon using focused ion beam. ACS Appl Electron Mater 2:677–682
Maynor MS, Nelson TL, O'Sullivan C, Lavigne JJ (2007) A food freshness sensor using the multistate response from analyte-induced aggregation of a cross-reactive poly(thiophene). Org Lett 9(17):3217–3220
McShan D, Ray PC, Yu H (2014) Molecular toxicity mechanism of nanosilver. J Food Drug Anal 22:116–127
Milan BZ, Mariza B, Jelena I, Mariza D, Jelena J, Jasna L, Tatjana B (2013) Nanotechnology and its potential application in meat industry. Tech Mess 54(2):168–175
Mills A, Hazafy D (2009) Nanocrystalline SnO_2-based, UVB-activated, colourimetric oxygen indicator. Sensors Actuators B Chem 136(2):344–349
Mittal A, Chisti Y (2013) Synthesis of metallic nanoparticles using plant extracts. Biotechnol Adv 31:346–356
Mohanpuria P, Rana NK, Yadav SK (2008) Biosynthesis of nanoparticles: technological concepts and future applications. J Nanopart Res 10:507–517
Muñoz-García J, Vázquez L, Cuerno R, Sánchez-García JA, Castro M, Gago R (2009) Self-organized surface nanopatterning by ion beam sputtering. In: Wang ZM (ed) Toward functional nanomaterials. Springer, Berlin, pp 323–398
Naraginti S, Kumari PL, Das RK, Sivakumar A, Patil SH, Andhalkar VV (2016) Amelioration of excision wounds by topical application of green synthesized, formulated silver and gold nanoparticles in albino Wistar rats. Mater Sci Eng C 62:293–300
Nayak D, Ashe S, Rauta PR, Kumari M, Nayak B (2016) Bark extract mediated green synthesis of silver nanoparticles: evaluation of antimicrobial activity and antiproliferative response against osteosarcoma. Mater Sci Eng 58:44–52
Neo YP, Ray S, Jin J, Gizdavic-Nikolaidis M, Nieuwoudt MK, Liu D, Quek SY (2013) Encapsulation of food grade antioxidant in natural biopolymer by electrospinning technique: a physicochemical study based on zein–gallic acid system. Food Chem 136:1013–1021
Newell DG, Koopmans M, Verhoef L, Duizer E, Aidara Kane A, Sprong H, Opsteegh M, Langelaar M, Threfall J, Scheutz F, van der Giessen J, Kruse H (2010) Food-borne diseases. The challenges of 20 years ago still persist while new ones continue to emerge. Int J Food Microbiol 139:S3–S15
Noori S, Zeynali F, Almasi H (2018) Antimicrobial and antioxidant efficiency of nanoemulsion-based edible coating containing ginger (*Zingiber officinale*) essential oil and its effect on safety and quality attributes of chicken breast fillets. Food Control 84:312–320
Ojha KS, Perussello CA, García CÁ, Kerry JP, Pando D, Tiwari BK (2017) Ultrasonic-assisted incorporation of nano-encapsulated omega-3 fatty acids to enhance the fatty acid profile of pork meat. Meat Sci 132:99–106
Ostermann R, Cravillon J, Weidmann C, Wiebcke M, Smarsly BM (2011) Metal–organic framework nanofibers via electrospinning. Chem Commun 47:442–444
Ozimek L, Pospiech E, Narine S (2010) Nanotechnologies in food and meat processing. Acta Sci Pol Technol Aliment 9(4):401–412
Pandey S, Zaidib MGH, Gururani SK (2013) Recent developments in clay-polymer nano composites. Sci J Rev 2(11):296–328. https://doi.org/10.14196/sjr.v2i11.1056
Panea B, Ripoll G, González J, Fernández-Cuello A, Alberti P (2013) Effect of nanocomposite packaging containing different proportions of ZnO and Ag on chicken breast meat quality. J Food Eng 123:104–112
Patel PD, Beveridge C (2003) In-line sensors for food process monitoring and control. In: Rapid and on-line instrumentation for food quality assurance, pp 215–239
Pathakoti K, Manubolu M, Hwang HM (2017) Nanostructures: current uses and future applications in food science. J Food Drug Anal 25(2):245–253
Patil MP, Kim GD (2018) Marine microorganisms for synthesis of metallic nanoparticles and their biomedical applications. Colloids Surf B Biointerfaces 172:487–495

Patil MP, Ngabire D, Thi HHP, Kim MD, Kim GD (2017) Ecofriendly synthesis of gold nanoparticles and evaluation of their cytotoxic activity on cancer cells. J Clust Sci 28:119–132

Patil N, Bhaskar R, Vyavhare V, Dhadge R, Khaire V, Patil Y (2021) Overview on methods of synthesis of nanoparticles. Int J Curr Pharm Res 13:11–16

Paul B, Bhuyan B, Purkayastha DD, Dhar SS (2016) Photocatalytic and antibacterial activities of gold and silver nanoparticles synthesized using biomass of Parkia roxburghii leaf. J Photochem Photobiol B Biol 154:1–7

Phull AR, Abbas Q, Ali A, Raza H, Zia M, Haq IU (2016) Antioxidant, cytotoxic and antimicrobial activities of green synthesized silver nanoparticles from crude extract of Bergenia ciliata. Fut J Pharm Sci 2:31–36

Picouet PA, Fernandez A, Realini CE, Lloret E (2014) Influence of PA6 nanocomposite films on the stability of vacuum-aged beef loins during storage in modified atmospheres. Meat Sci 96:574–580

Pimpin A, Srituravanich W (2012) Review on micro-and nanolithography techniques and their applications. Engl J 16:37–56

Pinelli JJ, de Abreu H, Martins H, Guimarães AS, Isidoro SR, Gonçalves MC, Junqueira de Moraes TS, Ramos EM, Piccoli RH (2021) Essential oil nanoemulsions for the control of clostridium sporogenes in cooked meat product: an alternative? LWT 143:111123

Pugazhendhi S, Sathya P, Palanisamy P, Gopalakrishnan R (2016) Synthesis of silver nanoparticles through green approach using Dioscorea alata and their characterization on antibacterial activities and optical limiting behavior. J Photochem Photobiol B Biol 159:155–160

Qi P, Zhang D, Wan Y (2013) Sulfate-reducing bacteria detection based on the photocatalytic property of microbial synthesized ZnS nanoparticles. Anal Chim Acta 800:65–70

Rajeshkumar S, Ponnanikajamideen M, Malarkodi C, Malini M, Annadurai G (2014) Microbe-mediated synthesis of antimicrobial semiconductor nanoparticles by marine bacteria. J Nanostructure Chem 4:1–7

Rajkuberan C, Prabukumar S, Sathishkumar G, Wilson A, Ravindran K, Sivaramakrishnan S (2017) Facile synthesis of silver nanoparticles using *Euphorbia antiquorum* L. latex extract and evaluation of their biomedical perspectives as anticancer agents. J Saudi Chem Soc 21:911–919

Ramachandraiah K, Han SG, Chin KB (2015) Nanotechnology in meat processing and packaging: potential applications – a review. Asian Australas J Anim Sci 28:290–30212

Ramesh P, Kokila T, Geetha D (2015) Plant mediated green synthesis and antibacterial activity of silver nanoparticles using *Emblica officinalis* fruit extract. Spectrochim Acta Part A 142:339–343

Rao NH, Lakshmidevi N, Pammi S, Kollu P, Ganapaty S, Lakshmi P (2016) Green synthesis of silver nanoparticles using methanolic root extracts of Diospyros paniculata and their antimicrobial activities. Mater Sci Eng C 62:553–557

Rashidi L, Kashravi-Darani K (2011) The application of nanotechnology in food industry. Crit Rev Food Sci Nutr 51:723–730

Ravichandran R (2010) Nanotechnology applications in food and food processing: innovative green approaches, opportunities and uncertainties for global market. Int J Green Nanotechnol 1(2): P72–P96

Ravichandran M, Hettiarachchy NS, Ganesh V, Ricke SC, Surendra S (2011) Enhancement of antimicrobial activities of naturally occurring phenolic compounds by nanoscale delivery against *Listeria monocytogenes*, *Escherichia coli* O157:H7 and *Salmonella Typhimurium* in broth and chicken meat system. J Food Saf 31:462–471

Rhim JW, Ng PKW (2007) Natural biopolymer-based nanocomposite films for packaging applications. Crit Rev Food Sci Nutr 47(4):411–433

Rhim JW, Park HM, Ha CS (2013) Bio-nanocomposites for food packaging applications. Prog Polym Sci 38:1629–1652

Saini P, Saha SK, Roy P, Chowdhury P, Babu SPS (2016) Evidence of reactive oxygen species (ROS) mediated apoptosis in Setaria cervi induced by green silver nanoparticles from Acacia auriculiformis at a very low dose. Exp Parasitol 160:39–48
Schaefer M (2005) Double tightness. Lebensmittelchemie 37:52–55
Shan D, Wang Y, Xue H, Cosnier S (2009) Sensitive and selective xanthine amperometric sensors based on calcium carbonate nanoparticles. Sensors Actuators B Chem 136(2):510–515
Sharma A et al (2010) Antibody immobilized cysteamine functionalized-gold nanoparticles for aflatoxin detection. Thin Solid Films 519(3):1213–1218
Sharma C, Dhiman R, Rokana N, Panwar H (2017) Nanotechnology: an untapped resource for food packaging. Front Microbiol 8:1735
Shi Z, Lian Y, Liao FH, Zhou X, Gu Z, Zhang Y et al (2000) Large scale synthesis of single-wall carbon nanotubes by arc-discharge method. J Phys Chem Solids 61:1031–1036
Shiri MSZ, Henderson W, Mucalo MR (2019) A review of the lesser-studied microemulsion-based synthesis methodologies used for preparing nanoparticle systems of the noble metals, Os, Re, Ir and Rh. Materials 12(12):1896
Silvestre C, Duraccio D, Cimmino S (2011) Food packaging based on polymer nanomaterials. Prog Polym Sci 36:1766–1782
Singh PK, Jairath G, Ahlawat SS (2015) Nanotechnology: a future tool to improve quality and safety in meat industry. J Food Sci Technol. https://doi.org/10.1007/s13197-015-2090-y
Sinha RS, Yamada K, Okamoto M, Ueda K (2002) Polylactide-layered silicate nanocomposite: a novel biodegradable material. Nano Lett 2(10):1093–1096
Smolander M, Hurme E, Koivisto M, Kivinen S (2004) PCT international patent application WO 102185: A1
Sofos JN (2008) Challenges to meat safety in the 21st century. Meat Sci 78:3–13
Soni N, Sonam P (2014) Green nanoparticles for mosquito control. Sci World J 2014:1–6
Soni A, Kandeepan G, Mendiratta SK, Shukla V, Kumar A (2016) Development and characterization of essential oils incorporated carrageenan based edible flm for packaging of chicken patties. Nutri Food Sci 46(1):82–95
Sorrentino A, Gorrasi G, Vittoria V (2007) Potential perspectives of bio nanocomposites for food packaging applications. Trends Food Sci Technol 18(2):84–95
Sozer N, Kokini JL (2009) Nanotechnology and its applications in the food sector. Trends Biotechnol 27(2):82–89
Sriram MI, Kalishwaralal K, Barathmanikanth S, Gurunathani S (2012) Size-based cytotoxicity of silver nanoparticles in bovine retinal endothelial cells. Nanosci Methods 1:56–77
Su SS, Chang I (2018) Review of production routes of nanomaterials. In: Brabazon D, Pellicer E, Zivic F, Sort J, Baró MD, Grujovic N, Choy K-L (eds) Commercialization of nanotechnologies – a case study approach. Springer, Berlin, pp 15–29
Tharanathan RN (2003) Biodegradable films and composite coatings: past, present and future. Trends Food Sci Technol 14:71–78
Thomas R, Janardhanan A, Varghese RT, Soniya E, Mathew J, Radhakrishnan E (2014) Antibacterial properties of silver nanoparticles synthesized by marine Ochrobactrum sp. Braz J Microbiol 45:1221–1227
Velusamy P, Das J, Pachaiappan R, Vaseeharan B, Pandian K (2015) Greener approach for synthesis of antibacterial silver nanoparticles using aqueous solution of neem gum (*Azadirachta indica* L.). Ind Crop Prod 66:103–109
Villamizar R, Maroto A, Xavier RF, Inza I, Figueras M (2008) Fast detection of *Salmonella Infantis* with carbon analytical nanotechnology for food analysis 17 nanotube field effect transistors. Biosens Bioelectron 24:279–283
Weiss J, Gibis M, Schuh V, Salminen H (2010) Advances in ingredient and processing systems for meat and meat products. Meat Sci 86:196–213
Wu C, Wang L, Hu Y, Chen S, Liu D, Ye X (2016) Edible coating from citrus essential oil-loaded nanoemulsions: physicochemical characterization and preservation performance. Royal Soc Chem 6:20892–20900

Yang L, Li Y (2005) Quantum dots as fluorescent labels for quantitative detection of *S. typhimurium* in chicken carcass wash water. J Food Prot 6:1241–1245

Yang L, Li Y (2006) Simultaneous detection of Escherichia coli O157:H7 and Salmonella Typhimurium using quantum dots as fluorescence labels. Analyst 131(3):394–401

Yotova L, Yaneva S, Marinkova D (2013) Biomimetic nanosensors for determination of toxic compounds in food and agricultural products. J Chem Technol Metall 48(3):215–227

Yusop SM, O'Sullivan MG, Preuß M, Weber H, Kerry JF, Kerry JP (2012) Assessment of nanoparticle paprika oleoresin on marinating performance and sensory acceptance of poultry meat. LWT Food Sci Technol 46:349–355

Zan G, Wu Q (2016) Biomimetic and bioinspired synthesis of nanomaterials/nanostructures. Adv Mater 28:2099–2147

Zhang C, Yin AX, Jiang R, Rong J, Dong L, Zhao T, Sun LD, Wang J, Chen X, Yan CH (2013) Time-temperature indicator for perishable products based on kinetically programmable Ag overgrowth on Au nanorods. ACS Nano 7(5):4561–4568

Zhang H, Liang Y, Li X, Kang H (2020) Effect of chitosan-gelatin coating containing nano-encapsulated tarragon essential oil on the preservation of pork slices. Meat Sci 166:108137

Zheng Z, Zhang X, Carbo D, Clark C, Nathan CA, Lvov Y (2010) Sonication-assisted synthesis of polyelectrolyte-coated curcumin nanoparticles. Langmuir 26:7679–7681

Zhou Y et al (2009) Colloidal gold probe-based immunochromatographic assay for the rapid detection of brevetoxins in fishery product samples. Biosens Bioelectron 24(8):2744–2747

Chapter 17
Current Techniques and Technologies of Meat Quality Evaluation

Ishamri Ismail and Nurul Huda

17.1 Introduction

Meat is the main component of animal carcasses used as human food. The percentage of edible meat that can be consumed from each type of animal varies depending on species, age, sex, genetics, rearing method, feeding and other factors such as the specific butchering techniques used, consumer preferences for certain cuts, extent of trimming to remove excess fat and connective tissue (Prache et al. 2022; Clinquart et al. 2022). Advancements in meat processing technology and techniques may influence these percentages over time. The term "percentage of edible meat or meat carcasses" typically refers to the ratio of usable meat obtained from an animal carcass after processing. On average, about 50–80% of the live weight of a livestock and poultry is consider as meat carcasses. Kušec et al. (2022) reported the percentage of pork carcass around 24–46% for natural reared outdoor and 46–61% under industrial condition. This includes popular cuts like ham, loin, shoulder, belly, and neck. Meanwhile Foraker et al. (2022) reported the percentage of beef meat carcass generally about 63.5% for conventional beef cattle, and 63.0% for beef x dairy crossbreeds.

Meat consumption is projected to rise by 14% by 2030, primarily driven by population growth. This surge in meat consumption is anticipated to be most significant in Africa (30%), trailed by Asia and the Pacific Region (18%), Latin America (12%), North America (9%), and notably less in Europe (0.4%)

I. Ismail
School of Food Industry, Faculty of Bioresources and Food Industry, Universiti Sultan Zainal Abidin, Kampus Besut, Terengganu, Malaysia

N. Huda (✉)
Department of Livestock Production, Faculty of Sustainable Agriculture, Universiti Malaysia Sabah, Sandakan, Malaysia
e-mail: drnurulhuda@ums.edu.my

S. A. Rather, F. A. Masoodi (eds.), *Hand Book of Processed Functional Meat Products*, https://doi.org/10.1007/978-3-031-69868-2_17

(Font-i-Furnols 2023). From 2000 to 2019, in general, global per capita meat consumption rose from 29.5 kg to 34.0 kg. Notably, countries with consumption levels below the 2000 world average, such as Peru, Russia and Vietnam, saw significant increases of nearly 2 kg/capita/year. Additionally, South American countries which had relatively high consumption in 2000 such as Argentina, Brazil, Chile, and Colombia, also experienced annual increases of over 1 kg/capita. Several countries such Canada, New Zealand, and Switzerland seemed to be approaching the peak consumption levels for certain types of meats (Whitton et al. 2021). Meat carries significantly larger environmental and climate footprints compared to plant-based foods and can also have adverse health effects. While technological advancements can enhance the sustainability of meat production, shifts in consumption patterns are necessary. Particularly in high-income countries with high average meat consumption, substantial reductions will be crucial (Parlasca and Qaim 2022). Therefore, efforts toward achieving greater sustainability in the production of meat, meat products, and meat alternatives are crucial. Equally important is offering consumers accurate and reliable information about advancements made in this regard, along with the benefits and drawbacks associated with consuming various products (Font-i-Furnols 2023).

17.2 Nutritional Composition

17.2.1 Protein in Meat

Meat is good source of protein. Yang et al. (2023) reported the protein content of different part of meat ranged between 26.06% (shoulder) to 32.43% (eye round) which higher than protein content of chicken (27.40%) or pork (28.21%). Meanwhile Drewnowski (2024) reported the protein content of beef, pork, lamb and poultry is 27.27%, 25.39%, 25.22% and 23.18% respectively. The higher content of protein in meat provides supporting various physiological functions essential for good health, growth, and overall well-being. Meat is recognized as a complete protein source, containing all nine essential amino acids that the body requires but cannot produce independently. These amino acids play vital roles in numerous bodily functions, such as muscle repair and growth, immune response, and hormone regulation. Yang et al. (2023) found that the proportion of essential amino acids (EAA) to total amino acids (TAA) in beef, chicken, and pork was approximately 44.76%, 44.78%, and 45.17% respectively. Additionally, the protein efficiency ratio (PER) for beef, chicken, and pork was 3.42, 3.33, and 3.44 respectively. PER is widely regarded as a reliable indicator of protein quality, reflecting how effectively the body utilizes protein. A PER score close to or exceeding 3 is considered indicative of high-quality protein.

17.2.2 Fat in Meat

Meat can indeed be a source of dietary fat, depending on the type of meat and its preparation method. Meat comprises fatty tissues, which are fat cells containing lipids, with fat levels typically ranging from 8% to 20% across animal carcasses (Ahmad et al. 2018). The fat content of Hanwoo breed sirloin varied between 5.94% and 25.39%, whereas for the Australian beef sirloin varied between 15.15% and 10.2% (Hwang and Joo 2017). Najar-Villarreal et al. (2019) reported the fat content of NR and Angus breed ground beef varied between 15.7% and 19.8%. meanwhile Yang et al. (2023) reported the fat content of beef tenderloin, shoulder, eye round and flank is 4.07%, 16.82%, 2.17% and 8.04% respectively. Nogoy et al. (2022) conducted a comparison of the fat content in beef from cattle fed either with grass or with grain-based feed. Their findings indicated that beef fed with grass generally contains lower total fat levels compared to beef fed with grain across all breeds studied. In term of fatty acid composition, there were minimal amounts of short-chain and medium-chain fatty acids detected in meat, and the fatty acid composition across different types of animals was generally similar. The primary fatty acids found included palmitic acid (C16:0), stearic acid (C18:0), oleic acid (C18:1n9c), and linoleic acid (C18:2n6c) (Yang et al. 2023). Eicosapentaenoic acid (C-20:5) and docosahexaenoic acid (C-22:6) are typically present in meat tissues at low concentrations (Ahmad et al. 2018). Grass-fed beef exhibited higher levels of trans-vaccenic acid and long-chain n-3 polyunsaturated fatty acids (PUFA) compared to grain-fed beef. Additionally, grass-fed beef contained an increased total n-3 PUFA level, resulting in a reduced n-6 to n-3 ratio, potentially offering more health benefits than grain-fed beef (Nogoy et al. 2022).

17.2.3 Mineral in Meat

Meat is rich in macro and micro mineral elements. It is also a good source of several essential minerals, including Fe, Zn, Se, P, Mg and K. Each of these minerals serves diverse functions in the growth, development, and upkeep of the human body. The mineral composition of meat may fluctuate depending on factors like the type of meat, the animal's diet, and cooking techniques. Potassium (K) is notably the most abundant mineral, followed by phosphorus (P), sodium (Na), and magnesium (Mg). Ahmad et al. (2018) reported the K content in meat varied between 335–369 mg/100 g, meanwhile in mutton varied between 244–303 mg/100 g. Yang et al. (2023) mentioned the Na content in meat varied between 48.40–63.47 mg/100 g. Meanwhile the Fe and Ca content varied between 1.34–2.46 mg/100 g and 4.25–7.47 mg/100 g respectively. It is evident that offal organs contain higher mineral concentrations, such as iron (Fe), zinc (Zn), and copper (Cu), compared to the minerals found in muscle tissues (Ahmad et al. 2018). The meat liver contains high amount of Fe (7.1–9.5 mg/100 g) compare than in muscular tissue (0.99–2.40 mg/100 g).

17.2.4 Vitamin in Meat

Meat serves as a significant source of various vitamins, including water-soluble vitamin B1 (thiamine), B2 (riboflavin), B3 (niacin), B5 (pantothenic acid), B6, B12, and fat-soluble vitamins A, D, and E (Mohammed et al. 2020). These vitamins plays important role in metabolism, brain development and function, immune function, nerve function, DNA synthesis, the synthesis of neurotransmitters and others. Ahmad et al. (2018) mentioned the vitamin B1 content in beef, mutton, veal and pork varied between 0.06–1.20 mg/100 g, meanwhile for B2, B3, B5, B6 and B12 varied between 0.21–0.26 mg/100 g, 4.99–7.11 mg/100 g, 0.20–0.40 mg/100 g, 0.50–0.60 mg/100 g and nil—0.02 mg/100 g respectively. Vitamin C is almost nil in different types of animal carcass. Mohammed et al. (2020) reported the vitamin A, D and E in camel, beef and mutton varied between 0.004–0.008 mg/100 g, 0.002–0.005 mg/100 g, 0.35–0.86 mg/100 g, respectively. Vitamins naturally occurring in meat can be depleted during processing, particularly with heat exposure, especially affecting vitamin B. Such losses range from approximately 15–40% when boiling, 40–50% with frying, 30–60% during roasting, and 50–70% with canning. In contrast, vitamin A demonstrates greater resilience and can withstand temperatures as high as 80 °C without significant loss (Ahmad et al. 2018).

17.2.5 Carbohydrate in Meat

Meat is generally low in carbohydrates, particularly when compared to plant-based products. However, there are small amounts of carbohydrates present in some parts of meat, primarily in the form of glycogen. Liver primarily serves as the main source of carbohydrates in meat, accounting for roughly half of the total carbohydrates present in the body (Ahmad et al. 2018). Glycogen, a carbohydrate stored in the liver, indirectly influences the color, texture, tenderness, and water retention capacity of meat. Before slaughter, stored glycogen is converted into glucose, and then further metabolized into lactic acid. This conversion of glycogen to lactic acid depleted the carbohydrate content in meat. Bobokulov et al. (2021) mentioned the carbohydrate content in Karakul lamb is 0.60%, meanwhile Cutroneo et al. (2022) reported that carbohydrate content in red and white meat is 0.00%.

17.3 Physical and Textural Quality

17.3.1 Meat Quality

Structural features like connective tissue, muscle fibres, and tendons are visible in joint meat, and water is the most significant component, making up to 75% of meat weight (Geletu et al. 2021). Meat is a crucial livestock product, serving as primary protein source for human consumption. Meat quality is influenced by both scientific factors and consumer preferences (Liu et al. 2022). Scientific factors affecting meat quality include composition, nutrition, colour, water-holding capacity (WHC), tenderness, functionality, flavours, spoilage, and contamination (Wu et al. 2022). Consumer preferences are closely linked to human senses like sight, smell, taste, and mouthfeel. Visual criteria encompass meat colour, drip or purge amount, meat texture, fat amount and fat distribution, and fat colour. Consumer choice for cooked meat is mostly determined by mouth-feel factors such as tenderness, flavour, and juiciness (Garmyn 2020).

The freshness of meat is essential for consumer satisfaction and acceptability in cooking. When fresh meat is slaughtered and stored at freezing temperatures, it goes through biochemical and physical changes. Discoloration in meat relates to biochemical changes, and tenderness is influenced by muscle characteristics and post-mortem changes. Key quality characteristics for fresh meat are colour, WHC, texture, and fat (intramuscular fat, intermuscular fat and subcutaneous fat) (Joo et al. 2013). Fresh meat must visually appeal to consumers during selection, emphasizing colour and textural characteristics closely linked to WHC (Geletu et al. 2021). For cooked meat, essential eating quality factors are tenderness, flavor, and juiciness, aligning with consumer expectations. The importance of these palatability traits may vary among consumers based on past experiences and cultural backgrounds, thus impacting preferences across countries (Garmyn 2020).

High-quality meat focuses on its raw and cooked appearance, whereas the quality of processed meat products depends on different functional responsibilities. These roles include WHC, fat-binding capacity, emulsification, gel formation, formability, viscosity, adhesion, dispersion, stabilisation, fibre and film formation, flavour development, and texture. The assessment of processed beef products' quality is highly dependent on their intended purpose.

17.3.2 Meat Colour

Colour plays a vital role in the visual quality of meat, acting as the primary indicator to customers of its freshness and quality. Unfavourable colour might result in dismissal prior to assessing other factors like as tenderness, flavour, texture, and taste. Meat colour significantly influences consumer judgements at the moment of sale, making it a crucial quality component that affects their preferences (de Araújo et al. 2022).

Meat colour is influenced by the animal's species, age, and muscle type. Different species have distinct colour ranges due to the myoglobin (Mb) content in the muscle. Mb directly affects the meat colour, intensifying it from white or pink to red. Beef appears darker than lamb or pork because of its higher Mb content (Bak et al. 2019). As animals age, Mb increases, resulting in darker meat. Mb accumulates in red muscle fibres, and muscles with a higher proportion of these fibres, like leg muscles, are darker. Factors like exercise, diet, genetics, and environment also impact Mb content in muscles (Joo et al. 2013).

Meat colour depends mainly on the Mb content in muscles and its chemical state, existing as DeoxyMb, OxyMb, or MetMb. DeoxyMb, also called 'myoglobin' or 'reduced Mb,' is purplish-red and lacks a ligand at the sixth coordinate position of the heam group. It's responsible for the initial meat colour after cutting or vacuum packing (Wu et al. 2020). OxyMb, a cherry-red form, quickly forms when DeoxyMb is exposed to oxygen. For oxygenation, OxyMb must be in the ferrous state with oxygen occupying the sixth binding site of the ferrous haem iron (Suman and Joseph 2013). Both DeoxyMb and OxyMb can oxidize into MetMb, where the haem iron is in the ferric state, and water occupies the sixth coordinate position. MetMb cannot bind oxygen and is physiologically inactive. While OxyMb gives a desirable fresh color to meat, MetMb imparts a brown color associated with a lack of freshness, making it unacceptable to consumers (Suman and Joseph 2013).

Discoloration of meat during processing and storage is influenced by various factors. The accumulation of MetMb depends on both intrinsic and extrinsic factors. Intrinsic factors include muscle pH, composition, age, breed, sex, and diet, while extrinsic factors involve pre-slaughter treatment, carcass processing methods, temperature, oxygen, lighting, microbial growth, and packaging atmosphere during distribution (Poveda-Arteaga et al. 2023). Ruedt et al. (2023) highlight that the speed of OxyMb oxidation or MetMb reduction primarily determines meat colour stability. The discoloration rate varies by muscle, with those having redder muscle fibre showing faster discoloration due to higher oxygen consumption (Suman and Joseph 2013).

The freshness of meat is significantly influenced by the pH of the muscle, impacting its colour, water-holding capacity (WHC), and texture. In the case of dark-cutting beef, the dark colour does not necessarily indicate high myoglobin levels. Instead, it is due to a high ultimate pH (above pH 6.0), significantly exceeding the isoelectric point of myofibrillar protein (Wu et al. 2020). This alters muscle structure, causing it to either reflect less or absorb more light, resulting in a dark appearance. This condition is referred to as DFD (dark, firm, and dry) meat, particularly in beef and sheep (Ponnampalam et al. 2017).

Unlike the dark colour in DFD, PSE (pale, soft, exudative) occurs when the pH of meat is less than 6 at 45 min after slaughter (Jankowiak et al. 2021). In PSE meats, acidification after slaughter happens faster than normal, and lower pH values are reached in the muscle while the carcass temperature is still high. The combination of low pH and high temperature in PSE meat leads to the denaturation of some muscle proteins, reducing their water holding capacity (Terlouw et al. 2021). Ruedt et al. (2023) proposed that the scattering of light from the meat surface is probably caused by variations in refractive indices between the sarcoplasm and myofibrils. The larger the difference, the more scattering happens, leading to a paler appearance of the meat.

To maintain desirable colour, it's important to keep retail meat consistently cool during distribution and display, preventing unwanted colour shifts caused by chemical reactions. High storage temperatures can also trigger colour changes due to microbial growth. When oxidation of Mb occurs in the presence of sulfhydryl groups and oxygen, sulfmyoglobin—a green pigment—is formed (Ruedt et al. 2023). Storage temperature significantly affects lipid oxidation, contributing to meat discoloration. The combined effects of Mb and lipid oxidation cause discoloration in retail meats (Wang et al. 2021a). These reactions generate products that mutually accelerate oxidation. Lipid oxidation breakdown products are linked to off-flavours, odours, and the loss of fresh meat colour.

Adding antioxidants to meat during processing helps delay lipid oxidation and preserve colour. Plant-based sources rich in bioactive phenolic compounds are effective alternatives to synthetic antioxidants (Aminzare et al. 2019). Vitamins E, C, B complex, and beta-carotene are widely researched dietary antioxidants. Fruits and vegetables offer vitamin C and carotenoids, while vitamin E is found in whole grains and high-quality green vegetables (Petcu et al. 2023). Farm animals get vitamin E from pasture, legumes, forage crops, silage, yeasts, and bioactive plants and herbs. Vitamin E, a potent lipid-soluble antioxidant, prevents fatty acid peroxidation in cell membranes. Studies show that supplementing vitamin E in livestock diets has a notable effect on beef, delaying discoloration and lipid oxidation (Ponnampalam et al. 2022; Pokorný et al. 2001; Faustman et al. 1998). Liu et al. (1994) examined lipid oxidation in cooked *Gluteus medius* samples from Holstein steers fed different doses of α-tocopheryl acetate for 42 or 126 days. α-tocopherol concentrations increased in both raw and cooked muscle. The study found that ingesting 500 mg of α-tocopheryl acetate per day for 126 days resulted in 3.4 g of α-tocopherol per gram of cooked *Gluteus medius*. Galvin et al. (2000) had similar findings in their study on dietary vitamin E supplementation and cholesterol oxidation in cooked beefsteaks. Vitamin E increased meat's oxidative stability, indicated by a significant decrease in TBARS after supplementation.

The discoloration of retail meat is influenced by the packaging, storage atmosphere, and storage temperature. Three primary packaging methods used for displaying beef in refrigerated cabinets are overwrapped packaging (OWP) with an oxygen-permeable film, vacuum skin packaging (VSP) utilizing impermeable bags, and modified atmosphere packaging (MAP) involving a gas mixture of 80% oxygen and 20% carbon dioxide (McMillin 2017). OWP, the simplest method, allows oxygen to diffuse across the permeable film, forming a thin layer of oxymyoglobin on the meat's surface (Millar et al. 1994). Consumers generally find the resulting bright red oxymyoglobin pigment on the cut surface desirable (Font-i-Furnols and Guerrero 2014). VSP entails placing the meat in an impermeable plastic pouch and creating a vacuum to eliminate air. Although this anaerobic environment provides extended shelf life for retail cuts (Lagerstedt et al. 2011), the myoglobin pigments remain in the reduced deoxymyoglobin state, resulting in a dark purple meat colour. High-oxygen MAP employs a gas mixture of 80% oxygen and 20% carbon dioxide (80% O_2, 20% CO_2), promoting the formation of a deeper layer of oxymyoglobin on the meat's surface (McMillin 2017). Consequently, in high-oxygen MAP packs, the meat surface appears as a bright cherry red, which is appealing to consumers (McMillin 2017).

Another phenomenon that sometimes occurs in cooked meat is persistent pink and premature browning. Persistent pinking in cooked beef, where the meat stays pink even when fully cooked, is not desirable. Despite reaching the right temperature, this pink appearance may be seen as undercooked. It's important to understand that persistent pinking is more about the look of the cooked meat than a safety issue, and it's unlikely to harm consumers beyond the objectionable colour (Suman et al. 2016). Several factors, with higher-than-normal meat pH being a key one, can cause pinking. Studies have shown a strong link between the internal cooked colour, different forms of myoglobin (Mb), pH, and Mb denaturation. When the pH is higher than normal (> 5.7), myoglobin is protected from heat-induced changes, causing the meat to remain pink even after cooking. High-pH beef, especially from dark-cutting beef or those with Bos indicus influence, is more prone to persistent pinking. Additionally, pinking can be influenced by incomplete combustion products from cooking methods like wood-fired barbecues and higher levels of nitrate/nitrite in water (Suman et al. 2016). Contrarily, premature browning (PMB) in cooked beef occurs when myoglobin denaturation happens at a temperature lower than 71 °C, giving a misleading appearance of thorough cooking. This can pose a food safety risk, especially with the potential survival of harmful bacteria like E. coli O157:H7. The occurrence of PMB is influenced by factors like meat source, cooking methods, equipment, and storage conditions (Suman et al. 2016).

17.3.3 Drip and Purge

Water-holding capacity (WHC) stands out as one of the key quality attributes in fresh meat, significantly impacting meat colour (Joo et al. 2013). Beyond its influence on colour, WHC plays a crucial role in various physical properties of raw meat, such as texture and firmness. In the context of cooked meat, WHC directly affects juiciness and tenderness. Given its multifaceted impact, WHC holds great importance for meat producers, sellers, and consumers alike. It not only influences consumer acceptance but also plays a role in the final weight loss of meat sold by weight (Watanabe et al. 2018).

WHC refers to the ability of meat to hold inherent or added moisture throughout fabrication, processing, and storage. Some moisture loss is inevitable during mild treatments due to the presence of free water in muscle tissue. When meats are cut, the wetness of the surfaces varies widely, with some being dry and others releasing a significant amount of liquid (drip) due to reduced WHC of muscle proteins (Bowker 2017). It is well recognized that water molecules within the meat tissue often play an important role in maintaining hydrogen bonds and hydrophobic interactions in the three-dimensional protein conformation. The net charge of the protein determines the amount of bound water, with repulsion enhancing binding (Zhang et al. 2022). Meanwhile, polar groups on amino acid side chains (approximately 76–80%) attract water molecules through van der Waals forces. These blocked polar groups reduce water-binding and decrease WHC. Hence, the solubility and state of myofibrillar and sarcoplasmic proteins play a crucial role in determining meat WHC (Puolanne and Halonen 2010; Ishamri et al. 2010).

Net charge effect: During the conversion of muscle to meat during process of glycolysis, the accumulation of lactic acid reduces the pH of the meat. Once the pH reaches the isoelectric point (pI) of major proteins, such as myosin (pI = 5.4), the net charge of the protein becomes zero. This implies that the positive and negative charges within the protein become essentially equal. Consequently, these balanced charges attract each other, leading to a reduction in the amount of water that the protein can attract and hold. Moreover, as the net charge approaches zero, repulsion of structures in the myofibril is diminished, allowing these structures to pack more closely together, resulting in reduced space within the myofibril (Huff-Lonergan and Lonergan 2005).

Steric effects: Myofibrils are a significant part of muscle cells, making up 82–87% of the cell volume. The majority of water in living muscle cells is found within these myofibrils, with as much as 85% of the cell's water held there. Capillary forces from the arrangement of thick and thin filaments within the myofibril help retain this water. Changes during rigor development, like the formation of cross-bridges and shortening of sarcomeres, reduce available space for water within the myofibril. The net charge of protein structures in the myofibril also affects the available space for water. Protein connections link myofibrils to each other and the cell membrane. Preserving these connections in postmortem muscle may transfer the reduction in myofibril diameter to the muscle cell, potentially impacting cell volume. Consequently, expelled water from the myofibril and muscle cell gathers in the extracellular space, leading to gaps between muscle cells and bundles after rigor development, acting as primary channels for purge flow, commonly known as "drip channels" (Huff-Lonergan and Lonergan 2005).

Protein denaturation and myofibrillar structural shrinkage, play pivotal roles in influencing the WHC of cooked meat (Hughes et al. 2014). The heating temperatures lead to varied protein denaturation processes, resulting in muscle tissue shrinkage. This shrinkage in the sarcomere structure diminishes the WHC of muscles. Heat-induced shortening can even cause the tearing of the myofiber structure due to the force generated by the contraction of cross-bridges, formed by myosin and actin filaments (Ertbjerg and Puolanne 2017). Poor WHC in meat translates to a significant loss of liquid (drip loss), contributing to increased weight loss (shrinkage) during storage.

17.3.4 Texture and Firmness

The determination of meat texture involves considering the contractile and metabolic types, size and number of muscle fibres, connective tissue content, composition and distribution, as well as the content and lipid composition of intramuscular fat (Listrat et al. 2016). A coarse texture on the cut surface is created by larger muscle fibres and bundles. These muscle fibres, typically ranging from 10 to 100 micrometre in diameter, contribute to the diverse texture of meat (Joo et al. 2013). Fibre diameter is influenced by factors such as species, age, nutrition, genetics, and muscle fibre types (Listrat et al. 2016). The size of muscle fibres affects the size of fibre bundles and muscle growth potential, resulting in noticeable coarseness in meat cross-sections (Arshad et al. 2018).

Meat surface coarseness is influenced by thickened connective tissue strands and larger muscle bundles (Joo et al. 2013). The intramuscular connective tissue network is categorized into endomysium (around muscle fibres), perimysium (around muscle fascicles), and epimysium (around the whole muscle) (Finni et al. 2023). The amount of connective tissue varies with species, age, nutrition, and muscle fibre types (Listrat et al. 2016). Increased connective tissue is associated with greater meat coarseness, often seen in older animals with a higher proportion of connective tissue (Stanisławczyk et al. 2021).

Meat firmness is characterized by a rigid structure, high water retention, and minimal fluid loss during processing or cooking. However, the presence of intramuscular fat deposits can create a sense of firmness without actually impacting fluid retention. Soft and watery meat often results from a rapid decline in pH and elevated postmortem temperatures, causing low fluid retention due to the denaturation of muscle proteins and resulting in a lack of firmness (Joo et al. 2013). This lack of firmness is a common issue, particularly in PSE meat. PSE meat results from extensive denaturation of myofibrillar and sarcoplasmic proteins, resulting in a pale colour, soft structure, and excessive drip (Mashood et al. 2023). The total lipid content of muscle, known as intramuscular fat (IMF), plays a role in the firmness and juiciness of cooked meat. IMF is commonly referred to as marbling fat, although strictly speaking, it consists of small pieces of adipose tissue primarily composed of neutral lipid. Since IMF is deposited between the perimysium surrounding muscle bundles, meat firmness is partially influenced by IMF concentration, which is affected by the composition of fatty acids and temperature. Therefore, a high tissue temperature, soft fatty tissue, and a minimal quantity of IMF contribute to a soft-textured meat (Geletu et al. 2021). The composite fatty acids in meat have melting points ranging between about 25 °C and 50 °C, with saturated fatty acids (SFA) melting at higher temperatures and polyunsaturated fatty acids (PUFA) at lower temperatures (Pintado et al. 2020).

17.3.5 Marbling

Marbling refers to small fat flecks found between individual muscle fibre bundles. Higher levels of intramuscular fat (IMF) typically lead to improvements in tenderness, juiciness, and mouthfeel (Pintado et al. 2020), though debates exist about its contribution to tenderness. Feeding a significant amount of high-calorie foods, such as cereal grains like corn or barley, changes the colour of carcass fat from yellowish to white, increasing the likelihood of achieving a higher quality meat grade. While some consumers may find yellow fat objectionable, it does not affect the taste of cooked meat (Arai 2014).

Factors influencing IMF deposition can be divided into genetic (including breed, sex differences, and heritability), management (covering weaning age, castration, slaughter weight and age, and environmental factors), and nutritional factors (such as fat metabolism, fat digestion and absorption, glucose/starch availability, roughage vs concentrate ratio, dietary energy and protein levels, and vitamin A, D, and C levels) (Park et al. 2018). The content of IMF is primarily determined by the number and size of intramuscular adipocytes, with the formation of these adipocytes being crucial as they provide sites for later marbling fat deposition (Li et al. 2020a). Most intramuscular adipocytes are placed between primary and secondary muscle bundles in the perimysium of beef cattle and pigs, while high-quality grade Japanese Black cattle may also have marbling adipocytes within muscle bundles (Li et al. 2020a). Nutrient partitioning, specifically lipogenesis, occurs in a sequence: visceral fat, subcutaneous fat, intermuscular fat, and IMF. Although variations exist among species, IMF generally increases with age once major muscle growth stages are complete. IMF deposition is heritable and positively correlated with overall body fatness in the animal (Pewan et al. 2020).

Carcass maturity and marbling levels are essential factors in evaluating meat quality. The United States Department of Agriculture (USDA) employs a grading system that categorizes carcasses into eight groups based on the combination of maturity and marbling level. Maturity signifies the physiological age of the animal, considering factors like ossification and dentition, rather than its chronological age (Liu et al. 2022). In most carcass grading systems, the amount and distribution of marbling in the longissimus dorsi play a crucial role due to their strong association with beef palatability. The USDA system assesses marbling by visually evaluating the amount and distribution of intramuscular fat on the cut surface at the 12th to 13th rib longissimus muscle (LL) interface. The marbling scores range from the highest amount (abundant) to the lowest amount (practically devoid) based on current USDA grading standards. While this interface is the primary point for marbling score determination, Acheson et al. (2018) noted that marbling amount can vary across different anatomical locations in a carcass. In contrast, the Japanese Meat Grading Association (JMGA system) conducts carcass grading at the rib site, specifically between the 5th and 6th ribs (Liu et al. 2022). A recent study in a French private meat plant found no significant difference in marbling scores between the 5th and 10th ribs, suggesting that marbling assessment could be conducted at the quarter carcass level (Liu et al. 2021). The summary of global beef grading systems, from several countries is provided in Table 17.1.

The marbling traits, including the size (fine or coarse) and spatial distribution of marbling particles (MPs), are crucial considerations alongside the degree of marbling itself (Konarska et al. 2017). Utilizing computer image analysis enables the measurement of both the size characteristics (fineness or coarseness) of MPs and the overall degree of marbling in the muscle (Konarska et al. 2017). Computer image analysis emerges as a valuable tool for evaluating marbling traits, particularly in assessing the size characteristics that determine finely and highly marbled beef.

Table 17.1 Summary of global beef grading of several countries

Country	Marbling Scores	Quality Grades	Basis of Quality Grading	Yield Grades	Basis of Yield Grading	Reference
South Korea	9: 1–9 (larger more abundant) BMS score of 8 or 9 (above abundant)	1++, 1+, 1, 2, and 3 QG1++	Marbling score, meat colour, fat colour, firmness and texture of lean meat, and maturity of Longissimus dorsi (LD) muscle at the 13th rib	A, B, C	Rib eye, back fat thickness and carcass weight	Jo et al. (2012), Chung et al. (2018)
Japan (JMGA)	12: 1–12 (larger more abundant)	1, 2, 3, 4, 5	Marbling, meat colour and brightness, meat firmness and texture, and fat colour, lustre, and quality	A, B, C	Thoracic longissimus muscle area, rib thickness, cold split carcass weight, and subcutaneous fat thickness	Motoyama et al. (2016)
USA (USDA)	Abundant to practical devoid	Prime, choice, select, and standard	Lean maturity, bone maturity, lean colour, marbling, loin eye area, and subcutaneous fat thickness at the 12th and 13th ribs	1, 2, 3, 4, 5	Hot carcass weight, back fat thickness, kidney, pelvic, and heart fat percentage, and longissimus muscle area	Jaborek et al. (2020), Hale et al. (2013)
Australia (MSA)	100 to 1190	MSA grade: 3, 4, 5 star MSA score: 46–63 (3 star), 64–75 (4 star), 76–100 (5 star)	Hot carcass weight, rib fat depth, hump height, marbling, pH, and ossification	–	–	MLA (2018), McGilchrist et al. (2019)

JMGA Japanese Meat Grading Association, USDA United States Department of Agriculture, MSA Meat Standards Australia

17.4 Traceability in Supply Chain

The food industry is focusing more on customers and needs faster responses to handle food issues. Good traceability systems help reduce the production and distribution of unsafe or low-quality products, minimizing the risk of negative publicity, liability, and recalls. The current food labelling system doesn't guarantee authenticity, quality, and safety. Therefore, traceability is used to ensure food safety and quality and build consumer confidence (Aung and Chang 2014). In the wake of alarming foodborne illness cases like mad cow disease in the 1990s, there has been a significant rise in concerns about food safety. In response, food manufacturers have adopted traceability systems to monitor and control the journey of food through the supply chain. These systems offer transparent information about food quality at every stage of the supply chain, becoming a crucial tool to ensure food safety and quality, aligning with existing regulations (Schmidt and Moreno 2022). Traceability in logistics management refers to the capability to track the movement of a food product across different levels of the food supply chain. It involves identifying the origin of the product (traced upward) and its destination (tracked downward) at any point in the supply chain. Traceability plays a crucial role in early hazard detection, managing quality issues, and facilitating effective product recalls (Schmidt and Moreno 2022).

Traceability systems have gained global prominence due to heightened concerns about safety, quality, transparency, and regulatory compliance (Bosona and Gebresenbet 2013). The need for efficiency in business operations, the reduction of errors, and the prevention of incidents and fraud further drive the adoption of these systems. In the meat industry, traceability proves vital for controlling and enhancing production, crisis management, and livestock identification. It plays a crucial role in preserving the identity of meat products, incorporating attributes like safety, farming conditions, and animal welfare. Additionally, traceability safeguards product reputation, prevents adulteration and fraud, and ensures the safety, quality, and authenticity of meat products for consumers (Botilias et al. 2023; Jose and Shanmugam 2020). The term 'traceability'encompasses both 'internal traceability' within an organization and 'external traceability' between businesses or across the entire value chain. Full chain traceability, which covers all stages of product development, aligns with consumers 'growing interest in the environmental and social aspects of the products they purchase.

Consumer interest in traceability varies globally, with Greek consumers particularly valuing pork traceability (Dima et al. 2022). In contrast, Muslim consumers are increasingly concerned about cross-contamination issues in their Halal food, prompting a stronger emphasis on monitoring products throughout the supply chain (Ab Rashid and Bojei 2020). To ensure the authenticity of Halal products and address environmental factors in the Halal industry, food companies must implement a systematic traceability system. This helps enhance the integrity of the Halal food supply chain and mitigate the risk of cross-contaminations (Ab Rashid and Bojei 2020). Traceability not only builds consumer confidence but also educates them about the quality of meat products. Despite its importance, the meat industry faces challenges in selecting and integrating traceability systems. The traceability process spans from animal delivery to consumption, requiring compliance with EU and state regulations. To address these challenges, modern technologies like cloud and IoT (Internet of Things) play a crucial role in efficient data archiving, storage, rapid recall, and cost-effectiveness (Botilias et al. 2023).

Traceability in food manufacturing has evolved since the 1970s with the introduction of the first Global Standards (GS1) barcodes. The core idea of traceability is assigning a unique identifier to each batch or lot of food material, tracking it at all stages through the food chain. GS1 has commonly used product barcodes, Electronic Product Codes (EPCs), or Radio Frequency Identification (RFID) tags. Quick response (QR) codes, capable of storing more data than standard barcodes, are gaining popularity, especially with the development of mobile warehouse management systems that allow reading QR codes with mobile devices (Caveen et al. 2021).

17.4.1 Types of Traceability

17.4.1.1 One Up One Down Traceability

Traditionally, supply chains have relied on the concept of 'one up one down' (OUOD) traceability. The OUOD approach, requiring identification of immediate suppliers and customers, is tedious and time-consuming. Investigators review records at the last known node, and the process may take days or weeks, disrupting businesses and affecting consumers. The OUOD system is not designed for quick contamination source identification, especially with real-life complexities like incomplete or erroneous data (Caveen et al. 2021).

17.4.1.2 External Traceability

External traceability necessitates the unique identification of all traceable items and the sharing of information among all relevant distribution channel participants. Products are identified for traceability purposes through the assignment of a unique product identification number and a batch/lot number. To uphold external traceability, these identification numbers must be communicated on product labels and associated paper or electronic business documents. This connection links physical products with the information required for traceability. External traceability enables both tracing back to suppliers (supplier traceability) and tracking forward to clients (client traceability) (Caveen et al. 2021).

17.4.1.3 Internal Traceability

Internal traceability involves maintaining processes within an enterprise to connect the identities of raw materials to those of the finished goods. When one material is combined with others and undergoes processing, reconfiguration, or repackaging, the new product must be assigned its own Unique Product Identifier. The connection between this new product and its original material inputs (including batters, breading, seasonings, marinades, salt, packaging materials, and various other inputs) must be sustained for traceability. A label displaying the lot number of the traceable input item should remain on the packaging until the entire traceable item is used up. This principle holds true even when the traceable item is part of a larger packaging hierarchy, such as cases, pallets, or shipment containers (Caveen et al. 2021).

17.4.1.4 Full Chain Traceability

Full chain traceability is a modern concept that utilizes technology to gain more visibility into a product's journey through the supply chain. This approach provides a complete view of the product from farm to fork (or sea to plate), reducing the potential for safety breakdowns in food production. Advanced Enterprise Resource Planning (ERP) systems, along with the Internet of Things (IoT), enable the collection of massive amounts of data across all aspects of a product's life cycle. This data, linked from electronic freshness sensors, digital farming, logistics, etc., can lead to the development of consumer-friendly apps offering detailed information about the food's origin, including harvest, processing and distribution methods, temperature history, freshness data, and environmental impact records (Caveen et al. 2021).

17.4.2 Traceability System from Farm to Folk

17.4.2.1 Controlling Authorities

A traceability system requires a central authority to handle key functions:

- Setting identification standards and providing codes to livestock producers for their animals.
- Defining standards for participants (farmers, transporters, traders, abattoir companies) and overseeing their accreditation.
- Recording all animal movements within the system and verifying them when necessary.

These functions can be managed by a single or multiple authorities to introduce checks within the system. Registering animal identification codes and linking them to owners and farms/holdings is crucial for effective traceability (FAO/WHO 2004).

17.4.2.2 Traceability at Farms/Holdings

A register of accredited farms or holdings is essential. These are farms whose management practices have been approved after a comprehensive inspection and reporting process by the scheme. Regular inspections by an authorized inspectorate are necessary, and the registry should be updated accordingly. Clear standards or a code of conduct must be established for these farms, and if they fail to adhere to these standards, their marketing privileges should be suspended until the issues are rectified. One key element, among others, in maintaining these standards is the consistent reporting by farmers of the movement of animals to and from their farms (FAO/WHO 2004).

17.4.2.3 Traceability of Animals

A centralized authority oversees an animal identification scheme, assigning codes and managing identification devices for the traceability system. Records are maintained for issued identification codes, including details of the animals. Identification devices must meet specific standards for readability, tamper-resistance, and fraud prevention.

There are two primary identification systems:

(a) Group identification, where all animals share the same code until slaughter.
(b) Individual identification, assigning a unique identity number to each animal throughout its life.

Farmers keep a record of unique numbers and descriptions for individual animal identification, informing the central authority for registration. To handle lost or malfunctioning identification, a backup system is crucial. In Europe, cattle have tamper-proof plastic tags in both ears, providing a cost-effective solution with a large readable primary tag and a small metal tag as a backup in case of primary tag loss or malfunction (FAO/WHO 2004). While ear tags and tattoos are affordable and straightforward for live animal identification, Radio Frequency Identification (RFID) tags and transponders are expensive and may be unreliable. Subcutaneous transponders raise concerns about animal welfare and the potential risk of entering the food chain. Innovative ideas, like injecting a unique antigen into pigs for readable antibodies in the meat, might not be well-received by consumers (Yordanov and Angelova 2006).

17.4.2.4 Traceability at Slaughter House

Upon arrival at the abattoir, the animal's identity and history are entered into system database. Abattoirs must keep records of all arrivals and inform the central authority about arrivals and slaughtering to update the system. They need to monitor animal identification to reject those from farms without accreditation. Rejections must be recorded and reported to the central authority (FAO/WHO 2004). Abattoirs must follow a code of conduct for animal welfare and maintain hygienic practices. Subsequently, after slaughter and hide removal, the animal is quartered into hindquarters and forequarters. Each quarter is labelled with a carcass label, including details like carcass number, ear tag number, farmer's name and address, country of origin, date of birth, factory of slaughter, slaughter date, sex & grade, and cold weight. All this information is meticulously recorded in database. Inside the slaughter and processing plant, the simpler options for identification include paper bar codes, RFID tags, batch markers, molecules bar codes, quantum dots, and microwave radar (Yordanov and Angelova 2006).

In the halal industry, the slaughterhouse plays a pivotal role in upholding stringent compliance standards. Exclusively committed to the slaughtering of halal animals, the abattoir operates with a dedicated focus on adhering to the highest halal practices. Traceability in this domain extends its purview across diverse dimensions, encompassing premises, animals, slaughtermen, halal inspectors, slaughter tools, slaughtering procedures, stunning processes, and detailed slaughter records. Consequently, the identification of the halal critical point (HCP) within these elements emerges as a matter of paramount importance (JAKIM 2020). The determination of HCP in the halal meat industry is intricately tied to the comprehensive process flow, thereby fostering effective traceability (Yaacob et al. 2021). This systematic approach ensures a meticulous examination of each facet, aligning with the industry's rigorous standards and facilitating a seamless trace of the entire halal meat production process.

17.4.2.5 Traceability at Processing Stage

After the animal is quartered, it proceeds to the boning hall, where it undergoes the deboning process, and each quarter is transformed into primal cuts. Following the cutting process, the individual cuts undergo weighing, vacuum packing, and labelling with the European Article Number (EAN) barcode. The EAN-128 label includes essential information such as the product code, batch code, country of origin, country of slaughter, factory of slaughter, factory of cutting, slaughter date, pack date, cut by date, and use by date (Yordanov and Angelova 2006). This barcode facilitates the traceability of primal cuts back to a group of animals slaughtered on a specific date. Subsequently, these primal cuts are scanned into the system storage facility, either in freezer storage or chill storage. Primal cuts remain in chill storage for few days, and a traceability record is meticulously maintained through scanning. This record precisely notes the entry of the product into the chiller or freezer room and when it was dispatched. This comprehensive traceability system ensures transparency and accountability in the handling and distribution of the primal cuts (GS1 2005).

17.4.2.6 Traceability at Distribution Level

The transportation of primal cuts to retail stores involves the utilization of refrigerated trucks, meticulously maintaining temperatures within the range of 0–3 °C. This stringent protocol is integral to the establishment of traceability mechanisms, achieved through the incorporation of unique barcodes. This sophisticated system facilitates efficient tracking and identification of each individual item. The comprehensive traceability process allows for both upstream and downstream tracking, encompassing the product's journey from the company to the store. Additionally, the system possesses the capability for product withdrawal or recall if circumstances necessitate such actions. In cases involving defective products, the efficacy of traceability becomes apparent, as the unique barcode enables tracking in both upstream (from store to company) and downstream (from company to store) directions. The developed traceability platform serves as a robust tool for administrators, enabling them to conduct searches based on the product's barcode. This functionality provides access to a wealth of detailed information pertaining to the specific product and its originating animal (Botilias et al. 2023).

17.4.2.7 Traceability at Store Level

When the primal cuts reach retail stores, the EAN-128 code on the pack is scanned to confirm beef receipt and transfer traceability details to the stores' database. Primals are scanned into storage and, when ready for cutting and packing, are scanned out (GS1 2005). This creates a record of the beef delivery and processing time, establishing a link between the primal cut and the batch of animals it came from. A Desktop PC-based database captures label information through scanning. For daily pre-packed product production, the computer system records traceability info from the EAN-128 label on the primal. It transfers this data to the pre-packed products via the retailer's scales, printing a batch number (daily lot number) on each item for unique reference (GS1 2005). After the product is registered, the administrator can print and mark the product with the corresponding QR code for identification purposes (Botilias et al. 2023). If consumers have queries about beef origin, quoting the QR code allows tracing back to the processing factory and the batch of animals. For serve-over products, scanning the EAN-128 label transfers information to the back-office PC. The system assigns a lot number to serve-over primal within a specific timeframe. When consumers buy a meat cut from the batch, traceability info is provided through the unique daily lot number on the price label. The system's software can generate a traceability report and a 'whiteboard' sheet for in-store display. A label copy from daily serve-over products is printed and stored in a manual logbook for batch records. The computer system can analyse all traceability data by date, farmer, batch, primal, etc., and supports immediate product recall (GS1 2005).

17.4.2.8 Traceability by the Consumer

When consumers decide to purchase a meat product and find it free from defects while on the store shelves, the process of scanning the QR code becomes an assurance of safety and quality. This engagement with the QR code allows consumers to delve into the intricate details of the product, unravelling its journey from the origin to the current retail environment. The QR code serves as a gateway to a comprehensive digital platform that tracks the origin of the product and its journey through the supply chain. Consumers gain insights into not only the product itself but also the farm where the animal was raised, crucial details such as the date of birth and slaughter, the type of feed the animal received, and various other relevant particulars. This transparency not only fosters consumer confidence but also aligns with the growing demand for visibility into the entire supply chain, providing a rich and informative experience for individuals seeking to make informed choices about the products they bring into their homes (Botilias et al. 2023).

17.4.3 Technological Advancement in Traceability Tools and Technology Solutions

Automated data collection significantly reduces the time and costs associated with data processing and maintenance. Manual gathering of information for large operations can be time-consuming, as workers need to record activities and then transmit or transcribe and input data into the computer system. This manual process increases the risk of recording errors, as demonstrated by a Grocery Manufacturers Association (GMA) study, which found a 36% error rate in consumer-packaged goods orders in the United States. Such errors can lead to inventory inaccuracies and stock shortages.

To address these challenges, traceability initiatives increasingly rely on technology to offer efficient and accurate tracking of products across the supply chain. This involves various technologies for product identification, data capture, analysis, storage, and transmission, as well as comprehensive systems integration. Hardware components, including measuring/sensing equipment, identification tags, and labels, are coupled with software to create robust traceability systems.

Technologies like barcodes and Radio Frequency Identification (RFID) are particularly effective for data collection, boasting an accuracy rate of over 99%. These tools swiftly scan and record essential information such as product codes, lot numbers, invoice data, and order numbers in less than a second.

17.4.3.1 Barcodes

Various hardware tools, such as barcode scanners and printers, play a pivotal role at diverse stages of the livestock process, making substantial contributions to the immediate and efficient management of information. The software platform functions as the central hub of the traceability system, facilitating precise data exchange between human and technological resources throughout the process (Botilias et al. 2023). A barcode, a visual representation of data readable by machines, furnishes information about the attached object. Barcodes convey data through variations in the widths and spacing of parallel lines (1D) or patterns like rectangles, dots, and hexagons in two dimensions (2D). Initially, barcodes were scanned using specialized optical scanners known as barcode readers (Wudhikarn et al. 2022). Over time, scanners and interpretive software became available on various devices, including desktop printers and smartphones.

Product identification methods encompass (a) one-dimensional barcodes, such as the European Article Number (EAN)/Universal Product Code (UPC) barcode family, including Global Standard (GS1-128) and Info Global Trade Item Number (ITF-14) barcodes, widely established and used in retail. (b) two-dimensional barcodes, forming a square with tiny dots, like the Quick Response (QR) Code commonly used in traceable labels containing product traceability information. The QR code is advantageous for its quick readability, ample data capacity, and compact size. (c) Multi-dimensional barcodes, storing information on the x-axis, y-axis, and colours, offering significantly more data than two-dimensional codes (Kafetzopoulos et al. 2020). Barcodes serve for tracking products between businesses and stores, while QR codes are employed for consumer-side product tracking (Botilias et al. 2023).

However, barcode deformation or damage can impede accurate scanning, resulting in severe data loss. This can lead to delays in construction site operations (Akshatha et al. 2017). In cases where barcode labels cannot be scanned, workers are required to manually input numeric codes, causing potential slowdowns and inaccuracies in information entry, further complicating material management processes (Islam Chowdhury et al. 2019). Suboptimal material management can negatively impact labour productivity and lead to inefficient use of work hours, causing a decline in construction productivity (Xin et al. 2021).

17.4.3.2 Radio Frequency Identification (RFID)

RFID tags offer a durable and potentially more cost-effective method than barcodes for tracking meat. They remain unaffected by contamination and can withstand various temperatures and cleaning chemicals. Transponders, whether injected or embedded in ear tags for animal identification, use a short-range, pulsed echo principle for remote activation. These devices primarily serve identification purposes and transmit information only upon request. A basic system includes a tamper-proof identifier permanently attached or implanted, an activating/reading device with antennas for both the identifier and reader, and software for accurate electronic recording and data transfer. Transponders should be small, lightweight, and robust, lasting throughout the animal's lifetime. The protective covering, like bioglass for injectable transponders or plastic for ear tags, is crucial, promoting encapsulation by connective tissue for injectable transponders (Yordanov and Angelova 2006).

RFID technology for identification presents advantages over competing technologies, such as not requiring an unobstructed line of sight like barcodes. Radio telemetry uses low-frequency radio signals, providing good penetration of most non-metallic or ferromagnetic materials. However, when animals with implanted transponders are slaughtered, recovery can be challenging. Proper injection and methods to link carcass identity with the hook holding it during slaughter are crucial. Special detectors at the end of the slaughter line ensure transponder recovery. To avoid transponders in the food chain, electronic ear tags can be used as an alternative. Conventional tagging, barcode labelling systems, and RFID tags may introduce errors and lack precision. DNA technology, tracing animals and by-products through their DNA code, offers a more precise solution, overcoming the challenges associated with other identification methods (Yordanov and Angelova 2006).

17.4.3.3 DNA-Based Traceability

DNA technology, as it advances, introduces molecular gene-level analysis for accurate traceability (Guan et al. 2018). DNA-based methods focus on polymorphic DNA markers, categorized into methods requiring the clone or sequence of the marker (e.g., Restriction Fragment Length Polymorphisms—RFLPs, microsatellites/short tandem repeats—STRs, Single Nucleotide Polymorphisms—SNPs). Sequencing animal DNA is accurate but not always feasible; PCR-RFLPs offer a more convenient and cost-effective alternative. Other methods for example Amplified Fragment Length Polymorphisms (AFLPs), Variable Number of Tandem Repeats (VNTRs), and Random Amplified Polymorphic DNA (RAPDs) are less desirable due to technical challenges (Laliotis et al. 2018).

Mitochondrial DNA (mtDNA) presents advantages in species discrimination, especially in detecting adulteration in meat. PCR protocols detect ovine and caprine sequences, allowing clear species identification (Prusakova et al. 2018). In the dairy sector, detecting adulteration practices in goat or sheep milk is crucial. Methods like PCR-RFLP analysis or mtDNA are effective, and newer methods like End-Point PCR and RT-PCR reveal higher fraud levels (Kemal Seçkin et al. 2017). While DNA methods may not discriminate at the breed level, they play a crucial role in traceability, especially for quality assurance in Protected Designation of Origin (PDO) or Protected Geographical Indication (PGI) products. Two DNA methods, microsatellites (STRs) and Single Nucleotide Polymorphisms (SNPs), are reliable, cost-effective, and repeatable. They address genetic traceability demands in small ruminant species and their products (Kannur et al. 2017).

17.4.3.4 Internet of Things (IoT)

The Internet of Things (IoT) is defined as a network connecting objects for management, data mining, and access to generated data (Dorsemaine et al. 2015). It encompasses wirelessly connected devices such as RFID tags, sensors, actuators, and mobile phones, facilitating data transfer through internet connectivity for storage on cloud servers. Managed cloud services play a crucial role in enhancing secure data sharing, improving business efficiency, and reducing waste and operational inefficiencies. This promotes the utilization of smart IoT applications, particularly in supply chains, where IoT contributes to increased transparency and traceability (Rejeb et al. 2021).

The 2013 horsemeat scandal in parts of Europe served as a catalyst for a loss of public trust in the food industry (Tripathi et al. 2023). Foods labelled as containing beef were found to contain undeclared or inaccurately declared horse meat, reaching up to 100% of the meat content in some instances (Barnard and O'Connor 2017). Furthermore, certain products were discovered to contain other undisclosed meats, such as pork. This erosion of trust, especially concerning halal considerations, underscores the importance of addressing consumer concerns related to food integrity (Rejeb et al. 2021).

To address these concerns and enhance food supply chain processes, food manufacturers should leverage emerging technologies like the Internet of Things (IoT) and blockchain (Rejeb et al. 2021). These technologies, involving interconnected devices, sensors, and advanced orchestration capabilities, offer innovative potential for the food supply chain (FSC). In the halal industry, IoT provides five significant benefits: product traceability, improved supply chain efficiencies, streamlined livestock management, authentication of halal status, and monitoring of halal certifications (Rejeb et al. 2021). Real-time monitoring of employee safety, productivity, and food safety is made possible through wearable IoT devices, ensuring comprehensive traceability throughout the entire food chain. Additionally, IoT enables the integration of various sensors, such as temperature, humidity, oil, salt, metal, colour, pH, and viscosity sensors, to detect and prevent food fraud (Bouzembrak et al. 2019).

In response to the challenges faced by IoT-based food traceability systems, particularly regarding data integrity, trust, authenticity, and security, Michelin et al. (2018) have proposed an efficient blockchain-based approach. This addresses the limitations associated with IoT devices, including their limited computing power and storage capacity, which can impede the timely tracing of food sources and batches (Tripathi et al. 2023).

17.4.3.5 Blockchain

The concept of Blockchain Technology (BCT) was introduced in 2008 by Satoshi Nakamoto for Bitcoin, providing a decentralized public transaction ledger. Over the years, BCT has evolved through different generations: digital currency (Blockchain 1.0), digital economy (Blockchain 2.0), and digital society (Blockchain 3.0) (Patel et al. 2023). Blockchain is a distributed ledger connecting blocks through cryptographic methods, with each block containing data and virtual signatures, forming a chain (Giungato et al. 2017). The information on a blockchain is immutable, making it suitable for various industries, including finance, cybersecurity, healthcare, and food supply chains (Radanović and Likić 2018). It employs a hash algorithm, linking blocks together to prevent tampering. In the food sector, BCT finds applications in traceability, fraud detection, and food safety. By recording real-time data on production, processing, transportation, and more, BCT ensures timely and tamper-proof sharing of information. The involvement of smart contracts further enhances food safety management (HACCP, GMP, GHP etc.,) ensuring adherence to standards and certifications (Galvez et al. 2018). Consumers can access detailed product information by scanning a QR code, fostering trust and transparency (Kamilaris et al. 2019). The participants in the food chain, including producers, processors, logistics suppliers, retailers, and consumers, act as nodes in a peer-to-peer network. BCT safeguards privacy by storing private details in a permissioned chain and releasing necessary information on a public chain (Li et al. 2020b).

Blockchain technology ensures food safety by monitoring the production, processing, and handling of food. The international trade of food presents challenges in maintaining quality and safety (Creydt and Fischer 2019). Safety concerns in animal products include hormones, antibiotic residues, adulteration, zoonosis, and microbial contamination (Bintsis 2018). Blockchain, combined with sensors, enables real-time monitoring of factors like animal handling and temperature, ensuring product safety. In the food supply chain (FSC), Blockchain identifies discrepancies in standards, allowing corrective actions and timely recalls, enhancing buyer trust (Patel et al. 2023).

Recording data on Blockchain includes information on feed, veterinary drugs, animal breeds, and quantities. Consensus mechanisms ensure accurate data, automatically rejecting inconsistent or illegal information. Recording genomic and proteomic data provides assurance of product origin (Xu et al. 2022c). Blockchain-driven traceability in FSC involves three layers: physical flow, digital record system using sensors and technologies, and a blockchain network verifying data with smart contracts. This immutable record-keeping platform prevents data tampering, addressing stakeholders' demand for integrity and transparency in the food supply chain (Kamilaris et al. 2019). For instance, in 2016, Walmart became the first U.S.-based retailer to employ Blockchain technology in the animal products supply chain. This innovation was introduced to monitor pork products in China, subsequently imported to the U.S. via Latin America (Patel et al. 2023). As a result, the tracing of pork products from their origin to U.S. stores, which used to take several days in the past, was significantly reduced to just a few minutes. Additionally, this implementation offered comprehensive details about the entire journey of pork, from the farm to Walmart's retail store in the U.S. (Wass 2017).

17.5 Preservatives and Additives

The burgeoning global population and the widespread reach of the food supply chain have spurred an escalating demand for large-scale meat production. This surge in demand presents a myriad of challenges for the meat industry as it endeavours to uphold quality and safety standards throughout the entire supply chain. Traditionally, the use of food additives and preservatives has been a common practice in preserving perishable commodities, including meat, aimed at maintaining product consistency, quality, and safety. However, in the contemporary landscape, there is a discernible shift in consumer preferences towards minimally processed foods. This shift has propelled the exploration of innovative approaches to meat preservation, focusing on mitigating microbial contamination, ensuring prolonged shelf life, and preserving the organoleptic qualities of the meat. Simultaneously, regulatory bodies across different countries have instituted standards to govern the permissible levels of additives and preservatives employed in meat processing (Bouzembrak et al. 2019).

To meet these regulatory standards, a plethora of analytical methods has been developed for both qualitative and quantitative detection of preservatives, additives, and contaminants in meat. This chapter delves into a comprehensive discussion on the current and emerging trends in meat preservatives and additives, elucidating the regulatory frameworks that oversee their utilization. Additionally, the discourse extends to an exploration of the safety implications associated with the use of meat preservatives and additives. By scrutinizing the safety aspects, this chapter provides a holistic understanding of the implications and considerations associated with the use of preservatives and additives in the meat industry. In essence, the chapter serves as a valuable resource for comprehending the intricate dynamics of meat preservation, regulatory compliance, and safety concerns in the context of evolving consumer preferences and global standards (Bouzembrak et al. 2019).

17.5.1 Preservatives

Traditional methods of preserving meat, such as curing, salting, and smoking, have been employed for centuries to enhance the shelf life of meat products (Lonergan et al. 2019). Among these, curing, primarily achieved through the use of salt, has been a prominent preservation technique. Salt acts by inhibiting the growth of microorganisms through water removal, leading to enlarged extracellular spaces. The increased salt concentration induces the formation of large ice crystals, resulting in decreased water holding capacity. However, it's worth noting that the specific grouping of salt can impact the characteristics of the preserved meat (Jiang et al. 2019).

Food preservatives are substances incorporated into food products to prevent and delay microbial spoilage. These compounds act by inhibiting, retarding, or halting processes like fermentation, acidification, or other forms of food decomposition (Sulieman et al. 2023). The main objective of preservatives is to impede the growth of microorganisms that might induce food spoilage and pose a risk of foodborne illnesses. Furthermore, these preservatives play a vital role in prolonging the shelf life of food items, facilitating their distribution and sale to consumers by ensuring an extended period of product viability (Sulieman et al. 2023). As defined by the FDA, a chemical preservative is any substance added to food to prevent or slow its deterioration (Surendran Nair et al. 2020). The suitability of a chemical preservative requires it to be nontoxic, easily soluble, free from imparting off-flavours, exhibit antimicrobial properties across varying pH levels, and be economically viable and practical (MacDonald and Reitmeier 2017).

A range of chemicals serves as food preservatives, such as nitrites or nitrates, butylated hydroxy anisol (BHA), butylated hydroxyl toluene (BHT), tert-butylhydroquinone (TBHQ), vinegar, citric acid, and calcium propionate. Consumer concerns have been raised regarding preservatives with complex chemical names. Changing consumer preferences have led to a demand for fresher, more natural, and nutritionally healthier foods, prompting interest in "bio-preservation," utilizing natural compounds with antioxidant and antimicrobial properties (Barberis et al. 2018).

Meat preservatives can be broadly categorized into artificial and natural preservatives. Artificial food preservatives can be categorized into various groups, distinguished by their chemical composition and mode of action. Some prevalent types are listed below:

17.5.1.1 Artificial Preservatives

Organic Acids and Derivatives

Organic acids and derivatives including sorbic acid, benzoic acid, lactic acid, citric acid, and acetic acid, find extensive use in acidic food products to inhibit microbial growth.

Sorbic acid, designated as E 200, functions as a general food preservative, while its derivative salts, potassium sorbate (E 202) and calcium sorbate (E 203), are permitted for incorporation into meat products (Bajcic et al. 2021). Originally sourced from plant origins, sorbic acid, identified as 2,4-hexadienoic acid, undergoes synthetic production and is commercially available in powdered or granulated form. Boasting a neutral taste and odour profile, sorbic acid proves advantageous for application without affecting the sensory properties of products (Shabir et al. 2011; Esquivel-Ferriño et al. 2012). Given their distinctive physical and chemical characteristics, sorbates find application across diverse food categories, encompassing aqueous solutions, fat emulsions, suspensions, gels, and other products with high water content. Typical usage involves the addition of sorbates to food within the range of 0.1–0.2%. While the recommended daily intake (RDI) varies by country regulations, it is generally up to 25 mg/kg for humans (Dehghan et al. 2018; EFSA 2015).

The antimicrobial mechanism of sorbates is not fully elucidated, but their weak lipid-soluble nature allows sorbic acid to dissolve microbial membranes, penetrate cells, alter pH within the cell, disrupt transport and metabolic functions, ultimately leading to cell death. Sorbates are mainly employed as inhibitors of mould and yeast growth, with additional effectiveness in reducing bacterial biofilm production. Their efficiency is heightened at lower pH levels, particularly in the range of pH 3–6.5. Numerous studies on laboratory animals have indicated minimal harmful effects on human nutrition (Bajcic et al. 2021).

Dating back to the sixteenth century, benzoic acid, a naturally occurring aromatic carboxylic acid found in prunes, cinnamon, and cloves, is recognized as a preservative (E 210), along with its salts, including sodium benzoate (E 211), potassium benzoate (E 212), and calcium benzoate (E 213) (Bajcic et al. 2021). Benzoic acid exhibits low toxicity within permissible limits and poses little risk to human health, although there is a concern about its conversion to benzene, a carcinogenic and toxic compound, but studies suggest a low likelihood of this reaction occurring in food (Zeece 2020).

Benzoates have antimicrobial properties, primarily inhibiting yeasts and moulds, and, like sorbates, they also have a bactericidal effect. Their use is most effective in products with a pH below 7, showing optimal effects within the pH range of 2.5–4 (Bajcic et al. 2021). In the food industry, they are added as additives in amounts ranging from 0.05% to 0.1% and used in minced meat (Turnbull et al. 2021). The recommended daily intake (RDI) of benzoate is up to 5 mg/kg body weight, and excessive consumption can lead to abdominal pain, liver and kidney enlargement, and diarrheal (Kalpana and Rajeswari 2019).

The antimicrobial activity of organic acids like lactic, acetic, and citric acids and/or their salts relies on reducing the pH level of a medium or foodstuff, hindering microbial growth (Hauser et al. 2016; da Costa et al. 2019; Flores and Toldrá 2021). The undissociated and uncharged organic acid molecules play a key role in antimicrobial activity, as their concentration increases at low pH (Theron and Lues 2007). Notably, organic acids exhibit optimal antimicrobial effects at low pH (Theron and Lues 2007). The undissociated molecules, believed to be lipophilic, can easily traverse the lipidic membrane of target microorganisms (bacteria, yeasts, and fungi), entering their cytoplasm and causing cell death (Hauser et al. 2016; McDermott et al. 2018).

In adherence to European food legislation, the application of lactic, citric, and acetic acids should follow the *quantum satis* principle, where no maximum level is specified, and the applied amount should not exceed what is necessary to achieve the intended technological purpose (Regulation 2008). Conversely, lactic and acetic acids at 2% demonstrated potent antimicrobial activity in decontaminating surfaces contaminated with *E. coli* O157:H7 on beef meat, as observed by Carpenter et al. (2011). These organic acid washes resulted in a 0.6 to 1 log/cm^2 reduction in the pathogen compared to control samples (no-wash meats) or water-wash meat samples, preventing residual growth.

Using combinations of various organic acids and their salts can be effective in enhancing the antimicrobial and preservative effects on treated meat products. For instance, Mbandi and Shelef (2001) observed a synergistic effect between sodium lactate and sodium diacetate on microbial growth in beef meat. A combination of 2.5% sodium lactate and 2% sodium diacetate showed bacteriostatic effects on *L. monocytogenes* and bactericidal effects on *Salmonella Enteritidis* after 20 days of beef storage at 10 °C. Another combination of 1.8% sodium lactate and 0.1% sodium diacetate displayed a listeriostatic effect, with *Salmonella* counts below 10 cells/g after 30 days of refrigerated storage at 5 °C. Additionally, applying 0.2% sodium acetate with 1.8% or 2.5% sodium lactate significantly boosted antimicrobial activity against both pathogens (Ben Braïek and Smaoui 2021).

Moreover, a spray washing method with combinations like 2% lactic acid +1.5% propionic acid or 1.5% acetic acid +1.5% propionic acid demonstrated a substantial decrease in total viable counts to undetectable levels during the storage of fresh sheep/goat meats compared to single acid treatments (Dubal et al. 2004). Additionally, the shelf-life of meat samples treated with these organic acid combinations increased to 11 days compared to 3 days in untreated meats.

Nitrites/Nitrates

Nitrites/nitrates are frequently used in cured meat products to prevent the growth of *Clostridium botulinum* and maintain colour stability. Cured meats boast a rich history as a means of preserving food, dating back to 3000 B.C. In the absence of refrigeration, people utilized a combination of nitrite/nitrate salt, common salt (NaCl), and spices to preserve fresh meat. Nitrite, a versatile chemical present in the biological nitrogen cycle, plays a crucial role in this preservation process. It is added to cured meat products in concentrations less than 150 ppm to prevent the growth of harmful microorganisms like *Clostridium botulinum* and enhance the colour, flavour, and texture of the meat (Shakil et al. 2022). The effectiveness of nitrite is influenced by factors such as pH, residual nitrite levels, salt concentration, iron content, and storage temperature, with acidic conditions enhancing its efficacy. Nitrite acts on multiple sites in bacteria, blocking metabolic enzymes, limiting oxygen absorption, and disrupting the proton gradient. It also binds to iron, reducing its availability for crucial bacterial activities (Tompkin et al. 2020). In cured meat products, nitrite primarily aims to prevent the growth of *C. botulinum*, with a small amount required for colour development. Higher nitrite levels lead to increased suppression of *C. botulinum* growth (Lim et al. 2016). Studies also indicate that nitrite inhibits starter cultures and bacteriocin production in fermented sausages, particularly when concentrations reach 100 ppm (Scannell et al. 2001). Additionally, pathogens like *Listeria monocytogenes*, *Staphylococcus aureus*, *Bacillus cereus*, and *E. coli* grow more slowly in the presence of nitrite at levels found in cured meats (Milkowski et al. 2010).

Sulphites

Sulphites are employed in the food industry not only for meat preservation but also to enhance its visual appeal, creating a more attractive colour that consumers associate with freshness.

Sulphites, including sulphur dioxide (SO_2) and related compounds like sodium, potassium, and calcium salts, are widely used in the meat industry as preservatives. They serve various functions, such as inhibiting microbial growth, reducing deterioration, enhancing colour, acting as bleaching agents, inhibiting enzymes, and scavenging oxygen (D'Amore et al. 2020). Each country has the authority to approve or ban the usage of specific sulfiting additives. For instance, potassium sulphite (INS-225) is not approved in Europe and the USA (FDA 2019) but is permitted in Australia and New Zealand (FSANZ 2020). Similarly, sodium thiosulfate (INS-539) is not approved in Europe but is widely used in other countries for its preservative and antioxidant properties. Sulphites demonstrate antimicrobial effects that are dependent on dosage, acting either by extending the microbial lag phase (microbiostatic) or causing microbial death (microbicidal). They interact with disulfide bonds and thiol groups in microbial proteins and enzymes, disrupting DNA synthesis. This leads to conformational changes in proteins and the inactivation of enzymes, affecting crucial processes in the microorganism life cycle (Simon and Kroneck 2013). With a broad spectrum of activity, sulphites are effective against both Gram-negative and Gram-positive bacteria, yeasts, moulds, and foodborne pathogens. The antimicrobial effectiveness varies depending on the type of microorganism. Sulphites inhibit the growth of various bacteria, including lactic acid-producing and acetic acid-producing bacteria, as well as other pathogenic bacteria. Additionally, they retard the growth of *Clostridium botulinum* cells and exhibit efficacy against mould and yeast species (D'Amore et al. 2020).

17.5.1.2 Natural Preservatives

Bio-preservation utilizes natural resources to prolong the shelf life and improve the safety of meat products. Many microorganisms, plants, animal and marine by-products and insect exhibit antimicrobial properties.

Essential Oils and Plant Extracts

Essential oils (EOs) and plant extracts emerge as viable alternatives to synthetic additives for preserving foods, especially meats and meat products, in alignment with the rising demand for safe and naturally processed options (Aguiar Campolina et al. 2023). Phenolic compounds, prevalent in plants across various categories, including herbs, spices, vegetables, fruits, wine, essential oils, olive oil, and oil seeds, have been harnessed by the food industry for their antioxidant and antimicrobial functions (Kalogianni et al. 2020; Yu et al. 2021). Plant-derived polyphenols encompass diverse classes, such as phenolic acids, flavones, flavanols, flavanones, flavonols, isoflavones, coumarins, anthocyanins, quinones, alkaloids, and terpenoids (Yu et al. 2021). Essential oils derived from aromatic plants, comprising a blend of phenolic compounds like terpenes, terpenoids, and phenylpropanoids, present varying concentrations. For example, origanum EO contains 30% carvacrol and 27% thymol as principal components (Bakkali et al. 2008). Rich in phenolic compounds, essential oils from plants like oregano, clove, coriander, ginger, rosemary, sage, thyme, and mint exhibit robust antimicrobial properties (Kalogianni et al. 2020). The antimicrobial efficacy of essential oils is attributed to the phenolic compounds they contain (Chouhan et al. 2017) However, the underlying mechanisms, though not entirely elucidated, are complex due to the diverse bioactive molecules in plants. Possible actions on bacterial cells include increased membrane permeability, decreased proton force, metabolite and ion leakage, effects on enzymes, and alterations in membrane fatty (Aguiar Campolina et al. 2023). The impact of essential oils can vary with bacterial species, with Gram-positive bacteria generally being more susceptible than Gram-negative bacteria (Nazzaro et al. 2013).

Peptides

Antibacterial peptides, such as lipopeptides (LPs) and antimicrobial peptides (AMPs), exhibit potential as preservatives for meats and meat products. LPs, recognized for their amphiphilic nature, function as biosurfactants with hydrophilic peptide and hydrophobic fatty acid chains, often produced by bacteria like *Bacilli* (Pilz et al. 2023). Plant AMPs, crucial for plant defence against pathogens (Lima et al. 2022), are ribosomally derived peptides sharing characteristics with counterparts in microbes, insects, and animals, possessing a positive charge and amphipathic nature (Tam et al. 2015). Insects have contributed to the identification of over 200 AMPs to date (Erdem Büyükkiraz and Kesmen 2022). Both LPs and AMPs from plants and insects display robust antibacterial properties against various pathogens, including those associated with foodborne illnesses. Their primary mode of action involves interactions with bacterial cell membranes, leading to depolarization. Some AMPs function intracellularly, impeding cellular processes and synthesis pathways (Lima et al. 2022; Pilz et al. 2023). Despite their promising antimicrobial properties, only a few AMPs, like daptomycin and natamycin, have clinical approval. Daptomycin, effective against Gram-positive pathogens, gained U.S. FDA approval in 2003 (Eisenstein 2004). While some LPs have undergone testing for antioxidant activity in the food industry, their potential as antibacterial preservatives remains an area that warrants further exploration (Jemil et al. 2020).

Bacteriophages

Phages, also known as bacteriophages, are viruses that specifically infect bacterial cells, making them the most abundant biological entities on Earth (Batinovic et al. 2019). They often have a narrow host range, targeting specific bacterial species or strains within a species (Koskella and Meaden 2013), although some phages show broader activity, like phage SS3e and phage PS5 that infect various bacteria including *Salmonella Enteritidis* and *E. coli* O157:H7 (Kim et al. 2018; Duc et al. 2020). Phages follow either lytic or lysogenic life-cycle strategies (Ofir and Sorek 2018). Lytic phages attach to a bacterium, introduce their genome, replicate using the host's machinery, and then release new phages by lysing the host cell. Lysogenic phages integrate their genome into the bacterial chromosome without causing cell lysis, forming prophages.

The emphasis is on lytic phages for screening against foodborne pathogens. Ideal candidates are lytic, polyvalent, thermo-stable, and lack resistance or virulence genes. When polyvalent phages are unavailable, a combination of phage strains, known as a phage cocktail, may be used for effective mitigation against diverse strains. Research has delved into phage roles in controlling Gram-positive and Gram-negative foodborne pathogens like *L. monocytogenes*, *S. aureus*, *Salmonella*, *E. coli* O157:H7, *C. jejuni*, and *Y. enterocolitica* (Ge et al. 2022; Elois et al. 2023). Recent examples in meats include *Salmonella* phage MSP1, effective against various serovars and stable across pH and temperature ranges, significantly reducing *S. Thompson* on chicken meat (Park et al. 2023). Another phage, Tequatrovirus EP01, showed a broad host range against *E. coli* and some *Salmonella* strains, with stability and no virulence or drug resistance genes. EP01 significantly reduced *E. coli* O157:H7, O114: K90, and O142: K86 on unspecified meat (Zhou et al. 2022b). The polyvalent phage PS5 and List-Shield bacteriophage cocktail also demonstrated efficacy against *Salmonella*, *E. coli*, and *L. monocytogenes* in various meat samples (Duc et al. 2020; Ishaq et al. 2020).

Animal and Marine by-Products

Lactoferrin, an iron-binding glycoprotein found in milk, exhibits antimicrobial activity against various foodborne microorganisms and viruses. Approved for use in meat products, it acts by chelating iron and destabilizing the outer membrane of Gram-negative bacteria, leading to increased membrane permeability. However, microorganisms with low iron needs, like LAB, are not sensitive to lactoferrin. Some studies suggest that lactoferrin provides iron to beneficial bacteria like bifidobacteria, inhibiting the growth of harmful microorganisms (Davidson et al. 2013; Davidson et al. 2012).

Chitosan, derived from crustacean shells and fungi, is a biodegradable polymer known for its antimicrobial properties. Approved as Generally Recognized as Safe (GRAS) for use in foods, including meat products, chitosan's effectiveness depends on pH levels. At lower pH, it carries positively charged amino groups that interact with bacterial cell membranes, disrupting their structure and causing leakage of cellular contents. It inhibits various microorganisms such as *B. subtilis*, *E. coli*, *Pseudomonas fragi*, and *S. aureus* (Chen and Mustapha 2008; Rabea et al. 2003). In practical use, a 3% chitosan solution demonstrated efficacy against *Clostridium perfringens* in cooked ground beef and turkey during prolonged cooling at abuse temperatures (Surendran Nair et al. 2020). Chitosan's application as a food preservative is limited by its insolubility at neutral and higher pH, but water-soluble chitosan derivatives and N-alkylated disaccharide chitosan derivatives show promise as alternatives (Barberis et al. 2018).

Compounds in milk, like casein and whey proteins, are not just nutritious—they're natural antimicrobial agents (Phelan et al. 2009). These peptides combat various harmful microorganisms, including bacteria and fungi. Produced through processes like enzymatic hydrolysis or fermentation, they're effective against a range of pathogens. Casein, the major milk protein, yields antibacterial peptides like casocidin and isracidin, targeting bacteria such as *Staphylococcus* and *Listeria* (Szwajkowska et al. 2011; Korhonen and Rokka 2011). Peptides from whey proteins, such as β-lactoglobulin and α-lactalbumin, show antimicrobial activity against pathogens like *S. aureus* and *E. coli* (Korhonen and Rokka 2011; Szwajkowska et al. 2011). These peptides, accumulating on cell membranes, enhance permeability, leading to cell death (Cheng et al. 2013b).

17.5.2 Additives

The term "food additive," as defined by the Food and Drug Administration (FDA), is quite broad. It covers any substance that directly or indirectly becomes a part of or affects the characteristics of any food. This includes substances used in various stages of food production, processing, treatment, packaging, transportation, or storage. Food additives play a crucial role in preserving or improving different aspects of food products, such as safety, freshness, nutritional value, taste, texture, and appearance. The use of these additives has become more prevalent in recent years, aligning with the increased production of prepared, processed, and convenient foods in our modern food landscape (FSIS 2015).

The FDA categorizes food additives into two main groups: "direct food additives" and "indirect food additives." Direct food additives are intentionally added to food for specific purposes. For example, phosphates are used in meat and poultry products to retain moisture and protect flavour. On the other hand, indirect food additives become part of the food in tiny amounts through processes like packaging, storage, or other handling. A simple example is the slight transfer of packaging materials into foods during storage. It's important to note that manufacturers of food packaging materials must demonstrate their safety to the FDA before approval for use in contact with food is granted. This careful oversight reflects the commitment to ensuring the safety and integrity of the food supply chain (FSIS 2015).

"Clean label" has become significant in the food industry, but its definition is unclear. Consumers associate clean label with products having clear front-of-pack claims like "free from...", "organic", or "no additives" (Asioli et al. 2017). In a stricter sense, clean-label products, seen on the back of the pack, have a short list of familiar, non-synthetic ingredients. The Institute of Food Technologists defines clean label as making a product with as few ingredients as possible that consumers recognize as wholesome (Delgado-Pando et al. 2021). The meat industry, part of the broader food industry, faces challenges in adopting clean-label options. Meat products need additives during processing for technological and taste reasons. Many additives used in meat processing are crucial for safety and shelf life. Challenges are significant in meat production, where synthetic-sounding ingredients often provide essential functions for meat quality (Delgado-Pando et al. 2021).

17.5.2.1 Texturizers

Processed meats are complex compositions of muscle and various non-muscle components. Achieving the ideal combination necessitates mutual supplementation, reddish-pink pigment during heat treatment. The presence of antioxidants and salt, along with adjustments in pH, influences colour development (Skibsted 2011). Residual nitrite levels are important for maintaining the cured colour, usually recommended at 10–15 ppm. Higher nitrite levels or low pH may cause undesirable discoloration in meats (Shakil et al. 2022). High nitrite concentrations pose risks, especially for infants, leading to a condition called methemoglobinemia (Brender 2020). Nitrite reacting with certain amines can produce cancer-causing nitrosamines (Oliveira et al. 2004). Processed meat has been classified as carcinogenic due to nitrite-related concerns, with links to colorectal cancer. Many countries have restricted nitrite use in processed foods due to these health risks (Shakil et al. 2022).

17.5.2.3 Flavouring Agents

Flavouring agents are additives used to modify the taste of food by enhancing or completely altering the final taste of the produced foodstuff. According to the FDA, spices and spice extractives can be labelled as "natural flavours," "flavours," or "natural flavouring" on meat and poultry products without individually specifying each component. This practice is permissible as these substances are primarily employed for their flavour-enhancing properties rather than their nutritional content (FSIS 2015). In contrast, "dried meat or poultry stocks," "dried broth," "meat extracts," and "dried beef plasma" cannot be categorised as "natural flavourings". Monosodium glutamate (MSG), classified as a flavour enhancer, must be explicitly mentioned on the label when added to a product. Additionally, MSG and hydrolysed protein are related, as the breakdown of proteins during the hydrolysis process may form free glutamate, which combines with sodium to create MSG. Hydrolysed animal or vegetable protein cannot be listed as "natural flavouring"; FSIS regulations mandate specific identification and disclosure of the protein source, such as "hydrolysed wheat protein" or "hydrolysed milk protein" (FSIS 2013).

Yeast extract is derived from *Saccharomyces* yeast, rich in glutamic acid and 5′-ribonucleotides, imparting a meat-like flavour. It contains non-volatile flavour precursors like sugars, amino acids, nucleotides, peptides, lipids, and thiamine. When heated, volatile compounds contribute to its meaty flavour (Zhang et al. 2018). Essential aroma-active compounds include 2-methyl-3-furanthiol, 2-furan methanethiol, 2,3-butanedione, and others formed through the Maillard reaction. Proteolytic enzymes in yeast extract, especially thermally treated, enhance its meaty flavour and umami taste (Liu et al. 2015).

Smoking is a widely used method especially to enhance the meat flavour. Smoking methods include traditional and industrial approaches, utilizing different smoke generation technologies (Ledesma et al. 2017). Smouldering smoke, maintained between 500 °C and 800 °C, is the common method for generating smoke (Sikorski and Sinkiewicz 2014). Traditional and industrial smouldering smoke are prevalent in producing smoked meat products. Traditional smoking involves wood smouldering beneath hanging meat in a smokehouse, requiring skilled control of factors like temperature and humidity (Ledesma et al. 2017). In industrial production, automated smokehouses use smouldering-type generators for efficient smoke generation. This reduces harmful compound formation, like polycyclic aromatic hydrocarbons (PAHs) in meat products (Sikorski and Sinkiewicz 2014). Liquid smoke, a modern method, is created by condensing wood smoke through controlled pyrolysis, offering a healthier alternative to traditional smoking (Yin et al. 2021).

Animal fat plays a key role in flavour by serving as a base and storage for flavour compounds, influencing how flavours are released over time. Unsaturated lipids and phospholipids contribute to the formation of volatile compounds, while lipid oxidation creates aroma molecules through complex reactions (Frank et al. 2016). Selecting and using animal fats correctly is crucial for optimal results in meat product manufacturing. Some animal fats are well-suited for this purpose, like pig fat, which is often preferred due to its higher quantity, availability, suitable tissue structure, composition, and mild taste (Heinz and Hautzinger 2007). In contrast, beef fat is less suitable for further processing due to its firmer texture, yellowish colour, and more intense flavour. Mutton fat from older animals is generally not preferred due to its unpleasant flavour, while lamb fat, with its neutral taste, is commonly enjoyed with lamb chops and suitable for Halal meat products. Chicken fat, neutral in taste, is ideal for pure chicken products, often adhering to chicken muscle tissue as intermuscular fat and processed without separating it from the lean meat. However, most chicken fat comes from chicken skin, which is minced and processed into a fat emulsion before being added during chopping (Heinz and Hautzinger 2007).

17.5.2.4 Antioxidants

Antioxidant terminates chain reactions caused by oxidation of unsaturated fats, preventing the production of aldehydes and ketones that lead to a rancid taste. Phenolic antioxidants, like BHA, BHT, TBHQ, and tocopherols, are antioxidants utilized to preserve the freshness and quality of meat. They work by slowing down the development of rancidity in fats, sausages, and dried meats. Additionally, these antioxidants contribute to the protection of essential nutrients, including vitamin A, in various food items. Propyl gallate is employed as an antioxidant to deter rancidity in products like rendered fats or pork sausage. It can be utilized in conjunction with other antioxidants, such as BHA and BHT (FSIS 2015).

Synthetic antioxidants are not practically useful for two main reasons. Firstly, synthetic antioxidants like BHA, TBHQ, and propyl gallate are generally not allowed in most meat products. In the EU, these synthetics are limited to specific uses in animal-derived products, such as lard and dehydrated meats, with a maximum dosage of 200 mg/kg (EC 2018). Notably, the FDA in the US currently prohibits the use of BHA and BHT as antioxidants in certain meat products (CFR 2020). Secondly, synthetic antioxidants are primarily lipophilic, acting in the lipid fraction, while natural antioxidants from plant extracts exhibit various polarities, with most acting in interfaces or the polar phase of emulsions or ground meat products. Consequently, the differing lipid-water behaviour of natural and synthetic antioxidants suggests that their effects would be distinct and potentially complementary when incorporated into a specific meat product (Estévez 2021).

Traditional methods of controlling lipid oxidation in meat, which involve adding antioxidants during processing, remain effective. Plant-derived antioxidants from fruits, herbs, and spices, such as phenolic compounds, carotenoids, and essential oils, are known for their antioxidant properties (Trautwein et al. 2015). Phenolic compounds, found widely in plant-based foods, are potent antioxidants due to their ability to absorb UV radiation and stabilize free radicals. Synthetic antioxidants like BHA and BHT have been compared with natural alternatives like rosemary extract in pork sausages, showing that rosemary extract was more effective in preventing an increase in thiobarbituric acid-reactive substance (TBARS) values (Sebranek et al. 2005). Legislation-wise, rosemary extract (E392) has been approved as a food additive in meat products, and its use is permitted in various countries. However, its application is subject to regulations to ensure safe usage (Kalogianni et al. 2020). Carotenoids, which include β-carotene and lycopene, act as antioxidants and food colorants, with polyisoprenoid structures and bilateral symmetry around the central double bond. They can scavenge peroxyl radicals, preventing lipid degradation. Despite their antioxidant benefits, they may alter the appearance of fresh raw meat or mask spoilage signs (Oswell et al. 2018). Essential oils, extracted from different plant components, contain phenols like carvacrol, eugenol, and thymol, known for their antioxidative activity. These oils interact with major and minor constituents, serving as free radical scavengers and hydrogen donors (Kalogianni et al. 2020).

Nitrite and sulphites are commonly used additives for preserving meat products. They have the antioxidant properties, which help delay lipid oxidation. Nitrite plays as an antioxidant in cured meats by preventing rancidity during storage and inhibiting lipid and protein oxidation. It stabilizes heme iron and chelates metallic ions, protecting meat molecules from oxidation (Feng et al. 2016). In cured flavour development, nitrite influences meat flavour by suppressing lipid oxidation and preventing the formation of undesirable flavours. It has both antioxidant and pro-oxidant properties, affecting the production of specific flavour compounds (Ursachi et al. 2020). While nitrite has significant benefits in preserving and enhancing cured meats, concerns about its potential health risks have led to regulations and increased research on natural alternatives. The use of nitrite is strictly regulated in many countries, and there is ongoing exploration of substitutes that provide similar functions without health risks. Despite challenges, nitrite remains a key ingredient in the hurdle technology of meat curing, ensuring safety and quality (Shakil et al. 2022).

Sulphites play a crucial role in inhibiting enzymatic browning (EB) and nonenzymatic browning (NEB) in meat. They stabilize and minimize these processes by preventing proteolytic breakdown, inhibiting lipid peroxidation reactions, and interfering with the synthesis of brown pigments. In EB, sulphites inhibit the breakdown of meat proteins and lipid peroxidation, indirectly stabilizing oxygen-binding forms in myoglobin and haemoglobin. This retention of meat colour and flavour delays the development of oxidative off-flavours (Kuijpers et al. 2013). In NEB, particularly in the Maillard reaction, sulphites reduce and inhibit the formation of melanoidins (brown nitrogenous polymers) by reacting with carbonyl groups and acting as terminators of polymer formation (D'Amore et al. 2020; van Boekel 2006). These additives are considered allergens and are regulated in the EU, with a maximum permitted dose of 450 mg/kg. They are allowed only in specific meat products like breakfast sausage, longaniza fresca, butifarra fresca, and burger meat with 4% cereal or vegetable (Delgado-Pando et al. 2021).

17.6 Freshness and Spoilage Identification

17.6.1 Meat Freshness Marker

Meat freshness serves as a critical indicator for assessing the quality and safety of meat products, garnering significant attention from producers, sellers, and consumers (Qiao et al. 2017). The relationship between microbial development and chemical transformations during meat storage serves as a crucial indicator for monitoring meat quality and freshness (Nychas et al. 2008). The evaluation of fresh meat quality typically involves observing microbiological and sensory changes over storage time. Chemical metabolites generated during microbiological deterioration, such as biogenic amines, ammonia gas, total volatile basic nitrogen (TVB-N), trimethylamine (TMA), and xanthine, serve as valuable indicators of meat freshness (Erna et al. 2021).

17.6.1.1 Biogenic Amines

Biogenic amines (BAs), resulting from the decarboxylation of enzymatic amino acids by microbial enzymes and tissue activity, play a significant role in indicating meat quality and freshness. High BAs concentrations indicate food spoilage, with common ones being cadaverine, putrescine, spermidine, spermine, β-phenylethylamine, tyramine, and histamine (Vasconcelos et al. 2021). Factors like pH, chemical composition, handling, and processing influence BAs presence in food (Dabadé et al. 2021). Decarboxylase-positive microorganisms, including *Enterobacteriaceae* and some lactic acid bacteria, contribute to BAs production (Schirone et al. 2022).

Meat, rich in protein and amino acids, is prone to BAs production, especially during prolonged storage or processing. Fresh meat initially contains spermine and spermidine, with other BAs increasing during storage due to protein breakdown (Triki et al. 2018). Proteolysis, influenced by factors like acidity and dehydration, promotes BAs accumulation, even in fermented meat products (De Mey et al. 2014). Due to health concerns, BAs serve as indicators of freshness, quality, and spoilage in meat. Various indices, such as Biogenic Amine Index (BAI) and specific amine ratios, help classify meat quality based on BAs concentrations. However, the effectiveness of these indicators varies based on processing methods, making them more suitable for fresh than fermented or heat-treated meat products (Schirone et al. 2022). Various analytical methods to identify BAs, including thin-layer chromatography (TLC), gas chromatography (GC), capillary electrophoresis (CE), and high-performance liquid chromatography (HPLC), have been employed to quantify biogenic amines in foods. Challenges such as effective separation, retention from interference compounds, and improving sensitivity are focal points in biogenic amine research (Erna et al. 2021).

17.6.1.2 Total Volatile Basic Nitrogen

Total volatile basic nitrogen (TVB-N), encompassing volatile amines like ammonia, dimethylamine (DMA), and trimethylamine (TMA), is a crucial factor in assessing meat freshness. These volatile amines result from amino acid decomposition and microbiological activities of *Pseudomonas* bacteria. TVB-N, primarily consisting of ammonia, is considered an indicator of meat spoilage. Analysing TVB-N, however, faces challenges such as complex sample treatments, time-consuming processes, low repeatability, and atmospheric contamination (Lee et al. 2018). To overcome these limitations, alternative approaches, including spectroscopy, hyperspectral imaging (HSI), and colorimetric sensor imaging, have been explored (Tang and Yu 2020).

17.6.1.3 Hypoxanthine

Postmortem nucleotide decomposition in meat leads to a significant increase in inosine concentration, subsequently converted into hypoxanthine and xanthine through bacterial metabolism. This degradation sequence, observed since the 1950s, correlates well with the degree of spoilage (Vanegas et al. 2014). Hypoxanthine, a major ATP catabolite, accumulates gradually in food products, particularly in seafood, making it a valuable freshness indicator. Detection of hypoxanthine and xanthine can be achieved by immobilizing the enzyme xanthine oxidase (XOx) on the platinum electrode surface, catalysing their oxidation (Vanegas et al. 2014). Due to its potential toxic effects, high concentrations of hypoxanthine in blood can contribute to an increase in reactive oxygen species (ROS) generation, leading to transfusion-related tissue damage (Casali et al. 2016). The levels of hypoxanthine are also associated with conditions like Lesch–Nyhan disease, characterized by high uric acid excretion concentrations (Torres et al. 2013).

17.6.2 Freshness Indicators

Efforts to enhance food safety and control involve the use of analytical methods to assess whether food meets regulatory requirements. The food industry continually seeks effective methods for identifying freshness markers, crucial for evaluating food freshness and preventing food poisoning. Freshness indicators, distinct from sensors, lack receptor or transducer components and communicate information through direct visual changes (Erna et al. 2021).

17.6.2.1 Time–Temperature Indicators (TTI)

TTIs serve as a simple quality recording system that demonstrates irreversible visual responses, providing critical parameter tracking across the supply chain. These indicators change irreversibly over time through mechanical, chemical, or enzymatic processes. TTIs are reliable tools for consumers to assess food quality, responding with visible changes correlating with the food's shelf life at the target temperature (Kerry et al. 2006). Studies by Kim et al. (2012) on ground beef quality using a microbial TTI demonstrated its predictive capacity for off-flavours and quality changes. Mataragas et al. (2019) developed a microbial TTI dependent on temperature and inherent properties of the growth medium, consistent with various organisms and beneficial for a wide range of foods.

17.6.2.2 Colour-Based pH-Sensitivity Indicators

Colour-based pH-sensitivity indicators using natural colours or synthetic dyes have shown promise for assessing microbial metabolites in packaged fish, shrimp, and broiler chicken. These indicators, such as bromocresol, curcumin, and purple bromocresol, undergo colour shifts in response to changes in pH triggered by volatile amines from meat and aquatic product spoilage (Ezati et al. 2019). Incorporating natural pigments like anthocyanins as indicators offers advantages such as biodegradability, structural conformation influence on colour, and sensitivity to pH changes. Studies by Yoshida et al. (2014) and Liu et al. (2017) on biodegradable freshness indicators based on chitosan, anthocyanin, and purple sweet potatoes demonstrated their effectiveness in tracking food quality degradation.

Recent developments in freshness indicators show potential for applications in meat packaging. Lee and Shin (2019) emphasized the success of intelligent packaging for evaluating microbial metabolites, specifically pH-sensitivity indicators for colour-based tracking. Indicator sensor packaging technology allows real-time and on-site monitoring of food spoilage without the need for expensive equipment, contributing to maintaining high-quality food fit for human consumption (Choi et al. 2017). As the food industry evolves to meet consumer demands for healthy and high-quality products, innovations in food-packaging systems become essential, helping reduce adverse effects on the environment (Han et al. 2018).

17.6.3 Recent Monitoring of Meat Freshness and Spoilage

Traditionally, the assessment of meat freshness has relied on two primary methodologies: sensory evaluations and chemical/biochemical analyses targeting specific bio-indicators (Gil et al. 2011). These traditional approaches have been effective in detecting physical and microbial contamination, assessing authenticity, and identifying potential adulteration in meat products. Sensory tests, chemical analysis, and viable bacterial counts are commonly utilized as part of these traditional methods.

Recent advancements in analytical techniques have broadened the array of methods used to assess meat freshness. These include electrochemical biosensors, potentiometric biosensors, amperometric biosensors, electronic nose and gas sensors, optical colorimetric and fluorometric sensors, and smartphone detection methods.

17.6.3.1 Electrochemical Biosensor

An alternative platform for detecting meat freshness is through electrochemical means. Electrochemical sensors, particularly potentiometric and amperometric sensors, have been extensively utilized in this context. These sensors, with various modifications, offer high sensitivity, desirable properties, and low detection limits (Johnson et al. 2019). An electrochemical sensor comprises a chemical recognition system (critical for sensor functionality) and a physicochemical transducer that converts chemical reactions into signals discernible by modern electrical devices (Kimmel et al. 2012).

Potentiometric Biosensor

Potentiometric sensors, relying on polymeric membrane ion-selective electrodes (ISE) and ion-selective field-effect transistors (ISFET), evaluate changes from one electrode to another (Lee et al. 2019). In meat industry applications, potentiometry is preferred due to its cost-effectiveness and convenience. For instance, Gil et al. (2010) utilized an electronic tongue with potentiometric electrodes to measure pork loin freshness over 14 days. Furthermore, electronic tongues, employing sets of electrodes for metal ions (Au, Ag, Cu, Pb, Zn, C), have been explored for potentiometric measurements over 10 days to evaluate physicochemical, microbial, and biochemical changes in refrigerated pork loin (Gil et al. 2011). These techniques demonstrate potential for qualitative or semi-quantitative assessment of freshness in various meat samples.

Amperometric Biosensor

Amperometric measurements record the cell's current flow at a constant applied potential, relying on enzymatic or bioaffinity reactions at the electrode surface. Dervisevic et al. (2015) developed a responsive amperometric biosensor using a nanocomposite host matrix with multi-walled carbon nanotubes for xanthine detection in fish. Sadeghi et al. (2014) created an amperometric biosensor for assessing xanthine concentrations in poultry freshness. Additionally, Albelda et al. (2017) proposed a microenvironment for direct electrochemistry of xanthine oxidase, utilizing a graphene-titanium dioxide nanocomposite for assessing hypoxanthine in pork loin. Teng et al. (2010) manufactured a disposable amperometric xanthine oxidase biosensor with manually printed carbon-working, carbon-counter, and Ag/AgCl reference electrodes on a polyethylene terephthalate substrate. The device, developed using traditional xanthine and hypoxanthine screen printing methods, showed promise in freshness assessment.

17.6.3.2 Electronic Nose and Gas Sensor

An electronic nose is a device designed to detect both simple and complex odours, utilizing a chemical sensor array system with partial specificity and a proficient pattern recognition system. This innovative tool has the capability to analyse volatile organic compounds (VOCs) emitted by microorganisms, making it a potential alternative method for the identification and classification of various chemicals and bacteria (Bonah et al. 2020). Electronic noses, integrated with gas sensors, have been instrumental in assessing meat freshness, detecting factors like pH, colour, and bacterial presence. In a study by Chen et al. (2019), both human sensory evaluation and an electronic nose successfully categorized pork, beef, and mutton samples based on storage times. The samples stored for 1–2 days, 3–4 days, and 5–7 days were identified as fresh, sub-fresh, and putrid, respectively. Discriminant factor analysis, using sensory evaluation as a reference, confirmed the E-nose's accurate assessment of freshness, achieving accuracies of 89.5%, 84.2%, and 94.7% for pork, beef, and mutton, respectively. According to Abdallah et al. (2013), electronic nose effectively detected total viable counts (TVC) in various beef and sausage samples, quantifying volatile organic compounds at concentrations from 50 ppb to >350 ppb. The concentrations of gases collected before and after contamination with these bacterial pathogens (*E. coli* O157: H7, *Salmonella typhimurium* 857, *Staphylococcus aureus* 29,213 and *Pseudomonas aeruginosa* 27,853) were highly correlated ($P < 0.005$).

Gas sensor technology, applied in areas like food safety, medical diagnosis, and environmental control, is increasingly crucial across various fields. Semiconductor gas sensors are efficient tools for gauging food freshness, detecting changes in toxic gas levels (e.g., oxygen, carbon dioxide, nitrogen, ammonia, and methane). A recent advancement by Galstyan et al. (2020) introduced a sensitive gas sensor for real-time monitoring of seafood quality, focusing on dimethylamine, using niobium-doped titanium dioxide nanotubes. Song et al. (2020) presented mesoporous film gas sensors for identifying hydrogen sulphide in fresh pork, based on SnO_2 hierarchical architectures from waste scallion root bio-templates. Shi et al. (2018) developed gas sensors with titanium dioxide-polyaniline/silk fibroin nanocomposites for ammonia detection and pork freshness assessment through in situ polymerization.

17.6.3.3 Optical Colorimetric and Fluorometric Sensor

Colorimetric sensor arrays have become a valuable tool in various fields, offering a rapid and cost-effective analytical approach. Inspired by the mammalian olfactory system, which can discern trillions of scents, these arrays use chemoresponsive dyes and solid supports to create unique patterns for different analytes. A cross-responsive sensor array was developed for real-time meat freshness detection. The design relies on colorimetric reactions between volatile compounds from meat samples (e.g., biogenic amines, sulphides) and selected dyes. A polyvinylidene difluoride (PVDF) membrane served as the flexible substrate, with colorimetric dyes immobilized through sol-gel processing. The sensor array successfully quantified representative volatiles in rotting meat, and the data were processed using hierarchical cluster analysis (HCA) and principal component analysis (PCA). The fabricated array effectively differentiated meat freshness at various storage intervals (Nie et al. 2022).

Moreover, optical techniques like fluorescence spectroscopy and light scattering have been explored for non-destructive monitoring of meat freshness. Fluorescence sensors convert weak spoilage signals into visible fluorescence enhancement or quenching, corresponding to different freshness levels (Zhuang et al. 2023). This transformation principle, akin to a "magnifier," amplifies the "small" into the "large" (Singh et al. 2022). Fluorescence sensors offer advantages such as digitization, miniaturization, intellectualization, multi-functionality, systematization, and networking compared to traditional methods. Meat spoilage, mainly caused by putrefying bacteria like *Pseudomonas*, is effectively monitored by fluorescence sensors using total viable bacteria count as a freshness index. Some sensors also target pathogenic bacteria like *Staphylococcus aureus* to prevent food poisoning incidents (Zhou et al. 2022a; Du et al. 2022). Despite these advancements, accurately predicting meat freshness remains a challenge.

17.6.3.4 Smartphone Detection

Smartphone spectroscopy, a method within smartphone-based sensing applications, involves capturing images with the built-in optical camera and analysing them using external attachments or third-party applications. Sensors based on smartphone spectroscopy can be categorized into micro-imaging, colorimetric, and luminescence sensors, each utilizing different types of light signals (Albelda et al. 2017). Recent advancements include smartphone-based sensors using the Janus concept for meat freshness detection. Franco et al. (2021) developed a real-time food freshness solution utilizing Janus colour-indicator labels that change colour from pink to purple when fresh to spoilage chicken passes through them. Furthermore, Seddaoui and Amine (2021) designed a colorimetric smartphone-based immunoassay for on-site detection of pork adulteration in meat, offering a promising approach for rapid and responsive real-time monitoring of food fraud.

17.7 Non-destructive Methods for Quality Evaluation

Meat quality assessments traditionally employ two major measurement methods: subjective and objective. Subjective methods rely on sensory evaluation, incorporating visual and eating experiences. However, these methods are highly dependent on evaluators' specific experiences, lack repeatability, and pose challenges in quantification. In contrast, objective evaluation methods, historically relying on laboratory tests, provide accurate results but often damage or destroy the meat product, resulting in cumbersome and time-consuming procedures (Shi et al. 2021).

Non-destructive detection techniques (NDDT) emerge as a promising field, incorporating physics, electronics, computer science, and artificial intelligence. NDDT describes the internal and external properties of a substance without altering its original state. This approach is particularly relevant in the realm of meat quality inspection due to its rapid, real-time, accurate, and non-destructive nature. The advent of artificial intelligence (AI) further revolutionizes non-destructive detection, providing technical support for online meat grading and evaluation. Common AI technologies utilized for non-destructive meat quality evaluation, including spectroscopic techniques (near-infrared spectroscopy (NIR), Raman spectroscopy (RS), and terahertz (THz) spectroscopy), imaging techniques (hyperspectral imaging (HSI), X-ray imaging, and thermal imaging (TI)) computer vision systems (digital camera or smartphone shooting, ultrasound imaging, nuclear magnetic resonance (NMR), computed tomography (CT), near-infrared spectral imaging (NIR)) (Wu et al. 2022).

17.7.1 Spectroscopic Techniques

Spectroscopy encompasses a range of wavelengths, from γ-rays to radio waves, providing a versatile tool for qualitative and quantitative analysis in the interdisciplinary field of physics and chemistry (Cheng et al. 2013a). This section focuses on three spectroscopic techniques: Near-Infrared Spectroscopy (NIR), Raman Spectroscopy (RS), and Terahertz (THz) Spectroscopy—integral to meat and meat product assessment.

17.7.1.1 Near-Infrared Spectroscopy

Near-infrared spectroscopy (NIR) uses electromagnetic radiation with wavelengths between 780 and 2526 nm. It detects distinct spectral curves for different chemicals, forming the basis for analysing correlations between spectral data and physical and chemical indices of samples (Alexandrakis et al. 2012). NIR spectroscopy is employed for detecting meat storage putrefaction and assessing properties like moisture, protein, and fat during meat processing (Collell et al. 2011). Recent studies focus on freshness detection, utilizing NIR for quantifying total volatile basic nitrogen (TVB-N) and assessing microbial contamination in meats. While NIR spectroscopy is effective for non-destructive meat quality detection, it primarily focuses on internal components. Combining NIR with computer vision technology could enhance the accuracy and comprehensiveness of meat quality assessment (Shi et al. 2021).

17.7.1.2 Raman Spectroscopy

Raman spectroscopy, based on Raman scattering, provides molecular vibration and rotation information. It complements infrared spectroscopy by offering unique Raman spectral signals for functional group molecules in meat. This method, requiring no sample pretreatment, excels in providing specific molecular signals (Wu et al. 2022). Applications of Raman Spectroscopy in meat quality assessment include predicting tenderness, cooking loss, and analysing chemical components like water, protein, and lipids (Wu et al. 2022). It has proven effective in evaluating safety quality by detecting spoilage and adulteration (Boyacı et al. 2014). Recent studies highlight Raman Spectroscopy's efficacy in assessing meat tenderness, juiciness, and texture, offering a non-invasive and direct approach (Schmidt et al. 2013; Fowler et al. 2018). Its application in analysing chemical components provides valuable insights into changes in chemical bonds. Raman Spectroscopy also contributes to safety quality assessment by distinguishing foreign materials and detecting boar taint (Liu et al. 2016).

17.7.1.3 Terahertz Spectroscopy

Terahertz (THz) falls between infrared light and microwave radiation, possessing both photon and electron properties (Liang et al. 2019). THz spectroscopy, operating in the far-infrared band, utilizes unique radiation properties for meat quality assessment (Han and Chen 2018). Biological macromolecules like amino acids, peptides, DNA, and small molecules such as nucleotides absorb THz waves due to rotational, vibrational, and overall molecular cluster vibrations (Liang et al. 2019). THz spectroscopy aids in identifying meat from different tissues, varieties, or brands, forming the basis for authenticity and adulteration detection. Applications include determining the appearance and internal components, assessing additives, and detecting foreign materials. Despite its promise, THz spectroscopy faces challenges such as limited theoretical research, water absorption interference, and the need for improved experimental platforms. Ongoing research focuses on addressing these challenges and exploring THz Spectroscopy's potential for scientific studies and industrial applications in meat processing (Wu et al. 2022).

17.7.2 *Imaging Techniques*

Imaging techniques give valuable spatial details, and spectroscopy provides abundant spectral data, both crucial for understanding the sensory, chemical, and physical aspects of meat. By analysing these features, we can objectively assess meat surfaces and qualities, including structural and textural features not easily noticed by humans (Xiong et al. 2017). Advancements in computer hardware and image processing have led to innovative imaging methods like hyperspectral imaging (HSI), X-ray imaging, and thermal imaging (TI) (Wu et al. 2022).

17.7.2.1 Hyperspectral Imaging

Hyperspectral imaging (HSI), derived from hyperspectral remote sensing imaging, covers a broad spectrum from ultraviolet to thermal infrared. It combines spectral detection and computer vision, offering both spectral and image details for efficient sample analysis (Xiong et al. 2015). Divided into visible/near-infrared (400–1000 nm) and near-infrared (900–1700 nm) techniques, HSI captures real-time information, distinguishing it from traditional spectroscopy (Shi et al. 2021). In meat quality evaluation, HSI proves effective for both sensory and chemical assessments. It accurately classifies tenderness, achieving significant accuracies for cooked-beef samples. Moreover, it predicts colour, pH value, and tenderness of beef, showing promising results (Cluff et al. 2013; Wu et al. 2012). In sensory attribute prediction, HSI in reflectance mode successfully modelled beef marbling with a high correlation coefficient of 95% (Aredo et al. 2017). Regarding chemical components, HSI detects moisture content in raw meat and estimates water content in meat products, demonstrating its versatility (Barbin et al. 2013; Achata et al. 2021). However, its application for online meat quality detection requires advanced algorithms to handle complex data efficiently. The future of HSI aims for high speed, low cost, and real-time monitoring systems (Wu et al. 2022).

17.7.2.2 X-Ray Imaging

X-rays have the ability to penetrate substances, diffract, and cause fluorescence in materials. By measuring how different substances absorb X-rays, we can obtain transmission and tomographic images for analysing internal structures (Shi et al. 2021). Computed tomography (CT), a prevalent X-ray imaging technique, is widely used in food quality and safety inspection, particularly for meat structure and component analysis (Wu et al. 2022). CT provides insights into heat-induced structural changes in meat, quantifying internal components like water, fat, connective tissue, and myofibrils (Miklos et al. 2015). It aids in estimating lean meat percentage in pig carcasses (Furnols et al. 2009) and distinguishing fresh meat from frozen meat through diffraction-enhanced imaging. X-ray imaging also proves effective in detecting foreign materials in meat, such as bone fragments, and qualitatively analysing food mould (Wu et al. 2022).

17.7.2.3 Thermal Imaging

Thermal imaging is a fast, non-invasive, and precise technique capable of capturing real-time images of moving targets and displaying temperature variations across a wide range. A typical setup includes a thermal camera, an infrared detector, and a signal processing unit. It's extensively used for ensuring meat safety and monitoring quality (Xiong et al. 2017). Compared to traditional methods, it offers faster and more accurate temperature readings without physical contact. Nanni Costa et al. (2016) utilized infrared thermography to assess pork and ham quality on a slaughter line. By analysing surface temperature differences, they could classify ham quality effectively, especially in hams with lower fat cover. On the other hand, thermal imaging is utilized to diagnose diseases in animals. Schaefer et al. (2012) explored its potential in detecting bovine respiratory disease complex (BRD) in cattle. They found that animals positive for BRD exhibited higher peak infrared thermal values compared to those negative for the disease. Furthermore, thermal imaging aids in cooking chicken to the right temperature and helps monitor the health of livestock during transportation (Ibarra et al. 2000). However, it can be affected by outside temperatures and may not be reliable for unstable objects (Wu et al. 2022).

17.7.3 Computer Vision Systems

Computer vision, or machine vision, utilizes image sensors instead of human eyes to capture target image data. It employs computer technology to analyse and process images, identifying, tracking, and detecting target objects. A typical machine vision system includes a computer, an industrial camera, an illumination system, and image processing software (Shi et al. 2021). Machine vision is used to extract features from meat-related images. It starts with illumination and image acquisition, using methods like digital camera or smartphone shooting, ultrasound imaging, nuclear magnetic resonance (NMR), and computed tomography (CT) (Wu et al. 2022).

17.7.3.1 Smartphone Shot

In this system, a digital camera takes the place of the human eye, while a learning algorithm acts as a substitute for the human brain. The camera captures consistent image data, reducing noise. Then, the learning algorithm correlates the image data with relevant quality levels (Jackman et al. 2011). Sun et al. (2016) developed a system to measure pork loin quality objectively, achieving high prediction accuracy using AI models by extracting colour features and marble patterns from meat images. Similarly, Liu et al. (2018) investigated the ability of machine vision to predict pork intramuscular fat percentage, while Arsalane et al. (2018) utilized machine vision to evaluate beef freshness with perfect prediction accuracy. Some studies have explored the use of machine vision to monitor meat defects, such as dark, firm, and dry (DFD) beef. Chmiel et al. (2012) found a significant relationship between colour components and pH, indicating the potential to detect DFD beef. Others, like Chmiel and Słowiński (2016), reported promising results in detecting meat defects using machine vision, particularly in industrial settings. While machine vision is effective in analysing external features like colour and texture for meat quality inspection, it falls short in assessing internal components such as moisture content and protein content. Therefore, it mainly focuses on detecting external properties like meat colour, marbling pattern, tenderness, freshness, and fat content, rather than internal characteristics (Taheri-Garavand et al. 2019).

17.7.3.2 Ultrasound Imaging

Ultrasonic technology comes in two main types: power ultrasonic waves and detection ultrasonic waves. Power ultrasonic waves, with low frequency and high energy, are used for tasks like food sterilization and thawing. Detection ultrasonic waves, with high frequency and low energy, analyse food quality (Wang et al. 2019). They help quickly and non-destructively assess meat quality without changing its internal traits (Zhang Xin et al. 2018). Studies have shown various uses of ultrasonic technology in meat quality assessment. For example, it can determine fat, moisture, and protein levels in meat mixtures by measuring ultrasonic velocity (Benedito et al. 2001). Ultrasonic imaging can identify fat content in pork loin accurately. It can even estimate beef marbling in live cattle (Fukuda et al. 2013). Ultrasonic technology also predicts salt content in brined pork muscles and helps assess carcass composition in live animals (De Prados et al. 2015). While ultrasonic detection has been used for non-destructive testing in meat quality assessment, it faces challenges such as irregularities in tested meat and variations in measurement sites and frequencies, leading to measurement errors (Jiang et al. 2017).

17.7.3.3 Nuclear Magnetic Resonance

Nuclear magnetic resonance (NMR) and magnetic resonance imaging (MRI) are techniques used for evaluating fresh meat. NMR generates an electromagnetic signal by interacting with sample nuclei in a magnetic field, using both high (HF) and low field (LF) systems. LF-NMR (30–100 MHz) is commonly used, while HF-NMR (>100 MHz) is less frequent (Antequera et al. 2021). LF-NMR, applied in pork and chicken studies, showed correlations between cooking conditions, water distribution, and sensory attributes (Bertram et al. 2005). In beef, LF-NMR determined moisture content with high precision (Pereira and Colnago 2012). LF-NMR also effectively classified beef and horse mixtures, demonstrating its sensitivity in species differentiation (Jakes et al. 2015). MRI serves as a valuable solution for the meat industry's challenges. It provides three-dimensional images with various contrasts, allowing insights into meat products. Its non-destructive nature allows continuous monitoring during processing. MRI reveals crucial components such as fat, connective tissue, water, and even salt distribution. MRI also reveals details at the cellular level, helping to understand muscle structure (Clerjon and Bonny 2017).

17.7.3.4 Computed Tomography

Computed Tomography (CT) is an imaging technology that uses X-ray beams to create non-invasive images of objects without the need for incisions (Lell and Kachelriess 2020). Recent attention has focused on X-ray computed tomography (CT) scanning in pigs, sheep, and beef cattle (Prieto et al. 2010). CT scanning offers non-invasive *in vivo* predictions of carcass composition, facilitating breeding programs in pigs and sheep (Prieto et al. 2010). Although not suitable for live beef cattle, CT scanning of primal cuts provides a cost-effective alternative to manual dissection for determining carcass composition (Navajas et al. 2010). Advancements like spiral CT scanning enable faster and more comprehensive assessments of carcass composition, muscle volume, and weight, with high accuracy reported in sheep and beef cattle (Prieto et al. 2010).

17.8 Physical Sensors and Biosensors for Quality Control During Processing

Currently, meat quality control relies heavily on manual analysis, which can be time-consuming. Assessing meat quality is a complex task that involves evaluating the molecular composition of meat using biological receptors (Zaukuu et al. 2020). Therefore, there's a need for simple, rapid, and reliable automated analytical methods to enhance meat quality control. In engineering, receptors are akin to sensors, which are man-made and adaptable for various purposes. While sensors encompass acoustic, chemical, electronic, optical, mechanical, and thermal types (Zaukuu et al. 2020). This section specifically addresses physical sensors such as bioimpedance, microwave, and UV-visible fluorescence, as well as biosensors like e-tongue, e-nose, e-eye, and optoelectronic sensors.

17.8.1 *Physical Sensors*

17.8.1.1 Bioimpedance

Low frequencies are commonly used to examine biological tissue and evaluate meat quality, offering versatility in various applications. This method, called bioimpedance, seems simple, requiring only a low-frequency generator, a few electrodes, and a view meter. At low frequencies, the cell membrane acts as non-conductive, while the intracellular fluid (ICF) behaves like an internal electrolyte, showing capacitance when exposed to alternating electric current. The electric current passes through the tissue, flowing through the extracellular fluid (ECF) or both the ECF and ICF compartments (Damez and Clerjon 2013).

Bioimpedance is highly effective in assessing the effects of freezing and refrigerated storage on meat quality post-slaughter. These processes can damage meat cells, causing changes in its electrical properties. Utilizing impedance spectroscopy with tetrapolar electrodes allows for precise differentiation between fresh and frozen-thawed pork. Parameters such as tetrapolar electrodes and phase angle demonstrate nearly perfect separation between fresh and frozen-thawed samples, even as the meat approaches the end of its shelf life. This suggests that bioelectrical impedance could serve as a rapid and reliable method for quality control in both fresh and frozen pork (Osen et al. 2022).

A study by Afonso et al. (2020) highlights the value of bioimpedance parameters for objectively measuring meat characteristics. These parameters correlate strongly with various meat quality traits, including intramuscular fat (IMF), total collagen, and shear force. For instance, a significant negative correlation exists between sarcomere length and shear force, indicating that shorter sarcomeres result in tougher meat. Moreover, the study demonstrates that bioimpedance can accurately predict IMF content in meat samples, even when the fat content is relatively low. However, the accuracy of prediction models varies depending on the IMF level within the samples. Models developed for samples with higher IMF content explain a larger proportion of the observed variation compared to those developed for samples with lower IMF content (Afonso et al. 2020).

17.8.1.2 Microwave

Microwaves and electromagnetic sensors offer a non-invasive solution for assessing meat quality attributes but are not widely used in the meat industry. These technologies can evaluate characteristics related to the structure, composition (especially water content), water state, water distribution, and water content profiles in meat products. Dielectric relaxation spectroscopy, for example, analyses how polar molecules respond to an external electric field, providing insights into water content and state. Microwave applications in the meat industry typically involve cavity, antenna, or probe measurements, including a method using transmission-reflection measurements with two antennas and a conveyor belt (Damez and Clerjon 2013).

Microwave spectroscopy on water activity, the observation a correlation between water activity and the spectra at various frequencies showed high for both high and low salt groups. A decrease in amplitude in the reflection spectra and transmission spectra corresponded to a decrease in water activity in the dry-cured meat samples across different salt groups. These results indicate that microwave spectroscopy can effectively differentiate between samples with varying water activity levels, even with minor differences among weight loss groups (Bjarnadottir et al. 2015).

Microwave polarimetry was used to track the decline in dielectric anisotropy during beef aging studied by Clerjon and Damez (2007). Measurements on muscle samples at 10 GHz indicated a slight decrease in anisotropy over time, although results were affected by external factors and imperfections in meat structure. Rectangular probe tests also showed consistent anisotropy in muscle properties, with a reduction observed during aging. These findings highlight the potential of microwave polarimetry for monitoring tissue changes during aging.

Dielectric spectroscopy in the radio frequency range of 100 Hz—100 kHz has shown promise in distinguishing lard from other animal fats. Results revealed that animal fats with high C18 composition had lower dielectric constants, while those with higher unsaturation degrees of C18 had higher dielectric constants. Statistical analysis effectively classified different groups of animal fats, with PCA showing graphical separation and achieving 85% accuracy in classifying dielectric spectra. This suggests that dielectric spectroscopy combined with quantitative analysis offers a rapid and cost-effective method for discriminating lard from other animal fats, requiring minimal sample preparation and low chemical usage (Amat et al. 2019).

17.8.1.3 UV-Visible Fluorescence

Fluorescence spectroscopy employs UV light to excite electrons in certain compounds, causing them to emit lower-energy light. Initially, a molecule absorbs a photon, moving from its ground to an excited electronic state. Subsequent collisions with other molecules reduce its vibrational energy until it returns to its ground state, emitting a photon. Different vibrational levels result in emitted photons with varying energies and frequencies, enabling analysis of molecular structure (Damez and Clerjon 2013). Tryptophan serves as a vital intrinsic fluorescent marker, particularly in muscle microenvironments. For opaque samples like meat, front-face fluorescence is used, often revealing structural details without added probes. Fluorescence is a selective technique, with excited species depending on the excitation wavelength. Recent studies have explored fluorescence spectroscopy's role in assessing food quality, offering valuable insights (Karoui and Blecker 2011).

Meat contains various fluorescent compounds like tryptophan residues in proteins, nicotinamide adenine dinucleotide (NADH), porphyrin, pyridinoline in collagen, lipid oxidation products, vitamin A, and riboflavin (Sahar and Dufour 2015). Researchers have used fluorescence spectroscopy to assess meat quality, focusing on detecting collagen content in different meat tissues like adipose and connective tissues. Egelandsdal et al. (2005) successfully quantified collagen in various muscles such as pork *glutens medius* and beef *masseter* using fluorescence spectroscopy, achieving prediction errors of 0.55%. Similarly, studies by Egelandsdal et al. (2002) and Allais et al. (2004) demonstrated the potential of fluorescence spectroscopy in measuring meat tenderness and texture by correlating fluorescent emission spectra with tenderness values and textural properties of meat products.

Moreover, fluorescence spectroscopy has been used to evaluate additional quality parameters beyond texture. Lee et al. (2018) examined total volatile basic nitrogen (TVB-N) content in pork during cold storage, noting an increase in fluorescence intensity as TVB-N content rose, indicating deteriorating meat quality. Fluorescence spectroscopy is effective in monitoring changes in meat caused by thermal treatments. Sahar et al. (2009) used synchronous fluorescence spectroscopy (SFS) to study cooking kinetics in bovine meat, identifying components related to Maillard reaction products and tryptophan residues. Similarly, Sahar et al. (2016) demonstrated the ability of SFS coupled with chemometrics to detect thermal changes during meat cooking rapidly and non-invasively. However, high-temperature cooking can produce harmful compounds like heterocyclic aromatic amines (HAA). Fluorescence spectroscopy has been used to detect HAA production in grilled meat. Sahar et al. (2010) successfully predicted concentrations of four HAA using SFS data and PARAFAC. Additionally, Li et al. (2011) employed derivative-SFS to rapidly determine benzo(a)pyrene, a carcinogenic compound, in processed meat and fish samples.

17.8.2 Biosensor

Basically, the biosensor system (including e-tongue, e-nose, and e-eye) consists of a sampling system, sensor array, and computer interface, as depicted in Fig. 17.1.

17.8.2.1 Electronic Tongue (e-Tongue)

Unlike the human tongue, the e-tongue demonstrates improvements in sensitivity, selectivity, and multiplexing capacity, akin to modern biosensors. It is a technology comprising a set of sensors that react when immersed in chemical solutions as shown in Fig. 17.2. Essentially, it consists of an electrochemical cell (sensor array), a measurement module, and an appropriate pattern recognition system capable of identifying simple and complex molecular systems that constitute taste (Munekata et al. 2023). Analysing liquid samples with the e-tongue poses no issues, but solid samples may require transformations that ideally should not alter the biochemical composition. Protocols for transforming meat, poultry, or fish samples into a liquid state without significant chemical alteration have posed challenges in the meat and fish industries (Zaukuu et al. 2020).

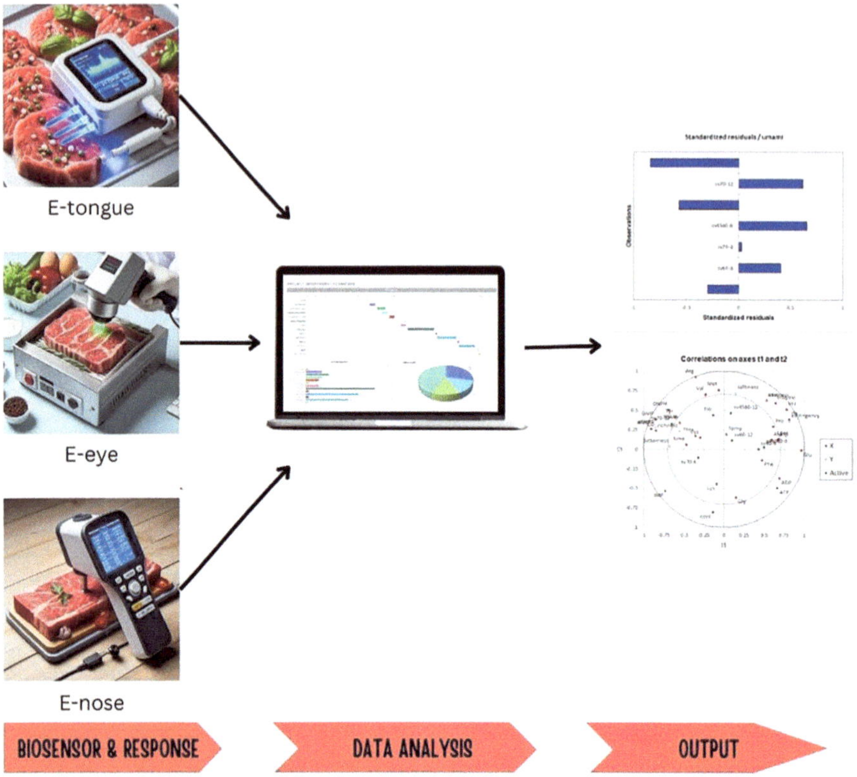

Fig. 17.1 Schematic illustration of the biosensor

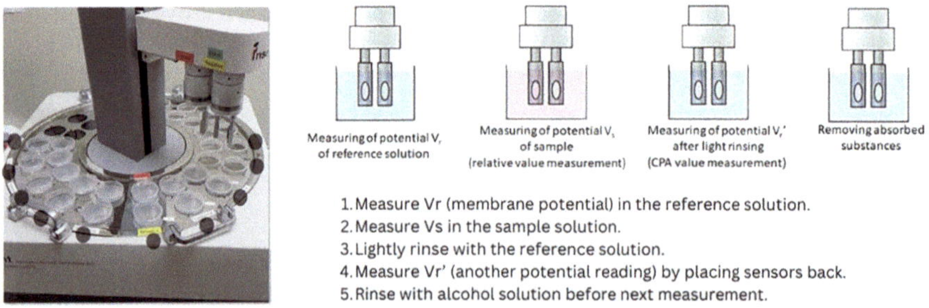

Fig. 17.2 Measurement procedure of an electronic tongue for identifying basic tastes in meat

Extraction and ultrafiltration methods have proven effective and cost-efficient for e-tongue analysis. For instance, in beef analysis, de-ionized water at 100 °C was used to homogenize the meat, followed by filtration and centrifugation (Zhang et al. 2012). While effective, the high temperature used in extraction procedures could potentially alter meat components and requires optimization (Zaukuu et al. 2020). Other effective extraction methods include microwave-assisted extraction, accelerated solvent extraction, and pressurized liquid extraction method (Tan et al. 2019; Chu and Metcalfe 2007). These methods have shown promise in various studies, offering less dense liquid extracts suitable for e-tongue analysis.

E-tongue has demonstrated acceptable correlations with human panellists' organoleptic results and exhibits higher sensitivity than the human tongue. Major categories for e-tongue meat analysis include process monitoring, freshness evaluation, shelf-life investigation, authenticity assessment, foodstuff recognition, quantitative analysis, and other quality control studies (Mimendia et al. 2010). In red meat analysis, the potentiometric e-tongue has been used to assess freshness, flavour, and shelf-life of beef, monitor pork storage, and discriminate between goat, beef, and lamb meat (Zaukuu et al. 2020). Additionally, the voltametric e-tongue has been utilized for detecting and quantifying putrescine and ammonia in beef, addressing challenges associated with biogenic amine formation in protein-rich foods (Apetrei and Apetrei 2016). The e-tongue, particularly the potentiometric variant, has been widely utilized due to its partially selective sensors, easy maintenance, and minimal technical requirements. Collaborative techniques with e-nose have also been employed to classify different types of meat effectively (Haddi et al. 2015).

However, sensor drift remains a concern in e-tongue applications as it can impact experimental results. To mitigate this issue, various interventions have been developed, including robust experimental design and mathematical approaches involving multivariate data analysis. Multivariate statistics optimize mathematical and statistical methods to process data effectively (Zaukuu et al. 2020).

17.8.2.2 Electronic Nose (e-Nose)

The e-nose addresses these challenges by offering enhanced sensitivity compared to the human nose for odour detection. It relies on electrochemical sensors and advanced algorithms to detect and recognize odours, providing a valuable tool for rapid quality evaluation in meat and fish products (Zaukuu et al. 2020). The e-nose converts sensory information from odorous substances into electrical signals through an array of gas sensors and artificial intelligence algorithms. These sensors interact with odorous compounds, resulting in changes in resistance and charge carrier concentration, which are then translated into odour profiles. Semiconductor metal oxides are commonly used due to their affordability, longevity, and high sensitivity, making them suitable for assessing meat and fish freshness (Loutfi et al. 2015).

In spoilage monitoring, e-nose technology has emerged as a valuable tool due to its ability to detect odour changes associated with the decomposition of food components like protein, fat, and carbohydrates by enzymes and bacteria. These changes, including discoloration and texture alterations, indicate spoilage (Papadopoulou et al. 2013). E-nose technology is also pivotal for distinguishing between different types of meat to ensure quality control and prevent adulteration. Studies have successfully used e-nose systems to differentiate between samples from sheep, cattle, poultry, and swine, showcasing its potential (Nurjuliana et al. 2011). Furthermore, e-nose analysis has been employed to assess the impact of dietary supplementation, such as vitamin E and selenium, on meat attributes, correlating aroma profiles with antioxidant status (Zhang et al. 2010). Monitoring meat production processes, including seasoning and ham ripening, has been achieved through e-nose technology, aiding in the detection of contaminants like ochratoxin A (Lippolis et al. 2016). Additionally, e-nose systems effectively detect sensory defects like "boar taint" in pork and the development of warmed-over flavour in pre-cooked meat (Tikk et al. 2008; Vestergaard et al. 2006). By correlating sensor responses with sensory and chemical analyses, e-nose technology provides a rapid and efficient means of identifying undesirable odours and flavours in meat products.

Although the e-nose effectively mimics human olfaction, challenges remain. Variations in sensor sensitivity to different gases and environmental conditions may affect detection accuracy (Berna 2010). Additionally, sample homogenization and headspace generation time can influence the concentration of volatile compounds and signal strength, thus impacting the reliability of e-nose analyses (González-Martín et al. 2000).

17.8.2.3 Electronic Eye (e-Eye)

E-eye technology, a visual information recognition and analysis system, plays a crucial role in evaluating food quality, particularly in terms of cost-effectiveness, portability, and large-scale applicability (YongXia et al. 2019). The human eye's ability to discern colours by wavelength differentiation is emulated through colour spaces such as HSI (hue, saturation, intensity), HSV (hue, saturation, value), HSL (hue, saturation, lightness) and HSB (hue, saturation, brightness), with e-eye leveraging colorimetry, spectrophotometry, or computer vision for its applications (Wu and Sun 2013). While e-eye excels at capturing appearance-related characteristics, it falls short in detecting flavour or aroma components. Hence, combining multiple technologies like e-nose and e-tongue has emerged as an alternative for comprehensive sample analysis (Wang et al. 2021b).

The e-eye technology has been extensively studied for quality evaluation in fresh meat and various meat products, focusing on colour, marbling level, quality prediction, defect detection, and sorting operations (Munekata et al. 2023). Colour is a crucial attribute in meat quality evaluation, with e-eye demonstrating good correlation with colorimeters in chicken and pork meat studies (Barbin et al. 2016). The selection of colour space and chemometric method significantly influences prediction accuracy, as observed in recent studies on pork meat (Sun et al. 2016). Marbling, reflecting intramuscular fat distribution, is vital for meat quality. E-eye has shown high accuracy in grading marbling in beef and pork, with data treatment methods affecting method precision (da Costa Barbon et al. 2017). Studies exploring e-eye's role in predicting physicochemical and sensory attributes of meat have shown varying accuracies among variables (Sun et al. 2018). Defect detection is another key application of e-eye, particularly in identifying issues like white stripes in poultry and pale, soft, and exudative characteristics in pork (Barbut et al. 2008). E-eye technology has been effective in sorting meat cuts, with high accuracy in classification achieved using algorithms like Linear Discriminant Analysis (Hassoun et al. 2023). E-eye technology has diverse applications, including monitoring changes in superficial meat colour during processing and evaluating intramuscular fat content in dry-cured ham. It has been instrumental in sorting meat products and discriminating dry-cured hams by geographical origin and quality grade (Muñoz et al. 2019).

While e-eye offers advantages like non-destructive sampling and easy operation, it requires controlled environments and careful sample handling. Calibration and background separation are essential for accurate data acquisition, highlighting considerations for effective use (Munekata et al. 2023).

17.8.2.4 Optoelectronic Sensor

An electronic nose, which relies on the combined response of an array of sensors, is a promising approach for sensing meat freshness. However, conventional electronic nose sensors have drawbacks like poor chemical specificity, sensor drift, and sensitivity to humidity changes, limiting their reliability for discriminating similar mixtures (Li and Suslick 2016). In recent years, colorimetric sensor arrays (CSAs) have emerged as a novel optoelectronic nose for detecting various analytes. Unlike traditional electronic noses, CSAs use chemically responsive dyes in hydrophobic matrices to probe a wide range of chemical reactivity. Digital imaging of the colour changes enables the identification of unique patterns for different odorants (Li and Suslick 2016). Additionally, CSAs are less susceptible to humidity and temperature fluctuations, with extended chemical activity retention (Xu et al. 2022b).

The colorimetric sensor arrays (CSAs) containing various pH-responsive dyes have been developed. These arrays utilize cross-response, where each dye responds differently to total volatile basic nitrogen (TVB-N) and other gases released during spoilage (Lin et al. 2023). The breakdown of beef proteins during storage leads to the production of TVB-N, altering the pH of the packaging environment (Xu et al. 2022a). pH-responsive colour indicators have emerged as promising tools for monitoring meat freshness due to their simplicity and visual colour changes. However, while pH-responsive dyes offer convenience for assessing meat quality because their colour changes are easily distinguishable by the naked eye, relying on a single type of pH-responsive dye makes it challenging to pinpoint the onset of the spoilage threshold (Hu et al. 2020).

For instance, a sensor array was developed to monitor fresh pork in Modified Atmosphere Packaging (MAP) (Salinas et al. 2014). It combines pH indicators and selective chromogenic reagents on different materials to detect meat spoilage through colour changes. It can distinguish between pork stored for up to 7 days based on colour patterns. Another developed tool, an optoelectronic nose, has been developed to monitor chicken meat aging (Salinas et al. 2012). It uses 16 pigments in various materials to detect colour changes indicating chicken aging in a modified packaging atmosphere. Statistical analyses like Principal Components Analysis (PCA) and Partial Least Squares (PLS) show its ability to differentiate aging at 2-day intervals and correlate colour data with meat aging. The sensor effectively monitors volatile emissions from meat samples, distinguishing between biogenic amines and sulphides over storage periods. Statistical methods like Hierarchical Cluster Analysis (HCA) and PCA cluster meat samples based on freshness, demonstrating high accuracy in identifying spoilage (Li and Suslick 2016).

17.9 Conclusions

In conclusion, advancements in technology play a pivotal role in ensuring the quality and safety of meat products. From enhancing traceability with tools like IoT, RFID, and blockchain to exploring innovative preservation methods such as essential oils and bacteriophages, the meat industry continues to adapt to evolving consumer preferences and regulatory standards. Moreover, the development of non-destructive assessment techniques, including spectroscopic, imaging methods, computer vision system and physical sensors, offers rapid and reliable means of evaluating meat quality attributes without altering its state. As these technologies continue to evolve, their integration holds immense potential to revolutionize meat quality control, providing stakeholders with the tools they need to meet consumer expectations while upholding industry standards for safety and excellence.

In this context, ongoing research and innovation are essential to address challenges such as regulatory scrutiny over additives and the need for continuous improvement in assessing meat freshness. By leveraging the capabilities of emerging technologies and embracing a culture of innovation, the meat industry can navigate these challenges effectively, ensuring the production of consistent high-quality meat products that meet the demands of today's discerning consumers.

References

24CFR (2020) Code of federal regulations. In: Food additives permitted for direct addition to food for human consumption. Electronic Code of Federal Regulations (eCFR). https://www.ecfr.gov/current/title-21/chapter-I/subchapter-B/part-172. Accessed 4 Mar 2024

Ab Rashid N, Bojei J (2020) The relationship between halal traceability system adoption and environmental factors on halal food supply chain integrity in Malaysia. J Islam Mark 11:117–142. https://doi.org/10.1108/JIMA-01-2018-0016

Abdallah SA, Al-Shatti LA, Alhajraf AF, Al-Hammad N, Al-Awadi B (2013) The detection of foodborne bacteria on beef: the application of the electronic nose. Springerplus 2:687. https://doi.org/10.1186/2193-1801-2-687

Achata EM, Esquerre C, Ojha KS, Tiwari BK, O'Donnell CP (2021) Development of NIR-HSI and chemometrics process analytical technology for drying of beef jerky. Innov Food Sci Emerg Technol 69:102611. https://doi.org/10.1016/j.ifset.2021.102611

Acheson R, Woerner DR, Walenciak CE, Colle MJ, Bass PD (2018) Distribution of marbling throughout the m. Longissimus thoracis et lumborum of beef carcasses using an instrument-grading system. Meat Muscl Bio 2. https://doi.org/10.22175/mmb2018.04.0005

Afonso J, Guedes C, Santos V, Morais R, Silva J, Teixeira A, Silva S (2020) Utilization of bioelectrical impedance to predict intramuscular fat and physicochemical traits of the beef longissimus thoracis et lumborum muscle. Food Secur 9. https://doi.org/10.3390/foods9060836

Aguiar Campolina G, das Graças Cardoso M, Rodrigues-Silva-Caetano A, Lee Nelson D, Mendes Ramos E (2023) Essential oil and plant extracts as preservatives and natural antioxidants applied to meat and meat products: a review. Food Technol Biotechnol 61:212–225. https://doi.org/10.17113/ftb.61.02.23.7883

Ahmad RS, Imran A, Hussain MB (2018) Nutritional composition of meat. Meat Science and Nutrition. Intech. https://doi.org/10.5772/intechopen.77045

Akshatha M, Alankrutha KP, Janitha Annet G, Monteiro L, George SV (2017) Student authentication and verification system using barcode scanner. Int J Int Things 6:71–74. https://doi.org/10.5923/j.ijit.20170602.13

Albelda JAV, Uzunoglu A, Santos GNC, Stanciu LA (2017) Graphene-titanium dioxide nanocomposite based hypoxanthine sensor for assessment of meat freshness. Biosens Bioelectron 89:518–524. https://doi.org/10.1016/j.bios.2016.03.041

Alexandrakis D, Downey G, Scannell AG (2012) Rapid non-destructive detection of spoilage of intact chicken breast muscle using near-infrared and Fourier transform mid-infrared spectroscopy and multivariate statistics. Food Bioprocess Technol 5:338–347. https://doi.org/10.1007/s11947-009-0298-4

Allais I, Viaud C, Pierre A, Dufour É (2004) A rapid method based on front-face fluorescence spectroscopy for the monitoring of the texture of meat emulsions and frankfurters. Meat Sci 67:219–229. https://doi.org/10.1016/j.meatsci.2003.10.009

Amat M, Abd Aziz S, Tan C, Mustafa S, Abd Gani SS, Rokhani FZ (2019) Lard classification from other animal fats using Dielectric Spectroscopy technique. Int Food Res J 26:773–782

Aminzare M, Hashemi M, Ansarian E, Bimkar M, Hassanzadazar H, Mehrasbi M, Daneshamooz S, Raeisi M, Jannat B, Afshari A (2019) Using natural antioxidants in meat and meat products as preservatives: A review. Adv Anim Vet Sci 7:417–426. https://doi.org/10.17582/journal.aavs/2019/7.5.417.426

Antequera T, Caballero D, Grassi S, Uttaro B, Perez-Palacios T (2021) Evaluation of fresh meat quality by Hyperspectral Imaging (HSI), Nuclear Magnetic Resonance (NMR) and Magnetic Resonance Imaging (MRI): A review. Meat Sci 172:108340. https://doi.org/10.1016/j.meatsci.2020.108340

Apetrei IM, Apetrei C (2016) Application of voltammetric e-tongue for the detection of ammonia and putrescine in beef products. Sens Actuators B Chem 234:371–379. https://doi.org/10.1016/j.snb.2016.05.005

Arai T (2014) The development of animal nutrition and metabolism and the challenges of our time. Front Vet Sci 1:23. https://doi.org/10.3389/fvets.2014.00023

Caveen A, Archer M, Plat M (2021) The impact of improved traceability on the safety of food In: RS Standards. Lloyd's Register Foundation, London. https://www.lrfoundation.org.uk/496616/siteassets/pdfs/impact_of_improved_traceability_on_the_safety_of_food___full_report.pdf. Accessed 21 Feb 20

Aredo V, Velásquez L, Siche R (2017) Prediction of beef marbling using hyperspectral imaging (HSI) and partial least squares regression (PLSR). Sci Agropecu 8:169–174. https://doi.org/10.17268/sci.agropecu.2017.02.09

Arsalane A, El Barbri N, Tabyaoui A, Klilou A, Rhofir K, Halimi A (2018) An embedded system based on DSP platform and PCA-SVM algorithms for rapid beef meat freshness prediction and identification. Comput Electron Agric 152:385–392. https://doi.org/10.1016/j.compag.2018.07.031

Arshad MS, Sohaib M, Ahmad RS, Nadeem MT, Imran A, Arshad MU, Kwon J-H, Amjad Z (2018) Ruminant meat flavor influenced by different factors with special reference to fatty acids. Lipids Health Dis 17:223. https://doi.org/10.1186/s12944-018-0860-z

Asioli D, Aschemann-Witzel J, Caputo V, Vecchio R, Annunziata A, Næs T, Varela P (2017) Making sense of the "clean label" trends: A review of consumer food choice behavior and discussion of industry implications. Food Res Int 99:58–71. https://doi.org/10.1016/j.foodres.2017.07.022

Aung MM, Chang YS (2014) Traceability in a food supply chain: safety and quality perspectives. Food Control 39:172–184. https://doi.org/10.1016/j.foodcont.2013.11.007

Bajcic A, Petronijevic RB, Sefer M, Trbovic D, Djordjevic V, Ciric J, Nikolic A (2021) Sorbates and benzoates in meat and meat products: importance, application and determination. In: 61st international meat industry conference, IOP conference series: earth and environmental science BC, Canada, IOP Publishing, 29 May – 2 June 2022

Bak KH, Bolumar T, Karlsson AH, Lindahl G, Orlien V (2019) Effect of high pressure treatment on the color of fresh and processed meats: a review. Crit Rev Food Sci Nutr 59:228–252. https://doi.org/10.1080/10408398.2017.1363712

Bakkali F, Averbeck S, Averbeck D, Idaomar M (2008) Biological effects of essential oils: a review. Food Chem Toxicol 46:446–475. https://doi.org/10.1016/j.fct.2007.09.106

Barberis S, Quiroga HG, Barcia C, Talia JM, Debattista N (2018) Chapter 20 – Natural food preservatives against microorganisms. In: Grumezescu AM, Holban AM (eds) Food safety and preservation. Academic, pp 621–658

Barbin DF, ElMasry G, Sun D-W, Allen P (2013) Non-destructive determination of chemical composition in intact and minced pork using near-infrared hyperspectral imaging. Food Chem 138:1162–1171. https://doi.org/10.1016/j.foodchem.2012.11.120

Barbin DF, Mastelini SM, Barbon S Jr, Campos GF, Barbon APA, Shimokomaki M (2016) Digital image analyses as an alternative tool for chicken quality assessment. Biosyst Eng 144:85–93. https://doi.org/10.1016/j.biosystemseng.2016.01.015

Barbut S, Sosnicki AA, Lonergan SM, Knapp T, Ciobanu DC, Gatcliffe LJ, Huff-Lonergan E, Wilson EW (2008) Progress in reducing the pale, soft and exudative (PSE) problem in pork and poultry meat. Meat Sci 79:46–63. https://doi.org/10.1016/j.meatsci.2007.07.031

Barnard C, O'Connor N (2017) Runners and riders: The horsemeat scandal, eu law and multi-level enforcement. Cambridge Law J 76:116–144

Batinovic S, Wassef F, Knowler SA, Rice DT, Stanton CR, Rose J, Tucci J, Nittami T, Vinh A, Drummond GR (2019) Bacteriophages in natural and artificial environments. Pathogens 8:100. https://doi.org/10.3390/pathogens8030100

Ben Braïek O, Smaoui S (2021) Chemistry, safety, and challenges of the use of organic acids and their derivative salts in meat preservation. J Food Qual 2021:6653190–6653120. https://doi.org/10.1155/2021/6653190

Benedito J, Carcel J, Rossello C, Mulet A (2001) Composition assessment of raw meat mixtures using ultrasonics. Meat Sci 57:365–370. https://doi.org/10.1016/S0309-1740(00)00113-3

Berna A (2010) Metal oxide sensors for electronic noses and their application to food analysis. Sensors (Basel, Switzerland) 10:3882–3910. https://doi.org/10.3390/s100403882

Bertram HC, Aaslyng MD, Andersen HJ (2005) Elucidation of the relationship between cooking temperature, water distribution and sensory attributes of pork—a combined NMR and sensory study. Meat Sci 70:75–81. https://doi.org/10.1016/j.meatsci.2004.12.002

Bintsis T (2018) Microbial pollution and food safety. AIMS Microbiol 4:377–396. https://doi.org/10.3934/microbiol.2018.3.377

Bjarnadottir SG, Lunde K, Alvseike O, Mason A, Al-Shamma'a AI (2015) Assessing quality parameters in dry-cured ham using microwave spectroscopy. Meat Sci 108:109–114. https://doi.org/10.1016/j.meatsci.2015.06.004

Foraker BA, Frink JL, Woerner DR (2022) Invited review: a carcass and meat perspective of crossbred beef × dairy cattle. Trans Anim Sci 6(2):txac027. https://doi.org/10.1093/tas/txac027

Bobokulov N, Khatamov A, Abduzoirova D, Yusupov A, Urimbetov A, Olmasov B (2021) Meat productivity of sheep in Uzbekistan and its relationship with different factors. E3S Web Conf 258:04020. https://doi.org/10.1051/e3sconf/202125804020
Bonah E, Huang X, Aheto JH, Osae R (2020) Application of electronic nose as a non-invasive technique for odor fingerprinting and detection of bacterial foodborne pathogens: a review. J Food Sci Technol 57:1977–1990. https://doi.org/10.1007/s13197-019-04143-4
Bosona T, Gebresenbet G (2013) Food traceability as an integral part of logistics management in food and agricultural supply chain. Food Control 33:32–48. https://doi.org/10.1016/j.foodcont.2013.02.004
Botilias G-P, Margariti SV, Besarat J, Salmas D, Pachoulas G, Stylios C, Skalkos D (2023) Designing and developing a meat traceability system: a case study for the greek meat industry. Sustain For 15:12162. https://doi.org/10.3390/su151612162
Bouzembrak Y, Klüche M, Gavai A, Marvin HJP (2019) Internet of Things in food safety: literature review and a bibliometric analysis. Trends Food Sci Technol 94:54–64. https://doi.org/10.1016/j.tifs.2019.11.002
Bowker B (2017) Chapter 4 – Developments in our understanding of water-holding capacity. In: Petracci M, Berri C (eds) Poultry quality evaluation. Woodhead Publishing, pp 77–113
Boyacı IH, Temiz HT, Uysal RS, Velioğlu HM, Yadegari RJ, Rishkan MM (2014) A novel method for discrimination of beef and horsemeat using Raman spectroscopy. Food Chem 148:37–41. https://doi.org/10.1016/j.foodchem.2013.10.006
Brender JD (2020) Human health effects of exposure to nitrate, nitrite, and nitrogen dioxide. In: Sutton MA, Mason KE, Bleeker A, Hicks WK, Masso C, Raghuram N, Reis S, Bekunda M (eds) Just enough nitrogen: perspectives on how to get there for regions with too much and too little nitrogen. Springer International Publishing, Cham, pp 283–294
Burgos-Díaz C, Wandersleben T, Marqués AM, Rubilar M (2016) Multilayer emulsions stabilized by vegetable proteins and polysaccharides. Curr Opin Colloid Interface Sci 25:51–57. https://doi.org/10.1016/j.cocis.2016.06.014
Carpenter C, Smith J, Broadbent J (2011) Efficacy of washing meat surfaces with 2% levulinic, acetic, or lactic acid for pathogen decontamination and residual growth inhibition. Meat Sci 88: 256–260. https://doi.org/10.1016/j.meatsci.2010.12.032
Casali E, Berni P, Spisni A, Baricchi R, Pertinhez TA (2016) Hypoxanthine: a new paradigm to interpret the origin of transfusion toxicity. Blood Transfus 14:555–556. https://doi.org/10.2450/2015.0177-15
CFR (2024) Listing of color additives exempt from certification. 21 CFR Part 73. In: https://www.ecfr.gov/current/title-21/chapter-I/subchapter-A/part-73. Accessed 4 Mar 2024
Chen M, Mustapha A (2008) Natural antimicrobial compounds as meat preservatives. Global science books. Foods 2:102–104
Chen L, Chen J, Ren J, Zhao M (2011) Modifications of soy protein isolates using combined extrusion pre-treatment and controlled enzymatic hydrolysis for improved emulsifying properties. Food Hydrocoll 25:887–897
Chen J, Gu J, Zhang R, Mao Y, Tian S (2019) Freshness evaluation of three kinds of meats based on the electronic nose. Sensors 19:605. https://doi.org/10.3390/s19030605
Cheng J-H, Dai Q, Sun D-W, Zeng X-A, Liu D, Pu H-B (2013a) Applications of non-destructive spectroscopic techniques for fish quality and safety evaluation and inspection. Trends Food Sci Technol 34:18–31. https://doi.org/10.1016/j.tifs.2013.08.005
Cheng X, Tang X, Wang Q, Mao XY (2013b) Antibacterial effect and hydrophobicity of yak κ-casein hydrolysate and its fractions. Int Dairy J 31:111–116. https://doi.org/10.1016/j.idairyj.2012.12.004
Chmiel M, Słowiński M (2016) The use of computer vision system to detect pork defects. LWT-Food Sci Technol 73:473–480. https://doi.org/10.1016/j.lwt.2016.06.054
Chmiel M, Słowiński M, Dasiewicz K, Florowski T (2012) Application of a computer vision system to classify beef as normal or dark, firm, and dry. J Anim Sci 90:4126–4130. https://doi.org/10.2527/jas.2011-5022

Choi YS, Choi JH, Han DJ, Kim HY, Lee MA, Kim HW, Jeong JY, Kim CJ (2011) Effects of rice bran fiber on heat-induced gel prepared with pork salt-soluble meat proteins in model system. Meat Sci 88:59–66. https://doi.org/10.1016/j.meatsci.2010.12.003

Choi I, Lee JY, Lacroix M, Han J (2017) Intelligent pH indicator film composed of agar/potato starch and anthocyanin extracts from purple sweet potato. Food Chem 218:122–128. https://doi.org/10.1016/j.foodchem.2016.09.050

Chouhan S, Sharma K, Guleria S (2017) Antimicrobial activity of some essential oils—present status and future perspectives. Medicines 4:58. https://doi.org/10.3390/medicines4030058

Chu S, Metcalfe CD (2007) Analysis of paroxetine, fluoxetine and norfluoxetine in fish tissues using pressurized liquid extraction, mixed mode solid phase extraction cleanup and liquid chromatography-tandem mass spectrometry. J Chromatogr A 1163:112–118. https://doi.org/10.1016/j.chroma.2007.06.014

Chung KY, Lee SH, Cho SH, Kwon EG, Lee JH (2018) Current situation and future prospects for beef production in South Korea – a review. Asian Australas J Anim Sci 31:951–960. https://doi.org/10.5713/ajas.18.0187

Clerjon S, Bonny J-M (2017) NMR imaging of meat. In: Webb GA (ed) Modern magnetic resonance. Springer International Publishing, Cham, pp 1–20

Clerjon S, Damez JL (2007) Microwave sensing for meat and fish structure evaluation. Meas Sci Technol 18:1038–1045. https://doi.org/10.1088/0957-0233/18/4/011

Clinquart A, Ellies-Oury MP, Hocquette JF, Guillier L, Santé-Lhoutellier V, Prache S (2022) Review: On-farm and processing factors affecting bovine carcass and meat quality. Animal 16:100426. https://doi.org/10.1016/j.animal.2021.100426

Cluff K, Naganathan GK, Subbiah J, Samal A, Calkins CR (2013) Optical scattering with hyperspectral imaging to classify longissimus dorsi muscle based on beef tenderness using multivariate modeling. Meat Sci 95:42–50. https://doi.org/10.1016/j.meatsci.2013.04.014

Collell C, Gou P, Arnau J, Comaposada J (2011) Non-destructive estimation of moisture, water activity and NaCl at ham surface during resting and drying using NIR spectroscopy. Food Chem 129:601–607. https://doi.org/10.1016/j.foodchem.2011.04.073

Commission E (2012) Commission regulation (EU) no 231/2012 of 9 March 2012 laying down specifications for food additives listed in annexes II and III to regulation (EC) no 1333/2008 of the European Parliament and of the Council. Off J Eur Communities 55:1–295

Creydt M, Fischer M (2019) Blockchain and more – Algorithm driven food traceability. Food Control 105:45–51. https://doi.org/10.1016/j.foodcont.2019.05.019

Cutroneo S, Angelino D, Tedeschi T, Pellegrini N, Martini D, SINU Young Working Group (2022) Nutritional quality of meat analogues: Results from the food labelling of Italian Products (FLIP) Project. Front Nutr 9:852831. https://doi.org/10.3389/fnut.2022

da Costa Barbon APA, Barbon S Jr, Campos GFC, Seixas JL Jr, Peres LM, Mastelini SM, Andreo N, Ulrici A, Bridi AM (2017) Development of a flexible computer vision system for marbling classification. Comput Electron Agric 142:536–544. https://doi.org/10.1016/j.compag.2017.11.017

da Costa RJ, Voloski FL, Mondadori RG, Duval EH, Fiorentini ÂM (2019) Preservation of meat products with bacteriocins produced by lactic acid bacteria isolated from meat. J Food Qual 2019:1–12. https://doi.org/10.1155/2019/4726510

Dabadé DS, Jacxsens L, Miclotte L, Abatih E, Devlieghere F, De Meulenaer B (2021) Survey of multiple biogenic amines and correlation to microbiological quality and free amino acids in foods. Food Control 120:107497

Damez J-L, Clerjon S (2013) Quantifying and predicting meat and meat products quality attributes using electromagnetic waves: an overview. Meat Sci 95:879–896. https://doi.org/10.1016/j.meatsci.2013.04.037

D'Amore T, Di Taranto A, Berardi G, Vita V, Marchesani G, Chiaravalle AE, Iammarino M (2020) Sulfites in meat: occurrence, activity, toxicity, regulation, and detection. A comprehensive review. Compr Rev Food Sci Food Saf 19:2701–2720. https://doi.org/10.1111/1541-4337.12607

Davidson PM, Taylor TM, Schmidt SE (2012) Chemical preservatives and natural antimicrobial compounds. In: Doyle MP, Buchanan RL (eds) Food microbiology: fundamentals and frontiers, 4th edn. Wiley: ASM Press, Washington, DC, pp 765–801

Davidson PM, Critzer FJ, Taylor TM (2013) Naturally occurring antimicrobials for minimally processed foods. Annu Rev Food Sci Technol 4:163–190. https://doi.org/10.1146/annurev-food-030212-182535

de Araújo PD, Araújo WMC, Patarata L, Fraqueza MJ (2022) Understanding the main factors that influence consumer quality perception and attitude towards meat and processed meat products. Meat Sci 193:108952. https://doi.org/10.1016/j.meatsci.2022.108952

De Mey E, De Klerck K, De Maere H, Dewulf L, Derdelinckx G, Peeters M-C, Fraeye I, Vander Heyden Y, Paelinck H (2014) The occurrence of N-nitrosamines, residual nitrite and biogenic amines in commercial dry fermented sausages and evaluation of their occasional relation. Meat Sci 96:821–828. https://doi.org/10.1016/j.meatsci.2013.09.010

De Prados M, García-Pérez JV, Benedito J (2015) Non-destructive salt content prediction in brined pork meat using ultrasound technology. J Food Eng 154:39–48. https://doi.org/10.1016/j.jfoodeng.2014.12.024

Dehghan P, Mohammadi A, Mohammadzadeh-Aghdash H, Ezzati Nazhad Dolatabadi J (2018) Pharmacokinetic and toxicological aspects of potassium sorbate food additive and its constituents. Trends Food Sci Technol 80:123–130. https://doi.org/10.1016/j.tifs.2018.07.012

Delgado-Pando G, Ekonomou SI, Stratakos AC, Pintado T (2021) Clean label alternatives in meat products. Food Secur 10. https://doi.org/10.3390/foods10071615

Dervisevic M, Custiuc E, Çevik E, Şenel M (2015) Construction of novel xanthine biosensor by using polymeric mediator/MWCNT nanocomposite layer for fish freshness detection. Food Chem 181:277–283. https://doi.org/10.1016/j.foodchem.2015.02.104

Dickinson E (2009) Hydrocolloids as emulsifiers and emulsion stabilizers. Food Hydrocoll 23: 1473–1482. https://doi.org/10.1016/j.foodhyd.2008.08.005

Dickinson E (2019) Strategies to control and inhibit the flocculation of protein-stabilized oil-in-water emulsions. Food Hydrocoll 96:209–223. https://doi.org/10.1016/j.foodhyd.2019.05.021

Dima A, Arvaniti E, Stylios C, Kafetzopoulos D, Skalkos D (2022) Adapting open innovation practices for the creation of a traceability system in a meat-producing industry in Northwest Greece. Sustain For 14:5111. https://doi.org/10.3390/su14095111

Ding Y, Chen L, Shi Y, Akhtar M, Chen J, Ettelaie R (2021) Emulsifying and emulsion stabilizing properties of soy protein hydrolysates, covalently bonded to polysaccharides: the impact of enzyme choice and the degree of hydrolysis. Food Hydrocoll 113:106519. https://doi.org/10.1016/j.foodhyd.2020.106519

Dorsemaine B, Gaulier JP, Wary JP, Kheir N, Urien P (2015) Internet of things: a definition & taxonomy. In: 9th international conference on next generation mobile applications, services and technologies, 72–77. Cambridge, UK, IEEE, 9–11 Sep 2015

Drewnowski A (2024) Perspective: the place of pork meat in sustainable healthy diets. Adv Nutr 15(5):100213. https://doi.org/10.1016/j.advnut.2024.100213

Du H, Ping T, Wu W, Yang Q (2022) Sandwich fluorescence detection of foodborne pathogen Staphylococcus aureus with CD fluorescence signal amplification in food samples. Food Secur 11:945. https://doi.org/10.3390/foods11070945

Dubal Z, Paturkar A, Waskar V, Zende R, Latha C, Rawool D, Kadam M (2004) Effect of food grade organic acids on inoculated S. aureus, L. monocytogenes, E. coli and S. Typhimurium in sheep/goat meat stored at refrigeration temperature. Meat Sci 66:817–821. https://doi.org/10.1016/j.meatsci.2003.08.004

Duc HM, Son HM, Yi HPS, Sato J, Ngan PH, Masuda Y, Honjoh K-i, Miyamoto T (2020) Isolation, characterization and application of a polyvalent phage capable of controlling Salmonella and Escherichia coli O157: H7 in different food matrices. Food Res Int 131:108977. https://doi.org/10.1016/j.foodres.2020.108977

EC (2008) Regulation, Council 1333/2008 of 16 December 2008. In: Food additives. Accessed 3 Mar 2024

EC (2018) European Commission, European Regulation (EU) 2018/1481. In: Off J Eur Union. 251
EFSA (2015) Scientific opinion on the re-evaluation of sorbic acid (E 200), potassium sorbate (E 202) and calcium sorbate (E 203) as food additives. EFSA 13:4144. https://doi.org/10.2903/j.efsa.2015.4144
Egelandsdal B, Wold JP, Sponnich A, Neegård S, Hildrum KI (2002) On attempts to measure the tenderness of Longissimus Dorsi muscles using fluorescence emission spectra. Meat Sci 60: 187–202. https://doi.org/10.1016/S0309-1740(01)00121-8
Egelandsdal B, Dingstad G, Tøgersen G, Lundby F, Langsrud Ø (2005) Autofluorescence quantifies collagen in sausage batters with a large variation in myoglobin content. Meat Sci 69:35–46. https://doi.org/10.1016/j.meatsci.2004.06.004
Eisenstein BI (2004) Lipopeptides, focusing on daptomycin, for the treatment of Gram-positive infections. Expert Opin Investig Drugs 13:1159–1169. https://doi.org/10.1517/13543784.13.9.1159
Elois MA, Silva R, Pilati GVT, Rodríguez-Lázaro D, Fongaro G (2023) Bacteriophages as biotechnological tools. Viruses 15:349. https://doi.org/10.3390/v15020349
Erdem Büyükkiraz M, Kesmen Z (2022) Antimicrobial peptides (AMPs): a promising class of antimicrobial compounds. J Appl Microbiol 132:1573–1596. https://doi.org/10.1111/jam.15314
Erna KH, Rovina K, Mantihal S (2021) Current detection techniques for monitoring the freshness of meat-based products: a review. J Packag Technol Res 5:127–141. https://doi.org/10.1007/s41783-021-00120-5
Ertbjerg P, Puolanne E (2017) Muscle structure, sarcomere length and influences on meat quality: a review. Meat Sci 132:139–152. https://doi.org/10.1016/j.meatsci.2017.04.261
Esquivel-Ferriño PC, Favela-Hernández JMJ, Garza-González E, Waksman N, Ríos MY, Camacho-Corona MR (2012) Antimycobacterial activity of constituents from foeniculum vulgare var. Dulce grown in Mexico. Molecules 17:8471–8482. https://doi.org/10.3390/molecules17078471
Estévez M (2021) Critical overview of the use of plant antioxidants in the meat industry: opportunities, innovative applications and future perspectives. Meat Sci 181:108610. https://doi.org/10.1016/j.meatsci.2021.108610
Ezati P, Tajik H, Moradi M (2019) Fabrication and characterization of alizarin colorimetric indicator based on cellulose-chitosan to monitor the freshness of minced beef. Sens Actuators B Chem 285:519–528. https://doi.org/10.1016/j.snb.2019.01.089
FAO (2011) Caramel colours. In: Monograph 11 prepared at the 74th JECFA 2011. https://www.fao.org/fileadmin/user_upload/jecfa_additives/docs/monograph11/additive-329-m11.pdf. Accessed 4 Mar 2024
FAO/WHO (2004) Section 4: traceability. In: Good practices for the meat industry. Switzerland. https://www.fao.org/3/y5454e/y5454e04.pdf. Accessed 21 Feb 2024
Faustman C, Chan WKM, Schaefer DM, Havens A (1998) Beef color update: the role for vitamin E. J Anim Sci 76:1019–1026. https://doi.org/10.2527/1998.7641019x
FDA (2019) Code of federal regulation title 21. In: Food and drugs. Silver Spring, MD. https://www.ecfr.gov/cgi-bin/textidx?SID=f8a2a8eadcdf4d12038eb7a20ffe8931&mc=true&tpl=/ecfrbrowse/Title21/21tab_02.tpl. Accessed 2 Mar 2024
Feng X, Li C, Jia X, Guo Y, Lei N, Hackman RM, Chen L, Zhou G (2016) Influence of sodium nitrite on protein oxidation and nitrosation of sausages subjected to processing and storage. Meat Sci 116:260–267. https://doi.org/10.1016/j.meatsci.2016.01.017
Finni T, de Brito FH, Maas H (2023) Force transmission and interactions between synergistic muscles. J Biomech 152:111575. https://doi.org/10.1016/j.jbiomech.2023.111575
Flores M, Toldrá F (2021) Chemistry, safety, and regulatory considerations in the use of nitrite and nitrate from natural origin in meat products-Invited review. Meat Sci 171:108272. https://doi.org/10.1016/j.meatsci.2020.108272
Font-i-Furnols M (2023) Meat consumption, sustainability and alternatives: an overview of motives and barriers. Food Secur 12:2144. https://doi.org/10.3390/foods12112144

Font-i-Furnols M, Guerrero L (2014) Consumer preference, behavior and perception about meat and meat products: an overview. Meat Sci 98:361–371. https://doi.org/10.1016/j.meatsci.2014.06.025

Fowler SM, Schmidt H, van de Ven R, Hopkins DL (2018) Preliminary investigation of the use of Raman spectroscopy to predict meat and eating quality traits of beef loins. Meat Sci 138:53–58. https://doi.org/10.1016/j.meatsci.2018.01.002

Franco MR, da Cunha LR, Bianchi RF (2021) Janus principle applied to food safety: an active two-faced indicator label for tracking meat freshness. Sens Actuators B Chem 333:129466. https://doi.org/10.1016/j.snb.2021.129466

Frank D, Joo S-T, Warner R (2016) Consumer acceptability of intramuscular fat. Korean J Food Sci Anim 36:699–708. https://doi.org/10.5851/kosfa.2016.36.6.699

FSANZ (2020) Food standards Australia New Zealand. In: Australia New Zealand food standards code—Standard 131—Food additives. Canberra, https://www.foodstandards.gov.au/code/Documents/Sched%2015%20Food%20additives%20v157.pdf. Accessed 28 Feb 2024

FSIS (2013) U.S. Department of Agriculture. In: Natural flavors on meat and poultry labels. https://www.fsis.usda.gov/food-safety/safe-food-handling-and-preparation/food-safety-basics/natural-flavors-meat-and-poultry. Accessed 5 Mar 2024

FSIS (2015) U.S. Department of Agriculture. In: Additives in meat and poultry products. https://www.fsis.usda.gov/food-safety/safe-food-handling-and-preparation/food-safety-basics/additives-meat-and-poultry. Accessed 5 Mar 2024

Fukuda O, Nabeoka N, Miyajima T (2013) Estimation of marbling score in live cattle based on ICA and a neural network. In: 2013 IEEE international conference on systems, man, and cybernetics, 1622–1627. Manchester, UK, IEEE, 13–16 Oct 2013

Furnols MFI, Teran MF, Gispert M (2009) Estimation of lean meat content in pig carcasses using X-ray computed tomography and PLS regression. Chemom Intell Lab Syst 98:31–37. https://doi.org/10.1016/j.chemolab.2009.04.009

Galstyan V, Ponzoni A, Kholmanov I, Natile MM, Comini E, Sberveglieri G (2020) Highly sensitive and selective detection of dimethylamine through Nb-doping of TiO2 nanotubes for potential use in seafood quality control. Sens Actuators B Chem 303:127217. https://doi.org/10.1016/j.snb.2019.127217

Galvez JF, Mejuto JC, Simal-Gandara J (2018) Future challenges on the use of blockchain for food traceability analysis. TrAC Trends Anal Chem 107:222–232. https://doi.org/10.1016/j.trac.2018.08.011

Galvin K, Lynch A-M, Kerry J, Morrissey P, Buckley D (2000) Effect of dietary vitamin E supplementation on cholesterol oxidation in vacuum packaged cooked beef steaks. Meat Sci 55:7–11. https://doi.org/10.1016/S0309-1740(99)00119-9

Garmyn A (2020) Consumer preferences and acceptance of meat products. Food Secur 9:708. https://doi.org/10.3390/books978-3-03943-081-9

Ge H, Fu S, Guo H, Hu M, Xu Z, Zhou X, Chen X (2022) Application and challenge of bacteriophage in the food protection. Int J Food Microbiol 380:109872. https://doi.org/10.1016/j.ijfoodmicro.2022.109872

Geletu US, Usmael MA, Mummed YY, Ibrahim AM (2021) Quality of cattle meat and its compositional constituents. Vet Med Int 2021:7340495–7340499. https://doi.org/10.1155/2021/7340495

Gil L, Barat JM, Baigts D, Martínez-Máñez R, Soto J, Garcia-Breijo E, Llobet E (2010) A potentiometric electronic tongue to monitor meat freshness. In: 2010 IEEE international symposium on industrial electronics, 390–395. Bari, Italy, IEEE, 4–7 Jul 2010

Gil L, Barat JM, Baigts D, Martínez-Máñez R, Soto J, Garcia-Breijo E, Aristoy MC, Toldrá F, Llobet E (2011) Monitoring of physical–chemical and microbiological changes in fresh pork meat under cold storage by means of a potentiometric electronic tongue. Food Chem 126:1261–1268. https://doi.org/10.1016/j.foodchem.2010.11.054

Giungato P, Rana R, Tarabella A, Tricase C (2017) Current trends in sustainability of bitcoins and related blockchain technology. Sustain For 9:2214. https://doi.org/10.3390/su9122214

Glorieux S, Goemaere O, Steen L, Fraeye I (2017) Phosphate reduction in emulsified meat products: impact of phosphate type and dosage on quality characteristics. Food Technol Biotechnol 55:390–397. https://doi.org/10.17113/ftb.55.03.17.5089

González-Martín I, Pérez-Pavón JL, González-Pérez C, Hernández-Méndez J, Álvarez-García N (2000) Differentiation of products derived from Iberian breed swine by electronic olfactometry (electronic nose). Anal Chim Acta 424:279–287. https://doi.org/10.1016/S0003-2670(00)01106-5

GS1 (2005) Beef traceability case study. In: GS1 Ireland, Dublin, Ireland. https://gs1ca.org/gs1ca-components/documents/can-trace/Beef_Traceabilty_Case_Study.pdf. Accessed 21 Feb 2024

Guan F, Jin Y-T, Zhao J, Xu A-C, Luo Y-Y (2018) A PCR method that can be further developed into PCR-RFLP assay for eight animal species identification. J Anal Methods Chem 2018: 5890140. https://doi.org/10.1155/2018/5890140

Haddi Z, El Barbri N, Tahri K, Bougrini M, El Bari N, Llobet E, Bouchikhi B (2015) Instrumental assessment of red meat origins and their storage time using electronic sensing systems. Anal Methods 7:5193–5203. https://doi.org/10.1039/C5AY00572H

Hale DS, Goodson K, Savell JW (2013) USDA beef quality and yield grades. In: Department of Animal Science Texas A&M AgriLife Extension Service College Station. https://meat.tamu.edu/beefgrading/. Accessed 20 Mac 2024

Han C, Chen Y (2018) Propagation modeling for wireless communications in the terahertz band. IEEE Commun Mag 56:96–101. https://doi.org/10.1109/MCOM.2018.1700898

Han JW, Ruiz-Garcia L, Qian JP, Yang XT (2018) Food packaging: a comprehensive review and future trends. Compr Rev Food Sci Food Saf 17:860–877. https://doi.org/10.1111/1541-4337.12343

Hassoun A, Aït-Kaddour A, Abu-Mahfouz AM, Rathod NB, Bader F, Barba FJ, Biancolillo A, Cropotova J, Galanakis CM, Jambrak AR (2023) The fourth industrial revolution in the food industry—Part I: industry 4.0 technologies. Crit Rev Food Sci Nutr 63:6547–6563. https://doi.org/10.1080/10408398.2022.2034735

Hauser C, Thielmann J, Muranyi P (2016) Organic acids: usage and potential in antimicrobial packaging. In: Antimicrobial food packaging. Academic, Elsevier, pp 563–580

Heinz G, Hautzinger P (2007) Meat processing technology. In: For small- to medium-scale producers. Food and Agriculture Organization of The United Nations Regional Office for Asia and The Pacific Bangkok. https://www.fao.org/3/ai407e/ai407e.pdf. Accessed 3 Mac 2024

Houben G, Abma P, Van den Berg H, Van Dokkum W, Van Loveren H, Penninks A, Seinen W, Spanhaak S, Vos J, Ockhuizen T (1992) Effects of the colour additive caramel colour III on the immune system: a study with human volunteers. Food Chem Toxicol 30:749–757. https://doi.org/10.1016/0278-6915(92)90076-W

Hu B, Sun D-W, Pu H, Wei Q (2020) A dynamically optical and highly stable pNIPAM@ Au NRs nanohybrid substrate for sensitive SERS detection of malachite green in fish fillet. Talanta 218: 121188. https://doi.org/10.1016/j.talanta.2020.121188

Huff-Lonergan E, Lonergan SM (2005) Mechanisms of water-holding capacity of meat: the role of postmortem biochemical and structural changes. Meat Sci 71:194–204. https://doi.org/10.1016/j.meatsci.2005.04.022

Hughes JM, Oiseth SK, Purslow PP, Warner RD (2014) A structural approach to understanding the interactions between colour, water-holding capacity and tenderness. Meat Sci 98:520–532. https://doi.org/10.1016/j.meatsci.2014.05.022

Hwang YH, Joo ST (2017) Fatty acid profiles, meat quality, and sensory palatability of grain-fed and grass-fed beef from Hanwoo, American, and Australian crossbred cattle. Korean J Food Sci Anim Resour 37:153–161. https://doi.org/10.5851/kosfa.2017.37.2.153

Ibarra JG, Tao Y, Xin H (2000) Combined IR imaging-neural network method for the estimation of internal temperature in cooked chicken meat. Opt Eng 39:3032–3038. https://doi.org/10.1117/1.1314595

Ishamri I, Nurul H, Fazilah A, Noryati I (2010) Effects of washing on the functional properties of duck meat. Int J Poult Sci 9:556–561. https://doi.org/10.3923/ijps.2010.556.561

Ishaq A, Ebner PD, Syed QA, Ur Rahman HU (2020) Employing list-shield bacteriophage as a bio-control intervention for Listeria monocytogenes from raw beef surface and maintain meat quality during refrigeration storage. LWT-Food Sci Technol 132:109784. https://doi.org/10.1016/j.lwt.2020.109784

Islam Chowdhury A, Rahman M, Sakib N (2019) A study on multiple barcode detection from an image in business system. Int J Comput Appl 181:30–37. https://doi.org/10.5120/ijca2019918340

Jaborek JR, Relling AE, Fluharty FL, Moeller SJ, Zerby HN (2020) Opportunities to improve the accuracy of the United States Department of Agriculture beef yield grade equation through precision agriculture. Transl Anim Sci 4:txaa033. https://doi.org/10.1093/tas/txaa033

Jackman P, Sun D-W, Allen P (2011) Recent advances in the use of computer vision technology in the quality assessment of fresh meats. Trends Food Sci Technol 22:185–197. https://doi.org/10.1016/j.tifs.2011.01.008

Jakes W, Gerdova A, Defernez M, Watson AD, McCallum C, Limer E, Colquhoun IJ, Williamson DC, Kemsley EK (2015) Authentication of beef versus horse meat using 60 MHz 1H NMR spectroscopy. Food Chem 175:1–9. https://doi.org/10.1016/j.foodchem.2014.11.110

JAKIM (2020) Malaysian Halal certification procedure manual. In: Department of Islamic Development Malaysia. https://www.halal.gov.my/v4/index.php?data=bW9kdWxlcy9uZXdzOzs7Ow==&utama=panduan&ids=gp4. Accessed 21 Feb 2024

Jankowiak H, Cebulska A, Bocian M (2021) The relationship between acidification (pH) and meat quality traits of polish white breed pigs. Eur Food Res Technol 247:2813–2820. https://doi.org/10.1007/s00217-021-03837-4

Jemil N, Ouerfelli M, Almajano MP, Elloumi-Mseddi J, Nasri M, Hmidet N (2020) The conservative effects of lipopeptides from Bacillus methylotrophicus DCS1 on sunflower oil-in-water emulsion and raw beef patties quality. Food Chem 303:125364. https://doi.org/10.1016/j.foodchem.2019.125364

Jiang Y, Cheng T, Wang J, Yu L (2017) Effect of ultrasonic thawing on the quality characteristics of frozen pork. Meat Res 31:14–19. https://doi.org/10.7506/rlyj1001-8123-201711003

Jiang Q, Jia R, Nakazawa N, Hu Y, Osako K, Okazaki E (2019) Changes in protein properties and tissue histology of tuna meat as affected by salting and subsequent freezing. Food Chem 271:550–560. https://doi.org/10.1016/j.foodchem.2018.07.219

Jiménez-Colmenero F, Cofrades S, Herrero AM, Fernández-Martín F, Rodríguez-Salas L, Ruiz-Capillas C (2012) Konjac gel fat analogue for use in meat products: comparison with pork fats. Food Hydrocoll 26:63–72. https://doi.org/10.1016/j.foodhyd.2011.04.007

Jo C, Cho SH, Chang J, Nam KC (2012) Keys to production and processing of Hanwoo beef: a perspective of tradition and science. Anim Front 2:32–38. https://doi.org/10.2527/af.2012-0060

Johnson J, Atkin D, Lee K, Sell M, Chandra S (2019) Determining meat freshness using electrochemistry: are we ready for the fast and furious? Meat Sci 150:40–46. https://doi.org/10.1016/j.meatsci.2018.12.002

Joo ST, Kim GD, Hwang YH, Ryu YC (2013) Control of fresh meat quality through manipulation of muscle fiber characteristics. Meat Sci 95:828–836. https://doi.org/10.1016/j.meatsci.2013.04.044

Jose A, Shanmugam P (2020) Supply chain issues in SME food sector: a systematic review. J Adv Manag Res 17:19–65. https://doi.org/10.1108/JAMR-02-2019-0010

Kafetzopoulos D, Stylios C, Skalkos D (2020) Managing traceability in the meat processing industry: principles, guidelines and technologies. In: Proceeding of the 9th international conference on information and communication technologies in agriculture, Food & Environment (HAICTA 2020)Thessaloniki, Greece, 24–27 September 2020

Kalogianni AI, Lazou T, Bossis I, Gelasakis AI (2020) Natural phenolic compounds for the control of oxidation, bacterial spoilage, and foodborne pathogens in meat. Food Secur 9:794. https://doi.org/10.3390/foods9060794

Kalpana VN, Rajeswari VD (2019) 1 – Preservatives in beverages: perception and needs. In: Grumezescu AM, Holban AM (eds) Preservatives and preservation approaches in beverages. Academic, pp 1–30

Kamilaris A, Fonts A, Prenafeta-Boldú FX (2019) The rise of blockchain technology in agriculture and food supply chains. Trends Food Sci Technol 91:640–652. https://doi.org/10.1016/j.tifs.2019.07.034

Kannur BH, Fairoze MN, Girish PS, Karabasanavar N, Rudresh BH (2017) Breed traceability of buffalo meat using microsatellite genotyping technique. J Food Sci Technol 54:558–563. https://doi.org/10.1007/s13197-017-2500-4

Karoui R, Blecker C (2011) Fluorescence spectroscopy measurement for quality assessment of food systems—a review. Food Bioprocess Technol 4:364–386. https://doi.org/10.1007/s11947-010-0370-0

Kemal Seçkin A, Yilmaz B, Tosun H (2017) Real-time PCR is a potential tool to determine the origin of milk used in cheese production. LWT–Food Sci Technol 77:332–336. https://doi.org/10.1016/j.lwt.2016.11.065

Kerry JP, O'Grady MN, Hogan SA (2006) Past, current and potential utilisation of active and intelligent packaging systems for meat and muscle-based products: a review. Meat Sci 74:113–130. https://doi.org/10.1016/j.meatsci.2006.04.024

Kim GH, Chin KB (2021) Characteristics of low-nitrite pork emulsified-sausages with paprika oleoresin solution during refrigerated storage. J Anim Sci Technol 63:394–439. https://doi.org/10.5851/kosfa.2021.e8

Kim H-Y, Lee E-S, Jeong J-Y, Choi J-H, Choi Y-S, Han D-J, Lee M-A, Kim S-Y, Kim C-J (2010) Effect of bamboo salt on the physicochemical properties of meat emulsion systems. Meat Sci 86: 960–965. https://doi.org/10.1016/j.meatsci.2010.08.001

Kim Y-A, Jung S-W, Park H-R, Chung K-Y, Lee SJ (2012) Application of a prototype of microbial time temperature indicator (TTI) to the prediction of ground beef qualities during storage. Korean J Food Sci Anim 32:448–457. https://doi.org/10.5851/kosfa.2012.32.4.448

Kim S, Kim S-H, Rahman M, Kim J (2018) Characterization of a Salmonella Enteritidis bacteriophage showing broad lytic activity against Gram-negative enteric bacteria. J Microbiol 56: 917–925. https://doi.org/10.1007/s12275-018-8310-1

Kimmel DW, LeBlanc G, Meschievitz ME, Cliffel DE (2012) Electrochemical sensors and biosensors. Anal Chem 84:685–707. https://doi.org/10.1021/ac202878q

Konarska M, Kuchida K, Tarr G, Polkinghorne RJ (2017) Relationships between marbling measures across principal muscles. Meat Sci 123:67–78. https://doi.org/10.1016/j.meatsci.2016.09.005

Korhonen HJ, Rokka S (2011) Properties and applications of antimicrobial proteins and peptides from milk and eggs. In: Bioactive food proteins and peptides, 1st edn. CRC Press, p 49

Koskella B, Meaden S (2013) Understanding bacteriophage specificity in natural microbial communities. Viruses 5:806–823. https://doi.org/10.3390/v5030806

Kuijpers TFM, Gruppen H, Sforza S, van Berkel WJH, Vincken J-P (2013) The antibrowning agent sulfite inactivates Agaricus bisporus tyrosinase through covalent modification of the copper-B site. FEBS J 280:6184–6195. https://doi.org/10.1111/febs.12539

Kušec G, Komlenić M, Gvozdanović K, Sili V, Krvavica M, Radišić Ž, Kušec ID (2022) Carcass composition and physicochemical characteristics of meat from pork chains based on native and hybrid pigs. PRO 10:370. https://doi.org/10.3390/pr10020370

Lagerstedt Å, Ahnström ML, Lundström K (2011) Vacuum skin pack of beef — a consumer friendly alternative. Meat Sci 88:391–396. https://doi.org/10.1016/j.meatsci.2011.01.015

Laliotis GP, Koutsouli P, Bizelis IA (2018) Implementation of contemporary DNA based techniques on traceability process of small ruminant species and products. J Adv Vet and Anim Res 5:255–264. https://doi.org/10.5455/javar.2018.e274

Ledesma E, Rendueles M, Díaz M (2017) Smoked food. In: Current developments in biotechnology and bioengineering. Elsevier, pp 201–243

Lee EJ, Shin HS (2019) Development of a freshness indicator for monitoring the quality of beef during storage. Food Sci Biotechnol 28:1899–1906. https://doi.org/10.1007/s10068-019-00633-5
Lee H, Kim MS, Lee W-H, Cho B-K (2018) Determination of the total volatile basic nitrogen (TVB-N) content in pork meat using hyperspectral fluorescence imaging. Sens Actuators B Chem 259:532–539. https://doi.org/10.1016/j.snb.2017.12.102
Lee K, Baek S, Kim D, Seo J (2019) A freshness indicator for monitoring chicken-breast spoilage using a Tyvek® sheet and RGB color analysis. Food Packag Shelf Life 19:40–46. https://doi.org/10.1016/j.fpsl.2018.11.016
Lell MM, Kachelriess M (2020) Recent and upcoming technological developments in computed tomography: high speed, low dose, deep learning, multienergy. Investig Radiol 55:8–19. https://doi.org/10.1097/RLI.0000000000000601
Li Z, Suslick KS (2016) Portable optoelectronic nose for monitoring meat freshness. ACS Sensors 1:1330–1335. https://doi.org/10.1021/acssensors.6b00492
Li N, Luo HD, Jia YZ, Zhou N, Li YQ (2011) Rapid determination of benzo(a)pyrene in processed meat and fish samples by second-derivative constant-energy synchronous fluorescence spectrometry. Food Addit Contam: Part A 28:235–242. https://doi.org/10.1080/19440049.2010.535857
Li X, Fu X, Yang G, Du M (2020a) Review: enhancing intramuscular fat development via targeting fibro-adipogenic progenitor cells in meat animals. Animal 14:312–321. https://doi.org/10.1017/S175173111900209X
Li X, Jiang P, Chen T, Luo X, Wen Q (2020b) A survey on the security of blockchain systems. Futur Gener Comput Syst 107:841–853. https://doi.org/10.1016/j.future.2017.08.020
Liang QI, Maocheng Z, Jie Z, Yuweiyi T (2019) Preliminary investigation of Terahertz spectroscopy to predict pork freshness non-destructively. Food Sci Tech 39:563–570. https://doi.org/10.1590/fst.25718
Lim S-C, Foster NF, Riley TV (2016) Susceptibility of Clostridium difficile to the food preservatives sodium nitrite, sodium nitrate and sodium metabisulphite. Anaerobe 37:67–71. https://doi.org/10.1016/j.anaerobe.2015.12.004
Lima AM, Azevedo MI, Sousa LM, Oliveira NS, Andrade CR, Freitas CD, Souza PF (2022) Plant antimicrobial peptides: an overview about classification, toxicity and clinical applications. Int J Biol Macromol 214:10–21. https://doi.org/10.1016/j.ijbiomac.2022.06.043
Lin Y, Ma J, Sun D-W, Cheng J-H, Wang Q (2023) A pH-Responsive colourimetric sensor array based on machine learning for real-time monitoring of beef freshness. Food Control 150: 109729. https://doi.org/10.1016/j.foodcont.2023.109729
Lippolis V, Ferrara M, Cervellieri S, Damascelli A, Epifani F, Pascale M, Perrone G (2016) Rapid prediction of ochratoxin A-producing strains of Penicillium on dry-cured meat by MOS-based electronic nose. Int J Food Microbiol 218:71–77. https://doi.org/10.1016/j.ijfoodmicro.2015.11.011
Listrat A, Lebret B, Louveau I, Astruc T, Bonnet M, Lefaucheur L, Picard B, Bugeon J (2016) How muscle structure and composition influence meat and flesh quality. Sci World J 2016:3182746–3182714. https://doi.org/10.1155/2016/3182746
Liu Q, Scheller KK, Schaefer DM, Arp SC, Williams SN (1994) Dietary α-Tocopheryl acetate contributes to lipid stability in cooked beef. J Food Sci 59:288–290. https://doi.org/10.1111/j.1365-2621.1994.tb06949.x
Liu J, Song H, Liu Y, Li P, Yao J, Xiong J (2015) Discovery of kokumi peptide from yeast extract by LC-Q-TOF-MS/MS and sensomics approach. J Sci Food Agric 95:3183–3194. https://doi.org/10.1002/jsfa.7058
Liu X, Schmidt H, Mörlein D (2016) Feasibility of boar taint classification using a portable Raman device. Meat Sci 116:133–139. https://doi.org/10.1016/j.meatsci.2016.02.015
Liu B, Xu H, Zhao H, Liu W, Zhao L, Li Y (2017) Preparation and characterization of intelligent starch/PVA films for simultaneous colorimetric indication and antimicrobial activity for food

packaging applications. Carbohydr Polym 157:842–849. https://doi.org/10.1016/j.carbpol.2016.10.067
Liu J-H, Sun X, Young J, Bachmeier L, Newman D (2018) Predicting pork loin intramuscular fat using computer vision system. Meat Sci 143:18–23. https://doi.org/10.1016/j.meatsci.2018.03.020
Liu J, Pogorzelski G, Neveu A, Legrand I, Pethick D, Ellies-Oury M-P, Hocquette J-F (2021) Are marbling and the prediction of beef eating quality affected by different grading sites? Front Vet Sci 8:611153. https://doi.org/10.3389/fvets.2021.611153
Liu J, Ellies-Oury M-P, Stoyanchev T, Hocquette J-F (2022) Consumer perception of beef quality and how to control, improve and predict it? Focus on eating quality. Foods 11:1732. https://doi.org/10.3390/foods11121732
Lonergan S, Topel D, Marple D (2019) Fresh and cured meat processing and preservation. In: Lonergan S, Topel D, Marple D (eds) The science of animal growth and meat technology. Academic, Iowa State University, pp 205–228
Loutfi A, Coradeschi S, Mani GK, Shankar P, Rayappan JBB (2015) Electronic noses for food quality: a review. J Food Eng 144:103–111. https://doi.org/10.1016/j.jfoodeng.2014.07.019
MacDonald R, Reitmeier C (2017) Chapter 6- Food processing. In: MacDonald R, Reitmeier C (eds) Understanding food systems. Academic, pp 179–225
Mashood Q, Karlsson AH, Wallenbeck A, Eriksson S, Johansson AM, Segerkvist KA (2023) A review on pale, soft, and exudative (PSE)-like zones in pork: current knowledge on underlying factors and identification of knowledge gaps for further research. Acta Agric Scand A Anim Sci 1-10:1–10. https://doi.org/10.1080/09064702.2023.2279079
Mataragas M, Bikouli VC, Korre M, Sterioti A, Skandamis PN (2019) Development of a microbial time temperature indicator for monitoring the shelf life of meat. Innov Food Sci Emerg Technol 52:89–99. https://doi.org/10.1016/j.ifset.2018.11.003
Mbandi E, Shelef LA (2001) Enhanced inhibition of Listeria monocytogenes and Salmonella enteritidis in meat by combinations of sodium lactate and diacetate. J Food Prot 64:640–644. https://doi.org/10.4315/0362-028X-64.5.640
McDermott A, Whyte P, Brunton N, Lyng J, Fagan J, Bolton D (2018) The effect of organic acid and sodium chloride dips on the shelf-life of refrigerated Irish brown crab (Cancer pagurus) meat. LWT - Food Sci Technol 98:141–147. https://doi.org/10.1016/j.lwt.2018.08.039
McGilchrist P, Polkinghorne RJ, Ball AJ, Thompson JM (2019) The meat standards Australia index indicates beef carcass quality. Animal 13:1750–1757. https://doi.org/10.1017/s1751731118003713
McMillin KW (2017) Advancements in meat packaging. Meat Sci 132:153–162. https://doi.org/10.1016/j.meatsci.2017.04.015
Michelin RA, Dorri A, Steger M, Lunardi RC, Kanhere SS, Jurdak R, Zorzo AF (2018) Speedychain: a framework for decoupling data from blockchain for smart cities. In: Proceedings of the 15th EAI international conference on mobile and ubiquitous systems: computing, networking and services, 144–154. New York, USA, ACM, 5–7 November 2018
Miklos R, Nielsen MS, Einarsdóttir H, Feidenhans R, Lametsch R (2015) Novel X-ray phase-contrast tomography method for quantitative studies of heat induced structural changes in meat. Meat Sci 100:217–221. https://doi.org/10.1016/j.meatsci.2014.10.009
Milkowski A, Garg HK, Coughlin JR, Bryan NS (2010) Nutritional epidemiology in the context of nitric oxide biology: a risk–benefit evaluation for dietary nitrite and nitrate. Nitric Oxide 22: 110–119. https://doi.org/10.1016/j.niox.2009.08.004
Millar S, Wilson R, Moss BW, Ledward DA (1994) Oxymyoglobin formation in meat and poultry. Meat Sci 36:397–406. https://doi.org/10.1016/0309-1740(94)90135-X
Mimendia A, Gutiérrez JM, Leija L, Hernández PR, Favari L, Muñoz R, del Valle M (2010) A review of the use of the potentiometric electronic tongue in the monitoring of environmental systems. Environ Model Softw 25:1023–1030. https://doi.org/10.1016/j.envsoft.2009.12.003
MLA (2018) Meat & livestock Australia. In: Meat standards Australia beef information kit. Meat & Livestock Australia Limited. https://www.mla.com.au/globalassets/mla-corporate/marketing-

beef-and-lamb/documents/meat-standards-australia/msa-beef-tt_full-info-kit-lr.pdf. Accessed 24 Mac 2024

Mohammed HHH, Jin G, Ma M, Khalifa I, Shukat R, Elkhedir AE, Zeng Q, Noman AE (2020) Comparative characterization of proximate nutritional compositions, microbial quality and safety of camel meat in relation to mutton, beef, and chicken. LWT-Food Sci Technol 118: 108714. https://doi.org/10.1016/j.lwt.2019.108714

Motoyama M, Sasaki K, Watanabe A (2016) Wagyu and the factors contributing to its beef quality: a Japanese industry overview. Meat Sci 120:10–18. https://doi.org/10.1016/j.meatsci.2016.04.026

Munekata PES, Finardi S, de Souza CK, Meinert C, Pateiro M, Hoffmann TG, Domínguez R, Bertoli SL, Kumar M, Lorenzo JM (2023) Applications of electronic nose, electronic eye and electronic tongue in quality, safety and shelf life of meat and meat products: a review. Sensors (Basel, Switzerland) 23. https://doi.org/10.3390/s23020672

Muñoz I, Gou P, Fulladosa E (2019) Computer image analysis for intramuscular fat segmentation in dry-cured ham slices using convolutional neural networks. Food Control 106:106693. https://doi.org/10.1016/j.foodcont.2019.06.019

Najar-Villarreal F, Boyle EAE, Danler RD, O'Quinn TG, Houser TA, Gonzalez JM (2019) Fatty acid composition, proximate analysis, and consumer sensory evaluation of United States retail grass-fed ground beef. Meat Muscle Biol 3:389–398. https://doi.org/10.22175/mmb2019.06.0018

Nanni Costa L, Stelletta C, Cannizzo C, Gianesella M, Lo Fiego D, Morgante M (2016) The use of thermography on the slaughter-line for the assessment of pork and raw ham quality. Ital J Anim Sci 6:704–706. https://doi.org/10.4081/ijas.2007.1s.704

Navajas EA, Glasbey CA, Fisher AV, Ross DW, Hyslop JJ, Richardson RI, Simm G, Roehe R (2010) Assessing beef carcass tissue weights using computed tomography spirals of primal cuts. Meat Sci 84:30–38. https://doi.org/10.1016/j.meatsci.2009.08.006

Nazzaro F, Fratianni F, De Martino L, Coppola R, De Feo V (2013) Effect of essential oils on pathogenic bacteria. Pharm 6:1451–1474. https://doi.org/10.3390/ph6121451

Nie W, Chen Y, Zhang H, Liu J, Peng Z, Li Y (2022) A novel colorimetric sensor array for real-time and on-site monitoring of meat freshness. Anal Bioanal Chem 414:6017–6027. https://doi.org/10.1007/s00216-022-04176-3

Nogoy KMC, Sun B, Shin S, Lee Y, Zi Li X, Choi SH, Park S (2022) Fatty acid composition of grain- and grass-fed beef and their nutritional value and health implication. Food Sci Anim Resour J 42(1):18–33. https://doi.org/10.5851/kosfa.2021.e73

Nurjuliana M, Che Man YB, Mat Hashim D, Mohamed AKS (2011) Rapid identification of pork for halal authentication using the electronic nose and gas chromatography mass spectrometer with headspace analyzer. Meat Sci 88:638–644. https://doi.org/10.1016/j.meatsci.2011.02.022

Nychas G-JE, Skandamis PN, Tassou CC, Koutsoumanis KP (2008) Meat spoilage during distribution. Meat Sci 78:77–89. https://doi.org/10.1016/j.meatsci.2007.06.020

Ofir G, Sorek R (2018) Contemporary phage biology: from classic models to new insights. Cell 172:1260–1270. https://doi.org/10.1016/j.cell.2017.10.045

Oliveira SM, Lopes TIMS, Rangel AOSS (2004) Spectrophotometric determination of nitrite and nitrate in cured meat by sequential injection analysis. J Food Sci 69:C690–C695. https://doi.org/10.1111/j.1365-2621.2004.tb09917.x

Osen DE, Abie SM, Martinsen ØG, Egelandsdal B, Münch D (2022) Bioimpedance-based authentication of defrosted versus fresh pork at the end of refrigerated shelf life. J Electr Bioimpedance 13:125–131. https://doi.org/10.2478/joeb-2022-0017

Oswell NJ, Thippareddi H, Pegg RB (2018) Practical use of natural antioxidants in meat products in the US: a review. Meat Sci 145:469–479. https://doi.org/10.1016/j.meatsci.2018.07.020

Papadopoulou OS, Panagou EZ, Mohareb FR, Nychas G-JE (2013) Sensory and microbiological quality assessment of beef fillets using a portable electronic nose in tandem with support vector machine analysis. Food Res Int 50:241–249. https://doi.org/10.1016/j.foodres.2012.10.020

Park SJ, Beak SH, Jung DJS, Kim SY, Jeong IH, Piao MY, Kang HJ, Fassah DM, Na SW, Yoo SP, Baik M (2018) Genetic, management, and nutritional factors affecting intramuscular fat deposition in beef cattle – a review. Asian Australas J Anim Sci 31:1043–1061. https://doi.org/10.5713/ajas.18.0310

Park H, Kim J, Kim H, Cho E, Park H, Jeon B, Ryu S (2023) Characterization of the lytic phage MSP1 for the inhibition of multidrug-resistant Salmonella enterica serovars Thompson and its biofilm. Int J Food Microbiol 385:110010. https://doi.org/10.1016/j.ijfoodmicro.2022.110010

Parlasca MC, Qaim M (2022) Meat consumption and sustainability. Annu Rev Resour Econ 14:17–41. https://doi.org/10.1146/annurev-resource-111820-032340

Parthasarathy DK, Bryan NS (2012) Sodium nitrite: the "cure" for nitric oxide insufficiency. Meat Sci 92:274–279. https://doi.org/10.1016/j.meatsci.2012.03.001

Patel AS, Brahmbhatt MN, Bariya AR, Nayak JB, Singh VK (2023) Blockchain technology in food safety and traceability concern to livestock products. Heliyon 9:e16526. https://doi.org/10.1016/j.heliyon.2023.e16526

Pereira FMV, Colnago LA (2012) Determination of the moisture content in beef without weighing using benchtop time-domain nuclear magnetic resonance spectrometer and chemometrics. Food Anal Methods 5:1349–1353. https://doi.org/10.1007/s12161-012-9383-9

Petcu CD, Mihai OD, Tăpăloagă D, Gheorghe-Irimia R-A, Pogurschi EN, Militaru M, Borda C, Ghimpețeanu O-M (2023) Effects of plant-based antioxidants in animal diets and meat products: a review. Food Secur 12:1334. https://doi.org/10.3390/foods12061334

Pewan SB, Otto JR, Huerlimann R, Budd AM, Mwangi FW, Edmunds RC, Holman BWB, Henry MLE, Kinobe RT, Adegboye OA, Malau-Aduli AEO (2020) Genetics of Omega-3 long-chain polyunsaturated fatty acid metabolism and meat eating quality in Tattykeel Australian white lambs. Genes 11:587. https://doi.org/10.3390/genes11050587

Phelan M, Aherne A, FitzGerald RJ, O'Brien NM (2009) Casein-derived bioactive peptides: biological effects, industrial uses, safety aspects and regulatory status. Int Dairy J 19:643–654. https://doi.org/10.1016/j.idairyj.2009.06.001

Pilz M, Cavelius P, Qoura F, Awad D, Brück T (2023) Lipopeptides development in cosmetics and pharmaceutical applications: a comprehensive review. Biotechnol Adv 67:108210. https://doi.org/10.1016/j.biotechadv.2023.108210

Pintado T, Ruiz-Capillas C, Jiménez-Colmenero F, Herrero AM (2020) Impact of culinary procedures on nutritional and technological properties of reduced-fat longanizas formulated with chia (Salvia hispanica L.) or Oat (Avena sativa L.) emulsion gel. Food Secur 9. https://doi.org/10.3390/foods9121847

Pokorný J, Yanishlieva N, Gordon M (2001) Antioxidants in food. Woodhead Publishing Limited, Cambridge

Ponnampalam EN, Hopkins DL, Bruce H, Li D, Baldi G, Bekhit AE-d (2017) Causes and contributing factors to "dark cutting" meat: Current trends and future directions: a review. Compr Rev Food Sci Food Saf 16:400–430. https://doi.org/10.1111/1541-4337.12258

Ponnampalam EN, Kiani A, Santhiravel S, Holman BWB, Lauridsen C, Dunshea FR (2022) The importance of dietary antioxidants on oxidative stress, meat and milk production, and their preservative aspects in farm animals: antioxidant action, animal health, and product quality-invited review. Anim 12:3279. https://doi.org/10.3390/ani12233279

Poveda-Arteaga A, Krell J, Gibis M, Heinz V, Terjung N, Tomasevic I (2023) Intrinsic and extrinsic factors affecting the color of fresh beef meat- comprehensive review. Appl Sci 13:4382. https://doi.org/10.3390/app13074382

Prache S, Schreurs N, Guillier L (2022) Factors affecting sheep carcass and meat quality attributes. Animal 16:100330. https://doi.org/10.1016/j.animal.2021.100330

Prieto N, Navajas EA, Richardson RI, Ross DW, Hyslop JJ, Simm G, Roehe R (2010) Predicting beef cuts composition, fatty acids and meat quality characteristics by spiral computed tomography. Meat Sci 86:770–779. https://doi.org/10.1016/j.meatsci.2010.06.020

Prusakova OV, Glukhova XA, Afanas'eva GV, Trizna YA, Nazarova LF, Beletsky IP (2018) A simple and sensitive two-tube multiplex PCR assay for simultaneous detection of ten meat species. Meat Sci 137:34–40. https://doi.org/10.1016/j.meatsci.2017.10.017
Puolanne E, Halonen M (2010) Theoretical aspects of water-holding in meat. Meat Sci 86:151–165. https://doi.org/10.1016/j.meatsci.2010.04.038
Qiao L, Tang X, Dong J (2017) A feasibility quantification study of total volatile basic nitrogen (TVB-N) content in duck meat for freshness evaluation. Food Chem 237:1179–1185. https://doi.org/10.1016/j.foodchem.2017.06.031
Rabea EI, Badawy MET, Stevens CV, Smagghe G, Steurbaut W (2003) Chitosan as antimicrobial agent: applications and mode of action. Biomacromolecules 4:1457–1465. https://doi.org/10.1021/bm034130m
Radanović I, Likić R (2018) Opportunities for use of blockchain technology in medicine. Appl Health Econ Health Policy 16:583–590. https://doi.org/10.1007/s40258-018-0412-8
Regulation EC (2008) No 1333/2008 of the European Parliament and of the Council of 16 December 2008 on food additives. Off J Eur Union 354:16–33. http://data.europa.eu/eli/reg/2008/1333/oj
Rejeb A, Rejeb K, Zailani S, Treiblmaier H, Hand KJ (2021) Integrating the internet of things in the halal food supply chain: a systematic literature review and research agenda. Internet of Things 13:100361. https://doi.org/10.1016/j.iot.2021.100361
Roldán M, Antequera T, Pérez-Palacios T, Ruiz J (2014) Effect of added phosphate and type of cooking method on physico-chemical and sensory features of cooked lamb loins. Meat Sci 97: 69–75. https://doi.org/10.1016/j.meatsci.2014.01.012
Ruedt C, Gibis M, Weiss J (2023) Meat color and iridescence: origin, analysis, and approaches to modulation. Compr Rev Food Sci Food Saf 22:3366–3394. https://doi.org/10.1111/1541-4337.13191
Ruslan NN, Tang JYH, Huda N, Ismail-Fitry MR, Ishamri I (2023) Effects of phosphate and two-stage sous-vide cooking on textural properties of the beef semitendinosus. Food Sci Anim Resour 43:491–501. https://doi.org/10.5851/kosfa.2023.e11
Sadeghi S, Fooladi E, Malekaneh M (2014) A nanocomposite/crude extract enzyme-based xanthine biosensor. Anal Biochem 464:51–59. https://doi.org/10.1016/j.ab.2014.07.013
Sahar A, Dufour É (2015) Classification and characterization of beef muscles using front-face fluorescence spectroscopy. Meat Sci 100:69–72. https://doi.org/10.1016/j.meatsci.2014.09.142
Sahar A, Boubellouta T, Portanguen S, Kondjoyan A, Dufour É (2009) Synchronous front-face fluorescence spectroscopy coupled with parallel factors (PARAFAC) analysis to study the effects of cooking time on meat. J Food Sci 74:E534–E539. https://doi.org/10.1111/j.1750-3841.2009.01365.x
Sahar A, Portanguen S, Kondjoyan A, Dufour É (2010) Potential of synchronous fluorescence spectroscopy coupled with chemometrics to determine the heterocyclic aromatic amines in grilled meat. Eur Food Res Technol 231:803–812. https://doi.org/10.1007/s00217-010-1323-6
Sahar A, Rahman U, Kondjoyan A, Portanguen S, Dufour E (2016) Monitoring of thermal changes in meat by synchronous fluorescence spectroscopy. J Food Eng 168:160–165. https://doi.org/10.1016/j.jfoodeng.2015.07.038
Salinas Y, Ros-Lis JV, Vivancos J-L, Martínez-Máñez R, Marcos MD, Aucejo S, Herranz N, Lorente I (2012) Monitoring of chicken meat freshness by means of a colorimetric sensor array. Analyst 137:3635–3643. https://doi.org/10.1039/C2AN35211G
Salinas Y, Ros-Lis JV, Vivancos J-L, Martínez-Máñez R, Marcos MD, Aucejo S, Herranz N, Lorente I, Garcia E (2014) A novel colorimetric sensor array for monitoring fresh pork sausages spoilage. Food Control 35:166–176. https://doi.org/10.1016/j.foodcont.2013.06.043
Scannell AGM, Schwarz G, Hill C, Ross RP, Arendt EK (2001) Pre-inoculation enrichment procedure enhances the performance of bacteriocinogenic Lactococcus lactis meat starter culture. Int J Food Microbiol 64:151–159. https://doi.org/10.1016/S0168-1605(00)00455-4
Schaefer AL, Cook NJ, Bench C, Chabot JB, Colyn J, Liu T, Okine EK, Stewart M, Webster JR (2012) The non-invasive and automated detection of bovine respiratory disease onset in receiver

calves using infrared thermography. Res Vet Sci 93:928–935. https://doi.org/10.1016/j.rvsc.2011.09.021
Schirone M, Esposito L, D'Onofrio F, Visciano P, Martuscelli M, Mastrocola D, Paparella A (2022) Biogenic amines in meat and meat products: a review of the science and future perspectives. Food Secur 11. https://doi.org/10.3390/foods11060788
Schmidt BV, Moreno MS (2022) Traceability optimization in the meat supply chain with economic and environmental considerations. Comput Ind Eng 169:108271. https://doi.org/10.1016/j.cie.2022.108271
Schmidt H, Scheier R, Hopkins DL (2013) Preliminary investigation on the relationship of Raman spectra of sheep meat with shear force and cooking loss. Meat Sci 93:138–143. https://doi.org/10.1016/j.meatsci.2012.08.019
Sebranek J, Sewalt V, Robbins K, Houser T (2005) Comparison of a natural rosemary extract and BHA/BHT for relative antioxidant effectiveness in pork sausage. Meat Sci 69:289–296. https://doi.org/10.1016/j.meatsci.2004.07.010
Seddaoui N, Amine A (2021) Smartphone-based competitive immunoassay for quantitative on-site detection of meat adulteration. Talanta 230:122346. https://doi.org/10.1016/j.talanta.2021.122346
Shabir G, Anwar F, Sultana B, Khalid ZM, Afzal M, Khan QM, Ashrafuzzaman M (2011) Antioxidant and antimicrobial attributes and phenolics of different solvent extracts from leaves, flowers and bark of Gold Mohar [Delonix regia (Bojer ex Hook.) Raf.]. Molecules 16:7302–7319. https://doi.org/10.3390/molecules16097302
Shakil MH, Trisha AT, Rahman M, Talukdar S, Kobun R, Huda N, Zzaman W (2022) Nitrites in cured meats, health risk issues, alternatives to nitrites: a review. Food Secur 11:3355. https://doi.org/10.3390/foods11213355
Shi Y, Li Z, Shi J, Zhang F, Zhou X, Li Y, Holmes M, Zhang W, Zou X (2018) Titanium dioxide-polyaniline/silk fibroin microfiber sensor for pork freshness evaluation. Sens Actuators B Chem 260:465–474. https://doi.org/10.1016/j.snb.2018.01.078
Shi Y, Wang X, Borhan MS, Young J, Newman D, Berg E, Sun X (2021) A review on meat quality evaluation methods based on non-destructive computer vision and artificial intelligence technologies. Food Sci Anim Resour 41:563–588. https://doi.org/10.5851/kosfa.2021.e25
Sikorski ZE, Sinkiewicz I (2014) Smoking traditional. In: Devine C, Dikeman M (eds) Encyclopedia of meat sciences. Elsevier, pp 321–327
Silva MM, Reboredo FH, Lidon FC (2022) Food colour additives: a synoptical overview on their chemical properties, applications in food products, and health side effects. Food Secur 11:379. https://doi.org/10.3390/foods11030379
Simon J, Kroneck PMH (2013) Chapter two – Microbial sulfite respiration. In: Poole RK (ed) Advances in microbial physiology, 62. Academic, pp 45–117
Singh H, Singh G, Kaur N, Singh N (2022) Pattern-based colorimetric sensor array to monitor food spoilage using automated high-throughput analysis. Biosens Bioelectron 196:113687. https://doi.org/10.1016/j.bios.2021.113687
Skibsted LH (2011) Nitric oxide and quality and safety of muscle based foods. Nitric Oxide 24:176–183. https://doi.org/10.1016/j.niox.2011.03.307
Song B-Y, Zhang M, Teng Y, Zhang X-F, Deng Z-P, Huo L-H, Gao S (2020) Highly selective ppb-level H2S sensor for spendable detection of exhaled biomarker and pork freshness at low temperature: mesoporous SnO2 hierarchical architectures derived from waste scallion root. Sens Actuators B Chem 307:127662. https://doi.org/10.1016/j.snb.2020.127662
Stanisławczyk R, Rudy M, Gil M, Duma-Kocan P, Żurek J (2021) Influence of horse age, marinating substances, and frozen storage on horse meat quality. Anim 11. https://doi.org/10.3390/ani11092666
Sulieman AME, Abdallah EM, Alanazi NA, Ed-Dra A, Jamal A, Idriss H, Alshammari AS, Shommo SAM (2023) Spices as sustainable food preservatives: a comprehensive review of their antimicrobial potential. Pharm 16:1451. https://doi.org/10.3390/ph16101451

Suman SP, Joseph P (2013) Myoglobin chemistry and meat color. Annu Rev Food Sci Technol 4: 79–99. https://doi.org/10.1146/annurev-food-030212-182623
Suman SP, Nair MN, Joseph P, Hunt MC (2016) Factors influencing internal color of cooked meats. Meat Sci 120:133–144. https://doi.org/10.1016/j.meatsci.2016.04.006
Sun XD, Arntfield SD (2012) Gelation properties of myofibrillar/pea protein mixtures induced by transglutaminase crosslinking. Food Hydrocoll 27:394–400. https://doi.org/10.1016/j.foodhyd.2011.11.001
Sun X, Young J, Liu JH, Bachmeier L, Somers RM, Chen KJ, Newman D (2016) Prediction of pork color attributes using computer vision system. Meat Sci 113:62–64. https://doi.org/10.1016/j.meatsci.2015.11.009
Sun X, Young J, Liu J-H, Newman D (2018) Prediction of pork loin quality using online computer vision system and artificial intelligence model. Meat Sci 140:72–77. https://doi.org/10.1016/j.meatsci.2018.03.005
Surendran Nair M, Nair DVT, Kollanoor Johny A, Venkitanarayanan K (2020) Chapter 12 – Use of food preservatives and additives in meat and their detection techniques. In: Biswas AK, Mandal PK (eds) Meat quality analysis. Academic, pp 187–213
Szwajkowska M, Wolanciuk A, Barlowska J, Król J, Litwińczuk Z (2011) Bovine milk proteins as the source of bioactive peptides influencing the consumers' immune system-a review. Anim Sci Paper Rep 29:269–280
Taheri-Garavand A, Fatahi S, Omid M, Makino Y (2019) Meat quality evaluation based on computer vision technique: a review. Meat Sci 156:183–195. https://doi.org/10.1016/j.meatsci.2019.06.002
Tam JP, Wang S, Wong KH, Tan WL (2015) Antimicrobial peptides from plants. Pharm 8:711–757. https://doi.org/10.3390/ph8040711
Tan J, Lu X, Fu L, Yang G, Chen J (2019) Quantification of Cl-PAHs and their parent compounds in fish by improved ASE method and stable isotope dilution GC-MS. Ecotoxicol Environ Saf 186:109775. https://doi.org/10.1016/j.ecoenv.2019.109775
Tang X, Yu Z (2020) Rapid evaluation of chicken meat freshness using gas sensor array and signal analysis considering total volatile basic nitrogen. Int J Food Prop 23:297–305. https://doi.org/10.1080/10942912.2020.1716797
Teng Y, Chen C, Zhou C, Zhao H, Lan M (2010) Disposable amperometric biosensors based on xanthine oxidase immobilized in the Prussian blue modified screen-printed three-electrode system. Sci China Chem 53:2581–2586. https://doi.org/10.1007/s11426-010-4038-4
Terlouw EMC, Picard B, Deiss V, Berri C, Hocquette JF, Lebret B, Lefèvre F, Hamill R, Gagaoua M (2021) Understanding the determination of meat quality using biochemical characteristics of the muscle: stress at slaughter and other missing keys. Food Secur 10. https://doi.org/10.3390/foods10010084
Theron MM, Lues JF (2007) Organic acids and meat preservation: a review. Food Rev Int 23:141–158. https://doi.org/10.1080/87559120701224964
Tikk K, Haugen J-E, Andersen HJ, Aaslyng MD (2008) Monitoring of warmed-over flavour in pork using the electronic nose—correlation to sensory attributes and secondary lipid oxidation products. Meat Sci 80:1254–1263. https://doi.org/10.1016/j.meatsci.2008.05.040
Tompkin RB, Bedale W, Milkowski A, Glass K, Sindelar JJ (2020) Nitrite. In: Davidson PM, Taylor TM, David JRD (eds) Antimicrobials in food, 4th edn. CRC Press, Boca Raton, pp 169–236
Torres AC, Ghica ME, Brett CM (2013) Design of a new hypoxanthine biosensor: xanthine oxidase modified carbon film and multi-walled carbon nanotube/carbon film electrodes. Anal Bioanal Chem 405:3813–3822. https://doi.org/10.1007/s00216-012-6631-1
Trautwein C, Friedman SL, Schuppan D, Pinzani M (2015) Hepatic fibrosis: concept to treatment. J Hepatol 62:S15–S24. https://doi.org/10.1016/j.jhep.2015.02.039
Triki M, Herrero AM, Jiménez-Colmenero F, Ruiz-Capillas C (2018) Quality assessment of fresh meat from several species based on free amino acid and biogenic amine contents during chilled storage. Food Secur 7:132. https://doi.org/10.3390/foods7090132

Tripathi AK, Akul Krishnan K, Pandey AC (2023) A novel blockchain and internet of things-based food traceability system for smart cities. Wirel Pers Commun 129:2157–2180. https://doi.org/10.1007/s11277-023-10230-9

Turnbull D, Jack MM, Coder PS, Picut CA, Rodricks JV (2021) Extended One-Generation Reproductive Toxicity (EOGRT) study of benzoic acid in Sprague Dawley rats. Regul Toxicol Pharmacol 122:104897. https://doi.org/10.1016/j.yrtph.2021.104897

Ursachi CŞ, Perţa-Crişan S, Munteanu F-D (2020) Strategies to improve meat products' quality. Food Secur 9:1883. https://doi.org/10.3390/foods9121883

van Boekel MAJS (2006) Formation of flavour compounds in the Maillard reaction. Biotechnol Adv 24:230–233. https://doi.org/10.1016/j.biotechadv.2005.11.004

Vanegas DC, Gomes C, McLamore ES (2014) Xanthine oxidase biosensor for monitoring meat spoilage. In: Cullum BM, McLamore ES (eds) Proceedings smart biomedical and physiological sensor technology XI, 91070V. Baltimore MD, USA, 22 May 2014

Vasconcelos H, de Almeida JM, Matias A, Saraiva C, Jorge PA, Coelho LC (2021) Detection of biogenic amines in several foods with different sample treatments: an overview. Trends Food Sci Technol 113:86–96. https://doi.org/10.1016/j.tifs.2021.04.043

Vestergaard JS, Haugen J-E, Byrne DV (2006) Application of an electronic nose for measurements of boar taint in entire male pigs. Meat Sci 74:564–577. https://doi.org/10.1016/j.meatsci.2006.05.005

Wang J, Zhou C, Wang C, Zheng Y, Ye K, Tang C, Zhou G (2019) The application of ultrasonic technology in meat processing and its influence on flavor precursors. Sci Technol Food Ind 40:320–323. https://doi.org/10.13386/j.issn1002-0306.2019.16.053

Wang Y, Domínguez R, Lorenzo JM, Bohrer BM (2021a) The relationship between lipid content in ground beef patties with rate of discoloration and lipid oxidation during simulated retail display. Food Secur 10:1982. https://doi.org/10.3390/foods10091982

Wang Y, Liu Y, Cui Q, Li L, Ning J, Zhang Z (2021b) Monitoring the withering condition of leaves during black tea processing via the fusion of electronic eye (E-eye), colorimetric sensing array (CSA), and micro-near-infrared spectroscopy (NIRS). J Food Eng 300:110534. https://doi.org/10.1016/j.jfoodeng.2021.110534

Wass S (2017) Food companies unite to advance blockchain for supply chain traceability. In: Global Trade Review. 22. https://www.gtreview.com/news/fintech/food-companies-unite-to-advance-blockchain-for-supply-chain-traceability/. Accessed 4 Mac 2024

Watanabe G, Motoyama M, Nakajima I, Sasaki K (2018) Relationship between water-holding capacity and intramuscular fat content in Japanese commercial pork loin. Asian Australas J Anim Sci 31:914–918. https://doi.org/10.5713/ajas.17.0640

Whitton C, Bogueva D, Marinova D, Phillips CJC (2021) Are we approaching peak meat consumption? analysis of meat consumption from 2000 to 2019 in 35 countries and its relationship to gross domestic product. Animals 11:3466. https://doi.org/10.3390/ani11123466

Wu D, Sun D-W (2013) Colour measurements by computer vision for food quality control—a review. Trends Food Sci Technol 29:5–20. https://doi.org/10.1016/j.tifs.2012.08.004

Wu D, Sun D-W, He Y (2012) Application of long-wave near infrared hyperspectral imaging for measurement of color distribution in salmon fillet. Innov Food Sci Emerg Technol 16:361–372. https://doi.org/10.1016/j.ifset.2012.08.003

Wu S, Han J, Liang R, Dong P, Zhu L, Hopkins DL, Zhang Y, Luo X (2020) Investigation of muscle-specific beef color stability at different ultimate pHs. Asian Australas J Anim Sci 33:1999–2007. https://doi.org/10.5713/ajas.19.0943

Wu X, Liang X, Wang Y, Wu B, Sun J (2022) Non-destructive techniques for the analysis and evaluation of meat quality and safety: a review. Food Secur 11:3713. https://doi.org/10.3390/foods11223713

Wudhikarn R, Charoenkwan P, Malang K (2022) Deep learning in barcode recognition: a systematic literature review. IEEE 10:8049–8072. https://doi.org/10.1109/ACCESS.2022.3143033

Xin TP, Kasim N, Zainal R, Musa SMS, Noh HM (2021) Study on barcoding implementation for improving materials tracking at construction site. Res Manage Technol Bus 2:722–733. https://doi.org/10.30880/rmtb2021.02.01.056

Xiong Z, Xie A, Sun D-W, Zeng X-A, Liu D (2015) Applications of hyperspectral imaging in chicken meat safety and quality detection and evaluation: a review. Crit Rev Food Sci Nutr 55: 1287–1301. https://doi.org/10.1080/10408398.2013.834875

Xiong Z, Sun D-W, Pu H, Gao W, Dai Q (2017) Applications of emerging imaging techniques for meat quality and safety detection and evaluation: a review. Crit Rev Food Sci Nutr 57:755–768. https://doi.org/10.1080/10408398.2014.954282

Xu MM, Kaur M, Pillidge CJ, Torley PJ (2022a) Microbial biopreservatives for controlling the spoilage of beef and lamb meat: their application and effects on meat quality. Crit Rev Food Sci Nutr 62:4571–4592. https://doi.org/10.1080/10408398.2021.1877108

Xu W, He Y, Li J, Deng Y, Xu E, Feng J, Ding T, Liu D, Wang W (2022b) Non-destructive determination of beef freshness based on colorimetric sensor array and multivariate analysis. Sens Actuators B Chem 369:132282. https://doi.org/10.1016/j.snb.2022.132282

Xu Y, Li X, Zeng X, Cao J, Jiang W (2022c) Application of blockchain technology in food safety control: current trends and future prospects. Crit Rev Food Sci Nutr 62:2800–2819. https://doi.org/10.1080/10408398.2020.1858752

Yaacob TZ, Abd-Razak SS, Ishak SZ (2021) An overview of halal critical control point risk management plan (HCCPRMP) in retail distribution of halal frozen meat supply chain. In: 3rd international halal management conference, 122–126. Universiti Sains Islam Malaysia, USIM, 23–24 Aug 2021

Yang Y, Zheng Y, Ma W, Zhang Y, Sun C, Fang Y (2023) Meat and plant-based meat analogs: nutritional profile and in vitro digestion comparison. Food Hydrocoll 143:108886. https://doi.org/10.1016/j.foodhyd.2023.108886

Yin X, Chen Q, Liu Q, Wang Y, Kong B (2021) Influences of smoking in traditional and industrial conditions on flavour profile of harbin red sausages by comprehensive two-dimensional gas chromatography mass spectrometry. Food Secur 10:1180. https://doi.org/10.3390/foods10061180

YongXia C, RuiXin L, ZhaoZhou L, PengJu C, LiLi W, YanLi W, SuiQing C (2019) Quality evaluation based on color grading: quality discrimination of the Chinese medicine Corni Fructus by an E-eye. Sci Rep 9:17006. https://doi.org/10.1038/s41598-019-53210-5

Yordanov D, Angelova G (2006) Identification and traceability of meat and meat products. Biotechnol Biotechnol Equip 20:3–8. https://doi.org/10.1080/13102818.2006.10817295Yoshida

Yoshida CMP, Maciel VBV, Mendonça MED, Franco TT (2014) Chitosan biobased and intelligent films: monitoring pH variations. LWT-Food Sci Technol 55:83–89. https://doi.org/10.1016/j.lwt.2013.09.015

Yu HH, Chin Y-W, Paik H-D (2021) Application of natural preservatives for meat and meat products against food-borne pathogens and spoilage bacteria: a review. Food Secur 10:2418. https://doi.org/10.3390/foods10102418

Załubska K, Klich D, Jackowska-Tracz A, Didkowska A, Bogdan J, Anusz K (2023) Dyes used in processed meat products in the polish market, and their possible risks and benefits for consumer health. Food Secur 12:2610. https://doi.org/10.3390/foods12132610

Zarringhalami S, Sahari M, Hamidi-Esfehani Z (2009) Partial replacement of nitrite by annatto as a colour additive in sausage. Meat Sci 81:281–284. https://doi.org/10.1016/j.meatsci.2008.08.003

Zaukuu JLZ, Bazar G, Gillay Z, Kovacs Z (2020) Emerging trends of advanced sensor based instruments for meat, poultry and fish quality– a review. Crit Rev Food Sci Nutr 60:3443–3460. https://doi.org/10.1080/10408398.2019.1691972

Zeece M (2020) Chapter Seven – Food additives. In: Zeece M (ed) Introduction to the chemistry of food. Academic, pp 251–311

Zhang Xin ZX, Song Lei SL, Gao Tian GT, Zhang Lin ZL, Jiang Yun JY, Li JiaoLong LJ, Gao Feng GF, Zhou GuangHong ZG (2018) Effect of ultrasonic thawing method on quality characteristics of chicken breast meat. Food Sci 39:135–140. https://doi.org/10.7506/spkx1002-6630-201805021

Zhang W, Xiao S, Samaraweera H, Lee EJ, Ahn DU (2010) Improving functional value of meat products. Meat Sci 86:15–31. https://doi.org/10.1016/j.meatsci.2010.04.018

Zhang MX, Wang XC, Liu Y, Xu XL, Zhou GH (2012) Isolation and identification of flavour peptides from Puffer fish (Takifugu obscurus) muscle using an electronic tongue and MALDI-TOF/TOF MS/MS. Food Chem 135:1463–1470. https://doi.org/10.1016/j.foodchem.2012.06.026

Zhang Z, Zhang L, Liu Y, Wang H, Yu C, Zeng H, Wang L-m, Xu B (2018) Interface-engineered Li7La3Zr2O12-based garnet solid electrolytes with suppressed Li-Dendrite formation and enhanced electrochemical performance. ChemSusChem 11:3774–3782. https://doi.org/10.1002/cssc.201801756

Zhang Y, Kim YHB, Puolanne E, Ertbjerg P (2022) Role of freezing-induced myofibrillar protein denaturation in the generation of thaw loss: a review. Meat Sci 190:108841. https://doi.org/10.1016/j.meatsci.2022.108841

Zhou B, Ye Q, Li F, Xiang X, Shang Y, Wang C, Shao Y, Xue L, Zhang J, Wang J, Ding Y, Chen M, Wu Q (2022a) CRISPR/Cas12a based fluorescence-enhanced lateral flow biosensor for detection of Staphylococcus aureus. Sens Actuators B Chem 351:130906. https://doi.org/10.1016/j.snb.2021.130906

Zhou Y, Li L, Han K, Wang L, Cao Y, Ma D, Wang X (2022b) A polyvalent broad-spectrum Escherichia phage Tequatrovirus EP01 capable of controlling Salmonella and Escherichia coli contamination in foods. Viruses 14:286. https://doi.org/10.3390/v14020286

Zhuang Q, Peng Y, Nie S, Guo Q, Li Y, Zuo J, Chen Y (2023) Non-destructive detection of frozen pork freshness based on portable fluorescence spectroscopy. J Food Compos Anal 118:105175. https://doi.org/10.1016/j.jfca.2023.105175

Chapter 18
Innovation in Sensory Assessment of Processed Meat Products

Tina Nurkhoeriyati and Nurul Huda

18.1 Introduction

In spite of negative insights and attitudes towards animal-source food, mainly processed meat, the global mean intake of processed meat increased by 152.8% between 1990 and 2018 (Miller et al. 2022). On the one hand, meat and meat products have a significant amount of protein (Pires et al. 2019), vitamins, and minerals (Jimènez-Colmenero et al. 2001) an excellent nutrition source for the general population. On the other hand, there are health concerns, such as fat, saturated fatty acid, cholesterol, and salt content in meat and meat products (Jimènez-Colmenero et al. 2001; Polizer Rocha et al. 2019). As a result, some researchers have developed new processed meat products with improved nutritional quality to address those issues (Pires et al. 2019; Barros et al. 2019; Paglarini et al. 2020). Developing new products like meat products requires sensory methods and consumer research.

Prinyawiwatkul et al. (2023) mentioned that the Institute of Food Technologists (IFT) defined sensory evaluation as follows:

> A scientific discipline used to evoke, measure, analyze, and interpret reactions to those characteristics of food and materials as they are perceived by the senses of sight, smell, taste, touch, and hearing. (Anonymous, 1975, quoted in Prinyawiwatkul et al. (2023)

T. Nurkhoeriyati
Food Technology Department, Faculty of Engineering, Bina Nusantara University, Jakarta, Indonesia
e-mail: tina.nurkhoeriyati@binus.ac.id

N. Huda (✉)
Livestock Production Department, Faculty of Sustainable Agriculture, Universiti Malaysia Sabah, Sandakan, Malaysia
e-mail: drnurulhuda@ums.edu.my

S. A. Rather, F. A. Masoodi (eds.), *Hand Book of Processed Functional Meat Products*, https://doi.org/10.1007/978-3-031-69868-2_18

Regarding important sensory attributes of meat and meat products, Cardona et al. (2023) found that taste or flavor, texture, appearance, color, and odor were the most significant intrinsic attributes affecting the products' acceptance.

This writing aims to overview novel sensory evaluation techniques applied to processed meat products. The methods discussed cover methods of sensory assessments in general, quick-fast descriptive and dynamic sensory techniques, and emotions in sensory meat science.

18.2 Methods of Sensory Assessments

Sensory methods can be divided into analytical (discrimination and descriptive tests) and affective (preferences and hedonic) methods (Aminah Abdullah 2000; Ruiz-Capillas et al. 2021). The purpose of discrimination tests or comparison tests is to register the discrimination abilities of sensory assessors. Meanwhile, descriptive tests assess the verbal expressiveness of sensory assessors' sensory perception. Last, among affective tests, preference tests and hedonic tests are performed to evaluate the acceptance of the product (DLG Sensory Analysis Committee 2017; Ruiz-Capillas et al. 2021). Summarized conventional and novel techniques can be seen in Table 18.1.

Traditional methods, however, have their drawbacks and sensory scientists work to improve them or develop new ones (Torrico 2021). For instance, the 'quick' or 'short-term methods' as alternatives to the QDA® or Spectrum™ and consensus profiling assessments (Schneider-Häder et al. 2016) to overcome the time and labor intensity in the conventional ones. Moreover, dynamic sensory tests accommodate the nature of the eating experience as a dynamic process (Dijksterhuis and Piggott 2001). In addition, sensory scholars also investigate new sensory techniques to dive into the emotions evoked by food consumption to capture consumer experience (Martins et al. 2021).

18.3 Quick-Fast Descriptive Sensory Techniques

Based on the sensory perception of appropriately trained panelists, conventional descriptive sensory analyses produce detailed product profiles but are also labor and time-consuming in training; hence, they are expensive (Rodríguez-Noriega et al. 2021). Sensory evaluation provides valid and reliable information for food manufacturers so that management can make informed business decisions (Meilgaard et al. 2016). As innovation costs and time pressures increase, decision-makers must have access to relevant information more quickly, and thus, newer approaches are used to waive the drawbacks of traditional methods (Schneider-Häder et al. 2016). Those approaches, but not limited to, are FCP (Free Choice Profiling), FP (Free

Table 18.1 Summary of conventional and novel sensory techniques

Sensory Test	Types		Subtypes	Assessors	Question?
Analytic	Discrimination	Difference	Duo-trio	Semi-trained/Trained, generally 20–40 subjects, can be used with as few as 5–8 subjects; brief training required	Does a sensory difference exist between products?
			Triangle		
			Sequential Test		
			Difference Test	Untrained/trained, generally 30 or more subjects, as few as 12–15 subjects; brief training required	
			Difference -from-control Test		
			PM (Projective mapping) aka Napping®	Naïve consumers/ trained	
		Sensitivity	Threshold Determination	Trained	
	Descriptive		Flavor Profile	Trained/greater Trained, a recommended minimum number is generally 5 subjects	How does a sensory difference exist between products?
			Texture Profile		
			QDA® (Quantitative Descriptive Analysis)		
			Spectrum™		
			TI (Time-intensity)		
			TDS (Temporal dominance of sensations)		
			FCP (Free Choice Profiling)	Untrained/semi-trained, generally 60–80 subjects	
			Flash Descriptive Analysis		
			CATA (Check-all-that-apply)		
			FP (Flash profiling)		
			RATA (Rate-all-that-apply)		

(continued)

Table 18.1 (continued)

Sensory Test	Types	Subtypes	Assessors	Question?
Affective	Preference	Pair-comparative	Naïve consumers, generally, 50–100 subjects	Which product is preferred?
		PSP (Polarized sensory positioning)		
	Acceptance			How acceptable is the product?

Profiling), Projective Mapping aka Napping®, CATA (Check-All-That-Apply), and RATA (Rate-All-That-Apply).

18.3.1 Free Choice Profiling (FCP)

It is a novel method of profile analysis where each assessor creates a unique profile of the products, describing them in words that they understand without having to define them. Furthermore, Generalized Procrustes Analysis (GPA) is used to rationalize the spatial configurations obtained from individual profiles to generate a consensus configuration that shows how the samples relate to one another for the panel (Williams and Langron 1984). Hence, it does not require highly trained assessors or the terminology agreement process as in conventional descriptive analysis, which results in more efficiency in time. In the procedure, quantitative attributes can be described using individual intensity scales or specified scales determined by the assessment leader (Schneider-Häder et al. 2016).

Several studies on processed meat have utilized this technique. A study by Li et al. (2022a) showed that FCP, along with olfactometry, showed that when added at the highest level, thiamine supplementation affected aroma in terms of toasted odor and that sulfur-derived compounds, which contribute to the "cured meat odor," were found in dry-cured loins. Moreover, Li et al. (2022b) suggested that the same methods could detect thiamine supplementation improved cured and savory odors in Thia-sausages.

18.3.2 Flash Profiling (FP)

FP combines FCP with a modified assessment system, the product ranking system; hence, rather than rating intensities on a scale, each attribute is ranked according to its intensity (Schneider-Häder et al. 2016). Some studies showed that flash profiling was an effective alternative to conventional profiling. In addition to dynamic methods, Lorido et al. (2018) evaluated various formulas of dry-cured loin that replaced NaCl with KCl using the FP approach. They concluded that this method provides quick, effective sample discrimination—more encouraging results from Albert et al. (2011) showed that FP with semi-trained panelists could be used as a quick alternative to QDA® in fish nuggets with different textural layers served hot.

18.3.3 Check-All-That-Apply (CATA)

CATA is a relatively new addition to food sensory evaluation, but it appears to be gaining popularity in product optimization, which is used to understand how consumers perceive products (Valentin et al. 2012). It has been applied to the optimization of processed meat products such as the partial replacement of beef with texturized soy protein along with flavor enhancement by the addition of nutritional yeast in meatballs (Grasso et al. 2019), the substitution of synthetic antioxidants with natural extract of pink pepper residue in chicken burgers (Saldaña et al. 2020), the replacement NaCl by $CaCl_2$ in chicken nuggets (Barros et al. 2019), the fat replacement with *Agaricus bisporus* mushroom in beef burger (Patinho et al. 2021), and also another fat replacement with polysaccharide-based emulsion gels in Frankfurter sausages (Fontes-Candia et al. 2023), the nitrite substitution with Japanese radish derivatives on restructured cooked hams (Guimarães et al. 2022), the enrichment of NaCl-reduced burger with long-chain PUFA (Rios-Mera et al. 2021), then enhanced antioxidants activity of beef burgers with dried *Hibiscus sabdariffa* L. leaves (do Nascimento et al. 2023) to name but a few. Figure 18.1 shows an example of CATA questionnaire.

Fig. 18.1 Example of CATA questionnaire

Please check all sensory characteristics that apply to the beef jerky.

	398	**183**	**765**
Juicy	O	O	O
Salty	O	O	O
Umami	O	O	O
Tender	O	O	O
Snack	O	O	O
…..			

18.3.4 Rate-All-That-Apply (RATA)

For improved sample description and discrimination, Ares et al. (2014) used attribute ratings as a variant (RATA) to CATA questions by applying a 3- or 5-point intensity scale. Merlo et al. (2021) utilized this method to characterize smoked bacon by the attributes of smoky, crunchy, salty, soft, and bright and later found that these sensory attributes are generated by volatile organic compounds, free fatty acids, and their interactions and at the end increasing its overall liking score. In another study, A RATA test was used to profile the food's sensory qualities from the consumers' viewpoints. Intense woody and smoky flavors were thought to be manufactured in the Western style, and they were less recognizable and inconsistent with bulgogi. In contrast, samples with weaker woody and smoky flavors were favored. According to this study, flavorings that match a food system can increase customer preference and a sense of familiarity (Seo et al. 2021).

18.3.5 Napping®

Napping® or projective mapping is an alternative to both profiling and (dis)similarity scalability. It is an assessment method for developing a two-dimensional perceptual map that involves each assessor placing products according to how they perceive similarities and differences (Risvik et al. 1994). Figure 18.2 shows the visualization of a Napping® test for various beef jerky.

Grossi et al. (2012) utilized Napping® to investigate the sensory properties of pork sausages with reduced salt content and the addition of carrot fiber and potato starch processed by high-pressure processing. The study concluded that this approach is promising in producing sausages with low salt content. This method's advantage is that it is a rapid alternative to QDA® with untrained panelists (Albert et al. 2011).

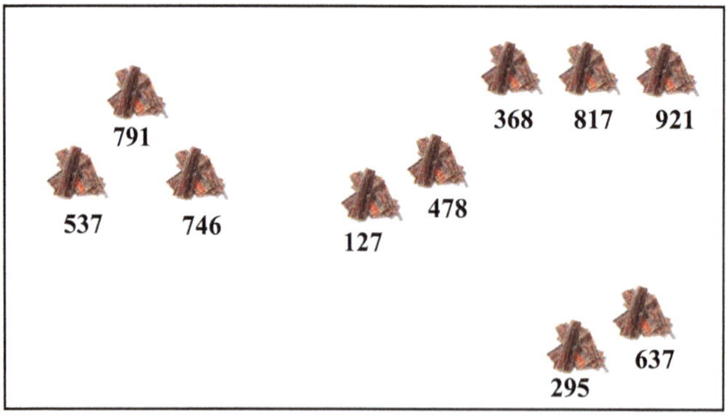

Fig. 18.2 Visualization of a Napping® test for various beef jerky

18.4 Dynamic Sensory Techniques

Most eating-related processes, including salivation and mastication, are dynamic; therefore, approaches that consider these characteristics are likely to yield more reliable results than those that ignore them (Dijksterhuis and Piggott 2001). There are several dynamic sensory methods, as follows:

18.4.1 Time-Intensity (TI)

It is practical for samples where the perceived flavor intensity changes as the product is tasted, such as the sweetness of artificial sweeteners or the bitterness of beer (Meilgaard et al. 2016). Dynamic sensory techniques are sometimes combined with others to get comprehensive sensory properties from processed meat. For example, Lorido et al. (2018) used FP in conjunction with TI and Temporal Dominance of Sensations (TDS) to assess the effect of replacing NaCl with KCl on the sensory attributes of dry-cured loins. The study suggested that TI revealed that samples with greater substitution levels of dry-cured loins were linked to high intensity values of pastiness and bitterness. On the other hand, the lowest degree of NaCl replacement was associated with the strongest and longest-lasting saltiness and high flavor intensity. However, TI focuses exclusively on one attribute. Hence, researchers have proposed other temporal methods that consider multiple attributes simultaneously (Le Quéré and Schoumacker 2023).

18.4.2 Temporal Dominance of Sensations (TDS)

The TDS methodology is a multi-attribute temporal method. Notably, there were variations in the temporal perception of the assessed flavor and texture qualities between the dry-cured loins with greater salt replacement levels and the control samples (Lorido et al. 2018). Several processed meat studies have performed TDS. Saldaña et al. (2019) found that TDS was a feasible technique for recording temporal sensory alterations in bacon smoked with different Brazilian woods. A similar result was found by Paglarini et al. (2020) that suggests dynamic sensorial methods are effective tools in characterizing reformulated Bologna sausage with the reduction of sodium and fat and the addition of functional emulsion gel.

18.4.3 Temporal CATA (TCATA)

TCATA has been used to explore the dynamic sensory (taste, flavor, and mouthfeel) perception induced by oral burn of complex liquid and solid foods, namely, tomato soup, curry rice, and beef patties (Lyu et al. 2024). They concluded that while oral burn suppressed the investigated sensory perceptions in selected samples, it minimally affected how these perceptions built up over time. It also lessens the lingering of these perceptions. Barker and McSweeney (2022) performed both static (CATA) and dynamic (TCATA) methods to characterize a hybrid burger containing yellow peas and ground chicken. Those techniques could identify the addition of yellow peas associated with off-flavors and reduce the consumers' liking.

18.5 Emotions in Sensory Meat Science

Investigating the emotions evoked by food consumption is necessary to understand consumer experiences during consumption better and improve food choice prediction compared to overall liking alone (Martins et al. 2021; Orr et al. 2023). Lorido et al. (2019) examined how consuming three types of Spanish dry-cured hams affected emotional responses by utilizing dynamic sensory assessment of Temporal Dominance of Emotions (TDE) along with TDS.

There are limited studies regarding emotional responses to processed meat consumption. Orr et al. (2023) developed emotion lexicons tailored to plant- and meat-based patties. According to the study, developing a consumer-led lexicon requires consideration of various case scenarios and product user levels. Meanwhile, Schouteten et al. (2016) studied the influence of ingredient information on emotional conceptualizations of burgers based on different raw materials of insects, plants, and meat. They suggested that different conditions (blind tasting, expected condition with no tasting, and informed condition with tasting) have little effect on emotional conceptualization. The investigators treated emotional data as CATA by only using the selection frequency.

18.6 Conclusions

Conventional descriptive sensory studies, which provide thorough product profiles based on the sensory perception of suitably trained personnel, are costly due to their time-consuming and labor-intensive nature. More recent methods achieve comparable outcomes by reducing or eliminating these problems. They do not, however, generate a completely accurate sensory profile of every distinct sample as the traditional ones. Dynamic sensory techniques and emotions in the sensory aspects of meat science may characterize processed meat sensory attributes more reliably and help gain an understanding of consumer experiences during consumption.

References

Abdullah A (2000) Prinsip Penilaian Sensori. Universiti Kebangsaan Malaysia, Bangi

Albert A, Varela P, Salvador A et al (2011) Overcoming the issues in the sensory description of hot served food with a complex texture. Application of QDA®, flash profiling and projective mapping using panels with different degrees of training. Food Qual Prefer 22:463–473. https://doi.org/10.1016/j.foodqual.2011.02.010

Ares G, Bruzzone F, Vidal L et al (2014) Evaluation of a rating-based variant of check-all-that-apply questions: Rate-All-That-Apply (RATA). Food Qual Prefer 36:87–95. https://doi.org/10.1016/j.foodqual.2014.03.006

Barker S, McSweeney MB (2022) Sensory characterization of yellow pea and ground chicken hybrid meat burgers using static and dynamic methodologies. J Food Sci 87:5390–5401. https://doi.org/10.1111/1750-3841.16380

Barros JC, Gois TS, Pires MA et al (2019) Sodium reduction in enrobed restructured chicken nuggets through replacement of NaCl with CaCl2. J Food Sci Technol 56:3587–3596. https://doi.org/10.1007/s13197-019-03777-8

Cardona M, Izquierdo D, Barat JM, Fernández-Segovia I (2023) Intrinsic and extrinsic attributes that influence choice of meat and meat products: techniques used in their identification. Eur Food Res Technol 249:2485–2514

Dijksterhuis GB, Piggott JR (2001) Dynamic methods of sensory analysis. Trends Food Sci Technol 11:284–290

DLG Sensory Analysis Committee (2017) Practice guide for sensory panel training, Frankfurt am Main

do Nascimento NP, Vinhal Borges ÍGA, de Castro JS et al (2023) Enhancing the quality and antioxidant properties of beef burgers with dried *Hibiscus sabdariffa* L. leaves: characterization and sensory evaluation. Food Biosci 103568:103568. https://doi.org/10.1016/j.fbio.2023.103568

Fontes-Candia C, Martínez-Sanz M, Gómez-Cortés P et al (2023) Polysaccharide-based emulsion gels as fat replacers in Frankfurter sausages: physicochemical, nutritional and sensorial evaluation. LWT 180:114705. https://doi.org/10.1016/j.lwt.2023.114705

Grasso S, Smith G, Bowers S et al (2019) Effect of texturised soy protein and yeast on the instrumental and sensory quality of hybrid beef meatballs. J Food Sci Technol 56:3126–3135. https://doi.org/10.1007/s13197-018-3552-9

Grossi A, Søltoft-Jensen J, Knudsen JC et al (2012) Reduction of salt in pork sausages by the addition of carrot fibre or potato starch and high pressure treatment. Meat Sci 92:481–489. https://doi.org/10.1016/j.meatsci.2012.05.015

Guimarães AS, Guimarães JS, Rodrigues LM et al (2022) Assessment of Japanese radish derivatives as nitrite substitute on the physicochemical properties, sensorial profile, and consumer acceptability of restructured cooked hams. Meat Sci 192:108897. https://doi.org/10.1016/j.meatsci.2022.108897

Jimènez-Colmenero F, Carballo J, Cofrades S (2001) Healthier meat and meat products: their role as functional foods. Meat Sci 59:5–13

Le Quéré JL, Schoumacker R (2023) Dynamic instrumental and sensory methods used to link aroma release and aroma perception: a review. Molecules 28:63

Li L, Perea-Sanz L, López-Díez JJ et al (2022a) Aroma enhancement in dry cured loins by the addition of nitrogen and sulfur precursors. Meat Sci 184:108698. https://doi.org/10.1016/j.meatsci.2021.108698

Li L, Perea-Sanz L, Salvador A et al (2022b) Understanding the impact of nitrogen and sulfur precursors on the aroma of dry fermented sausages. Meat Sci 192:108896. https://doi.org/10.1016/j.meatsci.2022.108896
Lorido L, Estévez M, Ventanas S (2018) Fast and dynamic descriptive techniques (flash profile, time-intensity and temporal dominance of sensations) for sensory characterization of dry-cured loins. Meat Sci 145:154–162. https://doi.org/10.1016/j.meatsci.2018.06.028
Lorido L, Pizarro E, Estévez M, Ventanas S (2019) Emotional responses to the consumption of dry-cured hams by Spanish consumers: a temporal approach. Meat Sci 149:126–133. https://doi.org/10.1016/j.meatsci.2018.11.015
Lyu C, Forde CG, Stieger M (2024) Effect of oral burn on dynamic taste, flavor and mouthfeel perception of tomato soups, curried rice and beef patties. Food Qual Prefer 115:105109. https://doi.org/10.1016/j.foodqual.2024.105109
Martins MM, Saldaña E, Teixeira ACB et al (2021) Going beyond sensory and hedonic aspects: a Brazilian study of emotions evoked by beef in different contexts. Meat Sci 180:108536. https://doi.org/10.1016/j.meatsci.2021.108536
Meilgaard MC, Civille GV, Carr BT (2016) Sensory evaluation techniques, 5th edn. CRC Press, Boca Raton
Merlo TC, Lorenzo JM, Saldaña E et al (2021) Relationship between volatile organic compounds, free amino acids, and sensory profile of smoked bacon. Meat Sci 181:108596. https://doi.org/10.1016/j.meatsci.2021.108596
Miller V, Reedy J, Cudhea F et al (2022) Global, regional, and national consumption of animal-source foods between 1990 and 2018: findings from the Global Dietary Database. Lancet Planet Health 6(3):e243–e256
Orr RE, Giezenaar C, Godfrey AJR, Hort J (2023) Development of a consumer-led emotion lexicon for meat and plant-based burger patties using digitally recreated eating contexts. J Sens Stud 38. https://doi.org/10.1111/joss.12824
Paglarini CS, Vidal VAS, dos Santos M et al (2020) Using dynamic sensory techniques to determine drivers of liking in sodium and fat-reduced Bologna sausage containing functional emulsion gels. Food Res Int 132:109066. https://doi.org/10.1016/j.foodres.2020.109066
Patinho I, Selani MM, Saldaña E et al (2021) Agaricus bisporus mushroom as partial fat replacer improves the sensory quality maintaining the instrumental characteristics of beef burger. Meat Sci 172:108307. https://doi.org/10.1016/j.meatsci.2020.108307
Pires MA, de Noronha RLF, Trindade MA (2019) Understanding consumer's perception and acceptance of bologna sausages with reduced sodium content and/or omega-3 addition through conjoint analysis and focus group. J Sens Stud 34. https://doi.org/10.1111/joss.12495
Polizer Rocha YJ, de Noronha RLF, Trindade MA (2019) Understanding the consumer's perception of traditional frankfurters and frankfurters with healthy attributes through sorting task and hard laddering techniques. Meat Sci 149:70–78. https://doi.org/10.1016/j.meatsci.2018.11.004
Prinyawiwatkul W, Tepper BJ, Hartel R (2023) Advances in sensory science: from perception to consumer acceptance. J Food Sci 88:2–4
Rios-Mera JD, Saldaña E, Patinho I et al (2021) Enrichment of NaCl-reduced burger with long-chain polyunsaturated fatty acids: effects on physicochemical, technological, nutritional, and sensory characteristics. Meat Sci 177:108497. https://doi.org/10.1016/j.meatsci.2021.108497
Risvik E, Mcewan JA, Colwill JS et al (1994) Projective mapping: a tool for Sensory Analysis and Consumer Research. Food Qual Prefer 5:263–269
Rodríguez-Noriega S, Buenrostro-Figueroa JJ, Rebolloso-Padilla ON et al (2021) Developing a descriptive sensory characterization of flour tortilla applying flash profile. Food Secur 10. https://doi.org/10.3390/foods10071473
Ruiz-Capillas C, Herrero AM, Pintado T, Delgado-Pando G (2021) Sensory analysis and consumer research in new meat products development. Food Secur 10:1–15
Saldaña E, Soletti I, Martins MM et al (2019) Understanding consumers' dynamic sensory perception for bacon smoked with different Brazilian woods. Meat Sci 154:46–53. https://doi.org/10.1016/j.meatsci.2019.04.006

Saldaña E, Serrano-León J, Selani MM, Contreras-Castillo CJ (2020) Sensory and hedonic impact of the replacement of synthetic antioxidant for pink pepper residue extract in chicken burger. J Food Sci Technol 57:617–627. https://doi.org/10.1007/s13197-019-04093-x

Schneider-Häder B, Derndorfer E, DLG Sensory Analysis Committee (2016) DLG Expert report 5/2016: sensory analysis: overview of methods and application areas classic descriptive tests & flash profiling. Frankfurt am Main

Schouteten JJ, De Steur H, De Pelsmaeker S et al (2016) Emotional and sensory profiling of insect-, plant- and meat-based burgers under blind, expected and informed conditions. Food Qual Prefer 52:27–31. https://doi.org/10.1016/j.foodqual.2016.03.011

Seo GG, Lee CL, Park SH et al (2021) Effect of chargrilled flavoring on the sensory perception and consumer acceptability of bulgogi (Korean barbecued beef). Food Sci Biotechnol 30:77–86. https://doi.org/10.1007/s10068-020-00848-x

Torrico D (2021) Novel techniques to measure the sensory, emotional, and physiological (biometric) responses of consumers toward foods and packaging. Foods 10(11):2620

Valentin D, Chollet S, Lelièvre M, Abdi H (2012) Quick and dirty but still pretty good: a review of new descriptive methods in food science. Int J Food Sci Technol 47:1563–1578

Williams AA, Langron SP (1984) The use of free-choice profiling for the evaluation of commercial ports. J Sci Food Agric 35:558–568. https://doi.org/10.1002/jsfa.2740350513

Chapter 19
Authentication of Processed Meat Products

Sadiya Sajad, Javed-ur-Rehman, Tahir Nazir, Sheikh Rafeh Ahmad, Humaira Fayaz, and Syed Arshad Hussain

19.1 Introduction

Processed meat products authentication is an important topic that should be addressed in order to promote fair trade and allow customers to make informed decisions. Authentication of processed meat products is crucial to ensure the quality, safety and compliance with regulatory standards (Ballin 2010). This process entails verifying the composition, origin and labelling details of processed meat to mitigate the risks of fraud, mislabelling or adulteration (Chuah et al. 2016). It is also a major concern for practicing religious people (Smith et al. 2008).

There are various methods and technologies employed to authenticate processed meat products. The analytical methods currently used for processed meat products authentication rely mainly on protein and DNA analysis (Orru et al. 2009; Rashid et al. 2014). Protein-based techniques include electrophoretic, chromatographic and spectroscopic approaches that have been shown useful to identify the species origin of meats. However, when assaying heat-treated material, these techniques are limited because of the denaturation of soluble proteins during food processing. Besides, cross-reactions occurring among closely related species often hinder analysis by immunoassays, which rely on the use of antibodies raised against a specific protein. Electrophoretic, chromatographic and immunological approaches based on protein fractions are typically ineffective in distinguishing closely related meat species in processed foods (Alikord et al. 2018). Using nucleic-acid-based analytical methods can overcome these difficulties because DNA is a very stable and long-lived biological molecule present in all tissues of all organisms.

S. Sajad (✉) · T. Nazir · S. R. Ahmad (✉) · H. Fayaz · S. A. Hussain
Division of Livestock Products Technology, FVSc & AH, SKUAST-Kashmir, Srinagar, India
e-mail: sh.rafeh@skuastkashmir.ac.in

Javed-ur-Rehman
Central Sericulture Research and Training Institute, Berhampore, West Bengal, India

S. A. Rather, F. A. Masoodi (eds.), *Hand Book of Processed Functional Meat Products*, https://doi.org/10.1007/978-3-031-69868-2_19

Molecular techniques for meat species identification and authentication of processed meat products have gotten a lot of interest in recent years because of their high sensitivity and specificity, as well as their short processing time and inexpensive cost (Chaudhary and Kumar 2022). The chapter deals with the most common approaches like protein and DNA based approaches, as well as microscopic techniques, Isotope Ratio Mass Spectrometry (IRMS), Nuclear Magnetic Resonance (NMR), Near-Infrared Spectroscopy (NIRS), Electronic Nose and Electronic Tongue used to validate the authenticity of processed meat products to date (Oh et al. 2014). Combining multiple authentication methods often provides a more comprehensive approach to ensure the accuracy and reliability of processed meat product authentication. It's essential to tailor the methods chosen based on the specific requirements and characteristics of the meat products in question.

19.2 Protein Based Analytical Techniques

Protein analysis techniques play a crucial role in authenticating processed meat products, helping to ensure that the labelled species matches the actual content of the product (Hsieh et al. 1995). The various protein based analysis techniques commonly used for the authentication of processed meats are tabulated (Table 19.1).

Table 19.1 Protein analysis-based techniques commonly used for the authentication of processed meat

Technique	Principle	Uses
Enzyme-linked immuno-sorbent assay (ELISA)	ELISA is based on the specific binding of antibodies to target proteins	Used for detecting species-specific proteins in processed meat, aiding in the identification of the presence or absence of specific species
Protein electrophoresis (2-dimensional electrophoresis)	Protein electrophoresis separates proteins in a gel based on their charge and size. Different species exhibit unique banding patterns	Protein electrophoresis is useful for comparing protein profiles among different meat samples, aiding in the identification of species-specific markers
Sodium dodecyl sulfate-polyacrylamide gel electrophoresis (SDS-PAGE)	SDS-PAGE is based on the separation of proteins based on their size and charge	PAGE is valuable for detecting protein patterns characteristic of specific species, helping in the identification and authentication of processed meats
Western blot Immuno-assay	Western blot immuno-assay is based on the transferring of proteins from a gel to a membrane, followed by detection using specific antibodies	This technique can be applied to identify and quantify specific proteins associated with certain animal species in processed meat samples

19.2.1 Enzyme-Linked Immunosorbent Assay (ELISA)

ELISA is an immunological technique in which an enzyme, a protein that catalyzes a biochemical reaction, is utilized to detect the presence of an antibody or an antigen in a sample. It can detect the presence of pork, horse, beef, chicken, turkey and goat meat adulterants to 0.1% (w/w) and is probably the most widely used technique for authentication purposes (Ha et al. 2019). Moreover, the two most commonly used ELISA techniques for food authentication are indirect and the sandwich ELISA (Nesic et al. 2017). The indirect ELISA uses two antibodies, one specific to the antigen and the other coupled to an enzyme. This second "enzyme-linked" antibody gives the assay its name, and causes a chromogenic or fluorogenic substrate to produce a signal. Sometimes, this second antibody can be linked to a protein such as avidin or streptavidin if the primary antibody is biotin labeled. In the sandwich ELISA, the antigen is bound between two antibodies: the capture antibody and the detection antibody. The detection antibody can be tied to an enzyme or can bind the conjugate (enzyme-linked antibody) that produces the biochemical reaction. This method produces qualitative or quantitative results. Qualitative ELISA detects the presence or absence of an analyte in the sample. The cut-off between positive and negative result is determined by the analyst and is statistically based. The quantitative ELISA, is used to detect amounts of the analyte, the optical density or fluorescent units of the sample is interpolated into a standard curve which is typically a serial dilution of the target (Asensio et al. 2008).

19.2.2 Protein Electrophoresis

Protein electrophoresis is the earliest method used for meat species identification. It is used for soluble proteins, especially sarcoplasmic proteins and is implemented on raw meat. Protein electrophoresis separates proteins in a gel based on their charge and size, thus, exhibiting unique banding patterns for different species. Two-dimensional electrophoresis makes it possible to separate thousands of proteins in one gel simultaneously. Within the first dimension, proteins are separated with the help of native technique depending on their iso-electric point while the second dimension uses the SDS-PAGE to separate proteins depending on their molecular weight. The various advantages of this method are: It's ability to simultaneously separate thousands of proteins, to separate proteins in the sample from contaminating substances by employing protein precipitation and to investigate the similarities and differences between proteins extracted from muscles of different animal species (Alikord et al. 2018).

19.2.3 Sodium Dodecyl Sulfate-Polyacrylamide Gel Electrophoresis (PAGE)

PAGE is the most commonly used gel electrophoretic system for analyzing proteins. This method is based on the separation of protein according to size and can also be used to determine the relative molecular mass of proteins. SDS is an anionic detergent which binds strongly to and denatures proteins to produce linear polypeptide chains. On average one SDS molecule will be present for every two aminoacids. The presence of β-mercaptoethanol assists in protein denaturation by reducing all disulfide bonds. The detergent binds to the hydrophobic region of the denatured protein in a constant ratio of about 1.4 g of SDS/gm of protein. The protein-SDS complex carries net negative charges, hence move towards the anode and the separation is based on the size of the proteins. Most SDS-PAGE gels are cast with a molar ratio of Bisacrylamide: Acrylamide of 1:29 which has been shown empirically to be capable of resolving polypeptides that differ in size by little as 3%. The Polyacrylamide gel is cast as a separating gel topped by a stacking gel. The stacking gel has properties that cause the proteins in the sample to be concentrated into a narrow band at the top of the separating gel. This is achieved by utilizing differences in ionic strength and pH between the resolving buffer and the stacking gel and involves a phenomenon known as isotachophoresis. Then it is analyzed quantitatively by determining the thickness of the protein band (intensity), volume and retention time (Hermanto et al. 2022).

19.2.4 Western Blot Immunoassay

Western blot immunoassay is a good technique due its high sensitivity and specificity to the species. In this technique, target proteins are transferred to a hydrophobic membrane after SDS-PAGE and detected using specific antibodies. After SDS-PAGE, a membrane is placed on the gel, to which the separated proteins in the gel are electrophoretically transferred. The membrane with transferred proteins is then probed with a primary antibody (an antibody specific for a target protein), washed and reacted with a secondary antibody labeled with an enzyme such as horseradish peroxidise (HRP). The bound enzyme activity is used to detect the target protein and visualized by a chemiluminescent or chromogenic method. E.g., Myosin binding protein C (MYBPC) is a muscle sarcoplasmic protein which has structural and regulatory roles. It has been identified as a potential mechanically deboned chicken meat (MDCM) biomarker in proteomics method for the detection of MDCM adulteration in meat mixtures. It is a suitable method for screening purposes and provides a rapid detection of fraudulent mixtures and it could also be quantitative (Deniz et al. 2017).

19.3 DNA Based Molecular Techniques

The protein based methods have been reported to be non-suitable for species identification in heated meat products due to denaturation of protein by intensive heating during food processing which in turn lead to modifications in the antigenic activity of molecules and their mobility after electrophoresis, consequently, changing the ability of antibody to identify its target protein, moreover, the possible cross-reaction between closely related species (Farag et al. 2015). Therefore, for these reasons protein based methods have been replaced by DNA-based ones. DNA is characterised by more stability under intensive heating, pressures and chemical processing. It also possesses a conserved structure in whole body cells, has a great identification power and potentially enables identical information to be obtained from the same animal, regardless of the tissue of origin (Rashid et al. 2014). Among the DNA based techniques, polymerase chain reaction (PCR) is the most employed, simple, time saving, sensitive and specific method that could identify the species of origin exposed to different processing conditions (Bottero and Dalmasso 2011; Floren et al. 2015).

19.3.1 Polymerase Chain Reaction (PCR)

Polymerase Chain Reaction (PCR) and its variants are powerful molecular biology techniques widely used for authenticating processed meat products by detecting and amplifying specific DNA sequences (Farag et al. 2015). When employing PCR-based techniques for meat authentication, it's important to consider factors such as primer design, DNA extraction methods and appropriate controls to ensure reliable and reproducible results. Additionally, validation of the chosen PCR assay is essential to confirm its specificity and sensitivity for the intended purpose (Saez et al. 2004; Ghovvati et al. 2009).

The various PCR-based techniques commonly employed for this purpose are:

- **Species-Specific PCR** which uses species-specific primers and is precise, sensitive, cheap and not a time consuming method which helps in novel identification of many mammalian and bird species in meat and meat products. The PCR method targets genomic and mitochondrial DNA for identification of meat species in large number of samples, even in cooked meat under different processing conditions without the need for further digestion of PCR products with restriction endonucleases. However, the most important requirement is that the nucleotide sequence of the gene used for species identification should be known for the purpose of primer designing (Mafra et al. 2008; Spychaj et al. 2009).
- **Multiplex PCR** techniques are highly favourable for the inspection of multiple targets in a single platform, which have gained huge popularity in food business (including meat authentication) based on the simplicity and low cost through

simple agarose gel analysis. The specificity of primers is a prerequisite for multiplex PCR assays. To obtain a specific multiplex PCR assay, the primers should specifically match the target species and have huge mismatches with non-targets. It has recently been used to locate animal DNA (Adenuga and Montowska 2023) and is regarded as the most efficient and trustworthy when compared to PCR-RFLP, RT-PCR, DNA barcoding and PCR based product DNA sequencing (Rajaei and Doosti 2023). However, mutual interference of components such as templates and primers would become more complex with increase of more primers and multiplicity of multiplex PCR reaction, which often cause lower efficiency and even the failure of amplification indicating that multiplex PCRs are often subjected to technological challenge.

- **Real time PCR** also known as qPCR where quantitative, is an advanced PCR technique that allows for the immediate detection of DNA product amplification while it is occuring in the reaction tubes in the thermal cycler (Meira et al. 2017). This real-time detection chemistry is enabled by fluorescent chemical compounds that when added to the PCR reaction mixture can monitor amplification of target DNA. SYBR Green 1, SYTO-9 and EvaGreen are simple inexpensive, nonspecific binding, reliable fluorescent detection methods. TaqMan probes, on the other hand are specific short-length amplicons carrying a reporter and a quencher dye that binds to the target DNA during amplification. The specific binding of the TaqMan probes increases the specificity of the assay (Dalsecco et al. 2018).
- **Reverse Transcription Polymerase Chain Reaction (RT-PCR)** is a relatively simple and inexpensive technique to determine the expression level of target genes. The principle technique in RT-PCR is total RNA isolation, reverse transcription (RT) and PCR. Reverse transcription involves the synthesis of DNA from RNA by using an RNA-dependent DNA polymerase. PCR can amplify a single or few copies of a piece of DNA across several orders of magnitude, generating thousands to millions of copies of a particular DNA sequence (Bachman 2013).
- **Nested PCR** is a sensitive technique to identify authentic meats and bloods. Identification of expensive and exotic crocodile products was done using sensitive nested-PCR method for very low DNA content products like meats and bloods. The amplification of targeted NADH dehydrogenase 5/6 genes with flanking by Cyt b and tRNA gene was done. The method also enabled sample DNA identification from dried animal blood. The mixture of raw materials and of DNA marker was subsequently determined to verify its specificity and sensitivity (Unajak et al. 2011).
- **Random Amplified Polymorphic DNA PCR (RAPD-PCR) and Arbitarily Primed PCR (AP-PCR)** are PCR based fingerprinting techniques for rapid detection of animal species in meat products. They follow a reproducible, rapid and simple method for simultaneous identification of multiple meat species in a single step DNA-based test dependent on the generation of species-specific fingerprintings by two different arbitrary DNA amplification approaches (RAPD-PCR and AP-PCR). RAPD-PCR fingerprintings allowed the

discrimination amongst pork, beef, lamb, chicken and turkey (Saez et al. 2004). Samples corresponding to each species were clustered together at similarity levels ≥75%. The DNA profiles consisted of a discrete but reproducible number of bands, interpreting the results by simple visual inspection. AP-PCR also allowed identification of the five tested species in every sample although more complex patterns were generated, including some low intensity bands. In both cases, a ramp time between annealing and extension temperatures was introduced to achieve good reproducibility (Calvo et al. 2001).

- **PCR-Restriction Fragment Length Polymorphism (RFLP)** is a widely used technique in meat authentication studies. Products from PCR amplifications are subjected to enzymatic restriction by specific endonucleases, resulting in a specific fingerprint in the product that is then used to identify the target species (Adenuga and Montowska 2023). PCR-RFLP technique has been use to genotype wild boar and domestic swine based on SNPs in the NR6A1 and MCR1 genes of each species. A clear distinction was seen between the two species and also to identify admixture based on the PCR-RFLP test (Koseniuk et al. 2021).
- **PCR-Forensically Informative Sequencing (FINS)** comprises of four steps, namely (1) DNA extraction from biological samples, (2) selection and amplification of specific DNA fragment, (3) determination of the sequence of the amplified DNA fragment and (4) cladistic analysis of the sample DNA sequence against a DNA database. Its success depends on the selection of DNA region and reference species (Li et al. 2011) (Table 19.2).

19.3.2 DNA Hybridization Techniques

DNA hybridization techniques are molecular biology methods that involve the binding of complementary DNA strands, allowing for the detection and authentication of processed meat products (Girish and Karabasanavar 2020). Several DNA hybridization techniques commonly used for detection and authentication of processed meat products are tabulated (Table 19.3).

- **Southern Blotting** is one of the oldest methods used to analyze DNA. DNA is first extracted from the sample and quantified by UV spectrometry. Then, a certain amount of DNA is digested by a specific restriction endonuclease and cut DNA samples are separated by agarose gel electrophoresis. Transfer and UV cross-linking of DNA fragments to a membrane is done along with hybridization with a specific nucleic acid probe, which will identify DNA target regions of interest (Girish and Karabasanavar 2020).
- **Dot Blot Hybridization** is a very potent technique that can be used for the identification of DNA and RNA of various species with varying degree of homology and for the estimation of relative amounts of nucleic acid with known homology (Bergmans and Gaastra 1988). In most cases, single stranded (ss) denatured DNA or RNA is bound to a filter support (e.g., nitrocellulose) and

Table 19.2 DNA analysis-based techniques commonly used for the authentication of processed meat and meat products

Technique	Principle	Application
Species-specific PCR	Species-specific PCR targets unique DNA sequences in the genome of a particular species. Primers are designed to amplify a region of DNA specific to the target species	This technique helps identify the presence or absence of a specific species in processed meat products
Multiplex PCR	Multiplex PCR is qualitative detection procedure which involves the simultaneous amplification of multiple target DNA sequences in a single reaction	It can be used to detect multiple species in a single assay. Multiplex PCR is useful for authenticating processed meat products that may contain a mixture of different species
Real-time PCR	Real-time PCR is a quantitative method which allows for the quantification of DNA during the amplification process. Fluorescent dyes or probes are used to monitor the accumulation of PCR products in real-time	qPCR is valuable for quantifying the amount of DNA from specific species in processed meat, providing both qualitative and quantitative information
Reverse transcription PCR (RT-PCR)	RT-PCR is used to amplify RNA sequences. It involves the reverse transcription of RNA into complementary DNA (cDNA) followed by PCR amplification of the cDNA	RT-PCR can be applied to detect and authenticate processed meat products that may undergo heat treatment, which degrades DNA but not RNA
Nested PCR	Nested PCR involves two sets of primers: The outer primers amplify a larger DNA fragment, and the inner primers amplify a smaller fragment within the first product	Nested PCR can enhance sensitivity and specificity, making it useful for detecting low concentrations of target DNA in processed meat samples
Random amplified polymorphic DNA PCR (RAPD-PCR) and Arbitarily primed PCR (AP-PCR)	RAPD-PCR and AP-PCR are based on the generation of species-specific fingerprinting and utilize short arbitrary primers to amplify random DNA sequences. It can generate unique banding patterns for different species	AP-PCR and RAPD can be employed for genetic fingerprinting and differentiation of species in processed meat products
PCR-restriction fragment length polymorphism (RFLP)	PCR-RFLP involves amplifying a target DNA fragment and digesting it with a restriction enzyme. The resulting fragment lengths are species-specific	PCR-RFLP is useful for distinguishing between closely related species in processed meat products

(continued)

Table 19.2 (continued)

Technique	Principle	Application
PCR-forensically informative sequencing (PCR-FINS)	PCR-FINS selects and amplifies a specific DNA fragment, determines its sequence and does cladistic analysis of the sample DNA sequence against a DNA database	PCR-FINS is useful for meat speciation of highly processed meat

Table 19.3 DNA hybridization techniques commonly used for the authentication of processed meat and meat products

Technique	Principle	Application
Southern blotting	Southern blotting involves the transfer of DNA fragments from a gel to a solid support (usually a membrane). The immobilized DNA is then probed with a labeled complementary DNA or RNA probe	Southern blotting can be used to detect specific DNA sequences in processed meat products, providing information about the presence or absence of particular genetic markers
Dot blot hybridization	Dot blot hybridization is a simplified version of southern blotting where denatured DNA samples are directly spotted onto a solid support, followed by hybridization with a labeled probe	Dot blot hybridization is a rapid and less labor-intensive method for detecting specific DNA sequences in processed meat
In situ hybridization	In situ hybridization involves the hybridization of a labeled probe directly to the DNA within intact cells or tissue sections	In situ hybridization can be used to detect and localize specific DNA sequences in processed meat products, providing information about the cellular distribution of target DNA
Fluorescence in situ hybridization (FISH)	FISH is a variant of in situ hybridization where fluorescently labeled probes are used for detection. This allows for visualization under a fluorescence microscope	FISH can be applied for the specific detection and localization of DNA sequences in processed meat samples, offering high sensitivity and spatial resolution
DNA Chip hybridization microarrays	PCR-based DNA chip hybridization microarrays consist of thousands of immobilized probes representing different DNA sequences. Fluorescently labeled DNA from the sample hybridizes with the immobilized probes	Microarrays enable the simultaneous detection of multiple DNA sequences, making them suitable for screening processed meat products for a broad range of genetic markers
DNA barcoding	Analyses of a specific region of DNA called DNA barcode is done. Its sequence is then compared to a reference library which contains information of many species linked to their barcodes	Food authentication by DNA barcoding is rooted in the concept of using short and standardized DNA sequences to differentiate between species

incubated with radioactively labelled ss DNA or RNA fragment (probe) complementary to the nucleic acids of interest. In subsequent washing steps the non-hybridized probe is removed. Hybridized probe will be retained, but may be washed off in further washing steps, depending on the ion concentration and temperature of the wash.

- **In Situ Hybridization** is a fast identification process based on the DNA hybridization which includes rapid extraction of DNA from meat samples and also use of selective oligonucleotide probes based on species-specific satellite DNA sequences. The DNA probes were radiolabelled and the detection was carried out by autoradiography (John et al. 1969). The probes were developed for the identification of meat from cattle, sheep/goat, horse, deer, pig, chicken and turkey. About 1–5% admixture of meat of one species in another could be detected. This method was also able to identify strongly heated meat products. The specific oligonucleotide can differentiate between closely related species like chicken and turkey or cattle and goat/sheep (Azad et al. 2023).
- **Fluorescence In Situ Hybridization (FISH)** is a cytogenetic technique used to detect and localise the presence of specific DNA sequences. In FISH, radioactivity is replaced with fluorescence. Hence, FISH protocols are based on the specific hybridization of fluorescently labelled probe with its complementary target sequence. Hybrids formed between the probes and their DNA targets are detected using fluorescence microscopy (Lecrenier et al. 2014).
- **DNA Chip Hybridization Microarrays** are very rapid and sensitive (0.1% admixtures of beef or chicken meat are detectable) and these primarily involve the use of oligonucleotide microarray hybridization analysis of PCR products from the mitochondrial cytochrome *b* gene DNA inorder to identify different animal species in meat samples. PCR products of cattle, pig, chicken, turkey, sheep and goat were unequivocally identified by hybridization with species-specific probe sequences immobilized on an oligonucleotide microarray (Peter et al. 2004).
- **DNA Barcoding** has received considerable interest due to its rapidity, accuracy and cost efficiency in authenticating food ingredients. The DNA barcoding is a detection, identification and diagnostic technique using a short standardized DNA marker. The genetic region known as a DNA barcode is composed of small part of genome. The gene fragment of target species located at the 5′ end of the barcode gene is amplified and sequenced to produce DNA barcodes that can be used as a master key for identifying species. The cytochrome c oxidase subunit 1 isolated from the mitochondrial DNA genome is considered as a universal barcode marker for identification of animal species (Dawan and Ahn 2022).

19.4 Spectroscopy Techniques

19.4.1 Mass Spectrometry (MS)

Mass spectrometry (MS) based proteomics methods also entered the field of meat authentication. Mass spectrometry has enabled the detection of peptide markers by liquid chromatography-mass spectrometry techniques (LC-MS) and this has enabled identification of specific proteins from soybean (Krager et al. 2022), fish (Carrera et al. 2013) and meat species (Montowska and Pospiech 2007; Prandi et al. 2017). These specific proteins or peptides (generated after tryptic digestion) are detected in mass spectrometric methods. Notably, the primary structure of proteins is, in general, quite stable against processing (Sentandreu et al. 2010). Furthermore, a limited degree of protein degradation is often less critical for the detection of species-specific peptides than DNA shearing for the PCR detection, as fragmentation events within the relatively short peptide sequences are less likely compared to degradation of the longer DNA templates (Montowska et al. 2015).

19.4.2 Near-Infrared Spectroscopy (NIRS)

Near-infrared spectroscopy is a non-destructive analytical technique that utilizes the absorption of near-infrared light by organic molecules to provide information about the chemical composition of a sample. NIRS has been employed for the authentication of processed meat products due to its ability to rapidly analyse samples without the need for extensive sample preparation. The procedure of NIRS used for processed meat product authentication purpose is given below:

Principle of NIRS
Near-infrared light, with wavelengths ranging from 700 to 2500 nm, is directed at the sample. The sample absorbs specific wavelengths of light based on the vibrational transitions of molecular bonds (such as C-H, N-H, and O-H bonds) present in the organic compounds. The transmitted or reflected light is then measured, and a spectrum is obtained. The resulting spectrum is analysed to identify characteristic absorption bands associated with different chemical components (Esteki et al. 2018).

Chemical Information Obtained
NIRS provides information about the chemical composition of the sample, including the presence and concentration of various compounds such as proteins, fats, moisture, and other organic constituents. Different meat species and their compositions exhibit unique spectral patterns, allowing for discrimination between species.

Applications
Species identification: NIRS can be used to distinguish between different meat species based on their unique spectral fingerprints. This is particularly useful for detecting mislabelling or adulteration in processed meat products.

Quality assessment: NIRS can assess the quality of meat products by analysing parameters such as fat content, protein content, and moisture levels. It can help verify whether a product meets specific quality standards.

Advantages of NIRS for Meat Authentication

Non-destructive: NIRS allows for rapid analysis without the need for sample destruction.

High throughput: It enables the analysis of a large number of samples in a relatively short time, making it suitable for quality control in industrial settings.

Reduced sample preparation: Compared to traditional methods, NIRS often requires minimal or no sample preparation, saving time and resources.

Instrumentation and Data Analysis

NIRS spectrometers: These instruments are designed to measure the absorbance of near-infrared light and generate spectra from the samples.

Calibration models: Calibration models are developed using a set of reference samples with known composition to correlate the spectral data with specific chemical attributes.

Data analysis: Multivariate statistical methods, such as Principal Component Analysis (PCA) and Partial Least Squares Regression (PLSR), are commonly used to analyze the complex spectral data obtained from NIRS. Calibration models are developed and validated to predict the concentration of specific components in unknown samples.

It's important to note that successful implementation of NIRS for meat authentication requires careful calibration and validation using representative samples. Additionally, factors such as sample homogeneity, instrument maintenance, and environmental conditions should be considered to ensure accurate and reliable results (Danezis et al. 2016).

19.4.3 Mid-Infrared Spectroscopy (MIR)

Mid-Infrared Spectroscopy (MIR) is a technique used for the analysis of molecular vibrations in the mid-infrared range (wavelengths typically between 2.5 and 25 micrometers or 4000–400 cm^{-1}). MIR spectroscopy has been applied for the authentication of processed meat products due to its ability to provide information about the chemical composition of samples (Mendes and Duarte 2021). Here's how MIR spectroscopy is utilized in this context:

Principle of MIR Spectroscopy

MIR spectroscopy relies on the absorption of infrared light by molecules, causing specific vibrational transitions. Different functional groups in organic compounds, such as C-H, O-H, and N-H bonds, exhibit characteristic absorption bands in the mid-infrared region. The resulting MIR spectra provide a fingerprint of the chemical composition of the sample (Meza-Márquez et al. 2010).

Applications

Species identification: MIR spectroscopy can distinguish between different meat species based on their unique spectral signatures. This is particularly useful for detecting mislabelling or adulteration in processed meat products. Pure chopped beef was differentiated from those contaminated with organ meats (kidney and liver) using MIR spectroscopy in combination with chemometrics where the SIMCA model differentiated pure and adulterated beef with 92% accuracy (Al-Jowder et al. 1999).

Quality assessment: MIR spectroscopy can assess the quality of meat products by analyzing various parameters, including protein, fat, moisture, and other components.

Advantages of MIR Spectroscopy for Meat Authentication

Non-Destructive: MIR spectroscopy allows for non-destructive analysis, meaning samples can be retained for further testing or quality control.

Rapid analysis: The technique provides rapid results, making it suitable for high-throughput analysis in industrial settings.

Quantitative and qualitative analysis: MIR spectroscopy can provide both quantitative and qualitative information about the composition of meat products.

Instruments Needed

MIR spectrometers: These instruments are designed to measure the absorption of mid-infrared light and generate spectra from the samples.

Data analysis

Chemometric models: Similar to Near-Infrared Spectroscopy (NIRS), MIR spectroscopy data is often analyzed using chemometric methods, including Principal Component Analysis (PCA) and Partial Least Squares Regression (PLSR).

Calibration models: Calibration models are developed using a set of reference samples with known composition to establish correlations between the spectral data and specific chemical attributes.

Quality Control and Process Monitoring

MIR spectroscopy can be employed for routine quality control in meat processing facilities. It can monitor key parameters, ensuring that products meet desired specifications.

Sample Preparation

MIR spectroscopy often requires minimal sample preparation. Solid samples may be ground into a powder, while liquid samples can be analyzed directly. Successful implementation of MIR spectroscopy for meat authentication involves careful calibration and validation of models using representative samples. Factors such as the homogeneity of samples, instrument calibration, and environmental conditions should be considered to ensure accurate and reliable results.

19.4.4 Ultraviolet-Visible Spectroscopy/UV-Vis Spectroscopy

Ultraviolet-visible spectroscopy is a technique commonly used for analysing the absorption of light in the ultraviolet and visible regions of the electromagnetic spectrum. While it may not be the primary method for meat authentication, UV-Vis spectroscopy can provide useful information related to the protein content and colour of processed meat (Patel et al. 2022).

Protein Content Analysis
UV-Vis spectroscopy can be used to estimate the protein content in meat samples based on the absorption of UV light by peptide bonds in proteins. Changes in protein content or discrepancies from expected levels can indicate potential issues, such as adulteration or mislabelling.

Colour Assessment
UV-Vis spectroscopy is often used for colourimetric analysis. The colour of meat is influenced by factors such as myoglobin content and its oxidation state. Monitoring changes in colour characteristics can provide insights into the freshness, processing, or preservation methods employed in meat products. UV-Vis spectroscopy can help assess colour parameters such as redness (related to myoglobin's absorption in the visible range).

Authentication Based on Absorption Bands
Different compounds in meat may have characteristic absorption bands in the UV-Vis spectrum. Monitoring these bands can help in the identification of specific components. UV-Vis spectroscopy can contribute to the authentication process by detecting the presence of certain compounds associated with particular meat species or indicating the presence of additives.

Quality Control
UV-Vis spectroscopy can be used for routine quality control in meat processing facilities, ensuring that products meet desired specifications. Monitoring parameters such as protein content, colour, or the presence of specific additives can help maintain consistent product quality.

Sample Preparation
UV-Vis spectroscopy may require minimal sample preparation. For liquid samples, it might involve dilution or extraction steps to obtain a suitable solution for analysis. The simplicity of sample preparation makes UV-Vis spectroscopy a relatively quick and straightforward method for analyzing meat samples. It's important to note that while UV-Vis spectroscopy can provide valuable information, its application in meat authentication is often part of a broader set of techniques. For a more comprehensive

understanding of processed meat authenticity, combining UV-Vis spectroscopy with other methods such as DNA analysis, mass spectrometry, or spectroscopic techniques like NIRS or MIR spectroscopy may be beneficial (Prieto et al. 2009). The choice of method depends on the specific goals of the analysis and the properties of interest. Additionally, proper calibration and validation procedures should be implemented to ensure the accuracy and reliability of the results.

19.4.5 Rapid Ambient Mass Spectrometry

Rapid ambient mass spectrometry is used for the direct as well as rapid analysis of compounds present in solid or liquid samples with the help of ionization sources in mass spectrometry (Cooks et al. 2006; Lu et al. 2018). It has also been used as a screening method in food safety analysis. The first ambient mass spectrometry technique developed was Desorption Electro Spray Ionization Mass Spectrometry (DESI-MS) {Kozlowski et al. 2015}. Afterwards, many other ambient mass spectrometric techniques were designed which include direct analysis in real time (DART), extractive electro-spray ionization (EESI), desorption atmospheric pressure chemical ionization (DAPCI), dielectric-barrier-discharge ionization (DABDI) or low temperature plasma (LTP) {Lu et al. 2018)}. Direct analysis in real time (DART) is one of the most popular ambient pressure ionization technique (Gupta and Samal 2022). In DART, the molecules are ionized by excited helium molecules and adducts, mainly protonated/deprotonated molecular ions are commonly formed. It has diverse applications such as food quality and safety control, drug screening, contamination analysis and environmental monitoring. The advantages of this technique are: the possibility of direct sampling at ambient conditions; to generate ions from liquid or solid samples; to be simple to operate; to have rapid analysis speed (acquisition time <60 s) compared to UPLC and GC-MS methods (Wells et al. 2008).

19.4.6 Fourier Transform Infrared Spectroscopy (FTIR)

Fourier transform infrared spectroscopy (FTIR) is a common type of Mid-Infrared spectroscopy used for its high sensitivity and ability to rapidly acquire spectra (Candogan et al. 2020). This spectroscopic technique is successfully utilized for detecting and characterization of different types of adulterants in meat and processed meat products. Fourier transform infrared spectroscopy (FTIR) is categorized under vibrational spectroscopic technique that employs interferometers to modulate the Fourier transform algorithm in the form of an electromagnetic signal to transform sample data into an optical spectrum obtained on a computer system. In the interferometer, light beams are scattered and then reassembled. Due to the simultaneous measuring of wavelengths, this approach is quick and has a better signal-to-noise ratio. Fourier Transform can be also used in optical, infrared, nuclear magnetic resonance, Raman, electron spin resonance spectroscopies as well as mass spectrometry. However, this method is generally applied to infrared spectroscopy. This spectroscopic technique is successfully utilized for detecting and characterization of different types of adulterants in meat and processed meat products (Meza-Márquez et al. 2010). Meza-Márquez et al. (2010) interrogated the presence of adulterants used in processed beef adulteration with different adulterants-horsemeat, trimmings of beef, and texturized plant protein by using Mid-Fourier Transform Infrared (MID–FTIR) spectroscopy and chemometrics. Principal component analysis (PCA) and partial least squares (PLS) regression models were used to detect and quantify the adulterants within the minced meat. The absorbance spectra had an excellent correlation with the concentration of constituents and the percentage of adulteration is measured (Alamprese et al. 2016).

19.5 Conclusion

Meat adulteration is a substantial problem in the food industry due to the increasing number of mislabelling incidents worldwide. Also, economically motivated adulteration or misrepresentation of meat products is more rampant in developing countries, however, being prevalent in the developed countries as well, thus, jeopardizing the processed meat market. The majority of the recent developments in this field have focused on domestic species, customer tastes and globalization of the meat business which has, thus, spurred the development of molecular (broadly protein and DNA based) identification methods for meat and meat products encompassing various single as well as amalgamated molecular techniques. Importantly, the molecular-based methods for meat authenticity are likely to continue to develop in potential as species coverage in the database grows, and in time these techniques may be embraced as complementary regulatory instruments to provide farm-to-fork traceability and prevent meat replacement in the market.

References

Adenuga BM, Montowska M (2023) A systematic review of DNA-based methods in authentication of game and less common meat species. Compr Rev Food Sci Food Saf 22:2112–2160. https://doi.org/10.1111/1541-4337.13142

Alamprese C, Amigo JM, Casiraghi E, Engelsen SB (2016) Identification and quantification of Turkey meat adulteration in fresh, frozen-thawed and cooked minced beef by FT-NIR spectroscopy and chemometrics. Meat Sci 121:175–181

Alikord K (2018) Species identification and animal authentication in meat products: a review. J Food Meas Charact 12:145–155

Al-Jowder O, Defernez M, Kemsley EK, Wilson RH (1999) Mid-infrared spectroscopy and chemometrics for the authentication of meat products. J Agric Food Chem 47:3210–3218

Asensio L, Gonzalez I, Garcıa T, Martın R (2008) Determination of food authenticity by enzyme-linked immunosorbent assay (ELISA). Food Control 19(1):1–8

Azad MAK, Dey M, Khanam F, Biswas B, Akhter S (2023) Authentication of meat and meat products using molecular assays: a review. J Agric Food Res 12:100586

Bachman J (2013) Reverse-transcription PCR (RT-PCR). Methods Enzymol 530:67–74

Ballin NZ (2010) Authentication of meat and meat products. Meat Sci 86(3):577–587

Bergmans HE, Gaastra W (1988) Dot-blot hybridization method. Methods Mol Biol 4:385–390

Bottero MT, Dalmasso A (2011) Animal species identification in food products: evolution of biomolecular methods. Vet J 190:34–38

Calvo JH, Zaragoza P, Osta R (2001) Random amplified polymorphic DNA fingerprints for identification of species in poultry. Poult Sci 80(4):522–524

Candogan K, Evrim AEG, Igci N (2020) Authentication and quality assessment of meat products by fourier transform infrared spectroscopy. Food Eng Rev 13: 13:66–91. https://doi.org/10.1007/s12393-020-09251-y

Carrera M, Cana B, Gallardo JM (2013) Proteomics for the assessment of quality and safety of fishery products. Food Res Int 54:972–979

Chaudhary P, Kumar Y (2022) Recent advances in multiplex molecular techniques for meat species identification. J Food Compos Anal 110:104581

Chuah L, He XB, Effarizah ME, Syahariza ZA, Shamila-Syuhada AK, Rusul G (2016) Mislabelling of beef and poultry products sold in Malaysia. Food Control 62:157–164

Cooks RG, Ouyang Z, Takats Z, Wiseman JM (2006) Detection technologies: ambient mass spectrometry. Science 311:1566–1570

Dalsecco LS, Palhares RM, Oliveira PC, Teixeira LV, Drummond MG, de Oliveira DAA (2018) A fast and reliable real-time PCR method for detection of ten animal species in meat products. J Food Sci 83(2):258–265

Danezis GP, Tsagkaris AS, Camin F, Brusic V, Georgiou CA (2016) Food authentication: techniques, trends & emerging approaches. TrAC Trends Anal Chem 85:123–132

Dawan J, Ahn J (2022) Application of DNA barcoding for ensuring food safety and quality. Food Sci Biotechnol 31(11):1355–1364

Deniz E, Karahisar S, Yildizhan H, Altuntas EG, Demiralp DO, Candogan K (2017) Western blot immunoassay in detecting mechanically deboned chicken meat in beef mixtures. In 63rd international congress of meat science and technology

Esteki M, Shahsavari Z, Simal-Gandara J (2018) Use of spectroscopic methods in combination with linear discriminant analysis for authentication of food products. Food Control 91:100–112

Farag MR, Alagawany M, El-Hack MEA, Tiwari R, Dhama K (2015) Identification of different animal species in meat ant meat products: trends and advances. Adv Anim Vet Sci 3(6):334–346

Floren C, Wiedemann I, Brenig B, Schutz E, Beck J (2015) Species identification and quantification in meat and meat products using droplet digital PCR (ddPCR). Food Chem 173:1054–1058

Ghovvati S, Nassiri MR, Mirhoseini SZ, Moussavi AH, Javadmanesh A (2009) Fraud identification in industrial meat products by multiplex PCR assay. Food Control 20(8):696–699

Girish PS, Karabasanavar NS (2020) Molecular techniques for speciation of meat. In: Meat quality analysis, pp 133–151

Gupta S, Samal N (2022) Application of direct analysis in real-time mass spectrometry (DART-MS) in forensic science: a comprehensive review. Egypt J Forensic Sci 12:17

Ha Y, Thienes CP, Agapov AA, Laznicka AV, Han S, Nadala C (2019) Comparison of ELISA and DNA lateral flow assays for detection of pork, horse, beef, chicken, Turkey and goat contamination in meat products. J AOAC Int 102(1):189–195

Hermanto S, Rudiana T, Zein MIHL, Wisudawati AW (2022) Methods validation of pork authentication in processed meat products (sausages) through densitometry analysis. Indones J Halal Res 4(1):35–44

Hsieh YP, Woodward BB, Ho S (1995) Detection of species substitution in raw and cooked meats using immunoassays. J Food Prot 58(5):555–559

John HA, Birnstiel ML, Jones KW (1969) RNA-DNA hybrids at the cytological level. Nature 223: 582–587

Koseniuk A, Smolucha G, Nantonek-Wisniewska M, Radko A, Rubis D (2021) Differentiating pigs from wild boars based on nr6a1 and mc1r gene polymorphisms. Animals 11(7):2123

Kozlowski RL, Mitchell TW, Blanksby SJ (2015) A rapid ambient ionization-mass spectrometry approach to monitoring the relative abundance of isomeric glycerophospholipids. Sci Rep 5(1): 9243

Krager J, Baumert JL, Downs ML (2022) Quantification of soy derived ingredients in model bread and frankfurter matrices with an optimized liquid chromatography-tandem mass spectrometry external standard calibratiom workflow. J Food Prot 85(2):311–322

Lecrenier MC, Ledoux Q, Berben G, Fumiere O, Saegerman C, Baeten V, Veys P (2014) Determination of the ruminant origin of bone particles using fluorescence in situ hybridization (FISH). Sci Rep 4:5730

Li M, Zhang KY, But PP, Shaw P (2011) Forensically informative nucleotide sequencing (FINS) for the authentication of Chinese medical materials. Chin Med 6:42

Lim J, Kim S, Kim S, Eo H, Kim C, Kee P, Kim W, Bhak J (2009) BioBarcode: a general DNA barcoding database and server platform for Asian biodiversity resources. BMC Genomics 10 (Suppl 3):S8

Lu H, Zhang H, Chingin K, Xiong J, Fang X, Chen H (2018) Ambient mass spectrometry for food science and industry. TrAC Trends Anal Chem 107:99–115

Mafra I, Ferreira I, Oliveira M (2008) Food authentication by PCR-based methods. Eur Food Res Technol 227(3):649–665

Meira L, Costa J, Villa C, Ramos F, Oliveira MBPP, Mafra I (2017) EvaGreen real-time PCR to determine horse meat adulteration in processed foods. LWT – Food Sci Technol 75:408–416

Mendes E, Duarte N (2021) Mid-infrared spectroscopy s a valuable tool to tackle food analysis: a literature review on coffee, dairies, honey, olive oil and wine. Food Secur 10(2):477

Meza-Márquez OG, Gallardo-Velázquez T, Osorio-Revilla G (2010) Application of mid-infrared spectroscopy with multivariate analysis and soft independent modeling of class analogies (SIMCA) for the detection of adulterants in minced beef. Meat Sci 86:511–519

Montowska M, Pospiech E (2007) Species identification of meat by electrophoretic methods. Acta Sci Pol Technol Aliment 6(1):5–16

Montowska M, Alexander MR, Tucker GA, Barrett DA (2015) Authentication of processed meat products by peptidomic analysis using rapid ambient mass spectrometry. Food Chem 187:297–304

Nesic K, Stojanovic D, Baltic MZ (2017) Authentication of meat and meat products vs. detection of animal species in feed – what is the difference? In: 59th international meat industry conference MEATCON earth and environmental science, vol 85, p 012043. https://doi.org/10.1088/1755-1315/85/1/012043

Oh JD, Song KD, Seo JH, Kim DK, Kim SH, Seo KS (2014) Genetic traceability of black pig meats using microsatellite markers. Asian-Australas J Anim Sci 27:926–931

Orru L, Catillo G, Napolitano F, De Matteis G, Scata MC, Signorelli F (2009) Characterization of a SNPs panel for meat traceability in six cattle breeds. Food Control 20:856–860

Patel D, Panchal D, Patel K, Dalwadi M, Upadhyay U (2022) A review on UV visible spectroscopy. Int J Creat Res Thoughts 10(10):2320–2882

Peter C, Brunen-Nieweler C, Cammann K, Borchers T (2004) Differentiation of animal species in food by oligonucleotide microarrayhybridization. Eur Food Res Technol 219:286–293

Prandi B, Lambertini F, Faccini A, Suman M, Leporati A, Tedeschi T, Sforza S (2017) Mass spectrometry quantification of beef and pork meat in highly processed food: application on Bolognese sauce. Food Control 74:61–69

Prieto N, Roehe R, Lavin P, Batten G, Andres S (2009) Application of near infrared reflectance spectroscopy to predict meat and meat products quality: a review. Meat Sci 83(2):175–186

Rajaei N, Doosti A (2023) Designing a novel method based on multiplex PCR for detecting various meat of birds in processed ground meat products. Food Chem 7:100177

Rashid PMA, Babashekh MO, Marouf AS, Amin KM (2014) Identification of animal species in meat broth by simplex and multiplex PCR. J Zankoy Sulaimani-Part A (JZS-A) 16(1):97–102
Saez R, Sanz Y, Toldra F (2004) PCR-based fingerprinting techniques for rapid detection of animal species in meat products. Meat Sci 66:659–665
Sentandreu MA, Fraser PD, Halket J, Patel R, Bramley PM (2010) A proteomic – based approach for detection of chicken in meat mixes. J Proteome Res 9:3374–3383
Smith G, Pendell D, Tatum J, Belk K, Sofos J (2008) Post-slaughter traceability. Meat Sci 80:66–74
Spychaj A, Paul EM, Edward P (2009) PCR methods in meat species identification as a tool for the verification of regional and traditional meat products. Acta Sci Pol Technol Aliment 8(2):5–20
Unajak S, Meesawat P, Anyamaneeratch K, Anuwareepong D, Srikulnath K, Choowongkomon K (2011) Identification of species (meat and blood samples) using nested-PCR analysis of mitochondrial DNA. Afr J Biotechnol 10(29):5670–5676
Wells JM, Roth MJ, Keil AD, Grossenbacher JW, Justes DR, Patterson GE, Barket DJ Jr (2008) Implementation of DART and DESI ionization on a fieldable mass spectrometer. J Am Soc Mass Spec 19(10):1419–1424

Chapter 20
Market, Regulatory and Consumer Challenges of Functional Meat Products

Martin Aduah, Frederick Adzitey, Adamu Yusuf Kabiru, and Nurul Huda

20.1 Introduction

A good understanding of the relationship between diet and health has a significant influence on how consumers approach nutrition. As a preventive health care technique, this will help achieve optimal health status by encouraging healthier food sources that will ensure general well-being and potentially lower the risk of diseases. According to Sunil et al. (2013), the concept of functional foods has shifted the focus of research from food's primary function as a source of energy and substances for body building, to the effects of biologically active dietary ingredients on human health. This category of foods has been described using different terms; including functional foods, nutraceuticals, designer foods, nutritional foods, medical foods, longevity foods, hypernutritional foods, superfoods, pharma foods, prescriptive foods, phyto-foods, therapeutic foods, fitness foods, and foodiceuticals (Sunil et al. 2013).

M. Aduah
Africa Centre of Excellence for Mycotoxin and Food Safety, Federal University of Technology, Minna, Nigeria
e-mail: aduah.ac2117284@st.futminna.edu.ng

F. Adzitey
Department of Animal Science, University for Development Studies, Tamale, Ghana

A. Y. Kabiru
Department of Biochemistry, Federal University of Technology, Minna, Nigeria
e-mail: adamu.kabiru@futminna.edu.ng

N. Huda (✉)
Livestock Production Department, Faculty of Sustainable Agriculture, Universiti Malaysia Sabah, Sandakan, Malaysia
e-mail: drnurulhuda@ums.edu.my

S. A. Rather, F. A. Masoodi (eds.), *Hand Book of Processed Functional Meat Products*, https://doi.org/10.1007/978-3-031-69868-2_20

According to El Sohaimy (2012), the European Food Information Council defined functional foods as those that include biologically active components that might enhance physical and mental well-being and potentially lower the likelihood of illnesses. Functional foods are "nutraceuticals" or "designer foods," according to Lang et al. (2009) that improve human health by fighting against diseases. The founder and chairman of the Foundation for Innovation in Medicine (FIM), Cranford, Stephen DeFelice created the phrase "nutraceutical" in 1989 by combining the words "nutrition" and "pharmaceutical". According to El Sohaimy (2012), a nutraceutical is "a food (or part of a food that provides medical or health benefits, including the prevention and/or treatment of a disease". Bech-Larsen and Grunert (2003) indicated that functional foods are foods that may offer health benefits beyond basic nutrition. They resemble conventional foods that are meant to be consumed as part of a regular diet but have been altered to fulfill physiological roles other than meeting basic nutrient requirements. A product that is labeled as functional has additional, scientifically advanced ingredients that have been shown to have certain health benefits (Niva 2007). The European Food Information Council states that biologically active ingredients with the ability to enhance both mental and physical health as well as lower the risk of illnesses are necessary for functional foods. According to El Sohaimy (2012), these foods include enhanced, enriched, fortified and healthy foods.

Recently, a lot of research has gone on to develop functional meat products. The focus has been on finding ways to improve the quantity of compounds that are good for human health and decrease the amount of substances that could be harmful to human health (Manassi et al. 2022).

Functional meat and meat products are products that are fortified or technologically or biotechnologically altered ingredients to increase their value and advance human health (Mukumbo et al. 2021). These products go beyond mere sustenance and offer additional health benefits. Some functional meat products include probiotic fermented meat products, prebiotic meat products, symbiotic fermented meat products, reduced-fat meat products, reduced-sodium meat products, natural additives, and bioactive peptides (Verruck and Marsico 2024). Functional meat products aim to provide both sustenance and health benefits, making them an interesting area of research and development (Decker and Park 2010). This chapter focuses on the market, regulatory and consumer challenges of functional meat products.

20.2 Regulations of Functional Meat Products

In Europe, EU Regulation 1924/2006 governs functional food claims in order to protect consumer safety (Lalor and Wall 2011). Based on the USA's Dietary Supplement Health and Education Act framework, a novel functional regulation system known as "Foods with Function Claims" was created in 2015 (Iwatani and Yamamoto 2019). According to Granato et al. (2020), there are several rules pertaining to functional meat products that include non-traditional meat constituents for specific health benefits. These regulations are in place to guarantee customer safety and proper product labeling.

For manufacturers to successfully introduce functional meat products to the market, they must comprehend and abide by these requirements (Fortin 2022). The Food and Drug Administration (FDA) and the United States Department of Agriculture (USDA) are principally responsible for the regulatory control of functional meat products in the United States (Fortin 2022). The USDA is in charge of the safety and labeling of meat and poultry products, while the FDA controls the use of specific functional ingredients, additives, and health claims (Pascall and Lee 2016). Iwatani and Yamamoto (2019) stated that in 2015 a fresh functional regulation framework, "Foods with Function Claims" was created. It was modelled after the Dietary Supplement Health and Education Act, which was implemented in the United States. In Japan, "foods for specified health uses" (FOSHU) is a functional food regulation that was implemented by the Ministry of Health, Labor, and Welfare in 1991 (Iwatani and Yamamoto 2019).

One of the key regulatory aspects for functional meat products is accurate and transparent labeling. Manufacturers must ensure that product labels provide clear and truthful information about the product's ingredients, nutritional content, and any health claims. Labeling requirements may vary depending on the specific claims made about the product's health benefits (Bagchi 2014).

The FDA Food Labeling guidelines contain the labeling requirements for foods under the Federal Food Drug and Cosmetic Act (Fortin 2022). In the marketplace, terminologies like "functional foods" and "nutraceuticals" are frequently employed. Under the auspices of the Federal Food, Drug and Cosmetic Act, the FDA regulates these foods and plays three key oversight functions when it comes to deciding whether health claims can be included on food labels (Hoadley and Rowlands 2014). Firstly, the issuance of authorizing regulations once there is substantial scientific agreement that the claim is supported by the available scientific data. The second method involves notifying the FDA of health claims that have been supported by a thoughtful evaluation of scientific data on the relationship between a food substance and disease, either by the National Academy of Sciences or another US government scientific agency. The third method allows the FDA to approve qualified health claims where the strength and caliber of the supporting scientific data are insufficient for the FDA to issue an enabling rule.

Claims about a food's nutritional content describe how much of a certain nutrient is present. FDA-authorizing regulations require the use of nutrient content claims on food labels (Bagchi 2014). For instance, the manufacturer of a functional meat product could have to get FDA permission for these claims and include the necessary disclosures on the product label if the product contains additional vitamins or minerals that support certain health advantages, such as immune support or bone health (FDA 2022). Strict regulations control the use of health claims on functional meat product labels to prevent making false or misleading claims. Scientific data proving a link between the functional components and the purported health benefits is required to back up health claims (Hahn 2005). Under the Food and Drug Administration Modernization Act (FDAMA) and the Nutrition Labeling and

Education Act (NLEA), the FDA in the US controls health claims on food labels. Either the notification procedure for qualifying health claims or the petition process for authorized health claims must be used by manufacturers to get FDA approval for their health claims (FDA 2018).

To guarantee the safety of any additional ingredients, including bioactive chemicals or functional additives used in functional meat products, extensive safety evaluations must be conducted. According to Pressman et al. (2017), manufacturers bear the responsibility of carrying out comprehensive safety assessments and furnishing regulatory bodies with proof to validate the safety of their products. Evaluation of any dangers related to the functional components employed entails toxicological investigations, allergy assessments, and other scientific evaluations (Pressman et al. 2017). Good manufacturing procedures (GMPs) are guidelines that manufacturers must follow during the production process to ensure the safety and quality of their products (FDA 2018).

For producers to export functional meat products to outside markets, they have additional requirements to take into account such as international laws. It is important to carefully adhere to each jurisdiction's legislation as different countries may have different criteria for product safety, health claims, and labeling (Fortin 2022). The goal of international trade agreements and harmonization initiatives, such as those led by the Codex Alimentarius Commission, is to encourage uniformity and alignment of national food laws (Fink 2023). For producers of functional meat products, navigating the regulatory environment across several jurisdictions is still a difficult task (Gordon et al. 2020).

20.2.1 Consumer Challenges of Functional Meat Products

The market for functional meat products is confronted with several challenges despite its bright future; such as regulatory barriers, customer distrust, and competition from other protein sources (Roberfroid 2007). Regional variations in the regulations on product safety, ingredient labeling, and health claims can be obstacles to new entrants and growth in the industry. Furthermore, variables including taste preferences, cost, and perceived health benefits may have an impact on how consumers see and embrace functional meat products (Bagchi 2014).

20.2.2 Market Challenges

The market environment for functional meat products is dynamic and multifaceted, offering producers both challenges and opportunities. The perception and acceptability of consumers present one of the major challenges. Even though people are becoming more interested in foods that improve health, some consumers could have doubts regarding the safety and effectiveness of functional meat products. For functional foods to succeed on the market, it is imperative that consumers are sensitized of their advantages and that any doubts regarding their legitimacy and safety are addressed (Katz et al. 2019). Furthermore, a lot of businesses are fighting for market dominance in the functional food industry, which is characterized by intense rivalry. Therefore, producers of functional meat products need to set themselves apart from the competition with creative recipes, persuasive marketing approaches, and unambiguous health benefits messaging (Frewer et al. 2003). In addition, the cost may be a challenge since functional meat products carry a higher price tag than traditional meat products, which keeps them out of the reach of some customer groups (Gomez-Luciano et al. 2019). Moreover, the worldwide reach of the meat sector poses logistical obstacles, concerning obtaining premium ingredients and guaranteeing uniformity in product excellence and effectiveness across various markets. The competitive landscape for functional meat products is further complicated by changing consumer trends and tastes, such as the emergence of plant-based alternatives (Nevalainen et al. 2023).

20.2.3 Regulatory Challenges

Another major obstacle for producers of functional meat products is navigating the regulatory environment (Elasrag 2016). Important regulatory organizations that provide guidelines for the production and distribution of functional meat to customers are the Food and Drug Administration (FDA) in the United States and the European Food Safety Authority (EFSA) in Europe (Owusu-Apenten and Vieira 2023). Labeling requirements compliance is one important regulatory factor to take into account. According to Gordon et al. (2020), manufacturers are required to truthfully convey the health benefits of their products without exaggerating their effectiveness or making deceptive claims.

Obtaining regulatory approval for health claims can be an expensive and time-consuming procedure that necessitates strong scientific proof of the benefits being suggested (Verbeke 2006). Concerns regarding the safety and appropriateness of functional substances for use in food products may arise from their use in meat products. To maintain customer safety and regulatory compliance, manufacturers must carry out thorough safety evaluations and adhere to laws controlling the use of additives, preservatives, and other functional components (Sardi et al. 2018). Manufacturers operating in several markets face additional challenges due to international trade agreements and harmonization initiatives aimed at matching regulatory standards across different jurisdictions.

20.2.4 *Consumer Challenges*

The success of functional meat products depends on comprehending and satisfying consumer tastes and expectations (Siró et al. 2008). Innovation in this field is being driven by consumer demand for foods that improve health, but producers need to balance taste and utility (Heasman and Mellentin 2001). Although health advantages are a major factor in purchasing intent, taste, texture, and appearance are still very important in determining consumer acceptance and recurring buying behaviour (Dean et al. 2009). Furthermore, buyers are becoming more critical of the sourcing and production methods used in the manufacturing of these products due to ethical, environmental, and animal welfare concerns (Bray et al. 2017). To gain customers' trust and loyalty, manufacturers should be open and honest about their sourcing and manufacturing methods, showcasing their dedication to ethical and sustainable food production (Aiking and de Boer 2019). Additionally, manufacturers must modify their product offerings and marketing strategies in response to shifting consumer preferences and lifestyles brought about by demographic shifts and changing lifestyles (Lin and Hsu 2015). For example, younger consumers may prioritize novelty and convenience, while older consumers may place a higher value on health and wellness (Loebnitz et al. 2020).

Industry players in the functional food area can however take advantage of these challenges to innovate and set themselves apart with their products. This can be done by creating new and enhanced functional meat products that satisfy consumer tastes and needs by utilizing developments in food science and technology (Abecassis et al. 2018). Furthermore, functional products can be made more trustworthy by using smart marketing and communication techniques that inform customers about their health benefits (Baker et al. 2022). To overcome these obstacles and finally realize the full potential of functional meat products in the global market, industry collaboration will be necessary to promote innovation, transparency, and consumer education.

20.3 Functional Meat Products Market

In recent years, growing consumer interest in health and wellbeing coupled with advancements in food technology and ingredient innovation have resulted in a major expansion of the market for functional meat products. The global market for dietary supplements and functional foods is huge; according to Bagchi (2014), the former is predicted to grow from over $30 billion in 2017 to over $440 billion in 2022, while the latter is forecast to rise from $133.1 billion in 2016 to $278.02 billion by 2024. The sales of functional food ingredients in Europe grew from 19,555 to 28,798 billion dollars between 2016 and 2023. Geographically, the market for functional meat products is expanding rapidly in all directions, with the rising major markets being North America, Europe, and Asia-Pacific. According to Fig. 20.1 (Wunsch

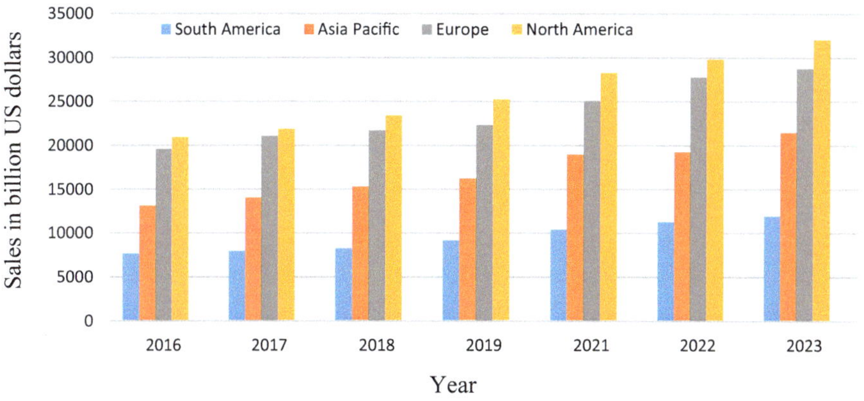

Fig. 20.1 Functional foods sales by region (Wunsch 2020)

2020), North America has been leading the market for functional food products since 2016 and this is projected to continue.

The market for functional meat products is expanding due to a number of significant trends and factors. First, the demand for foods that provide additional health advantages beyond basic nutrition is being driven by consumers' growing health consciousness (Kumar et al. 2021). Health-conscious consumers looking for convenient solutions to support their wellness goals are drawn to functional meat products, which are enriched with vitamins, minerals, antioxidants, and other bioactive substances (Lau et al. 2013).

Furthermore, interest in functional foods that address certain health issues like heart health, joint support, and immune function has increased due to the aging population and the rising prevalence of lifestyle-related disorders (Drewnowski and Gomez-Carneros 2000). Functional meat products formulated with ingredients targeting these health areas are well-positioned to capitalize on this demographic trend (Stanton et al. 2001). Advancements in food technology and ingredient innovation have also played a significant role in driving market growth. Manufacturers are increasingly leveraging novel ingredients and formulation techniques to enhance the functional properties of meat products while maintaining their taste, texture, and sensory appeal (Talens et al. 2022). This includes the use of encapsulation technologies to improve ingredient stability and bioavailability, as well as the incorporation of plant-based ingredients to enhance nutritional profiles.

Due to the aging population and the strong over-the-counter market in several nations, the global functional food and nutraceutical business is predicted to increase from $175 billion in 2013 to $424 billion by 2017 (Daliri and Lee 2015; Ken 2014). The market for nutraceutical beverages is anticipated to develop at the fastest rate possible, with a compound annual growth rate (CAGR) of 8.8%. In 2016, the industry is projected to be valued close to $87 billion. As the market for nutritional food grows at a compound annual growth rate (CAGR) of 6.4%, it is predicted to reach $67 billion in 2016. From 2022 to 2028, the global nutraceuticals market is projected to expand at a compound annual growth rate (CAGR) of 8.47% or from 425.57 to 693.06 billion USD (Ken 2014). The value of the global nutraceutical industry was estimated to be around US$233.9 billion in 2020 and is expected to rise to US$358.5 billion by 2027. China and Japan are the next biggest users of nutraceuticals after North America. When combined, they account for over 78% of the Asia-Pacific market (Jayaweera 2023). According to a study by Shimizu (2019), China is the second-biggest consumer of nutraceutical products in the globe due to its larger population and more health-conscious citizens. Japan, on the other hand, is the greatest consumer of nutraceuticals in Asia. According to Rajagopal (2014), the large population base and busy lifestyles of the urban middle-class population are expected to propel the Indian nutraceutical market to grow to US $4 billion by 2018 at a CAGR of 17%. Due to their growing middle-aged populations, the Russian and Hungarian nutraceutical markets are predicted to have market shares of 24.4% and 20.2%, respectively in 2017 (Daliri and Lee 2015).

20.3.1 Functional Meat and Meat Products

Research to create meat products with lower fat, sodium or cholesterol content has gained prominence by the desire to create meat products with better and healthier compositions (Pogorzelska-Nowicka et al. 2018; Bis-Souza et al. 2019). Furthermore, improved unsaturated fatty acid composition and addition of components has been encouraged for usage to enhance functional meats. While satisfying customer expectations for meat products with increased nutrition, these useful components may also have positive health consequences (Ursachi et al. 2020; Thøgersen and Bertram 2021). An increasing research interest is pointing to the health benefits of certain ingredients in meat and animal products. Recently, there has been a lot of focus on developing meat and meat products with physiological activities to enhance health and reduce the risk of diseases (Kumar et al. 2013).

Animal products account for almost 60% of total lipids, 70% of saturated fats, and 100% of total cholesterol in the diets of industrialized nations (Jiménez-Colmenero et al. 2001). The meat industry can explore new avenues for using food for health rather than nutrition. Some of these avenues include controlling the composition of raw and processed materials by reformulating the fatty acid content or adding dietary fiber, antioxidants, probiotics, and many other ingredients (Scollan et al. 2005). Additionally, during processes like fermentation, curing and aging, and enzymatic hydrolysis, other chemicals, particularly peptides can be produced from meat products to increase their functional value for customers (Zhang et al. 2010). A few functional meat components and their possible health benefits are displayed in Table 20.1.

Table 20.1 Some functional meat components and their potential health benefits

Functional components	Source	Potential benefits	Reference(s)
Linoleic acid	Chicken, beef, pork dairy and egg products	Reduce incidence of colon cancer (57%) and rectal cancer (59%), for weight loss, heart health, reduce inflammation and enhance eye health	Hunter (2012); Djoussé et al. (2001)
Long-chain omega-3 fatty acids	Salmon, cod liver oil, herrings, oysters, mackerel, fish and fish oil	Immune system booster, improves digestion and fertility	Deckelbaum and Torrejon (2012)
Conjugated linoleic acid (CLA)	Meat of ruminants and dairy products (cheese)	Improve body composition and reduces the risk of certain cancers, weight loss supplementation	Klein et al. (2000); Rainer and Heiss (2004)
Lactic acid bacteria	Naturally fermented meats including sausages, salami, and other cured meat products	Improve the quality of intestinal microflora and gastrointestinal health	Mathur et al. (2020); Kaveh et al. (2023)
Vitamin E	Goose meat, poultry meat, fish and dairy foods, Kielbasa sausage	Preventing heart disease and eye disorders, protecting against some cancers and promoting skin health	Mocchegiani et al. (2014); Rizvi et al. (2014)
Peptides	Dairy products, meat and seafood	Have anti-inflammatory, antioxidant, antihypertensive, and anticancer properties	Elias et al. (2008); Jillian (2023).
Selenium	Eggs, some fish, seafood	Selenium supports cognition, immune system function, and fertility	Benstoem et al. (2015); Rayman (2012)
Antioxidants (carosine, anserine)	Meat of herbivores (such as cows and sheep), poultry meat (native chicken meat, in particular)	Prevents body tissue damage by protecting cell membranes from oxidation., anti-glycation activity	Xing et al. (2019)
Probiotics	Yogurt, soft cheese, meat, fish, dairy products	Boost health of the gut microbiome, digestive balance, diarrhea treatment and prevention, mental health boost and improve heart health	Abatenh et al. (2018); Rijkers et al. (2011)

20.3.2 *Factors That Contribute to the Need for Functional Meat Development*

Predanócyová et al. (2023) stated that sustainable living, healthy lifestyles, and the COVID-19 pandemic's aftermath have a significant impact on current food market trends. As a result of their health advantages, functional meat products (FMP) are being consumed at an increasing rate. The demand for diets with additional health advantages has increased due to changes in the socio-economic and demographic characteristics of the population. A 2015 report from the U.S. State Department and the United Nations Department of Economic and Social Affairs stated that 8% of the world's population is over 65. This age group is predicted to account for 13% of the population by 2030, with a majority in developing countries. Thus, a higher quality of life is required as life expectancy rises (United Nations 2015).

In addition to stakeholders, governments, researchers, health professionals, and the food sector, the rising cost of healthcare has forced these groups to look for ways to manage these costs more effectively. The primary drivers of the nutraceutical and functional food business include the increasing prevalence of obesity, diabetes, eye health issues, cardiovascular disease, and a growing need for preventative medicine (Zawistowski 2008). Additionally, convenience and products with a better major component bioavailability are what consumers are looking for (Bourne 2013).

The global functional food and nutraceutical market is driven by consumers' growing awareness of the nutritional advantages of foods for potential illness prevention and health enhancement (Daliri and Lee 2015). Due to the rising demand for functional foods, the USA, which now has the largest market for these products worldwide, is predicted to expand by 21% over the next several years, reaching a valuation of USD 8.62 billion by 2015 (Daliri and Lee 2015). Iwatani and Yamamoto (2019) stated that Japanese people are renowned for living long lives and that the country has a long history of utilizing foods high in health benefits. As of 2007, Japan's market value for "foods for specified health uses" (FOSHU) goods was 6.2 billion dollars. Sales of functional foods, which included FOSHU products, totaled $8 billion in 2018 (Iwatani and Yamamoto 2019). Hence the market for innovative functional food products is continuously growing.

20.4 Functional Meat Products Launch Strategy

A comprehensive approach that includes product development, marketing, and distribution is necessary for the successful launch of functional meat products (Granato et al. 2020). Manufacturers can increase product awareness, consumer acceptance, and market penetration by carefully planning and carrying out each step of the launch process (Cooper 2019).

20.4.1 Market Research and Product Development

In-depth market research is necessary to comprehend customer preferences, industry trends, and the competitive environment before the launch of a new functional meat product (Guiné et al. 2020). In addition to evaluating the current functional meat products on the market, market research should analyze consumer demographics, dietary preferences, and purchasing patterns (Scholderer and Grunert 2005). Mozaffarian and Wu (2011) indicated that producers can create functional meat products that satisfy consumer preferences and demands by using market knowledge. To address common health issues or dietary trends, this may entail creating products with certain health-improving constituents, such as omega-3 fatty acids, vitamins, minerals, or antioxidants. To guarantee consumer acceptance and pleasure, additional consideration should be given to product features such taste, texture, packaging and convenience.

20.4.2 Regulatory Compliance

Launching functional meat products requires careful attention to regulatory compliance (Bagchi 2014). According to Fortin (2022), to stay out of legal hot waters and safeguard the public's health, manufacturers need to abide by rules regarding ingredient safety, labeling, and health claims. To achieve this, safety evaluations must be carried out. Regulatory clearances for health claims must be obtained, and product packaging must be appropriately labeled (Fortin 2022; Food and Drug Administration 2019). To maintain compliance with relevant laws and standards and manage the complicated regulatory landscape, regulatory experts and legal advisers should collaborate (Bagchi 2014). Manufacturers can reduce regulatory risks and expedite the launch process by proactively addressing regulatory requirements during the product development phase (Algorri et al. 2023).

20.4.3 Marketing and Promotion

To increase consumer awareness, interest, and demand for functional meat products, a strong marketing and promotion plan is necessary (Siró et al. 2008). A multichannel strategy including influencer alliances, social media, digital marketing, traditional advertising, and experiential marketing events may be used to achieve this (Grunert and Wills 2007). The marketing plan's essential components are shown in Fig. 20.2 (Shankar et al. 2012).

From the Fig. 20.2, higher sales and increased consumer interest will be achieved through the process of clearly communicating the health benefits, taste, and qualities of the functional meat product; providing information to educate consumers about the health benefits and overall well-being; organizing sampling and tasting events to allow consumers to experience the product firsthand, encourage trial, and purchase; leveraging digital platforms to reach target consumers, and solicit feedback and collaborating with retailers to secure prominent shelf placement to drive visibility and sales.

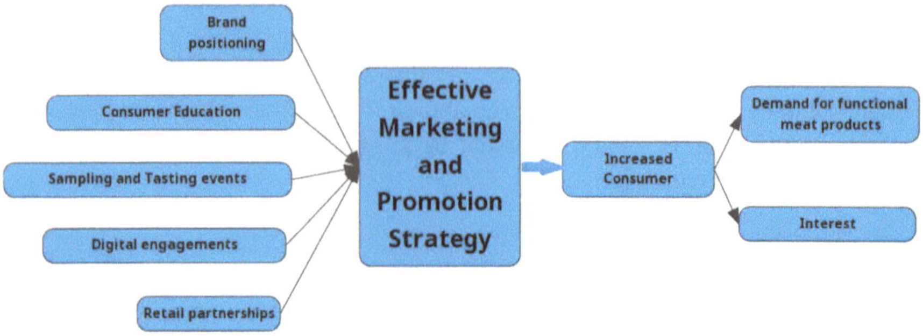

Fig. 20.2 Key elements of marketing strategy. (Source Shankar et al. 2012)

20.4.4 Distribution Strategy

For functional meat products to be widely available, it is essential to choose the appropriate distribution channels (Aaker and Moorman 2023). It is possible to leverage channels including direct-to-consumer sales, food service outlets, internet retailers, supermarkets, specialty food stores, health food stores, and traditional retail.

Manufacturers can effectively distribute products to target markets by partnering with well-established distributors and merchants, who can take advantage of the networks and infrastructure already in place (Gibbs and Humphries 2009). Additionally, investing in cutting-edge distribution channels like subscription services and e-commerce platforms can present chances to connect with tech-savvy customers and profit from dynamic consumer purchasing patterns (Marcus 2021). A thorough and well-executed plan that takes into account product development, marketing, distribution, and regulatory compliance is necessary for the launch of functional meat products (Brody and Lord 2007). Manufacturers can achieve a strong market position and maximize the success of their launch of a functional meat product by choosing the right distribution channels, adhering to regulatory requirements, executing efficient marketing strategies and performing in-depth market research.

20.5 Future Trends in the Development of Functional Meat Products

The growing demand from consumers for foods that provide health advantages beyond basic nutrition has sparked an evolution in the field of developing functional meat products (Granato et al. 2020). Future trends are anticipated to influence the development of functional meat products as science and technology progress together with our understanding of nutrition and health (Bigliardi and Galati

2013). According to Mont and Neuvonen (2014), these patterns are a reflection of changing consumer choices, technology advancements, and cultural shifts toward sustainability and wellness.

20.5.1 Personalized Nutrition

One trend that will probably be seen in the creation of functional meat products in the future is personalized nutrition; which is tailoring food items to individual dietary needs, tastes, and genetic profiles (Gibney et al. 2018). The potential for personalized nutrition approaches; which can recognize each person's specific nutritional needs and suggest tailored dietary interventions is being made possible by developments in genetic testing, biomarker analysis, and artificial intelligence. Personalized nutrition in the context of functional meat products refers to the creation of unique meat compositions enhanced with certain nutrients or bioactive chemicals to address genetic predispositions or individual health concerns (Panagoulias et al. 2022). Depending on a person's specific nutritional requirements, meat products fortified with specific vitamins, minerals, or phytonutrients may be designed to enhance immunological, cognitive, or cardiovascular health (Arshad et al. 2021).

20.5.2 Plant-Based Meat Alternatives

The demand for plant-based meat substitutes is being driven by the growing acceptance of plant-based diets and worries about environmental sustainability. Plant-based meat alternatives created from materials including legumes, grains, and vegetables are gaining popularity despite the continued popularity of traditional meat products (Hartmann and Siegrist 2017). According to Tyndall et al. (2024), incorporating plant-based components into meat products to improve their nutritional profile and sustainability may be a trend in the creation of functional meat products in the future. Tyndall et al. (2024) further noted that it is possible to include plant-based proteins, fibers, and antioxidants with meat-like flavour and texture while lowering the amount of saturated fat, increasing fiber consumption, and enhancing antioxidant activity.

20.5.3 Clean Label and Natural Ingredients

Cleaner, more transparent ingredient lists are becoming increasingly popular in functional meat products due to consumer demand for clean-label foods manufactured with natural, minimally processed components (Asioli et al. 2017). Customers are reading food labels more carefully and are looking for items without artificial substances, preservatives, or additives (Chen et al. 2022). To meet consumer expectations for transparency and authenticity and to improve product appeal, future trends in the manufacture of functional meat products must prioritize the use of natural, plant-based ingredients and clean label formulations. This can be achieved by using plant-derived colours and flavours; reducing the use of synthetic processing aids, and substituting natural additives for artificial ones (Rai et al. 2023).

20.5.4 Sustainable Production Practices

The need for more sustainable methods of producing meat is being driven by worries about the effects of meat production on the environment, including deforestation, greenhouse gas emissions, and water consumption (Alexander et al. 2017). To satisfy consumer demand for wholesome and tasty products, future efforts for the development of functional meat products will try to minimize the environmental impact of meat production (Ponnampalam et al. 2019).

According to Caporgno and Mathys (2018), adopting resource-efficient production techniques, obtaining meat from sustainably managed farms, and investigating substitute protein sources like cultured meat and insect-based ingredients can all help accomplish sustainable production. Furthermore, components with established environmental advantages, like omega-3 fatty acids produced from algae or plant-based proteins having a smaller environmental impact than conventional meat sources, may be included in functional meat products.

20.5.5 Technological Innovation

Innovation in the creation of functional meat products is being fueled by developments in food technology, such as cutting-edge processing methods, ingredient encapsulation, and functional ingredient delivery systems (Tomar et al. 2022). Modern technology may be used in the future to improve the bioavailability, stability, and effectiveness of functional additives in meat products.

According to Enrico (2019), delivery methods based on nanotechnology may make it possible to distribute bioactive substances to particular tissues or organs with precision, hence augmenting their therapeutic effects. Additionally, the development of meat products suited to each person's nutritional requirements and preferences may be made possible by advances in cell culture and three-dimensional (3D) printing (Eswaran et al. 2023).

20.6 Conclusion

The future development of functional meat products is expected to be shaped by emerging trends in personalized nutrition, plant-based alternatives, clean-label ingredients, sustainable production practices, and technological innovation. By embracing these trends and leveraging advancements in science and technology, manufacturers can create functional meat products that meet evolving consumer demands for health, sustainability, and convenience while delivering delicious and nutritious eating experiences. Regulations in the future must make sure that the obligation to shield consumers from deceptive claims is balanced with the need to give the food business access to markets.

The meat industry will successfully enter the functional food market by adhering to the necessary regulations, focusing on the right market, and developing functional meat products that are at the same time safe, flavoursome, and reasonably priced. Several studies conducted on processed meat products have demonstrated that it is possible to manufacture meat products that are healthier and more appealing to customers.

References

Aaker DA, Moorman C (2023) Strategic market management. Wiley

Abatenh E, Gizaw B, Tsegay Z, Tefera G, Aynalem E (2018) Health benefits of probiotics. J Bacteriol Infect Dis 2(1):17–27

Abecassis J, Cuq B, Escudier JL, Garric G, Kondjoyan A, Planchot V, Salmon JM, de Vries H (2018) Food chains; the cradle for scientific ideas and the target for technological innovations. Innov Food Sci Emerg Technol 46:7–17. https://doi.org/10.1016/j.ifset.2017.09.011

Aiking H, de Boer J (2019) Protein and sustainability–the potential of insects. J Insects Food Feed 5(1):3–7. https://doi.org/10.3920/JIFF2018.0011

Alexander P, Brown C, Arneth A, Dias C, Finnigan J, Moran D, Rounsevell MD (2017) Could consumption of insects, cultured meat or imitation meat reduce global agricultural land use? Glob Food Secur 15:22–32. https://doi.org/10.1016/j.gfs.2017.04.001

Algorri M, Cauchon NS, Christian T, O'Connell C, Vaidya P (2023) Patient-centric product development: a summary of select regulatory CMC and device considerations. J Pharm Sci 112(4):922–936. https://doi.org/10.1016/j.xphs.2023.01.029

Arshad MS, Khalid W, Ahmad RS, Khan MK, Ahmad MH, Safdar S, Kousar S, Munir H, Shabbir U, Zafarullah M, Nadeem M, Asghar Z, Suleria HAR (2021) Functional foods and human health: an overview. IntechOpen. https://doi.org/10.5772/intechopen.99000

Asioli D, Aschemann-Witzel J, Caputo V, Vecchio R, Annunziata A, Næs T, Varela P (2017) Making sense of the "clean label" trends: a review of consumer food choice behavior and discussion of industry implications. Food Rev Int 99:58–71. https://doi.org/10.1016/j.foodres.2017.07.022

Bagchi D (2014) Nutraceutical and functional food regulations in the United States and around the world. Elsevier

Baker MT, Lu P, Parrella JA, Leggette HR (2022) Consumer acceptance toward functional foods: a scoping review. Int J Environ Res Public Health 19:1217. https://doi.org/10.3390/ijerph19031217

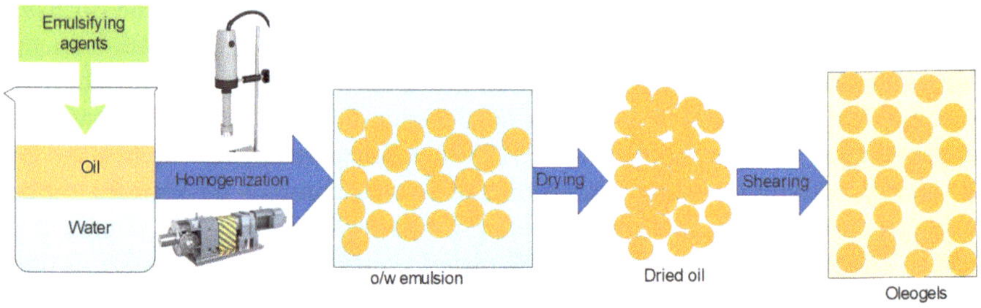

Fig. 21.3 Oleogel preparation method

addition to their functional properties as fat replacers, oleogels can provide nutritional benefits. The use of oleogels can help reduce the saturated fat content in food products by replacing solid fats with healthier liquid oils, which are rich in unsaturated fatty acids (Martins et al. 2017). Furthermore, oleogels can be formulated with functional ingredients, such as antioxidants, vitamins, and nutraceuticals, to improve the nutritional profile of food products (Rogers et al. 2008).

The application of oleogels as oil-based meat suspensions stabilized by adding organogelators was studied by Lupi et al. (2012). The stabilizing effects of added additives were evaluated by destabilizing the modified suspensions via centrifugation and comparing their rheological characteristics and oil loss to unmodified systems. Results indicated that both additives provided stabilization, with monoglycerides of fatty acids proving more effective even at low concentrations. The rheological approach was instrumental in determining the suitable type and amount of organogelator for use as an oil suspension stabilizer. In another study, the use of ethylcellulose as an organogelator for vegetable oils, aiming to produce oleogels with the potential to replace saturated fats in frankfurters was tested (Zetzl et al. 2012). Mechanical properties of oleogels made from canola, soybean, and flaxseed oils (10% ethylcellulose, 90% oil) were assessed, with higher unsaturation levels resulting in harder gels. The feasibility of replacing saturated animal fat with canola oil oleogels in frankfurters showed no significant differences in chewiness or hardness compared to controls. Furthermore, the substitution of beef fat (BF) with regular and structured canola oil (organogel produced with ethylcellulose (EC), with sorbitan monostearate (SMS)) in frankfurters was investigated (Barbut et al. 2016). Regular oil substitution doubled the frankfurters' hardness relative to BF. Higher concentrations of EC resulted in higher sensory hardness than BF. Oil-based frankfurters had smaller fat globules and were lighter in color than BF ones but showed higher smokehouse yields. The findings suggest the potential of organogels to replace BF in frankfurters, enabling manipulation of textural properties to resemble traditional products whilst reducing saturated fat content. In conclusion, the application of oleogels as oil-based meat suspensions can be effectively stabilized through the addition of organogelators. These oleogels have demonstrated potential as replacements for saturated fats in food products, without significantly altering key sensory attributes like chewiness and hardness.

21.5 Emulsion Gels as a Fat Replacement

Emulsion gels, semi-solid structures resulting from the trapping of liquid oil within a hydrocolloid matrix, are garnering interest as fat substitutes owing to their capability to mimic the texture and mouthfeel of fat (McClements and Gumus 2016). Emulsion gels are a complex colloidal material with both emulsion and gel states, formed by droplets of one liquid dispersed in another insoluble liquid. The fabrication of emulsion gels is a two-step process: the formation of a stable emulsion and the subsequent gelation. An emulsifier, water, and oil are amalgamated to create stable emulsions. The ratio of oil to water and the protein content are crucial parameters in stabilizing the oil-water interface. The transformation of the emulsion into a solid-like emulsion gel can be achieved either through thermal treatment or the incorporation of gelation agents in a cold-set gel. The schematic representation of emulsion gel preparation can be seen in Fig. 21.4. Heat treatment involves the denaturation and agglomeration of protein molecules to form new intra- and interchain disulfide bonds, thereby constituting a three-dimensional gel network structure. The preparation of a cold-set gel involves the introduction of gelation agents such as salt, acid, or enzymes into the emulsion.

The utilization of emulsion gels as fat substitutes in meat products is on the rise, attributed to their competency to replicate the physical attributes of animal fats. The production of such gels typically involves the creation of a protein-stabilized emulsion, potentially supplemented with a hydrocolloid stabilizer or additional ingredients post-emulsion. In a study conducted by Serdaroğlu et al. (2017), complete substitution of beef fat with emulsion gels in chicken patties led to the lowest sensory evaluation scores. However, patties with 25% and 50% fat substitution mirrored the sensory scores of samples containing full fat. Nacak et al. (2021) devised composite emulsion gels harnessing peanut and linseed oil as beef fat replacements. The sensory evaluations demonstrated these samples had superior oil scores and overall acceptability compared to control specimens with full fat. This outcome may be linked to the emulsion gel‘s ability to encapsulate oil beads and their inherent solid-like structures, thereby delivering the desired taste and

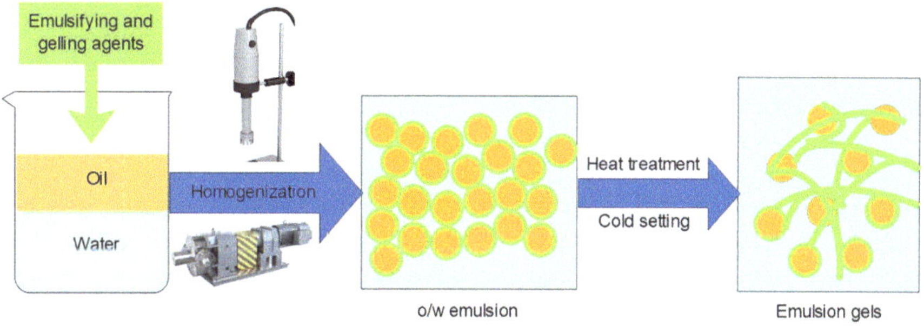

Fig. 21.4 Preparation method of emulsion gels

effectively mimicking fat characteristics. Pintado et al. (2016) utilized emulsion gels as carriers for health-promoting bioactive plant compounds, formulating gels from chia powder and olive oil as fat substitutes in low-fat Frankfurters. The incorporation of chia powder emulsion gel reduced the fat content by 40% and the energy intake by 30%, yielding a product that could be labeled as "reduced fat content". The reformulated Frankfurters were mineral-rich, containing elements like magnesium, manganese, and calcium, and exhibited enhanced lipid-protein interactions. The emulsion gel preserved the sensory characteristics within the accepted range, and the product exhibited oxidative stability during storage. This substitution resulted in a reduction in fat and energy content, an improved fatty acid ratio, reduced cooking loss, and elevated concentrations of certain minerals and amino acids. Furthermore, oat emulsion gels were employed as fat replacements to infuse meat products with β-glucan and monounsaturated fatty acids (MUFA) (Pintado et al. 2016a). In conclusion, research into the application of emulsion gels as fat substitutes in food products presents encouraging results. The studies highlighted here manifest that the substitution of animal fats with emulsion gels can successfully diminish fat and energy content, enhance nutritional value, and retain desirable sensory characteristics. The versatility of emulsion gels facilitates a variety of applications, including the creation of reduced-fat meat products and plant-based seafood analogs. These findings suggest that emulsion gels could serve as a valuable instrument in the development of healthier, low-fat food options without compromising taste or texture.

21.6 Protein-Based Hydrogelled Emulsions as Fat Replacers

A multitude of techniques have been recently devised to structure liquid oils for utilization in reformulated products to preserve the desirable attributes of the original product (Jiménez-Colmenero 2007). One specific technique involves the employment of protein-based gelled emulsions in meat products, which has been shown not to influence their hardness and water-holding capacity (Paglarini et al. 2022). Protein-based hydrogelled emulsions, semi-solid gels constituted by the entrapment of liquid oil within a protein matrix, have emerged as a promising alternative to traditional fat replacements due to their capability to offer a texture and mouthfeel akin to fat (Jimenez-Colmenero et al. 2015). The physical properties of the emulsion gels, such as structure and rheological behavior, are determined by factors like the role of the fat as an active or inactive filler, the droplet sizes, oil content, and processing conditions. The gel strength is governed by hydrogen and covalent bonds, as well as electrostatic and hydrophobic interactions. Incorporating other ingredients such as polysaccharides also influences the final properties of the gels due to their gelling/thickening properties. These protein-based emulsion gels are formed by oil droplets stabilized by a layer of protein adsorbed at the oil–water

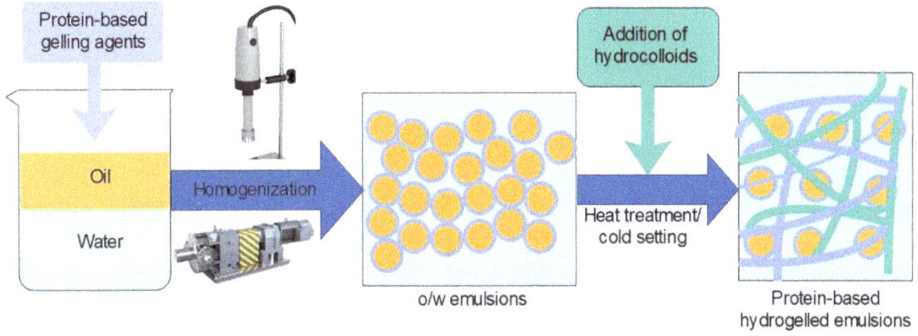

Fig. 21.5 Preparation method of protein-based hydrogelled emulsions

interface and entrapped in a protein gel network. The formation of a hydrogelled emulsion involves the creation of a protein-stabilized emulsion followed by the gelation of this emulsion. The resultant gel can be formed by inducing the aggregation of oil droplets and/or by gelling the continuous phase. The preparation of protein-based hydrogelled emulsions is illustrated in Fig. 21.5.

The characteristics of gelled emulsions are fundamentally reliant on the type and quantity of hydrocolloids and oil incorporated into the gelled emulsion. In terms of texture, products akin to the conventional product have been realized when carrageenan, soy protein isolate, or oat bran were integrated as gelling, emulsifying, or stabilizing agents (Pintado et al. 2018; de Souza et al. 2019). However, the replacement of animal fat with inulin-based gelled emulsions has resulted in a decrease in the hardness of meat products (Serdaroğlu et al. 2017; Glisic et al. 2019). From a nutritional standpoint, products formulated with gelled emulsions have demonstrated an increase in monounsaturated fatty acids (MUFA) and polyunsaturated fatty acids (PUFA) such as alpha-linolenic acid content and a decline in saturated fatty acids (SFA). Depending on the polysaccharide source harnessed to construct the gelled emulsion, final products with health claims could be accomplished.

Huang et al. (2022) prepared an emulsion from an aqueous soy protein isolate solution homogenized with coconut oil, to which konjac glucomannan was introduced to form a 3D cube emulsion gel with distinct hardness and strength via the TG cross-linking effect. Used as a fat substitute, the influence of varying protein and konjac concentrations on the emulsion gel systems was examined. The emulsion gel exhibited the potential to mimic a solid cube of fat, resembling pork fat in appearance, and demonstrating desirable functional attributes in terms of both mechanical and oral tribological properties. Consumers preferred the emulsion gel with an added protein content of 1% and konjac content of 4%. The 3D structures of fat substitutes can furnish the product with the desired visible cubic fat appearance, for which a protein–polysaccharide composite matrix offers a potential approach. Ran et al. (2022) amplified the cross-linking between soy protein and konjac glucomannan to mimic the texture of fish balls. The results indicated that the addition of konjac glucomannan significantly impacted the textural and rheological properties of plant-based fish balls, increasing hardness, chewiness, and gel strength at higher concentrations. The inclusion of konjac glucomannan facilitated the formation of a tighter gel network and denser cross-linking. Lower konjac concentrations increased the porosity and density of the structure, while an appropriate concentration enhanced elasticity and gel strength. In summary, protein-based hydrogelled emulsions represent a feasible strategy for reducing the fat content in meat products and enhancing their nutritional profiles. Future research should concentrate on optimizing these emulsions to achieve the optimal balance.

21.7 Microparticulated Powders and Microgels as Fat Replacers

Microparticulated powders are engineered to mimic the sensory and textural properties of fat. Microparticulated powders, like Simplesse®, are usually composed of proteins or carbohydrates that are processed into tiny particles, typically ranging from 0.1 to 100 micrometers in size (Singer et al. 1988). The small size of these particles allows them to mimic the sensation of fat on the tongue, providing a creamy and smooth texture similar to full-fat products (McClements and Li 2010). Proteins or carbohydrates are selected as the base material. Common choices include whey proteins, casein, or plant-based proteins for proteins, and maltodextrin or starch for carbohydrates. The material is then dissolved in water. The protein or carbohydrate solution is heated to denature the proteins or gelatinize the carbohydrates. This process allows the molecules to unfold and interact with each other. The heated solution is then passed through a high-pressure homogenizer, which physically breaks the material down into micro-sized particles. The process is usually repeated multiple times to achieve the desired particle size (Kew et al. 2020). Microgels are tiny, cross-linked polymer networks that swell in water. They typically range from tens of nanometers to a few micrometers in size. When these particles absorb water, they swell and form a gel-like structure, which can provide a texture similar to fat (Zhang and Cremer 2006). The preparation method of protein microgels is depicted in Fig. 21.6.

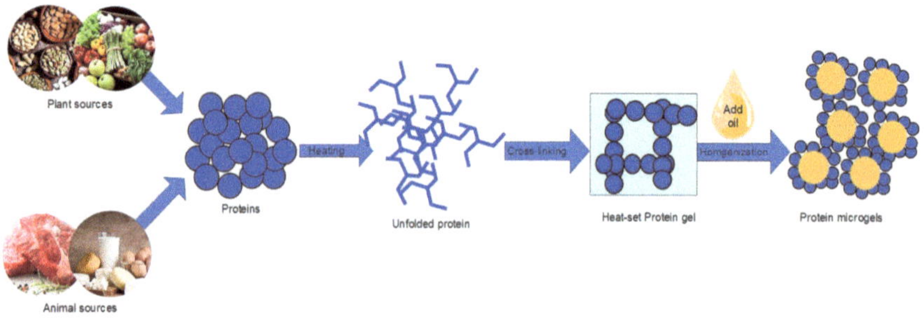

Fig. 21.6 Preparation method of protein microgels

One of the most commonly used microparticulated powders is Simplesse®, a microparticulated whey protein. It has been used as a fat mimetic in a variety of food applications due to its ability to provide a creamy texture similar to fat (McClements and Li 2010). However, its use in meat products is relatively unexplored and has the potential to be a promising area for future research (Jiménez-Colmenero 2013). The use of microparticulated proteins as fat replacers in meat products can help reduce the overall fat content while maintaining the desirable sensory properties typically associated with fat. This is particularly important as consumers continue to demand healthier options without sacrificing taste and texture. Incorporating microgels in meat products can help reduce the overall fat content while maintaining the desirable sensory properties typically associated with fat. This is particularly important as consumer demand for healthier options continues to grow. Notably, Pintado et al. (2016b) used chia in microgels as a fat replacer in frankfurters. Their study found that the frankfurters prepared with microgels showed similar textural properties to their full-fat counterparts. This suggests that microgels could provide a comparable mouthfeel to fat, making them a promising fat substitute in similar meat products. While the use of microgels shows promise, there are still challenges to overcome. One of the main issues is ensuring that the microgels remain stable during cooking and storage. The interaction of these microgels with other ingredients in the meat matrix also needs to be considered, as it can impact the overall product quality. Further, consumer acceptance is a critical factor. Even if the technology can produce low-fat meat products with similar texture and taste as traditional products, the ultimate success depends on consumer perception and acceptance of these modified products. In conclusion, while the use of microgels as fat replacers in meat products is a promising area of research, it is still in its early stages. Further research is required to fully understand their application and effects on the overall quality, safety, and acceptability of the meat products.

21.8 Conclusions

The production of low-fat meat products is a complex process that requires understanding various techniques and approaches to reduce fat content while maintaining product quality and consumer acceptability. Advances in lean meat selection, fat reduction methods, fat replacers, and formulation techniques have led to the development of a wide range of low-fat meat products that are both tasty and nutritious. As consumer demand for healthier meat products continues to grow, further research and innovation in this area will remain crucial for the development of novel, high-quality low-fat meat products.

References

Augustin MA, Hemar Y (2009) Nano-and micro-structured assemblies for encapsulation of food ingredients. Chem Soc Rev 38(4):902–912

Barbut S, Wood J, Marangoni A (2016) Potential use of organogels to replace animal fat in comminuted meat products. Meat Sci 122:155–162

de Souza PC, Martini S, Pollonio MAR (2019) Using emulsion gels made with sonicated soy protein isolate dispersions to replace fat in frankfurters. LWT 99:453–459

Domínguez R et al (2017) Effect of the partial replacement of pork backfat by microencapsulated fish oil or mixed fish and olive oil on the quality of frankfurter type sausage. J Food Sci Technol 54(1):26–37

Furlán LTR, Padilla AP, Campderrós ME (2014) Development of reduced fat minced meats using inulin and bovine plasma proteins as fat replacers. Meat Sci 96(2):762–768

Gharsallaoui A et al (2007) Applications of spray-drying in microencapsulation of food ingredients: an overview. Food Res Int 40(9):1107–1121

Glisic M et al (2019) Inulin-based emulsion-filled gel as a fat replacer in prebiotic-and PUFA-enriched dry fermented sausages. Int J Food Sci Technol 54(3):787–797

Heck RT et al (2018) Oxidative stability of burgers containing chia oil microparticles enriched with rosemary by green-extraction techniques. Meat Sci 146:147–153

Heck RT et al (2019) Volatile compounds and sensory profile of burgers with 50% fat replacement by microparticles of chia oil enriched with rosemary. Meat Sci 148:164–170

Hoffman JR, Falvo MJ (2004) Protein–which is best? Journal of sports science & medicine 3(3): 118–130

Huang L et al (2022) Create fat substitute from soybean protein isolate/konjac glucomannan: the impact of the protein and polysaccharide concentrations formulations. Front Nutr 9:843832

Jalal H et al (2013) Development of low fat meat products. International Journal of Food Nutrition and Safety 4(3):98–107

Jiménez-Colmenero F (2007) Healthier lipid formulation approaches in meat-based functional foods. Technological options for replacement of meat fats by non-meat fats. Trends Food Sci Technol 18(11):567–578

Jiménez-Colmenero F (2013) Potential applications of multiple emulsions in the development of healthy and functional foods. Food Res Int 52(1):64–74

Jiménez-Colmenero F et al (2010) Technological and sensory characteristics of reduced/low-fat, low-salt frankfurters as affected by the addition of konjac and seaweed. Meat Sci 84(3):356–363

Jimenez-Colmenero F et al (2015) Novel applications of oil-structuring methods as a strategy to improve the fat content of meat products. Trends Food Sci Technol 44(2):177–188

Kew B et al (2020) Review on fat replacement using protein-based microparticulated powders or microgels: a textural perspective. Trends Food Sci Technol 106:457–468

Kommuru TR (1999) Formulation development and in vitro/in vivo characterization of self-emulsifying drug delivery systems of coenzyme Q (10). Northeast Louisiana University, Monroe, LA, USA

Lorenzo JM et al (2016) Healthy Spanish salchichón enriched with encapsulated n− 3 long chain fatty acids in konjac glucomannan matrix. Food Res Int 89:289–295

Lupi FR et al (2012) Stabilization of meat suspensions by organogelation: a rheological approach. Eur J Lipid Sci Technol 114(12):1381–1389

Martins AJ et al (2017) Fortified beeswax oleogels: effect of β-carotene on the gel structure and oxidative stability. Food Funct 8(11):4241–4250

Martins A et al (2018) Edible oleogels: an opportunity for fat replacement in foods. Food Funct 9(2):758–773

McClements DJ, Gumus CE (2016) Natural emulsifiers – biosurfactants, phospholipids, biopolymers, and colloidal particles: molecular and physicochemical basis of functional performance. Adv Colloid Interf Sci 234:3–26

McClements DJ, Li Y (2010) Structured emulsion-based delivery systems: controlling the digestion and release of lipophilic food components. Adv Colloid Interf Sci 159(2):213–228
Mozaffarian D (2016) Dietary and policy priorities for cardiovascular disease, diabetes, and obesity: a comprehensive review. Circulation 133(2):187–225
Nacak B et al (2021) Peanut and linseed oil emulsion gels as potential fat replacer in emulsified sausages. Meat Sci 176:108464
Ozturk-Kerimoglu B et al (2022) Chemical, technological, instrumental, microstructural, oxidative and sensory properties of emulsified sausages formulated with microparticulated whey protein to substitute animal fat. Meat Sci 184:108672
de Souza Paglarini C et al (2022) Protein-based hydrogelled emulsions and their application as fat replacers in meat products: a review. Crit Rev Food Sci Nutr 62(3):640–655
Pintado T, Herrero AM, Ruiz-Capillas C et al (2016) Effects of emulsion gels containing bioactive compounds on sensorial, technological, and structural properties of frankfurters. Food Sci Technol Int 22(2):132–145
Pintado T, Herrero AM, Jiménez-Colmenero F et al (2016a) Emulsion gels as potential fat replacers delivering β-glucan and healthy lipid content for food applications. J Food Sci Technol 53: 4336–4347
Pintado T, Herrero AM, Jiménez-Colmenero F et al (2016b) Strategies for incorporation of chia (*Salvia hispanica* L.) in frankfurters as a health-promoting ingredient. Meat Sci 114:75–84
Pintado T et al (2018) Chia and oat emulsion gels as new animal fat replacers and healthy bioactive sources in fresh sausage formulation. Meat Sci 135:6–13
Ran X et al (2022) Improving the texture and rheological qualities of a plant-based fishball analogue by using konjac glucomannan to enhance crosslinks with soy protein. Innovative Food Sci Emerg Technol 75:102910
Rogers MA, Wright AJ, Marangoni AG (2008) Engineering the oil binding capacity and crystallinity of self-assembled fibrillar networks of 12-hydroxystearic acid in edible oils. Soft Matter 4(7):1483–1490
Serdaroğlu M, Nacak B, Karabıyıkoğlu M (2017) Effects of beef fat replacement withgelled emulsion prepared with olive oil on quality parameters of chicken patties. Korean J Food Sci Anim Resour 37(3):376–384
Simopoulos AP (2002) The importance of the ratio of omega-6/omega-3 essential fatty acids. Biomed Pharmacother 56(8):365–379
Singer NS, Dunn JM (1990) Protein microparticulation: the principle and the process. J Am Coll Nutr 9(4):388–397
Singer NS, Yamamoto S, Latella J (1988) Protein product base-US4734287A. US patent (Ed.) [preprint]
Singer NS, Latella J, Shoji Y (1990) Fat emulating protein products and process-US4961953A. US patent (Ed.) [preprint]
Solomando JC, Antequera T, Perez-Palacios T (2020) Evaluating the use of fish oilmicrocapsules as omega-3 vehicle in cooked and dry-cured sausages as affected by their processing, storage and cooking. Meat Sci 162:108031
Torres IC et al (2011) Effect of microparticulated whey protein with varying content of denatured protein on the rheological and sensory characteristics of low-fat yoghurt. Int Dairy J 21(9): n645–n655
Zetzl AK, Marangoni AG, Barbut S (2012) 'Mechanical properties of ethylcelluloseoleogels and their potential for saturated fat reduction in frankfurters. Food & Function 3(3):327–337. Edible oleogels: structure and health implications. AOCS press, Urbana 295–312
Zhang Y, Cremer PS (2006) Interactions between macromolecules and ions: the Hofmeister series. Curr Opin Chem Biol 10(6):658–663

Chapter 22
3D Printing in the Development of Meat Based Products

Kurnia Ramadhan and Nurul Huda

22.1 Introduction

The rise of three-dimensional (3D) printing technology has catalyzed transformative shifts in various industries. This innovative process, known as additive manufacturing, involves the sequential deposition of material to create three-dimensional objects from digital models. Using 3D printing technology, it is possible to create customized, complex shapes and geometries in a shorter amount of time.

The development of 3D printing technology has been ongoing since the 1980s, particularly after the patent for stereolithography was granted in 1986 (Fig. 22.1). Subsequent advancements have expanded the range of printable materials. Notably, researchers from Cornell University were the first to utilize 3D printing for edible materials in 2006, heralding the inception of 3D food printing—a significant milestone nearly two decades after the technology's introduction. This novel technique facilitates the precise, layer-by-layer construction of edible items, revolutionizing the fabrication of sophisticated edible structures (Nachal et al. 2019; Mantihal et al. 2020).

In 2010, a groundbreaking endeavor in 3D meat printing was undertaken, utilizing turkey meat, celery, and the enzyme transglutaminase to produce printed meat that could undergo traditional cooking techniques such as frying, baking, and stewing. The interest in developing meat products, including meat analogues, through 3D printing has surged, leading to more pioneering research. Notably, the

K. Ramadhan
Department of Food Science and Technology, Universitas Bakrie, Jakarta, Indonesia
e-mail: kurnia.ramadhan@bakrie.ac.id

N. Huda (✉)
Livestock Production Department, Faculty of Sustainable Agriculture, Universiti Malaysia Sabah, Sandakan, Malaysia
e-mail: drnurulhuda@ums.edu.my

S. A. Rather, F. A. Masoodi (eds.), *Hand Book of Processed Functional Meat Products*, https://doi.org/10.1007/978-3-031-69868-2_22

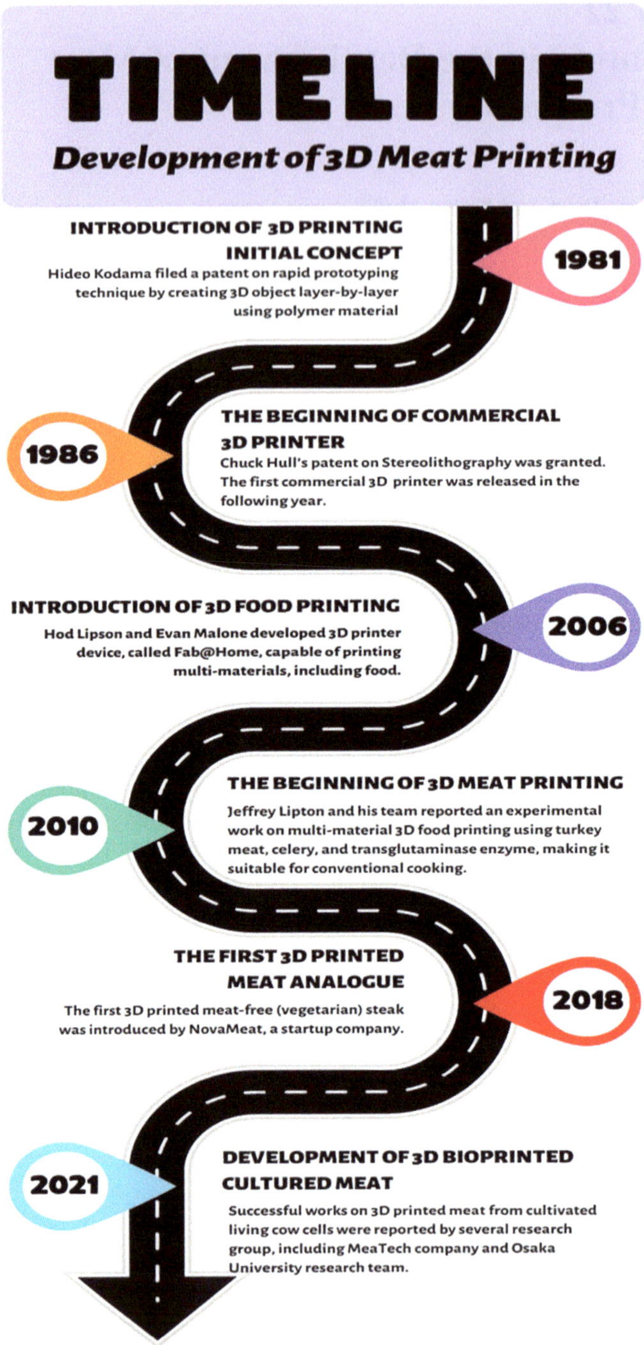

Fig. 22.1 Chronological timeline of 3D meat printing development

first 3D printed plant-based meat substitute, resembling a steak, was introduced in 2018. Advancements continued with the successful 3D bioprinting of beef steak using cultivated bovine muscle cells reported in 2021 (Kang et al. 2021; Pereira et al. 2021).

The advancements in 3D food printing are highly sought after to enhance the health and wellbeing of communities by enabling the personalization of nutrition and providing solutions for health issues, such as designing specialized diets for individuals with dysphagia (Jayaprakash et al. 2020; Dick et al. 2021b). The development of 3D-printed food that incorporates protein, dietary fibers, and micronutrients has been explored to achieve personalized nutrition goals (Chen et al. 2021; Huang et al. 2022; Wilson et al. 2023). Additionally, 3D food printing is being utilized to address sustainability challenges in the food system. In particular, the 3D printing of meat analogues is being investigated as a means to reduce carbon emissions from extensive livestock farming. Moreover, the 3D printing of traditional meats aims to improve the utilization of undervalued trimmings and off-cut meats, promoting more efficient resource use (Bhat et al. 2021; Dong et al. 2023). This chapter explores the use of 3D printing in developing meat-based products, an evolving field with considerable potential to revolutionize meat production and consumption.

22.2 3D Printing Process

Three-dimensional (3D) food printing is a complex process that involves several key components: food materials, a digital model and a 3D printer (Fig. 22.2). The food materials, serving as "inks", must be carefully selected and formulated to meet the requirements of the printing process. These materials should be easily processable into a form suitable for deposition by the printer, such as a paste or gel. They should

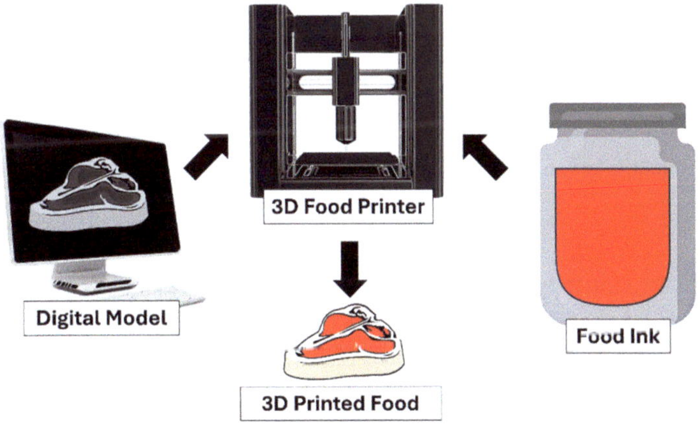

Fig. 22.2 Components of three-dimensional (3D) food printing

retain their shape after deposition, withstand the weight of subsequent layers, and possess the necessary properties for achieving the desired texture. Additionally, they should be capable of cooking or solidifying as needed. The taste, color, and nutritional content of these materials are also crucial considerations (Nachal et al. 2019).

The digital model of the object to be printed is another vital element. This model, created using Computer-Aided Design (CAD) software or derived from scanning an existing object, serves as the geometric blueprint for the intended structure. The model is converted into a stereolithography file (stl. file) and sent to slicing software, where the model's layers are generated and the printing parameters, such as speed, temperature, infill, and layer height, are set. A G-code file, a programming language used to instruct industrial machine controller computers, is then created with the dimensional and printing instructions and sent to the printer. The printer constructs the desired structure by depositing layers of food materials according to the pattern defined by the model (Jayaprakash et al. 2020).

The type of 3D printer and the food material used determine the printing technique employed, which could be extrusion, inkjet printing, or laser sintering. The printing conditions, including temperature and time, are crucial to the 3D food printing process. The temperature must be optimized to ensure the food material has the right consistency for printing and behaves as expected after deposition. The printing speed, which determines the duration of the process, should be adjusted to balance the resolution of the printed object and the total printing time. These conditions may need fine-tuning based on the specific food material and the desired characteristics of the final product. After printing, post-processing steps such as cooking, cooling, or setting may be necessary to solidify the printed structure and ensure it has the desired texture and taste. In some cases, additional ingredients or decorations may be added after printing. These post-processing steps should be carefully planned and executed to preserve the shape of the printed object and enhance its sensory attributes (Mantihal et al. 2020).

Among the types of 3D food printer, the extrusion-based 3D printing method is the most commonly used for 3D meat printing (Dick et al. 2019a). This method involves ejecting the material, i.e. meat paste or slurry, through the nozzle in a digitally controlled way. The stream is deposited in layers according to a pre-designed pattern, until a 3D structure is obtained. This method is suitable for meat because it allows for the use of viscous, paste-like materials, which is often the form that meat takes in this context.

22.3 Formulating the Meat for Enhancing Its Printability

The successful 3D printing of meat hinges on its printability, which is defined by the meat paste's ability to be extruded through the printer nozzle and retain its shape post-extrusion. During the extrusion process, the food material must exhibit low viscosity and yield stress to facilitate smooth flow. Once extruded, it requires adequate viscosity and yield stress to maintain its form and support subsequent layers. Moreover, the ink must possess robust mechanical support and gel strength to preserve the integrity of the structure, as noted by; thereby preventing collapse over time.

Meat, inherently a complex colloidal system, lacks natural printability. Enhancing the rheological and gel properties of meat paste during ink formulation is crucial for achieving optimal 3D printing outcomes. Currently, the printability of meat-based materials is primarily improved through pre-treatment methods that enhance colloidal interactions within minced meat or by incorporating additional hydrocolloids (Dick et al. 2021a). It is essential to consider the ink's resilience to post-treatment processes, aiming for a product that can evenly distribute mass, retain its complete shape, and minimize steaming losses during maturation. The rheological attributes of the raw materials, such as shear-thinning behavior, viscosity, and yield stress, significantly impact the printing process and the final product's quality. Proteins, fats, and carbohydrates, vital for a balanced diet, exhibit distinct rheological and gelation properties. Understanding these properties is imperative for effective 3D printing. Adjusting the ratio of meat to non-meat ingredients in ink formulations can enhance both rheology and printability, facilitating the development of nutrient-dense products.

The preparation of meat paste typically begins with the trimming of the meat to eliminate connective tissues and excess fat. The meat is subsequently finely ground and pulverized, usually in conjunction with salt at a concentration of approximately 1.5%. This mixture is often combined with water, generally around 15%, and hydrocolloids, which typically do not exceed a concentration of 1%. The presence of connective tissue within the meat paste can be mitigated by using a larger diameter printer nozzle, although this may lead to less accurate and precise shapes in the printed product (Dick et al. 2019a; Dong et al. 2023).

Fat can also enhance the flow behavior of meat materials, reduce viscosity, and improve ink flow and extrudability. They play a crucial role in gel formation, partnering with proteins to produce emulsion gels that exhibit a uniform structure, formability, and self-supporting characteristics. However, an excess of fat can result in a fluid-like behavior in the meat paste, leading to increased cooking loss and deformation (Dick et al. 2019b; Park et al. 2023). Fat can also be prepared separately from the meat paste. For instance, Park et al. (2023) prepared a fat paste from finely ground beef tallow and xanthan gum, which was used as a multi-filament filling in the ink cartridge to mimic marbled meat. Similarly, Dick et al. (2019b) used lard to create a composite multi-layer 3D printed meat product by placing a fat layer between layers of meat paste. Several plant-based oils have also been reported to enhance the 3D printability of meat paste prepared from chicken meat (Yin et al. 2024).

Hydrocolloids, a diverse array of long-chain polymers predominantly comprising polysaccharides and proteins, enhance the water retention and mechanical strength of gels, and stabilize hybrid dispersion systems in meat products. Within the context of 3D food printing, hydrocolloids interact with meat macromolecules, enhancing the colloidal stability and rheological properties of the meat. Serving primarily as thickening and gelling agents, hydrocolloids influence the extrusion of the ink by affecting its continuity and homogeneity. Their molecular structure, rich in hydrophilic groups such as hydroxyl and carboxyl, imparts viscosity to the solution, creating a stable colloidal system and optimizing the extrudability of meat materials. The gelling function of hydrocolloids primarily enhances the self-supporting capabilities of the ink. For instance, starch exhibits high viscosity and shear stability. At elevated concentrations, starch molecules interact more intensely, leading to a denser gel network and improved self-supporting properties of the ink (Ji et al. 2023). Consequently, starch is often added to other food materials to make them printable (Feng et al. 2019).

Transglutaminase, an enzyme that naturally occurs in nature, catalyzes the formation of isopeptide bonds between proteins or peptides. This enzyme essentially functions as a protein adhesive and is extensively employed in the food processing industry to enhance texture and fortify the binding of food products. This enzyme was instrumental in the groundbreaking work on 3D printed meat, as chronicled by Lipton et al. (2010). Their innovative research demonstrated the practicality of 3D printing turkey mash for sous-vide cooking. Furthermore, the application of this enzyme has been reported in 3D printing studies on surimi (Dong et al. 2020; Zhu et al. 2023). These studies concluded that the incorporation of transglutaminase can significantly enhance the 3D printability, viscoelastic properties, mechanical properties, and microstructure of surimi gels.

22.4 3D Printing of Traditional Meat Materials

3D printing technology for food has piqued the interest of numerous researchers, prompting them to explore its application across a variety of edible materials. With its potential to address sustainability issues, 3D printing is viewed as a promising approach to mitigate food waste by enhancing the utilization of undervalued meat cuts. Furthermore, the development of 3D printed meat for specific health conditions, such as dysphagia, has also captured the attention of many researchers. A summary of several studies on 3D meat printing, encompassing various sources of meat, can be found in Table 22.1.

Technological advancements in 3D food printing have facilitated the exploration of various meat types in diverse experimental scenarios. Studies focusing on beef have demonstrated the necessity of adding water and salt to produce a printable meat paste, which results in increased moisture content and reduced viscosity. Similarly, incorporating oil into the meat paste yields comparable effects, as reported in the studies on 3D printed chicken meat and fish. Notably, the self-supporting characteristic of the printed meat paste structure is crucial and can be maintained by ensuring a high yield stress at low shear rates. This property is typically achieved through the addition of hydrocolloids, as indicated in numerous studies. In addition to hydrocolloids, the transglutaminase enzyme has been shown to enhance self-supporting properties via a protein-crosslinking mechanism. The nozzle diameters employed in

Table 22.1 Recent studies on 3D printed meat products

Meat types	Other ingredients	Specific condition	Nozzle diameter size	Post processing	Reference
Beef	Water, salt, starch	Meat paste containing native and modified potato starch were evaluated for its printability	0.75 mm	Cooked in oven at 120 °C for 10 min	Liu et al. (2024)
Beef	Meat paste: salt, water, xanthan gum, gelatin Fat paste: tallow, xanthan gum	Fat paste was positioned at the core of the meat paste within the ink cartridge to produce a printed product that mimics marbled meat	1.5 mm	Cooked in oven at 200 °C for 10 min	Park et al. (2023)
Beef	Water, salt, starch	The effects of incorporating various types of modified starch into meat paste were evaluated with the aim of creating a meat product suitable for individuals with dysphagia	4 mm	Cooked in oven at 120 °C for 10 min	Liu et al. (2022)
Beef	Water, salt, hydrocolloids (xanthan gum, guar gum, carrageenan, or locust bean gum)	The meat was cooked before it was mixed with other ingredients. The objective of this study is to develop a 3D printed meat product that is suitable for individuals with dysphagia	1.2 mm	Cooked in oven at 120 °C for 15 min	Dick et al. (2021b)
Beef	Meat paste: salt, water, guar gum Lard was used separately	Lard was printed as layer between printed meat paste. The aim of the study is to develop a tailored meat product that contains composite ingredients arranged in a multi-layered structure	2 mm (meat); 1 mm (lard)	Sous-vide at 75 °C for 30 min	Dick et al. (2019b)
Pork	Water, salt, hydrocolloids (xanthan gum and/or guar gum)	Effects of hydrocolloids were evaluated to create 3D printed pork for individuals with dysphagia	1.2 mm	Steamed or cooked in oven at 100 °C for 7 min	Dick et al. (2020)
Chicken meat	Water, salt, chitosan, and plant-based oil (soybean, peanut, or olive oil)	Inclusion of plant-based oils into meat paste were evaluated	1.3 mm	Cooked in oven at 160 °C for 15 min, or microwaved at 280 W for 80 s	Yin et al. (2024)

(continued)

Table 22.1 (continued)

Meat types	Other ingredients	Specific condition	Nozzle diameter size	Post processing	Reference
Chicken meat	Salt	Optimum printing conditions were determined using response surface method: nozzle diameter size (1–1.5 mm), filament diameter (8–12 mm), nozzle temperature (25–45 ° C), salt addition (0–4%)	1.3 mm	Steamed or boiled 100 °C for 10 min	Yang et al. (2022)
Chicken meat	Sunflower oil, egg yolk, ice	Raw lean meat and cooked meat highly containing connective tissue were used at different ratios	2 mm	Not reported	Herold et al. (2021)
Chicken meat	Water and composite flour (mixture of grains)	Incorporation of composite flour was intended to enhance dietary fiber content of 3D printed meat product	0.82 mm	Not reported	Wilson et al. (2023)
Cod fish	Water, soybean dietary fiber, inulin, flaxseed oil	Fish flesh was steamed prior to mixing with other ingredients	0.84 mm	Not reported	Xie et al. (2022a)
Cod fish	Water	Fish flesh was steamed and subsequently washed in 0.1% NaCl solution, prior to mixing with water	Not reported	Not reported	Xie et al. (2022b)
Surimi	Potato starch and water (separated from surimi paste)	3D printing involves two separate fluids extruded through two co-axially assembled cartridges to create an inner fluid (surimi) and an outer fluid (potato starch gel)	1 mm (inner fluid) and 1.6 mm (outer fluid)	Steamed for 20 min	Kim et al. (2021)
Surimi	Water, glycerol, salt (NaCl, $CaCl_2$), and transglutaminase enzyme	Salts and transglutaminase was sprayed to the 3D printed meat	0.75–2 mm	Air fried, steamed, microwaved, heated in oven	Zhu et al. (2023)

printing vary, ranging from 0.75 mm to 4 mm. While larger diameters improve the printability of the meat paste, they concurrently diminish the shaping precision of the printed object.

The post-processing of printed meat commonly referred to as cooking, plays a pivotal role in preserving the shape of the printed meat and simulating the actual conditions under which the printed meat is ready for consumption. Documented post-processing methods include heating in an oven, steaming, air-frying, microwaving, and sous-vide cooking. These techniques ensure that the 3D printed meat retains its shape post-cooking and is safe for consumption.

22.5 3D Printing of Cultured Meat Products

Among the pioneering projects in the field of cultured meat was the introduction of a burger patty made from lab-grown meat in 2013. Cultured meat is considered as a more sustainable alternative to traditional meat production, requiring less land and water, producing lower greenhouse gas emissions, and leaving nearly zero pollutants in the environment. Additionally, it has the potential to mitigate the risk of zoonotic diseases by eliminating the direct consumption of animal meats (Handral et al. 2022; Guo et al. 2023).

Cultured meat production is grounded in the principles of tissue engineering and typically involves four steps: isolation and cultivation of animal cells, preparation of a culture medium, construction of a scaffold, and cell maturation in a bioreactor. Both embryonic and specific adult animal cells, such as myosatellite cells (the precursors to skeletal muscle) and adipose-derived cells, may be utilized. These adult cells are deemed more efficacious for proliferation and differentiation. The culture medium is a nutrient-rich liquid specifically designed to support cell growth on scaffolds or matrices. Scaffolds provide a framework for cells to adhere, grow, and achieve tissue maturity. Ideal scaffolds for cultured meat production should exhibit biological activity, ample surface area, contractility, and porosity to facilitate tissue maturation, while being non-toxic. Various biomaterials, including soy protein, collagen, cellulose, alginate, and chitosan, have been identified as effective scaffolds in cultured meat production. However, conventional scaffolding methods have yet to meet consumer expectations regarding the sensory profiles of meat (Handral et al. 2022; Guo et al. 2023; Tibrewal et al. 2023).

3D printing technology, particularly 3D bioprinting, is poised to revolutionize the production of cultured meat. The key distinction in 3D bioprinting is the use of cultured animal cells as "ink," which must remain viable during and after the printing process. To mimic the structure of traditional meat, it is crucial to incorporate cells that contribute to muscle and adipose tissue formation. These cells are typically encapsulated within hydrogels composed of selected biomaterials, such as gelatin, alginate, pea protein, and soy protein, which provide the necessary support for cell growth and differentiation (Handral et al. 2022; Guo et al. 2023; Tibrewal et al. 2023).

While the architecture of 3D bioprinted cultured meat can be extensively customized through digital modeling, achieving a resemblance to conventional meat requires further refinement. Additional efforts, including the integration of fabricated vascular structures, are essential to enhance the texture and functionality of the bioprinted meat, thereby bringing it closer to its traditional counterpart (Guo et al. 2023).

22.6 Nutritional and Sensory Properties of 3D Printed Meat

The nutritional properties of 3D printed meat products can be tailored; however, the formulation of the meat paste prioritizes printability. This necessitates a focus on the techno-functionality of the meat and other ingredients, especially in achieving the desired rheological properties. Most studies on 3D meat printing stipulate the addition of salt (NaCl) at approximately 1.5%, which inevitably contributes to an elevated sodium content. Conversely, the incorporation of hydrocolloids, particularly indigestible polysaccharides such as guar gum, xanthan gum, and locust bean gum, results in an increase in dietary fibers, though their concentration is typically limited to 1%. The inclusion of additional fat, intended to improve meat paste printability, is another factor that influences the nutritional properties of 3D printed meat (Dong et al. 2023). Furthermore, the integration of additional ingredients, designed to tailor the nutritional characteristics of 3D printed meat, is indeed feasible. However, it is crucial that these modifications do not compromise the printability of the meat paste.

In the realm of 3D bioprinting for cultured meat, it's fascinating to note that the nutritional composition closely mirrors that of meat derived from traditionally slaughtered animals. However, this similarity can fluctuate based on the specific types of cells utilized in the process, including myogenic cells, which are muscle-forming, and adipose cells, known for their richness in fat. Additionally, the biomaterials employed as scaffolding during cell cultivation, such as gelatin and alginate, also play a significant role in determining the nutritional composition. Intriguingly, these factors offer the potential for modulation, allowing for the customization of nutritional properties to meet specific requirements (Guo et al. 2023).

The sensory characteristics of 3D printed meat are significantly influenced by the composition of the raw materials and the printing method employed. The visual aspects are greatly determined by the design of the digital model. The printability of the meat paste, along with the size of the nozzle diameter which affects the precision of printing, plays a crucial role in shaping the appearance of the 3D printed meat. Moreover, the color of the 'meat ink' can be tailored using food-grade coloring agents, allowing for a customizable appearance. To achieve a multicolored effect in the printed meat, various colored meat pastes, fat pastes, or specially formulated food inks can be blended (Park et al. 2023). It is possible to utilize separate food ink cartridges to construct layers with distinct colors. Additionally, a single-nozzle 3D printer equipped with a dual co-axial food ink cartridge can produce meat with contrasting colors on the inner and outer surfaces, enhancing the visual appeal and simulating the natural variegation found in traditional meat cuts (Dick et al. 2019b). Multiple layers created in 3D meat printing can also led to the uniqueness of meat textural properties. Flavors can also be potentially manipulated together with the modulated meat texture.

22.7 Market Potential of 3D Printed Meat

3D food printing is rapidly emerging as a topic of significant interest and remains ripe for exploration. This innovative technology is still relatively unfamiliar to many consumers and even within the food industry itself. A recent study by Jayaprakash et al. (2020) suggests that 3D food printing could be effectively implemented in three distinct business models: customized design food, personalized snacks in semi-public spaces, and 3D food printing within hospital kitchens. Of these, the latter two models, which involve creating personalized meals and snacks for broader consumer segments, present more substantial business opportunities due to their potential for scalability and customization. These two models seem to be more suitable for 3D printed meat.

Within the domain of 3D printed traditional meat, the market segmentation reveals two primary consumer groups: those seeking cost-effective meat options, and individuals afflicted with dysphagia. The former demographic could be engaged as prospective purchasers of enhanced meat products, ingeniously restructured from less valued cuts of meat, offering improved quality at a reduced cost. For the latter category, particularly those with dysphagia, the integration of 3D food printing technologies in hospital kitchens or specialized dietary facilities presents a viable application, potentially revolutionizing meal preparation for this specific dietary need.

Regarding 3D bioprinted cultured meat, the target market encompasses conventional meat consumers who are notably conscious of environmental sustainability. This group is likely to expand, including a burgeoning segment of organic meat product consumers and those adopting a flexitarian diet (Siegrist et al. 2018; van Dijk et al. 2023). Crucially, the focus should extend beyond merely educating the populace on the technological intricacies of 3D printed meat. It is equally imperative to elucidate the role of this innovative technology in fostering sustainable living practices.

22.8 Current Opportunities and Challenges

3D meat printing technology, which includes both traditional and cultured meat, is an emerging frontier with vast potential yet to be fully realized. In the realm of traditional meat, 3D printing aims to optimize the use of less popular cuts, thereby enhancing food efficiency and minimizing waste. This innovative approach not only promotes sustainable consumption but also opens avenues for creating hybrid meat products. Hybrid meats, which combine a reduced amount of meat with functionally bioactive ingredients, offer nutritious alternative and present significant opportunities for further research and development. The exploration of these hybrid varieties could lead to advancements in food science and a new wave of sustainable meat products.

Simultaneously, the field of 3D bioprinting cultured meat is making strides in replicating the sensory and nutritional profiles of traditional meat. This technology provides a guilt-free alternative for meat enthusiasts by striving to match the taste and texture of conventional meat without the associated ethical and environmental concerns. The customization capabilities of 3D meat printing technology allow for unprecedented control over the final product's attributes, such as flavor profiles and texture. This level of customization paves the way for gourmet meats tailored to individual preferences, potentially revolutionizing the culinary landscape with personalized nutrition and taste experiences.

Despite the promising opportunities, there are significant challenges that must be addressed to advance 3D meat printing technology. One of the primary concerns is the food safety aspect of 3D printed meat. As we move towards mass production, it is imperative to establish rigid guidelines to ensure the safety and quality of the final products. These guidelines must encompass all stages of the production process, including the cleaning of 3D printer equipment for meat products, handling and storing the raw materials, etc. Additionally, from a business perspective, the economic feasibility of 3D printed meat is a critical factor. It is essential to explore the affordability of this product to ensure it can reach a wide consumer base. Achieving a balance between cost-effectiveness and high-quality production will be key to the widespread adoption of 3D printed meat and its contribution to sustainability goals.

22.9 Conclusion

3D meat printing, encompassing both traditional and cultured meat varieties, represents a frontier yet to be fully explored. Beyond the technological advancements, there is a vast potential for development in various facets related to 3D meat printing. This innovative approach offers a promising solution to ethical concerns and environmental challenges associated with traditional meat production. As this technology continues to evolve, it has the potential to transform the food industry profoundly. By offering sustainable meat alternatives, 3D printed meat could meet the increasing global meat demand while significantly reducing the ecological footprint of livestock farming. This innovation could lead to a substantial decrease in food waste, contributing to a more efficient and responsible food system worldwide.

References

Bhat ZF, Morton JD, Kumar S, Bhat HF, Aadil RM, Bekhit AEDA (2021) 3D printing: development of animal products and special foods. Trends Food Sci Technol 118:87–105. https://doi.org/10.1016/j.tifs.2021.09.020

Chen Y, Zhang M, Phuhongsung P (2021) 3D printing of protein-based composite fruit and vegetable gel system. LWT-Food Sci Technol 141:110978. https://doi.org/10.1016/j.lwt.2021.110978

Dick A, Bhandari B, Prakash S (2019a) 3D printing of meat. Meat Sci 153:35–44. https://doi.org/10.1016/j.meatsci.2019.03.005

Dick A, Bhandari B, Prakash S (2019b) Post-processing feasibility of composite-layer 3D printed beef. Meat Sci 153:9–18. https://doi.org/10.1016/j.meatsci.2019.02.024

Dick A, Bhandari B, Dong X, Prakash S (2020) Feasibility study of hydrocolloid incorporated 3D printed pork as dysphagia food. Food Hydrocoll 107:105940. https://doi.org/10.1016/j.foodhyd.2020.105940

Dick A, Dong X, Bhandari B, Prakash S (2021a) The role of hydrocolloids on the 3D printability of meat products. Food Hydrocoll 119:106879. https://doi.org/10.1016/j.foodhyd.2021.106879

Dick A, Bhandari B, Prakash S (2021b) Printability and textural assessment of modified-texture cooked beef pastes for dysphagia patients. Future Foods 3:100006. https://doi.org/10.1016/j.fufo.2020.100006

Dong X, Pan Y, Zhao W, Huang Y, Qu W, Pan J, Qi H, Prakash, S (2020) Impact of microbial transglutaminase on 3D printing quality of Scomberomorus niphonius surimi. Lwt 124:109123. https://doi.org/10.1016/j.lwt.2020.109123

Dong H, Wang P, Yang Z, Xu X (2023) 3D printing based on meat materials: challenges and opportunities. Curr Res Food Sci 6:100423. https://doi.org/10.1016/j.crfs.2022.100423

Feng C, Zhang M, Bhandari B (2019) Materials properties of printable edible inks and printing parameters optimization during 3D printing: a review. Crit Rev Food Sci Nutr 59:3074–3081. https://doi.org/10.1080/10408398.2018.1481823

Guo X, Wang D, He B, Hu L, Jiang G (2023) 3D bioprinting of cultured meat: a promising avenue of meat production. Food Bioprocess Technol:1–22. https://doi.org/10.1007/s11947-023-03195-x

Handral HK, Tay SH, Chan WW, Choudhury D (2022) 3D printing of cultured meat products. Crit Rev Food Sci Nutr 62(1):272–281. https://doi.org/10.1080/10408398.2020.1815172

Herold M, Morick S, Hensel O, Grupa U (2021) Influence of raw meat content on 3D-printing and rheological properties. Int J Food Stud 10:195–202. https://doi.org/10.7455/ijfs/10.1.2021.a6

Huang Y, Zhang M, Pattarapon P (2022) Reducing freeze-thaw drip loss of mixed vegetable gel by 3D printing porosity. Innov Food Sci Emerg Technol 75:102893. https://doi.org/10.1016/j.ifset.2021.102893

Jayaprakash S, Paasi J, Pennanen K, Ituarte IF, Lille M, Partanen J, Sozer N (2020) Techno-economic prospects and desirability of 3D food printing: perspectives of industrial experts, researchers and consumers. Food Secur 9:1725. https://doi.org/10.3390/foods9121725

Ji S, Zeng Q, Xu M, Li Y, Xu T, Zhong Y, Liu Y, Wang F, Lu B (2023) Investigation of the mechanism of different 3D printing performance of starch and whole flour gels from tuber crops. Int J Biol Macromol 241:124448. https://doi.org/10.1016/j.ijbiomac.2023.124448

Kang DH, Louis F, Liu H, Shimoda H, Nishiyama Y, Nozawa H, Kakitani M, Takagi D, Kasa D, Nagamori E, Irie S (2021) Engineered whole cut meat-like tissue by the assembly of cell fibers using tendon-gel integrated bioprinting. Nat Commun 12:5059. https://doi.org/10.1038/s41467-021-25236-9

Kim SM, Kim HW, Park HJ (2021) Preparation and characterization of surimi-based imitation crab meat using coaxial extrusion three-dimensional food printing. Innov Food Sci Emerg Technol 71:102711. https://doi.org/10.1016/j.ifset.2021.102711

Lipton J, Arnold D, Nigl F, Lopez N, Cohen D, Norén N, Lipson H (2010) Multi-material food printing with complex internal structure suitable for conventional post-processing. 21st Solid Freeform Fabrication Symposium. https://doi.org/10.26153/tsw/15245

Liu P, Dang X, Woo MW, Chattha SA, An J, Shan Z (2022) Feasibility study of starch-based biomass incorporated 3D printed beef. Starch-Stärke 74:2200030. https://doi.org/10.1002/star.202200030

Liu P, Zhou Z, Zhou W, Liu C, Wang N, Zhang F (2024) Spray-drying assisted 3D printing to manufacture beef products. Starch-Stärke:2300286. https://doi.org/10.1002/star.202300286

Mantihal S, Kobun R, Lee BB (2020) 3D food printing of as the new way of preparing food: a review. Int J Gastron Food Sci 22:100260. https://doi.org/10.1016/j.ijgfs.2020.100260

Nachal N, Moses JA, Karthik P, Anandharamakrishnan C (2019) Applications of 3D printing in food processing. Food Eng Rev 11:123–141. https://doi.org/10.1007/s12393-019-09199-8

Park JW, Lee SH, Kim HW, Park HJ (2023) Application of extrusion-based 3D food printing to regulate marbling patterns of restructured beef steak. Meat Sci 202:109203. https://doi.org/10.1016/j.meatsci.2023.109203

Pereira T, Barroso S, Gil MM (2021) Food texture design by 3D printing: a review. Food Secur 10:320. https://doi.org/10.3390/foods10020320

Siegrist M, Sütterlin B, Hartmann C (2018) Perceived naturalness and evoked disgust influence acceptance of cultured meat. Meat Sci 139:213–219. https://doi.org/10.1016/j.meatsci.2018.02.007

Tibrewal K, Dandekar P, Jain R (2023) Extrusion-based sustainable 3D bioprinting of meat & its analogues: a review. Bioprinting 29:e00256. https://doi.org/10.1016/j.bprint.2022.e00256

van Dijk B, Jouppila K, Sandell M, Knaapila A (2023) No meat, lab meat, or half meat? Dutch and Finnish consumers' attitudes toward meat substitutes, cultured meat, and hybrid meat products. Food Qual Prefer 108:104886. https://doi.org/10.1016/j.foodqual.2023.104886

Wilson A, Anukiruthika T, Moses JA, Anandharamakrishnan C (2023) Preparation of fiber-enriched chicken meat constructs using 3D printing. J Culin Sci Technol 21:127–138. https://doi.org/10.1080/15428052.2021.1901817

Xie Y, Yu X, Wang Y, Yu C, Prakash S, Zhu B, Dong X (2022a) Role of dietary fiber and flaxseed oil in altering the physicochemical properties and 3D printability of cod protein composite gel. J Food Eng 327:111053. https://doi.org/10.1016/j.jfoodeng.2022.111053

Xie Y, Yu X, Wang Z, Yu C, Prakash S, Dong X (2022b) The synergistic effects of myofibrillar protein enrichment and homogenization on the quality of cod protein gel. Food Hydrocoll 127:107468. https://doi.org/10.1016/j.foodhyd.2021.107468

Yang G, Tao Y, Wang P, Xu X, Zhu X (2022) Optimizing 3D printing of chicken meat by response surface methodology and genetic algorithm: feasibility study of 3D printed chicken product. LWT-Food Sci Technol 154:112693. https://doi.org/10.1016/j.lwt.2021.112693

Yin Y, Wang Y, Fang Q, Xiang M, Zhao X, Xu X, Li C (2024) Effects of pre-formulation and post-cooking method on the rheological and gelation properties of 3D printed chicken products. Food Chem 446:138857. https://doi.org/10.1016/j.foodchem.2024.138857

Zhu J, Cheng Y, Ouyang Z, Yang Y, Ma L, Wang H, Zhang Y (2023) 3D printing surimi enhanced by surface crosslinking based on dry-spraying transglutaminase, and its application in dysphagia diets. Food Hydrocoll 140:108600. https://doi.org/10.1016/j.foodhyd.2023.108600

Chapter 23
Valorization of Meat Waste and By-Products

Mohamed Tagrida

23.1 Introduction

The meat industry plays a crucial role in fulfilling global food demands which arise yearly, however, this industry is also responsible for the generation of substantial waste that poses environmental challenges if not managed properly (Roy et al. 2023). Meat waste includes unused animal parts, processed byproducts, and expired products. Improper disposal of this waste can lead to pollution, greenhouse gas emissions, and public health hazards (Akmal and Jamil 2021). Nevertheless, sustainable and innovative strategies to reduce such waste and reuse them can lie in the concept of industrial symbiosis. Industrial symbiosis is a collaborative approach where industries exchange resources, waste, and energy, fostering a circular economy and reducing the negative impact on the environment (Cecchin et al. 2020). Besides the proper management of meat industry wastes, such strategies can also contribute to reducing the current dilemma of food shortage which was exacerbated due to many factors including the augmented population globally, the demographic change in many of these populations, the rising conflicts in some parts of the world, and the climate change and its drawbacks on food production (Amicarelli and Bux 2021). By 2050, the worldwide need for protein will increase twofold, such rise can be attributed not only to population growth but also to a greater acknowledgment of the vital role protein plays in maintaining a healthy diet, particularly for children and the growing elderly population (Aspevik et al. 2018).

Approximately 40–60% of the overall weight of high protein containing animals and fish consists of residuals. Depending on the species, these residuals can include the heads, blood, skin, viscera, bones, hooves, and feathers (Chuck-Hernández and Ozuna 2019). Such materials can hold considerable benefits for being utilized in

M. Tagrida (✉)
Department of Food Technology, Engineering and Science, University of Lleida—Agrotecnio CERCA Center, Lleida, Spain

S. A. Rather, F. A. Masoodi (eds.), *Hand Book of Processed Functional Meat Products*, https://doi.org/10.1007/978-3-031-69868-2_23

different ways forfood and feed applications. The proper utilization of by-products necessitates a well-structured approach. Recently it was postulated that the immediate management of the by-products from the moment that they are produced until their eventual use through a well-planned industrial symbiosis, is crucial for the maximized transformation, or extraction of active components (Gooding and Meeker 2016). When incorporating these by-products into foods, understanding the interactions between them and the foodstuff becomes vital to monitor the possible change in their functionality. Efficient use of by-products may improve the gross income of different meat industries (Irshad and Sharma 2015).

The meat industry produces a high amount of by products including blood, bones, skin, fatty tissues, horns, hooves, and internal organs. Properly managing and environmentally disposing of these by products can be costly (Bux and Amicarelli 2022). Such expenses can be reduced through innovative methods by creating value added products that increase profitability. Mishandling or insufficiently treating these by products can result in many environmental and health problems. To address these concerns many regulations have been taken by many organizations in order to minimize the quantity of such wastes and what follows of environmental impacts (Hubalek et al. 2022). Therefore, the correct management of these by products can lead to the reduction of processing expenses which will eventually help to balance out the costs of disposal (McCabe et al. 2020).

Meat by-products have the potential to be a valuable resource when managed effectively to create additional value-added substances or products as the efficient utilization of these by-products can lead to a significant increase in the overall income from different meat sources (Toldrá et al. 2016). These by-products encompass a wide range of materials that can be regarded as essential nutrients such as vital amino acids, minerals, and vitamins. This presents a promising opportunity for the meat industry to extract value from these resources (Mazhangara et al. 2019). Innovative technologies have been implemented for processing meat by-products, leading to various applications. These include the production of edible food ingredients, animal feed and pet food industries (Irshad and Sharma 2015). Meat by-products can also serve as raw materials for generating bioactive molecules like protein hydrolysates with their significant bioactivities and extracts such as gelatin with valuable functional properties (Baldi et al. 2021). Additionally, there are non-food applications, such as fertilizers, chemical or pharmaceutical supplements, and energy production. Energy generation from meat wastes is based on producing biodiesel from waste animal fats or developing a second generation of bioderived diesel fuel, commonly known as biogas oil (Suresh et al. 2021).

23.2 Industrial Symbiosis

Industrial symbiosis is an approach that promotes the effective use of energy and materials by connecting traditionally separate industrial sectors. It fosters cooperation and synergy among several establishments, leading to environmental, economic, and social benefits in different areas including waste management (Tabassum et al. 2023). By applying industrial symbiosis among the different industrial bodies, a sustainable approach to waste management and resource utilization can be achieved. Applying industrial symbiosis in the meat industry can minimize the expected loss resulting from wastes and the inappropriate methods undertaken in the preparation of meat and manufacturing of other meat products, thus maximizing the profits of the meat manufacturers (Kowalski et al. 2023). The expected benefits of applying industrial symbiosis in the meat industry can include the reduction of wastes and preserving the raw materials more efficiently; this can take place by enabling the sharing of resources and by-products among different producing companies, hence the reduction of waste generation and promotion of resources efficiency (Kowalski et al. 2023). The by-products of one company can become valuable inputs for another; this would minimize the need for new raw materials and thus reduce environmental impact. The organic waste from slaughterhouses can be used as a feedstock for biogas production, reducing the reliance on fossil fuels (Baldi et al. 2021).

Industrial symbiosis can contribute to environmental sustainability by reducing greenhouse gas emissions and minimizing the use of non-renewable resources, this can take place by optimizing material and energy flows, therefore, contributing to the reduction of waste disposal and landfilling, and reducing the industry's ecological footprint (Sporchia et al. 2023). The implementation of industrial symbiosis can also result in the adoption of cleaner production technologies or green processes that can further enhance environmental sustainability. Collaboration and cooperation among different entities involved in meat and other meat products manufacturing, can lead to cost savings and so increase their profitability and improve the overall efficiency of the production process (McCabe et al. 2020).

23.3 Current Status of Meat Industry Wastes

Due to the expanded population worldwide, the demand for meat and other meat products is expected to grow. Meat production globally has grown by 1% in 2022 (approximate to 347 MT CWE), and the production of meat from different sources (poultry, beef, pork, and sheep meat) is expected to further grow by 10–15% by 2032 (OECD and FAO 2023). The drastic impact from such increased consumption raised concerns among environmentalists. Greenhouse gas emissions, deforestation, and water contamination are examples of the negative effects of the increased usage of meat products (Ryder et al. 2016). The meat processing industry generates approximately150 MT of by-products, yet only a small quantity of that are utilized to create high value products. By products from meat production can be regarded as either edible or inedible by-products. Edible by-products including lungs, liver, spleen, meat trimmed from head, fats around some organs, etc., are prepared and packaged to be consumed (Irshad and Sharma 2015). Inedible by-products such as bones, viscera, blood, hides are generally regarded as wastes and their accumulation can lead to negative impacts on the environment. The unsuitable utilization or disposal of such wastes can cause severe environmental and economical drawbacks, for instance, the slow decomposition of bones and hides release greenhouse gas emissions, additionally such wastes will end up in the landfills and water sources causing more pollution (Okoro et al. 2017). The augmented volumes of inedible wastes will require more energy and disposable methods that will eventually lead to resource depletion and add to the overall CO_2 footprint generated from the meat industry (Eldesouky et al. 2018). To reduce such impacts resulted from the accumulation of meat waste technological approaches based on the concept of waste-to-energy, can be implemented, as most of these wastes after proper processing, can produce biogases that are involved in clean energy generation (Badgett and Milbrandt 2020). Moreover, the utilization of inedible meat wastes can be a promising solution since these wastes contain valuable compounds that can be involved in many applications. Bones, fats, blood, and residual muscles are the most wastes that have been extensively utilized, and each has its unique valorization system that yields an added-value product (Pinto et al. 2022).

23.4 Valorization Systems for Meat Production Wastes

Valorization systems differ according to the end product targeted, and the type of waste used as starter material. Fats resulted from slaughterhouses or meat processing plants that cannot be consumed as an edible material can be used as a raw material to produce soap or other detergents (Skoronski et al. 2016). Recently, utilization of such fats for the production of biodiesel was suggested as a more profitable alternative. Biodiesel from fats can be produced via chemical catalysis, using acid or base followed by triglyceride hydrolysis then by transesterification (Andreo-Martínez et al. 2022). Alternatively, microalgae and enzymatic transesterification of fats into the monoalkyl esters to produce biodiesel can be regarded as cleaner alternative to minimize the production of hazardous waste and other by-products during the process of biodiesel production (Skoronski et al. 2016). Immobilized enzyme systems for biodiesel production from wasted fats, offers additional advantages as it can be easily separated from the catalyst to avoid any undesired contamination of the biodiesel. Additionally, using immobilized enzymes can reduce the production costs and enhance enzyme stability (Christopher et al. 2014). Moreover, using such technique would allow the usage of hydrate alcohol as substrate, which was not possible during normal chemical transesterification reactions, thus this can lower the costs of the process. Bones are other major wastes resulting from meat processing. Different methods were proposed for the better utilization of these bones, transformation into liquid phosphorus fertilizers through microbiological or chemical pathways was used widely (Saeid et al. 2014). The use of bones as fertilizers can be a valuable valorizing strategy as bones are regarded as the most concentrated form of organic phosphorus (Hong et al. 2022). Pyrolysis and vacuum assisted pyrolysis can be used for the transformation of bones into valuable products such as animal feed, cement, and fertilizers (Piccolla et al. 2021; Duvallet and Jewell 2023).

23.5 Utilization of By-Products into Value Added Products

The meat industry generates various by-products throughout its production processes. These include animal bones, blood, skin, fat, and internal organs, among others. Historically, these by-products were often discarded. However, with increasing concerns about environmental sustainability and resource efficiency, there has been a shift towards exploring more innovative and valuable uses for these by-products. Meat by-products contain many essential nutrients. Some can be used as pharmaceutical supplements because of their content of numerous components such as amino acids, minerals, fatty acids, hormones, and vitamins (Skoronski et al. 2016). The utilization of such by-products can offer a variety of compounds that are involved in many applications.

Blood has a high content of protein and heme iron, and for long was an important edible by-product in many parts around the world (Karimi et al. 2018). Recently, it can be used as an emulsifier, stabilizer, clarifier, colorant agent, and as a nutritional component in many food applications, it was also used as fertilizer, animal meal and binders in other non-food applications (Sayeed and Tiwari 2020). Blood plasma has a gel forming ability, because of its content of albumin therefore, it can be used as water and fat binder in many applications, moreover it also has an excellent foaming capacity thus, can be used as alternative to egg whites in the baking industry (Fernández et al. 2021). Blood can be separated into several fractions including fibrinogen, fibrinolysin, serotonin, albumin, immunoglobulins, and plasminogen which have their use in the medicinal sector, furthermore, many blood products can be used as important components for tissue culture and microbiological media (Sayeed and Tiwari 2020).

Hides and skins mainly usage is in clothing and furniture, however they also can be used for extraction of collagen and gelatin which have many food and industrial applications, collagen can act as an emulsifier in many meat products because it can bind to fat, thus making it a useful additive or filler for these meat products (Usman et al. 2023). Bones constitute a considerable portion of meat waste; it can be used as food, animal feed and a source of gelatin and other valuable minerals like calcium and phosphorus (Sporchia et al. 2023). It can be also used as an ingredient in the concrete industry and as a cement replacement (Gautam et al. 2022). Organs and glands can usually be discarded in many parts of the world; however, they have a high nutritional value and can offer a wide variety of flavors and textures thus can be consumed as foods in other parts of the world (Alao et al. 2017). Glands and some organs can be valuable sources for many enzymes and hormones which can be used to produce important pharmaceutical compounds and drugs (Jayathilakan et al. 2012). Cholesterol can be extracted from brains and spinal cords which is the starter material for the synthesis of vitamin D3 (Hirsch 2011). Melatonin, a hormone that is extracted from the pineal gland is being studied for its possible involvement to cure schizophrenia, insomnia, and other mental diseases (Bastos Jr et al. 2019). Bile extracted from gallbladder consist of many bio-compounds including pigments, proteins, acids, and cholesterol which can be used to treat indigestion, constipation, and gastrointestinal tract problems, additionally it can enhance the secretion and the overall performance of the liver (Jayathilakan et al. 2012). Liver can be source of vitamin B12 which is considered as treatment of anemia, and heparin which is used as anticoagulant (Murawska et al. 2016). Pancreas is a source of insulin, glucagon, chymotrypsin and trypsin which have functions in sugar metabolism regulation and improve cells healing (Cade and Hanison 2017). Progesterone, estrogen and relaxin are hormones that can be extracted from ovaries and have many medical uses (Hirpessa et al. 2020).

Different conversion methods have been used to produce value-added products from meat waste for different applications. The high moisture content in meat waste provides beneficial effects in some of the conversion methods such as anaerobic digestion, fermentation, microbial hydrolysis, etc. (Roy et al. 2023). Additionally, meat wastes might need size reduction or pretreatment (ultrasonication, microwave, etc.) to improve the conversion process and increase the yield of the value-added products.

Anaerobic digestion is an effective conversion method mainly for alternate energy production, as the rich content of meat waste is exhibiting a high potential for biogas production (Hamawand 2015). Typically, anaerobic digestion is a multi-beneficial microbiological process in which a consortium of different micro-organisms degrades or digest various organic materials in meat waste to produce biogas such as methane plus a stabilized end-product which is the digestate that can be used as fertilizer or crops growth promoter (Luste 2011). Despite the ability of this method to utilize meat processing waste into value added products, some limitations might arise particularly from using the meat waste themselves, as these substrates are considered as difficult materials to be anaerobically digested, majorly due to their complex structure, high protein and fat contents which can negatively affect the proliferation of the microorganisms thus, the digestion process (Hejnfelt and Angelidaki 2009). Ammonia released from the protein degradation during anaerobic digestion of waste has an inhibition effect on the microorganisms which can result in slowing the digesting process (Hansen et al. 1998). Fat content in meat waste on the other hand can yield to intermediates such as long-chain fatty acids which their accumulation can lead to microbial inhibition (Broughton et al. 1998).

Composting is a controlled, aerobic process that can transform meat wastes into nutrient-rich soil fertilizers through natural decomposition (Storino et al. 2016). The process creates a product that can be used to improve soil nutrients for agriculture applications, thus minimizing the accumulation of meat waste. Composts obtained from meat waste showed remarkable physical and chemical properties in addition to their rich content of different organic matter (Onwosi et al. 2017). Thermochemical conversion techniques including pyrolysis, incineration, hydrothermal carbonization, etc., were used to utilize meat waste into novel products used in different applications including energy production, animal feed, fertilizers, and pharmaceutical applications (Roy et al. 2023).

Rendering is a common process used to convert meat by-products into other products such as animal feed. Usually, the process involves heating meat by-products at high temperatures to separate the fat and moisture from the protein and other solids (Jayathilakan et al. 2012). The resulting product is known as rendered meat meal or meat and bone meal based on the protein content. Different types of meals can be produced depending on the specific by-product used (Leon et al. 2018). This includes bone meal, blood meal, or poultry offal meal. Rendered meals are widely used in livestock, poultry, and aquaculture diets (Jayathilakan et al. 2012).

Drying and grinding are methods employed to process different food waste into powdered or granular forms suitable as animal feed or fertilizers (Noori et al. 2022). By using heat or freeze-drying techniques, food waste is dried to minimize its moisture. Subsequently, the resulting dried materials are ground into fine particles or pellets, to serve as either animal meal or soil fertilizer. These products are rich in nutritional components including proteins, amino acids, vitamins, and other essential nutrients (Murugesan et al. 2021).

Hydrolysis and enzymatic processing methods are used to extract proteins, peptides, and amino acids from meat by-products (Angulo and Márquez 2023). These processes involve the use of enzymes or acid hydrolysis to break down complex proteins into smaller, more digestible components. The resulting hydrolysates are highly palatable and can be involved in many applications including industrial, pharmaceutical, and cosmetic applications (Skripsts et al. 2022).

Extrusion can be used to process a variety of meat waste as it is considered as an energy-efficient process (Dey et al. 2021). Extrusion process can be regarded as simultaneous application of different unit operations, including blending, kneading, cooking, and formation of extruded end product under high temperature, high shear, and short processing time (Offiah et al. 2019). Extrusion processing is favorable by many manufacturers particularly food producers due to the minimization of nutritional losses of the food ingredients compared to other conventional heat processing techniques (Leonard et al. 2020). Furthermore, the high temperature used during the process break enzymes that can have drawbacks on the final product such as lipases and lipoxidases in addition to inhibition of several antinutritional compounds such as phytates and trypsin inhibitors and food pathogens (Dey et al. 2021).

23.6 Applications of Meat Waste and By-Products

23.6.1 Animal Feed and Pet Food

One of the primary uses of meat by-products is in the formulation of animal feed and pet food. By employing proper processing techniques, by-products can be transformed into nutritious and cost-effective feed ingredients. This not only reduces waste but also adds value to the by-products by creating a new product with market demand (Jayathilakan et al. 2012). Animal bones obtained during meat processing are rich in minerals and proteins. Previously, bones were processed and transformed into various forms, including bone meal, bone flour, or bone char, which are valuable sources of calcium, phosphorus, and other essential minerals for animal nutrition (Gizzi et al. 2004). However, due to the potential transmission of diseases among animals consuming these types of feeds, bone meals were banned by many organizations worldwide particularly if the bones come from animals affected by transmissible diseases such as transmissible spongiform encephalopathies (TSE) (Gizzi et al. 2004).

Blood is a nutrient-rich by-product of the meat industry. It contains proteins, amino acids, and vitamins (Sayeed and Tiwari 2020). Blood meal is produced by drying and processing blood, resulting in a high-protein feed ingredient. It is often used in livestock, poultry, and aqua-culture diets as a source of protein and amino acids (Karimi et al. 2018). The use of blood as animal feed is a practice that has been employed in various contexts for its nutritional value and potential benefits. Blood meal was included in poultry diets as a source of highly digestible protein. As it can contribute to the growth and development of different poultry species. Blood meal is typically incorporated into their feeds at appropriate amounts to meet the protein requirements of the poultry (Beski et al. 2015). Blood meal can also be used in swine diets. It acts as valuable protein and amino acid sources. It was usually included in starter diets for piglets to support their early growth and development (King'ori et al. 1998). Additionally, blood meal has been utilized in aquaculture feeds for fish, shrimp, and other seafoods. It enhances the essential amino acids content of these species and helps promote their growth (Kirimi et al. 2017). Blood meal can be used as a component in feed formulations for various fish species, such as salmon, trout, and catfish, as well as shrimp. Kirimi et al. (2016) reported that replacement of normal fish meal with 50% and 100% dietary based on blood meal was found to reduce the incidence cost and increase the profit index of Nile tilapias fed on blood meal indicating an economic and nutritional benefits on feeding these fish with such dietary. Supplemented diets based on boiled cow blood meal and donkey blood meal were found to enhance the performance of *Clarias gariepinus* juveniles (Ogunji et al. 2020). The blood meals fed to these fish had no harmful effect on the digestive enzyme secretion and the overall metabolic activities. Blood meal was used to replace common fishmeal in practical diets of red tilapia juveniles, the proposed feed was found to have no drastic effects on the growth of the fish samples with decreasing production costs to up to 5% (Karaket et al. 2022). Spray-dried animal plasma utilized from blood waste was widely used in different types of pet foods to improve the textural properties of the feed by strengthening the cohesion between the various components in the feed, because of the excellent water binding and emulsifying capacity of the blood plasma (Ofori and Hsieh 2014). The use of spray-dried animal plasma in pet food enhanced the palatability of these pet feeds and was reported to be preferred by carnivores such as cats (Polo et al. 2005). While animal meals based on blood showed prospective characteristics, proper processing and handling are very important to ensure their safety and prevent the transmission of diseases. The risk of transmitting TSE diseases such as bovine spongiform encephalopathy (BSE) commonly known as mad cow disease reduced the preferability of blood as source of animal feed, in addition, the banning of animal feeds based on the animal blood by many countries and organizations makes feeds based on them an economical futility (Gizzi et al. 2004).

Skin and fat obtained from meat processing can be utilized in the production of animal feed. They are processed to extract edible fats and oils, which are valuable sources of energy for animals. Fats from different resources were investigated as aquaculture feed (Trushenski and Lochmann 2009), the feeds showed satisfying results in improving the aquaculture nutritional and physiological properties. Types and nature of fats used as animal feed should be considered, as fats susceptible to oxidation show negative effects on the growth and health of livestock consuming these types of feeds (Dibner et al. 1996). An approximate reduction of 11.4% and 8.8% in growth rate and feed intake, respectively was observed for pigs that were fed isocaloric diets added with peroxidized lipids compared to diets free from this type of lipids (Shurson et al. 2015). Furthermore, vitamin E content in serum of the fed specimens was significantly reduced, while the TBARS content was amplified after the intake of feeds containing peroxidized lipids, indicating that these feeds negatively affected the metabolic oxidative activities of pigs (Shurson et al. 2015). Apart from animal feed, fats can be used as part of human diets or used as cooking materials after the required pretreatments, they can be obtained after minimally processed treatments to make them convenient source of energy and micronutrients (BELL et al. 1997). Animal fats were found to acquire high levels of oleic acid and specific micronutrients including vitamins and essential fatty acids (Barendse 2014); thus, they can be considered as of immense benefit in human nutrition.

Internal organs (offal), such as liver, kidneys, spleen, lungs, and intestines are nutrient-dense by-products that can be utilized in animal feed production as well as human consumption (Baldi et al. 2021). They are often processed and transformed into meat by-product meals, which are protein-rich ingredients used in both human and animal diets. Typically, the offal undergo a rendering process at approximately 100 °C to eliminate a portion of the fat. The residual substance is then dehydrated and pulverized, resulting in a meal with 10% moisture and fat content ranging from 5% to 13%. Based on the protein content of the resulted meal, 55% or higher, is labeled as a meat meal. If the protein content ranged between 40 and 55%, it is categorized as a meat and bone meal, while a protein content of less than 40% is regarded as a bone meal (Manso et al. 1998). The different variations in the chemical composition of the final product are countered by blending to create a more uniform product regarded as suitable for animal feed (Cooke and Pugh 1980). Poultry offal was used as meal and introduced as diet for broiler chickens. Weight gain, feed conversion, carcass yield and noble cuts were markedly improved for specimens fed on these diets (Silva et al. 2014). Additionally, poultry's performance and carcass physical characteristics were enhanced as well as an observed increase in the abdominal fat.

23.6.2 *Food Applications*

23.6.2.1 Gelatin and Collagen

The extraction of gelatin and collagen from animal bones and skin is another important application of meat by-products. Collagen is the protein that lies between the connective tissues such as bones, tendons, and cartilage. It can be used in different cosmetic and medical applications such as wound dressings and bone grafts (Holmes et al. 2013). Gelatin on the other hand, is the protein that is derived from collagen via hydrolysis methods and have a variety of food applications such as confectionery, industrial applications such as adhesives, glues and matches, and medical applications such as soft gels, capsules, wound dressings, hemostatic, and ophthalmic agents (Usman et al. 2023). It also can be used for the production of packaging materials and biodegradable films and coating that can be used as bags or wraps for foodstuff protection, thus reducing the dependance on plastic materials and consequently reducing their harmful impact on the environment (Tagrida et al. 2022). The physicochemical properties of gelatin depend on the source, extraction conditions, and pretreatment methods (Usman et al. 2023). The application of more recently developed techniques can be used for the extraction of gelatin to overcome several limitations of the traditional process including low yield and quality, high energy and water consumption, and environmental issues that might arise due to the unwise production (Noor et al. 2021), thus, such product without the usage of toxic solvents and less time for extraction can fulfill the global trend of sustainable and green economy.

23.6.2.2 Functional Ingredients

By-products of the meat industry can be a valuable source of bioactive peptides, which are sequences containing between 2 to 20 amino acids that can have different biological activities such as antioxidant, immunomodulatory, hypocholesterolemic, and antithrombotic activities in the human physiological systems (Toldrá et al. 2016). Bioactive peptides can be released before food consuming particularly during several food processing stages such as fermentation or curing, others can be generated after consuming the food during gastrointestinal digestion (Chakrabarti et al. 2018). The naturally generated active peptides usually found to show weak activity which is related to the different set of enzymes used by the human body system during digestion resulting in a wide range of peptides with different amino acids chain sizes that have different characteristics (Daffre et al. 2008). To avoid the generation of inactive peptides, controlled hydrolysis conditions using known enzymes have been proposed to obtain more homogeneous peptides with better bioactivities (Toldrá et al. 2016). Blood from slaughterhouses and meat processing facilities is considered a good source of active peptides as it contains substantial amounts of proteins including hemoglobin. Blood based hydrolysates were reported to contain bioactive peptides which have antioxidant, antimicrobial, and opioid activities (Resende et al. 2023). The quality and activity of the generated bioactive peptides from blood depends on the enzymes used for the hydrolysis process and also on the type and source of the blood used (Toldrá et al. 2020). Furthermore, the hydrolysis degree determines the activity of the peptides generated in which conditions such as temperature, pH, and time are very important parameters to obtain them. The characteristics of the active peptides including size of peptide chain and amino acid sequences determines the bioactivity of the generated peptides (Borrajo et al. 2019). Collagen has a limited nutritional value as it deprives the essential amino acids, however after certain hydrolysis processes, it can be a useful source of active peptides (Zhao et al. 2021). Enzymes such as alcalase, pepsin, flavourzyme, trypsin, chymotrypsin, and papain were used for the hydrolysis process of collagen to generate active peptides that had numerous bioactivities including antioxidant, antimicrobial, and angiotensin-converting enzyme (ACE) inhibitory activities (Nuñez et al. 2020).

Meat by-products can be a valuable source of ingredients that can enhance the physicochemical properties of other foods such as color enhancer, flavoring agents, and textural consolidation supplements, in this context meat waste derivatives such fibrinogen, which can be converted to insoluble fibrin when reacted with thrombin and forms fibers by aggregation (Litvinov et al. 2021), immunoglobulins, and serum albumin were used as gelation agents and emulsifiers in other food and industrial applications (Li and Nie 2016). Other proteins, particularly plasma proteins were used as protein cross-linking or foaming agents (Samsalee and Sothornvit 2017). These proteins were reported to contain many bioactivities like antioxidant and antimicrobial activities (Toldrá et al. 2016).

23.6.2.3 Biofuels and Biochemicals

The fats and oils obtained from meat by-products can be processed to produce biofuels such as biodiesel. Additionally, they can serve as feedstock to produce various biochemicals, including fatty acids, glycerol, and bio-based plastics (Okoro et al. 2017). Biodiesel can be produced from various waste sources, including animal fats which are co-produced from slaughterhouses, meat processing facilities, and cooking places (Toldrá-Reig et al. 2020a). The production of biodiesel from animal fat waste has several advantages, including low cost, high availability, and reduced environmental impact. Additionally, biodiesel production from animal fat waste can reduce the cost of waste treatment, thus providing a valid renewable source of energy (Zhang et al. 2022). The production of biodiesel from animal fat involves a series of chemical reactions called transesterification, which transforms fats into biodiesel and glycerol as a byproduct of the process the glycerol itself can be used in the production of soap (Kwon et al. 2014). Fats obtained from rendering during meat processing are collected as the feedstock for biodiesel production, these fats undergo a pretreatment stage to remove impurities such as water, free fatty acids, and contaminants in order to improve the efficiency of the transesterification process

(Andreo-Martínez et al. 2022). After the required pretreatment processes, the fats go through the main transesterification step which involves reacting the pretreated fats with an alcohol in the presence of a catalyst, which is often a strong base, such as sodium hydroxide or potassium hydroxide (Kwon et al. 2014). The result of this reaction is fatty acid methyl or ethyl esters (biodiesel) depending on the alcohol used and glycerol. Subsequently, the mixture is let to separate into two phases which are biodiesel and glycerol. Glycerol is removed and the biodiesel is washed repeatedly to eliminate any remaining impurities, catalyst, or glycerol (Toldrá-Reig et al. 2020b). The biodiesel is then dried completely form water as it can affect the quality and stability of the biodiesel. Recent advancement in the production of biodiesel including hydro-oxygenation, hydro-isomerization in tubular reactors, ultrasounds assisted transesterification, and supercritical transesterification has been studied to improve biodiesel profitability (Babadi et al. 2022).

The resulting materials from the usage of the meat products such as energy production leftovers, ashes can contain high content of phosphorus, nitrogen and other macromolecules and minerals thus, such materials can be used as fertilizers (Toldrá et al. 2016). Meat waste as fertilizer presents a dual opportunity to address environmental concerns and enhance agricultural productivity, as they can be repurposed as a nutrient-rich fertilizer. When properly processed, these by-products can be transformed into organic fertilizers, promoting sustainable agriculture.

23.6.2.4 Pharmaceutical and Neutraceutical Products

Valuable components found in meat wastes such as active peptides, amino acids, minerals etc., can have potential pharmaceutical and nutraceutical applications. These compounds can be extracted and purified to develop functional food ingredients, dietary supplements, or pharmaceutical formulations (Timorshina et al. 2022). Bio-compounds such as collagen, extracted from skins and hides can be used as dressing for burns or skin ulcers (Shi et al. 2023). Glands and organs constitute edible byproducts with good nutritive value that are consumed in different parts of the world, these byproducts can also be a source of compounds that have medical use such as melatonin from the pineal gland, bile from the gall bladder, progesterone and estrogen from ovaries, heparin from the liver, and insulin from pancreas (Socas-Rodríguez et al. 2021). Protein hydrolysates from gelatin and collagen have a positive effect on bones and joints and they also can be used as health booster ingredients (Mora et al. 2018).

23.7 Conclusion

The utilization of by-products from the meat industry to produce value-added products offers significant economic, environmental, and social benefits. From animal feed to biofuels, pharmaceuticals to nutraceuticals, there is a diverse range of applications that can be explored. The transformation of meat by-products into valuable resources not only reduces waste but also contributes to a more sustainable and efficient meat industry, supporting the principles of the circular economy and promoting a greener future with healthier environment.

References

Akmal T, Jamil F (2021) Assessing health damages from improper disposal of solid waste in metropolitan Islamabad–Rawalpindi, Pakistan. Sustainability 13:2717

Alao BO, Falowo AB, Chulayo A, Muchenje V (2017) The potential of animal by-products in food systems: production, prospects and challenges. Sustainability 9:1089

Amicarelli V, Bux C (2021) Food waste measurement toward a fair, healthy and environmental-friendly food system: a critical review. Br Food J 123:2907–2935

Andreo-Martínez P, Ortiz-Martínez VM, Salar-García MJ, Veiga-Del-Bano JM, Chica A, Quesada-Medina J (2022) Waste animal fats as feedstock for biodiesel production using non-catalytic supercritical alcohol transesterification: a perspective by the PRISMA methodology. Energy Sustain Dev 69:150–163

Angulo M, Márquez MC (2023) A green technology approach using enzymatic hydrolysis to valorize meat waste as a way to achieve a circular economy. Appl Sci 13:8763

Aspevik T, Oterhals Å, Rønning SB, Altintzoglou T, Wubshet SG, Gildberg A, Afseth NK, Whitaker RD, Lindberg D (2018) Valorization of proteins from co-and by-products from the fish and meat industry. Chem Chem Technol Waste Valorization:123–150

Babadi AA, Rahmati S, Fakhlaei R, Barati B, Wang S, Doherty W, Ostrikov KK (2022) Emerging technologies for biodiesel production: processes, challenges, and opportunities. Biomass Bioenergy 163:106521

Badgett A, Milbrandt A (2020) A summary of standards and practices for wet waste streams used in waste-to-energy technologies in the United States. Renew Sust Energ Rev 117:109425

Baldi G, Soglia F, Petracci M (2021) Valorization of meat by-products. Food waste recovery. Elsevier

Barendse W (2014) Should animal fats be back on the table? A critical review of the human health effects of animal fat. Anim Prod Sci 54:831–855

Bastos MAV Jr, De Oliveira Bastos PRH, Portella RB, Soares LFG, Conde RB, Rodrigues JRPMF, Lucchetti G (2019) Pineal gland and schizophrenia: a systematic review and meta-analysis. Psychoneuroendocrinology 104:100–114

Bell SJ, Bradley D, Forse RA, Bistrian BR (1997) The new dietary fats in health and disease. J Am Diet Assoc 97:280–286

Beski SS, Swick RA, Iji PA (2015) Specialized protein products in broiler chicken nutrition: a review. Anim Nutr 1:47–53

Borrajo P, Pateiro M, Barba FJ, Mora L, Franco D, Toldrá F, Lorenzo JM (2019) Antioxidant and antimicrobial activity of peptides extracted from meat by-products: a review. Food Anal Methods 12:2401–2415

Broughton MJ, Thiele JH, Birch EJ, Cohen A (1998) Anaerobic batch digestion of sheep tallow. Water Res 32:1423–1428

Bux C, Amicarelli V (2022) Material flow cost accounting (MFCA) to enhance environmental entrepreneurship in the meat sector: challenges and opportunities. J Environ Manag 313:115001
Cade JE, Hanison J (2017) The pancreas. Anaesth Intensive Care Med 18:527–531
Cecchin A, Salomone R, Deutz P, Raggi A, Cutaia L (2020) Relating industrial symbiosis and circular economy to the sustainable development debate. In: Industrial symbiosis for the circular economy: operational experiences, best practices and obstacles to a collaborative business approach. Springer Nature, Cham, pp 1–25
Chakrabarti S, Guha S, Majumder K (2018) Food-derived bioactive peptides in human health: challenges and opportunities. Nutrients 10:1738
Christopher LP, Kumar H, Zambare VP (2014) Enzymatic biodiesel: challenges and opportunities. Appl Energy 119:497–520
Chuck-Hernández C, Ozuna C (2019) Protein isolates from meat processing by-products. In: Proteins: sustainable source, processing and applications. Elsevier
Cooke B, Pugh ML (1980) Slaughter waste in animal feeding, vol 3. BSAP Occasional Publication, pp 79–83
Daffre S, Bulet P, Spisni A, Ehret-Sabatier L, Rodrigues EG, Travassos LR (2008) Bioactive natural peptides. Stud Nat Prod Chem 35:597–691
Dey D, Richter JK, Ek P, Gu BJ, Ganjyal GM (2021) Utilization of food processing by-products in extrusion processing: a review. Front Sustain Food Syst 4:603751
Dibner J, Atwell C, Kitchell M, Shermer W, Ivey F (1996) Feeding of oxidized fats to broilers and swine: effects on enterocyte turnover, hepatocyte proliferation and the gut associated lymphoid tissue. Anim Feed Sci Technol 62:1–13
Duvallet TY, Jewell RB (2023) Recycling of bone ash from animal wastes and by-products in the production of novel cements. J Am Ceram Soc 106:3720–3735
Eldesouky A, Mesias F, Elghannam A, Escribano M (2018) Can extensification compensate livestock greenhouse gas emissions? A study of the carbon footprint in Spanish agroforestry systems. J Clean Prod 200:28–38
Fernández CL, Romero MC, Rolhaiser F, Fogar RA, Doval MM (2021) Fat substitutes based on bovine blood plasma and flaxseed oil as functional ingredients. Int J Gastron Food Sci 25: 100365
Gautam L, Jain JK, Jain A, Kalla P (2022) Recycling of bone China ceramic waste as cement replacement to produce sustainable self-compacting concrete. Structures, Elsevier 37:364–378
Gizzi G, Von Holst C, Baeten V, Berben G, Van Raamsdonk L (2004) Determination of processed animal proteins, including meat and bone meal, in animal feed. J AOAC Int 87:1334–1341
Gooding CH, Meeker DL (2016) Comparison of 3 alternatives for large-scale processing of animal carcasses and meat by-products. Profess Anim Sci 32:259–270
Hamawand I (2015) Anaerobic digestion process and bio-energy in meat industry: a review and a potential. Renew Sust Energ Rev 44:37–51
Hansen KH, Angelidaki I, Ahring BK (1998) Anaerobic digestion of swine manure: inhibition by ammonia. Water Res 32:5–12
Hejnfelt A, Angelidaki I (2009) Anaerobic digestion of slaughterhouse by-products. Biomass Bioenergy 33:1046–1054
Hirpessa BB, Ulusoy BH, Hecer C (2020) Hormones and hormonal anabolics: residues in animal source food, potential public health impacts, and methods of analysis. J Food Qual 2020:1–12
Hirsch AL (2011) Industrial aspects of vitamin D. In: Vitamin D. AGD Nutrition LLC, Lewisville
Holmes C, Wrobel JS, Maceachern MP, Boles BR (2013) Collagen-based wound dressings for the treatment of diabetes-related foot ulcers: a systematic review. Diabetes Metab Syndr Obes 6:17–29
Hong MH, Lee JH, Jung HS, Shin H, Shin H (2022) Biomineralization of bone tissue: calcium phosphate-based inorganics in collagen fibrillar organic matrices. Biomater Res 26:42
Hubalek S, Post MJ, Moutsatsou P (2022) Towards resource-efficient and cost-efficient cultured meat. Curr Opin Food Sci 47:100885

Irshad A, Sharma B (2015) Abattoir by-product utilization for sustainable meat industry: a review. J Anim Prod Adv 5:681–696
Jayathilakan K, Sultana K, Radhakrishna K, Bawa A (2012) Utilization of byproducts and waste materials from meat, poultry and fish processing industries: a review. J Food Sci Technol 49: 278–293
Karaket T, Seel-Audom M, Yuangsoi B (2022) Partial replacement of fish meal with blood meal in diet on growth performance for nursing red tilapia (*Oreochromis* spp.). Burapha Sci J 27:801–814
Karimi S, Mahboobi Soofiani N, Mahboubi A, Taherzadeh MJ (2018) Use of organic wastes and industrial by-products to produce filamentous fungi with potential as aqua-feed ingredients. Sustainability 10:3296
King'ori A, Tuitoek J, Muiruri H (1998) Comparison of fermented dried blood meal and cooked dried blood meal as protein supplements for growing pigs. Trop Anim Health Prod 30:191–196
Kirimi JG, Musalia LM, Magana A, Munguti JM (2016) Performance of Nile tilapia (*Oreochromis niloticus*) fed diets containing blood meal as a replacement of fish meal. J Agric Sci 8(8):79
Kirimi J, Musalia LM, Munguti J (2017) Effect of replacing fish meal with blood meal on chemical composition of supplement for Nile tilapia (*Oreochromis niloticus*). East Afr Agric For J 82:1–9
Kowalski Z, Kulczycka J, Makara A, MondelloG SR (2023) Industrial symbiosis for sustainable management of meat waste: the case of Śmiłowo Eco-Industrial Park, Poland. Int J Environ Res Public Health 20:5162
Kwon EE, Jeon EC, Yi H, Kim S (2014) Transforming duck tallow into biodiesel via noncatalytic transesterification. Appl Energy 116:20–25
Leon M, Garcia AN, Marcilla A, Martinez-Castellanos I, Navarro R, Catala L (2018) Thermo-chemical conversion of animal by-products and rendering products. Waste Manag 73:447–463
Leonard W, Zhang P, Ying D, Fang Z (2020) Application of extrusion technology in plant food processing byproducts: an overview. Compr Rev Food Sci Food Saf 19:218–246
Li JM, Nie SP (2016) The functional and nutritional aspects of hydrocolloids in foods. Food Hydrocoll 53:46–61
Litvinov RI, Pieters M, De Lange-Loots Z, Weisel JW (2021) Fibrinogen and fibrin. In: Macromolecular protein complexes III: structure and function. Springer, pp 471–501
Luste S (2011) Anaerobic digestion of organic by-products from meat-processing industry: the effect of pre-treatments and co-digestion. Itä-Suomen yliopisto
Manso T, Mantecón A, Giráldez FJ, Lavın P, Castro T (1998) Animal performance and chemical body composition of lambs fed diets with different protein supplements. Small Rumin Res 29: 185–191
Mazhangara IR, Chivandi E, Mupangwa JF, Muchenje V (2019) The potential of goat meat in the red meat industry. Sustainability 11:3671
Mccabe BK, Harris P, Antille DL, Schmidt T, Lee S, Hill A, Baillie C (2020) Toward profitable and sustainable bioresource management in the Australian red meat processing industry: a critical review and illustrative case study. Crit Rev Environ Sci Technol 50:2415–2439
Mora L, Toldrá-Reig F, Reig M, Toldrá F (2018) Meat by-products: new insights into potential technical and health applications. In: Novel proteins for food, pharmaceuticals, and agriculture: sources, applications, and advances, p 101
Murawska N, Fabisiak A, Fichna J (2016) Anemia of chronic disease and iron deficiency anemia in inflammatory bowel diseases: pathophysiology, diagnosis, and treatment. Inflamm Bowel Dis 22:1198–1208
Murugesan K, Srinivasan KR, Paramasivam K, Selvam A, Wong J (2021) Conversion of food waste to animal feeds. In: Current Developments in Biotechnology and Bioengineering. Elsevier
Noor NQIM, Razali RS, Ismail NK, Ramli RA, Razali UIIM, Bahauddin AR, Zaharudin N, Rozzamri A, Bakar J, Shaarani SM (2021) Application of green technology in gelatin extraction: a review. Processes 9:2227
Noori AW, Royen MJ, Medveďová A, Haydary J (2022) Drying of food waste for potential use as animal feed. Sustainability 14:5849

Nuñez SM, Guzmán F, Valencia P, Almonacid S, Cárdenas C (2020) Collagen as a source of bioactive peptides: a bioinformatics approach. Electron J Biotechnol 48:101–108
OeCD & FAO (2023) OECD-FAO Agricultural Outlook 2023–2032. OECD Publishing, Paris
Offiah V, Kontogiorgos V, Falade KO (2019) Extrusion processing of raw food materials and by-products: a review. Crit Rev Food Sci Nutr 59:2979–2998
Ofori JA, Hsieh YHP (2014) Issues related to the use of blood in food and animal feed. Crit Rev Food Sci Nutr 54:687–697
Ogunji JO, Iheanacho SC, Abe GA, Ikeh OR (2020) Assessing effects of substituting dietary fish meal with boiled donkey and cow blood meal on growth performance and digestive enzyme activities of *Clarias gariepinus* juvenile. J World Aquacult Soc 51:1066–1079
Okoro OV, Sun Z, Birch J (2017) Meat processing waste as a potential feedstock for biochemicals and biofuels–a review of possible conversion technologies. J Clean Prod 142:1583–1608
Onwosi CO, Igbokwe VC, Odimba JN, Eke IE, Nwankwoala MO, Iroh IN, Ezeogu LI (2017) Composting technology in waste stabilization: on the methods, challenges and future prospects. J Environ Manag 190:140–157
Piccolla CD, Hesterberg D, Muraoka T, Novotny EH (2021) Optimizing pyrolysis conditions for recycling pig bones into phosphate fertilizer. Waste Manag 131:249–257
Pinto J, Boavida-Dias R, Matos HA, Azevedo J (2022) Analysis of the food loss and waste valorisation of animal by-products from the retail sector. Sustainability 14:2830
Polo J, Rodríguez C, Saborido N, Ródenas J (2005) Functional properties of spray-dried animal plasma in canned petfood. Anim Feed Sci Technol:122331–122343
Resende D, Pereira R, Domínguez D, Pereira M, Pereira C, Pintado M, Valente LM, Velasco C (2023) Stress response of European seabass (*Dicentrarchus labrax*) fed plant-based diets supplemented with swine blood hydrolysates. Aquacult Rep 30:101600
Roy P, Mohanty AK, Dick P, Misra M (2023) A review on the challenges and choices for food waste valorization: environmental and economic impacts. ACS Environ Au 3:58–75
Ryder K, Bekhit AED, Mcconnell M, Carne A (2016) Towards generation of bioactive peptides from meat industry waste proteins: generation of peptides using commercial microbial proteases. Food Chem 208:42–50
Saeid A, Labuda M, Chojnacka K, Górecki H (2014) Valorization of bones to liquid phosphorus fertilizer by microbial solubilization. Waste Biomass Valorization 5:265–272
Samsalee N, Sothornvit R (2017) Modification and characterisation of porcine plasma protein with natural agents as potential cross-linkers. Int J Food Sci Technol 52:964–971
Sayeed R, Tiwari P (2020) Wealth from meat industry by-products and waste: a review. In: Sustainable food waste management: concepts and innovations. Springer, Singapore, pp 191–208
Shi S, Wang L, Song C, Yao L, Xiao J (2023) Recent progresses of collagen dressings for chronic skin wound healing. Collagen Leather 5:31
Shurson GC, Kerr BJ, Hanson AR (2015) Evaluating the quality of feed fats and oils and their effects on pig growth performance. J Anim Sci Biotechnol 6:1–11
Silva EPD, Rabello CBV, Lima MBD, Ludke JV, Arruda EMFD, Albino LFT (2014) Poultry offal meal in broiler chicken feed. Sci Agric 71:188–194
Skoronski E, De Oliveira DC, Fernandes M, Da Silva GF, Magalhães MDLB, João JJ (2016) Valorization of agro-industrial by-products: analysis of biodiesel production from porcine fat waste. J Clean Prod 112:2553–2559
Skripsts E, Mezule L, Klaucans E (2022) Primary sludge from dairy and meat processing wastewater and waste from biomass enzymatic hydrolysis as resources in anaerobic digestion and co-digestion supplemented with biodegradable surfactants as process enhancers. Energies 15: 4333
Socas-Rodríguez B, Álvarez-Rivera G, Valdés A, Ibáñez E, Cifuentes A (2021) Food by-products and food wastes: are they safe enough for their valorization? Trends Food Sci Technol 114:133–147

Sporchia F, Galli A, Kastner T, Pulselli FM, Caro D (2023) The environmental footprints of the feeds used by the EU chicken meat industry. Sci Total Environ 886:163960
Storino F, Arizmendiarrieta JS, Irigoyen I, Muro J, Aparicio-Tejo PM (2016) Meat waste as feedstock for home composting: effects on the process and quality of compost. Waste Manag 56:53–62
Suresh T, Sivarajasekar N, Balasubramani K (2021) Enhanced ultrasonic assisted biodiesel production from meat industry waste (pig tallow) using green copper oxide nanocatalyst: comparison of response surface and neural network modelling. Renew Energy 164:897–907
Tabassum Z, Mohan A, Mamidi N, Khosla A, Kumar A, Solanki PR, Malik T, Girdhar M (2023) Recent trends in nanocomposite packaging films utilising waste generated biopolymers: industrial symbiosis and its implication in sustainability. IET Nanobiotechnol 17:127–153
Tagrida M, Nilsuwan K, Gulzar S, Prodpran T, Benjakul S (2022) Fish gelatin/chitosan blend films incorporated with betel (*Piper betle* L.) leaf ethanolic extracts: characteristics, antioxidant and antimicrobial properties. Food Hydrocoll 108316
Timorshina S, Popova E, Osmolovskiy A (2022) Sustainable applications of animal waste proteins. Polymers 14:1601
Toldrá F, Mora L, Reig M (2016) New insights into meat by-product utilization. Meat Sci 120:54–59
Toldrá F, Gallego M, Reig M, Aristoy MC, Mora L (2020) Recent progress in enzymatic release of peptides in foods of animal origin and assessment of bioactivity. J Agric Food Chem 68:12842–12855
Toldrá-Reig F, Mora L, Toldrá F (2020a) Developments in the use of lipase transesterification for biodiesel production from animal fat waste. Appl Sci 10:5085
Toldrá-Reig F, Mora L, Toldrá F (2020b) Trends in biodiesel production from animal fat waste. Appl Sci 10:3644
Trushenski JT, Lochmann RT (2009) Potential, implications and solutions regarding the use of rendered animal fats in aquafeeds. Am J Anim Vet Sci 4:108–128
Usman M, Ishaq A, Regenstein JM, Sahar A, Aadil RM, Sameen A, Khan MI, Alam A (2023) Valorization of animal by-products for gelatin extraction using conventional and green technologies: a comprehensive review. Biomass Convers Biorefinery:1–13
Zhang Y, Duan L, Esmaeili H (2022) A review on biodiesel production using various heterogeneous nanocatalysts: operation mechanisms and performances. Biomass Bioenergy 158:106356
Zhao X, Zhang X, Liu D (2021) Collagen peptides and the related synthetic peptides: a review on improving skin health. J Funct Foods 86:104680

Chapter 24
Environmental Issues of Meat

Saeid Jafari, Mahdi Ebrahimi, Khursheed Ahmad Shiekh, and Kitipong Assatarakul

24.1 Introduction

The effects of meat production and consumption on the environment have recently received urgent attention. The adverse effects on our planet are becoming more obvious as global meat consumption rises. The environmental problems connected to the production of meat are serious and require our attention. They range from deforestation and habitat destruction to greenhouse gas emissions and water contamination (Winders and Ransom 2019).

Deforestation is one of the main environmental issues connected to the production of meat (Walker et al. 2013). Big areas of forest are destroyed, especially in places like the Amazon rainforest, to make room for the grazing of animals and the production of crops used as animal feed, including corn and soybeans. In addition to

Saeid Jafari, Mahdi Ebrahimi, Khursheed Ahmad Shiekh and Kitipong Assatarakul contributed equally with all other contributors.

S. Jafari · K. Assatarakul (✉)
Department of Food Technology, Faculty of Science, Chulalongkorn University, Bangkok, Thailand
e-mail: saeid.J@chula.ac.th; Kitipong.A@chula.ac.th

M. Ebrahimi
Department of Veterinary Preclinical Sciences, Faculty of Veterinary Medicine, University Putra Malaysia, Serdang, Malaysia
e-mail: Mehdiebrahimii@gmail.com

K. A. Shiekh
Department of Food Technology, Faculty of Science, Chulalongkorn University, Bangkok, Thailand

Department of Food Science, College of Agriculture and Life Sciences, Cornell University, Ithaca, NY, USA
e-mail: khursheedahmad.s@chula.ac.th

S. A. Rather, F. A. Masoodi (eds.), *Hand Book of Processed Functional Meat Products*, https://doi.org/10.1007/978-3-031-69868-2_24

destroying important ecosystems and species, this deforestation exacerbates climate change. When forests are destroyed, large amounts of stored carbon are released into the atmosphere, increasing greenhouse gas emissions. Forests serve as carbon sinks, absorbing large amounts of carbon dioxide from the environment.

Another serious issue related to the production of beef is greenhouse gas emissions. Methane is a powerful greenhouse gas that is produced by livestock, particularly ruminants like cattle (Ebrahimi et al. 2017; Jafari et al. 2020). It has a far larger warming potential than carbon dioxide. The production and delivery of animal feed as well as the processing and refrigeration of meat products all add to the meat industry's overall carbon footprint.

Addressing the unsustainable nature of meat production and consumption is crucial given the severity of these environmental problems. The environmental effect of eating meat can be considerably reduced by switching to a more sustainable, plant-based diet. Adopting agro-ecological methods that put emphasis on crop rotation, regenerative agriculture, and biodiversity preservation can also help to create a more resilient and sustainable food supply.

In this chapter, we emphasize the problems associated with consuming meat and other meat-related foods, human health, climate change, viral infections, as well as some ethical questions and problems relating to the production of livestock.

24.2 Meat Consumption and Dietary Habits

Meat has been an essential component of human diets since the dawn of time, providing not only essential nutrients but also cultural significance. However, changes in eating habits have raised concerns about the morality, health, and sustainability of meat consumption. Meat consumption has dramatically increased over the past few decades in many countries, with a trend that is most pronounced in the last 40 years. There is widespread consensus that meat consumption has increased over the past few decades, despite the fact that different research have produced differing statistics (González et al. 2020). This alteration in eating behaviors has had a profound impact on our nutritional choices. In order to put contemporary dietary behaviors into historical context, it is crucial to understand how meat intake has changed over time. Due to their reliance on hunting and gathering, the early humans needed meat to survive (Alt et al. 2022). During the shift to agricultural societies, meat consumption patterns changed as a result of better food availability and cultural norms.

Meat is a great source of high-quality protein, essential amino acids, vitamins (including vitamin B12), and minerals (including iron and zinc). However, consuming too much red and processed meat has been linked to a variety of diseases, including obesity, a number of cancers, and cardiovascular conditions (Shi et al. 2023). A study found a link between eating a lot of meat and non-communicable diseases. The numerous factors that underlie these connections between consuming meat and poor health outcomes are discussed in epidemiological studies. Regardless of body mass index, diets heavy in animal products and low in nutritious plant-based foods increase the risk of type 2 diabetes (Adeva-Andany et al. 2019).

Culture has a significant influence on food preferences, particularly the desire for meat eating. Cultural dietary behaviors are frequently influenced by religious beliefs, traditions, and culinary practices. Examining these factors reveals information about the variety of eating practices and the difficulties in changing them (Milford et al. 2019). The production of meat has a significant negative impact on the environment since it uses more resources, land, and emits more greenhouse gases (Kwasny et al. 2022). Examining the consequences of meat consumption on the environment and looking into sustainable alternatives are essential to halting climate change and protecting natural resources. Dietary limitations, financial worries, and cultural resistance are obstacles to limiting meat intake. However, cutting-edge strategies like meat alternatives, nutritional suggestions, and awareness campaigns offer paths to more sustainable and health-conscious eating habits. Altering dietary habits and reducing meat consumption are equally as vital as using less hazardous production techniques, especially in wealthy Western nations. A high-meat diet emits around 7.19 $kgCO_2e$/day, while a vegetarian diet emits just 3.81 $kgCO_2e$/day, which is nearly a 50% reduction in GHG emissions. The reduction is far more pronounced when following a vegan diet, which emits only 2.89 $kgCO_2e$ / day. Vegetarianism, veganism, and other plant-based diets are becoming more popular as a result of growing health, moral, and environmental concerns (Krizanova et al. 2021). Consuming meat has a significant impact on the environment, culture, and health of people. It is closely related to dietary habits. The significance of making thoughtful, knowledgeable food choices that match nutritional requirements with moral and ethical considerations (Kwasny et al. 2022). Researchers, governments, and the food industry must work together to promote dietary habits that are healthier and more sustainable.

Despite strong scientific data supporting the environmental advantages of cutting back on meat consumption, consumer dietary choices have not yet been dramatically impacted by these results. For instance, a survey conducted in North American countries found that consumption of red and processed meat is high in most of the country (Frank et al. 2021). These results highlight the fact that, despite strong scientific support, consumers' eating habits are not yet significantly changing in a way that would improve the environment and the public's health.

24.3 Meat Consumption and Its Impact on Human Health

The intake of meat is a crucial component of the human diet since it offers vital minerals and proteins. Concerns have been raised about the effects of excessive and unbalanced meat consumption on human health. Throughout human history, meat has played a key role in meals, providing vital elements such protein, iron, zinc, and vitamin B12. However, concerns about the potential negative impacts on health have been raised by the increased availability and consumption of meat, particularly in high-income countries. Consuming processed meats (such as bacon, sausages, and hot dogs) has been linked in a number of studies to an increased risk of some cancers, including colon cancer (IARC 2015). An increased risk of cardiovascular illnesses, including coronary heart disease, has been associated with high consumption of red meat, particularly processed and unprocessed red meat (Shi et al. 2023). According to several research, eating a lot of red meat increases your risk of developing type 2 diabetes (Adeva-Andany et al. 2019).

There is rising interest in alternative protein sources, such as plant-based diets and meat substitutes, due to the issues with excessive meat consumption. These substitutes might lessen some of the associated health hazards while still supplying the necessary nutrients. Consuming meat has both nutritional advantages and health hazards, making it a complicated topic. It's important to consume meat in moderation and to be informed of the different types and cooking techniques available. In order to improve general health, public health recommendations frequently urge cutting back on processed and red meat consumption and switching to a more diversified and plant-based diet. Consuming red meat or processed meat increased the risk of cancer, notably colon cancer, according to substantial epidemiological evidence (Farvid et al. 2021). One of the primary subjects of the studies discovered in the literature is colorectal cancer. According to study by (Rose and Strombom 2019) eating a vegan diet provided statistically significant protection against the occurrence of both male and female colon cancer. According to a recent review by (Farvid et al. 2021) that focused on different cancer, eating red meat increased the risk of developing the breast cancer.

Studies that concentrate on the impact of meat intake on the development of other diseases, such as metabolic disorders, which can also be enhanced by red meat consumption (Hidayat et al. 2022). The consumption of red meat, but not poultry, was found to be positively connected with a greater incidence of diabetes in prospective research carried out in China (Du et al. 2020). The consumption of more red meat was also found to be positively connected with other disorders, such as chronic renal disease (Mirmiran et al. 2020).

It's interesting to note that eating red meat can shorten life expectancy. A study by Wu et al. (2023) aimed on establishing if red meat consumption could affect life expectancy used data from 204 nations and results from international organizations. Although the use of red meat today has a detrimental effect on life expectancy in high- and middle-income nations, there was no correlation between it and life expectancy in low-income countries, it was noted (Ranabhat et al. 2020). It takes constant research and a comprehensive analysis of dietary preferences, personal health situations, and cultural considerations to comprehend the connection between meat consumption and human health. It is vital to promote healthy eating habits that prioritize both human health and environmental sustainability, and to base dietary decisions on a thorough understanding of the available information.

24.4 Meat Consumption and Climate Change

There are major effects on climate change from the production and consumption of animal meat. It is widely acknowledged that climate change may have an impact on the safety and quality of meat (Soliman and Safwat 2020). One of the greatest threats facing the world today is climate change, and the food system is a major contributor to greenhouse gas (GHG) emissions. Due to a number of variables, including the release of methane during enteric fermentation, changes in land use, and energy-intensive production techniques, meat production—in particular the production of beef and lamb—is linked to significant GHG emissions (Moumen et al. 2016). Methane is a strong GHG that is produced in substantial volumes during enteric fermentation in ruminant animals (such as cattle and sheep). Methane has a far greater short-term potential for global warming than carbon dioxide (CO_2). It is crucial to keep in mind that raising cattle is one of the main causes of global warming (Bačėninaitė et al. 2022). In especially when using fossil fuels, the energy-intensive nature of meat production, including feed production, transportation, and processing, results in significant emissions. Because of the land needed for grazing and the methane emissions linked to cattle digestion, beef production has one of the highest carbon footprints per protein unit (Veeramani et al. 2017). Due to their differing digestive systems and quicker development rates than beef and lamb, poultry and pork have lower carbon footprints than those of the latter two.

The effects of meat consumption on the environment must be measured and taken into account. One of the most popular methods for doing this is life cycle evaluation (LCA). Environmental effects of climate change, such as acidification, land and water consumption, and eutrophication, are evaluated through LCA. The first step is to determine which eating practices are most environmentally beneficial. Bruno et al. (2019) used LCA to calculate the carbon footprint of four food scenarios in a study done in Denmark (standard, carnivore, vegetarian and vegan). From farm production to consumption, all activities were considered. Carnivore diets had the highest environmental impact, as evidenced by the findings that they produce more CO_2 (1.83 t CO_2eq/person/year). Other diets, including vegetarianism or veganism, however, exhibited lower emissions (0.89 and 1.37 t CO_2eq /person/year, respectively) (Bruno et al. 2019). The majority of findings generally support that meat products, followed by dairy products, are the commodities with the highest environmental impact. This is because different agronomic and zootechnical activities are involved (Notarnicola et al. 2017).

Silva et al. (2023) also studied LCA for the production systems of chicken and pork. Per kilogram of live weight, values between 1.40 and 5.78 kgCO_2-eq and 1.05 to 9.35 kgCO_2-eq, respectively, are recorded in the production of pigs and chickens. According to Silva et al. (2023) the manufacture of feed was the major source of emissions. However, compared to the production of red meat, white meats like those of rabbit or fowl have a reduced carbon impact (Silva et al. 2023). It is possible to drastically cut GHG emissions linked to food production by switching to a plant-based diet and reducing or eliminating animal consumption (Hemler and Hu 2019).

Reduced use of synthetic fertilizers and rotational grazing are two sustainable and regenerative farming techniques that can reduce the carbon footprint of meat production (Teague and Kreuter 2020). Environmentally friendly alternatives to traditional meat are being developed and adopted, such as plant-based and cultured meats. Since meat consumption contributes to GHG emissions, it is closely tied to climate change. Reduced consumption of meat, especially beef and lamb, and the adoption of sustainable production methods are crucial measures for reducing the negative environmental effects of the meat industry. The issues created by meat consumption and its effects on climate change must be addressed by public education, regulatory changes, and consumer decisions. To achieve a more climate-resilient food system, greater research and innovation in sustainable agriculture and alternate protein sources are essential.

24.5 Meat Consumption and Viral Infections: Understanding the Connection

Meat consumption is a widespread dietary practice around the world, providing essential nutrients and proteins. However, the production, handling, and consumption of meat products can also be associated with certain risks, including the potential transmission of viral infections. In this section, we explore the relationship between meat consumption and viral infections, shedding light on the underlying factors and providing insight into preventive measures (Table 24.1). Zoonotic diseases are those that can be transmitted from animals to humans. Several viral infections, such as avian influenza (bird flu), swine influenza (swine flu), and SARS-CoV-2 (COVID-19), have zoonotic origins and are linked to the handling or consumption of meat products. Hepatitis E is one of the most serious viral illnesses that can be contracted by eating meat. The virus is mostly spread through undercooked meat, with pigs, wild boars, and deer serving as the main reservoirs. Fortunately, boiling the meats at temperatures above 60 °C can inactivate up to 80% of the virions (González et al. 2020). According to the most recent statistics on the epidemiology of hepatitis E virus (HEV) in the United States, eating meat frequently (more than 10 times per month) can raise the likelihood of an outbreak (Ditah et al. 2014). Rivero-Juarez et al. (2017) discussed a familial HEV outbreak involving nine family members in Cordoba, Spain, that was linked to eating wild boar meat and was supported by phylogenetic research. Based on those findings, detailed research on HEV in farm animals, game, and pork-derived products was recommended in order to evaluate the severity of this infectious disease and to implement control and prevention strategies to reduce the spread of this newly emerging infectious disease via this route. On the other hand, the norovirus is a very contagious virus that can cause gastroenteritis, also called the stomach flu. It is one of the main global

Table 24.1 Various meat consumption patterns and probable links to microbial/viral infections

	Meat consumption type	Potential microbial/viral infections
1	Raw or undercooked meat	*Salmonella* spp.
		Campylobacter spp.
		Escherichia coli
		Toxoplasma gondii
		Trichinella spp.
		Hepatitis E
2	Processed meats	*Listeria monocytogenes*
		Clostridium botulinum
		Hepatitis E
		Norovirus
3	Poultry consumption	Avian influenza (bird flu)
		Salmonella spp.
		Campylobacter spp.
		Clostridium perfringens
		Staphylococcus aureus
		Escherichia coli
4	Pork consumption	Hepatitis E
		Hepatitis A
		Nipah virus
		Swine influenza (swine flu)
		Taenia solium
		Japanese encephalitis
5	Red-meat (Beef) consumption	Hepatitis E
		Bovine spongiform encephalopathy (BSE)
		Coxiella burnetii (Q fever)
		Escherichia coli O157:H7
		Brucella spp.
		Foot-and-mouth disease (FMD)
6	Wild game consumption	*Brucella* spp.
		Trichinella spp.
		Rabies
		Hantavirus
		African swine fever (ASF)

Reference: González et al. (2020) and Tasiame et al. (2022)

contributors to foodborne diseases. When these meats are handled and processed, norovirus can spread. While it is mainly spread by direct contact between people, it can also contaminate food, notably processed meats, when it comes into touch with infected food handlers. As well as nausea, vomiting, diarrhea, and stomach cramps, norovirus infections can also cause fever and body aches. An investigation found rabies virus RNA in dogs slaughtered for human consumption in several regions of Ghana. 144 dogs were sampled, and three (2.08%) of those tested positive for the rabies virus (Tasiame et al. 2022). This study was necessary in terms of epidemiology of rabies transmission in Ghana because the discovery of rabies virus in slaughtered dogs processed without protective gear, sold, and consumed by the general public represented a potential source of rabies transmission and a threat to the fight against human rabies. These kinds of viral infections may also arise when humans come into close contact with infected animals or their contaminated meat. Intensive animal farming, commonly employed in meat production, can increase the risk of viral infections. Animals reared in crowded and unsanitary conditions are more susceptible to viral diseases. Viruses can rapidly spread among animals in confined spaces, creating an environment conducive to viral mutation and cross-species transmission. As an example, one of the most pressing issues at the moment is the discovery of potentially novel coronavirus in samples from bats and rats in Vietnam in 2018. It was found that coronavirus RNA was present in 22% of the bat fecal samples. Therefore, if security precautions are not implemented, zoonosis caused by eating bat meat is likely to occur. The SARS-CoV-2 new coronavirus has been a worldwide problem since December 2019. The main theory is that this RNA virus may have its origins in a seafood market from Wuhan in China, where vertebrate and invertebrate—wild and farmed—animal were sold (González et al. 2020).To mitigate the risk of viral infections, proper food safety measures are crucial. This includes maintaining hygiene and sanitation standards in slaughterhouses, processing plants, and marketplaces. Adequate cooking of meat products, especially poultry and pork, is necessary to eliminate or inactivate potential pathogens, reducing the risk of transmission to humans. The One Health approach emphasizes the interconnection between human, animal, and environmental health. It recognizes that the health of humans is closely linked to the health of animals and the environment. Adopting this holistic perspective enables the development of comprehensive strategies to prevent and control viral infections associated with meat consumption. The encroachment of human activities into natural habitats can lead to the emergence of novel viral diseases. As deforestation and habitat destruction continue, humans are more likely to come into contact with previously undisturbed animal species, potentially exposing themselves to new viruses. Thus, addressing the root causes of viral outbreaks requires sustainable land-use practices and the preservation of biodiversity. Consumers play a significant role in shaping the demand for meat products and influencing production practices. By making informed choices, such as opting for meat from sustainable and well-regulated sources, consumers can encourage responsible farming practices that prioritize animal health and reduce the risk of viral infections. Enhancing surveillance systems and early detection mechanisms is crucial for effective response to viral outbreaks. Strengthening international collaboration, knowledge sharing, and research on viral infections can improve our understanding of transmission patterns, host range, and preventive measures, facilitating swift and coordinated actions during public health emergencies. By addressing these factors collectively, we can strive for a healthier and more secure future for both humans and animals.

24.6 Social Aspects of Livestock Production

Significant societal implications of livestock production touch on many facets of society. Keeping animals so meets a variety of social needs. These range from highly visible ones like ensuring food security, providing proteins, and generating cash for farmers to more covert ones like carrying out cultural or religious functions, supplying ecosystem services, and gratifying their owners' passion for maintaining and caring for animals, to mention a few. When considering the future roles of farm animals in food systems and civilizations, one must take into account their negative effects on the sustainability of food production in addition to these beneficial features of animal farming. Here are some significant societal factors connected to the livestock production: (1) Livelihoods and rural communities: Livestock production is essential to millions of people's ability to support their way of life globally, especially in rural areas. For farmers and other people working in the livestock value chain, it offers employment opportunities, and food security. Livestock farming frequently supports local communities by economic growth and social cohesion in rural areas, (2) Cultural Heritage and traditional customs: Traditional agricultural methods, nomadic lifestyles, and indigenous civilizations have all included livestock keeping as an essential component. It frequently relates to cultural identity, rituals, and ceremonies, representing the wide variety of global human traditions and heritage, (3) Food security and nutrition: Livestock production is essential in resolving the problems associated with food security and nutrition. Meat is an example of foods derived from animals and is significant source of high-quality protein as well as macro- and micronutrients. Livestock farming can help improve diets and fight malnutrition in areas with limited access to a variety of food sources, especially for vulnerable populations like youngsters and expectant mothers, (4) Social networks and the cohesion of communities: Livestock farming has the potential to strengthen social networks and community cohesion. Activities involving livestock frequently entail community participation, and cooperation. Traditional methods like communal grazing, group herding, and cooperative breeding can improve social ties and resilience by encouraging social interactions, mutual aid, and social capital, (5) Consumer attitudes and social acceptability: Consumer preferences and social acceptability have a big impact on how livestock production is done. Consumer preferences and purchasing choices have been influenced by shifting societal values, rising concerns about sustainability, animal welfare, and health effects. These elements influence market changes that encourage the adoption of more moral and ethical methods for raising animals and the last but not the least is (6) Animal Welfare and Ethical Issues: Concerns for animal welfare and ethical issues are included in the social elements of livestock production. The promotion of techniques that put animal welfare first is becoming more and more popular, as is understanding of the need for humane treatment of animals. The efforts made to ensure adequate housing, access to food and water, and the avoidance of cruel or inhumane activities are among these factors.

Šrédl et al. (2021) concluded that the relationship between meat production and consumption, along with changes in the price of this commodity on the market, was what primarily determined the socioeconomic aspect of the sustainable development of livestock production in the Czech Republic. Furthermore, significant environmental factors including excessive water consumption for meat production, greenhouse gas emissions, and the potential for global warming had an additional impact on the sustainable growth of animal industry in Czechia. The social aspects of livestock production bring attention to how interdependent people, animals, and communities are. Promoting sustainable and responsible livestock production systems that are in line with societal demands, cultural values, and ethical considerations requires acknowledging and addressing these social factors.

24.7 Ethical Concerns and Animal Welfare Issues Concerned to Livestock Production

Many attentions have given to ethical issues surrounding the production of livestock as a result of expanding knowledge about animal welfare, environmental effect, and societal values. These issues are a reflection of a larger movement towards responsible, humane, and ecological methods of food production (Petherick 2005). Here, we'll review some of the most important moral issues surrounding the raising of animals: (1) Animal welfare: Treating and ensuring the wellbeing of animals used for food production—meat, dairy, and eggs—is of the utmost ethical importance. Numerous intensive farming methods, including stifling housing conditions, overcrowding, and confinement systems, can cause animals great stress and suffering. Animals' access to natural behaviors, proper nutrition, and medical treatment are among the top ethical concerns. There is a push for higher standards for animal welfare and the implementation of farming methods that put animals' health and quality of life first, (2) Factory farming is a system that stresses great production efficiency but frequently sparks ethical questions. It is sometimes referred to as industrial animal agribusiness. Large numbers of animals are kept in close quarters, which can cause stress on the body and mind, health problems, and a general decline in quality of life. Critics contend that industrial farming damages the environment and prioritizes profit over animal welfare (Table 24.2).

To address these issues, calls for switching to more ethical and environmentally friendly farming practices, including pasture-based or free-range farming, have evolved. (3) Environmental impact: The production of livestock has a considerable negative impact on the environment, contributing to resource depletion, water pollution, greenhouse gas emissions, and deforestation. The destruction of habitat and the conversion of land for grazing and the production of animal feed endangers biodiversity. Livestock emissions of the powerful greenhouse gas methane contribute to climate change. The need to lessen these environmental effects and encourages more sustainable practices, such regenerative agriculture and agroecology, as well as

Table 24.2 Key issues relating to animal welfare in livestock production

S. No.	Animal welfare issues	Description
1	Confinement systems	Animals are frequently confined in small quarters under intensive agricultural operations, limiting their normal movements and activities, such as battery cages for hens or gestation crates for sows
2	Overcrowding	Due to lack of space and social interactions, animals that are crowded together in small areas may experience stress, increased disease transmission, and decreased welfare
3	Lack of outdoor access	Some livestock production methods limit the availability of animals to outside spaces, preventing them from engaging in natural activities, grazing, or taking in sunlight and fresh air
4	Painful husbandry	If conducted without using appropriate pain management or alternatives, many husbandry practices, including as dehorning, debeaking, tail docking, and castration, can cause pain and discomfort to animals
5	Inadequate veterinary care	Animals' well-being and health may be harmed by undiagnosed or untreated illnesses and injuries caused by a lack of access to veterinary treatment and insufficient surveillance
6	Long-distance transport	Animals transported across long distances, frequently in confined spaces, may experience stress, tiredness, injury, and a lack of proper food, drink, and rest
7	Lack of environmental enrichment	Boredom, frustration, and deviant behaviors can result from repetitive environments without suitable enrichment, such as room to walk around, things to examine, or materials to manipulate
8	Inhumane slaughter	It is crucial to receive correct instruction and follow humane handling practices since poor stunning or slaughter methods might result in additional suffering and pain throughout the slaughter process

Reference: Petherick (2005) and Sinclair et al. (2019)

lowering the overall use of animal products, are ethical concerns, (4) Use and resistance of antibiotics: The unethical use of antibiotics in livestock production raises serious ethical questions. Antibiotic-resistant bacteria can emerge as a result of widespread usage of antibiotics in intensive farming systems as growth promoters and illness preventatives, endangering human health. The responsible use of antibiotics, the promotion of alternatives to routine antibiotic use (e.g., natural alternatives), and the advocacy for tougher controls to maintain the potency of these essential medications for both humans and animals are the main ethical concerns, (5) Social justice and labor conditions: Ethical considerations in the production of livestock go beyond issues of animal welfare and environmental effect. Important factors to take into account are worker rights in the industry, safe working conditions, and fair remuneration. Unfair labor practices and exploitation may be particularly prevalent among marginalized communities and migrant workers. Key ethical requirements include promoting fair trading procedures, protecting worker rights, and supporting sustainable livelihoods for people engaged in animal production, (6) Alternative approaches: The introduction of substitute meat products, such as meat derived from plants and meat that has been cultured or developed in a lab, has raised new ethical questions. These alternatives address issues with resource consumption, environmental effect, and animal welfare while offering viable solutions to lessen the reliance on conventional livestock production. The usage of genetically modified organisms (GMOs), the sustainability of the production processes, and the acceptance of these substitutes in various cultural and socioeconomic contexts all raise ethical issues, though.

A multifaceted strategy involving consumers, farmers, policymakers, and industry stakeholders is necessary to address these ethical issues. Supporting and promoting animal welfare standards that are strict, improving supply chain transparency and traceability, and encouraging responsible consumption choices are all part of it (Six et al. 2017). To promote a more ethical and sustainable future for livestock production, stakeholders must work together, engage in research and innovation, and change existing policies (Sinclair et al. 2019).

24.8 Conclusion

In conclusion, the environmental consequences of raising and consuming meat are serious and demand our attention. To accommodate the rising demand for meat, intensive farming methods have been used, which have led to deforestation, habitat loss, greenhouse gas emissions, water pollution, and the depletion of natural resources. These elements lead to environmental degradation overall, biodiversity loss, and climate change. Making the shift to a plant-based diet, encouraging regenerative farming methods, and encouraging creative alternatives to conventional meat production are essential steps in reducing the environmental impact of meat production. We can work toward a healthier planet and a more sustainable future for the next generations by making more environmentally responsible decisions and adopting a holistic approach to food production.

Acknowledgments This study is supported by the Second Century Fund (C2F), Chulalongkorn University.

References

Adeva-Andany MM, Rañal-Muíño E, Vila-Altesor M, Fernández-Fernández C, Funcasta-Calderón R, Castro-Quintela E (2019) Dietary habits contribute to define the risk of type 2 diabetes in humans. Clin Nutr ESPEN 1(34):8–17

Alt KW, Al-Ahmad A, Woelber JP (2022) Nutrition and health in human evolution–past to present. Nutrients 14(17):3594

Bačėninaitė D, Džermeikaitė K, Antanaitis R (2022) Global warming and dairy cattle: how to control and reduce methane emission. Animals 12(19):2687

Bruno M, Thomsen M, Pulselli FM, Patrizi N, Marini M, Caro D (2019) The carbon footprint of Danish diets. Clim Chang 156:489–507

Du H, Guo Y, Bennett DA, Bragg F, Bian Z, Chadni, M (2020) China Kadoorie Biobank collaborative group. Red meat, poultry and fish consumption and risk of diabetes: a 9 year prospective cohort study of the China Kadoorie Biobank. Diabetologia, 63:767–779

Ditah I, Ditah F, Devaki P, Ditah C, Kamath PS, Charlton M (2014) Current epidemiology of hepatitis E virus infection in the United States: low seroprevalence in the National Health and Nutrition Evaluation Survey. Hepatology 60(3):815–822

Ebrahimi M, Rajion MA, Adeyemi KD, Jafari S, Jahromi MF, Oskoueian E, Meng GY, Ghaffari MH (2017) Dietary n-6: n-3 fatty acid ratios alter rumen fermentation parameters and microbial populations in goats. J Agric Food Chem 65(4):737–744

Farvid MS, Sidahmed E, Spence ND, Mante Angua K, Rosner BA, Barnett JB (2021) Consumption of red meat and processed meat and cancer incidence: a systematic review and meta-analysis of prospective studies. Eur J Epidemiol 36:937–951

Frank SM, Jaacks LM, Batis C, Vanderlee L, Taillie LS (2021) Patterns of red and processed meat consumption across North America: a nationally representative cross-sectional comparison of dietary recalls from Canada, Mexico, and the United States. Int J Environ Res Public Health 18(1):357

González N, Marquès M, Nadal M, Domingo JL (2020) Meat consumption: which are the current global risks? A review of recent (2010–2020) evidences. Food Res Int 1(137):109341

Hemler EC, Hu FB (2019) Plant-based diets for personal, population, and planetary health. Advances. Nutrition 10(Supplement 4):S275–S283

Hidayat K, Zhu WZ, Peng SM, Ren JJ, Lu ML, Wang HP, Xu JY, Zhou H, Yu LG, Qin LQ (2022) The association between meat consumption and the metabolic syndrome: a cross-sectional study and meta-analysis. Br J Nutr 127(10):1467–1481

International Agency for Research on Cancer (IARC) (2015) IARC monographs evaluate consumption of red meat and processed meat. https://www.iarc.who.int/featured-news/media-centre-iarc-news-processed-meat-cancer/

Jafari S, Meng GY, Rajion MA, Ebrahimi M (2020) The use of plant by-products as non-conventional feedstuff for livestock feeding with reference to rumen methanogenesis. Agrofor Syst 94:1491–1500

Krizanova J, Rosenfeld DL, Tomiyama AJ, Guardiola J (2021) Pro-environmental behavior predicts adherence to plant-based diets. Appetite 1(163):105243

Kwasny T, Dobernig K, Riefler P (2022) Towards reduced meat consumption: a systematic literature review of intervention effectiveness, 2001–2019. Appetite 1(168):105739

Milford AB, Le Mouël C, Bodirsky BL, Rolinski S (2019) Drivers of meat consumption. Appetite 1(141):104313

Mirmiran P, Yuzbashian E, Aghayan M, Mahdavi M, Asghari G, Azizi F (2020) A prospective study of dietary meat intake and risk of incident chronic kidney disease. J Ren Nutr 30(2): 111–118

Moumen A, Azizi G, Chekroun KB, Baghour M (2016) The effects of livestock methane emission on the global warming: a review. Int J Glob Warm 9(2):229–253

Notarnicola B, Tassielli G, Renzulli PA, Castellani V, Sala S (2017) Environmental impacts of food consumption in Europe. J Clean Prod 1(140):753–765

Petherick JC (2005) Animal welfare issues associated with extensive livestock production: the northern Australian beef cattle industry. Appl Anim Behav Sci 92(3):211–234

Ranabhat CL, Park MB, Kim CB (2020) Influence of alcohol and red meat consumption on life expectancy: results of 164 countries from 1992 to 2013. Nutrients 12(2):459

Rivero-Juarez A, Frias M, Martinez-Peinado A, Risalde MA, Rodriguez-Cano D, Camacho A, García-Bocanegra I, Cuenca-Lopez F, Gomez-Villamandos JC, Rivero A (2017) Familial hepatitis E outbreak linked to wild boar meat consumption. Zoonoses Public Health 64(7): 561–565

Rose S, Strombom A (2019) Colorectal cancer prevention with a plant-based diet. Cancer Ther Oncol Int J 15(2):50–67
Shi W, Huang X, Schooling CM, Zhao JV (2023) Red meat consumption, cardiovascular diseases, and diabetes: a systematic review and meta-analysis. Eur Heart J 2:ehad336
Silva JP, Giehl CJ, Cardinal KM, Andretta I, Ribeiro AM (2023) Different life cycle assessment methods and causes of variation in estimates of global warming in chicken and pork production systems: a critical systematic review. Livest Sci 23:105320
Sinclair M, Fryer C, Phillips CJ (2019) The benefits of improving animal welfare from the perspective of livestock stakeholders across Asia. Animals 9(4):123
Six L, De Wilde B, Vermeiren F, Van Hemelryck S, Vercaeren M, Zamagni A, Masoni P, Dewulf J, De Meester S (2017) Using the product environmental footprint for supply chain management: lessons learned from a case study on pork. Int J Life Cycle Assess 22:1354–1372
Soliman A, Safwat AM (2020) Climate change impact on immune status and productivity of poultry as well as the quality of meat and egg products. In: Climate change impacts on agriculture and food security in egypt: land and water resources – smart farming – livestock, fishery, and aquaculture. Springer, pp 481–498
Šrédl K, Prášilová M, Severová L, Svoboda R, Štěbeták M (2021) Social and economic aspects of sustainable development of livestock production and meat consumption in the Czech Republic. Agriculture 11(2):102
Tasiame W, El-Duah P, Johnson SA, Owiredu EW, Bleicker T, Veith T, Schneider J, Emikpe B, Folitse RD, Burimuah V, Akyereko E (2022) Rabies virus in slaughtered dogs for meat consumption in Ghana: a potential risk for rabies transmission. Transbound Emerg Dis 69(4): e71–e81
Teague R, Kreuter U (2020) Managing grazing to restore soil health, ecosystem function, and ecosystem services. Front Sustain Food Syst 157
Veeramani A, Dias GM, Kirkpatrick SI (2017) Carbon footprint of dietary patterns in Ontario, Canada: a case study based on actual food consumption. J Clean Prod 20(162):1398–1406
Walker NF, Patel SA, Kalif KA (2013) From Amazon pasture to the high street: deforestation and the Brazilian cattle product supply chain. Trop Conserv Sci 6(3):446–467
Winders B, Ransom E (2019) Global meat: social and environmental consequences of the expanding meat industry. MIT Press
Wu Z, Wang W, Zhang K, Fan M, Lin R (2023) The global burden of disease attributable to diet high in red meat in 204 countries and territories, 1999–2019: an updated analysis of the Global Burden of Disease Study. Mol Nutr Food Res 6:2300144

Index

S. Rather, F. A. Masoodi (eds.), *Hand Book of Processed Functional Meat Products*, https://doi.org/10.1007/978-3-031-69868-2